IET MANUFACTURING SERIES 15

Composites Assembly for High Performance Fastener-less Structures

Other volumes in this series:

Volume 23 **Optimal Design Exploiting 3D Printing and Metamaterials** Paolo Di Barba and Sławomir Wiak (Editors)

Composites Assembly for High Performance Fastener-less Structures

Edited by
Hamed Yazdani Nezhad and
Vijay Kumar Thakur

The Institution of Engineering and Technology

Published by The Institution of Engineering and Technology, London, United Kingdom

The Institution of Engineering and Technology is registered as a Charity in England & Wales (no. 211014) and Scotland (no. SC038698).

First published 2022

The Institution of Engineering and Technology
Futures Place
Kings Way, Stevenage
Hertfordshire, SG1 2UA.

www.theiet.org

British Library Cataloguing in Publication Data
A catalogue record for this product is available from the British Library

ISBN 978-1-83953-149-1 (hardback)
ISBN 978-1-83953-150-7 (PDF)

Typeset in India by Exeter Premedia Services Private Limited

Contents

Preface xvii
About the Editors xxiii

1 Overview on design and manufacturing of assembled composite aerostructures 1
Marco Barile, Giuseppe Barile, Leonardo Lecce, and Giuseppe Petrone

1.1 Introduction 1
1.2 General philosophy for strength analysis of primary aerospace composite structures 1
1.2.1 Overview on structural analysis methods 2
1.3 Fastened joints 6
1.3.1 Load distribution in composite joints 6
1.3.2 Composite material failure at joint 6
1.4 Bonded joints 7
1.4.1 Types of failure in adhesive bonded joints 8
1.4.2 Stress analysis of adhesive bonded lap joints 9
1.5 Fabrication of bonded joints 10
1.5.1 Joints in space industry 15
1.5.2 Joints in aviation industry 20
1.6 Joint assembly 24
1.6.1 Significance of bond-line control 25
1.6.2 The need for bond-line control 25
1.6.3 Advantages and disadvantages of adhesive bonding 26
1.7 Conclusions 26
References 27

2 Processing of polymer composites: autoclave and microwave energy approaches 29
T Lachana Dora and Radha Raman Mishra

2.1 Introduction 29
2.2 Fundamentals 31
2.2.1 Basic principles 31
2.2.2 Mathematical models 34
2.2.3 Experimental details 36
2.3 Challenges in autoclave and microwave curing 39
2.4 Concluding remarks 40
References 40

3 Industry 4.0 for composites manufacturing **43**
Miroslav Stojkovic and Cristian Lira

3.1 Introduction 43
3.2 Composites context of Industry 4.0 44
3.3 Literature review 45
3.3.1 Composites technology 46
3.3.2 Business challenges 49
3.3.3 Industry 4.0 51
3.4 Critical assessment 55
3.4.1 Key trends 55
3.4.2 Gaps and research aim refinement 56
3.5 Concluding remarks 57
References 57

4 Development of fibre-reinforced polymer composites through direct digital manufacturing **61**
Ramesh Kumar Nayak and Hamed Yazdani Nezhad

4.1 Introduction 62
4.2 Fiber-reinforced AM composites 62
4.3 Fused deposition modeling 68
4.4 Selective laser sintering (SLS) 71
4.5 Mechanical properties 74
4.5.1 Tensile strength 74
4.5.2 Fatigue strength 75
4.5.3 Creep deformation 76
4.5.4 Fracture toughness 77
4.5.5 Machine learning 77
4.6 Applications 78
4.7 Sustainability 79
4.8 Challenges and future prospective 82
References 85

5 Joining and repair of resin-infused, continuous fibre-reinforced, thermoplastic acrylic-matrix composites for extended applicability **93**
Winifred Obande and Dipa Ray

5.1 Introduction 93
5.2 Liquid acrylic resin-based composites 94
5.2.1 Manufacturing 94
5.2.2 Comparison with conventional thermoset composites 96
5.3 Joining of acrylic-matrix composites 96
5.4 Thermoplastic acrylic-matrix fibre-metal laminates 102
5.5 End-of-life opportunities with acrylic-based composites 103
5.5.1 Applicability of recyclate matrix 106
5.5.2 Repairability of acrylic-matrix composites 107
5.5.3 Reshapability of acrylic-matrix composites 107

5.6 Future opportunities 109
5.7 Conclusion 110
References 111

6 Aerospace composites' repair: integrated processes' feasibility 117
Dileep Yaswanth Pasupuleti, Gopalakrishnan Kamalakannan, Guillermo Garcia Del Valle, Larissa Reuter, Laurine Dehée, Lucas Mestre, and Hamed Yazdani Nezhad

6.1 Introduction 118
6.1.1 Aim of the project 120
6.2 Structure and methodology 120
6.2.1 Methodology 121
6.2.2 Literature review 123
6.2.3 Damage in composite structures 123
6.2.4 Main causes and associated costs of damages 125
6.2.5 Typical damages in composites 127
6.3 NDI techniques 128
6.3.1 Ultrasonic testing (UT) 129
6.3.2 Thermography 130
6.3.3 Shearography 132
6.3.4 Combination of thermography and shearography 133
6.3.5 Other NDI techniques 134
6.3.6 Post repair inspection 135
6.3.7 Material removal 135
6.3.8 Surface preparation 137
6.3.9 Surface treatments 138
6.4 Material and conditions 140
6.4.1 Hard and soft patch 141
6.4.2 To bond or to bolt? 145
6.4.3 Quality control 146
6.4.4 Structural health monitoring 147
6.4.5 Repair certification 149
6.5 Needs assessment 151
6.5.1 Available systems 151
6.5.2 Gap analysis 152
6.5.3 Situation of the composite industry in the UK 152
6.6 Feasibility 153
6.6.1 NDI system 153
6.6.2 Commercial options 154
6.6.3 Machinable area 155
6.6.4 Material removal 156
6.6.5 Conventional machining 157
6.6.6 Laser system 157
6.6.7 3D scanning 158
6.6.8 Commercial options 159
6.6.9 Patch 160

6.6.10 Patch design 161
6.6.11 Patch production 162
6.6.12 Adhesives 163
6.6.13 On-site curing 163
6.7 Implementation 165
6.7.1 Experimental testing 165
6.8 Conclusion 186
References 189

7 Augmented reality-equipped composites bonded repair 197
Xi Wang, Hamed Yazdani Nezhad, Samuel Court, Bikram Thapa, and John Erkoyuncu

7.1 Introduction 198
7.1.1 Project background 198
7.1.2 Previous group project outcome 198
7.1.3 Gap analysis 198
7.1.4 Aim and objectives 199
7.2 Methodology 199
7.3 Literature review 200
7.3.1 Defects and damages in composite materials and structures 200
7.3.2 Manufacturing defects 200
7.3.3 Common service-life damages 200
7.3.4 Alternative classification of composite impact damage 202
7.3.5 Composite material repairing technique 202
7.3.6 Classification of composite material repairing method 202
7.3.7 Scarf-based repair process 202
7.4 Augmented reality 210
7.4.1 Augmented reality technique 210
7.4.2 Application of augmented reality into aircraft industry 211
7.5 Concept design 212
7.5.1 Project scope selection 212
7.5.2 Reparation method 213
7.5.3 Target process 213
7.5.4 Scenario design 214
7.5.5 Panel 214
7.5.6 Patches 214
7.5.7 Bagging layers design 214
7.5.8 Other assisting materials and equipment 215
7.5.9 Overall vacuum bagging system 216
7.5.10 Functionality design 216
7.6 Implementation and result 217
7.6.1 Hardware configuration 217
7.6.2 Developing environment configuration 217
7.6.3 Coding language 218

7.6.4 Developing process 218
7.6.5 Program validation 221
7.7 Discussion 224
7.7.1 Strengths 224
7.7.2 Challenges 224
7.7.3 Future work 225
7.7.4 Prospect 225
7.8 Conclusion 226
References 227

8 3D printing of multi-material polymer composite systems 231
Feroz Khan and Hamed Yazdani Nezhad

8.1 Introduction 231
8.2 Theoretical background 233
8.2.1 Material science 240
8.2.2 Polymer blends 241
8.2.3 Fillers and reinforcements 242
8.2.4 Particulates 245
8.2.5 Metallic polymer composites 248
8.2.6 Ceramic polymer composites 252
8.2.7 Carbon polymer composites 254
8.2.8 Fibres and whiskers 254
8.2.9 Final impressions 260
8.3 Objectives 261
8.4 Numerical simulation 262
8.4.1 Schematic 262
8.4.2 Parameters 264
8.4.3 Boundary conditions 266
8.4.4 Mesh discretisation 272
8.4.5 Estimating thermo-physical properties 273
8.5 Results and discussion 275
8.5.1 Estimated thermo-physical properties of specimens 275
8.5.2 Slicing 275
8.6 Conclusion 277
References 278

9 3D printing of composites for space applications 283
Sung Wook Paek and Sivagaminathan Balasubramanian

9.1 Introduction 283
9.2 3D-printed structures 284
9.2.1 Heat shields for suborbital flight 284
9.2.2 Radiation shields in low-Earth orbits or deep space 287
9.2.3 Issues in printing and assembling structural parts 291

9.3 3D-printed electronics 298
9.3.1 Traces and substrates 298
9.3.2 Passive components 300
9.3.3 Active components 303
9.4 3D-printed batteries 304
9.5 3D-printed devices for medical and life support purposes 307
9.6 Conclusion 309
Acknowledgements 310
References 310

10 Development and manufacturing of thermoplastic composite booms for drag augmentation system of a small satellite 321
Chun Ho Yau, Zaria Serfontein, and Hamed Yazdani Nezhad

10.1 Introduction 322
10.1.1 Background 322
10.1.2 Cranfield University's de-orbit mechanism 322
10.1.3 Project's aims and objectives 324
10.2 Designs of booms 325
10.2.1 TRAC boom 325
10.2.2 STEM boom 326
10.2.3 CTM booms 327
10.3 Additive manufacturing 333
10.3.1 Markforged's FDM printer and materials 334
10.4 Design of the CTM boom 335
10.4.1 Onyx and CuBe 336
10.4.2 Cross-section geometry 336
10.4.3 Finite element analysis of CTM boom 339
10.4.4 Three-point bending test methodology 343
10.5 Manufacturing of the CTM booms 349
10.5.1 Horizontal, half profile 350
10.5.2 Vertical, full profile 351
10.5.3 Horizontal, full profile 351
10.6 Conclusion and future work 353
References 353

11 Adhesively bonded polymer composite joints 355
E. A. S. Marques, R. J. C. Carbas, A. Akhavan-Safar, and L. F. M. da Silva

11.1 Introduction 355
11.2 Bonding of composite materials 356
11.3 Composite adherend modifications 359
11.3.1 Transverse toughened and reinforced adherends 359
11.3.2 Functionally graded substrates 363
11.4 Bi-adhesive joints 365
11.4.1 Manufacturing techniques 366

11.4.2 Bi-adhesive SLJs 367
11.4.3 Other types of joints 370
11.5 Composite adhesives with natural fibres 374
11.5.1 Surface preparation 374
11.5.2 Cork particles 375
11.5.3 Date palm tree fibres 376
References 377

12 Design principles and recent developments in adhesively bonded joints of fibre-reinforced plastic composite structures 385
Roya Akrami, Sakineh Fotouhi, Ali Tabatabaeian, Saeid Lotfian, Mahdi Bodaghi, and Mohammad Fotouhi

12.1 Introduction 385
12.2 Design considerations and effective parameters in ABJs in FRP composites 387
12.2.1 Basics of adhesive bonding design and manufacturing considerations 388
12.2.2 Joint configuration: geometries and dimensions 389
12.2.3 Stress distribution and failure modes 389
12.2.4 Adhesive type 390
12.2.5 Surface preparation 392
12.2.6 Environmental parameters 394
12.2.7 Ease of assembly and costs 394
12.3 Characterisation methods for adhesively bonded composite-to-composite joints 395
12.3.1 Tensile test 395
12.3.2 Shear test 395
12.3.3 Peel test 396
12.3.4 Wedge cleavage test 396
12.3.5 Double cantilever beam (DCB), end-notched flexure (ENF) and mixed-mode tests 396
12.4 Analytical and finite element models for analysis of adhesively bonded composite-to-composite joints 396
12.4.1 Analytical analysis 396
12.4.2 Finite element analysis 400
12.5 Design principles and recent developments in high-performance composite-to-composite ABJs 402
12.6 Conclusions 406
References 407

13 Mechanical degradation of composite bonded joints subjected to environmental effects 413
Sonia Ruth Rincon Urbina, David Ayre, and Hamed Yazdani Nezhad

13.1 Introduction 414
13.1.1 Objectives 414

13.2 Literature review 415
13.2.1 Adhesive joints 415
13.2.2 Design aspects 415
13.2.3 Manufacture process 419
13.2.4 Durability 421
13.2.5 Non-destructive testing 425
13.2.6 Damage prediction 427
13.2.7 Overview 434
13.3 Methodology, materials and methods 435
13.3.1 Research methodology 435
13.3.2 Adhered manufacturing 437
13.3.3 SJL manufacturing 437
13.3.4 Hygrothermal cycles 439
13.3.5 Bulk adhesive specimens manufacturing 439
13.3.6 Materials' characterization 444
13.3.7 Mechanical properties 447
13.4 Experimental results 451
13.4.1 Introduction 451
13.4.2 Moisture characterization 451
13.4.3 Adhesive bonded SLJ test 463
13.5 Degradation modelling of adhesively bonded SLJ 474
13.5.1 Finite-element analysis 474
13.6 Discussion 487
13.6.1 Introduction 487
13.6.2 Moisture absorption 487
13.6.3 Moisture effect adhesive mechanical properties 491
13.6.4 Mechanical test results 492
13.6.5 Non-destructive inspection adhesive bonded joints 494
13.6.6 Degradation modelling 494
13.7 Conclusions and contribution 495
References 497

14 Performance of aerospace composites in the presence of process-induced defects 507
Dimosthenis Stratakis, David Ayre, and Hamed Yazdani Nezhad

14.1 Introduction 507
14.1.1 Aim 508
14.1.2 Objectives 509
14.2 Materials and manufacturing 509
14.2.1 Materials 509
14.2.2 Manufacturing 510
14.3 Testing 523
14.3.1 Lap shear test 523
14.3.2 NDI 525
14.4 Results and discussion 527
14.4.1 Lap shear test 527

14.4.2 NDI 532
14.4.3 Optical microscopy 533
14.5 Conclusions 534
References 535

15 Interleaving in composites **537**
Mehdi Yasaee

15.1 Introduction 537
15.2 Measuring interleaving properties 539
15.3 Interleaving types and performance 541
15.3.1 Self-same matrix resin 541
15.3.2 Summary 548
15.3.3 Interleaving applied to stiffeners and adhesive joints 549
15.3.4 Selective interface interleaving 551
15.3.5 Summary 554
References 555

16 A deep learning-based tool to predict delamination induced interlaminar stresses in composite structures **559**
Marco Zacaria Wahaba, Hamed Yazdani Nezhad, Yazhini Chitra Pradeep, and Sathiskumar Anusuya Ponnusami

16.1 Introduction 559
16.2 Methodology 561
16.2.1 Design of experiments 561
16.2.2 Database development 563
16.2.3 Machine learning modelling 564
16.3 Results 569
16.3.1 Stress distributions within the composite panel 569
16.3.2 Performance measure plot 569
16.3.3 Training state plot 569
16.3.4 Error histogram 570
16.3.5 ML predictions and regression plot 570
16.4 Discussion 571
16.5 Conclusions 573
References 573

17 Damage assessment of composites based on non-destructive pulsed thermographic inspection **575**
Yifan Zhao, Weixiang Du, Lawrence Tinsley, Sri Addepalli, Jörn Mehnen, and Rajkumar Roy

17.1 Introduction 575
17.2 Experimental data 577
17.2.1 Specimen 577
17.2.2 Data collection 579
17.2.3 Temperature decay curve 580

17.3 Coefficient clustering analysis 581
17.4 Results and discussion 585
17.4.1 Damage detection 585
17.4.2 Damage measurement 586
17.5 Conclusions 590
Acknowledgements 590
References 591

18 Augmented reality-equipped composite monitoring 593
Idriss Ben Jamaa and Hamed Yazdani Nezhad

18.1 Introduction 594
18.2 Theory: measurement, simulation, and visualization 595
18.2.1 Review of literature and research 595
18.2.2 Stress and strain calculation through surface data 597
18.2.3 Image data processing and 3D modelling 599
18.3 Methodology 600
18.3.1 Deformation type 600
18.3.2 Composite laminates specifications 603
18.3.3 Photogrammetry measurements 603
18.3.4 LIDAR measurements 605
18.3.5 Point cloud denoising and distance measurements 606
18.3.6 Abaqus and MATLAB® simulation 607
18.3.7 AR Visualization 608
18.4 Results and discussion 609
18.4.1 Photogrammetry results 610
18.4.2 C2C from LIDAR and Abaqus results 612
18.4.3 Difficulties with MATLAB® code 615
18.4.4 Recommendations and future work 620
18.4.5 Scope of application 622
Appendix A 622
References 623

19 Energy harvesting and self-sensing multi-functional polymer composites 625
Tasnuva Khaleque, Hamed Yazdani Nezhad, and Marcin Słoma

19.1 Energy harvesting and self-sensing 626
19.2 Piezoelectric effect 627
19.2.1 Piezoelectric materials 630
19.3 Energy harvesting polymer composites 639
19.3.1 Energy harvesting fibre-reinforced polymer composites 639
19.3.2 Energy harvesting polymer nanocomposites 643

19.4 Self-sensing and health monitoring polymer composites 649
19.5 Conclusions 652
References 653

20 Tailoring thermo-mechanical properties of hybrid composite-metal bonded joint 667
Xiaolong Zhang, Vijay Kumar Thakur, Adrianus Indrat Aria, and Hamed Yazdani Nezhad

20.1 Introduction 668
20.1.1 Aim 669
20.1.2 Objectives 669
20.2 Literature review 670
20.2.1 Carbon nanotube 670
20.2.2 Composite-metal adhesively bonded joint 671
20.2.3 Effects of CNTs on thermo-mechanical performance 674
20.2.4 CNTs dispersion 676
20.2.5 Mechanism of joint failure by CTE mismatch 681
20.2.6 Analytical calculations 682
20.2.7 Methods of thermal strain measurement 683
20.3 Specimens manufacturing 685
20.3.1 Material 685
20.3.2 Specimens manufacturing 686
20.4 Testing 692
20.4.1 Thermal strain measurement 692
20.4.2 Scanning electron microscopy 694
20.4.3 Raman spectroscopy 695
20.5 Results and discussion 695
20.5.1 Coefficient of thermal expansion 695
20.5.2 Scanning electron microscopy 700
20.5.3 Raman spectroscopy 701
20.5.4 Bond deficiency on specimens 702
20.6 Conclusion and further work 703
References 704

21 High-performance nanocomposites for strain self-sensing applications in composite joints 709
Jim Nourry, Tadej Bregar, Marek Burda, Vijay Kumar Thakur, and Hamed Yazdani Nezhad

21.1 Introduction 710
21.1.1 Structure of carbon nanotubes 710
21.1.2 Electromechanical properties 711
21.1.3 CNTs for strain sensing 711

21.2 Materials and methods 714
21.2.1 CNT fabrication 714
21.2.2 Sample manufacturing 728
21.2.3 Tensile and electrical resistance testing 732
21.2.4 Scanning electron microscopy 734
21.2.5 Raman spectroscopy 735
21.3 Results and discussion 735
21.3.1 Carbon nanotubes characterisation 735
21.3.2 Strain sensing 738
21.4 Conclusion 744
21.5 Future work 745
Appendix A: Tensile tests 746
Appendix B : Load vs displacement curves 749
References 751

Index 757

Preface

An urgent situation for collecting knowledge, skills, challenges and state-of-the-arts motivated the editors and contributed authors to embark on the development of the current book at the wake of substantial need for multi-material and highly integrated polymer composite materials for high-performance applications driven by rapid and modern societal, industrial and economical establishment towards advanced, multifunctional, eco-friendly and sustainable components across a hierarchy of scales and technologies. Cross-disciplinary integrations (material, chemistry, electronic and digitalisation) are present on the establishment within the last three decades with distant disciplines mapped across electronic devices to materials sciences in sustainable and functional composites developments, across human error and operational managements in composites assembly, joining and integration, and across data-analytics driven approaches to composites' performance, their process digitalisation technologies and socio-economic challenges in addressing the ongoing industrial revolution and evolution. Due to very multidisciplinary nature of the subject of this book, *Composites Assembly for High Performance Fastener-less Structures*, together with our distinguished authors in their respective fields, we have provided chapters from pioneering and active researchers, managers and entrepreneurs across the globe and in world-leading research and development organisations, industries and universities to pull together a broad variety of disciplines involved in the subject that would benefit both the academic and industrial frameworks under continuous developments, and furthermore would act as a training setting benefitting current and future generation of trainees in the two.

The book contents are extensive however it is mainly divided into three major parts of (i) scientific and technical processes and state-of-the-arts involved in composites assembly, (ii) assurance of reliable processes and assembled composites using detection and instrumentation technologies, and (iii) structural integrity assessments of the composites. Last but not least, the book is wrapped up with the provision of scientific researches and innovation in the area of multifunctional, multi-material composite systems and their novel processes.

The editors have been overwhelmed by the supports, guidance and patience received from our dedicated authors, the IET office, and numerous colleagues, experts and pioneers in the field of high performance composites, and are grateful to all who commented, advised and corrected our course of actions prior and during the developmental journey of our comprehensive book within the last few years.

Hamed Yazdani Nezhad, Editor
Vijay Kumar Thakur, Co-Editor

List of Authors As They Appear Throughout Chapters

No.	Chapter	Author	Affiliation(s)	Country
1	1	Marco Barile	NOVOTECH Aerospace Advanced Technology s.r.l.	Italy
2	1	Giuseppe Barile	NOVOTECH Aerospace Advanced Technology s.r.l.	Italy
3	1	Leonardo Lecce	NOVOTECH Aerospace Advanced Technology s.r.l.	Italy
4	1	Giuseppe Petrone	Department of Industrial Engineering – Aerospace section, Università degli Studi di Napoli Federico II	Italy
5	2	T. Lachana Dora	Department of Mechanical Engineering, Birla Institute of Technology and Science Pilani	India
6	2	Radha Raman Mishra	Department of Mechanical Engineering, Birla Institute of Technology and Science Pilani	India
7	3	Miroslav Stojkovic	Airborne UK	UK
8	3	Cristian Lira	National Composites Centre	UK
9	4	Ramesh Kumar Nayak	Department of Materials and Metallurgical Engineering, Maulana Azad National Institute of Technology	India
10	4	Hamed Yazdani Nezhad	Aeronautics & Aerospace Research Centre Department of Mechanical Engineering & Aeronautics City, University of London	UK
11	5	Winifred Obande	School of Engineering, Institute for Materials and Processes, The University of Edinburgh	UK
12	5	Dipa Roy	School of Engineering, Institute for Materials and Processes, The University of Edinburgh	UK
13	6	Dileep Yaswanth PASUPULETI	School of Aerospace, Transport and Manufacturing, Cranfield University	UK
14	6	Gopalakrishnan KAMALAKANNAN	School of Aerospace, Transport and Manufacturing, Cranfield University	UK

No.	Chapter	Author	Affiliation(s)	Country
15	6	Guillermo GARCÍA DEL VALLE	School of Aerospace, Transport and Manufacturing, Cranfield University	UK
16	6	Larissa REUTER	School of Aerospace, Transport and Manufacturing, Cranfield University	UK
17	6	Laurine DEHÉE	School of Aerospace, Transport and Manufacturing, Cranfield University	UK
18	6	Lucas MESTRE	School of Aerospace, Transport and Manufacturing, Cranfield University	UK
19	7	Xi Wang	School of Aerospace, Transport and Manufacturing, Cranfield University	UK
20	7	Samuel Court	School of Aerospace, Transport and Manufacturing, Cranfield University	UK
21	7	Bikram Thapa	School of Aerospace, Transport and Manufacturing, Cranfield University	UK
22	7	John Erkoyuncu	School of Aerospace, Transport and Manufacturing, Cranfield University	UK
23	8	Feroz Khan	Aeronautics and Aerospace Research Centre, Department of Mechanical Engineering and Aeronautics, City, University of London	UK
24	9	Sung Wook Paek	City, University of London	UK
25	9	Sivagaminathan Balasubramanian	Samsung R&D Institute India	India
26	10	Chun Ho (Craig) Yau	Aeronautics and Aerospace Research Centre, Department of Mechanical Engineering and Aeronautics, City, University of London	UK
27	10	Zaria Serfontein	School of Aerospace, Transport and Manufacturing, Cranfield University	UK

No.	Chapter	Author	Affiliation(s)	Country
28	11	E.A.S. Marques	Institute of Science and Innovation in Mechanical and Industrial Engineering (INEGI)	Portugal
29	11	R.J.C. Carbas	Institute of Science and Innovation in Mechanical and Industrial Engineering (INEGI)	Portugal
30	11	A. Akhavan-Safar	Institute of Science and Innovation in Mechanical and Industrial Engineering (INEGI)	Portugal
31	11	L.F.M. da Silva	Departamento de Engenharia Mecânica, Faculdade de Engenharia, Universidade do Porto	Portugal
32	12	Roya Akrami	Faculty of Engineering, University of Strathclyde	UK
33	12	Sakineh Fotouhi	School of Engineering, University of Glasgow	UK
34	12	Ali Tabatabaeian	School of Engineering, University of Glasgow	UK
35	12	Saeid Lotfian	Faculty of Engineering, University of Strathclyde	UK
36	12	Mahdi Bodaghi	Department of Engineering, School of Science and Technology, Nottingham Trent University	UK
37	12	Mohammad Fotouhi	Microlab, Faculty of Civil Engineering and Geosciences, Delft University of Technology	The Netherlands
38	13	Sonia Ruth Rincon Urbina	Aerospace Technology Research Center, Colombian Air Force, Valle del Cauca	Colombia
39	13	David Ayre	School of Aerospace, Transport and Manufacturing, Cranfield University	UK
40	14	Dimosthenis Stratakis	School of Aerospace, Transport and Manufacturing, Cranfield University	UK
41	15	Mehdi Yasaee	School of Aerospace, Transport and Manufacturing, Cranfield University	UK

No.	Chapter	Author	Affiliation(s)	Country
42	16	Marco Wahaba Zacaria	Aeronautics and Aerospace Research Centre, Department of Mechanical Engineering and Aeronautics, City, University of London	UK
43	16	Yazhini Chitra Pradeep	Humanising Autonomy Ltd., London	UK
44	16	Sathiskumar Anusuya Ponnusami	Aeronautics and Aerospace Research Centre, Department of Mechanical Engineering and Aeronautics, City, University of London	UK
45	17	Yifan Zhao	Through-life Engineering Services, School of Aerospace, Transport and Manufacturing, Cranfield University	UK
46	17	Weixiang Du	Through-life Engineering Services, School of Aerospace, Transport and Manufacturing, Cranfield University	UK
47	17	Lawrence Tinsley	Through-life Engineering Services, School of Aerospace, Transport and Manufacturing, Cranfield University	UK
48	17	Sri Addepalli	Through-life Engineering Services, School of Aerospace, Transport and Manufacturing, Cranfield University	UK
49	17	Jörn Mehnen	Through-life Engineering Services, School of Aerospace, Transport and Manufacturing, Cranfield University	UK
50	17	Rajkumar Roy	School of Mathematics, Computer Science and Engineering, City, University of London	UK
51	18	Idriss Ben Jamaa	Aeronautics and Aerospace Research Centre, Department of Mechanical Engineering and Aeronautics, City, University of London	UK

No.	Chapter	Author	Affiliation(s)	Country
52	19	Tasnuva Khaleque	Aeronautics and Aerospace Research Centre, Department of Mechanical Engineering and Aeronautics, City, University of London	UK
53	19	Marcin Słoma	Faculty of Mechatronics, Warsaw University of Technology, Warsaw	Poland
54	20	Xiaolong Zhang	Enhanced Composites and Structures Centre, School of Aerospace, Transport and Manufacturing, Cranfield University	UK
55	20	Vijay Kumar Thakur	Biorefining and Advanced Materials Research Centre, SRUC, Edinburgh	UK
56	20	Indrat Aria	Surface Engineering and Precision Centre, Cranfield University	UK
57	21	Jim Nourry	Enhanced Composites and Structures Centre, School of Aerospace, Transport and Manufacturing, Cranfield University	UK
58	21	Tadej Bregar	Heating and Process Engineering, Faculty of Mechanical Engineering, University of Ljubljana, Ljubljana	Slovenia
59	21	Marek Burda	Ampashield Ltd., Cambridge House Camboro Business Park, Cambridge, Cambridgeshire	UK

About the Editors

Hamed Yazdani Nezhad is an associate professor in aerospace structures at the Aeronautics and Aerospace Research Centre at City, University of London, with expertise in composite materials, multifunctional composites and composites assembly. He is a principal investigator for a number of the UK Engineering & Physical Sciences Research Council projects, and co-investigator in the European Commission ones, for research on multifunctional composites and electromagnetic field aided composites tailoring, sensing and manufacturing. He is a fellow of UK Higher Education Academy and a Fellow of IMechE and a member of the European Society for Composite Materials, with publications in the area of composites assembly and electromagnetic field-materials interaction for multifunctional composites.

Vijay Kumar Thakur is a professor and the founding head of the Biorefining and Advanced Materials Research Centre at SRUC, Edinburgh, UK. He also holds an adjunct professor position in Jiangsu University, Riga Technical University, University of Petroleum & Energy Studies (UPES), Chandigarh University, and is a visiting professor at Shiv Nadar University and visitor at Cranfield University, UK. Before commencing his tenure at SRUC, Vijay has previously held faculty positions at Cranfield University, Washington State University, USA, and Nanyang Technological University, Singapore. He received PhD in polymer chemistry (2009) and did his post-doctoral study in materials science & engineering at Iowa State University (USA). He has published over 350 SCI journal articles; 52 Books; 40 Book Chapters and 2 Patents. He sits on the editorial board of several SCI journals.

Chapter 1

Overview on design and manufacturing of assembled composite aerostructures

Marco Barile[1], Giuseppe Barile[1], Leonardo Lecce[1], and Giuseppe Petrone[2]

1.1 Introduction

The aim of this chapter is to provide an overview on the methodologies used in the design and best practices applied in the manufacturing of assembled composite primary structures in the aerospace field.

Section 2 provides an overview on the general philosophy applied in the strength analysis, how its objectives are covered by the main product design phases and approach used for un-notched and notched composite laminates.

Sections 3 and 4 deal with the analysis process of fastened and bonded joints; in particular, they illustrate how the load is distributed in different configurations, an overview on how the stress analysis can be performed and typical modes of failure at joint. Section 5 introduces the processes used for assembly composite structures; in particular, the three main categories of joining of composites with thermoset matrices as co-curing, co-bonding and secondary bonding. Furthermore, the key factors to control adhesively bonded joint, together with relevant test cases in the space and aviation industry, have been illustrated.

1.2 General philosophy for strength analysis of primary aerospace composite structures

Usually, a structure is classified into various grades, according to the consequences of its failure. A primary structure element is a critical part, the failure of which could have a catastrophic effect upon the aircraft and/or for which characteristics have been identified as critical for other safety reasons. It can be any detail, element or assembly that significantly contributes to carrying flight, ground, pressure or control

[1]NOVOTECH Aerospace Advanced Technology srl, via G.Pascoli 7, 80026 Casoria (NA), Italy
[2]Department of Industrial Engineering – Aerospace section, Università degli Studi di Napoli Federico II, via Claudio 21, 81025 Napoli, Italy

loads. More in detail, the primary structures are normally classified into different levels of criticality as a function of the following:

- consequences of their failure
- experience of the loading environment (e.g. high or low vibration, manoeuvre)
- stress level
- in-service failures of similar design structure
- type of load path (single or multiple load path)
- location where flaw detection would be difficult
- items/areas where flaw detection would be difficult
- corrosion sensitive areas

Typical strength analyses can be divided into four main phases, as summarized below.

1. **Conceptual phase**: This phase of the project deals with the preliminary sizing of the structural elements.
2. **Development phase**: During this phase, a more detailed analysis shall be performed; the associated load loop shall be an updated and revised (more reliable) issue.
3. **Certification phase**: This phase covers the final loop, where the primary structural element must be carefully analysed.
4. **Sustaining phase**: During this phase, the strength analysis activity has to cover mainly 'non-conformities' due to production and support of the in-service issues.

The general philosophy for each design phase to cover the strength analysis required is summarized in the flowchart shown in Figure 1.1.

According to the flowchart reported in Figure 1.1, the objectives of the strength analysis are listed below :

- fulfilment of certification requirements
- weight optimization
- basic reference for damage tolerance and durability analyses
- drawing release and approval reference for non-conformity analysis
- qualification of finite method element
- preparation of structural tests

1.2.1 Overview on structural analysis methods

1.2.1.1 Section properties

In stress analysis, the section 'properties' is expressed by equations that relate applied loads and section geometry to the strength and stiffness. The characteristics that are associated with the cross section of the part under analysis are known as

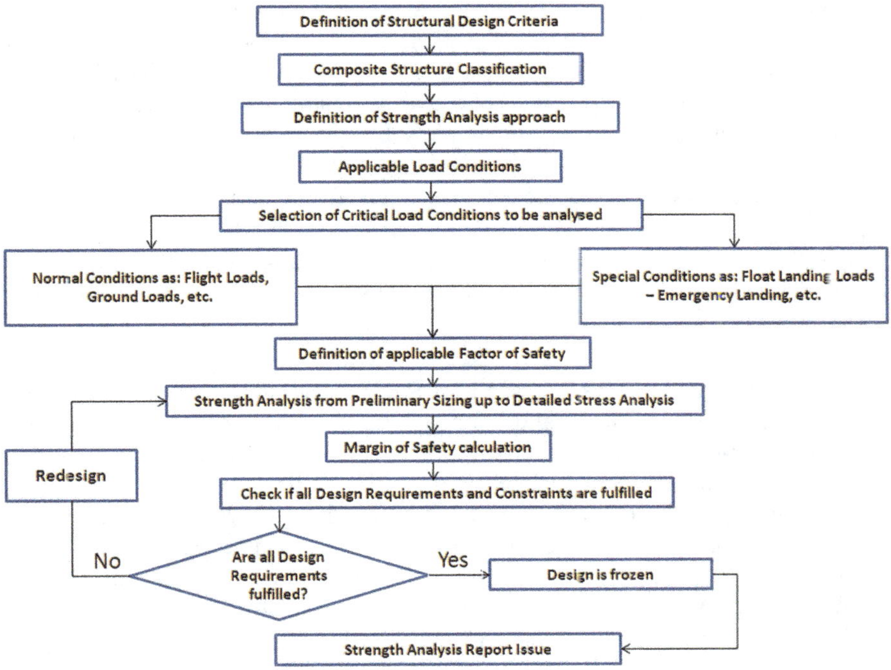

Figure 1.1 *Strength analysis general philosophy*

section properties. The most common properties that significantly influence the section property calculation are the following:

- geometry to be used for the strength analysis
- effective skin width estimation
- principal axes of rotation of unsymmetrical sections
- beam curvature effect

1.2.1.2 Theories of material failures for composite structures

This section presents an approach to the margin of safety calculation at the limit and ultimate loads of composite structures not subjected to failure of buckling or crippling (local and global instability).

The complexity of the stress distribution throughout the fibre and the matrix regions of the plies of a composite laminate means that precise analysis methods are not possible. Instead, the strength analysis of composites has developed generally into two methods.

- **Lamina level.** Average values of the stress (or strain) components within each ply are calculated using lamination theory (plane stress assumptions) and then compared against either the lamina allowables or failure criteria. This approach is called the 'First Ply Failure' (FPF), and it is where the failure of any lamina is considered as the complete failure of the laminate (a conservative approach).

- **Laminate level**. Average values of the stress or strain components in the laminate are compared to the laminate allowable (instead of lamina allowable as in the lamina level method). The laminate allowable must be derived from an experimental test campaign covering all applicable materials and laminates.

The accuracy of lamina failure criteria is very variable, and all of them have limitations. The most common lamina failure criteria may be grouped into two main categories:

1. Failure mode based (Max Stress or Max Strain and Hashin): This criterion treats each identifiable physical failure mode, such as fibre direction tensile failure and matrix-dominated transverse failure, separately. This criterion identifies clearly the laminate failure mode.
2. Empirical (Hill, Tsai-Wu and Hoffman): This criterion consists of a polynomial combination of the three stress (σ_x, σ_y, τ_{xy}) or strain (ε_x, ε_y, γ_{xy}) in a ply and attempts to combine the effects of several different failure mechanisms into one function. They are therefore less physically representative because they do not identify the failure mode of the laminate.

These failure criteria are used to find the first lamina to fail in the laminate, an approach called the FPF. The calculation of the residual strength of the remaining plies 'Sequential Ply Failure' is not normally used in the composite design because it may degenerate into an unreliable or un-conservative estimation of the laminate strength.

1.2.1.3 Analysis of un-notched and notched composite laminates

The methodology for the strength analysis of un-notched composite laminates is reported in Figure 1.2.

With regard to the notched composite laminates, the methodology proposed to cover the strength analysis is presented below and includes the following cases:

Laminates with stress concentrations: The presence of significant discontinuities (mainly holes) in a composite structure introduces stress concentrations (as shown in Figure 1.3). These stress concentrations are significantly influenced not only by the geometry but also by laminate composition and by plies orientation (laminate stiffness). As known, composites are brittle materials and thus they are unable to yield and redistribute loads near to stress concentration when close to failure. Thus, the high stress concentration at the edge of a hole will always initiate the failure. Usually, it is assumed that the failure will occur when the maximum tensile stress at the edge of the hole equals the strength of the un-notched material. The calculation of the maximum stress at the hole edge should be performed using a dedicated FE model.

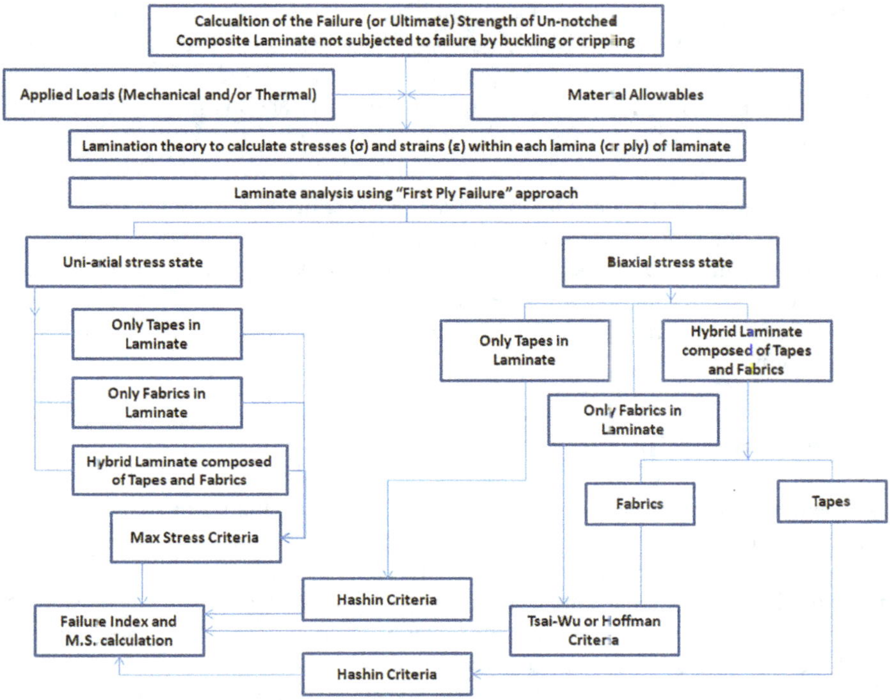

Figure 1.2 Strength analysis of un-notched laminates

This method is slightly conservative because it does not take into account the relief of the stress concentration effect at the edge of the hole due to matrix damage effects. This localized damage reduces the material stiffness and diminishes and spreads the stress concentration effects.

Laminates with impact damage: Impact damage can be created during manufacturing operations, foreign objects, maintenance procedures, etc. These types of impacts are normally classified as barely visible impact damage (BVID) and clearly visible impact damage (CVID). More in detail, the definitions are as follows:

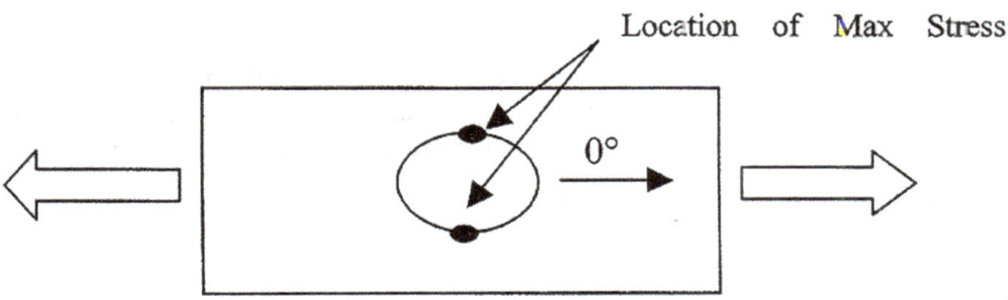

Figure 1.3 Maximum stress at the hole edge

- BVID is a type of damage that is so small that the probability of it being detected is very low. and thus it is considered present in the structure for its entire life
- CVID is a type of impact damage that is sufficiently large that there is a high probability of it being detected and thus adequate actions may be taken (i.e. repair of damage)

Impact damages are normally included in the certification strength tests to demonstrate the capability of the impacted structure to withstand the design loads. If during the design phase, impacted test data are not available, the following approach is suggested to cover the damage tolerance of the composite part. This approach is based on the experience of the majority of aerospace companies, and it is applicable to BVID only.

1.3 Fastened joints

Fastened joints are probably the most common source of failure in airframe structure; therefore, it is most important that all aspects of the design are given consideration when making the structural analysis. The weight of a fastened joint is usually small in relation to its importance; hence, conservative analysis should be used in its design. In this section, fastened joints for composite materials will be covered, and it will include joints involving, solid and blind rivets, nuts, bolts, inserts and lugs. The following flowchart (Figure 1.4) shows the analysis process of a fastened joint.

1.3.1 Load distribution in composite joints

In general, the load distribution in joint made from composite materials is more complex with respect to metallic structures, and this is due to the nature of the material, as shown in Figure 1.5. As known, in metallic joints, at ultimate loads, the material will deform plastically and redistribute the load equally to the fasteners, while composite joints have very little capacity to redistribute the loads due to their brittle nature. Thus, the majority of the load will go to the end of fasteners or to a fastener that is tighter in its hole.

The yielding of metallic materials permits ultimate strengths to be so high that the metal fatigue of the joint becomes the designing factor. However, in composites, the residual stress concentrations at failure prevent high static joint efficiencies. Usually, joints in metallic materials can have strengths up to 70% of the basic material allowable, whereas even the very best fastened joints in composite materials will still impose a loss of strength of about half the basic material allowable (although there is some evidence to indicate that interference-fit fasteners can alleviate this reduction) (Figure 1.6).

1.3.2 Composite material failure at joint

The composite designer must consider the following relevant stresses in mechanical joints as reported in Table 1.1.

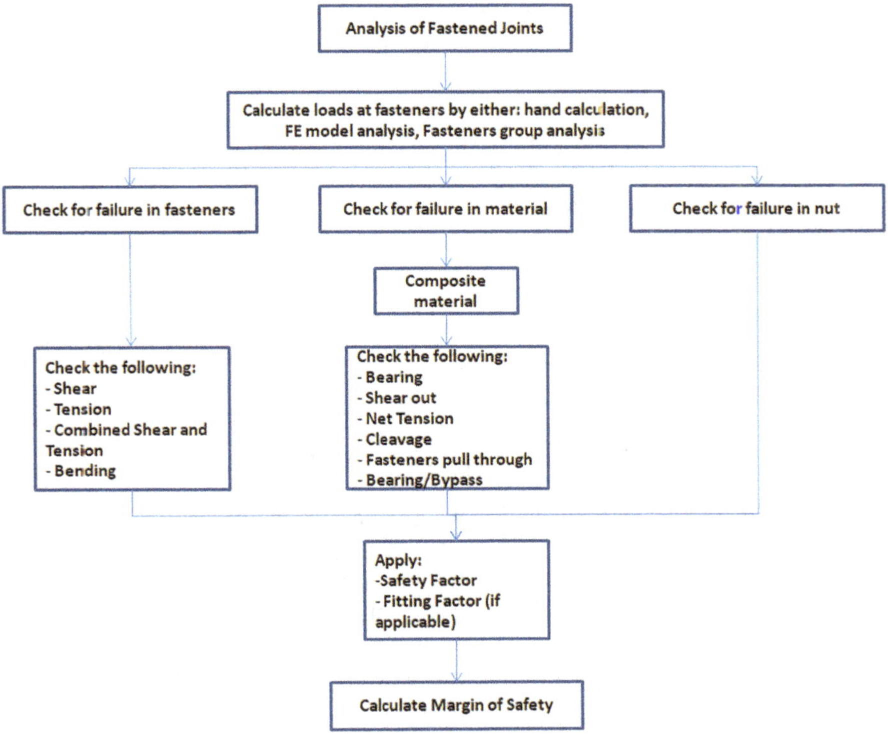

Figure 1.4 Analysis of a fastened joint

1.4 Bonded joints

Adhesive bonding of composites is often the best way to permanently join two components together but in practice is often backed up with rivets. For maximum structural efficiency, bonded joints should be designed in accordance with the following general principles.

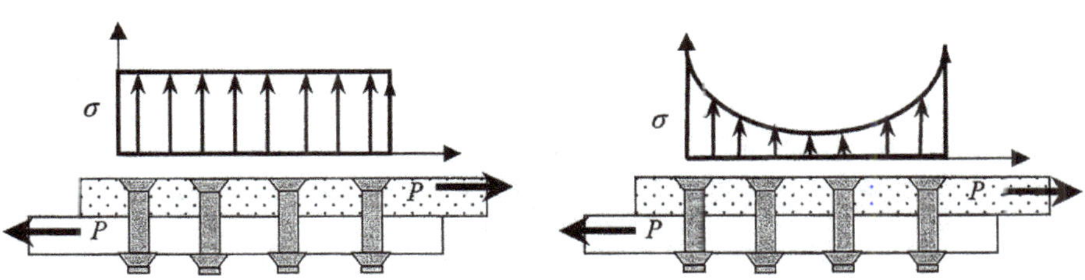

Figure 1.5 Comparison of load distribution in metallic (a) and composite (b) joints

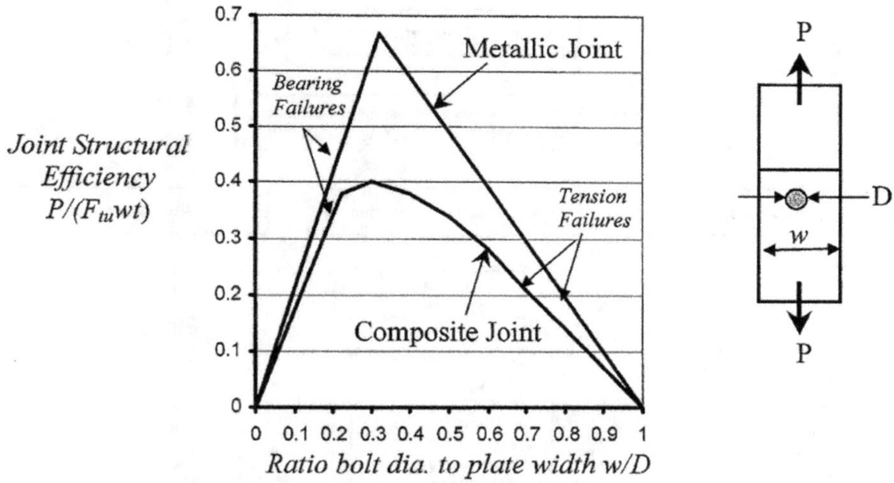

Figure 1.6 *Typical joint efficiency for metallic and composite joints [1]*

- Double-lap joints are more efficient from the structural point of view with respect to single-lap joints (Figure 1.7), mainly because they increase the bonded area and minimize the eccentricities.
- Joints with short overlap lengths are more efficient for reducing peak shear stress.
- Tests have generally shown that the highest strength is reached when (*Et*) value of the two adherents are similar ($E_1t_1 \approx E_2t_2$).
- Minimize joint eccentricity.
- Bonding dissimilar adherents with wide variation in their material coefficients of linear expansion should be avoided whenever possible.
- Adhesive bond strength decreases drastically with an increase in temperature.
- Adhesive ductility improves the shear load distribution by reducing stress peaking at the end of the joint.
- To reduce the effect of peeking and eccentricity, taper the end of adherents as follows (Figure 1.8).

1.4.1 Types of failure in adhesive bonded joints

The primary function of a bonded joint is to transfer load by shear or tension because bonded joints are inherently weak in peel and cleavage. However, some joints will develop peel stresses either because of eccentricities in the load path or due to externally applied peel loads. Figure 1.9 shows the possible failure modes of a bonded joint which in reality may occur in various combinations.

Table 1.1 Failure modes in bolted joints

Failure modes in bolted joints

Bearing: The bearing stress, σ_b, is the load P divided by the projected transverse cross-sectional area of the holes $\sigma_b = \frac{P}{Dt}$ – where D represents the nominal hole diameter and t is the material thickness. Bearing failures typically consist of crushing of a localized region of the composite laminate that is in contact with the fastener.

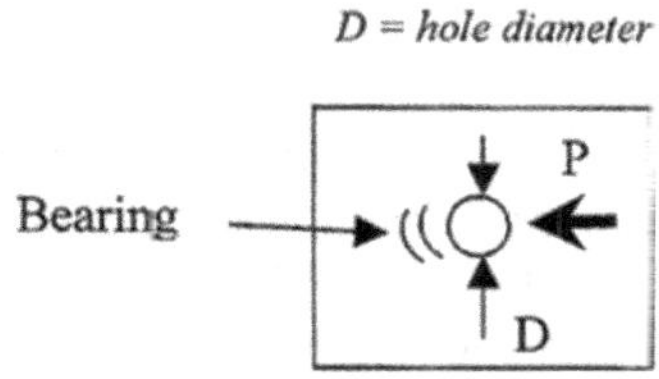

Shear out: The shear out stress is determined by the longitudinal shear surfaces and is given as $\sigma_{SO} = \frac{P}{2ted}$ – where *ed* is the edge distance, i.e. distance from the edge of the plate to a point on the hole edge located at 40° from the centre of the hole and t is the material thickness.

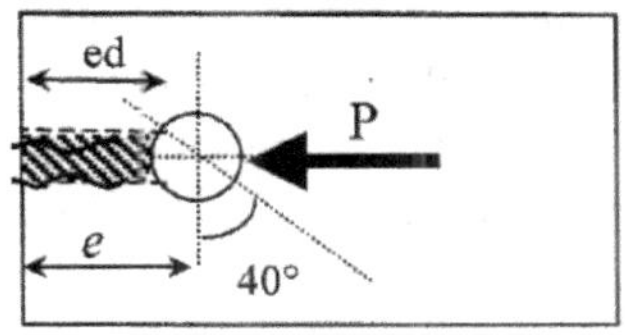

Net tension: The net section stress is $\sigma_N = \frac{P}{2tS}$ – where S is the minimum fastener side distance (see figure where $S < W$) and t is the material thickness.

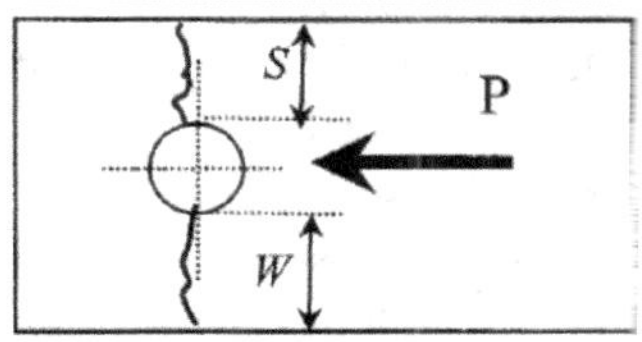

Fastener pull through: It is possible especially for composite materials with countersunk fasteners. This type of failure is common in structures subjected to internal pressure or diagonal tension after buckling.

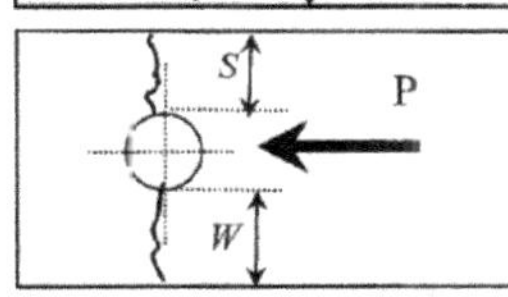

Cleavage: The shear out failure can also be sometimes characterized by a single cleavage (combined tension and shear) failure, where the apparent laminate transverse tensile strength is less than the corresponding in-plane shear strength.

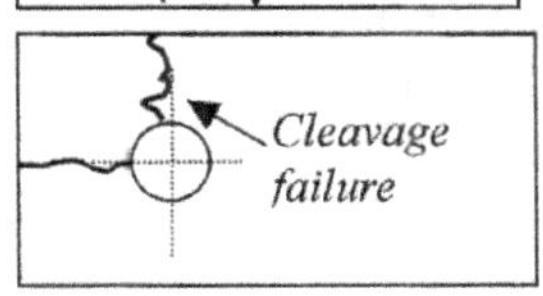

Bearing bypass: In composite joints, the stress concentration due to the combination of the bearing stress and bypass stress is more critical than just the bearing stress alone. This is due to the brittle nature of composite materials.

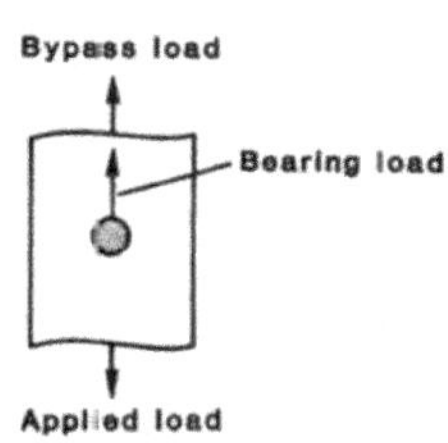

1.4.2 Stress analysis of adhesive bonded lap joints

Single-lap joints are the most common type of joint to be encountered in connecting two components. The bond area of a lap joint is equal to the lap length times the width of the joint. However, it should be noted that the bond area is not necessarily the load-carrying area of the joint. For example, Figure 1.10 shows that for

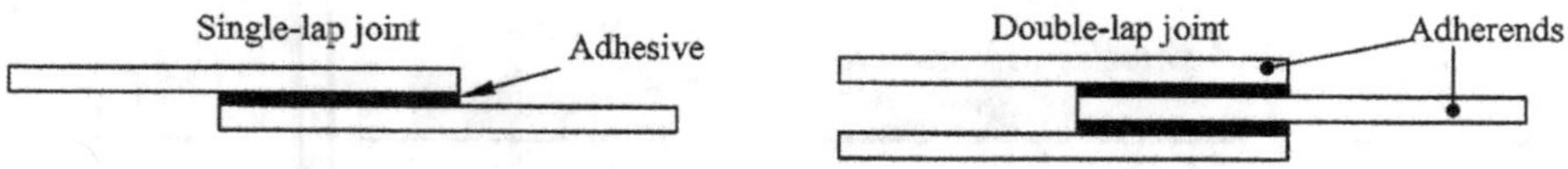

Figure 1.7 (a) Single- and (b) double-lap joint

intermediate and long overlaps, the load distribution is concentrated at the ends of the bond and that increasing the width of the joint will increase the bond strength, but increasing the lap length greater than 25 mm will not.

The failure initiation modes, displayed in Figure 1.11, describe the most important origins of potential failures of bonded joints. There are different root causes for these initial failure modes, and only major effects will be discussed in this paragraph.

A disbond is an initial area within a bonded joint without connection between adherend and adhesive. Typical root causes are massive contaminations of the adherend surface (e.g. with release agent) or failures within the adhesive application process (e.g. gaps within the adhesive layer). A disbond is detectable with non-destructive inspection (NDI) technologies as ultrasonic inspection within the individual limits of the detection threshold.

A weak bond is a bonded joint with a reduced strength between adherend and adhesive. It is characterized by an adhesive failure mode. The root cause is an insufficient adhesion of the adherend interfaces due to small contamination of the surface or unfavourable process conditions. A weak bond is not detectable by means of NDI due to the absence of a detectable interface layer. Research addresses the problem, but results are not expected short- to mid-term for industrial usage. Impact events within manufacturing and in service can lead to initial damages of the adherend and the adhesive. Damages resulting from impact are detectable by NDI within the individual limits of the detection threshold.

1.5 Fabrication of bonded joints

This section introduces the best practices used in composite assembly from the manufacturing point of view. Figure 1.12 shows the three main categories of joining of composites with thermoset matrices.

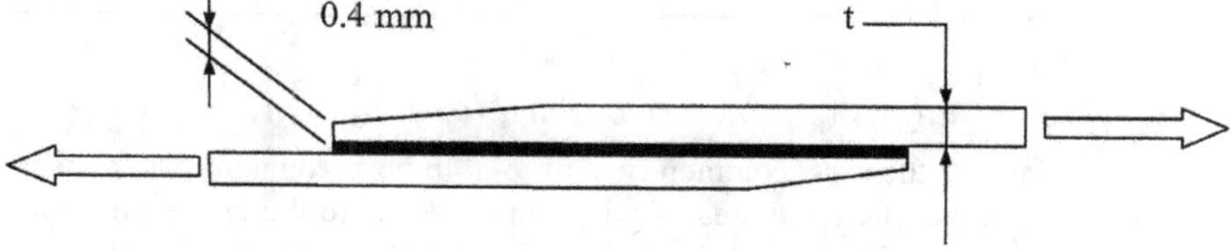

Figure 1.8 How to reduce the effect of peeking and eccentricity

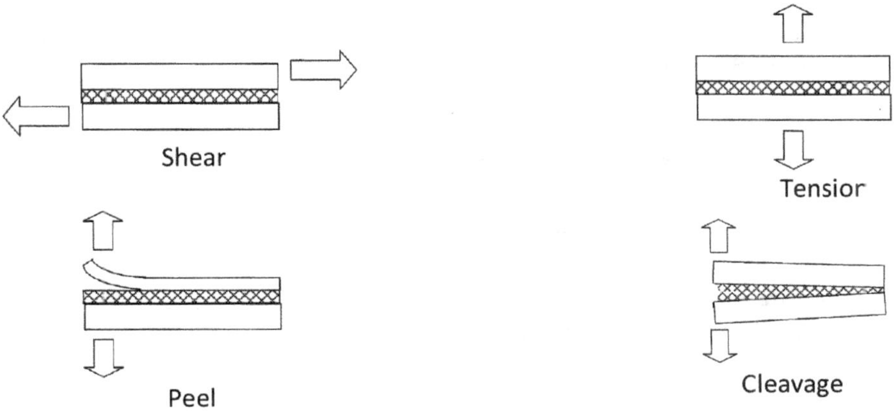

Figure 1.9 Possible failure modes of a bonded joint

In the **co-bonding** method, an uncured part is joined with one or more cured parts, typically with an additional layer of adhesive. The joining mechanism between the adhesive and the cured part is adhesion. Between the un-cured part and the adhesive, chemical cross-linking is taking place.

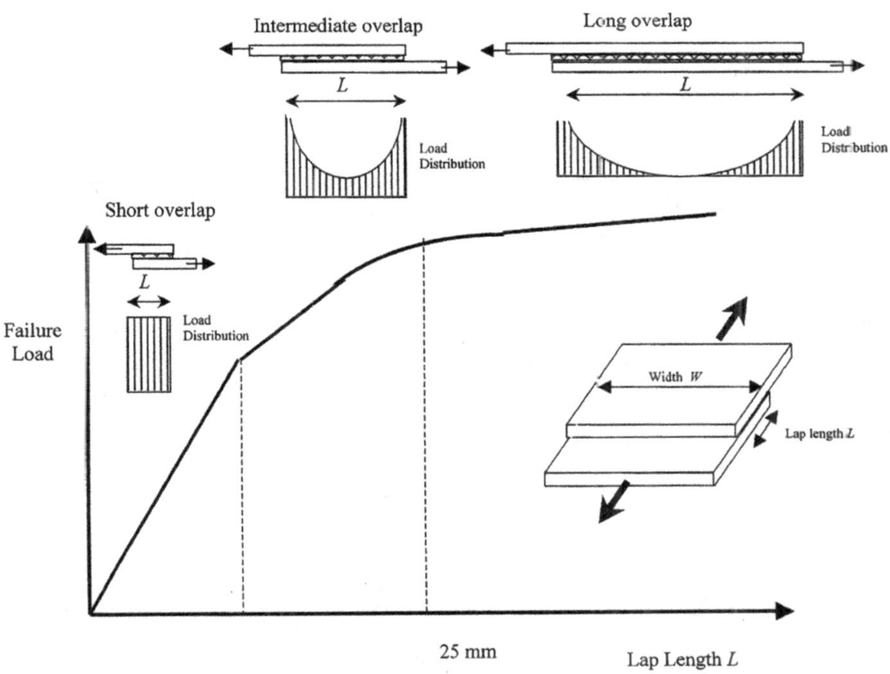

Figure 1.10 Failure load of a single-lap joint as a function of lap length [2]

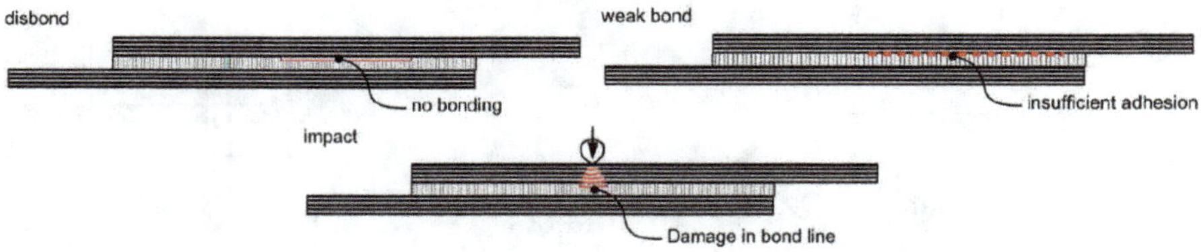

Figure 1.11 Typical failure initiation modes for bonded joints

Secondary bonding foresees that two previously cured parts are joined by a film or paste adhesive. The joining mechanism between adhesive and adherend is adhesion.

Adhesive bonding is the most common joint type of composite material. Conventionally, in this type of joining, two separately manufactured parts are integrated by using adhesives, which is also called secondary bonding. A proper surface preparation is mandatory to ensure the performance and process safety of a secondary or co-bonded joint. The most common surface preparation/protection is the application of peel ply for co-bonded and secondary bonded joints. This technology holds some challenging aspects, especially the activation of the surface and remaining contaminations on the surface without further cleaning or activation processes. The main arguments for application of peel ply as the baseline surface preparation are the reproducibility and simple application as well as the protection aspect during handling and storage of the single part. Additionally, surface cleaning and activation

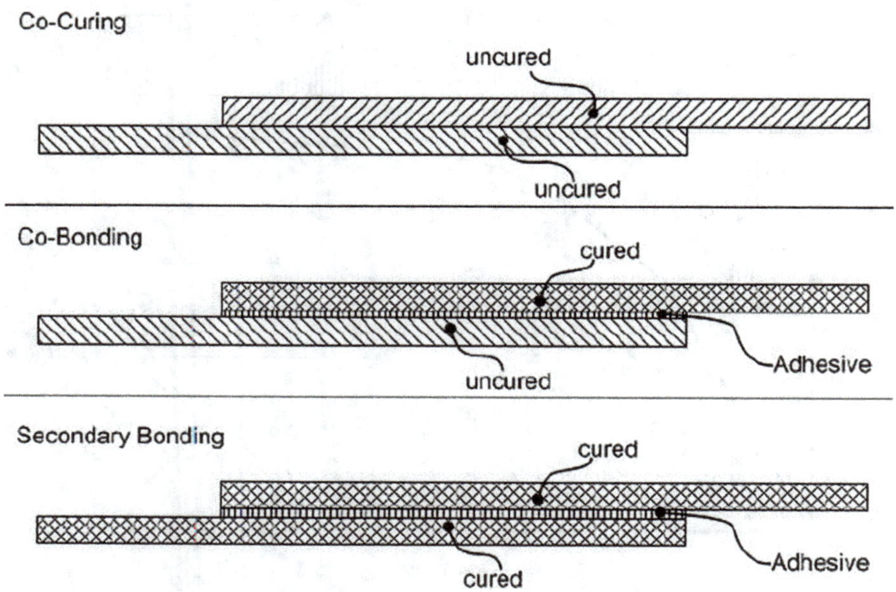

Figure 1.12 Classification of composite bonded joints

Figure 1.13 Schematic of a typical co-curing process

technologies as atmospheric plasma are applied more and more in industrial applications. Other novel surface pre-treatment technologies such as laser cleaning or vacuum blasting are under development. Adhesive bonding has significant advantages over mechanical fastening. First, the load at the joint interface distributes over an area rather than a concentrated point, which results in uniform stress distribution. As a result of uniform stress distribution, the joint becomes more resistant to mechanical loads, particularly repeated loadings. Furthermore, the additional weight that comes from adhesive is negligible compared to that from mechanical fastening. Adhesives not only bond the two surfaces but also seal the joint. Because of this characteristic, adhesive can also prevent galvanic corrosion between carbon fibre composites and metals. Plus, adhesives adapt complex surfaces very well and create a negligible small change in part dimensions.

The **co-curing** joining mechanism is chemical cross-linking in which composite subcomponents are produced in a single-shot manufacturing. It is a process wherein uncured composite parts are cured and bonded simultaneously during the same cure cycle. By integrally co-curing structures, one can greatly reduce the part count and eliminate both metal fasteners and extra processing steps, which in turn cut down the final assembly costs. Co-curing is commonly used in aircraft applications, where a large amount of small parts is integrated to form a large structure, thereby making the assembling process highly challenging and expensive. By unifying small parts into large structures simultaneously, co-curing is a process that promises to cut down both assembling time and cost. Figure 1.13 shows a typical process of co-curing.

Sub-structures of aircraft structures mainly consist of stiffened shells such as fuselage frames, ribs and multi-cell box beams. Conventionally, these stiffened shells are manufactured through a process wherein shells and stiffeners are fabricated separately and then are integrated through either mechanical fastening and/or secondary bonding. The main motivation of co-curing is to reduce assembly costs. Three cost-down mechanisms exist for co-curing: (1) eliminating extra processing step, (2) lowering processing time and (3) reducing metal fasteners count. Using secondary bonding to produce stiffened shells, such as skin-stiffener or skin-web connections, skins and stiffening members or spars are separately fabricated, and another curing operation to assemble the skins with stiffeners or spars is required. This additional curing operation is eliminated in the co-curing process, which is illustrated in Figure 1.14.

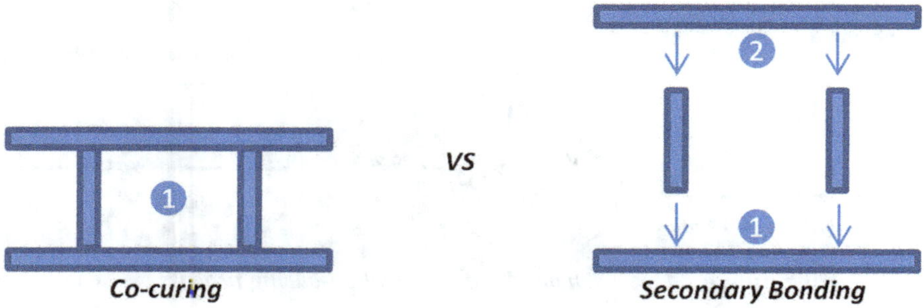

Figure 1.14 Illustration of Co-curing and secondary bonding

It is a trade-off in the case of decreasing processing time. Since lay-up work – placing resin mixtures on a mould at desired angles and places – is complicated in co-curing process, it requires additional lay-up time. Therefore, the advantage of skipping additional curing operations must be weighed against the additional lay-up time. This gets co-curing more feasible in manufacturing large and complex assemblies.

Co-curing provides an opportunity to lower the number of metal fasteners. One of the significant advantages of co-curing is the reduction in the part count. In both mechanical fastening and secondary bonding, materials should be in contact with each other properly and intimately, which is hard to achieve for complex structures. As a consequence of improper contact, undesirable assembly stresses would be introduced in the structure, which is challenging to account for in the design. The assembly stresses are deleterious for composite structures and affect the life of structure severely, particularly carbon epoxy composites by virtue of its brittle nature. By co-curing composite substructures, these assembly stresses can be eliminated. Moreover, structural adhesives that have higher density compared to resins are typically used to adhere sub-structures in secondary bonding. Hence, additional weight in assembly can be presented, which may cause a weight penalty in such applications. Plus, a high amount of structural adhesive is used in practice compared to that of the resin system in a co-cured integration, which also adds more weight to the assembly. Metal fastening integration generates galvanic corrosion risks for carbon fibre composites. Carbon fibre composites are electrochemically very noble and electrically conductive. Hence, when a metal is in contact with a carbon fibre composite, galvanic corrosion takes place. This issue becomes even more serious when a large surface of carbon fibre composites is coupled to small metallic parts, such as metal fasteners. In co-curing integration, resin does not only bond two surfaces but also seals them and consequently lessens the risk of galvanic corrosion for carbon fibre composites. So far, a variety of manufacturing techniques have been employed in co-curing technology. Co-curing processes can be categorized according to manufacturing techniques (Figure 1.15).

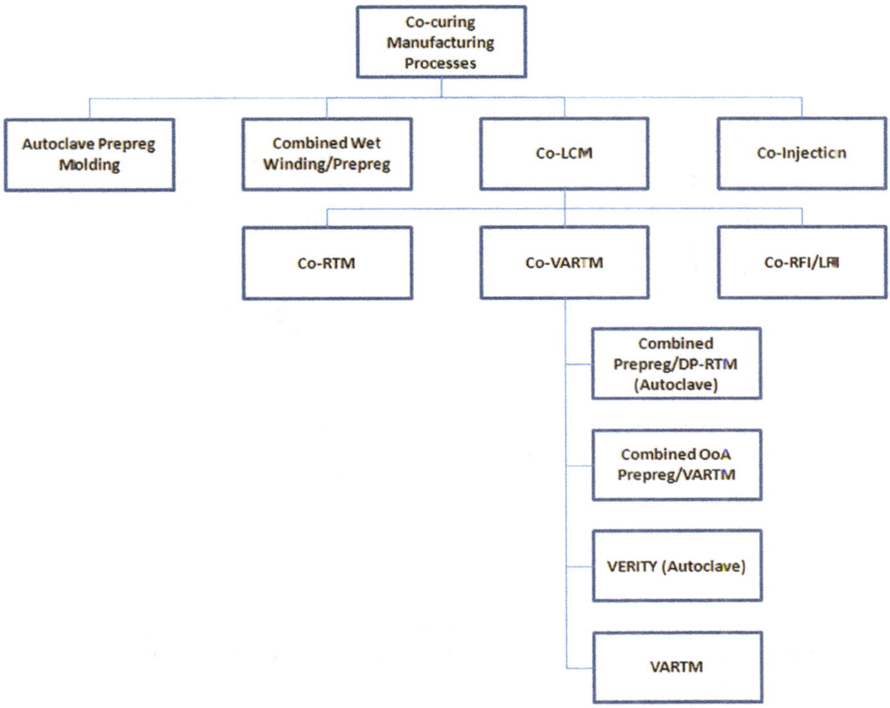

Figure 1.15 *Co-curing manufacturing processes*

1.5.1 Joints in space industry

Most larger parts of spatial structures are already made of CFRP, but the achieved weight saving may be 'jeopardized' by the use of metallic brackets as joining elements. Moreover, current metallic brackets and joints typically used in composite sandwich space structures may not be the most efficient solution from the following points of view: weight, thermal behaviour and punctual loads transference as well as their assembly process turn out to be expensive and invasive (need of bolts, inserts, reinforced cores, etc.) [3–5]. This section illustrates three case studies of different bonding techniques in the space industry.

Case study 1: Ariane 5 ACY 5400 – adhesive bonding

Case study 1 refers to ACY5400 that is an external structure of the upper part of the Ariane 5 launcher, as shown in Figure 1.16 [5].

The assembly comprises cylinders of different heights, i.e. 0.5, 1, 1.5 or 2 m, depending on the launcher configuration. This enables more flexibility for the Ariane 5 payload volume.

The shell is made of CFRP skins (1.1 mm thick) with an aluminium honeycomb core of thickness 22.8 mm. The sandwich cylinder is adhesively bonded to the upper and lower aluminium alloy attachment rings.

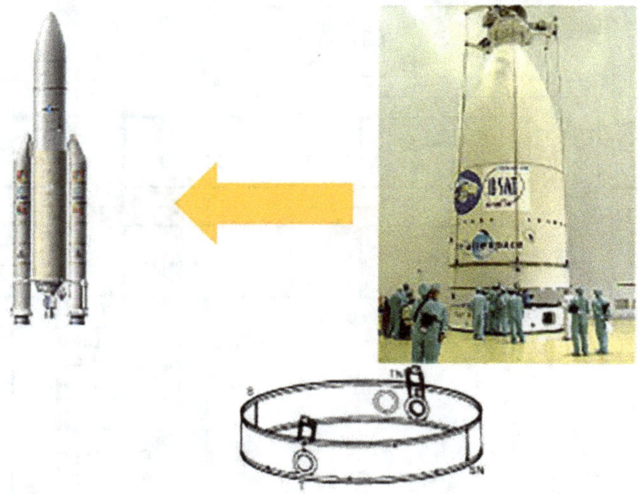

Figure 1.16 Ariane 5 ACY 5400

The interface ring-to-shell joints are solely bonded using an epoxy-based adhesive cured at room temperature. This reduced costs and avoided the need for extra equipment and processes.

The ACY 5400 design-development programme showed that bonded joints are feasible for large, highly loaded space structures and, in comparison with a conventional riveted assembly, they demonstrate mass and cost savings. In fact, the total cost of the bonded assembly is almost 10% approx. less than riveted one, and, also in terms of mass, it is more or less 10% lighter than riveted one (182.5 kg for the adhesive bonded configuration instead of 206.5 kg of riveted configuration). In conclusion, adhesive bonding demonstrates to be the lightest, fastest and most economical way of producing joints, although the preparation and quality assurance effort remained a significant cost factor.

Case study 2: LVA ring/CFRP thrust cone

Case study 2 refers to an adhesive connection between a cylindrical CFRP-skinned aluminium alloy honeycomb sandwich and a metallic attachment ring of assemblies destined for launch by Space Shuttle.

The thrust cone was manufactured from three 120° segments of CFRP-faced, aluminium alloy honeycomb. These were joined axially by bonded CFRP butt straps (as shown in Figure 1.17).

Four aluminium alloy rings were connected by various means to the cone, the lower of and the four being the interface flange for attaching the spacecraft to the launch vehicle. During the launch, a large bending moment is induced on the LVA frame-to-cone joint that is reacted by differential axial compression and tension loads across the diameter of the joint. These loads are reacted by the bond between the CFRP sandwich segments and the machined alloy frame.

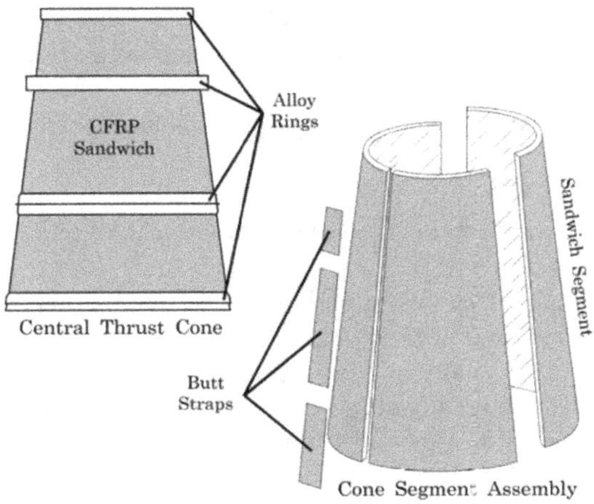

Figure 1.17 *LVA ring/CFRP thrust cone – basic construction of thrust cone assembly.*

The maximum ultimate running load was ±150 N/mm from a combination of static vibration loading. A baseline LVA frame joint was assessed, consisting of a simple twin lap shear bond, as shown in Figure 1.18. This proved inadequate, and the two realistic options shown were compared, by both finite-element analysis and mechanical testing of representative sections.

The general features used in both options to reduce the peak adhesive shear stresses at the edge of the bond were as follows:

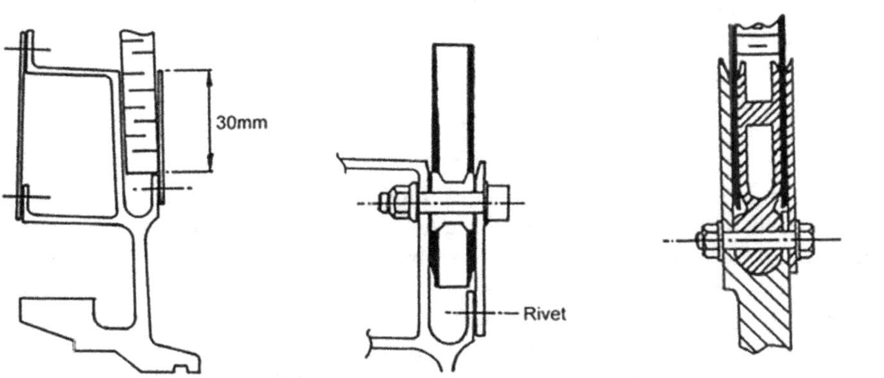

Figure 1.18 *LVA ring/CFRP thrust cone – joint configurations for LVA ring to thrust cone connection*

- increasing the thickness of the alloy closing plate and the CFRP skin in the region of the joint in order to balance the stiffness of the adherends, i.e. $E1t1 = E2t2$
- chamfering the edges of the adherends, so thickening the adhesive at the ends of the joint.

The chamfering was seen to increase the failure load of representative joints by 56%.

The twin double-lap face configuration was selected to balance the joint. This gave much lower shear and peel stresses than the single-lap shear of the baseline configuration. The mass penalty of this, over the baseline joint, was 3.2 kg.

For the bonded and bolted concept, the peel stresses are minimized by the constraint imposed by the bolts, adherend stiffness balancing and chamfering. The mass penalty of this joint over the baseline design was 1.4 kg.

The failure loads in tension, for preliminary tests on two specimens of each type, were as follows:

- four-lap shear face: 667 N/mm and 737 N/mm
- bolted (no adhesive): 244 N/mm
- bolted and bonded: 340 N/mm

These figures demonstrate the efficiency of the all-bonded construction. However, the bonded and bolted version was chosen because it had adequate strength, was lighter and easier to manufacture and was fail-safe. As part of the development testing programme, representative joint test samples were loaded in tension and compression. The results obtained from the tests on joint samples proved, with a high level of confidence, that the chosen design was capable of withstanding the ultimate flight loads imposed on the joint. This was verified by the following:

- the development testing of a complete cone
- qualification testing of the complete spacecraft structure
- proof testing of each flight cone

No failures occurred during any of these tests.

In conclusion, the case study shows the process undertaken to select a joint design strong enough to take the desired running load. The final design met this criterion whilst at the same time satisfying the need for mass control, ease of manufacture and fail-safe characteristics. Inspectability is one area where the joint configuration can cause problems, unless each manufacturing stage was closely monitored.

Case study 3: Payload attach fitting (PAF) conical structure

Case study 3 concerns investigations carried out in the project Composite Technology for Exploration (CTE) that was funded by the Game Changing Development Programme in the Space Technology Mission Directorate and the Spacecraft Payload Integration and Evolution Office in the Space Launch System (SLS) Programme aiming to advancing composite technologies and providing

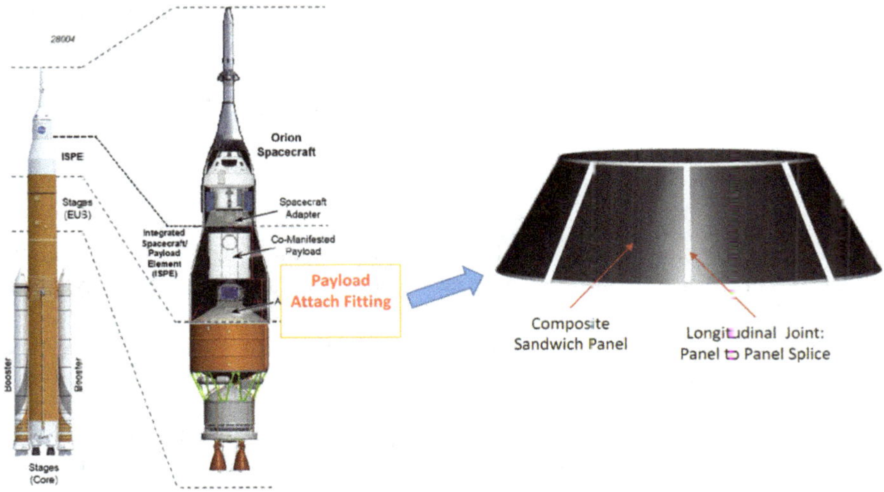

Figure 1.19 Exploded view of SLS showing PAF

lightweight structures to support future NASA exploration missions (Figure 1.19). In particular, the main scope of the CTE project was to demonstrate weight-saving, performance-enhancing composite bonded joint technology for SLS-scale composite hardware [6].

Sleight *et al.* [6] reported in their paper the details of the development including materials, design and analysis and manufacturing of a composite sandwich bonded longitudinal joint for a generic space launch vehicle structure called the CTE Point Design (Figure 1.20). The CTE Point Design is a conical sandwich structure with a 35° angle consisting of eight composite sandwich segments joined together with a double-lap composite joint. This last was selected as the most advantageous

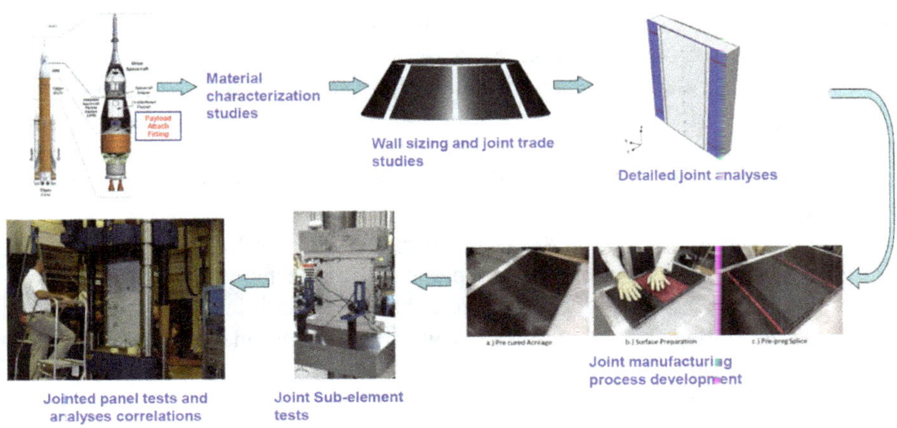

Figure 1.20 CTE technology plan for joint development

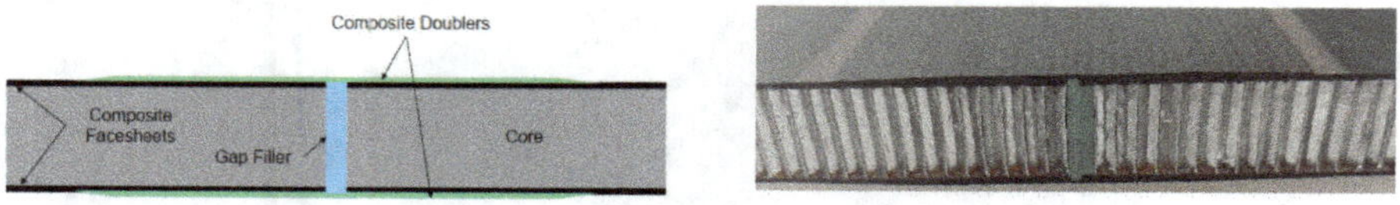

Figure 1.21 Cross-sectional view of panel joint [7]

longitudinal joint to advance for the CTE project. The joint selection was based on an assessment of figures of merit for mass, damage tolerance, inspectability, cost, design/analysis uncertainty and producibility.

The segmented panels are a sandwich construction with 3.1-lb/ft3, 1-in.-thick perforated Plascore aluminium core (5056) with (3/16-in.) hexagonal cells and 8-ply quasi-isotropic IM7/88552-1 face sheets with a layup of $[45/90/-45/0]_S$ (Figure 1.21). The material for the doubler plates is a T650/5320-1 epoxy and carbon fibre woven prepreg fabric with a layup of $[45_4]$. Doubler plates were bonded to the sandwich panels with FM209-1M adhesive. The doubler plates were 4.2 in. wide. During the joining process, an adhesive paste (EA 9396 6MD) was injected to fill a 0.1-in. gap between the two panel segments.

Longitudinal bonded joint sub-element test articles were fabricated and tested for several loading conditions to test the capability of the joint design for the as-designed loads. The test results show that test failure loads for the composite longitudinal bonded joint design significantly exceed the joint design loads with a 2.0 factor of safety for both pristine and impact-damage coupons. The analysis pre-test failure predictions for all sub-element bonded joint test coupons were all within 10% of the average test coupon failure load. This testing and analysis provide confidence in the potential use of composite bonded joints for future launch vehicle structures.

1.5.2 Joints in aviation industry

Vacuum-assisted resin transfer moulding (VARTM) is an affordable technique for co-curing, so it requires neither autoclave nor prepreg. Co-curing using VARTM was first proposed by Mahruz *et al.* [8] to fabricate skin stringers. In the technique, rigid soluble foam mandrels were utilized to create cavities. Moulds were also used as tools on which dry fibres are draped. Following the lay-up process, the conventional VARTM procedure was carried out and the whole skin-stringer structure was made in one curing cycle (Figure 1.22). After curing, foam mandrels were dissolved using acetone.

Co-liquid resin infusion (LRI) is another affordable technique for co-curing, since it requires neither autoclave nor prepreg. This technique is a well-known out-of-autoclave (OOA) production process, based on the infusion of liquid resin, mainly thermosetting, into the voids present in a laminate, made by layers of dry reinforcement (unidirectional or woven fabric plies).

LRI is widely used in the nautical sector to fabricate hulls, but today, it is gaining a growing space also in the aviation field, thanks to its advantages summarized below:

(a) (b)

Figure 1.22 *Co-cured skin stringer assembly (a) before foam removal and (b) after foam removal*

- low cost of the initial investment (slightly higher than that required to acquire equipment for the manual layup);
- ability to produce aerospace grade parts with a high degree of integration (reduction/elimination of fasteners and adhesives);
- large structures (with integrated details) can be infused in a single-shot process resulting in cost savings;
- almost infinite shelf-life of the non-binderized reinforcement and the possibility of using the same fabric with different resin systems, as long as they are compatible;
- technicians do not have to handle prepreg materials and the exposure time/contact with the resin is reduced;
- dry fibres/fabrics do not require refrigeration (as opposed to pre-pregs)
- no need for autoclave processing, process to be executed in oven or integrated heating tools resulting in cost savings (both capital and running costs);
- resins that require refrigeration (typically one part) occupy relatively small space (compared to pre-pregs) and thus do not require large freezers;
- reduction of costs for toolings, if compared to those made with more expensive production processes (toolings for autoclave, press).

Thanks to the introduction of aerospace-grade resins systems (e.g. EP2400 from Solvay, RTM six from Hexcel) and increasing usage of automated preforming processes, which could be both pick and place or automated tape/fibre placement, several examples of flying aerostructures are available on the market. Premium Aerotech (Germany) uses a patented infusion process, the vacuum-assisted process to produce the pressurized rear bulkheads of the Boeing 787; Boeing Aerostructures

(Australia) is using an infusion process called controlled atmospheric pressure resin infusion to produce the moving parts of the trailing edge of the Boeing 787 (ailerons, flaperons, inboard and outboard flaps and seven spoilers); Bombardier (Canada) is using an infusion process followed by autoclaving to make the wing of its C-Series aircraft. At last, Aerocomposit (Russia), with the aim of obtaining a reduction in weight and production costs of about 45% compared to the current ones, is developing an out-of-autoclave LRI process (oven cure), for the production of wing of the new Irkut MS-21 single-aisle commercial transport aircraft [9].

As opposed to LRI used in the nautical field, in aviation considering that resins need higher performance (e.g. higher Tg), dry fibres preforms are impregnated by resin in liquid form in oven, under the application of vacuum only [10, 11]. No external pressure is applied on the final part. The resin is heated and injected into a vacuum bag using special means. The part is allowed to cure under vacuum and temperature following a curing schedule according to the recommendation/instructions from the resin system manufacturer. The cure cycle results then in a fibre-reinforced structure. Within this context, it is worth to illustrate the main results of the SWING ('Smart WIng for New Generation of UAVs for civil use') project, an R&D action completed on October '21, that received partial funding from the Italian Ministry of Economic Development – MISE (PON 'Imprese e competitività' 2014–2020 – Grandi Progetti PON – Industria Sostenibile. F/090035/01–05/X36).

One of the SWING goals was to adapt automated OOA composites processes – including AFP and LRI – for cost-effective and sustainable production of smaller aircraft, using Piaggio's P180 Avanti forward wing as a demonstration platform (Figure 1.23). The idea developed by NOVOTECH, responsible for the manufacturing together with PROTOM (Italy), as project coordinator and responsible for the design, was to produce the forward wing of the P180 in just two main parts: an upper panel and a lower panel, both moulded with integrated stringer and spars. The moulded upper and lower panels have been joined using structural adhesive.

Raw materials used in SWING project for AFP preforming process followed by LRI are as follows:

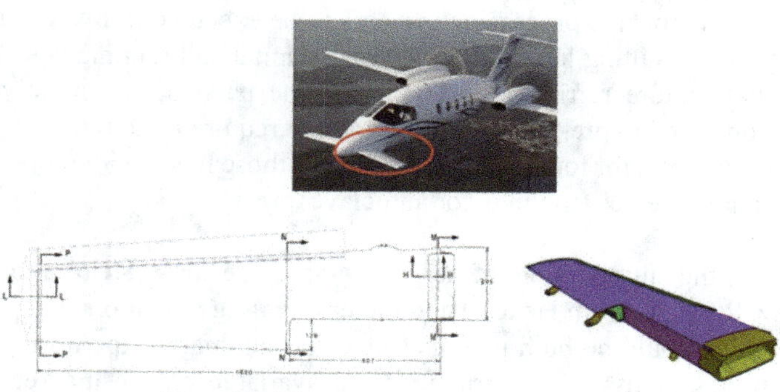

Figure 1.23 Piaggio's P180 Avanti forward wing as a demonstration platform

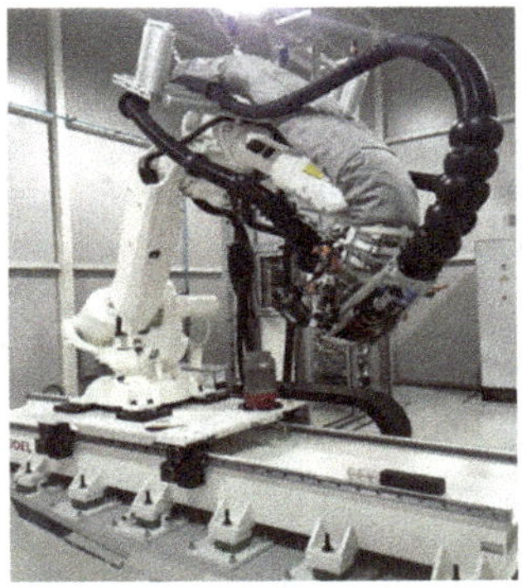

Figure 1.24 AFP robot available at Novotech's Avetrana plant

- PRISM TX1100 IMS 65-24K-UD196-6.35, as a dry fibre
- PRISM EP2400 Epoxy liquid resin system.

Materials were provided by Solvay (more information on www.solvay.com).

The preforming process has been performed by using a laser-assisted fibre placement robot developed by Coriolis Composites (France) and available at NOVOTECH's plant (Avetrana, Italy) (Figure 1.24).

The production process development has been carried out according to a bottom-up approach, starting from a small-scale demo up (approx. size 800 × 300 mm²) to the production of the final full-size wing box (approx. size 1 700 × 300 mm²). The details of different phases of the production process are depicted in Figure 1.25.

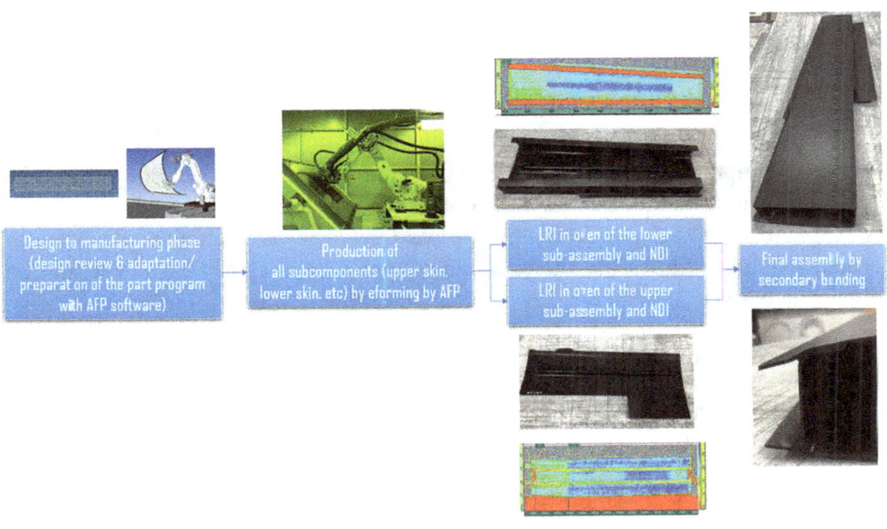

Figure 1.25 Swing Project: chain of manufacturing processes

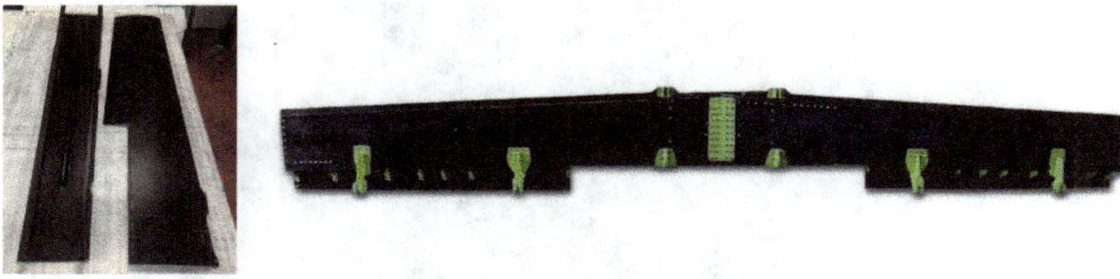

Figure 1.26 Final demonstrator (source: protom.com)

Figure 1.26 shows the final demo before assembly and final forward wing of the P180 before the testing activity. The other wing box has been developed by the project partner LAER company (Airola, Italy) with prepreg material.

According to the results of the activities performed during the project, the following conclusions can be given.

- In terms of material waste, the usage of AFP results in a reduction of about 20% of the material with respect to HLU (30% vs 4%).
- Regarding the ply placement, the HLU layup speed is much more affected by the number of ply segments composing the layer and decreases with bigger size ply segments, while for the AFP, the layup speed is constant, regardless of the ply size (intermediate compactions are not required).
- The HLU vs AFP laydown rate ratio is about 1:4.

1.6 Joint assembly

In general, tooling or assembly equipment in the form of jigs and fixtures is required to achieve a high-quality, reproducible adhesively bonded joint. The function of assembly equipment is to locate and retain components together throughout the bonding process, i.e. adhesive application, structure assembly and adhesive cure. As the majority of adhesive systems exist as liquids or pastes before they cure or harden, jigs and fixtures are there to control the following key factors within an adhesively bonded joint:

- bond-line thickness (BLT)
- joint alignment
- fillet profile

These factors in turn have a direct impact on mechanical performance, cosmetic appearance and assembly time.

1.6.1 Significance of bond-line control

Assembly tooling and related materials, in the context of adhesive bonding, can be defined as those systems/materials that retain components in a precise position and orientation during adhesive application and subsequent curing. With the exception of pressure-sensitive adhesive systems (i.e. adhesive tapes), all adhesives must exist in a liquid or semiliquid state prior to hardening/curing if an intimate bond is to be generated between the adhesive and the adherend. The surface of the adherend needs to be fully wetted by the adhesive to maximize the bonding area and enable intimate physical and chemical bonding to take place. This can only be achieved with a liquid phase; however, unconstrained liquids have no direct load-bearing properties and tend to be displaced and flow away from an area under pressure, such as between two adherends. The rate and level of displacement depend on the pressure exerted and the viscosity of the adhesive. In addition, any joint that can move (e.g. through gravity because it is not lying flat) will tend to do so if it is not held in place prior to adhesive cure/hardening. It is, therefore, necessary to use some sort of mechanical aid to locate and support the structure to be bonded until the adhesive in the structure has developed sufficient green strength or handling strength. Once this stage has been achieved, the structure can be self-supporting and tolerate the removal of, or from, jigs and fixtures.

Confusion often exists about how much pressure to apply to a joint during the adhesive curing process. In some instances, guidelines supplied by manufacturers indicate a certain minimum pressure; in other cases, no magnitude is indicated, and this is often interpreted as 'as much pressure as possible'. However, as adhesives exist as liquids or partial liquids for a period of time prior to curing/hardening, care must be taken to control the application of pressure in order to avoid squeezing the entire adhesive from the joint. Nevertheless, pressure is required to hold the adherends together, to encourage wetting of the surface and, on some occasions, to address poor fit-up or tolerances by forcing the adherend surfaces together. In order to address this issue, jigging or tooling is required to control the spacing between the adherends and therefore the BLT. This can be either internal, i.e. filler particles, glass beads, wires, carrier films, joint details, etc., or external, i.e. tooling, external shims, joint details, etc. In summary, pressure is required for adhesively bonded joints but only when some type of BLT control can balance it.

1.6.2 The need for bond-line control

For virtually all adhesive systems, there exists an optimum BLT range for achieving the desired mechanical properties of high bond strength and resistance to creep. If there is insufficient adhesive within the joint, it will be very weak and highly susceptible to defects such as bubbles or poorly wetted areas. If there is too much adhesive present, the joint may also be of lower strength since the properties of the bulk adhesive may dominate the structure due to inefficient load transfer from one adherend to the other. When the BLT is within the optimum range, load transfer is maximized and creep is minimized. It is, therefore, very important that the user takes this into

Table 1.2 Advantages and disadvantages of adhesive bonding

Advantages	Disadvantages
Provision of large, stress-bearing area – especially relevant for composites	Some type of surface preparation will be required – composite surface preparation will either require some type of cleaning, followed by abrasion/peel-ply removal and then a subsequent cleaning stage prior to bonding
Excellent fatigue strength	Significant cure times may be required
Good shock absorption – this property is dependent upon the type of adhesive selected and the design of the joint	Heat and pressure may be needed to cure an adhesive
Reduced galvanic corrosion – of particular importance where carbon fibre systems require to be joined to metals such as aluminium and steel	Jigs and fixtures may be required to locate components whilst the adhesive cures
No need to have access to the far side of adherend to make joint	Rigid process control is required to obtain consistent results
Bonding is possible on dissimilar materials	Conventional non-destructive inspection of bonded joints is difficult
Provision of smooth contours and sections around joint areas	Adhesives have a finite shelf-life prior to curing and may require special storage conditions before use
Weight reduction – in contrast to other joining technologies where fasteners are required and additional composite is required to maintain performance around holes, the only added mass is that of the adhesive which is not normally significant	Adhesives have poor resistance to peel and cleavage stresses in relation to shear and tension loading – ideally, the best loading direction is in compression

account during the design phase and then makes use of appropriate assembly aids during fabrication.

1.6.3 Advantages and disadvantages of adhesive bonding

Adhesives can provide many benefits over other types of joining technology and composites are no exception. However, it is important to provide a balanced view where disadvantages, both real and perceived, are described in Table 1.2.

1.7 Conclusions

The design and manufacturing of assembled composite structures for the aerospace field is still a challenge, and an overview of this topic, with a focus on the bonding technologies, was given in this chapter.

Recent advances in material and manufacturing technologies reveal that, at present, there is no feasible way to implement a structural bonded joint on aircraft/space primary structures without additional features, such as fasteners. The main reason is due to the lack of technologies able to give a secured certification. In recent years, in view to increase process safety and to eliminate the risk for the occurrence of a weak bond, intense research has been performed, and it comes out that major fields of development are process automation, extended NDT, understanding of bonding parameter influence and surface pretreatment for bonding.

A new technology solution to enable the fully bonded aircraft will only be established within a strong cooperation of industry and academics pushing, on the one hand, to a stable process with the needed performance within the right cost and lead time framework, on the other hand, to perform the needed research to be able to clearly understand the utilized physical or chemical mechanisms and potential influencing factors and environmental conditions.

Since this kind of assembly is strongly under investigation and development, a section on different bonding techniques in the aerospace/space industry has been provided through different case studies.

References

[1] Niu C. *Composite airframe structures: practical design information and data.* Hong Kong: Conmilit Press; 2010.

[2] ESDU 80039 *Elastic adhesive stresses in multistep lap joints loaded in tension.* ESDU; 1995.

[3] Makowski Z.S. *Analysis, design and construction of double layer grids.* London: Applied Science Publishing Ltd; 1981.

[4] Makowski Z.S. *Analysis, design and construction of braced domes.* London: Granada Publishing Corporation; 1984.

[5] Gonzalez R., Vogt P., Gonzalez A. 'Implementation of high-performance and reliable structural bonded joints'. Proceedings of the 3rd European Conference on Launcher Technology; Strasbourg, 2001.

[6] Sleight D.W., Segal K.N., Guin W.E, *et al.* 'Development of composite sandwich bonded longitudinal joints for space launch vehicle structures'. AIAA Scitech 2019 Forum; San Diego, CA, 2019.

[7] Cox S.B., Danley S.E., Wright J.R., Nettles A.T., Guin W.E., Segal K.N. 'Composite repair process for adhesively bonded joints'. CAMX 2019; Anaheim, 2019.

[8] Mahruz H., Majumdar P., Saha M., Shamery F., Jeelani S. 'Integral manufacturing of composite skin-stringer assembly and their stability analysis'. *Applied Composite Materials*. 2004.

[9] *Infused wing sheds light on aerocomposites future* [online]. 2018. Available from https://www.compositesworld.com/articles/infused-wing-sheds-light-on-aerocomposites-future

[10] Stokes V.K. 'Joining methods for plastics and plastic composites: an overview'. *Polymer Engineering and Science.* 1989;29:1310–24.
[11] Kenney W.E. *Designing plastic parts for ultrasonic assembly.* Vol. 64. Machine Design; 1992.

Chapter 2

Processing of polymer composites: autoclave and microwave energy approaches

T Lachana Dora[1] and Radha Raman Mishra[1]

In this chapter, autoclave and microwave applicator-based processing of polymer composites have been discussed. First, the principles and physics involved in the curing of polymer composites in both processes have been presented. The equipment details, process parameters, and processing steps have been discussed for both processes. The autoclave curing was compared with the microwave curing process for fabricating a fiber-reinforced polymer composite. The challenges in autoclave and microwave processing of polymer composites have been highlighted and possible future research directions have been discussed.

2.1 Introduction

In recent years, the polymer matrix composites (PMC) have emerged as the popular materials for manufacturing of components such as airplane wings and bodies, wind turbine blades, rail cars, and truck bodies [1]. These components are used in various applications, including aerospace, wind energy, automotive, and infrastructure due to their improved specific stiffness, specific strength, fatigue resistance, and corrosion performance as compared to the metallic materials. The properties of the PMC significantly depend upon the curing technique used for preparing the composite. The curing methods that are used for fabricating the PMCs can be broadly classified into radiation curing and thermal curing [2]. The detailed classification of curing methods has been presented in Figure 2.1.

In the radiation curing process using the energy sources gamma ray, X-ray, and electron beam, the unique energy absorption mechanism provides deep penetration inside the PMC, rapid curing speed, higher energy efficiency, improved resin stability, and freedom to handle complex and large structures as compared to the thermal curing processes. However, the need for radiation-sensitive materials, lower cost

[1]Department of Mechanical Engineering, Birla Institute of Technology and Science Pilani, Pilani, India

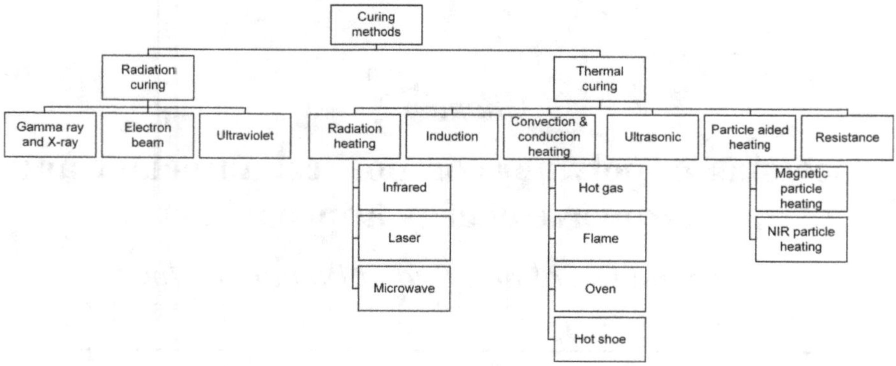

Figure 2.1 Different curing methods for the polymer matrix composites

efficiency, and challenges in handling highly radioactive nature limit the use of these processes. However, ultraviolet radiation is unable to cure the PMC efficiently due to small penetration in semi- and non-transparent polymer composites.

On the other hand, thermal curing techniques are frequently utilized due to the availability of more mature material systems and improved curing processes; however, they lack process efficiency. The thermal energy sources such as infrared/laser heating, microwave heating, and convection and conduction heating methods face issues such as surface heating, limited coupling with polymer composites, and high-temperature gradient, respectively. Additionally, the low heat conductivity of polymers and reinforcements (lead fibers and natural fibers) lead to longer cure cycles. Some of the emerging techniques for curing the PMC such as induction, ultrasonic, resistance, and thermal additive heating methods offer potential alternatives for the fabrication of industrial components; however, further investigations are required for understanding the curing mechanisms involved in these techniques and for fully utilizing the potential of the processes.

The thermal curing processes are more popular than radiation-based curing processes due to issues in radiation-based curing such as unavailability of specific materials, limited handling flexibility, issues in control, and uneconomical processing [2]. Most of the conventional thermal curing processes are time-consuming and uneconomical due to low thermal conductivity of the polymers (e.g. epoxy resin has 0.19 W/m K at 25°C); consequently, their composites show poor response to heat transfer. Long-curing cycle time and higher energy consumption are involved in very slow heating rates (1–2°C min^{-1}) to avoid degradation of the developed PMC parts. Moreover, the thermal curing of thick PMCs (thickness up to 100 mm) results in the non-homogeneous curing of the PMC and it induces large thermal stresses.

Autoclave curing is one of the most important thermal curing methods that have gained popularity for the curing of the high-performance industrial parts. The major challenges associated with autoclave curing while processing parts having huge size and higher thickness are long-curing cycles with high energy consumption, non-uniform curing of parts, high-induced residual stresses, large capital investment, and operating expense [3]. Availability of an oven or an autoclave that can accommodate larger size industrial components (e.g. the part of the Boeing-787 was cured inside an autoclave which was developed by ASC Process Systems, USA with working area: φ 9 × 25 m) is limited [4]. Therefore, it is necessary to explore and develop economical and highly efficient methods for curing the PMC.

In recent years, microwave energy has been used as an efficient energy source for more uniform curing of the PMCs [5]. Microwaves interact with the material in the waveform and try to dissipate energy inside the PMC depending upon the dielectric loss factor of the matrix or reinforcements and penetration depth of microwaves inside the PMC. The microwave energy gets dissipated in the high dielectric loss factor component of the PMC and volumetric heating of the PMC occurs rapidly up to the penetration depth. Beyond the penetration depth, the curing of the thin part depends upon the efficient heat conduction inside the PMC. However, the curing of thick parts has been reported with non-uniform heating, thermal stresses, loss of matrix or reinforcements, and generation of the hot spots [5, 6]. The size of a specific microwave applicator limits the size and thickness of the PMC parts that need to be cured. Moreover, the non-uniform distribution of microwaves inside the applicator cavity enhances the number of hot spots formed during the curing process. Therefore, further investigations are needed to explore the full potential of the microwave heating-based curing process.

In this chapter, the principles and physics involved in autoclave and microwave energy-based curing of the PMC have been presented. The machine architecture, process parameters, and recent advancements have been discussed. The challenges in autoclave and microwave processing of polymer composites have been highlighted, and possible future research directions have been discussed.

2.2 Fundamentals

2.2.1 Basic principles

Autoclave technology utilizes pressure and temperature cycles to cure the PMC. The main mechanism of curing in the autoclave method is based on forced convection (between the laminate system and the autoclave fluid) and conduction (between the bag and the tool). Figure 2.2 shows the possible heat transfer phenomenon during autoclave curing [7]. The main sources of heat are autoclave, mold, and lay-up. The main mechanisms of heat transfer are conduction, convection (force and natural), and radiation between different components, as indicated in Figure 2.2.

The heating of a PMC inside the microwave applicator occurs due to the conversion of microwave energy into heat energy. The wave energy dissipates at the atomic level depending upon the dielectric properties of the constituent materials (matrix and reinforcement), penetration depth of the microwaves in the PMC, crystallinity of the polymer, dipole structure, frequency of processing, and temperature of the PMC [5, 6]. A brief overview of different heating modes in microwave curing has been shown in Figure 2.3. The main sources of heat are energy conversion and heat transfer during the microwave curing of the PMC. The heat dissipation occurs

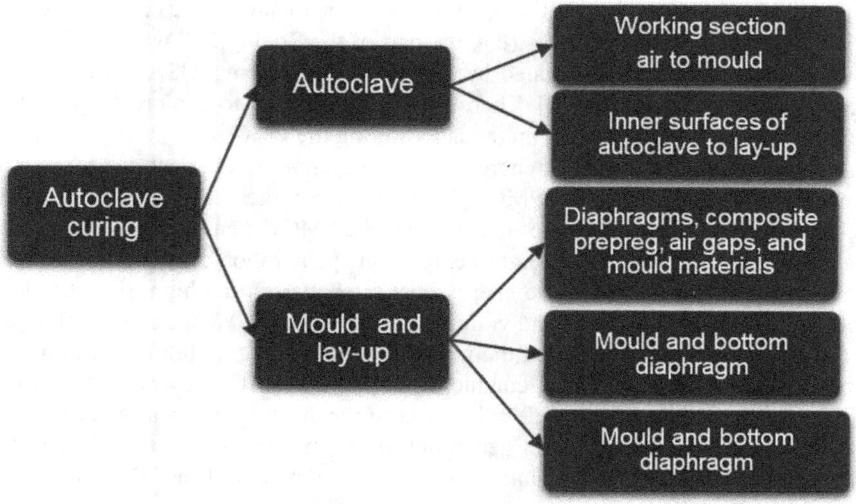

Figure 2.2 Different heat transfer modes during autoclave curing process

when microwaves interact with matrix or fiber. Heat conduct takes place between heated matrix–matrix, heated matrix–fiber, or heated fiber–matrix.

The heating behavior of the PMCs having short or long fibers in microwave/autoclave curing is shown in Figure 2.4. In the case of short fibers, the polymer matrix gets heated first and the heat is transferred to the core of the PMC while using the autoclave heating method. The conduction of heat takes place from the heated portion of the matrix to the unheated matrix and short fibers. Consequently, the temperature of the outer layers (T_1) is always higher as compared to the core ($T_2 < T_1$)

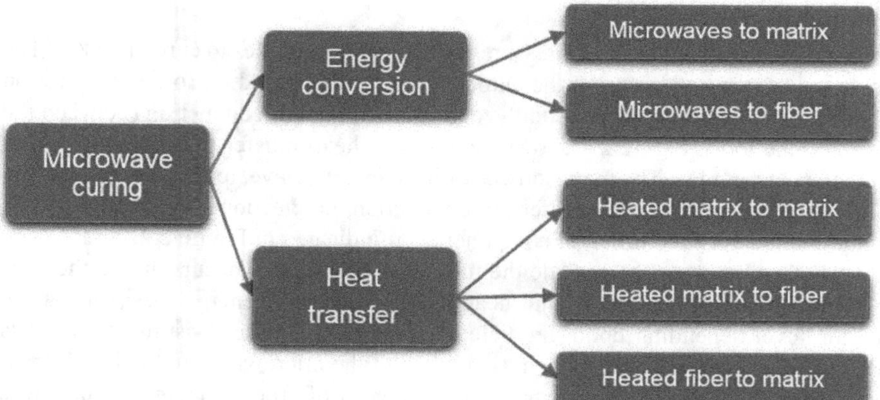

Figure 2.3 Heat dissipation and heat transfer modes during microwave curing process

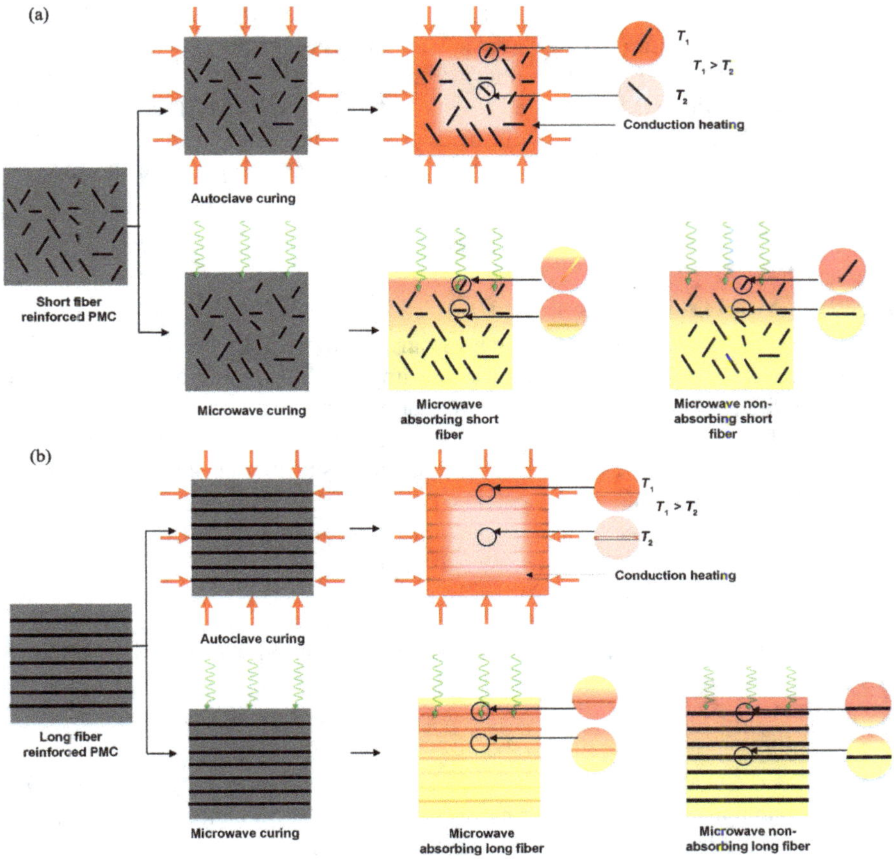

Figure 2.4 *Heating pattern of the PMCs having (a) short fiber and (b) long fiber during autoclave and microwave curing*

of the PMC (Figure 2.4a). On the other hand, when the short fiber-reinforced PMC is exposed to microwave radiations, heat dissipates in the matrix or short fibers, depending on the dielectric loss factor. If the dielectric loss factor of the polymer matrix is higher than that of fibers and couples with microwaves, it gets heated first and further transfers heat to the unheated matrix and fibers. Most of the polymers are transparent to microwaves and hence the dielectric loss factor of the short fibers plays an important role during curing. Microwaves absorbing short fibers quickly get heated and act as a secondary source of curing for matrix (Figure 2.4a).

Heating patterns of the long fiber-reinforced PMCs during autoclave curing and microwave curing have been shown in Figure 2.4b. In the case of the autoclave method, the heat gets conducted from the outer layer of the matrix to the core of the matrix and the long fibers present between the core and outer layer of the matrix get heated via conduction. In this mode of heating, the outer layer (T_1) has a higher temperature as compared to the core ($T_2 < T_1$). Whereas, while exposing the long fiber-reinforced PMC to microwaves, the heat gets dissipated in the high dielectric loss factor constituent (matrix or fiber). Furthermore, the transfer of heat occurs from heated constituents to unheated constituents via conduction. It is indicated by the pattern of heating in the PMCs during microwave and autoclave curing methods that heating in autoclaves will be more non-uniform as compared to microwave curing.

2.2.2 *Mathematical models*

2.2.2.1 Autoclave curing model

The heat transfer model of the autoclave curing process has been reported with assumptions that the convective heat transfer impact of the resin flow is considered to be insignificant and the resin and fiber are at the same temperature at any specific times [8]. The model will take into account the energy transfers within the layers, and in particular, the thermal evolution of the resin-fiber curing system can be described by the following energy balance equation [9]:

$$\rho c \frac{\partial T}{\partial t} = \frac{\partial}{\partial x}\left(K_x \frac{\partial T}{\partial x}\right) + \frac{\partial}{\partial x}\left(K_y \frac{\partial T}{\partial x}\right) + \frac{\partial}{\partial x}\left(K_z \frac{\partial T}{\partial x}\right) + \dot{q} \tag{2.1}$$

where T is the temperature, ρ is the density, c is the specific heat, and K_i is the thermal conductivity. The term $\dot{q}$ is internal heat generation per unit volume and is expressed as:

$$\dot{q} = \rho H_u \tfrac{d\alpha}{dt} \tag{2.2}$$

where H_u is the heat of reaction generated during dynamic scanning. The term $d\alpha/dt$ is cure rate that describes the curing of resin. An energy balance equation can be coupled with the resin kinetic model for solving the cure kinetics [8, 9]:

$$\frac{d\alpha}{dt} = \frac{1}{H_u}\left(\frac{dQ}{dt}\right)_T = \frac{H_T}{H_u}\frac{d\beta}{dt} = \frac{H_T}{H_u}\left(K_1 + K_2\beta^m\right)(1-\beta)^n \tag{2.3}$$

where H_T is the heat of reaction during the isothermal scanning, $(dQ/dt)_T$ is the rate of heat generation during the isothermal scanning, and $d\beta/dt$ the isothermal rate of cure. The value of K, m, n, and H_T can be estimated using the following equations [8, 9]:

$$K_i = A_i e^{\left(-\frac{A}{RT}\right)} \quad (i = 1, 2, \ldots) \tag{2.4}$$

$$m = C_1 e^{(-C_2 T)} \tag{2.5}$$

$$n = C_3 e^{(-C_4 T)} \tag{2.6}$$

$$H_T = D_0 + D_1 T \tag{2.7}$$

where A_1 and A_2 are pre-exponential factors, E_1 and E_2 are activation energies, R is universal gas constant, and C_1, C_2, C_3, C_4, D_0, and D_1 are constants.

2.2.2.2 Microwave curing model

Electromagnetic fields in the microwave frequency range are mostly used to create heat inside materials through microwave–material interactions. Permittivity and magnetic permeability of the material are responsible for microwave heat absorption in the presence of time-harmonic electric and magnetic fields. These characteristics and fields are regulated by a set of mathematical correlations known as Maxwell's equations and are given as [5, 6]:

$$\nabla \times \mathrm{E} = \mathrm{j}\mu\omega\mathrm{H} \tag{2.8}$$

$$\nabla \times \mathrm{H} = -\mathrm{j}\omega\varepsilon_0\varepsilon^*\mathrm{E} \tag{2.9}$$

$$\nabla.\left(\varepsilon\mathrm{E}\right) = 0 \tag{2.10}$$

$$\nabla.\mathrm{H} = 0 \tag{2.11}$$

where **E** and **H** are the time-harmonic electric and magnetic fields, ε^* is the complex permittivity of the material, ε is the permittivity, ε_0 is the permittivity in air, μ is the magnetic permeability, ω is the angular frequency equal to $2\pi f$, f is the frequency of microwaves, and j is the imaginary unit.

The material reaction to an applied electromagnetic field is based on dielectric, electric, and magnetic properties of matter.

The microwave coupling of material is described by Maxwell's equation. The major metric used to define the amount of heat generated by microwave contact with a material is its complex permittivity [5, 6, 10], which can be described as:

$$\varepsilon^* = \varepsilon' - j\varepsilon'' \tag{2.12}$$

where ε^* is the complex permittivity, ε' is the real part of complex permittivity, and ε'' is the dielectric loss factor (imaginary part). The ratio of imaginary part and real part of complex permittivity was termed as dissipation factor or loss factor ($\tan \delta$), an essential dielectric parameter that serves as a measure of the material's ability to produce heat, which is given as:

$$\tan\delta = \frac{\varepsilon''}{\varepsilon'} \tag{2.13}$$

The amount of energy absorbed by a material is directly proportional to the penetration depth (the distance from the surface of the material at which the magnitude of field strength drops by a factor of $1/e$). Microwaves, on the other hand, are not able to permeate all materials in the same way. Polymer and PMC are the mixed absorber of microwave radiation. The rate of microwave penetration into a material is determined the penetration depth [5, 6]:

$$D_p = \frac{1}{2\pi f\sqrt{\left[0.5\mu_0\mu'\varepsilon_0\varepsilon'\left\{\sqrt{\left(1+\left(\frac{\varepsilon''_{eff}}{\varepsilon'}\right)^2\right)}-1\right\}\right]}} \tag{2.14}$$

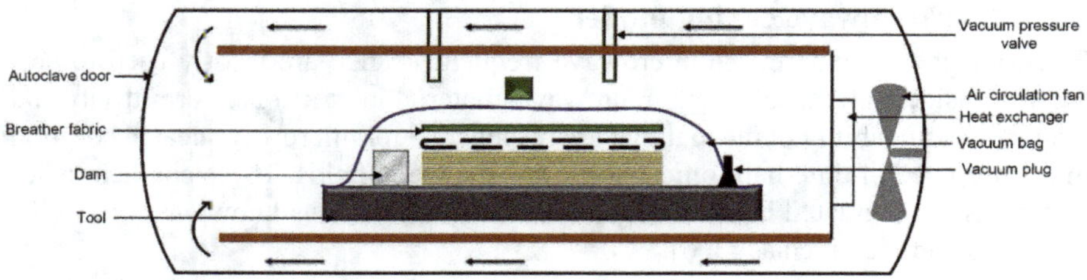

Figure 2.5 Schematic diagram of the experimental setups of autoclave

For dielectric heating, the electric field heating determines the power loss per unit volume, which can be expressed as [5, 10]:

$$P = \omega \varepsilon''_{eff} \varepsilon_0 E_{rms}^2 \tag{2.15}$$

where E_{rms} is the root mean square of the electric field, ε''_{eff} denotes to the effective relative dielectric loss factor, and μ''_{eff} denotes to the effective relative magnetic loss factor.

2.2.3 Experimental details

The schematic diagram of an autoclave that was used by various researchers for curing of the PMC is shown in Figure 2.5. The major components of an autoclave are as follows: a cylindrical pressure vessel with domed end, a fan in the dome end, and equipment for heating, cooling, vacuum, and instrumentation [11]. A concentric inner cylinder with one uncapped end is placed inside the outer cylinder. The prepreg (fiber impregnated with partially cured resin) covered with bleeder and breather is placed inside a vacuum bag, as shown in Figure 2.5. Generally, the excess resin is absorbed by the bleeders to utilize the optimal fiber volume fraction. Breathers allow escape of air and volatile gasses during the curing of the parts. The vacuum bag is connected to the vacuum system. A pressurized gas (usually, air or nitrogen) is circulated through the fan inside the inner cylinder, as indicated in Figure 2.5. Figure 2.6

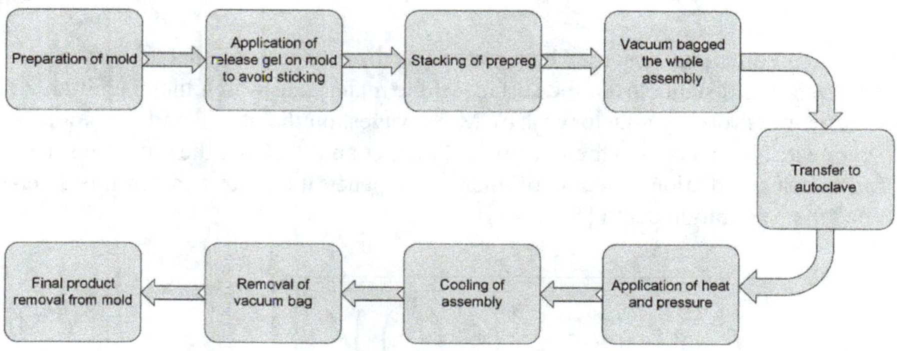

Figure 2.6 Processing steps involved in the autoclave curing of the PMC

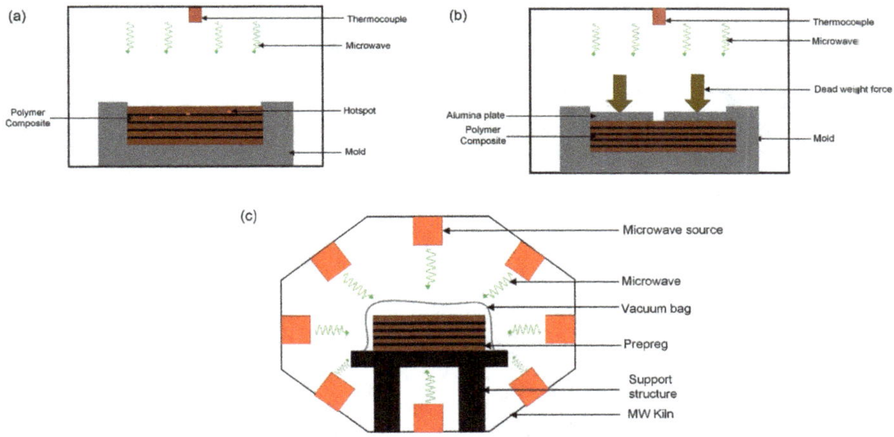

Figure 2.7 *Schematic diagrams of the experimental setups for (a) non-pressurized microwave cavity, (b) pressurized microwave cavity, and (c) multi-magnetron microwave cavity*

provides over view of the different steps involved in the autoclave curing of the PMC. The tool and laminate assembly are kept inside the autoclave. The pressure and temperature are applied simultaneously to cure the PMC.

Different experimental setups developed by researchers for microwave curing are shown in Figure 2.7. The basic experimental setup for microwave curing of the PMC is shown in Figure 2.7a. It consists of a mold of ceramic materials for holding the PMC [12, 13]. A thermocouple is used to monitor the temperature of the part during curing. Generally, the curing process is carried out in the multimode microwave applicators that include uneven distribution of field intensities. This experimental setup is unable to apply pressure and also lacks uniform heating. Another experimental setup (Figure 2.7b) was developed to apply pressure during microwave curing using dead weights [14]. This setup was useful in improving the mechanical properties of the PMCs; however, the issue of non-uniform heating could not be addressed. A new design of microwave applicator was developed for achieving more uniform heating (Figure 2.7c). It includes sixteen microwave sources symmetrically distributed on each side [15]. The applicator provides more flexibility in designing the heating pattern of the PMC during curing and hence achieving higher uniform heating of the PMC. Different processing steps involved in the microwave curing of the PMC are shown in Figure 2.8. The stacked PMC is placed in the mold and it is exposed to microwaves with/without pressure for curing the PMC.

The influencing parameters of autoclave and microwave curing processes of the PMC can be divided in two categories as follows: process-specific and material-specific [16–19]. These parameters need to be controlled and optimized for fabricating the required quality of the PMC parts. Some influencing parameters of autoclave and microwave curing with their effect on process output have been presented in Table 2.1.

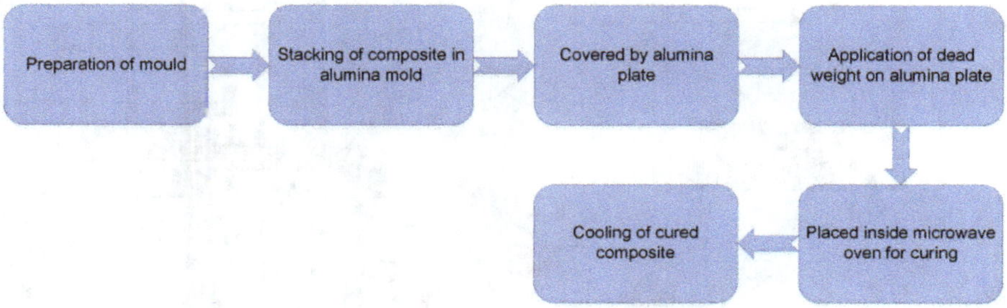

Figure 2.8 Processing steps involved in the autoclave curing of the PMC

A comparison between the autoclave and microwave curing methods has been presented in Table 2.2 while developing carbon fiber-reinforced polymer (CFRP) composites. A significant reduction in curing time (approximately three times) has been reported [20]. It is evident from the data that microwave curing induces higher voids in the cured CFRP composites as compared to autoclave curing due to the absence of pressure. The tensile properties of the autoclave-cured CFRP composite improve over microwave-cured CFRP. However, identifiable improvement in compressive strength, shear strength, and flexural strength has been reported.

Table 2.1 The influencing parameters for autoclave and microwave curing processes

Parameters	Curing method		Characteristics affected
	Autoclave	**Microwave**	
Process-specific	• Curing pressure • Curing temperature • Curing time • Vacuum pressure • Air flow rate	• Frequency • Power • Curing pressure • Curing temperature • Curing time • Vacuum pressure • Cavity shape and size	• Curing kinetics • Distribution of field intensity
Material-specific	• Thermal conductivity • Heat capacity • Density • Reinforcement effects • Additive effects	• Thermal conductivity • Heat capacity • Density • Reinforcement effects • Additive effects • Dielectric properties	• Penetration depth • Heat transfer • Heat dissipation

Table 2.2 Comparison of autoclave and microwave curing of carbon fiber-reinforced polymer

Characteristics	Curing method		References
	Autoclave	**Microwave**	
Curing time (min)	180	30	[20]
Void content (%)	0.1	19.3 and 8.9	[18]
Tensile strength (MPa)	750 ± 50 (cured at 120°C)	450 ± 50 (cured at 120°C) 400 ± 50 (cured at 180°C)	[20]
Compressive strength (MPa)	403 ± 9	422 ± 26	[21]
Shear strength (MPa)	47 ± 8	50 ± 4	[21]
Flexural strength (MPa)	650	700	[22]
Relative cost index (USD)	500–1000	200–1400	[18]
Production rate (per unit)	0.2–0.7	0.3–1.3	[18]

2.3 Challenges in autoclave and microwave curing

Autoclave curing and microwave curing have been explored by researchers and are well-understood. The processes have been employed for curing industrial components made up of advanced composite materials and satisfactory outcomes have been achieved. However, there are many challenges that are yet to be addressed [5, 6, 19]. Different challenges associated with autoclave and microwave curing are highlighted in Figure 2.9. Non-uniform heating of thick sections and limited flexibility of part shape and size are the major issues that need to be addressed in the case of both processes. Another challenge is due to the lower conductivity of the resins which severely affects the heat transfer. Most of the resins are transparent to microwaves and provide better chances of microwave absorption in conduction fibers such as carbon fiber. However, dissipated energy does not get transmitted during

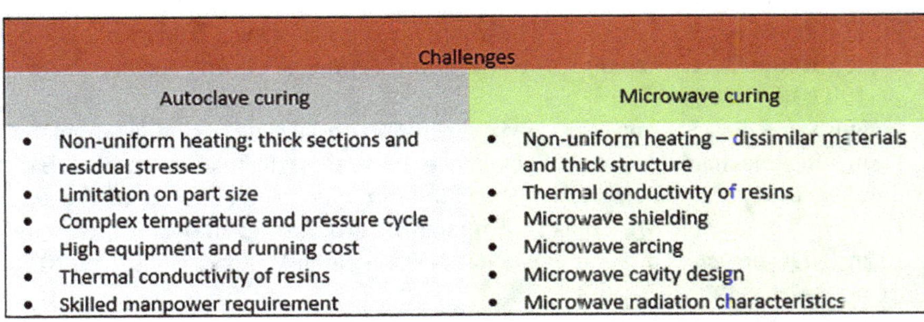

Figure 2.9 Challenges in autoclave and microwave curing

curing due to the lower conductivity of the polymer matrix. The situation gets worse with the combination of a resin and non-conducting fibers such as natural fibers.

2.4 Concluding remarks

Autoclave and microwave curing processes have demonstrated the potential to fabricate the PMCs at an industrial scale. Both the processes are capable of reducing the manufacturing cost with significant improvement in part quality. However, to realize such potential of these processes, the processes need to be explored for an in-depth understanding of the process mechanisms. On the other hand, the mechanical performances of the developed PMCs need to match the properties of conventionally cured components; this seems challenging as to date such targets seem unachievable. Achieving higher mechanical properties requires control of parameters, careful monitoring, and control of process parameters. The highlighted challenges indicate possibilities of research to improve the capabilities of the processes for realizing fabrication of the large-scale real-world components instead of relatively small test coupons. Furthermore, research on process control and optimization to improve cure quality and repeatability would be advantageous.

References

[1] Dang Z.-M., Yuan J.-K., Zha J.-W., Zhou T., Li S.-T., Hu G.-H. 'Fundamentals, processes and applications of high-permittivity polymer–matrix composites'. *Progress in Materials Science*. 2012;57(4):660–723.

[2] Abliz D., Duan Y., Steuernagel L., Xie L., Li D., Ziegmann G. 'Curing methods for advanced polymer composites - a review'. *Polymers and Polymer Composites*. 2013;**21**(6):341–48.

[3] Abrams F., Tolle T.B. 'An analysis of E-beam potential in aerospace composite manufacturing'. *Evolving Technologies for the Competitive Edge*. 1997;42:548–57.

[4] *ASC completes "world's largest autoclave" [online]*. 2006. Available from http://www.aschome.com/index.php/en/asc-completes-worlds-largest-autoclave [Accessed 16 Aug 2022].

[5] Thostenson E.T., Chou T.-W. 'Microwave processing: fundamentals and applications'. *Composites Part A: Applied Science and Manufacturing*. 1999;30(9):1055–71.

[6] Mishra R.R., Sharma A.K. 'Microwave–material interaction phenomena: heating mechanisms, challenges and opportunities in material processing'. *Composites Part A: Applied Science and Manufacturing*. 2016;81(2):78–97.

[7] Davé R.S., Henne M., Louis B., Ermanni P. 'Processing: autoclave modeling' in *Wiley encyclopedia of composites*. NY, US: John Wiley & Sons, Inc; 2011. pp. 1–13.

[8] Chanda M. *Introduction to polymer science and chemistry: A problem-solving approach*. Boca Raton, FL: CRC Press; 2006.

[9] Abdelal G.F., Robotham A., Cantwell W. 'Autoclave cure simulation of composite structures applying implicit and explicit FE techniques'. *International Journal of Mechanics and Materials in Design*. 2013;9(1):55–63.

[10] Das S., Mukhopadhyay A.K., Datta S., Basu D. 'Prospects of microwave processing: an overview'. *Bulletin of Materials Science*. 2009;32(1):1–13.

[11] Hubert P., Fernlund G., Poursartip A. 'Autoclave processing for composites' in *Manufacturing techniques for polymer matrix composites (pmcs)*. Philadelphia: Woodhead Publishing; 2012. pp. 414–34.

[12] Das S., Mukhopadhyay A.K., Datta S., Basu D. 'Prospects of microwave processing: an overview'. *Bulletin of Materials Science*. 2009;32(1):1–13.

[13] Rabby M.R.E., Jeelani S., Rangari V.K. 'Microwave processing of sic nanoparticles infused polymer composites: comparison of thermal and mechanical properties'. *Journal of Applied Polymer Science*. 2014;**132**(12):41708.

[14] Singh M.K., Zafar S. 'Influence of microwave power on mechanical properties of microwave-cured polyethylene/coir composites'. *Journal of Natural Fibers*. 2020;17(6):845–60.

[15] Li D., Li Y., Zhou J., Zhao Z. 'A novel method to improve temperature uniformity in polymer composites microwave curing process through deep learning with historical data'. *Applied Composite Materials*. 2020;27(1):1–17.

[16] Stefaniak D., Kappel E., Spröwitz T., Hühne C. 'Experimental identification of process parameters inducing warpage of autoclave-processed CFRP parts'. *Composites Part A: Applied Science and Manufacturing*. 2012;43(7):1081–91.

[17] Boey F.Y.C., Lye S.W. 'Effects of vacuum and pressure in an autoclave curing process for a thermosetting fibre-reinforced composite'. *Journal of Materials Processing Technology*. 1990;23(2):121–31.

[18] Galos J. 'Microwave processing of carbon fibre polymer composites: a review'. *Polymers and Polymer Composites*. 2021;29(3):151–62.

[19] Mgbemena C.O., Li D., Lin M.-F., *et al.* 'Accelerated microwave curing of fibre-reinforced thermoset polymer composites for structural applications: a review of scientific challenges'. *Composites Part A: Applied Science and Manufacturing*. 2018;115(1):88–103.

[20] Balzer B.B., McNabb J. 'Significant effect of microwave curing on tensile strength of carbon fiber composites'. *Journal of Industrial Technology*. 2008;24(3):1–9.

[21] Li Y., Cheng L., Zhou J. 'Curing multidirectional carbon fiber reinforced polymer composites with indirect microwave heating'. *The International Journal of Advanced Manufacturing Technology*. 2018;97(1-4):1137–47.

[22] Shimamoto D., Tominaga Y., Hotta Y. 'Effect of microwave irradiation on carbon fiber/epoxy resin composite fabricated by vacuum assisted resin transfer molding'. *Advanced Composite Materials*. 2016;25(1):71–9.

Chapter 3
Industry 4.0 for composites manufacturing

Miroslav Stojkovic[1] and Cristian Lira[2]

3.1 Introduction

This chapter reviews current theoretical and empirical studies related to the impact of Industry 4.0 on composites manufacturing. The aim is to identify industry trends and the gaps in the professional practice, making them available to the research community for further investigation.

The research scope is defined at an intersect of the three areas of interest: composites technology, business challenges, and Industry 4.0. The literature search is conducted by keyword search relevant to the areas of interest in academic databases. By further filtering of literature, the relevant papers were established, which were critically reviewed to establish key trends and gaps in professional practice. The critical review explored appropriate theoretical approaches for the area of interest, examined ethical position, and suggested a conceptual framework for investigating Industry 4.0 within composites industrial context.

Findings identified by reviewing literature predict the growth of the composites sector, dependent on the paradigm shift in the way composites are manufactured. Automation and more horizontally and vertically integrated business models are seen as trends in achieving this. However, before end-to-end automation is enabled, a knowledge base of complex process interactions needs to be established and be made available to manufacturing entities. Design for Manufacturing (DfM) technology is identified to enable across the company integration and capture of information by applying system theory and enabling knowledge base generation. However, there is a gap in composites theory and practice related to the model for DfM implementation that enables linking of departments and integration of the composites manufacturing business. The literature revealed that implementation of DfM requires management commitment and access to necessary resources. Review of Industry 4.0 literature demonstrated the ability to deliver on automation, integration, and DfM of composites manufacturing, by employing technologies such as autonomous robots, enabled by sensors, digital twinning, and the industrial Internet of Things (IIoT).

[1]Airborne UK, Membury Airfield, Lambourn, Berks, UK
[2]School of Science & Technology, City, University of London, London, England, UK

The reviewed literature investigates integration efforts needed to achieve DfM, however, fails to address the value proposition or organisational resistance in establishing such integration, as the industry does not understand the long-term gains and payback of implementing Industry 4.0 technologies.

The literature review indicated the key gap in professional practice as a lack of the model for structuring and implementing Industry 4.0 technology into composites business. The exposed gap covers an array of identified topics that need further investigation. The existence of identified gaps evidenced by the lack of literature and available knowledge reinforces the need for further research.

3.2 Composites context of Industry 4.0

This section introduces opportunities presented by the digital and automation technologies combined into Industry 4.0 in resolving challenges of composites manufacturing and supporting the innovation and management practices used to commercialise composites technology. Composites and their challenges are briefly introduced, followed by Industry 4.0 take on resolving listed challenges.

Composite materials with their specially engineered properties, formed by adding two or more existing materials to exploit their benefits, present significant opportunities to a variety of sectors, ranging from aerospace and automotive to renewable energy, marine, and construction [1]. Unusual geometries, non-uniform weight distributions, directional strength, and stiffness are the main advantages that composites can offer in existing or new products [1, 2]. Although the industry exists and delivers to the aforementioned markets, it is still reported to be in its infancy and needing further maturation when compared with the metallic counterparts [2–4].

The large-scale examples of composites use in the aerospace industry (Boeing-787, Airbus-A350) demonstrate not only the demand for composites due to their ability to offer a reduction in the structural weight but also the challenges presented by the lack of standardisation and the amount of tacit experience that this industry relies and is built upon [5–7]. Those challenges have led to significant delays and cost over-runs in mentioned programmes. However, literature does not provide sufficient explanation as to why the industry is still largely relying on tacit knowledge and practices that lack standardisation.

On the other hand, Black [8] and Maier [9] suggest that the composites could benefit from the current trends of the Industry 4.0 in combating standardisation and tacit knowledge. Existing research on this topic includes reports of industry attempts to standardise production by transforming challenging manual manufacturing steps into the automated ones, using technologies such as advanced fibre placement (AFP) or automatic tape laying (ATL) for automated deposition of the composite materials [1]. The research from Reference 2 (p. 4) contends that the final solution is not the mechanisation or automation alone and should include the development of integrated knowledge base around composites technology, although establishing such knowledge base and overall value proposition of such a system for the real composites business are not addressed.

Therefore, the main research question that guides this investigation is as follows:

How can manufacturing benefit from Industry 4.0 to lower the cost of composites and make them easier to develop and produce, enabling widespread adoption?

The following sub-questions are formulated to support the investigation:

1. What is the value proposition of Industry 4.0 to composites manufacturing?
2. What are the obstacles (barriers and constraints) and attractions (incentives or drivers) to this change?
3. What are the skills needed to enact this change?
4. Can Industry 4.0 help knowledge capture and re-use?

3.3 Literature review

As proposed by Reference 10 (pp. 10, 16), literature review within the research context aims to create new dimension or fresh perspective that makes a distinct contribution, and review should take a critical approach, which means recognising strengths and weaknesses of existing research and the ability to articulate why and how these ideas and theories might be improved. The current research scope is defined at the intersection of the three areas of interest, as indicated in Figure 3.1.

1. **Composites technology**: This topic was considered first to provide an overview of the technological challenges still existing in industry.
2. **Business challenges**: This topic was considered to provide an overview of the business challenges related to composites.
3. **Industry 4.0**: This topic was considered, with the view of providing information as to how Industry 4.0 can alleviate some of the previously identified challenges.

Literature search was carried out using the following:

1. External databases Scopus, Science Direct, Sage, and Google Scholar, where journal and conference articles are found by applying keywords such as 'Composites, Manufacturing', 'Cost-effective manufacturing', 'DfM', or 'Smart Factory'.
2. References and citations made within interesting papers being investigated as potential sources of relevant material.

Literature was filtered by excluding nonrelevant sectors and non-English language. Further relevance was established by reviewing titles and reading abstracts. Finally, the relevant papers are established by reading the full text.

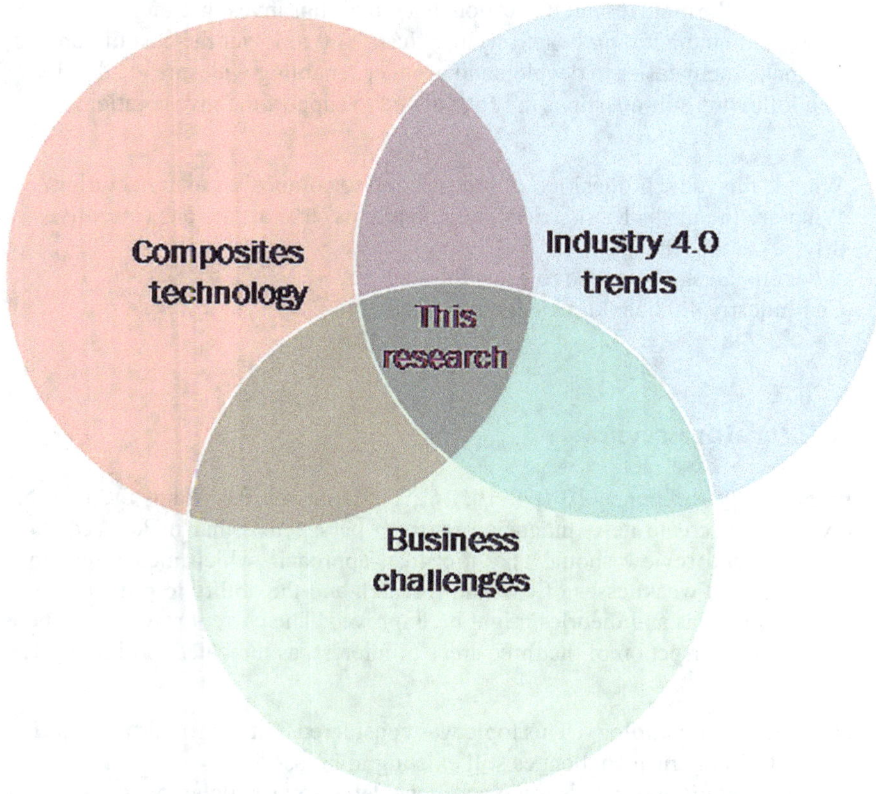

Figure 3.1 Research scope

3.3.1 Composites technology

This section provides the reader with an understanding of composites technology trends. The research identified the following three main topics:

1. Design for composites manufacturing
2. Manufacturing method
3. Material state as a function of the production process

3.3.1.1 Design for composites manufacturing

Composites suffer if design, engineering, and manufacturing are separated as the composites product and process are highly interdependent [11–13]. Despite these characteristics, the sector followed the principle of division of labour that enabled growth of traditional metal industries in the industrial revolution, which means disconnection of design, engineering, and production from physical craftsmanship skills [2].

Potter [14] explores this interdependency and highlights the importance of generalising manufacturing development knowledge that feeds into DfM. Potter further proposes that DfM enables interactions with the design and production, therefore, enhancing the chances of the right-first-time manufacture. Andersson *et al.* [15] report on Svenska Aeroplan Aktie Bolag implementation of DfM to its Carbon Fibers and Carbon Fiber-Reinforced plastic (CFRP) aircraft production, highlighting the need to implement DfM in all departments and develop proprietary DfM methods.

Youssef [16] explores the general application of DfM and proposes that DfM is a time- and computer-based technology that promotes advantages in terms of quality, flexibility, responsiveness, and cost by applying system theory in linking departments and enterprise integration. However, implementation of DfM requires management commitment to deliver company-wide changes as well as the set of computer-based tools and technologies to enable deployment.

Reviewed literature exposes sensitivity of composites to the division of design, development, and manufacture, proposing utilisation of DfM to link design and manufacturing through knowledge obtained in development. Industry typically applies system theory principles in realising DfM technology. However, a gap has been identified in availability of theoretical and practical guidelines related to the implementation of DfM technology not only to the technical or technological processes of composites but also across the entire enterprise.

3.3.1.2 Manufacturing method

From the early days of composites manufacturing, using the process called "bucket-and-brush" (repeat manual deposition of the dry fabric layers followed by the brush application of wetting resin from the bucket), through to more sophisticated and standardised utilisation of prepreg materials (pre-combined fabric and resin with the standardised quantity of each), the composites manufacturing process heavily relies on skilled humans craftsmanship skill [17, 18]. This self-taught skill acquired through the many years of experience is passed on to the next generations by directly working with the master laminator, causing significant challenges in expanding the workforce and scaling up the production.

Elkington *et al.* [19] suggest that the reason for this is that humans have the dextrous and problem-solving capabilities to tackle craftsman-like tasks such as composites lamination and can successfully adapt their skills to form a wide variety of complex shapes. These existing traditions and cultural norms of manufacturing practice are the direct cause of the tacit knowledge still existing in the industry [2, 19, 20].

Elkington *et al.* [19] propose three potential avenues to improve on the current state: automation, the evolution of the manual approach, and improvements in training.

Potter *et al.* [21] contend that the manual processes still dominate the manufacturing of very many, if not most, composite products, adding that even if AFP is used for material deposition, all the other manufacturing process steps may still be manual. Potter further proposes that before automating manufacturing, a clear understanding of the complex interactions between the parameters governing the manufacturing process needs to be developed, and information built into a knowledge base and fed back to support the design process.

The reviewed literature proposes automation as an ultimate goal, preceded by creating a knowledge base of complex process interactions, however, omits to explore the knowledge base constitution, enabling generalisation and transfer of captured knowledge.

3.3.1.3 Material state as a function of production process

Chatzimichali and Potter [2] indicate that the main difficulty in utilising composite materials is a requirement for a very deep understanding of the material behaviour during manufacturing. Talreja [22] proposes that a given manufacturing process results in a composite part with the process-specific material state as the material in its final form that is created at the same time as the component manufacture.

Hinton *et al.* [23] benchmark various polymer composites failure theories, and the work exposes the scale of a challenge when trying to predict the composites failure. The work of Lu [5] further states that it remains very difficult to predict fracture of composites when compared to metals.

To overcome the deficiencies of available composite failure theories, Talreja [22] proposes in-process characterisation of the material state by measuring introduced defects as the function of the manufacturing process parameters.

The proposal to enrich the design process by the in-process measurements of characteristics influencing the quality and hence the material state has also been made by Potter *et al.* [21].

An example of the process parameters and material state interactions is demonstrated by the study of Mgbemena *et al.* [24], which evaluates microwave curing in order to accelerate production process whilst maintaining controllability. Major factors influencing microwave curing of thermoset–matrix composites are identified and recommendations for efficient cure cycle design are provided. An importance of accelerated cure is exemplified in aerospace maintenance, repair and overhaul (MRO) where the average duration for in-service composite repair to perform successfully, at the flight line, could take up to 15 hours according to Commercial Aircraft Composite Repair Committee. Any variation in curing of thermoset polymers can potentially generate a defect precursor and dramatically lower down the adhesion properties. Ogonna Mgbemena, Yazdani Nezhad *et al.* [24] addressed the state-of-the-art in academic research and industrial applications and identified the major challenges for energy-saving high volume microwave processing techniques for thermoset composites. The lack of a reliable curing state measurement technique *in situ* has been identified as a major barrier towards designing controlled efficient microwave processes for composites. Another relevant challenge for accelerated microwave curing of fibre-reinforced polymer composites is that geometric features must be taken into account along with material constants for designing an optimum microwave curing cycle for a composite structure. The literature data collated in the review highlight a challenge of this process to be the development of the uniform temperature field throughout the cured part and hence the homogeneous cure. However, a coupled material process approach should provide a predictive tool to achieve optimum microwave process and mechanical performance overcoming current challenges in accelerated microwave curing of fibre-reinforced polymer composites. This should overcome the wrong implementation of recommended curing cycles used in conventional heating for microwave curing, which is currently a source of degradation of the final microwaved composite product and poor mechanical properties of the structural part cured via microwave heating. Once developed, the commercial solution should satisfy the requirement for fully controllable and timely efficient curing, provide essential information about the process induced material state and allowing to certify the repair or the structural component manufactured via accelerated microwave curing.

Reviewed literature proposes in-process measurement of the material and process parameters as input into the knowledge base of complex process interactions. However, architecture, implementation, and techno-economic viability of technology that enables critical information capture and reuse in either design or manufacturing stage in real commercial settings remains under investigation.

3.3.2 Business challenges

This section provides the reader with a better understanding of the existing research that focuses on business challenges faced by the composites industry. The research identified the following three main topics:

1. Composites market
2. Business models
3. Operational management

3.3.2.1 Composites market

Composites material usage is expected to grow globally [25], and the UK composites strategy developed by the Composite Leadership Forum [26] declared that the UK has the opportunity to grow its 2015 composite product market of £2.3 to £12 billion by the year 2030. The condition to achieve growth is the paradigm shift in composites manufacturing. The original UK composites strategy document [27] states that by investing in automation equipment companies can produce cost-effective, world-class composite products that are fit to compete with the best in the market. O'Dea [25] further suggests that the composites manufacturing challenge is represented in the lack of high volume processes for making structural parts, and industry report by Brosius [28] highlighted that the trends in a run-up to 2019 period were manufacturing process simulation, automated layup and moulding, as well as the IIoT, all of which should drive down the cost of composites.

Growth drivers are influenced by advantageous relative strength and stiffness to achieve light-weighting, gains in fuel efficiency, reduction in parts and maintenance cost as well as governmental policies, such as reducing CO_2 and fuel efficiency in the automotive sector, enabling competitive energy rates compared to the fossil fuels in the offshore renewable sector, and sustainability in construction [25, 26, 29].

According to Witten *et al.* [30], barriers are material standardisation, the cost of CFRP products restricting current usage in lower and mass volume production, relative immaturity of the manufacturing process, and the elaborate certification process.

The reviewed literature reveals an opportunity for growth and the general trends towards Industry 4.0 as a way to mature technology. However, literature was not found that addresses at which point automation and digitalisation satisfy the business case to be implemented.

3.3.2.2 Business models

According to Maine and Garnsey [12], Pisano [31], and Lubik and Garnsey [32], composite manufacturers execute science-based innovation business model by commercialising translation of the technological advances from the development into the commercial offers. Harms *et al.* [11] add that this model suffers from the challenges of technology and market uncertainty.

Maine and Garnsey [12] propose a business model for advanced material ventures that takes into account the key variables such as radical technology driving the innovation and value creation within the multiple markets, the position of the company within the value chain, and their influence on the ability to create value. However, this model does not address the complete ecosystem within which the composite companies operate.

Lubik and Garnsey [32] combine the resource-based theory [33] and ecosystem analysis, which are bridged by the business model concept. This theory claims that the advanced material venture cannot realise the potential of its innovation unless it can identify and participate in a business environment that can meet its critical requirements for co-innovation and adoption.

The resource-based theory and ecosystem analysis provide a good basis for establishing composites manufacturing business models through exploring the lines of value creation and needed resources. Proposed theory, however, needs further validation in case of composites, as it covers generic advanced materials ventures and concepts of value add based on the division of labour principle (that is deemed incompatible with the composites), amongst participants in the ecosystem.

3.3.2.3 Operation models

Fabris and Poursartip [3] used Technology Modularity–Technology Maturity framework of Pisano and Shih [13] and classified composites as high maturity in commercialisation but low at implementation. To implement a new or modified process typically requires process development which in itself is a research and low maturity. Modularity is defined as low, meaning that product development process, R&D, and manufacturing are intimately interconnected, which agrees with the views of

Maine and Garnsey [12], Chatzimichali and Potter [2], and Harms *et al.* [11]. Tang *et al.* [34] report that Boeing's 787 Dreamliner suffered from this when outsourcing design and manufacturing of parts, as integrated operations and body of knowledge were largely missing due to being distributed and not integrated within the wider supply chain.

As reported by Witten *et al.* [30], to overcome the issues of value creation and supply chain management, the companies operating within the composites market look to integrate horizontally (aiming at expanding their portfolios on a defined product) and vertically (to exploit shares of the up- or downstream value chains).

The review of the operational management literature reveals that vertical and horizontal business integration by acquisition is proposed to protect and manage value creation. Literature, however, does not address the model that benefits from the specialisation and risk-sharing of the supply chain, as attempted in case of Boeing-787, and benefits from the operational system integration of companies within this chain, aiding transparency, quality, and value creation.

3.3.3 Industry 4.0

This section provides the reader with an understanding of Industry 4.0 in its composite context. Nine underpinning technologies, building blocks of the Industry 4.0 [9, 35], are considered: autonomous robots, simulation, horizontal and vertical system integration, the IIoT, cybersecurity (CS), the cloud, additive manufacturing (AM), augmented reality (AR), and big data and analytics.

3.3.3.1 The industrial Internet of Things

Irving and Soutis [1] describe Industry 4.0 as the capability of automation and data exchange in manufacturing technologies. In the essence of Industry 4.0, the cyber-physical system is a term describing the connectivity of the physical assets (automated and sensorised manufacturing equipment) with the cyberspace (digital twins of physical assets) using resources such as the IIoT. Gerbert [35] claims that the IIoT is a vital part of the Industry 4.0 representing the connectivity between equipment and product with digital resources. Black [8] reports on the importance of digital asset tracking enabled by the IIoT where composite manufacturers lose 10–15% of the inventory without this technology and save up to 80% on material spend when digital tracking is in place.

3.3.3.2 Simulation

Gerbert [35] reports that simulations are used more extensively in plant operations to leverage real-time data and mirror the physical world in a virtual model, including machines, products, supply chains, and humans, allowing operators to test and optimise machine settings in the virtual world before the physical actions take place.

Industry 4.0 refers to a variety of simulations across the product development chain as digital twins, encompassing design, analysis, process development, and production [8]. According to Irving and Soutis [1] with tools that enable virtual verification, the 'right-first-time' approach for composites is more realistic as products could be virtually tested at any stage of the development.

However, Tyrrell [36] contends that there is still work to be done to convince the composites manufacturing industry of the value brought by simulation technologies, and Richardson [37] poses a question as to how much digitalisation and automation are also seen as 'dark arts' as the customers do not always understand the long-term gains and payback.

3.3.3.3 Horizontal and vertical system integration

Deuter and Pethig [38] propose that digital twin technology enables both the vertical and horizontal integration of the value chains, naming them the key value-add elements of the product lifecycle management (PLM).

The paper of Perez-Lara *et al.* [39] proposes the generic company structure for vertical and horizontal integration, and Black [8] reports that to make this a reality, a company needs a digital enterprise platform able to integrate data from different sources and with different protocols, such as PLM software by Siemens used to offline test and develop composite products before real manufacture takes place. Irving and Soutis [1] claim that with this technology, products reach the market as much as 50% faster, with at least the same level of quality achieved without PLM.

According to Snudden [40], system integration provides opportunities for composites manufacturing to provide services, without owning infrastructure, but offering complete transparency enabled by system integration and IIoT. This is demonstrated by Airborne on-demand manufacturing portal [41]. However, this business model requires the adoption of a wider ecosystem and changes in business processes.

The reviewed literature reveals the possibility of achieving a new manufacturing paradigm by enacting vertical and horizontal system integration, however, fails to address the implementation model or organisational resistance in establishing such integration.

3.3.3.4 Autonomous robots

The third industrial revolution introduced robots in the 60 seconds, and the composites industry started experimenting with robots in manufacturing about 30 years ago [42]. Richardson [37] states that industrial automation ensures reliability of composites manufacturing. However, Industry 4.0 brings together automated machinery, including robots, with intelligence and connectivity which allow smart production decisions to be made in reaction to changing requirements. Snudden [40] mentions robotic solutions in composites such as pick and place, automated tape laying (ATL), automated fibre placement (AFP), filament winding or pultrusion, suffering from requiring large capital investments, and requiring the vision of the long-term programmes to guarantee the capital investment and personnel training payback. Björnsson [43] proposes tackling high investment costs by utilising Commercial-Off-The-Shelf solutions able to meet rigorous quality standards and handle a variety of materials and products.

Snudden [40] continues that traditional robotics generally require monitoring and inspection since there are no feedback loops to confirm the automated operation performed satisfactorily. Jay [42] claims that AFP automation technology is only active for 25–35% of the time available to deposit material on to the mould, with an inspection making the rest of the time. As those solutions are developed by the companies traditionally operating in the automation, but not composite specific, the material behaviour during the production is often neglected, leading to serious issues with process reliability and productivity [44]. Inability to capture the expert skills and develop automated technologies relevant to composites results in limiting the composite production capability [2].

Papadopoulos [45] investigates the concern of anticipated lowering in employment by the rise of automation, reporting that the Massachusetts Institute of Technology (MIT) study carried out in France, revealed that every 20% increase in robot use from 2010 to 2015 led to a 3.2% decrease in employment; however for firms adopting robots, total employee hours increased by 10.9% and wages rose as well. Fuscaldo [46] claims that the automation results in the net gain for employment. Küpper *et al.* [47] report that robotics and computerisation reduce the number of jobs in production; however, the number of manufacturing jobs requiring IT and data science skills increases. The report adds that the future employee skills are a major challenge and mechatronics skills (combined mechanical, electronic, and IT skills) will be more relevant, rather than purely mechanical skills.

3.3.3.5 Additive manufacturing

Gartner and Fink [48] report that AM includes sheet lamination, and Snudden [40] states that advanced composites are AM processes. Di Francesco [49] states the final composite part consists of material layers built up during production, so automated deposition of fibre or tape can be considered a large-scale 3D printer. Daminabo *et al.* [50] consider multiple techniques and polymer materials systems. This review provides a good overview of existing AM techniques, highlighting how AM techniques are still significantly limited by the range of printable functional material systems that meet the requirements of a broad range of industries. Daminabo *et al.* [50] state that AM is capable of cost-effectively improving innovation, production, and service lead times, whilst delivering a high level of flexibility in manufacturing. However, the rapid development of AM technologies has been a key limitation to its development as this is leading to the lack of specific design principles, manufacturing guidelines, and standards that guides AM. A challenge that needs to be tackled strategically is related to the fact that advancement in AM, for example, advancements in material processing and optimisation, introduces new changes within the ecosystem, thereby making guidelines, principles, and standards more difficult to define. In this respect, the potential of AM as a sustainable and decentralised manufacturing system fails to be exploited by missing to capture existing tacit knowledge and transferring it into guidelines. Guidelines about process cost-efficiency and wider range of materials processability and environmental sustainability like reusability, recyclability, recovery, and disposal of products at their end-of-life are still missing. The implementation of digital twins of the AM process will be potentially capable of supporting a variety of standards or customised product and services quality improvement via transfer of the tacit knowledge into digital twins and having the digital twin as the repository of the currently missing guidelines. This approach will be particularly meaningful when used to identify innovative materials and recent research activities that may support breakthroughs for AM and its sustainability.

3.3.3.6 Big data and analytics

Richardson [37] reports of digital engineering within the composites context aiming to capture the knowledge of experienced personnel and translate it to the tools that people with less experience can use, and end-to-end automation might not be the right answer to all the parts and sectors, specifically when the value proposition is not fully understood. However, tailored solutions that technologically and economically make sense coupled up with in-process monitoring would be of practical value. Black [8] reports the Industry 4.0 is the marriage of advanced manufacturing techniques with information technology, data, and analytics, and Maier [9] proposes that smart factories can collect data during the manufacturing process and link measured parameters with virtual models that replicate that particular stage of the development, enabling the live feedback and corrective action, as well as the knowledge capture and deeper insight of the process.

The reviewed literature proposes that capturing data from manufacture is in the essence of the digital proposition, and an indication is given that selective rather than the end-to-end implementation of Industry 4.0 could deliver the practical value. However, the literature fails to examine and propose methodology by which the value-add level of Industry 4.0 is determined.

3.3.3.7 The cloud

Black [8] reports that ‘cloud’ enables large data storage and computing power for data analysis enabling rapid computing scalability and flexibility, seamless data sharing across sites, and utilising data analytics and artificial intelligence as a service. All these activities provide a deeper insight into the manufacturing process based on the historical data or ideally real-time feedback, allowing adaptive manufacturing.

3.3.3.8 Cybersecurity

Butt [51] reports that CS is rapidly becoming a major concern for manufacturers and consumers. With the increased connectivity and use of standard communication protocols that come with Industry 4.0, the need to protect critical industrial systems and manufacturing lines from CS threats increases dramatically. As a result, secure, reliable communications, as well as sophisticated identity and access management of machines and users, are essential.

3.3.3.9 Augmented reality

Snudden [40] reports that virtual reality immerses users in a computer-generated world, and AR overlays digital information on to the physical world. As composites manufacturing is a dominantly manual process, these devices can be used to train operators as well as in production to inspect the location and orientation of the placed material. Nezhad *et al.* [52] developed an AR equipped composite bonded assembly and repair for aerospace application. Nezhad described the enormous opportunities and interests of the global aircraft MRO market where a competitive industry solution which bears the merit of efficiency and also provides accessibility to explicitly instruct the operations of MRO engineers and operators is in highly demanded. Nezhad identified the use of visual representation and discussed the gap between existing AR industry applications and the composite bonded repair process with digital cameras embedded in the HoloLens. With the use of HoloLens, the synthetic scene presented directly in front of the operator's view enables close scrutiny of the current step and the steps to be followed to complete successfully the repair-like orientation of fibrous reinforcement and optimisation of the fibre-to-resin ratio, which are key parameters to obtain strength-to-weight superiority of the repair and avoid non-uniform stress transfer and ineffective bonding.

3.4 Critical assessment

The reviewed literature was used to establish the key trends and gaps in professional practice, define the ethical position, and elicit opportunities for further research.

3.4.1 Key trends

The key trends within the professional practice are established by analysing and comparing trends identified for each area of interest.

The identified key trends at the intersection of the area of interest for this research, as presented in Figure 3.1, are activities, realisation of which overcomes barriers and constraints of the traditional composites industry, enabling composites market growth, greater value generation, and waste reduction via reduction of manufacturing uncertainty and variability whilst enabling production scaling by:

- Developing and deploying DfM (or other appropriate process excellence methodologies) including product and process knowledge serving as an architectural plan for Industry 4.0 implementation
- Adopting level of automation and autonomous robots as appropriate to tackle inconsistencies, uncertainties, and variability
- Enhancing use of cyber connectivity through IIoT to tackle sensor data capture, departmental information exchange, and feedback
- Digital twinning of operations for optimisation and future states prediction

and via Reduction of business model uncertainties by:

- Adequately utilising the wider business ecosystem
- Vertical and horizontal system integration to overcome the issues of market and technology uncertainties, value creation, and supply chain management

However, the literature suggests that the adoption of mentioned trends is constrained by:

- Understanding the techno-economic viability (cost–benefit analysis) of implementing Industry 4.0 technology to the composites manufacturing companies when delivering to different sectors with varied requirements
- Change of cultural norms, skills, business processes, and organisational structure within each company of the ecosystem, as this may be met with resistance from the stakeholders
- CS issues posed by the introduction of Industry 4.0 technologies
- Adoption of Industry 4.0 by the wider composites business ecosystem

3.4.2 Gaps and research aim refinement

The professional practice gap analysis is carried out by evaluating the trends in the reviewed literature. By attempting to evaluate how investigated literature answered research question and supplementary questions, the key gap in professional practice is formulated as:

A lack of the model for structuring and implementing the value-add level of Industry 4.0 technology into the composites business delivering to a diverse range of sectors.

The above-elicited gap in professional practice is echoed by the recent publication of Butt [51], who claims that implementation of Industry 4.0 enabling technologies is a difficult task and becomes even more challenging without any standardised approach. Butt [51] adds that the barriers include, but are not limited to, lack of knowledge, inability to realistically quantify the return on investment, and lack of a skilled workforce. Although [51] is valuable as it proposes an implementation model based on recognised by manufacturer's six sigma process excellence methodology, it is not directly addressing composites and needs further justification in composites manufacturing practice.

Therefore, the challenge of addressing Industry 4.0 implementation model for composites manufacturing sector remains, and further research in topics mentioned below is needed to deliver desired changes, which are as follows:

1. Cost of implementation and payback of Industry 4.0
2. Operational systems integration of business functions through Industry 4.0
3. Methods for knowledge capture, generalisation and transfer through Industry 4.0
4. Business model adaptation to protect competitiveness
5. Establishing a new combined skill from the commercial company point of view

3.5 Concluding remarks

This review provides indications of the Industry 4.0 impact on composites manufacturing. Although the reviewed literature proposes that new trends can reduce manufacturing and business model uncertainties, evidence also suggests that the adoption of Industry 4.0 requires an understanding of techno-economic viability of this technology, change of cultural norms, skills, business processes and organisational structure, CS improvements to match Industry 4.0 technologies, and adoption by the wider composites business ecosystem.

The absence of literature investigating the model for structuring and implementing Industry 4.0 technology into the composites sector identifies that there is scope for further research.

On the highest level, why businesses should invest in the technology must be justified.

The next level to the challenge is where the technology should be applied. It could be business-wide, or it could be on a manufacturing process level, depending on the requirements of the technology.

How technology is applied requires a different set of capabilities. These could include software development, robotics, or data science. A manufacturing business may not have these capabilities, so the question arises of whether to build the capabilities in house or outsource.

A driver in this decision would be the need to retain or capture competitive advantage, and this may require broad knowledge capture or re-examining the business model.

Further research is needed to address the above considerations and contribute to establishing an industry-relevant model for the implementation of Industry 4.0 technologies within composites manufacturing settings.

References

[1] Irving P., Soutis C. *Polymer composites in the aerospace industry*. 2nd ed. Woodhead Publishing; 2020.

[2] Chatzimichali A.P., Potter K.D. 'From composite material technologies to composite products: a cross-sectorial reflection on technology transitions and production capability'. *Translational Materials Research*. 2015;2(2):026001.

[3] Fabris J., Poursartip A. 'Using process modelling as a job-aid to reduce composites manufacturing risk'. *SAMPE Conference Proceedings*; WA, 2017.

[4] Sloan J. Boeing offers insights on 787 composites lessons, composites world [online]. 2014. Available from https://www.compositesworld.com/blog/post/despite-787-boeing-not-sold-oncomposites [Accessed 17 Mar 2020].

[5] Lu K. 'The future of metals'. *Science*. 2010;328:319–20.

[6] Marsh G. 'Airbus takes on boeing with composite A350 XWB'. *Reinforced Plastics*. 2007;51:26–27.

[7] Slayton R., Spinardi G. 'Radical innovation in scaling up: Boeing's dreamliner and the challenge of socio-technical transitions'. *Technovation.* 2016;47:47–58.

[8] Black S. Composites and Industry 4.0: where are we? [online]. 2017. Available from https://www.compositesworld.com/articles/composites-and-industry-40-where-arewe [Accessed 29 Mar 2020].

[9] Maier J. *Made smarter review*. London, UK: Department for Business, Energy & Industrial Strategy; 2017. Available from https://assets.publishing.service.gov.uk/government/uploads/system/uploads/attachment_data/file/655570/20171027_MadeSmarter_FINAL_DIGITAL.pdf

[10] Jesson J., Matheson L., Lacey F.M. *Doing your literature review: traditional and systematic techniques*. 1st ed. London: Sage; 2011.

[11] Harms R., Marinakis Y., Walsh S.T. 'Lean startup for materials ventures and other science-based ventures: under what conditions is it useful?'. *Translational Materials Research.* 2015;2(3):035001.

[12] Maine E., Garnsey E. 'Commercializing generic technology: the case of advanced materials ventures'. *Research Policy.* 2006;35:375–93.

[13] Pisano G.P., Shih W.C. 'Does America really need manufacturing?'. *Harvard Business Review*. 2012;90:94–102.

[14] Potter K.D. *Introduction to composite products—design, development and manufacture*. London/Berlin: Chapman and Hall/Springer; 1997.

[15] Andersson F., Hagqvist A., Sundin E., Björkman M. 'Design for manufacturing of composite structures for commercial aircraft – the development of a DFM strategy at SAAB aerostructures'. *Proceedings of the 47th CIRP Conference on Manufacturing Systems*; Linköping, 2014.

[16] Youssef M.A. 'Design for manufacturability and time-to-market, part 1: theoretical foundations'. *International Journal of Operations & Production Management*. 1994;14(12):6–21.

[17] Bloom L.D., Elkington M., Chatzimichali A., Potter K. 'On prepreg properties and manufacturability'. *19th International Conference on Composite Materials, ICCM 19*; Montreal, 2013.

[18] Paton R. *Forming technologies for thermoset composites, composites forming technologies*. 1st ed. Long A.C. (ed.). Cambridge: Woodhead; 2007.

[19] Elkington D. 'Understanding the lamination process'. *ICCM 19*; Montreal, Canada, 2013.

[20] Fabris J., Roughley D., Poursartip A., Maine E. 'Managing the technological and market uncertainty of composites innovation: a case study of composites manufacturers in Western Canada and interventions by a translational research centre'. *Translational Materials Research.* 2017;4(4).

[21] Potter K., Ward C., Bloom D., Crowley D., Elkington M., Mahadik Y. *Design for manufacture: advancing lamination technologies towards right first time every time*. Bristol: University of Bristol; 2016.

[22] Talreja R. 'Manufacturing defects in composites and their effects on performance' in *Polymer composites in the aerospace industry*. Vol. 1; 2020. pp. 83–97.

[23] Hinton P.M., Kaddour D.A.S., Soden M.P.D. *Failure criteria in fibre reinforced polymer composites: the world-wide failure exercise*. 1st ed. Elsevier: Oxford; 2004.

[24] Mgbemena C.O., Li D., Lin M.-F, *et al*. 'Accelerated microwave curing of fibre-reinforced thermoset polymer composites for structural applications: a review of scientific challenges'. *Composites Part A: Applied Science and Manufacturing*. 2018;115(1):88–103.

[25] O'Dea N. *Global outlook for the composites industry*. Birmingham: Lucintel; 2018.

[26] Composite Leadership Forum *The 2016 UK composites strategy*. Hemel Hempstead: Composite Leadership Forum; 2016.

[27] BIS, Department for Innovation & Skills *The UK composites strategy*. s.l: Department for Business, Innovation & Skills; 2009.

[28] Brosius D. Looking ahead at composites trends in 2019 [online]. 2019. Available from https://www.compositesworld.com/articles/looking-ahead-in-2019 [Accessed 24 May 2020].

[29] Mathes V. 'The composites industry: plenty of opportunities in heterogeneous market'. *Reinforced Plastics*. 2018;62(1):44–51.

[30] Witten E., Mathes V., Sauer M., Kühnel M. *Composites market report 2018*. Frankfurt: Industrievereinigung Verstärkte Kunststoffe; 2018.

[31] Pisano G.P. 'The evolution of science-based business: innovating how we innovate'. *Industrial and Corporate Change*. 2010;19(2):465–82.

[32] Lubik S., Garnsey E. 'Early business model evolution in science-based ventures'. *Long Range Planning*. 2016;49(3):393–408.

[33] Penrose E.T. *The theory of the growth of the firm*. 4th ed. Oxford: Oxford University Press; 1959.

[34] Tang C.S., Zimmerman J.D., Nelson J. 'Managing new product development and supply chain risks: the boeing 787 case'. *Supply Chain Forum: An International Journal*. 2009;10(2):74–86.

[35] Gerbert P. Industry 4.0: the future of productivity and growth in manufacturing industries [online]. 2015. Available from https://www.bcg.com/publications/2015/engineered_products_project_business_industry_4_future_productivity_growth_manufacturing_industries.aspx [Accessed 30 Mar 2020].

[36] Tyrrell M. 'Simulate to innovate' in *Cim composites in manufacturing*; 2016. pp. 35–37.

[37] Richardson M. 'A digitally bright future for composites' in *Composites in manufacturing*; 2020. pp. 18–20.

[38] Deuter A., Pethig F. 'The digital twin theory – a new view on a buzzword'. *Industry 4.0 Management*. 2019;35(1):27–30.

[39] Perez-Lara M., Saucedo J., Marmolejo J.A., Salais T., Vasant P. 'Vertical and horizontal integration systems in Industry 4.0'. *Wireless Networks*. 2018;24(8).

[40] Snudden J. 'Progression to the next industrial revolution: Industry 4.0 for compsites'. *Reinforced Plastics*. 2019;63(3):136–42.

[41] Van Mourik A. The need for speed: why the composites industry should radically change now [online]. 2018. Available from https://www.airborne.com/the-need-for-speed-why-the-composites-industry-shouldradically-change-now/ [Accessed 6 Jun 2020].

[42] Jay M.L. Rise of the robots [online]. 2017. Available from http://compositesmanufacturingmagazine.com/2017/09/composites-industry-willbenefit-from-robotics-and-automation/ [Accessed 2 Apr 2020].

[43] Björnsson A. *Enabling automation of composite manufacturing through the use of off the shelf solutions*. Linköping: Licentiate dissertation; 2014.

[44] Lukaszeewicz D., Ward C., Potter K. 'The engineering aspects of automated prepreg layup: history, present and future'. *Composites Part B*. 2012;43(3):997–1009.

[45] Papadopoulos L. More robots lead to hiring more people, study shows [online]. 2020. Available from https://interestingengineering.com/more-robots-lead-to-hiring-more-people-studyshows [Accessed 6 Jun 2020].

[46] Fuscaldo D. No robots won't take our jobs but we need new skills [online]. 2019. Available from https://interestingengineering.com/no-robots-wont-take-our-jobs-but-we-need-newskills [Accessed 6 Jun 2020].

[47] Küpper D., Kuhlmann K., Köcher S., Dauner T., Burggraaff P. The factory of the future [online]. 2016. Available from https://www.bcg.com/publications/2016/leaning-manufacturing-operations-factoryof-future.aspx [Accessed 17 Jan 2020].

[48] Gartner J., Fink M. 'The magic cube: towards a theoretical framework to explain the disruptive potential of additive manufacturing'. *Translational Materials Research*. 2018;5(2).

[49] Di Francesco M. 'Digital manufacturing: turning high-end composites manufacturing systems into 3D printers'. *SAMPE Conference*; Nantes, 2019.

[50] Daminabo S., Goel S., Grammatikos S., Nezhad H.Y., Thakur V.K. 'FDM-based additive manufacturing (3D printing): techniques for polymer material systems'. *Materials Today*. 2020;16(1).

[51] Butt J. 'A strategic roadmap for the manufacturing industry to implement Industry 4.0'. *Designs*. 2020;4(2):11–30.

[52] Nezhad H.Y., Wang X., Court S.D., Thapa B., Erkoyuncu J.A. 'Development of an augmented reality equipped composites bonded assembly and repair for aerospace applications'. *IFAC-PapersOnLine*. 2020;53(3):209–15.

Chapter 4

Development of fibre-reinforced polymer composites through direct digital manufacturing

Ramesh Kumar Nayak[1] and Hamed Yazdani Nezhad[2]

Direct digital manufacturing (DDM) is nothing but additive manufacturing (AM) digitally of three-dimensional (3D) printing. It has changed the way things are manufactured traditionally. The term DDM process undergoes from a digital blueprint of a product to a finished physical product. The technology is used to generate an object by computers and manufacture digitally to create real-world objects ranging from simple toy pieces to complex fighter jet parts. It is extensively adopted by many industries for product development such as toys, construction, medical devices, engineering components for automotive, marine, aerospace, space, and even human organs. One of the most important advantages of DDM is to manufacture the products with the high-dimensional accuracy where the cost and lead times can be reliably predicted. It is major breakthrough in manufacturing technology because this process can directly manufacture components without the need for expensive tooling, molding or machining to create complex geometries. In general, fibers have been traditionally used in many manufacturing processes for various reasons. However, using conventional methods, there are obstacles in obtaining the desired complex geometries and low setup costs. AM offers possible avoidance of these limitations. Bridging AM with fiber-reinforced materials offers a vast opportunity for lightweight and strong parts. Depending on the affinity, fibers with different structures can be mixed with different matrix materials and, thus, create stronger parts with improved mechanical properties. Process parameters like raster angle, infill speed, layer thickness, and nozzle temperature also strongly impact physical properties of fiber-reinforced AM products and are considered carefully. The products are used in many industries such as aerospace, motorsports, and biomedicine, where the weight, strength, and complexity of parts are critical. Fiber-reinforced polymer composites (FRP) are the leading engineering materials in twenty-first century. The manufacturing of FRP composites through DDM is a challenge. Therefore,

[1]Department of Materials and Metallurgical Engineering, Maulana Azad National Institute of Technology, Bhopal, India

[2]Aeronautics & Aerospace Research Centre Department of Mechanical Engineering & Aeronautics City, University of London, London, UK

this chapter focuses on the principle, manufacturing process, and its application to different products, and limitations are discussed.

4.1 Introduction

Manufacturing has a history of several thousand years and is driven by several approaches. Each such process results in the well-defined changes in the physical and chemical characteristics of the materials. What really separates DDM is the freedom of design it offers. In conventional manufacturing, the design team needs to consider the manufacturing constraints first and then design accordingly. Let us say a designer has an innovative idea about creating a curved steering wheel for an automobile that is extremely ergonomic. It can be extremely beneficial for drivers who need to undertake a long journey. Since driver fatigue is one of the main causes of road accidents, this new design can prove to be a boon by increasing the comfort level of the driver and reducing fatigue. However, if the production team says that they just do not have any tool sets that will produce a curved steering wheel, it simply cannot be manufactured. Traditionally, the fiber-reinforced polymer composites, nanocomposites, and hybrid nanocomposites are manufactured by hand lay-up methods. The effect of fiber hybridization on mechanical and tribological properties is studied by different researchers [1–8]. DDM frees designers to design exactly the way they want, without having to worry about the manufacturing concerns. If they can create a 3D-printing file, it can be manufactured. Design is no longer constrained by the limitations of conventional manufacturing processes. This frees the product development team to focus completely on designing a part exactly the way they envisage it. An added advantage this freedom offers is that products not only function better; they look better as well. Just like traditional manufacturing, there are different 3D-printing technologies. And just like traditional manufacturing, each technology has advantages and disadvantages. The present 3D-printing technologies include stereo lithography (SLA), digital light processing, fused deposition modeling (FDM), selective laser sintering (SLS), selective laser melting, electronic beam melting, laminated object manufacturing and variants of these basic technologies. Which 3D printer to choose depends entirely upon the manufacturer's objectives. Fiber-reinforced plastics (FRP) have recently gained attention for their potential use in 3D printing to enhance the mechanical strength and elasticity of manufactured parts [3, 4]. These composite materials can match the strength of many metals but at much lighter weight. In particular, the addition of carbon fibers to plastic resins is becoming a widely used strategy to enhance the mechanical properties of 3D-printed parts.

4.2 Fiber-reinforced AM composites

Parts fabricated with fiber-reinforced AM (FRAM) consist of a matrix and reinforcement elements, as shown in Figure 4.1. The chart presented in Figure 4.1 is a collection of materials used in recent FRAM studies. There are numerous research projects

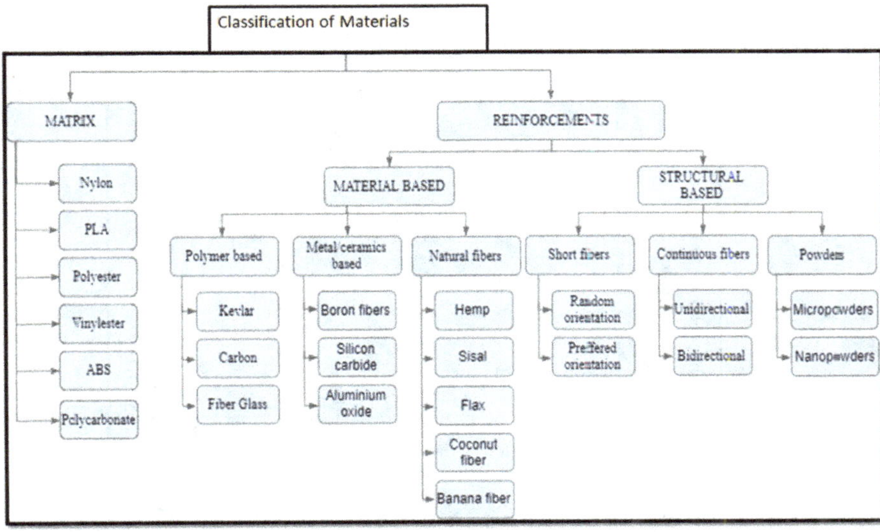

Figure 4.1 Classification of FRAM materials [9]

under development to produce lightweight, low-cost, and sustainable matrices by selecting the appropriate combination of matrix and reinforcement materials. A collection of FRAM materials and their mechanical properties is shown in Table 4.1 [9].

Fibers in FRAM can be used in the form of short, continuous, and powder based. In the short fibers group, the most common ones are carbon and glass fibers [10]. Short fibers are easier and cheaper to produce than continuous ones. Until 2014, continuous fibers were used with conventional manufacturing methods. In 2014, continuous fibers started to be used in AM, as well. Several studies have shown that short (discrete or chopped) fibers extruded as one filament successfully integrated with matrix material are applicable in several AM techniques [11, 12]. Newly developed FRAM machines printing the fibers and matrix materials continuously layer-by-layer have presented successful research, development, and industrial results. Similar to short fiber technologies, matrix and filament materials could also be reinforced in micro- and nano-level using powder bed technologies' fiber binding with laser or adhesives [13]. The schematic representation of the FRAM process is illustrated in Figure 4.2.

Matrices, in this case polymers, play a vital role in composite materials (CMs) by (1) keeping fibers in place, (2) transferring stress between fibers, (3) protecting fibers from adverse environmental conditions, (4) preventing surface abrasion, and (5) supporting fibers under compression loading [14]. The matrix materials also strongly influence shear and compressive properties of CMs. The matrix material should be thermally, physically, and chemically (matrix should not react with fibers) compatible with the reinforcements. Polyester is widely used in biomedical

Table 4.1 Mechanical properties of materials used in FRAM [9]

Reinforced fibers	Density (g/cm³)	Tensile modulus (GPa)	Tensile strength (MPa)	Thermal conductivity (W/m*k)	Linear coefficient of thermal expansion (10^{-6}/°C)	Diameter (μm)
			Polymer fibers			
Carbon	1.8	230–270	3600	21–180	4–8	5–10
Kevlar	1.4	70.5–112.4	3620	0.04	–4.86 to –3.96	11.9
Glass	2.54	50–90	3000	3.21	4–9	10
			Metal ceramics fibers			
Boron	2.45	450	3500	3.22	4.5	140
Al_2O_3	3.4–3.9	215–413	2100–3100	25.08	8.1	10–12
SiC	2.55–3.08	450	2750–3440	120	2.77	14.5–140
			Natural fibers			
Jute	1.3–1.45	39.4	320–800	0.059	–0.6	20–200
Flax	1.4–1.5	62.4	345–2000	2	–0.8	12–600
Sisal	1.33–15	21.9	373–700	0.25–0.38	–3.9	8–200
			Matrix materials			
Nylon	1.15–1.85	2–4	40–60	0.2	80	14–24
ABS	1.01–1.21	1.4–3.1	25–50	0.15–2	72–108	50–53
PLA	1.252	1.28	59	0.11–0.19	85	60
Polyester	1.2–1.5	3–3.5	40–90	0.2	124	12–25
Vinyl ester	1.12–1.32	3–3.5	73–81	0.3–0.4	16–22	240
Polycarbonate	1.06–1.2	2.6	45–70	0.2	65–76	150

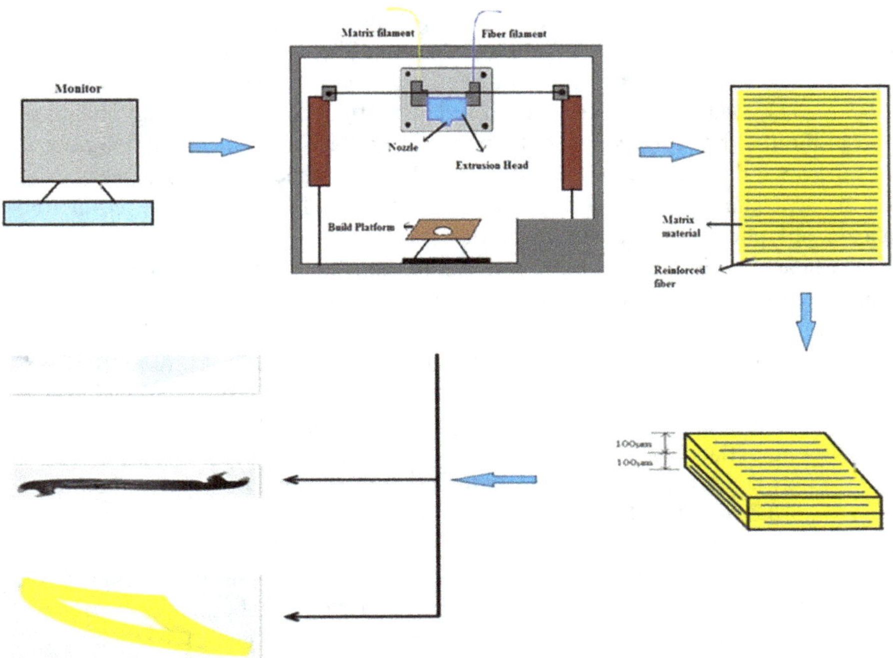

Figure 4.2 Schematic layout of FRAM [9]

applications due to its low viscosity, gravity, cost, and ease of use. Glass fiber-reinforced polyester is also popular in the boating industry [15].

The primary limitations of polyester are its unrestrainable volumetric shrinkage and consequent depression on the surface, which is caused by the difference between the shrinkage values of matrix and fiber materials [14]. Pereira *et al.* successfully built porous cubes of poly (3-hydroxybutyrate), a natural thermoplastic polyester produced by microorganisms via SLS with minimal dimensional discrepancies [16]. Shape and dimensions of the resulting model are very close to the virtual model. Further, as an example for biomedical applications, Oliveira *et al.* performed experimental analysis to build scaffolds of polyester and nylon using SLS [17].

Acrylonitrile butadiene styrene (ABS) is a versatile thermoplastic polymer that exhibits good mechanical characteristics under a varying temperature range (75–85°C) [18]. The material is generally associated with impact resistance, rigidity, low creep, low weight, good dimensional stability, and toughness, hence considered suitable for CMs [18]. Several experiments were performed on ABS to further improve toughness, impact resistance, and heat resistance. Boparai *et al.* performed a comparison analysis of tribological properties of ABS and Nylon 6-Al-Al2O3 parts made with FDM [19]. The experiment used three different compositions of Al and Al_2O_3 for tribological investigations and concluded that wear resistance behavior, thermal stability, and stiffness were more for CMs prepared with different proportions compared to parts produced with only ABS. Mori *et al.* developed a 3D printer to manufacture die-less forming of carbon fiber-reinforced polymer (CFRP) parts [20]. CFs are sandwiched and heated between the upper and lower plates of ABS made by 3D printing. Results showed that fatigue and static strength increased due to thermal bonding between the fibers and the reinforcements. Ajinjeru *et al.* identified valid processing conditions for high-performance thermoplastics on large format AM and compared the rheological behavior of high-performance thermoplastic, polyphenylsulfone with that of low-temperature polymer (ABS) [21].

Polylactic acid (PLA) is a thermoplastic and biodegradable material usually considered to be the most suitable material for FDM. Due to low glass transition temperature, it is considered unsuitable for high-temperature applications. To enhance its mechanical properties, PLA can be annealed [22–24], reinforced with fibers, nanoparticles [25–27] or undergo chain extending [28]. PLA has a wide array of applications including medical devices where PLA scaffolds have shown biocompatibility potentials [29] with controlled and interconnected porosity [30].

Carbon nanofibers are popular additives to CMs. Incorporating carbon nanofibers into thermoset and thermoplastic polymers can improve material properties. High strength and modulus is achieved by incorporating carbon nanofibers in thermoplastics, but small improvements are seen by incorporating them in a thermosetting polymer [14]. Carbon nanotubes, the allotrope of carbon, have outstanding thermal, mechanical, optical, and electrical properties and their elastic modulus is much higher than that of carbon fibres (CFs) [14]. Incorporating them in polymer composites improves one or more material property such as modulus, strength, thermal conductivity, and electrical conductivity. Carbon nanotubes are active participants as reinforcements in polymer matrices due to (1) more dispersion in polymer matrices, (2) strong bonding with the matrix material, and (3) higher solubility in solvents with functional groups [14]. Ivanova *et al.* proposed a theory to introduce nanoparticles (metal, ceramic, and carbon) to AM [31]. Adding nanoparticles helps improve thermal and electric conductivities, mechanical properties, etc. The authors also mentioned the limitations of using nanoparticles in AM such as nozzle clogging, rough surface finish of printed parts, etc. Zheng *et al.* investigated distribution of nanoparticles inside a matrix material [32]. The authors suggested that emulsion polymerization should be done to avoid agglomeration of nanoparticles and to produce good interfacial bonding between the reinforcements and the matrix material. Kim *et al.* compared homogeneous and heterogeneous dispersion of BaTiO3 nanoparticles in polyvinylidene fluoride matrix materials [33]. The parts produced by FDM have a low degree of agglomeration, porosities, and cracks compared to traditional ways of building composites.

Zhang *et al.* worked on characterizing deformations and residual stresses induced after fabricating a layer-by-layer structure of polymer-based CMs [34]. ABS, carbon nanotube-reinforced ABS, and short CF-reinforced ABS were used for this study. Displacement fields, shrinkage, porosity, and microstructure analysis were performed to study the effect of processing parameters like raster angle and printing speed. Results showed that as printing speed increases, porosity, shrinkage, and residual stresses increase for the final product. 45° raster angles showed the minimum shrinkage and porosity compared to 0° and 90°. Matsuzaki *et al.* performed 3D printing of a CFRP circle to examine the effects of set radius of curvature and fiber bundle size on the precision of the radius of curvature. It was demonstrated that with a larger fiber bundle size or a smaller set radius, the printed radius decreased from the set value. Also, analytical models to predict the actual printed radius were constructed and validated based on the experimental study [35]. Ning *et al.* studied the effects of FDM process parameters on tensile properties of CFRPs [12]. Raster angle, infill speed, layer thickness, and nozzle temperature were considered as major process parameters and their effects on tensile strength, Young's modulus, yield strength, ductility, and toughness were investigated. Tian *et al.* studied performance characteristics and interfaces of 3D-printed composite parts by investigating the effect of process parameters on pressure and temperature [36]. At 200–230°C, good impregnation of fibers and plastics was achieved, and bond strength also improved by taking the layer thickness from 0.4 to 0.6 mm and hatched a gap of 0.6 mm. The authors also concluded that maximum flexural strength and flexural modulus were achieved at 27% fiber content.

Papon *et al.* studied the effect of nozzle geometries and nanofiber concentrations on void contents of CMs [37]. Square nozzle geometry showed increased tensile strength and reduced void formation in cellulose nanofiber/PLA CMs. Transmission electron microscopy (TEM) analysis showed that there was uniform dispersion of carbon nanofibers with aligned orientation. Eichenhofer *et al.* introduced continuous lattice fabrication (CLF) for manufacturing fiber-reinforced thermoplastic composites and demonstrated its ability to exploit anisotropic material properties [38]. The authors proved that CLF enables orientation in all the X, Y, Z coordinates and continuous deposition of high fiber volume fraction along the building trajectory. Farina *et al.* employed 3D printing to design reinforced particles in the cement mortar [39]. Rough and smooth reinforced particles were selected to compare the load carrying capacity of the final product. Results of this study showed that the matrix reinforced with rough fibers (Ti6Al4V) exhibited shear failure and high interfacial bond strength compared to the matrix reinforced with smooth fibers (photopolymeric), which eventually increases its load carrying capacity. Hill *et al.* analyzed the dependency of material properties on raster angle [40]. Tensile strength, yield strength, elastic modulus, hardness, and density of the material were experimentally measured as functions of raster angle. AM is probably the most versatile technology for manufacturing complex structures, as it is difficult to fabricate complex structures of 3D-printed composite products using conventional methods like hand layup, spray up, pultrusion, filament winding, compression and injection molding, etc. [41]. Since AM does not require any special tooling, there is a potential for less cost, labor, and production time. In this regard, FRAM produced parts are likely to be lighter, sustainable, and lower cost than traditionally made parts [42]; however, most AM-produced parts require post-processing like sanding, trimming, infiltrating, and polishing.

Previous research has compared the physico-mechanical properties of conventional methods and AM methods of making CMs. Kuldeep Agarwal *et al.* performed a comparison study between conventional composite processes and composite filament fabrication (CFF) [41]. Elastic modulus, tensile strength, and fatigue life were compared for the different processes and the effect of process parameters like fiber volume fraction, fill pattern, etc. was also studied in this work. Results in this study showed that parts produced with CFF have better mechanical properties compared to conventional techniques. Van Der Klift *et al.* fabricated CFRTPs with a Mark One 3D printer using DDM [43]. The aim of this study was to compare tensile properties of two-layer and six-layer CFRTPs made by DDM and by conventional methods. The 2CF specimens came close to the rule of mixing for composites made by conventional methods due to the formation of many voids in the 6CF composites. Sugiyama *et al.* used a continuous carbon fiber (CCF) 3D printer to manufacture CFRTP sandwich structures with various core shapes and surface as a single piece. The study investigated increasing the flexibility of core shape design. It was demonstrated that CCF 3D printers could be applied for flexible design of core shapes matching the required strength and stiffness [44]. Justo *et al.* reported experimental characterization of fully made nylon-based additive layer manufactured (ALM) CMs [45]. Results of this study showed that mechanical properties of ALM-produced CMs were not yet comparable with autoclave manufacturing of CMs due to high porosities and low fiber volume. Dhariwala *et al.* designed a scaffold using hydrogels combined with cell-encapsulation to provide mechanical strength and access to nutrients for new tissue using SLA, which was not quite achievable by traditional cell-seeding techniques [46]. Results showed that SLA is highly useful in producing complex constructs and the cells were viable and proliferating within the constructs.

4.3 Fused deposition modeling

Conventional fused deposition modeling (FDM) follows the additive paradigm of depositing material in successive layers in order to produce a 3D part. Contrary to powder- or ink-based deposition processes, FDM extrudes flattened lines of molten material to form layers that rapidly solidify upon extrusion from the nozzle. Commonly available FDM materials, such as ABS and PLA, are thermoplastics; that is, they become molten when heat is applied to them and cool rapidly in air after the heating source is removed. Moreover, their molten forms are viscoelastic fluids that enable these materials to sustain their own weight and shape upon release from the nozzle, resulting in printed filaments with ~1:1 aspect (height-to-width) ratio. By overlapping two-dimensional (2D) patterns printed with filaments of ~1:1 aspect ratio, the FDM process can create very complicated 3D structures that are challenging to produce with traditional manufacturing techniques such as lithography or micromachining.

Although FDM parts have attracted attentions in various industries, they are mostly used as prototypes rather than end-user functional products. Lack of mechanical properties of built parts is the main weakness of FDM process that restricted its functional applications [47]. FDM parts lack strength, toughness, and consistency in mechanical properties, hence the end-use parts have been unreliable for functional engineering applications [48, 49]. Typically, injection molded parts have stronger

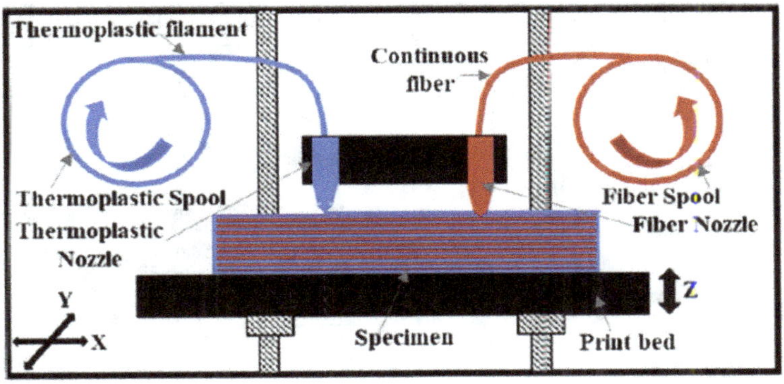

Figure 4.3 Schematic diagram of continuous fiber-reinforced additively manufactured (CFRAM) printing process [61]

mechanical properties than 3D-printed parts due to the high injection pressure. Several research studies are conducted to improve mechanical properties of FDM polymers by the optimization of printing process [50, 51]. FDM has been researched extensively in recent years to improve the product properties for new applications including auto industry [52], medical applications [53], telecommunications [54], and more [55–57]. Nanomaterials, micro-particles, short and long fibers have been added to polymers to improve mechanical properties [9]. The inclusion of reinforced fibers in printed parts has shown promising solutions to manufacture stronger FDM parts [58]. Different fibers have been used to improve mechanical properties of FDM parts [59, 60].

Figure 4.3 shows the schematic diagram of continuous fiber-reinforced additively manufactured printing process [61]. The carbon fiber-reinforced specimens showed the highest resistance to failure and had the highest fiber packing density. Studying mechanical and structural analysis sheds light on potential application of these materials in aerospace, automotive, and biomedical industries to replace conventional polymer composites. It is demonstrated that glass fibre-reinforced polymer (GFRP) composites can show progressive failure and pseudo ductility using 3D printing technology. This is achieved by integrating a combination of discrete fiber orientations and stacking sequence using monotype fibers [61].

Currently, FRP printing techniques are unable to align fibers along a contoured surface, therefore preventing them from creating shell structures with similar strength-to-weight ratios as those made with traditional layering/mold approaches. The FRP printing system detailed in this manuscript provides the first demonstration of 3D-printed FRP patterns with the fibers aligned along the contour profile. This represents a significant step toward customized 3D printing of functional contoured structures. Additionally, advancements in the print head design will address the need for a faster, more flexible and robust printing system [62].

Case 1: The use of advanced materials within performance cycling has led to outstanding results, and specialist bicycle components are now becoming lighter

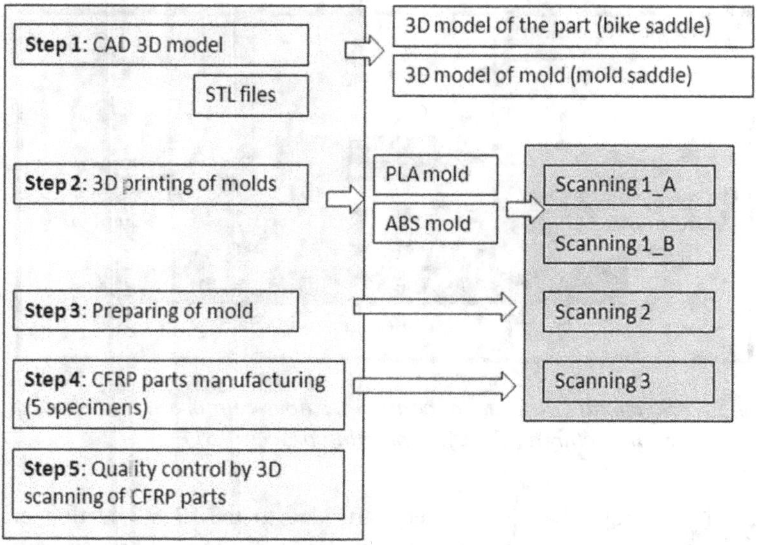

Figure 4.4 Novel method for manufacturing complex carbon fiber-reinforced polymer (CFRP) composites parts [63]

whilst retaining desired special mechanical properties. Traditional molds (metal or epoxy resin block, composite molds) used for composite materials are often very robust; however, they are only cost-effective when used in conjunction with mass production. For production of small batches or customized parts, there is a growing need for molds that can be produced quickly and at a relatively low cost. 3D printing through the use of FDM/fused-filament-fabrication (FFF) processes, utilizing low-price filaments, such as PLA or ABS, could be an approach that will significantly reduce production time and costs to produce molds for CM. Two materials were selected for evaluation, PLA and ABS, which were used to create the mold(s) using FDM. The design of the novel mold and CFRP bike saddle was then presented within this paper, where the newly manufactured CFRP parts and mold(s) were then scanned using noncontact 3D scanning methods in order to perform dimensional evaluation of the results; these results were then presented and compared against the original CAD model. It was found that the manufacturing technology used to create the mold and CFRP parts allowed for good results to be achieved. The proposed method consists of several steps as shown in Figure 4.4. The 3D CAD model of the mold is converted to a STL (standard triangulation) file format. Within the second phase of the method, the mold is 3D printed by FDM using two different materials on two different 3D printers. The molds are prepared for vacuum bag forming methodology and oven curing, within the third step. Following the manufacture of CFRP, specimens are subjected to a quality inspection, using noncontact 3D scanning techniques, in order to perform the dimensional evaluation of the results. Three

Table 4.2 Estimate of mold cost, lead time, and the mold weight [63]

Manufacturing method/material type	Cost	3d print/matching time	Mold weight/raw material
CNC milling/AI block	1500 EUR	18 hours	2800/6000 g
CNC milling/epoxy block	1300 EUR	14 hours	1200/2600 kg
FDM process/PLA	150 EUR	18 hours 37 minutes	180 g
FDM process/ABS	200 EUR	14 hours 11 minutes	150 g

3D scans were performed as follows for the 3D-printed mold, the final mold, and the CFRP parts.

The mold cost, lead time, and the mold weight are discussed in Table 4.2. The manufacture of CFRP parts in 3D-printed molds is a variable process that depends on parameters such as the shape and structure of the printed mold, uniform distribution of mold material, mold coating material and its thickness, constant thickness of the mold walls, uniform distribution of the vacuuming bag, vacuum pressure applied, oven-baking temperature, cooling and demolding of the part. A future direction of research for these types of mold(s) would be to use them in the curing autoclave process of FRP. The same temperature can be used for the mold(s); however, various additional pressures that may rise up in the range of four to five bars will be needed. Another important study would be to investigate the design of these types of mold(s), so that they do not deform under the external pressures applied, whilst also retaining conditions resembling a low wall thickness. Additionally, studies would also need to investigate the use of different types of thermoset polymer on the mold surfaces in order to improve the stiffness and the thermal expansion qualities of the mold.

4.4 Selective laser sintering (SLS)

The basic fundamentals of the AM and 3D-printing technologies were already pointed out by the Battelle Memorial Institute (Ohio, USA) in the late 1960s, where first attempts took place to create solid objects using photopolymer resin processed by two intersecting laser beams [64]. Since that time, the principle of a solid "formless" or rather "freeform" manufacturing technology to generate complex 3D near-net-shaped components is gain in gainer-greater significance [65, 66]. Thus, based on the recent development of material science and the numerous amount of new AM process technologies, the scope could be expanded from the initially, mainly area of fabricating haptic conceptual models due to their weak mechanical and thermal properties (so-called rapid prototyping) [66, 67] particularly to the direct manufacturing of smaller, not excessively stressed structural components [68]. The difficulties of these state-of-the-art technologies can be overcome by a technological and procedural conception of a hybrid AM process in shape of sandwich structures fabricated by SLS and combined with partly and/or globally cladded fiber-reinforced

Table 4.3 *Qualitative evaluation and comparison of commercially most proven AM technologies according to assorted limiting conditions [68]*

	SL	FLM	SLS	LBM
Mechanical properties	••	••	•••	••••
Thermal properties	•	••	••	••••
Component accuracy	••••	••	•••	•••
Surface quality	••••	•	••	••
Finishing and cleaning	•••	•••	••	•
Long-term stability	•	••	••	••••
Processing time	••	•	•••	••
Production rate*	••	•	••••	••
Intended use†	•	••	•••	••••
Total score	20	16	24	26

*Space and economic efficiency.
†• Model making, …, •••• industrial purpose.
FLM, fused layer modeling; LBM, laser beam melting; SL, stereo lithography; SLS, selective laser sintering.

composites. At this, the main objective of the so-called "SLS/FRP" hybrids is to produce components with notably complex structures, especially concerning high-strength, multifunctional and individually tailored lightweight engineering parts. The qualitative evaluation and comparison of commercially most proven AM technologies according to assorted limiting conditions are reported in Table 4.3 and the qualitative evaluation and comparison of SLS, VARI, and their combination regarding "SLS/FRP" hybrids are reported in Table 4.4.

Table 4.4 *Qualitative evaluation and comparison of SLS, VARI, and their combination regarding "SLS/FRP" hybrids [69]*

SLS		**VARI**	**Hybrid**
Mechanical properties	•••	••••	••••
Thermal properties	••	••	••
Component accuracy	•••	••	•••
Surface quality	••	•	••
Finishing and cleaning	••	•	•
Long-term stability	••	••••	•••
Production rate*	••••	•	••••
Reduced complexity costs	••••	•	•••
Light weight potential	•••	••••	••••
Component distortion	••	••••	•••
Total score	**27**	**24**	**29**

*Space and economic efficiency (even with small lot sizes as low as 1 and with more variants).
*SLS=selective laser sintering. VARI, vacuum assisted resin infusion.

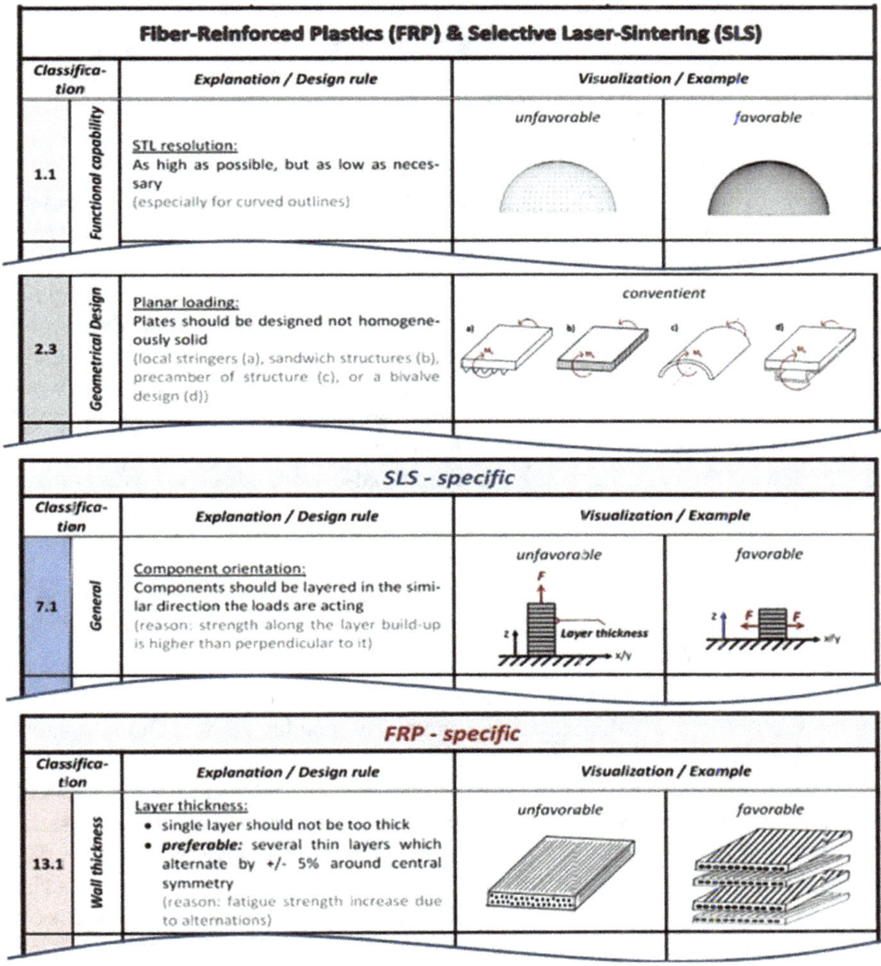

Figure 4.5 Extract of the "SLS/FRP"-hybrids design guideline catalogue [69]

Consequently, to fulfill the systematic support of the design engineer's tasks, Figure 4.5 defines specific design guidelines of "SLS/FRP" hybrids regarding design modifications on both the combined and individual consideration of fiber-reinforced plastic and SLS aspects. Thereafter, SLS-specific issues like the component build orientation according to the individually trending load paths need to be focused, apart from the aspects of wall thicknesses, inner and outer radii, edges, and overhangs. In the end, the correct spatial layout and choice of the composite laminate is made.

Starting from a broad range of AM technologies and state-of-the-art approaches in terms of additive hybrids, this contribution presents a technological and procedural conception of a hybrid production process in terms of fiber-reinforced sandwich structures based on the SLS technology. Moreover, and in accordance with this approach, the existing influences on the design process (e.g. design guidelines) as well as the holistic development methodology are described and further on presented within a refined systematic "hybrid additive lightweight engineering" (HALE) development process. Thus, the link of both practices – meaning the AM of support geometries of complex structural components combined with an ensuing completion by local or even though global composite reinforcements – and herewith the combination of the assets, while largely eliminating the drawbacks of both, paves the way for high-strength, multifunctional and individually tailored lightweight structures. As a next step, further application examples are intended in the field of medicine (e.g. prostheses). Depending on the corresponding patient needs (specific data such as form, loads or connection points) an individually tailor-made prosthesis may thus be conceptualized and finally fabricated. In future methodological work, a special view of a detailed carbon fiber-reinforced plastics (CFRP) development process will be pursued based on the presented HALE development process, but also in view of the basic lightweight and material-oriented design (LMOD) methodology [69].

4.5 Mechanical properties

4.5.1 Tensile strength

Mechanical properties of FRAM have been thoroughly investigated. Starting with tensile properties, different studies have been conducted on different factors. Generally, testing was done to investigate improving the maximum carrying load capacity and the effects of different printing parameters on this aspect. Van der Klift et al. conducted a brief study on CF-reinforced nylon specimens using Japanese standard JIS K 707 [43]. Dickson *et al.* extended the study by investigating the effect of volumetric fraction and fiber orientation angles on tensile properties of Kevlar, CF, and F Grein forced nylon specimens [47]. Results showed that CF-reinforced specimens performed 6.3 times better than non-reinforced nylon specimens in tensile testing. Melenka *et al.* evaluated and predicted tensile properties of Kevlar reinforced nylon by using volume average stiffness [70]. The elastic moduli for specimens with fiber volume of 4.04, 8.08, and 10.1% were 1.76, 6.92, and 9.00 GPa, respectively. Model results differed by 57.5, 6.2, and 0.1% for the 4.04, 8.08, and 10.1% fiber volume fraction specimens, respectively. Li *et al.* manufactured a polymer-based CM using FDM by taking PLA as the matrix material and CCF as the reinforcement and investigated the mechanical strength and thermodynamic properties of the final product [71]. Chen *et al.* and Imeri investigated the effect of fiber orientation to tensile properties by changing the number of rings to the specimens [72, 73]. From these results, CFs with two rings isotropic infill showed the best results with 419 MPa ultimate strength. In general, FRAM parts performed much better than plain 3D-printed parts with ABS, PLA, nylon, and others. In addition, the effect of fiber orientations is thoroughly understood, but there is still more work to be done in attempting to predict tensile properties of these parts. Specifically, parts

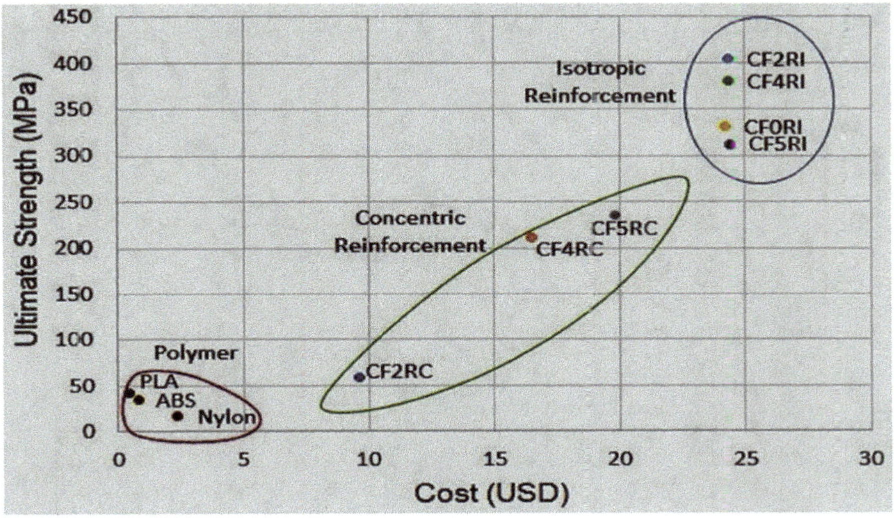

Figure 4.6 *UTS-cost comparison of FRAM and conventional 3D-printing material [73]*

can be assumed to consist of laminas and then analyzed using classical laminate theory [74]. Still the model will not take into account surface and internal defects that come from the tool path of the printer. This remains an open problem for researchers to solve. On the other hand, the better tensile properties lead to a higher cost. Thus, FRAM is more expensive than PLA and ABS, as shown in Figure 4.6.

4.5.2 Fatigue strength

Fatigue failure constitutes 50–90% of all mechanical failures [75]. Hence, it is important to know how a part would react under cyclic loading. In FRAM parts, it is also important to know how the parts can be made stronger with varying fiber volume, direction, and other significant parameters. On the subject of fatigue properties, Kuchipudi investigated the effect of volume fraction of fiber to the number of cycles up to failure [76]. Specimens of 25 and 50% fiber volume with 0°, 45°, and 90° fiber orientations, respectively, were fatigue-tested at different loads. In terms of strength, 0° specimens proved to be the strongest, while 90° ones were the weakest. In terms of ductility, compared to 0° and 90° orientations, 45° specimens elongated more. Imeri *et al.* investigated the effect of fiber orientation on fatigue properties [77]. Specimens were tested at different loads with a load ratio $R = 0.1$. Fiber orientation was divided into two groups: concentric and isotropic. Concentric specimens with two and three rings failed in the first load of 3.33–0.33 kN and were much weaker than the isotropic ones. Isotropic infill implies the way the fiber is laid down, not mechanical properties. CF specimens with zero and one rings proved to be the strongest. Imeri *et al.* performed analysis of variance (ANOVA) on the experimental data to understand the effect of load, material, number of rings, and the interaction of

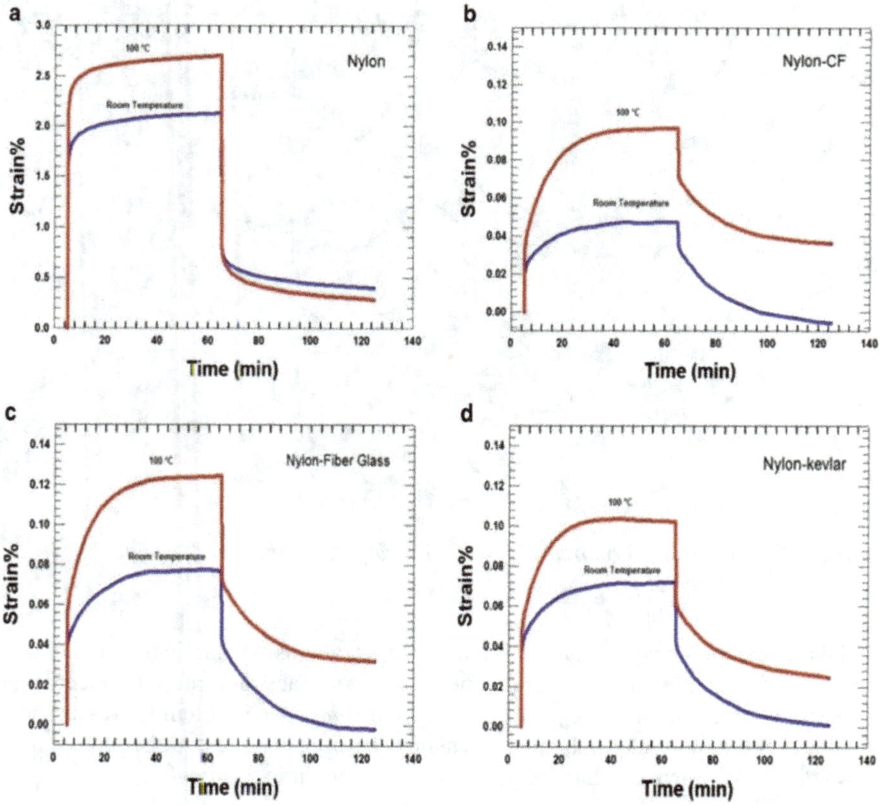

Figure 4.7 Strain of FRAM specimens at two different temperatures [78]

load and material. From ANOVA, it was found that the interaction of load and material was not significant. However, load, material, and rings with very small *p*-values, proved to have a significant effect on the number of cycles [77].

4.5.3 Creep deformation

Creep is a continuing deformation of material under constant stress over a period of time. Studying FRAM properties is of great importance due to many applications of FRAM above the transition temperature. However, creep properties of FRAM have not been investigated thoroughly. Mohammadizadeh *et al.* intended to increase the knowledge base of creep behavior of fiber-reinforced (CF, Kevlar, FG) specimens at room temperature and 100°C [78].

Figure 4.7 shows strain of FRAM specimens at two different temperatures [78]. It is observed that nylon specimens without reinforcement deformed the most while CF-reinforced specimens deformed the least. Statistical analysis with a 98% coefficient of determination verified there was difference in the creep models for different fibers. Strain of 3D-printed polymer composites at two different temperatures is shown in Figure 4.7 for each FRAM specimen.

4.5.4 Fracture toughness

Fracture toughness is the ability of a component to resist fracture. In FRAM, the fiber orientation and interlayer bonding are crucial to the crack propagation of the part. There are a few studies that have attempted to increase fracture toughness of FRAM parts. Recently, Papon *et al.* have investigated inter-laminar fracture toughness of short carbon fiber filled PLA composites with different layer orientations (0°/ 90° and 45°/–45°) [79]. The authors have shown that square nozzle geometry improves the bond strength and enhances the contact surface areas between adjacent beads by conducting fracture toughness tests on the printed specimens. Furthermore, Young *et al.* have investigated the interlayer fracture toughness of additively manufactured unreinforced and carbon fiber-reinforced ABS. The fracture toughness of the reinforced specimens was reduced due to poorly wetting of carbon fiber [80]. Since there are not many studies in the literature, this property of FRAM materials can be investigated further by varying fiber composition, fiber alignment, and process parameters.

4.5.4.1 Topology optimization

Topology optimization (TO) is a mathematical method of structural optimization in which material is distributed depending on constraints of the design volume or area [81]. With the ability of AM to create complex parts, the possibility to bridge AM with TO rises for the purpose of reducing material use and mass [82]. In 3D printing of single materials, there have been applications of TO for AM [83–85]. However, when it comes to FRAM, applications of TO be fewer in number. Among various reasons, it could be that in addition to boundary conditions, design volume, and other constrains, optimizing for FRAM has more barriers. First, modeling of the part has to be for multi-materials. Next, fiber angle optimization has to be conducted, as well. Even though these constrains exist, there are some studies conducted in the literature. Jacome *et al.* have developed an analysis framework for TO of 3D-printed reinforced composites [86]. The approach developed is a multistep optimization. First, the structure is optimized in a design volume. The second step is dependent on manufacturability criteria and the design is explored in the fiber angle space. Since Jacome *et al.* do not have a specific FRAM hardware system, their focus is on the first step of this multidisciplinary design optimization framework. Hoglund *et al.* presented a TO method for continuous fiber angle optimization approach for fused filament fabrication parts [87]. Optimization is done in a 2D area and the fiber angle after was shown to align along with the axis of the structural members formed after optimization. After testing, parts showed to be stiffer compared to the structures designed without accounting for fiber direction.

4.5.5 Machine learning

Machine learning (ML) is a branch of artificial intelligence, which perceives or recognizes patterns from complex data sets [88]. This method has proved to be valuable in various research fields. Researchers have just recently started to apply ML in the field of FRAM, as well. Gu *et al.* conducted a study designing bio inspired composite materials using ML [89]. Designs were simulated, 3D printed, and tested for tensile strength. This study showed that new materials with bio inspired microstructures can be designed using ML and their properties can also be predicted. These new microstructures can be printed using AM. ML is not only applicable in AM research but also in AM industry. AREVO Inc., based in California, has established a machine capable of printing fiber-reinforced composites [90]. AREVO Inc. uses ML in the digital design and process software of FRAM with parameters including temperature, print path, warpage, shrinkage of the printed material, and residual stresses. Although ML is still in its early days, it is envisaged that its potential in FRAM could soon be realized. This may include improving efficiency, real-time control, and predictive maintenance of machines.

4.6 Applications

FRAM has become the choice of a wide range of applications in automotive, aerospace, biomedical, and other industries [91]. In the automotive industry, Divergent 3D introduced one of the world's first 3D-printed CF-made lightweight cars [92]. This culminated in PSA Group's strategic partnership with Divergent 3D to build carbon-based vehicle structures [93]. Reduced weight and high strength are imperative in aerospace industries [94]. NASA accomplished numerous studies in order to reduce emission and fuel burn and make lighter turbofans for business jets [95]. Another application of composite-based AM is a jet engine that was developed by CFM with GE Aviation. This next generation engine called the "LEAP Engine" with durable and lightweight features offers 15% increased fuel efficiency, 50% reduced emissions, and 75% lower noise effect [96]. Successful utilization of this process has relatively popular aircraft applications in various Airbus and Boeing airplane models [95].

Lightweight, high strength, and individual parts like steering rack extensions, shown in Figure 4.8, wheel suspension, structural brackets, and other parts have been applied to Formula SAE and F1 cars, respectively [97]. Apart from cars, AREVO Inc. prints composites using a robotic arm and has announced the world's first 3D-printed CF e-Bike frame, which could be a promising alternative for bike manufacturers [90]. Successful results of these implementations have proven the functionality of FRAM in robust engineering applications. In general, conventionally machined aluminum parts could be replaced with FRAM ones out of CF with nylon, based on the evidence presented in this review [97]. Successful implementation of FRAM-made wrenches (Figure 4.9) has provided ergonomic, lightweight, affordable solutions to making tools for industrial applications. Furthermore, Hofstätter *et al.* replace injection molding inserts with FRAM ones and showed that lifetime and crack-propagation velocity were successfully improved [98]. Finally, FRAM has been applied to various industries and is growing day by day due to its advantages over traditional methods.

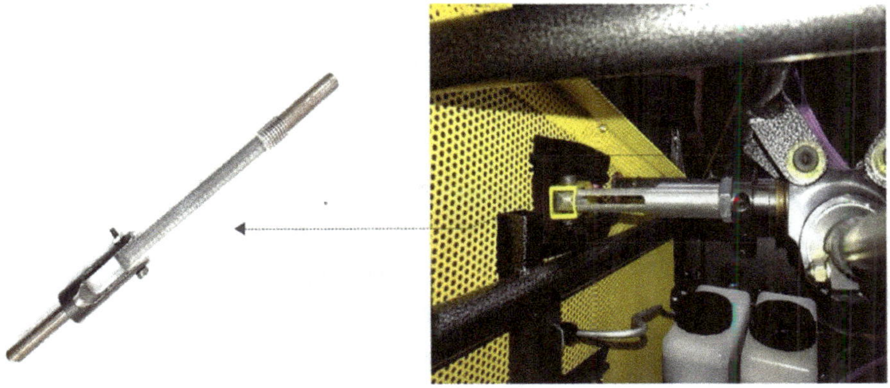

Figure 4.8 Steering rack and its location on Formula SAE car [97]

4.7 Sustainability

During a product development, many questions need to be considered such as how to design a product, how to manufacture a product, how to use a product, how to dispose (including recycle, reuse, and remanufacturing) a product and in the best way how society will be effected by it. This means a product have different phases: design, manufacturing, use, and disposal. All of these phases have different sustainability perspectives and indicators that need to be considered. In the end, final product and all these four phases will eventually affect society somehow. From society perspective, one of the most common definitions of sustainability and sustainable development was provided by the Brundtland [99] commission: "Sustainable development is development that meets the needs of the present without compromising

Figure 4.9 Aluminum, FRAM, and PLA wrenches [97]

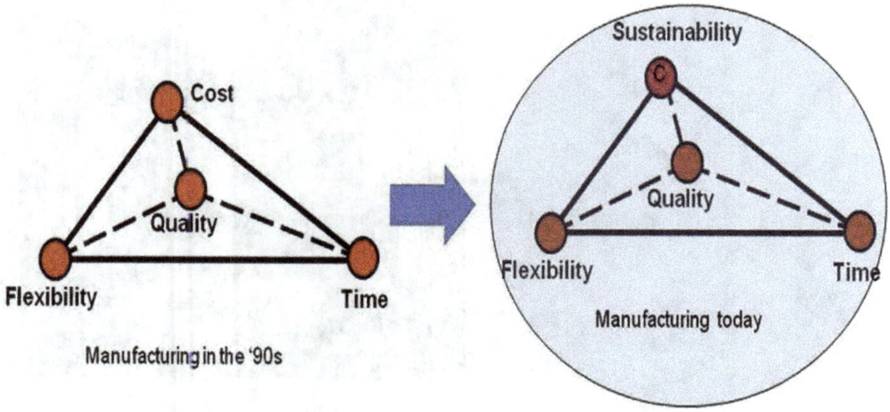

Figure 4.10 Manufacturing decision-making attributes in 1990s and at present time [102]

the ability of future generations to meet their own needs" [99]. The aspects of sustainability are many, and National Institute of Standards and Technology has categorized them to environmental stewardship, economic growth, social well-being, technological advancement, and performance management [100]. But generally, sustainability have three common dimensions: social, economic, and environmental dimensions. Where traditionally, economic aspects dominate decision-making, with respect to sustainability, these dimensions should be considered simultaneously and equally [101, 102]. Traditionally in the manufacturing phase, the performance of a production system is assessed by monitoring four main classes of manufacturing attributes, namely, cost, time, quality, and flexibility. Nowadays, however, additional issues must be considered, such as energy and resources efficiency that are apart of sustainability [102], which is shown in Figure 4.10. It is evident that sustainability has evolved to be a key manufacturing decision attribute that incorporates cost. In order to support later discussions on sustainability-related implications of DDM, the next section provides an overview of the related sustainability indicators [103].

Having in mind the evolution of manufacturing paradigms and the growing importance of sustainability, the question arises: which are the differences for DDM in comparison to traditional manufacturing paradigms? In relation to the indicators provided in the previous section, Table 4.5 summarizes an overview for the sustainability implications of DDM relative to the other manufacturing paradigms based on existing literature/studies. It is also indicated whether the specific aspects of DDM tend to be rather: beneficial (+), unfavorable (−), or whether it cannot clearly be assessed for the general case (±).

Table 4.5 Implication of DDM toward sustainability

Economic	Environmental	Social
• Higher material utilization (+) • Simpler, more efficient supply chains with less transportation efforts (+) • Less materials and energy losses due to less inventory (+) • Less waste and better waste management through possibility of direct recycling (+) • User-oriented manufacturing less-over production in stocks (+) • No molds, etc. necessary (+) • Higher specific energy demand (−) • Quality issues are not finally solved, thus risk of bad parts and rework (−) • Potentially higher profit due to customer-specific solutions (+) • Profitability could be proved in selected cases (±) • Longer manufacturing time (−)	• Ambivalent studies in terms of an environmental impact or eco-efficiency (±)	• Equal possibilities to all participants in markets and societies (+) • Bridge technological, educational, and cultural gaps between developing and developed countries (+) • User-oriented products, more customer satisfaction (+) • Potential benefits on human/worker health (+) • Unclear impact on an employment situation of industry (±)

4.8 Challenges and future prospective

Numerous industries and institutions are investigating ways to utilize FRAM to print lightweight and sustainable products to compete with conventional techniques. Among these are Xi'an Jiaotong University (China), Eskisehir Osmangazi University (Turkey), Georgia Institute of Technology (USA), Rochester Institute of Technology (USA), and Tennessee Technological University (USA) [93]. Furthermore, researchers at Oak Ridge National Laboratory have built car frames using AM- and CF-reinforced ABS to replace serial production [104, 105]. Boeing and Ford Motor Company provided funding for Stratasys' Robotic Composite 3D Demonstrator to print lightweight and low-volume parts [106]. Another potential application of FRAM could be in biomedical engineering for load-bearing composite orthopedic implants and prosthetic limbs [107]. Moreover, fiber-filled plastics are modified to have enhanced thermal and electrical conductivity, thus making it possible to utilize their full potential in product design [108].

There are other novel FRAM processes. One of them that was invented by continuous composites utilizes fast UV light-curing thermosets reinforced with continuous fibers to print 3D structures by means of a robotic arm [109]. Similarly, Eichenhofer *et al.* conduct extensive research on CLF, which is not based on layer by layer build up, but rather, extruding fiber in 3D space, which allows engineers to fully utilize load bearing capabilities of composites [110]. All in all, latest developments in FRAM will expand its application field to different industries such as concrete printing, dentistry, and architecture.

The utilization of FRAM fabricated parts is growing in various manufacturing fields; however, the current studies report a certain number of barriers, which are related to processes, materials, and machines. It is predicted that the upcoming research studies will focus on the improvements of parameters related to these factors. The authors of this chapter identify the following core challenges provided in Table 4.6 that the researchers will be investigating to improve FRAM.

The present paper examines the advanced studies held in the field of FRAM and proves that this technology has a potential to be used in various engineering fields successfully. The research studies conducted are mainly on evaluating and improving the mechanical properties of the FRAM parts. It has been shown that mechanical properties like strength, thermal and electrical conductivity, fatigue life, and creep resistance are greatly improved by using FRAM. Moreover, FRAM processes also reduce the time and cost associated in manufacturing the industrial complex work pieces.

Despite these improvements, FRAM faces challenges in the area of post-processing and porosities. Post-processing needs to be done in order to remove the support materials and maintain dimensional accuracies. The presence of porosities inside or on the surface of the part leads to more stress concentration and eventually reduces mechanical strength. Many researchers put significant efforts into experimental analyses to discover ways to reduce porosity.

Table 4.6 Future challenges

Challenge	Clarification
Enhancement of interfacial bonding	High interfacial bonding between the matrix and reinforcements is desirable in order to enhance the mechanical properties of the composite materials.
Minimization of porosities	Preventing the formation of voids inside the specimen by selecting the appropriate process parameters like inert atmosphere, nozzle temperature, fan cooling speed, printing speed, etc.
Refractory behavior of materials	Improving the thermal performance of the polymer matrix composite in order to use the min high-temperature applications, i.e., automotive and aerospace industries.
Electric and magnetic properties	Improvement of electrical conductivity and magnetic properties of polymer matrix composites by selecting the appropriate fiber type and composition.
Flexibility in fiber placement	Increasing the degrees of freedom of the fiber placement inside the matrix material in such a way that the mechanical properties get enhanced in all three directions.
Cost reduction of machine sand materials	Due to the limited availability of the industrial-scale FRAM printers, it is necessary to develop new materials and printers in order to make them available at lower cost.
Limited adaptation of the technology	Considering the advantages of FRAM fabricated parts in making light weight and sustainable products, integration of the technology to their field of studies will have to grow.
Generative design based on fiber orientation	TO software tools for FRAM parts along with optimization of fiber placement angle should be developed.

Raster angle, infill speed, layer thickness, and nozzle temperature have a significant effect on the porosity level, shrinkage, and microstructure of the material. These process parameters have direct effect on the ductility, toughness, Young's modulus and strength of the material. Careful consideration of these parameters will broaden the application of FRAM in several industries.

Furthermore, linkage of FRAM with TO and ML is on its origin, but very promising. Compared to single material AM, where the TO is concentrated in the structure mainly, fiber alignment in FRAM structures needs to be optimized. Not only research institutions but also industrial companies are starting to use the generative design for making stronger parts. However, significant efforts need to be done in the future to ease the implementation of TO and ML in FRAM processes. Smart manufacturing system of FRAM will be established by a fusion of current manufacturing system and artificial intelligence technology in the near future. Thus, smart FRAM printers will predict defects of products and troubles of printing in advance and adjust process parameters to perform normal operation. Therefore, the field of TO, ML, and big data analytics will be significant and be expanded rapidly. Current possible applications of FRAM in multiple industries such as aerospace, automotive, and biomedicine are presented in this review. In conclusion, FRAM is a promising field with many challenges to be tackled and different areas to be explored more in the future. It is clear that the next generation of FRAM-related research will be closely tied to the advancement of computers, resources, and analytical tools.

Compared to traditional manufacturing, AM has the ability to create composites with shorter cycle times and less waste due to near net shape printing. Current trends in composite AM are moving toward the ability to include continuous fiber as a filament, in order to obtain material properties that are comparable with traditional manufacturing methods. Also, the ability to reorient the reinforcing material during the printing process is currently gaining traction due to its ability to customize part strength. AM contains the capacity to create a fully optimized product, with the ability to align fibers multiple ways as well as control the fiber volume throughout the part processes that would be unfeasible with traditional composite manufacturing methods. The technology is able to create customizable products without having to invest in tooling for each product iteration. However, to compete with the strength of traditional methods, interfacial properties must be better understood. In the future, the difference between parts made with traditional composite manufacturing technology and AM will continue to decrease. The merging of the manufacturing methods will yield higher quality parts, with more customization and less cost and variability in the manufactured part. While there are many advantages of DDM/3D printing, there are still some constraints that the 3D-printing industry needs to overcome. Today's 3D printers, especially from pioneers like Stratasys, print 3D objects faster than traditional manufacturing. However, it is ideal for small batch customized productions rather than volume manufacturing as of date.

Another constraint they have is small printing area. Both these constraints have so far prevented 3D printers from being used for mainstay manufacturing. However, as technology progresses, there is a measurable change in the speed of printing. Each newer generation of 3D printers is faster and more accurate than its predecessor.

There is progress in the field of large scale or big area 3D printing as well. With this new printing technology, it is now possible to 3D print tools like patterns for foundry industry, jigs, and fixtures to aircraft wings, turbines, submarine hulls, or even entire shells of small vehicles. As more and more companies enter the 3D printers' market due to the huge benefits DDM offers and the coming to age of industry 4.0, there are exciting times ahead for the 3D-printing sector in India and the world.

References

[1] Jesthi D.K., Nayak R.K., Nanda B.K., Das D. 'Assessment of abrasive jet machining of carbon and glass fiber reinforced polymer hybrid composites'. *Materials Today: Proceedings*. 2019;18(7):3116–21.

[2] Jesthi D.K., Mandal P., Rout A.K., Nayak R.K. 'Effect of carbon/glass fiber symmetric inter-ply sequence on mechanical properties of polymer matrix composites'. *Procedia Manuf.* 2018;20:530–35.

[3] Jesthi D.K., Mandal P., Rout A.K., Nayak R.K. 'Enhancement of mechanical and specific wear properties of glass/carbon fiber reinforced polymer hybrid composite'. *Procedia Manufacturing*. 2018;20:536–41.

[4] Jesthi D.K., Mohanty S.S., Nayak A., Panigrahi A., Nayak R.K. 'Improvement of mechanical properties of carbon/glass fiber reinforced polymer composites through inter-ply arrangement'. *IOP Conference Series: Materials Science and Engineering*. 2018;377(1):012182.

[5] Jesthi D.K., Nayak A., Mohanty S.S., Rout A K., Nayak R.K. 'Evaluation of mechanical properties of hybrid composite laminates reinforced with glass/carbon woven fabrics'. *IOP Conference Series: Materials Science and Engineering*. 2018;377(1):012157.

[6] Jesthi D.K., Nayak R.K. 'Influence of glass/carbon fiber stacking sequence on mechanical and three-body abrasive wear resistance of hybrid composites'. *Materials Research Express*. 2020;7(1):015106.

[7] Jesthi D.K., Nayak R.K. 'Evaluation of mechanical properties and morphology of seawater aged carbon and glass fiber reinforced polymer hybrid composites'. *Composites Part B: Engineering*. 2019;174:106980.

[8] Jesthi D.K., Nayak R.K. 'Sensitivity analysis of abrasive air-jet machining parameters on machinability of carbon and glass fiber reinforced hybrid composites'. *Materials Today Communications*. 2020;25(1980–2015):101624.

[9] Fidan I., Imeri A., Gupta A, *et al.* 'The trends and challenges of fiber reinforced additive manufacturing'. *International Journal of Advanced Manufacturing Technology*. 2019;102(5):1801–18.

[10] Unterweger C., Brüggemann O., Fürst C. 'Synthetic fibers and thermoplastic short-fiber-reinforced polymers: properties and characterization'. *Polymer Composites*. 2014;35(2):227–36.

[11] Ivey M., Melenka G.W., Carey J.P., Ayranci C. 'Characterizing short-fiber-reinforced composites produced using additive manufacturing'. *Advanced Manufacturing: Polymer & Composites Science*. 2017;3(3):81–91.

[12] Ning F., Cong W., Qiu J., Wei J., Wang S. 'Additive manufacturing of carbon fiber reinforced thermoplastic composites using fused deposition modeling'. *Composites Part B: Engineering*. 2015;80(1):369–78.

[13] Gu D., Wang H., Chang F, *et al*. 'Selective laser melting additive manufacturing of tic/alsi10mg bulk-form nanocomposites with tailored microstructures and properties'. *Physics Procedia*. 2014;56:108–16.

[14] Mallick P.K. *Fiber-reinforced composites: materials, manufacturing, and design*. 3rd ed. Boca Raton, FL: CRC Press; 2007.

[15] Chalmers D.W. 'The potential for the use of composite materials in marine structures'. *Marine Structures*. 1994;7(2):441–56.

[16] Pereira T.F., Oliveira M.F., Maia I.A., Silva J.V.L., Costa M.F., Thiré R.M.S.M. '3D printing of poly (3-hydroxybutyrate) porous structures using selective laser sintering'. *Macromolecular Symposia*. 2012;319(1):64–73.

[17] Oliveira M., Maia I., Noritomi P, *et al*. 'Building tissue engineering scaffolds utilizing rapid prototyping'. *Matéria (Rio de Janeiro)*. 2006;12:373–82.

[18] Campo E.A. *Complete part design handbook*. Cincinnati: Hanser; 2006.

[19] Boparai K., Singh R., Singh H. 'Comparison of tribological behaviour for Nylon6-Al-Al_2O_3 and ABS parts fabricated by fused deposition modelling'. *Virtual and Physical Prototyping*. 2015;10(2):59–66.

[20] Mori K. -ichiro, Maeno T., Nakagawa Y. 'Dieless forming of carbon fibre reinforced plastic parts using 3D printer'. *Procedia Engineering*. 2014;81(2):1595–600.

[21] Ajinjeru C., Kishore V., Liu P, *et al*. 'Determination of melt processing conditions for high performance amorphous thermoplastics for large format additive manufacturing'. *Additive Manufacturing*. 2018;21(5):125–32.

[22] Nugroho P., Mitomo H., Yoshii F., Kume T. 'Degradation of poly (L-lactic acid) by γ-irradiation'. *Polymer Degradation and Stability*. 2001;72(2):337–43.

[23] Urayama H., Kanamori T., Fukushima K., Kimura Y. 'Controlled crystal nucleation in the melt-crystallization of poly (l-lactide) and poly (l-lactide) / poly (d-lactide) stereocomplex'. *Polymer*. 2003;44(19):5635–41.

[24] Tsuji H., Ikada Y. 'Properties and morphologies of poly (L-lactide): 1. annealing condition effects on properties and morphologies of poly (L-lactide)'. *Polymer*. 1995;36(14):2709–16.

[25] Urayama H., Ma C., Kimura Y. 'Mechanical and thermal properties of poly (L-lactide) incorporating various inorganic fillers with particle and whisker shapes'. *Macromolecular Materials and Engineering*. 2003;288(7):562–68.

[26] Trimaille T., Pichot C., Elaissari A., Fessi H., Briançon S., Delair T. 'Poly (D, L-lactic acid) nanoparticle preparation and colloidal characterization'. *Colloid & Polymer Science*. 2003;281(12):1184–90.

[27] Hu X., Xu H.-S., Li Z.-M., Zhong G. 'Morphology and properties of poly (L-lactide) (PLLA) filled with hollow glass beads'. *Macromolecular Materials and Engineering*. 2007;292(5):646–54.

[28] Di Y., Iannace S., Di Maio E., Nicolais L. 'Reactively modified poly(lactic acid): properties and foam processing'. *Macromolecular Materials and Engineering*. 2005;290(11):1083–90.

[29] Lam C.X.F., Olkowski R., Swieszkowski W., Tan K.C., Gibson I., Hutmacher D.W. 'Composite PLDLLA/TCP scaffolds for bone engineering: mechanical and in vitro evaluations scaffolds for bone engineering: mechanical and in vitro evaluations' in Lim C.T., Goh J.C.H. (eds.). *13Th international conference on biomedical engineering*. Berlin, Heidelberg: Springer; 2009. pp. 1480–83.

[30] Bose S., Vahabzadeh S., Bandyopadhyay A. 'Bone tissue engineering using 3D printing'. *Materials Today*. 2013;16(12):496–504.

[31] Ivanova O., Williams C., Campbell T. 'Additive manufacturing (am) and nanotechnology: promises and challenges'. *Rapid Prototyping Journal*. 2013;19(5):353–64.

[32] Zheng H., Zhang J., Lu S., Wang G., Xu Z. 'Effect of core–shell composite particles on the sintering behavior and properties of nano-Al_2O_3/polystyrene composite prepared by SLS'. *Materials Letters*. 2006;60(9):1219–23.

[33] Kim H., Fernando T., Li M., Lin Y., Tseng T.-L.B. 'Fabrication and characterization of 3D printed $BaTiO_3$/PVDF nanocomposites'. *Journal of Composite Materials*. 2018;52(2):197–206.

[34] Zhang W., Wu A.S., Sun J, *et al*. 'Characterization of residual stress and deformation in additively manufactured ABS polymer and composite specimens'. *Composites Science and Technology*. 2017;150(6):102–10.

[35] Matsuzaki R., Nakamura T., Sugiyama K, *et al*. 'Effects of set curvature and fiber bundle size on the printed radius of curvature by a continuous carbon fiber composite 3D printer'. *Additive Manufacturing*. 2018;24(3):93–102.

[36] Tian X., Liu T., Yang C., Wang Q., Li D. 'Interface and performance of 3D printed continuous carbon fiber reinforced PLA composites'. *Composites Part A: Applied Science and Manufacturing*. 2016;88(4):198–205.

[37] Papon E.A., Haque A. 'Tensile properties, void contents, dispersion and fracture behaviour of 3D printed carbon nanofiber reinforced composites'. *Journal of Reinforced Plastics and Composites*. 2018;37(6):381–95.

[38] Eichenhofer M., Wong J.C.H., Ermanni P. 'Continuous lattice fabrication of ultra-lightweight composite structures'. *Additive Manufacturing*. 2017;18:48–57.

[39] Farina I., Fabbrocino F., Carpentieri G, *et al*. 'On the reinforcement of cement mortars through 3D printed polymeric and metallic fibers'. *Composites Part B: Engineering*. 2016;90(2):76–85.

[40] Hill N., Haghi M. 'Deposition direction-dependent failure criteria for fused deposition modeling polycarbonate'. *Rapid Prototyping Journal*. 2014;20(3):221–27.

[41] Agarwal K., Kuchipudi S.K., Girard B., Houser M. 'Mechanical properties of fiber reinforced polymer composites: a comparative study of conventional and additive manufacturing methods'. *Journal of Composite Materials*. 2018;52(23):3173–81.

[42] Additive manufacturing of polymer matrix composites | Intechopen [online]. Available from https://www.intechopen.com/chapters/60211 [Accessed 8 Jan 2022].

[43] Der Klift F.V., Koga Y., Todoroki A., Ueda M., Hirano Y., Matsuzaki R. '3D printing of continuous carbon fibre reinforced thermo-plastic (CFRTP) tensile test specimens'. *Open Journal of Composite Materials*. 2016;06(1):18–27.

[44] Sugiyama K., Matsuzaki R., Ueda M., Todoroki A., Hirano Y. '3D printing of composite sandwich structures using continuous carbon fiber and fiber tension'. *Composites Part A: Applied Science and Manufacturing*. 2018;113(2):114–21.

[45] Justo J., Távara L., García-Guzmán L., París F. 'Characterization of 3D printed long fibre reinforced composites'. *Compos Struct*. 2018;185(5):537–48.

[46] Dhariwala B., Hunt E., Boland T. 'Rapid prototyping of tissue-engineering constructs, using photopolymerizable hydrogels and stereolithography'. *Tissue Engineering*. 2004;10(9–10):1316–22.

[47] Dickson A.N., Barry J.N., McDonnell K.A., Dowling D.P. 'Fabrication of continuous carbon, glass and kevlar fibre reinforced polymer composites using additive manufacturing'. *Additive Manufacturing*. 2017;16(1):146–52.

[48] Nakagawa Y., Mori K. -ichiro, Maeno T. '3D printing of carbon fibre-reinforced plastic parts'. *International Journal of Advanced Manufacturing Technology*. 2017;91(5):2811–17.

[49] Justo J., Távara L., García-Guzmán L., París F. 'Characterization of 3D printed long fibre reinforced composites'. *Compos Struct*. 2018;185(5):537–48.

[50] Song Y., Li Y., Song W., Yee K., Lee K.-Y., Tagarielli V.L. 'Measurements of the mechanical response of unidirectional 3D-printed PLA'. *Materials & Design*. 2017;123(1):154–64.

[51] Wu W., Geng P., Li G., Zhao D., Zhang H., Zhao J. 'Influence of layer thickness and raster angle on the mechanical properties of 3D-printed peek and a comparative mechanical study between peek and ABS'. *Materials*. 2015;8(9):5834–46.

[52] Ilardo R., Williams C.B. 'Design and manufacture of a formula SAE intake system using fused deposition modeling and fiber-reinforced composite materials'. *Rapid Prototyping Journal*. 2010;16(3):174–79.

[53] springerprofessional.de Enhancing dimensional accuracy of FDM based biomedical implant replicas by statistically controlled vapor smoothing process. Available from https://www.springerprofessional.de/en/enhancing-dimensional-accuracy-of-fdm-based-biomedical-implant-r/10993642 [Accessed 8 Jan 2022].

[54] Yeole S., Madhav C., Regula S. 'Importance and utilization of 3D printing in various applications'. *International Journal of Modern Engineering Research*. 2016;02:24–29.

[55] Aw Y.Y., Yeoh C.K., Idris M.A., Amali H.K., Aqzna S.S., Teh P.L. 'A study of tensile and thermal properties of 3D printed conductive ABS – ZnO composite'. *AIP Conference Proceedings*. 2017;1835(1):020008.

[56] Wang X., Jiang M., Zhou Z., Gou J., Hui D. '3D printing of polymer matrix composites: a review and prospective'. *Composites Part B: Engineering*. 2017;110(2):442–58.

[57] Niaza K., Senatov F.S., Stepashkin A., Anisimova N.Y., Kiselevsky M. 'Long-term creep and impact strength of biocompatible 3D-printed PLA-based scaffolds'. *Nano Hybrids and Composites*. 2017;13:15–20.

[58] Brooks H., Molony S. 'Design and evaluation of additively manufactured parts with three dimensional continuous fibre reinforcement'. *Materials & Design*. 2016;90(1):276–83.

[59] Caminero M.A., Chacón J.M., García-Moreno I., Rodríguez G.P. 'Impact damage resistance of 3D printed continuous fibre reinforced thermoplastic composites using fused deposition modelling'. *Composites Part B: Engineering*. 2018;148:93–103.

[60] Goh G.D., Dikshit V., Nagalingam A.P, *et al.* 'Characterization of mechanical properties and fracture mode of additively manufactured carbon fiber and glass fiber reinforced thermoplastics'. *Materials & Design*. 2018;137(6):79–89.

[61] Vemuganti S., Soliman E., Reda Taha M. '3D-printed pseudo ductile fiber-reinforced polymer (FRP) composite using discrete fiber orientations'. *Fibers*. 2020;8(9):53.

[62] Tse L.Y.L., Kapila S., Barton K. *Contoured 3D Printing of Fiber Reinforced Polymers*. Austin, TX: University of Texas at Austin; 2016.

[63] Bere P., Neamtu C., Udroiu R. 'Novel method for the manufacture of complex CFRP parts using FDM-based molds'. *Polymers*. 2020;12(10):2220

[64] History of additive manufacturing: science & engineering book chapter | IGI global [online]. Available from https://www.igi-global.com/chapter/history-of-additive-manufacturing/182410 [Accessed 8 Jan 2022].

[65] Huang S.H., Liu P., Mokasdar A., Hou L. 'Additive manufacturing and its societal impact: a literature review'. *International Journal of Advanced Manufacturing Technology*. 2013;67(5):1191–203.

[66] Gausemeier J., Echterhoff N., Kokoschka M., Wall M., Pasquier P. Thinking ahead the future of additive manufacturing – analysis of promising industries [online]. 2011. Available from https://www.semanticscholar.org/paper/Thinking-ahead-the-Future-of-Additive-Manufacturing-Gausemeier-Echterhoff/1316e48a0ce7e4d0dd9f1728f71947ef20c6f960 [Accessed 8 Jan 2022].

[67] Campbell T., Williams C., Ivanova O., Garrett B. *Could 3D Printing Change the World? Technologies, Potential, and Implications of Additive Manufacturing*. Atlantic Council; 2011.

[68] Gibson I., Rosen D., Stucker B. *Additive Manufacturing Technologies: 3D Printing, Rapid Prototyping, and Direct Digital Manufacturing*. 2nd ed. Springer; 2015.

[69] Kaspar J., Häfele T., Kaldenhoff C., Griebsch J., Vielhaber M. 'Hybrid additive design of FRP components – fiber-reinforced sandwich structures based on selective laser sintering technology'. *Procedia CIRP*. 2017;60(5):235–40.

[70] Melenka G.W., Cheung B.K.O., Schofield J.S., Dawson M.R., Carey J.P. 'Evaluation and prediction of the tensile properties of continuous fiber-reinforced 3D printed structures'. *Composite Structures*. 2016;153:866–75.

[71] Li N., Li Y., Liu S. 'Rapid prototyping of continuous carbon fiber reinforced polylactic acid composites by 3D printing'. *Journal of Materials Processing Technology*. 2016;238(4):218–25.

[72] Chen Y., Rios C.O., Imeri A., Russell N.A., Fidan I. 'Investigation of the tensile properties in fibre-reinforced additive manufacturing and fused filament fabrication'. *International Journal of Rapid Manufacturing*. 2020;9(2–3):251–67.

[73] Investigation of the mechanical properties for fiber reinforced additively manufactured components-ProQuest [online]. Available from https://www.proquest.com/openview/97af19d504ddf7cc85749d86ce9d9010/1?pq-orig site=gscholar&cbl=18750&diss=y [Accessed 9 Jan 2022].

[74] Li L., Sun Q., Bellehumeur C., Gu P. 'Composite modeling and analysis for fabrication of FDM prototypes with locally controlled properties'. *Journal of Manufacturing Processes*. 2002;4(2):129–41.

[75] Wiley.com Metal fatigue in engineering [online] | wiley. Available from https://www.wiley.com/en-us/Metal+Fatigue+in+Engineering%2C+2nd+Edition-p-9780471510598 [Accessed 9 Jan 2022].

[76] Kuchipudi S.C. *The Effects of Fiber Orientation and Volume Fraction of Fiber on Mechanical Properties of Additively Manufactured Composite Material*. Minnesota State University, Mankato: All Graduate Theses, Dissertations, and Other Capstone Projects; 2017.

[77] Imeri A., Fidan I., Allen M., Perry G. 'Effect of fiber orientation in fatigue properties of FRAM components'. *Procedia Manufacturing*. 2018;26(9):892–99.

[78] Mohammadizadeh M., Fidan I., Allen M., Imeri A. 'Creep behavior analysis of additively manufactured fiber-reinforced components'. *The International Journal of Advanced Manufacturing Technology*. 2018;99(5):1225–34.

[79] Papon E.A., Haque A. 'Fracture toughness of additively manufactured carbon fiber reinforced composites'. *Additive Manufacturing*. 2019;26(5):41–52.

[80] Young D., Wetmore N., Czabaj M. 'Interlayer fracture toughness of additively manufactured unreinforced and carbon-fiber-reinforced acrylonitrile butadiene styrene'. *Additive Manufacturing*. 2018;22(4):508–15.

[81] Andreassen E., Clausen A., Schevenels M., Lazarov B.S., Sigmund O. 'Efficient topology optimization in MATLAB using 88 lines of code'. *Structural and Multidisciplinary Optimization*. 2011;43(1):1–16.

[82] Gao W., Zhang Y., Ramanujan D, *et al*. 'The status, challenges, and future of additive manufacturing in engineering'. *Computer-Aided Design*. 2015;69(2):65–89.

[83] Kwok T.-H., Li Y., Chen Y. 'A structural topology design method based on principal stress line'. *Computer-Aided Design*. 2016;80(47):19–31.

[84] Zegard T., Paulino G.H. 'Bridging topology optimization and additive manufacturing'. *Structural and Multidisciplinary Optimization*. 2016;53(1):175–92.

[85] Wu J., Aage N., Westermann R., Sigmund O. 'Infill optimization for additive manufacturing-approaching bone-like porous structures'. *IEEE Transactions on Visualization and Computer Graphics*. 2018;24(2):1127–40.

[86] Bahamonde L.G., Tatting B., Harik R, *et al*. 'An analysis framework for topology optimization of 3D printed reinforced composites '. *AIAA/ASCE/AHS/ASC Structures, Structural Dynamics, and Materials Conference*; Florida, 2018.

[87] Jiang D., Hoglund R., Smith D. 'Continuous fiber angle topology optimization for polymer composite deposition additive manufacturing applications'. *Fibers*. 2019;7(2):14.

[88] Michalski R., Carbonell J., Mitchell T. *Machine Learning. An Artificial Intelligence Approach*. Vol. 2. Springer; 1983.

[89] Gu G., Chen C.-T., Richmond D.J., Buehler M.J. 'Bioinspired hierarchical composite design using machine learning: simulation, additive manufacturing, and experiment'. *Materials Horizons*. 2018;5(5):939–45.

[90] AREVO launches battery-assisted 3D printed carbon-fiber frame bike [online]. Available from https://www.compositesworld.com/news/arevo-launches-battery-assisted-3d-printed-carbon-fiber-frame-bike [Accessed 9 Jan 2022].

[91] A new composite-manufacturing approach takes shape [online]. Available from https://str.llnl.gov/june-2017/lewicki [Accessed 9 Jan 2022].

[92] Luxury CAR maker PSA group puts 3D printing to use for CAR chassis-3dprint.com | the voice of 3D printing / additive manufacturing [online]. Available from https://3dprint.com/159574/peugeot-3d-printing-chassis/ [Accessed 9 Jan 2022].

[93] Wohlers associates [online]. Available from https://wohlersassociates.com/2018report.htm [Accessed 9 Jan 2022].

[94] Sculpteo *3D Printing Applications: Benefits of 3D Printing in Every Field*. Sculpteo; 2022.

[95] Hofstätter T., Pedersen D.B., Tosello G., Hansen H.N. 'State-of-the-art of fiber-reinforced polymers in additive manufacturing technologies'. *Journal of Reinforced Plastics and Composites*. 2017;36(15):1061–73.

[96] CFM International Leap engines – CFM international jet engines [online]. Available from https://www.cfmaeroengines.com/engines/leap/ [Accessed 9 Jan 2022].

[97] Imeri A., Russell N., Rust J., Sahin S., Fidan I. 'Maker: 3D printing as an alternative to fabricate the motorsports parts'. *ASEE Annual Conference Proceedings*; Columbus, OH, 2017.

[98] Hofstätter T., Pedersen D.B., Tosello G., Hansen H.N. 'Applications of fiber-reinforced polymers in additive manufacturing'. *Procedia CIRP*. 2017;66(3):312–16.

[99] Brundtland G.H. 'Our common future—call for action'. *Environmental Conservation*. 1987;14(4):291–94.
[100] Joung C.B., Carrell J., Sarkar P., Feng S.C. 'Categorization of indicators for sustainable manufacturing'. *Ecological Indicators*. 2013;24(2):148–57.
[101] Seliger G., Bayat N., Consiglio S, *et al*. *Sustainability in manufacturing: recovery of resources in product and material cycles*. Springer; 2007.
[102] Salonitis K., Stavropoulos P. 'On the integration of the cax systems towards sustainable production'. *Procedia CIRP*. 2013;9(3):115–20.
[103] Chen D., Heyer S., Ibbotson S., Salonitis K., Steingrímsson J.G., Thiede S. 'Direct digital manufacturing: definition, evolution, and sustainability implications'. *Journal of Cleaner Production*. 2015;107(9):615–25.
[104] Talagani F., DorMohammadi S., Dutton R, *et al*. 'Numerical simulation of big area additive manufacturing (3D printing) of a full size CAR'. *SAMPE Journal*. 2015;51:27.
[105] Curran S., Chambon P., Lind R, *et al*. 'Big area additive manufacturing and hardware-in-the-loop for rapid vehicle powertrain prototyping: a case study on the development of a 3-D-printed shelby cobra'. *SAE Technical Papers*. 2016;1302890:01–328.
[106] Royal Aeronautical Society 3D printing takes-off. Available from https://www.aerosociety.com/news/3d-printing-takes-off/ [Accessed 9 Jan 2022].
[107] Matsuzaki R., Ueda M., Namiki M, *et al*. 'Three-dimensional printing of continuous-fiber composites by in-nozzle impregnation'. *Scientific Reports*. 2016;6:23058.
[108] Mulholland T., Goris S., Boxleitner J., Osswald T., Rudolph N. 'Process-induced fiber orientation in fused filament fabrication'. *Journal of Composites Science*. 2018;2(3):45.
[109] 'Additive manufacturing industry got interested in fiber reinforced composites'. *Basalt. Today*. 2017.
[110] Eichenhofer M., Maldonado J., Klunker F., Ermanni P. 'Analysis of processing conditions for a novel 3D-composite production technique'. *20th International Conference on Composite Materials*; Copenhagen, 2015.

Chapter 5

Joining and repair of resin-infused, continuous fibre-reinforced, thermoplastic acrylic-matrix composites for extended applicability

Winifred Obande[1] and Dipa Ray[1]

5.1 Introduction

Nowadays, there is renewed interest in thermoplastics (TP) as matrices for fibre-reinforced polymer (FRP) composites because their use opens up a world of post-processing opportunities, which are desirable for end-of-life waste management. Already, reactive, *in situ* polymerisable thermoplastic matrix systems have been demonstrated as viable alternatives to traditional melt-processible thermoplastics (as shown in Table 5.1). Importantly, these reactive monomers and oligomers offer a route to effectively bridge the processibility/post-processibility gap between traditional thermoplastic and thermoset matrices.

Innovative, liquid, thermoplastic acrylic resins such as Arkema's flagship family of Elium® resins have come to the fore in the last decade for composite applications due to their suitability for liquid resin infusion of continuous fibre-reinforced thermoplastic composites at ambient temperatures under vacuum. They exhibit full retention of thermoplasticity after *in situ* polymerisation and have shown great promise as thermoplastic matrices contributing to a circular economy with attractive assembly, reuse and repair opportunities that must be explored [9]. It is worth noting that studies on the reactive processibility of acrylic resins by bulk polymerisation date back as far as the 1940s [10].

Owing to room-temperature processibility, these acrylic resins are ideally suited to the fabrication of large-scale structures. This makes them advantageous over early-generation reactive thermoplastic resins, which cannot be readily applied for the production of larger-scale components [2–4]. This is an important consideration because it demonstrates the extended practicability of liquid acrylic resin technology for applications such as those in the wind energy sector [5, 11, 12] where blade lengths continue to increase to meet the growing global demand. Already, a rich

[1]School of Engineering, Institute for Materials and Processes, The University of Edinburgh, Edinburgh, Scotland, UK

Table 5.1 Important considerations on the processibility and post-processibility of thermoplastic (traditional and novel reactive) and thermosets

	Traditional thermoplastic FRP matrices	Traditional thermoset FRP matrices	Reactive thermoplastic FRP matrices
Examples	PEEK PP PBT PA-6 or -12	Epoxy Polyester Vinylester Benzoxazine	Acrylics CBT [1–5] Caprolactam [6, 7] Laurolactam [1, 8]
Processing	Suited to melt-processing; typically requiring elevated temperatures (150°C–400°C) and pressures	Suited to a broad range of mature processing technologies including liquid resin infusion from ambient temperatures to <200°C	Suited to liquid resin infusion; processible from ambient temperatures to <200°C
Post-processing	Weldable, thermoformable, recyclable and repairable	Repairable; not readily weldable; not thermoformable or recyclable	Weldable, thermoformable, recyclable and repairable

CBT, cyclic butylene terephthalate; PA-, polyamide; PBT, polybutylene terephthalate; PEEK, polyether ether ketone; PP, polypropylene.

body of literature [9] demonstrates the growing research interest in this area, with some scholarly attention on the reuse/recycling and joining potentials with thermoplastic acrylic-matrix composites. These attributes are essential to meeting environmental targets associated with waste management, as summarised in Figure 5.1.

In this chapter, the recent trends in developing the maturity of thermoplastic acrylic-matrix composites towards extended applicability are summarised, with insights for prospects and future opportunities also provided.

5.2 Liquid acrylic resin-based composites

5.2.1 Manufacturing

Liquid resin infusion has been used widely for manufacturing liquid acrylic resin-based composites [13–18], although resin transfer moulding (RTM) [15, 19–21], press moulding and wet lay-up [22] have also been used.

Auxiliary materials are used to aid the liquid resin infusion process and the subsequent demoulding process. A schematic has been shown in Reference [9] where a flexible bag is used in conjunction with a rigid tool base. A typical resin degassing process is carried out for the traditional thermoset resins such as epoxy during the resin infusion process to remove dissolved gases and micro-bubbles [23]. However, degassing is not that common with liquid acrylic resins due to their short pot life

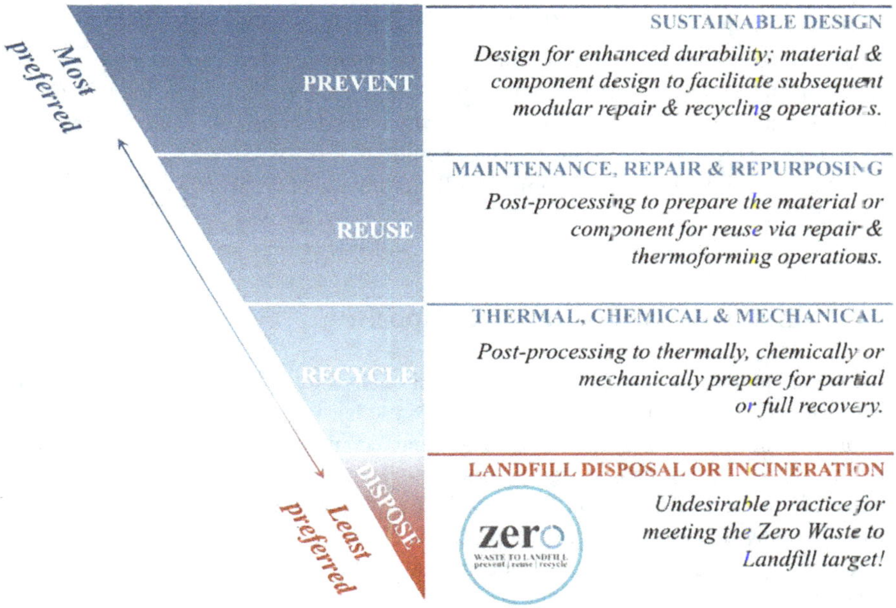

Figure 5.1 *Graphical representation of the most commonly adopted waste management practices and their relative desirability in terms of fit formeeting the* Zero Waste to Landfill *target*

and short infusion window (e.g. 20–35 minutes for Elium® 150 [24]). Degassing was attempted on acrylic resin E-280, but no significant benefits were reported [14]. Apart from Cousins *et al.* [25], no other authors reported a resin degassing step for liquid acrylic resin-based composites. A conventional liquid resin infusion configuration with a flexible membrane is the most commonly employed technique.

From a manufacturing perspective, these acrylic resins are attractive drop-in substitutes for room-temperature infusible epoxy resins systems particularly because their use does not necessitate mould heating or specialised tooling designs. Nevertheless, it is important to consider one major reported challenge associated with acrylics. They are known to be susceptible to boiling when peak exothermic temperatures during bulk free-radical polymerisation exceed their boiling temperature (100°C) [26, 27]. This, however, is more problematic where laminate thicknesses approach or exceed 12 mm; e.g. the production of an 11.3-mm-thick laminate is associated with a 21.5°C higher peak exotherm temperature than a 3.8-mm-thick laminate [26]. Further increases in thickness would typically require the use of exothermic control agents to mitigate the risks posed by the Trommsdorf (or auto-acceleration) effect [5].

5.2.2 Comparison with conventional thermoset composites

A considerable body of literature has been published on the comparative performance of thermoplastic acrylic composites with respect to their traditional thermoset counterparts as summarised in the review by Obande *et al.* [9]. Table 5.2 presents an overview of the key properties from the authors' review. Overall, acrylic-matrix composites exhibit good comparability with composites comprising non-aerospace-grade epoxies.

5.3 Joining of acrylic-matrix composites

In terms of joining methodologies, most of the well-known techniques employed for thermoset-matrix composites are readily applied to their thermoplastic counterpart. The use of mechanical fasteners and adhesive bonding techniques is considerably mature for traditional thermoplastic-matrix composites. Additionally, thermoplastic-matrix composites afford the added benefit of being suited to newer bonding technologies such as fusion bonding or welding [28]. Of course, this covers a broad range of techniques as shown in Figure 5.2, all of which can negate the use of fasteners and as such constitute economical and weight-reduction measures [29].

As the demand for thermoplastic-matrix composites continues to rise, improved potentials for repair and application of automation principles to joining methodologies would facilitate more successful adoptions of thermoplastic-matrix-based structural materials as drop-in replacements for both their thermosetting-matrix-based counterparts and metals. This would facilitate the construction of large structures such as the wind turbine blade, where a modular approach may be adapted for assembly operations. The potential for assembling structures in an industrial environment adds to its cost-effectiveness. Various welding techniques are used for thermoplastic composites such as induction, ultrasonic and resistance welding, where heat is generated at the joining interface and fusion bonding takes place at the softened interface under pressure [19, 30]. Upon softening, the contacting surfaces allow polymeric chain entanglements under pressure leading to a bonded interface. Advancements in the ultrasonic welding of thermoplastic composites have been reported in a recent review paper by Bhudolia *et al.* [30].

As liquid acrylic resin-based composites reinforced with continuous reinforcement fibres are relatively new in the field of structural composites, not many joining or welding works have been reported yet. Only a few published works exist where authors have welded acrylic/glass or acrylic/carbon fibre composites. Murray *et al.* [19] reported findings on the fusion joining of acrylic/glass composites for wind turbine blade applications [19]. The authors used three different joining methods for acrylic/glass composites: conventional adhesive bonding, resistance welding and induction welding. Details of the materials and bonding/heating details are presented in Table 5.3 [19].

The infrared thermal images of the glass fibre-reinforced acrylic composites during welding are shown in Figure 5.3.

Table 5.2 Overview of comparative performance of carbon and glass fibre-reinforced acrylic and epoxy composites as reviewed by Obande et al. [9]

	Fibre volume fraction (%)	Tensile strength (MPa)	Young's modulus (GPa)	Flexural strength (MPa)	Flexural modulus (GPa)	Glass transition temperature (°C)
CF/Acrylic	48–60	831–1004	53–72	590–813	51–103	110–120
CF/Epoxy	60	1,217	67	835	51	–
GF/Acrylic	47–54	860–1241	40–53	879–1006	37–40	106–122
GF/Epoxy	44–50	860	40	809–869	35–38	119

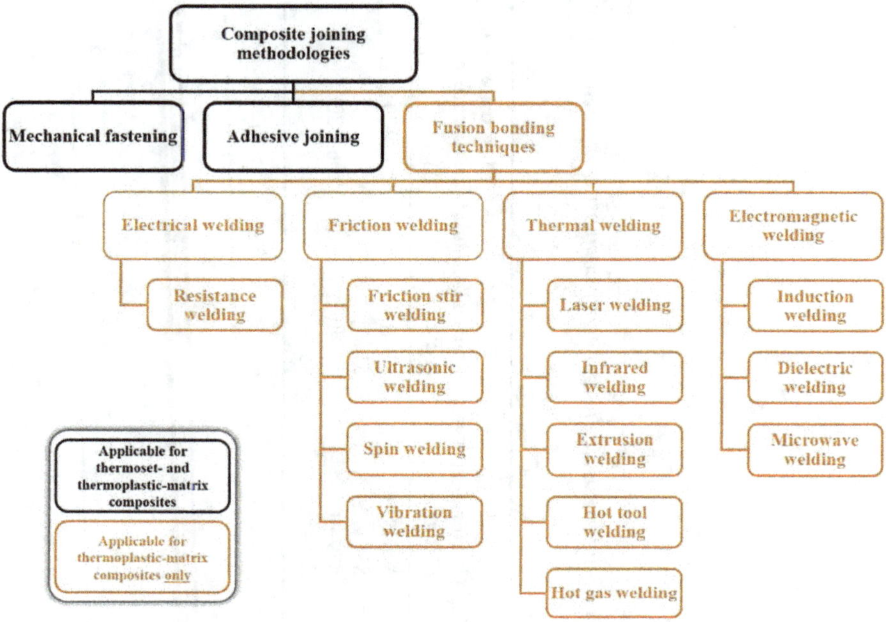

Figure 5.2 An overview of bonding techniques for polymer-matrix composites

Fusion-welded glass fibre-acrylic composites can exhibit up to 100% higher static lap-shear strengths (~20 MPa) than their adhesively bonded counterparts, which had strengths of 10 and 17 MPa [19]. The lap shear stress of welded acrylic/glass composites in comparison to their adhesively bonded counterparts is presented in Figure 5.4. Overall, fusion welding can offer considerable time- and cost-saving for glass fibre-reinforced acrylic components with high bond strength [19]. However, in selecting the welding process for field applications, care must be taken in weld preparation as possible implications of integrated conductive elements on lightning strike vulnerability may need to be considered for the renewable energy sector [19, 31]. The use of integrated lightning protection systems may effectively prevent catastrophic lightning damage [31] . When applied within a 5-m-long blade tip comprising a carbon fibre heating element, simulated 1.5-MV experiments only produced outer skin damage, which could be readily thermally repaired [31].

The ultrasonic welding technique can also be applied to acrylic-based composites [30, 32]. For the welding study, acrylic laminates with the integrated energy director (ED) of semi-circular configuration have been reported in the literature [20, 30, 32]. A constant-time mode may be used during the ultrasonic welding by varying four parameters: weld time, hold time, weld pressure and amplitude. These weld parameters can significantly influence the resin melt and resin flow at the welding interface. In fact, an increase in weld strength has been reported with an increase in the weld time at a specific pressure. Interestingly, beyond an optimal weld time, weld strength can begin to decrease due to various phenomena (e.g.: fibre breakage and

Table 5.3 Summary of the bonding and heating information from a study by Murray et al. [19] (Copyright permission from Elsevier [19])

Bond method	Materials	Bond preparation
Adhesive bonds	Acralock SA10-60 epoxy-based adhesive	A bead of adhesive was applied, along with 1% by weight glass beads, and cured at room temperature under clamp pressure for 48 h
	Plexus 590 acrylic-based adhesive	
	Plexus 310 acrylic-based adhesive	
Resistance-welded bonds	Stainles-steel 304 mesh with a wire diameter of 0.094 mm and 53% opening area heating element with acrylic film on either side (from McMaster Carr)	A current of 16 A and voltage of 2.5 V was applied to the heating element for 5 min
	Fiberglast 4.0-oz unidirectional carbon-fiber heating element infused with Elium 188; multipurpose polyester and epoxy sizing on fibres	A current of 7 A and volage of 7 V was applied to the heating element for 5 min
	Fibreglast 3.5-oz plain-weave carbon-fiber heating element infused with Elium 188 multipurpose polyester and epoxy sizing on fibres	A current of 5.5 A and voltage of 7 V was applied to the heating element for 5 min
	Vectroply C-BX 0450 biaxial carbon-fibre heating element infused with Elium 188; multipurpose polyester and epoxy sizing on fibers	A current of 6 A and volage of 4.5 V was applied to the heating element for 10 min
Induction-welded bonds	Fibreglast 3.5-oz plain-weave carbon-fiber heating element infused with Elium 188 multipurpose polyester and epoxy sizing on fibres	The induction coil was located 10 mm above the heating element and operated at 4.25kW (85% of maximum power) for 5 min

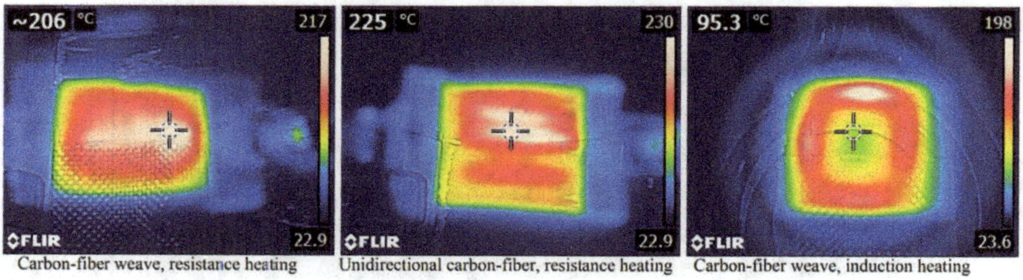

Figure 5.3 *Infrared thermal images of heating using various elements: resistance heating of a carbon fibre weave (left), resistance heating of unidirectional carbon fibre (middle) and induction heating of a carbon fibre weave (right). Note, the induction heating element was placed 3.5 mm above the top surface of the induction coil [19] (Copyright permission from Elsevier [19]).*

delamination) [32]. A combination of higher weld time and pressure, may result in resin being squeezed out of the interface, resulting in damage to the adherends and dramatic weld-strength reduction [32]. Examples of ED design strategies used in the preparation of composites for ultrasonic welding are shown in Figures 5.5–5.7 [32].

The maximum lap shear strength of the welded laminate was 23.2% higher than the adhesively bonded Elium® laminates. Significant plastic deformation of the acrylic resin and the shear cusp formation near resin-rich sites were observed under

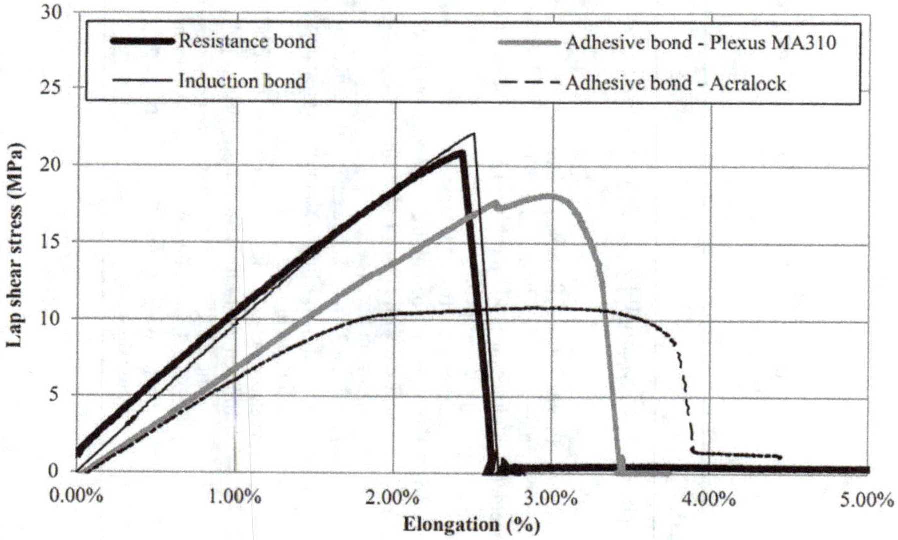

Figure 5.4 *Elongation as a function of lap-shear stress for various bond types as reported by Murray et al. [19] (Copyright permission from Elsevier [19])*

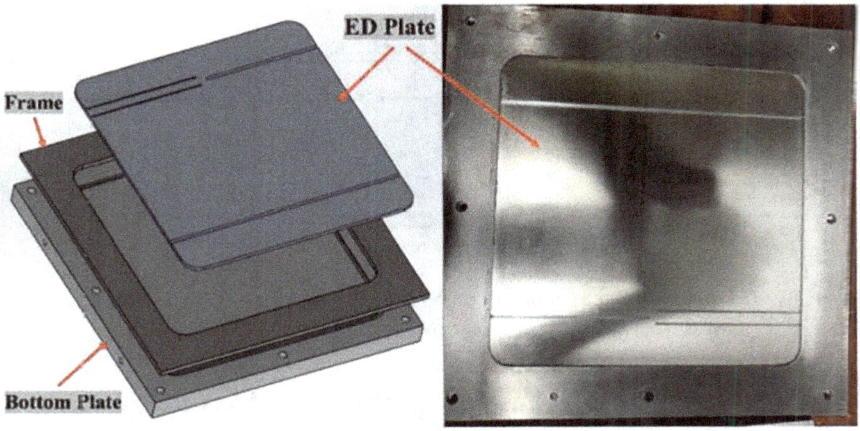

Figure 5.5 Mould design for the production of composite laminates with energy directors (EDs) for ultrasonic welding [32]

SEM. These observations were typical of the optimised weld condition and had a direct relationship with strong interfacial bonding.

In the first section of this chapter, the need for the prevention of landfill waste via intelligent design was highlighted. While durable joints are ideal, the ability to subsequently separate joints for repair and other recycling operations is equally important. Disassembly of welded joints can be possible under certain thermal and loading conditions, as reported in Reference [29]. Successful disassembly was reported at 130 and 150°C, and a heat-affected zone was found to exist in both the adherends. Although the presence of this zone could have implications on subsequent reuse of the adherends, it may be possible to mitigate its size and severity by following the authors' recommendations.

Figure 5.6 A bubble-free energy director (ED) on a twill-woven, carbon fibre-reinforced acrylic adherend prepared for ultrasonic welding by Bhudolia et al. [32]

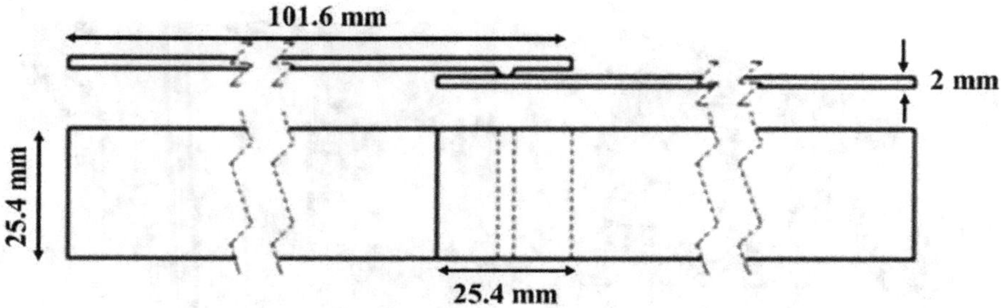

Figure 5.7 Dimensioned schematic of a semi-circular ED configuration employed in weld preparation as reported by Bhudolia et al. [32]

5.4 Thermoplastic acrylic-matrix fibre-metal laminates

Hybridisation offers attractive mass- and cost-reduction strategies where two or more material classes can be combined synergistically to realise a new material with highly tailorable and desirable properties [33, 34]. While a plethora of classifications and compositions may be considered, by far the most commonly encountered hybrids of fibre-reinforced composites are fibre-metal laminates (FML). These hybrids typically combine alternating fibre-reinforced composite and metal layers [9] and are often classified based on the constituents and configuration employed in their fabrication, as shown in Figure 5.8. They were originally developed to exploit the weight-saving benefits of composites in a relatively cost-effective manner with possible enhancements in fire resistance, damage tolerance, fatigue and impact properties [35, 36] making them highly attractive materials.

Fibrous reinforcements were first incorporated into metal-bonding adhesives in the 1970s in a bid to address the poor fatigue performance of monolithic metals which had catastrophic in-service repercussions on the aviation industry. Thereafter, more advanced FMLs such as GLARE, ARALL and CARALL, which are glass fibre-reinforced, aramid fibre-reinforced and carbon fibre-reinforced epoxy-based aluminium FMLs were developed for commercial use between the late 1970s and the early 1990s. Nowadays, more combinations of metallic alloys and advanced FRP composites have contributed to the wealth of literature and overall maturity of these hybrids [37]. Table 5.4 presents a summary of some of the employed techniques and materials in recent works on acrylic-matrix FMLs.

In Table 5.4, an array of metal surface treatments can be seen. Some notable differences arise from the use of strategically drilled holes as resin flow paths for radial, through-thickness resin infusion by Mamalis *et al.* [38]. On the other hand, other authors [40–43] did not report similar measures to enhance flow. Some commonly cited advantages of FMLs include the potential for strategic performance enhancements (fatigue, impact, formability, etc.) [37, 38, 40, 44]. However, the implementation of FML design is not without its challenges; fibre-matrix and fibre-metal adhesion play key roles in the extent of improvement realised. Consequently, appropriate surface treatments and adhesion promoters such as those described

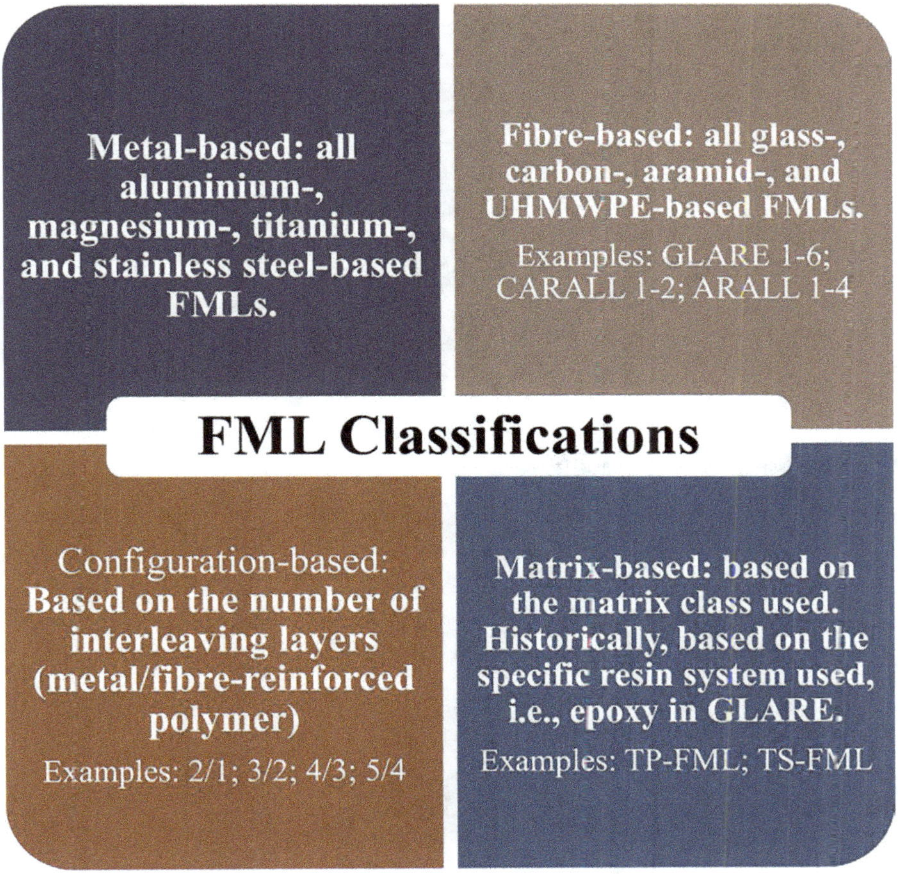

Figure 5.8 Constituent- and configuration-based classifications of FMLs

in Table 5.4 may be required [38, 39, 44], . From the properties summarised in Table 5.5 , it can be observed that the use of treated alloys within FMLs can confer marked superiority relative to equivalent fibre-reinforced polymer laminate with the same constitutents when compared on the basis of flexural strengths and moduli, shear strength, and mode-I interlaminar fracture toughness.

5.5 End-of-life opportunities with acrylic-based composites

One of the most common routes to recycling thermoplastic composites takes the form of a primary mechanical process involving cutting or grinding the bulk component into smaller pieces and a secondary thermal process by which the resized material is compression moulded into the desired shape *as-processed* or with added

Table 5.4 Overview of recent works on acrylic-based FMLs

Metal alloy	Reinforcement	Matrix	Metal surface treatment	Manufacturing process	Ref.
6082-T6 Al Aluminium 0.71-mm-thick	Glass	E-180	Various surface treatments and adhesion promoters	Vacuum-assisted resin infusion	[38, 39]
Ti-6Al-4V Titanium 1.5-mm-thick	UHMWPE *treated with PDA and PDA + CNT*	E-188	Anodising, etching, annealing	Vacuum-assisted resin infusion	[40]
Ti-6Al-4V Titanium 0.4-mm-thick	UHMWPE *treated with PDA and PDA + CNT*	E-188	Annealing	Vacuum-assisted resin infusion	[41]
Ti-6Al-4V Titanium 0.4-mm-thick	UHMWPE *treated with PDA and PDA + CNT*	E-188	Sand-blasting, etching, anodising, annealing	Vacuum-assisted resin infusion	[42]
304-2B-SS Stainless steel 0.48-mm-thick	Basalt	E-150	Sand-blasting	Vacuum-assisted resin infusion	[43]

Table 5.5 Some mechanical properties (rounded values) of acrylic-based and epoxy-based FMLs comprising various fibre types and metal alloys

FML type	Flexural strength (MPa)	Flexural modulus (GPa)	Shear strength (MPa)	Mode-I fracture toughness (kJ/m^2)	Ref.
Glass-aluminium-acrylic FML	231–631	33–41	19–37	0.4	[38, 39]
Glass-acrylic FRP	553	28	40	2.1	[38]
UHMWPE-titanium-acrylic FML	700–1650	75–130	3–10	0.3–2.5	[40–42]
UHMWPE-acrylic FRP	300–400	5–6	10–15	2–3	[45]
Basalt-stainless steel-acrylic FML	538	88	–	–	[43]
Basalt-acrylic FRP	257	21	–	–	[43]

Table 5.6 Overview of some common recycling strategies [47]

	Advantages	Disadvantages
Mechanical + reconsolidation shredding/ grinding and reforming	• Relatively cost-effective compared to thermal and chemical processes	• Exposure to fibre dust • Down-cycling • Poor resource recovery
Thermal pyrolysis; fluidised bed process	• By-products can be valuable feedstock for other processes • Can retain good fibre length scales	• Can degrade the fibre sizing • Not economical • Poor resource recovery • The matrix cannot be recovered
Chemical solvolysis; dissolution	• Good resource recovery • Lower temperatures • Less detrimental to fibres	• Exposure to toxic chemicals • Not economical

materials [46]. The associated factors of these and other routes such as pyrolysis and solvolysis are summarised in Table 5.6.

Dissolution can be effectively used to separate the fibrous reinforcement from the consolidated acrylic polymer matrix, as shown in Figure 5.9. Further post-processing of these recyclates can constitute an effective solution from waste and raw material supply perspectives.

5.5.1 Applicability of recyclate matrix

On the applicability of acrylic (Elium®) resin recyclate, it is believed that once thermally depolymerised, it can be purified and should perform on par with its virgin counterpart [48]. In a study by Gebhardt *et al.* [49], it was shown that up to 7.5 wt% recyclate could be dissolved into virgin monomer for subsequent resin infusion. However, both the viscosity and rate of polymerisation increased dramatically at this composition. It was suggested instead that 5 wt% could be readily employed to avoid these issues. Nonetheless, with a matrix recovery potential of up to 85% and

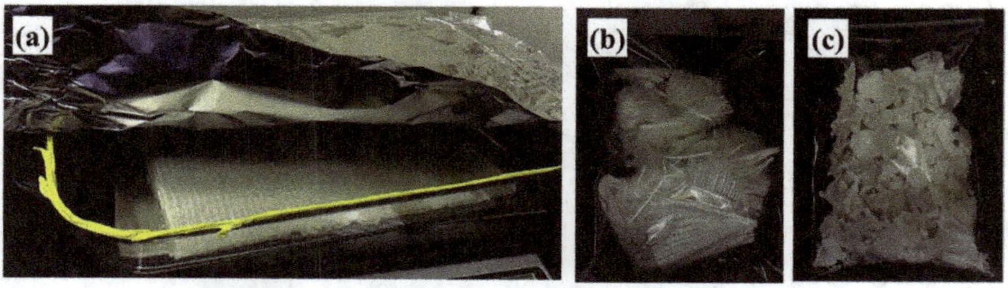

Figure 5.9 Fibre and polymer matrix recovery using a dissolution approach for pre-consolidated, in situ polymerised glass fibre-reinforced acrylic laminate (Copyright permission from Elsevier [25])

comparable flexural performance across the unreinforced reference virgin acrylic and recyclate-containing acrylic materials, this work provides an essential foundation for further development towards circularity. Similar findings were reported by Cousins et al. [25] who dissolved up to 30 wt% of acrylic polymer in monomer and reported significant increases in viscosity and reaction rates. In the same work by Cousins *et al.* [25], chloroform (4 L or 5.96 kg) was employed in the dissolution of a large section of a spar cap, weighing 0.924 kg. The sequential process required a total of 108 hours from immersion to separation and precipitation. Precipitation also required 8 L of methanol. Although technically viable, the use of solvents in the recycling process makes it undesirable in terms of its environmental-friendliness and viability for industrial applications. The time and materials required in this work and another study by Gebhardt *et al.* [50] clearly highlight that the approach of dissolving acrylic polymer in acrylic monomer offers simplicity and time savings.

Another major consideration in the adoption of recyclate-based composite systems into routine applications across diverse sections is the effect of recycling methodologies on performance. For example, successive thermocompression recycling of flax fibre-reinforced acrylic composites can give rise to progressive increases in density (up to +33% after five recycling cycles) [51]. Similarly, repeated recycling operations can result in reductions in tensile strength and strain from Cycle 1: 15.6 MPa and 0.72% to Cycle 5: 13 MPa and 0.23%, respectively. These changes occur predominantly due to the inevitable shortening of fibres, loss of fibre orientation and of course, the weakening effects of polymer chain scission [51].

5.5.2 Repairability of acrylic-matrix composites

Although a clear case has been made for the importance of repair in the first section and Figure 5.1 of this chapter, with the exception of the aforementioned works on welding, no dedicated studies have explored the repairability of acrylic-based composites. Undoubtedly, laboratory-scale and field-based studies focusing on structural repair can go a long way towards addressing this critical gap, possibly increasing component durability and minimising landfill disposal of acrylic-based materials. When fully developed and employed with other techniques such as welding, it may be possible to avoid complete decommissioning and replacement of fractured and damaged components [52].

Some promising work in this area has been from a feasibility study carried out at the Composites Group at the University of Edinburgh, where impact-damaged coupons measuring 150 mm × 45 mm were successfully repaired, as shown in Figure 5.10, using a LAB 450 PEI hot press. After repair operations, the impact-damaged zone, which had previously been visible using backlighting, can no longer be seen (Figure 5.10). By exploiting the thermoformability of acrylic-matrix composites, this work demonstrates new repair opportunities for these materials.

5.5.3 Reshapability of acrylic-matrix composites

Reshaping is another viable route to effectively reuse end-of-life acrylic-based composite materials. The unique ability to repeatedly flatten and curve thermoplastic

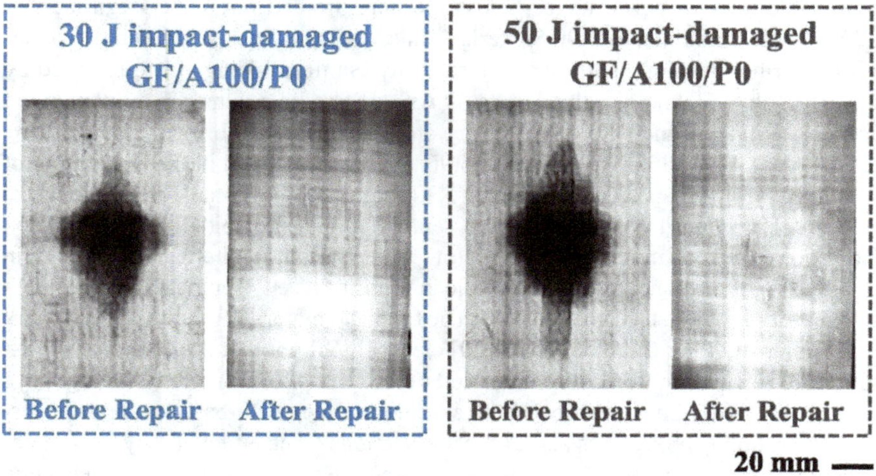

Figure 5.10 *High-contrast back-lit images of impact-damaged coupons before and after hot-press repair (temperature: 200°C; pressure: 29.5 bar; duration: 3 minutes)*

components not only offers desirable design flexibility but can make the repurposing process easier for end-of-life components. Perhaps, even more importantly, larger-scale parts can be readily reshaped and employed for new applications, while maintaining a favourable fibre length scale. This was demonstrated by Cousins *et al.* [25] who flattened a thick-section spar cap comprising glass fibre-reinforced thermoplastic acrylic matrix using a hot press as shown in Figure 5.11.

Further processing (planing) facilitated the extraction of thinner sheets from the flattened sections, from which the authors thermoformed a skateboard deck. Such post-processing can be vastly beneficial from waste management hierarchy and circularity perspectives. Finally, an important consideration for future research

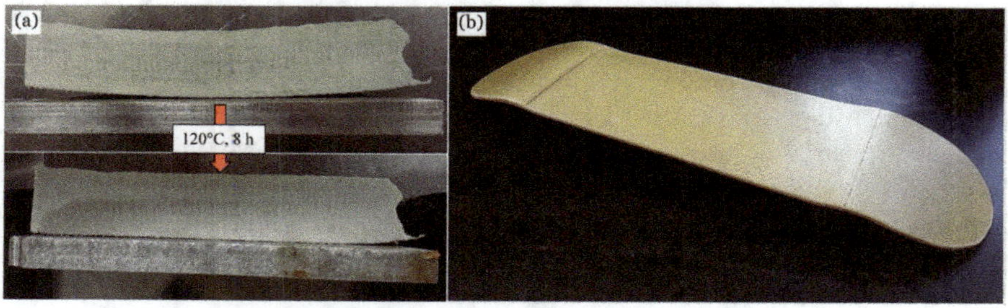

Figure 5.11 *The thermoformability of glass fibre-reinforced acrylic laminates can be exploited for (a) flattening an end-of-life curved section to produce (b) a new curved component with new design requirements (Copyright permission from Elsevier [25])*

efforts building on this area is that acrylics are amorphous materials with no partial crystallinity and as such, no specialised cooling regime is required to tune their microstructure.

5.6 Future opportunities

A key driver of the recent research activities and contributions described in this chapter is the need to address the ongoing end-of-life waste management challenges. Already, the threats posed by the current practices from ecological impact and resource depletion perspectives are clearly understood. The correct identification of open research problems and implementation of targeted innovation can effectively build on the foundation laid by current research efforts to extend the applicability of acrylic-based composites. The inherent reshapability, weldability and recyclability of thermoplastic acrylic matrix-based materials may be developed to facilitate scalable and practical solutions towards sustainability. Herein, we present insights into the current trends and unique opportunities in these areas.

Cousins *et al.* [25] showed that it was possible to reshape end-of-life acrylic-matrix composites components in two ways: (i) a spar cap section was flattened using a hot press and (ii) a skateboard demonstrator was thermoformed from a thinner section obtained by planing. This demonstrates the feasibility of serial production by hot-pressing large-scale, end-of-life, thermoplastic acrylic composites. However, the thermal stability of the acrylic matrix to sustain repeated thermal reprocessing or thermoforming needs further investigation. The reinforcement fabric architecture is another factor to consider for establishing reshaping and thermoforming successfully for continuous fibre-reinforced acrylic composites. Moreover, the exploration of possible routes towards material and process data gathering and the harmonisation of standard practices may help to build a robust database and to identify key areas where reuse could be readily adopted. Building a performance database for reclaimed and repurposed components can increase the much-needed confidence for market uptake and adoption in the future. As research advances in this area, tailored protocols for assessing and reassigning materials for use will be essential, taking into consideration the potential changes to end-user requirements and new design envelopes. There is also a huge cross-technological opportunity that may be readily exploited; advances in smart composites with embedded health monitoring systems could play a key role in these efforts.

Perhaps, a redefinition of the term '*end-of-life*' may be required as more and more repurposed parts become adopted across different sectors.

The research contributions to joining featured in this chapter highlight considerable growth and application potentials in this area. Some important factors must be considered for further development and the demonstration of industrial/commercial viability. Laboratory-based studies have shown both weldability [20, 27, 30, 32, 53] and disassembly [29] possibilities with thermoplastic acrylic-matrix composites. However, the scalability and verification of field-based applicability remain unexplored. Furthermore, the identification of a through-thickness heat-affected zone by Frederick *et al.* [29] highlights an area of concern that necessitates further investigation of the impact of welding parameters on the structural integrity, durability and reuse potential of welded acrylic-based components. As microstructural changes are likely to occur in the heat-affected zone, researchers may consider in-depth analyses to assess such changes. *In addition, are current sizing technologies optimised for thermal welding acrylic-based composites? Can welding-induced defects be readily mitigated by strategic optimisation of welding parameters? Can additional lightning strike protection be effective in addressing the risks posed by the use of conductive heating elements for welding?* These open problems warrant further investigations.

The findings reported in studies by Gebhardt *et al.* [49, 50] and Cousins *et al.* [25] show promising alternatives to conventional pyrolytic and solvolytic recycling methodologies. These works demonstrated a viable route towards attaining supply chain sustainability and landfill-free management schemes for end-of-life acrylic-matrix composites by employing acrylic monomer as a reactive solvent to dissolve pre-consolidated acrylic polymers and composites (e.g.: end-of-life components) for subsequent composite material production. . For relatively thin laminates, the use of some recyclate acrylic feedstock will considerably reduce the burden and demand for virgin acrylic monomer, which invariably translates into reduced depletion of fossil fuel resources. Research efforts building on this knowledge could, therefore, seek to develop similar stretegies for thicker and larger components, considering considering the associated increase in viscosity and polymerisation rates. Perhaps novel and scalable alternative methodologies hold the key to the commercial-readiness of this recycling route. Of course, a material-based approach could also prove impactful; innovative initiators and additives may be developed to tune polymerisation rates as required or bicomponent co-monomer systems may be investigated.

Repairability has been relatively unexplored in this area. Not much is understood about laboratory-scale and field-based repair of acrylic-matrix composites. It must be seen that *in situ*, field-based repair under ambient conditions is desirable for repair without high-cost decommissioning campaigns.

Future work can benefit from life-cycle and techno-economic analyses to assess the environmental and cost impacts of second-generation acrylic-matrix composite materials and structures (repurposed, reclaimed and reused), paying particular attention to their durability and short-, mid- and long-term impacts of full implementation. An understanding of direct scalability will also be important.

5.7 Conclusion

With the growing demand for sustainable, high-performance materials, a greater need arises to exploit the numerous post-processing advantages that the use of thermoplastic acrylic matrices affords. Acrylic-matrix composites offer advantageous room-temperature processibility and remain recyclable, thermoformable and weldable even after consolidation. There are many attractive material- and process-based opportunities for addressing the open problems surrounding resource depletion and waste management. Owing to growing concerns over the environmental impacts associated with material/resource demand and waste management, repairability, recyclability and weldability are some of the most important material selection criteria within the composites industry. The ability to repair and recycle can considerably change the way end-of-life components are handled, ultimately providing alternatives to landfilling of waste. Joining can yield great weight-saving benefits (avoiding the use of fasteners) and enhance design flexibility for large-scale components and multi-material systems. Acrylic matrices as reshapable, recyclable and joinable materials, create a world of possibilities that must be explored. Moreover, cross-technological and inter-disciplinary research activities can contribute to the success of these explorations, which will undoubtedly have positive environmental and cost implications.

References

[1] Zingraff L., Michaud V., Bourban P.-E., Månson J.-A.E. 'Resin transfer moulding of anionically polymerised polyamide 12'. *Composites Part A: Applied Science and Manufacturing*. 2005;36(12):1675–86.

[2] Ó Brádaigh C.M., Doyle A., Doyle D., Feerick P.J. 'Electrically-heated ceramic composite tooling for out-of-autoclave manufacturing of large composite structures'. *International SAMPE Tech Conference*; Long Beach, CA, 2011.

[3] Doyle A., Feerick P., Mallon P., Ó Brádaigh C., Doyle D. 'A heated mould for moulding polymeric composites, A method for making such mould and its use'. 2012.

[4] Gardiner G. *Thermoplastic wind blades: to be or not?* Composites World; 2012. pp. 30–35.

[5] Murray R.E., Snowberg D., Berry D., Beach R., Rooney S., Swan D. 'Manufacturing a 9-meter thermoplastic composite wind turbine blade'. *32nd Technical Conference American Society for Composites*; West Lafayette, Indiana, 2017. pp. 29–43.

[6] Rijswijk Kvan., Teuwen J.J.E., Bersee H.E.N., Beukers A. 'Textile fiber-reinforced anionic polyamide-6 composites. Part I: the vacuum infusion process'. *Composites Part A: Applied Science and Manufacturing*. 2009;40(1):1–10.

[7] Pillay S., Vaidya U.K., Janowski G.M. 'Effects of moisture and UV exposure on liquid molded carbon fabric reinforced nylon 6 composite laminates'. *Composites Science and Technology*. 2009;69(6):839–46.

[8] Ó Máirtín P., McDonnell P., Connor M.T., Eder R., Ó Brádaigh C.M. 'Process investigation of a liquid PA-12/carbon fibre moulding system'. *Composites Part A: Applied Science and Manufacturing*. 2001;32(7):915–23.

[9] Obande W., Ó Brádaigh C.M., Ray D. 'Continuous fibre-reinforced thermoplastic acrylic-matrix composites prepared by liquid resin infusion – a review'. *Composites Part B: Engineering*. 2021;215:108771.

[10] van Rijswijk K., Bersee H.E.N. 'Reactive processing of textile fiber-reinforced thermoplastic composites – an overview'. *Composites Part A: Applied Science and Manufacturing*. 2007;38(3):666–81.

[11] Arkema. "Elium resin: a disruptive innovation in the world of composites," iinnovative webzine. 2019. Available from https://www.arkema.com/global/en/webzine/post/elium-resin-a-disruptive-innovation-in-the-world-of-composites/

[12] Murray R.E., Beach R., Barnes D., *et al.* 'Structural validation of a thermoplastic composite wind turbine blade with comparison to a thermoset composite blade'. *Renewable Energy*. 2021;164(4):1100–7.

[13] Matadi Boumbimba R., Coulibaly M., Khabouchi A., Kinvi-Dossou G., Bonfoh N., Gerard P. 'Glass fibres reinforced acrylic thermoplastic resin-based tri-block copolymers composites: low velocity impact response at various temperatures'. *Composite Structures*. 2017;160:939–51.

[14] Bhudolia S.K., Perrotey P., Joshi S.C. 'Optimizing polymer infusion process for thin ply textile composites with novel matrix system'. *Materials*. 2017;10(3):293.

[15] Bhudolia S.K., Joshi S.C. 'Low-velocity impact response of carbon fibre composites with novel liquid methylmethacrylate thermoplastic matrix'. *Compos Struct*. 2018;203(24):696–708.

[16] Bhudolia S.K., Perrotey P., Joshi S.C. 'Experimental investigation on suitability of carbon fibre thin plies for racquets'. *Proceedings of the Institution of Mechanical Engineers, Part P: Journal of Sports Engineering and Technology*. 2016;230(2):64–72.

[17] Lorriot T., El Yagoubi J., Fourel J., Tison F. 'Non-conventional glass fiber NCF composites with thermoset and thermoplastic matrices'. *20th International Composite Materials Conferences*; Copenhagen, 2015.

[18] Obande W., Mamalis D., Ray D., Yang L., Ó Brádaigh C.M. 'Mechanical and thermomechanical characterisation of vacuum-infused thermoplastic- and thermoset-based composites'. *Materials & Design*. 2019;175:107828.

[19] Murray R.E., Roadman J., Beach R. 'Fusion joining of thermoplastic composite wind turbine blades: Lap-shear bond characterization'. *Renewable Energy*. 2019;140(1):501–12.

[20] Bhudolia S.K., Gohel G., Kah Fai L., Barsotti R.J. 'Fatigue response of ultrasonically welded carbon/Elium® thermoplastic composites'. *Materials Letters*. 2020;264(2):127362.

[21] Boufaida Z., Farge L., André S., Meshaka Y. 'Influence of the fiber/matrix strength on the mechanical properties of a glass fiber/thermoplastic-matrix plain weave fabric composite'. *Composites Part A: Applied Science and Manufacturing*. 2015;75:28–38.

[22] Baley C., Lan M., Bourmaud A., Le Duigou A. 'Compressive and tensile behaviour of unidirectional composites reinforced by natural fibres: influence of fibres (flax and jute), matrix and fibre volume fraction'. *Materials Today Communications*. 2018;16(76):300–6.

[23] Afendi M., Banks W.M., Kirkwood D. 'Bubble free resin for infusion process'. *Composites Part A: Applied Science and Manufacturing*. 2005;36(6):739–46.

[24] Arkema. ELIUM®150 technical datasheet – liquid thermoplastic resin for glass- reinforced composite. 2017. Available from https://cstjmateriaux-composites.files.wordpress.com/2017/11/elium_150_tech_data_sheet_grp_160908.pdf

[25] Cousins D.S., Suzuki Y., Murray R.E., Samaniuk J.R., Stebner A.P. 'Recycling glass fiber thermoplastic composites from wind turbine blades'. *Journal of Cleaner Production*. 2019;209(13):1252–63.

[26] Han N., Baran I., Zanjani J.S.M., Yuksel O., An L., Akkerman R. 'Experimental and computational analysis of the polymerization overheating in thick glass/Elium® acrylic thermoplastic resin composites'. *Composites Part B: Engineering*. 2020;202(3):108430.

[27] Murray R.E., Penumadu D., Cousins D., *et al*. 'Manufacturing and flexural characterization of infusion-reacted thermoplastic wind turbine blade sub-components'. *Applied Composite Materials*. 2019;26(3):945–61.

[28] Panneerselvam K., Aravindan S., Noorul Haq A. 'Study on resistance welding of glass fiber reinforced thermoplastic composites'. *Materials & Design*. 2012;41(1):453–9.

[29] Frederick H., Li W., Palardy G. 'Disassembly study of ultrasonically welded thermoplastic composite joints via resistance heating'. *Materials*. 2021;14(10):2521.

[30] Bhudolia S.K., Gohel G., Leong K.F., Islam A. 'Advances in ultrasonic welding of thermoplastic composites: a review'. *Materials*. 2020;13(6):1284.

[31] Murray R.E., Plumer A., Beach R., Broome P. 'Validation of a lightning protection system for a fusion-welded thermoplastic composite wind turbine blade tip'. *Wind Engineering*. 2022;46(1):260–72.

[32] Bhudolia S.K., Gohel G., Leong K.F., Barsotti R.J. 'Investigation on ultrasonic welding attributes of novel Carbon/Elium® composites'. *Materials*. 2020;13(5):1117–5.

[33] Safri S.N.A., Sultan M.T.H., Jawaid M., Jayakrishna K. 'Impact behaviour of hybrid composites for structural applications: a review'. *Composites Part B: Engineering*. 2018;133(4):112–21.

[34] Bunea M., Cîrciumaru A., Buciumeanu M., Bîrsan I.G., Silva F.S. 'Low velocity impact response of fabric reinforced hybrid composites with stratified filled epoxy matrix'. *Composites Science and Technology*. 2019;169:242–8.

[35] Vlot A., Gunnik J.W. (eds.). *Fibre Metal Laminates: An Introduction*. Springer Science & Business Media; 2011.

[36] Vogelesang L.B., Vlot A. 'Development of fibre metal laminates for advanced aerospace structures'. *Journal of Materials Processing Technology*. 2000;103(1):1–5.

[37] Ding Z., Wang H., Luo J., Li N. 'A review on forming technologies of fibre metal laminates'. *International Journal of Lightweight Materials and Manufacture*. 2021;4(1):110–26.

[38] Mamalis D., Obande W., Koutsos V., Blackford J.R., Ó Brádaigh C.M., Ray D. 'Novel thermoplastic fibre-metal laminates manufactured by vacuum resin infusion: the effect of surface treatments on interfacial bonding'. *Materials & Design*. 2019;162:331–44.

[39] Robert C., Mamalis D., Obande W., Koutsos V., Ó Brádaigh C.M., Ray D. 'Interlayer bonding between thermoplastic composites and metals by in-situ polymerization technique'. *Journal of Applied Polymer Science*. 2021;138(40):51188.

[40] Shanmugam L., Kazemi M.E., Rao Z., Yang L., Yang J. 'On the metal thermoplastic composite interface of Ti alloy/UHMWPE-Elium® laminates'. *Composites Part B: Engineering*. 2020;181(7):107578.

[41] Shanmugam L., Kazemi M.E., Qiu C., Rui M., Yang L., Yang J. 'Influence of UHMWPE fiber and Ti6Al4V metal surface treatments on the low-velocity impact behavior of thermoplastic fiber metal laminates'. *Advanced Composites and Hybrid Materials*. 2020;3(4):508–21.

[42] Kazemi M.E., Shanmugam L., Chen S., Yang L., Yang J. 'Novel thermoplastic fiber metal laminates manufactured with an innovative acrylic resin at room temperature'. *Composites Part A: Applied Science and Manufacturing*. 2020;138(4):106043.

[43] Yaghoobi H., Taheri F. 'Mechanical performance of a novel environmentally friendly basalt-Elium ® thermoplastic composite and its stainless steel-based fiber metal laminate'. *Polymer Composites*. 2021;**42**(9):4660–72.

[44] Kazemi M.E., Shanmugam L., Yang L., Yang J. 'A review on the hybrid titanium composite laminates (HTCLs) with focuses on surface treatments, fabrications, and mechanical properties'. *Composites Part A: Applied Science and Manufacturing*. 2020;128(4):105679.

[45] Kazemi M.E., Shanmugam L., Lu D., Wang X., Wang B., Yang J. 'Mechanical properties and failure modes of hybrid fiber reinforced polymer composites with a novel liquid thermoplastic resin, Elium®'. *Composites Part A: Applied Science and Manufacturing*. 2019;125(1):105523.

[46] Allagui S., El Mahi A., Luc R.J., Beyaoui M., Bouguecha A., Haddar M. 'Thermoplastic Elium recycling: mechanical behaviour and damage mechanisms analysis by acoustic emission'. *Applied Condition Monitoring*. 2021;17:53–61.

[47] Pegoretti A. 'Towards sustainable structural composites: a review on the recycling of continuous-fiber-reinforced thermoplastics'. *Advanced Industrial and Engineering Polymer Research*. 2021;4(2):105–15.

[48] Sekhar Burgula R., Jeelani Basri D., Chandra Mohan K. 'Elium composite resin - exuberant innovation which is easily processed by RTM, infusion and pultrusion'. *International Journal of Advanced Research Engineering and Technology*. 2021;12:580–5.

[49] Gebhardt M., Manolakis I., Chatterjee A., *et al*. 'Reducing the raw material usage for room temperature infusible and polymerisable thermoplastic CFRPs through reuse of recycled waste matrix material'. *Composites Part B: Engineering*. 2021;216(2):108877.

[50] Gebhardt M., Chakraborty S., Manolakis I., Meiners D. 'Closed-Loop room temperature recycling of Elium CFRPs and its influence on the 2nd generation composite properties'. *Zeitschrift Kunststofftechnik*. 2020;1:179–210.

[51] Allagui S., El Mahi A., Rebiere J.-L., Beyaoui M., Bouguecha A., Haddar M. 'Effect of recycling cycles on the mechanical and damping properties of flax fibre reinforced Elium composite: experimental and numerical studies'. *Journal of Renewable Materials*. 2021;9(4):695–721.

[52] Welder S.M., Lause H.J., Fountain R. 'Structural repair systems for thermoplastic composites'. *Materials & Design*. 1986;7(3):147–9.

[53] Bhudolia S.K., Gohel G., Kantipudi J., Leong K.F., Barsotti R.J. 'Ultrasonic Welding of Novel Carbon/ Elium® Thermoplastic Composites with Flat and Integrated Energy Directors: Lap Shear Characterisation and Fractographic Investigation'. *Materials*. 2020;13(7):1634.

Chapter 6

Aerospace composites' repair: integrated processes' feasibility

Dileep Yaswanth Pasupuleti[1], Gopalakrishnan Kamalakannan[1], Guillermo Garcia Del Valle[1], Larissa Reuter[1], Laurine Dehée[1], Lucas Mestre[1], and Hamed Yazdani Nezhad[1]

Repairs of aerospace composite structures are demanding: many different complex processes are performed while maintaining a strict process control and considering the stringent regulatory framework. Composite materials are becoming a standard for aircraft structure design, so the maintenance, repair and overhaul (MRO) industry needs to be ready to face the new associated challenges.

Different stages are considered in a composite repair: inspection, material removal, surface preparation, patch design, manufacturing and application and post-repair inspection. Many technologies can be applied to each of these tasks; consequently, there is not any agreement on a single solution to guarantee a fast, reliable and efficient repair.

The aim of this research was to analyse all the available technologies in order to suggest an integrated system capable of addressing all the involved steps in a cost-efficient way. The proposed system includes a two-stage non-destructive inspection (NDI), combining pulsed thermography and ultrasonic testing. Its execution is automated, carried out by a robotic arm, which also performs other tasks such as material removal or patch application to increase reliability and repeatability. A laser system is integrated for both micro-machining and surface preparation; this is a future-proofing feature, as laser technology is expected to become a standard for composites processing.

To validate the system feasibility, three main areas are covered. First, NDI experiments are performed to analyse if the information obtained is enough to ensure a fast and reliable damage assessment. Then, CAD modelling is done to integrate all

[1]Enhanced Composites and Structures Centre, School of Aerospace, Transport and Manufacturing, Cranfield University, Bedford, UK

the equipment in a portable platform. Finally, cost analysis guarantees the economic viability of the design and technologies used.

6.1 Introduction

Few inventions have changed the way people live and experience the world as much as the invention of the aeroplane. Since its humble origins in 1914, with the first commercial flight in history (St. Petersburg-Tampa Air Line in the US), commercial aviation has evolved and changed in many ways, representing today a vast industry, generating $664 billion of gross domestic product per year, carrying nearly 3.6 billion passengers (in 2015) and directly employing 10 million people worldwide [1].

Nevertheless, this is an extremely complex industry for many reasons:

- capital intensive: requiring extremely expensive and sophisticated equipment and facilities
- labour intensive and high unionization
- thin profit margin
- seasonal industry with marked peaks during summer
- intensely competitive market
- lofty barriers to entry
- government stake in ownership
- duopoly in suppliers (Boeing and Airbus) and oligopoly in both jet engine manufacturers and airlines
- strong regulatory presence

Among all of the complexities stated before, one is extremely important: aviation is one of the most heavily regulated markets. Government bodies such as the Federal Aviation Administration (FAA) or the European Aviation Safety Agency (EASA) are in charge of providing the regulatory framework for the industry, establishing rules influencing personnel licensing and formation, airline licensing and operation, airport regulations, airspace management, security and, of course, aircraft design, registration, operation and maintenance [2].

A key term of the aircraft regulation is airworthiness, defined as fitness for flight operations, in all possible environments and foreseeable circumstances for which the aircraft or device has been designed.

Two types of airworthiness can be considered [3]:

1. Initial airworthiness, meaning that the aircraft design and manufacturing meets the minimum conditions laid down in its type certificate.
2. Continuing airworthiness, defined as all the processes ensuring that, at any time in its life, an airplane complies with the technical conditions linked to the issue of the Certificate of Airworthiness and is in a condition for safe operation.

Figure 6.1 Aircraft in a heavy maintenance hangar [4]

Continuing airworthiness is ensured by carrying out all the required maintenance tasks stated in the aircraft documentation, executed by an approved MRO provider (Figure 6.1).

Figure 6.2 shows the global MRO market forecast for the next ten years, predicting a 50% increase in the global market size, with growth in the four MRO segments: line maintenance, disassembled components, engines and airframe.

MRO companies are considered, in general, as not very innovative and resistant to changes [5]. However, a major change is currently happening in the aviation industry, with the extensive use of composite materials in aircraft structures. MRO providers are currently in need of a revolutionary transformation of their procedures, equipment and know-how in order to adapt to the new generation of aircrafts.

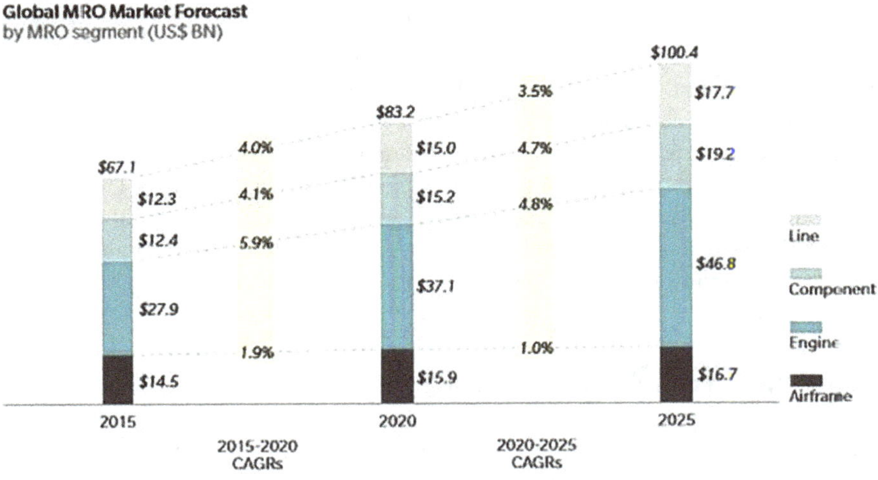

Figure 6.2 Global MRO market forecast by MRO segment [5]

6.1.1 Aim of the project

The Aerospace Composite Repair project aimed to develop a state-of-the-art, highly efficient, rapid and reliable technology for accelerated processing and monitoring of repairs for composite restoration in aerospace primary structures. The project has the potential to underpin all composite activities in the UK and Europe by introducing a series of knowledge-based pioneer scientific and high technology readiness level (TRL) activities. Hence, it enhances the competitiveness with North American advanced composites research, enabling manufacturers to create products with higher added value. In addition, the UK has a strong aircraft industry that requires new manufacturing and joining techniques and training to fabricate, assembly and repair new composite aircraft (e.g. B787, A350 and Bombardier C-series) to remain competitive. The current project/system will highly contribute to such objectives.

Therefore, this project is aimed to develop a rapid, reliable and integrated repair system for aircraft composite structures, at the lowest possible cost. At the end of this report, an innovative, cost-efficient and state-of-the-art design will be provided, integrating and automating (as much as possible) the required steps for carrying out the repair: NDI, material removal, surface preparation, patch manufacturing and measurement of the bond integrity of the repaired structure, in accordance with airworthiness regulations.

Consequently, two key aspects are the main drivers of the project: integration and automation. There are many standalone systems able to carry out the required processes for a repair in a composite structure. However, the underlying idea of the project is to develop a unique system comprising all of the aforementioned capabilities, in order to save in terms of cost, setup and processing times.

Regarding the automation, it is important to mention that one of the major maintenance challenges with composite material is the lack of reliability and repeatability of the bonded repairs. The main objective is to develop an automated system, capable of performing the whole repair in a fast, cheap and repeatable manner. Hence, the dependence on the technician's skills decreases, which will notably ease problems regarding the certification of the repair.

To ensure that state-of-the-art processes and techniques will be analysed and applied: in any case, these will be relatively mature technologies, with a TRL of 7–8. However, due to the academic nature of this project, promising technologies with lower TRL classification (5–6) will also be considered. The different categories of TRL can be seen in Figure 6.3.

6.2 Structure and methodology

This section provides information about two key aspects related to the report: how is the report structured and what is the methodology of the project, in order to get consistent and coherent results.

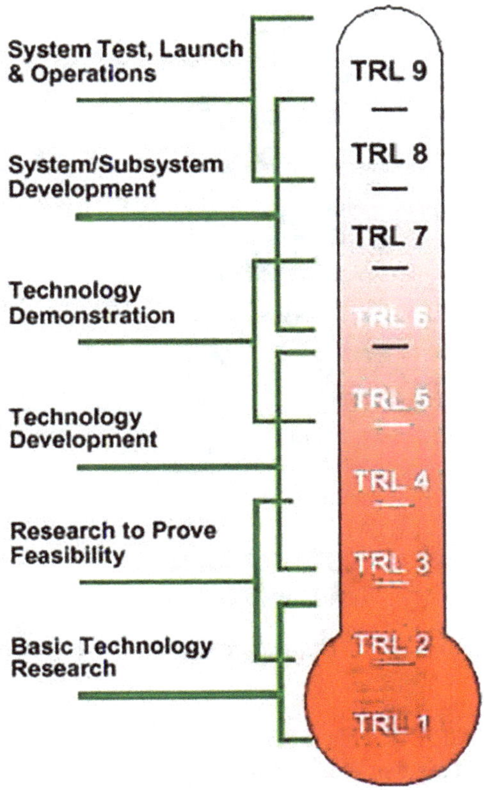

Figure 6.3 NASA TRL scale [6]

6.2.1 Methodology

Several milestones are set to ensure a correct and efficient development of the project, based on the different stages considered: literature review, needs assessment, feasibility study, implementation (covering multiple areas) and final product proposal. All these stages are shown in Figure 6.4.

Therefore, an analysis of state-of-the-art technologies for aerospace composite repair has been carried out. With all this information, altogether with a brief market research, the technology challenges and gaps (knowledge gaps) are identified. In addition, damage mapping activities on the Cranfield University's Boeing 737 helped to analyse the most vulnerable areas of an aircraft structure.

The next logical step is a brainstorming, with an eye kept on the project goals, to come up with ideas about a system being able to cover the identified gaps.

The feasibility study starts afterwards: an in-depth analysis of the available technologies determines how to proceed with future stages of the project. A number of inputs are used during this stage, coming from different sources, as indicated in

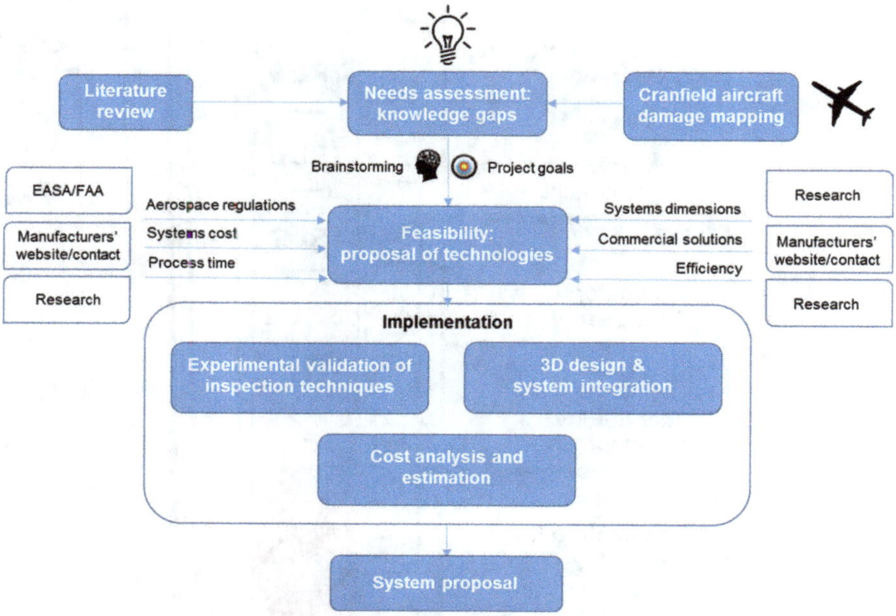

Figure 6.4 Project methodology overview

Figure 6.4 (e.g. EASA/FAA). This is done in order to make the decisions taken reliable enough, guaranteeing that the final design is going to be competitive.

The next step is the implementation stage, with several different activities and studies carried out within it. The main objectives of this stage are to:

- validate the proposed NDI techniques with laboratory scale tests
- carry out the integration of different technologies in one single system: a virtual experiment including the required 3D modelling
- execution of a cost analysis of the system, in order to check its competitiveness

Finally, a proposal for the system is presented. It is evaluated and summarized, including a future outlook for the project. However, providing a work plan and schedule for a project chartered for manufacturing the system proposed here will not be the focus of this report as it requires further refinement within a consortium of the University, OEM and MRO.

A B737-400 analysis consists of a damage and repair mapping activity, carried out in the airplane. The objective of this activity is to identify the most susceptible areas to suffer in service damage in a typical airliner, in order to design a system able to cope with those damages. The study comprised a visual inspection, damage measurement, location of the damages in an aircraft plan and a document summarizing all the information gathered, including images. It is noteworthy that the B737 primary structure does not contain composite; however, the type and location of damages are likely to be similar in composite aircraft.

6.2.2 Literature review

The literature review was carried out to identify current/future trends and techniques applied to composite repair. It is structured in three subsections:

1. damage in composite structures, providing an overview of the composite material usage in commercial aircrafts and highlighting the typical sources and types of damage
2. evaluation of the NDI techniques, used both to detect damages and to assess the quality of the final repair
3. study of the repair process, presenting different procedures for each repair stage (material removal, surface preparation, repair patch application, etc.)

6.2.3 Damage in composite structures

6.2.3.1 Composite material in the aerospace sector

The use of composite materials in the aviation industry has skyrocketed in recent years, as they allow significant weight savings. This is due to their excellent strength-to-weight ratio. As an example, more than 50% of the structural weight of a B787 (and 80% of the volume) corresponds with composite materials, as seen in Figure 6.5.

Many different materials are combined to create aerospace-grade composite materials: fibres made of carbon, glass, aramid, boron or quartz embedded in thermoset (epoxy, polyester, phenolic, etc.) or, more recently, thermoplastic matrixes [8]. However, the most widely used composite material in aerospace structures is carbon fibre reinforced plastic (CFRP) monolithic laminate, built with plies made of carbon fibres embedded in an epoxy (thermoset) matrix.

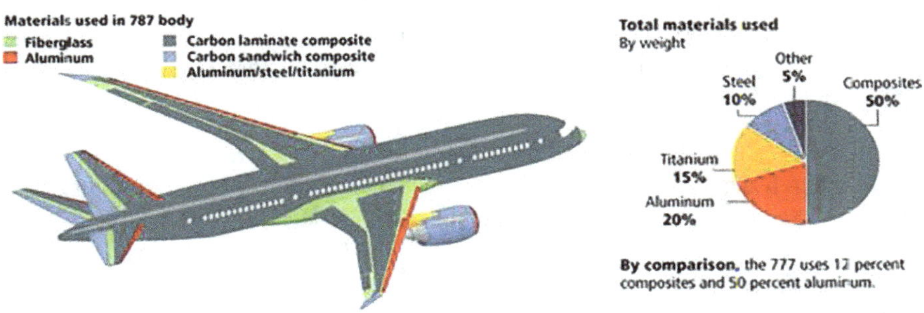

(a) (b)

Figure 6.5 Material usage in a B787 Dreamliner [7]

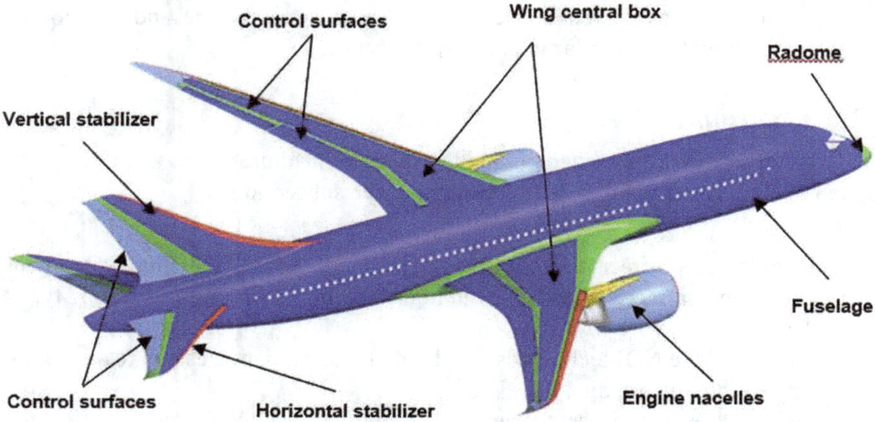

Figure 6.6 Parts manufactured in composite materials in modern aircraft [10]

The use of composite materials in commercial aircraft started in 80 seconds: aircrafts such as the B767, B737 and A320 incorporated the first parts made of polymer composites, generally in the form of sandwich panels and in secondary, non-critical structure. Nowadays, these materials are replacing metal alloys in primary structure (i.e. the structure that bear flight loads and is critical to flight) in aircrafts such as the Airbus A380, A350 and Boeing B787. The main parts manufactured with composite materials are [9] as follows:

- Fuselage: both skin and stringers, made of CFRP solid laminates of variable thickness and co-cured or -bonded
- Wing structure: spars, ribs, stringers and skins, made of CFRP and GFRP solid laminates; the same is applied for horizontal and vertical stabilizers
- Radome: usually made of glass or quartz fibre sandwich panels. It is a secondary structure
- Engine nacelles: usually manufactured with glass fibre sandwich panels, mainly receive impact damage; it is a secondary structure
- Control surfaces: considered secondary structure, they are made from honeycomb panels

Figure 6.6 depicts a schematic breakdown of a modern airliner structure and points out that these parts are made of composite materials.

Despite its extremely interesting mechanical and chemical properties, composite materials present a major drawback: they are susceptible to damage, potentially reducing their performance in a significant way. Impact damages are especially dangerous, able to cause delamination and matrix cracking while being usually difficult to see with the naked eye.

6.2.4 Main causes and associated costs of damages

This study is mainly focused on in-service damage, i.e. the damage that randomly occurs while the aircraft is in operation. Therefore, no pre-existing conditions, which are described in Figure 6.1, are taken into account.

There are two main categories of in-service damage [11]:

- Ageing damage ('wear and tear'): fatigue and degradation processes, which are leading to matrix micro-cracking and progressive delamination
- Accidental damage, which comprises impact damage, impact delamination and lightning strikes

These damages will be caused by one, or a combination of, the following sources [12]:

- in-service loads
- impact: both airborne (hail, lightning, wildlife) and on ground (hail, runway debris, collision with vehicles and structures)
- maintenance errors
- heat and ultraviolet light exposure
- surface coating removal and erosion

Eighty-seven percent of composite damages are caused by impacts from 10 to 100 J (low-energy impacts), according to industry data. The lower the energy associated, the higher the probability to happen, at least when considering debris impact during take-off. The probability level at different energy levels and take-off speeds is shown in Figure 6.7. Obviously, these events have been associated with repair costs, which sum up to the approximate value of $200,000 per aircraft [12]. These low-energy impacts usually lead to undetectable, barely visible impact damage (BVID) [13].

It is also important to consider that a fuselage damage is likely to happen once every 1,000 flights for narrow-body aircrafts and once every 4,600 flights for wide-body aircrafts [12]. Typically, the most affected areas are the external surfaces (fuselage, wing) and front-facing parts of the aircraft.

The effect of these impacts will be different depending on the impacted part as well as impactor type: soft body (flows with a loss of structural cohesion) – hail during flight or bird strike and hard body (solid material) – tool drop, runway debris or hail on ground, according to European studies [14]. Figure 6.8 summarizes the typical impact load cases for commercial airlines, with figures for the typical masses and energies involved.

Bird strikes are historical causes of damage: about 90,000 impacts, leading to a cost of 614–1,280 million $US in terms of repairs, delays and cancellations of flights, only in the US in the past 20 years [15]. This is equivalent to a declared cost of about 160,000 US$ per incident, between direct repair cost and associated costs (passenger's re-allocation, accommodation, etc.), as can be seen in Figure 6.9

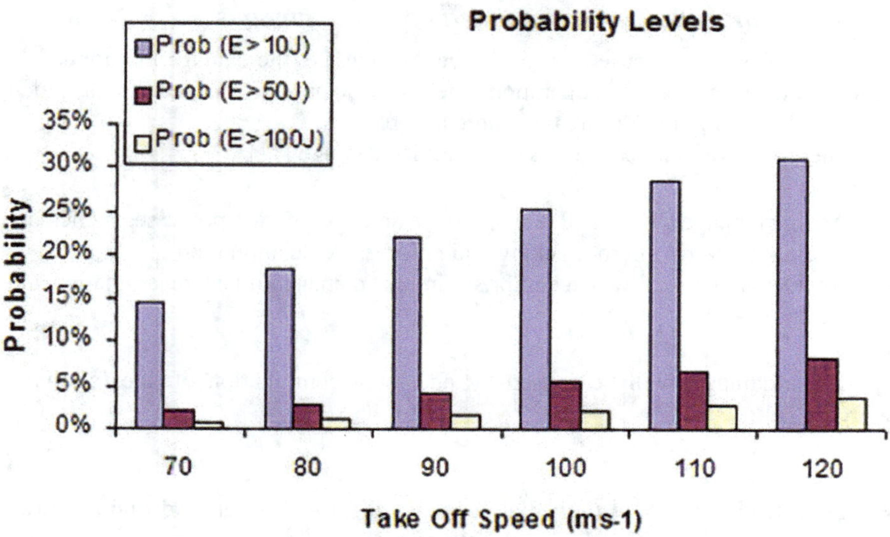

Figure 6.7 Probability of impact damage caused by debris during taking off [12]

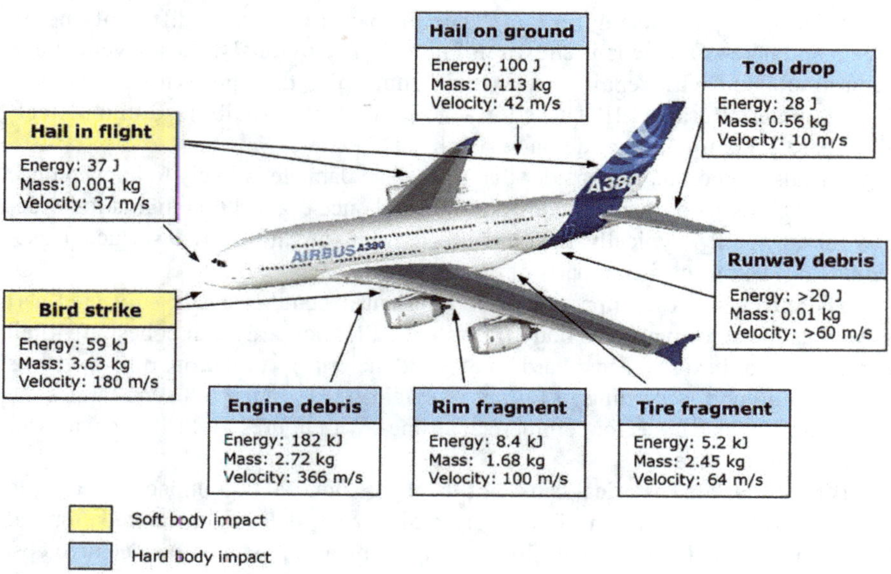

Figure 6.8 Typical examples of impact load cases in a commercial [14]

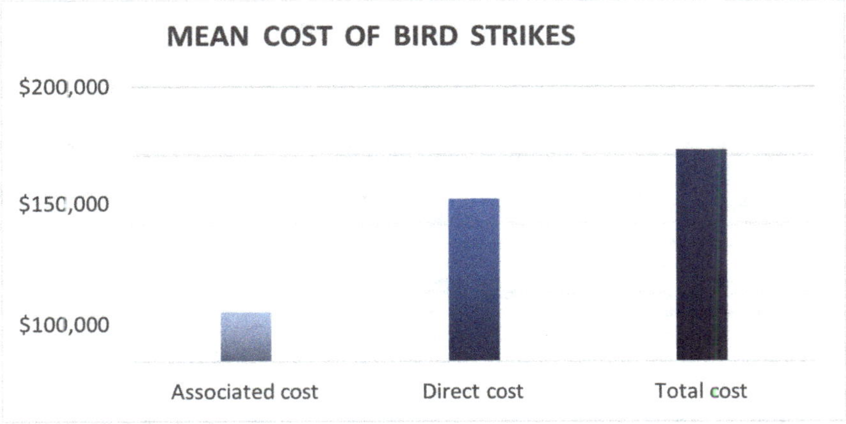

Figure 6.9 Mean costs associated with bird strikes (data from Reference 15)

6.2.5 Typical damages in composites

The type and extent of damage are crucial regarding its influence on the structural integrity and its detectability. The main types of damages are summarized in Table 6.1.

As some damages may become undetectable (e.g. BVID), composite structures must be designed to be damage tolerant [8]. Apart from that, the type of damage considered to be the most critical for aerospace structures is inter-laminar delamination. Airworthiness rules state that no delamination at all is allowed in the primary structure; this is because they reduce structural stability and strength [8]. In fact, inter-laminar delamination becomes the dominant mechanism of energy absorption

Table 6.1 Most common damage types in composite materials

Defect type	Geometry/appearance
Matrix imperfections	Spherical (porosities) Linear along interface (cracks)
Delamination	Surface defect Discontinuity between plies
Fibre breakage	Discontinuity within long fibres
Scratch	Elongated and deep groove of material removed by a sharp object
Dent	Volumetric depression caused by an external impact
Damaged fastener hole	Distorted shape of the hole
Erosion	Rough surface, exposing interior fibres Removal of resin layers at the edges of laminates
Heat damage	Dis-colourization Magnitude depends on intensity of heat source
Lightning burn	Localized burnt surface
Fluid ingestion	Slight volume increase Progressive with time

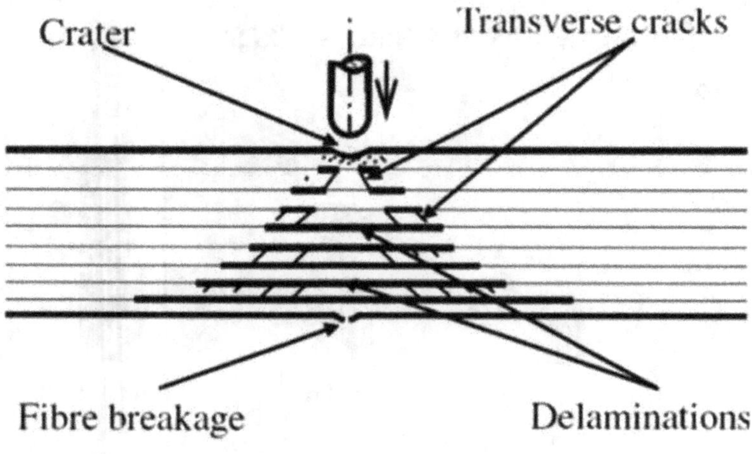

Figure 6.10 Scheme of damage mechanisms caused by BVID [11]

after impact, accounting for up to 60% of the total energy absorption in thermosetting matrix CFRP [13, 16].

Composite materials are also prone to a particular damage that is not critical for metallic structures: low-velocity impact damage. In contrast with metallic parts, a composite one, subjected to low energy impact, may show no marks at all on the surface (causing a BVID, as mentioned before). This impact leads to a delamination in the composite layout, greatly reducing the component compression strength [11]. The damage is magnified from the impacted surface towards the opposite one and has the potential to damage reinforcing elements such as stringers. A schematic image of various damage mechanisms caused by BVID can be seen Figure 6.10

Therefore, due to the great risk that these damages cause to aircraft airworthiness and its high frequency of occurrence, a main focus of this project will be the development of a testing system able to detect this type of damage, as well as others.

All these in-service damages explained before are detected during inspections by using one or a combination of NDI methods, as can be seen in the next part.

6.3 NDI techniques

To perform a proper repair, a thorough inspection of the part is required to determine the location and extension of damages. NDI is the key to achieve this. However, damages in composite materials are difficult to assess: there is not a single technique able to detect all kinds of damages accurately. Hence, the applicability of each technique (ultrasonic, thermography, shearography, etc.) will be discussed, and a combination of two or more will be considered for the final system.

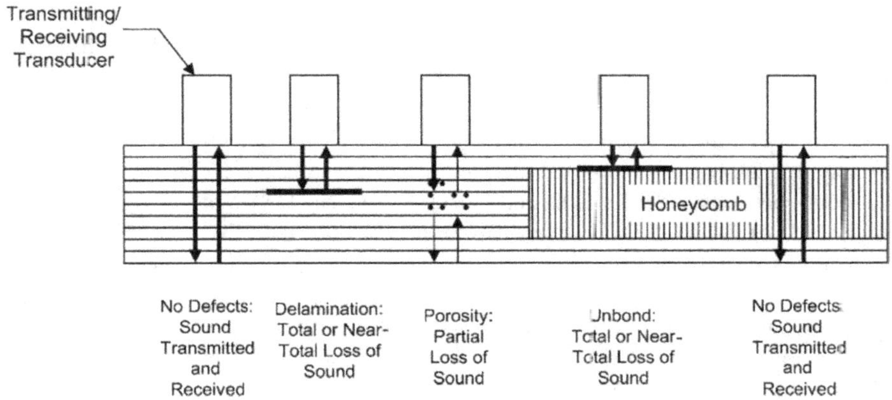

Figure 6.11 Schematics of the pulse-echo ultrasonic inspection [19]

6.3.1 Ultrasonic testing (UT)

6.3.1.1 Overview

This technique is based on the propagation of ultrasonic waves within the object tested. Normal beam inspection (also known as pulse-echo) is the most used process [17], based on detecting the time required for an ultrasonic pulse to travel through the thickness of the material. This process also allows determining the exact thickness of a laminate, requiring only one side access [18]. Figure 6.11 shows how pulse-echo works.

This technique is widely used for composite materials testing [20], competing with other procedures such as thermography or shearography [21]. Table 6.2 summarizes its most important benefits and drawbacks [17].

Table 6.2 Benefits and drawbacks of ultrasonic testing

Benefits	**Drawbacks**
High sensitivity and penetrating power	Sensitive to surface quality and thickness (scattering effects)
High location accuracy	Linear defects aligned with the ultrasound beam may be undetected
Size and shape detection	Difficult and very expensive for large areas
Through-thickness defect location	Part accessibility is required
Single-side access	Long calibration process
High portability	Manual operation (slow, certification required)
Low cost	Requires a coupling medium

6.3.1.2 Techniques

Contact UT can be divided into two modes: A-scan and C-scan, which have different accessibility requirements. The combination of both techniques allows to locating the defect in top view (C-scan) and along the thickness with a time-dependent signal (A-scan), providing a three-dimensional (3D) model of the defect. However, contact UT requires the use of a couplant to ease energy transmission (acoustic impedance reduction) between the transducer and the tested material [18]. Without the couplant, only 0.04% of the input sound energy goes into the specimen from the air [22].

An interesting contactless UT is the laser UT (LUT), which generates ultrasonic waves by means of a laser pulse normal to the surface to be analysed. Another laser pulse permits to read reflections/transmissions [18]. Fast and easily automatable, LUT enables to scan large areas. However, equipment is expensive, and special conditions are required [23].

6.3.1.3 Applicability

This technique is especially suited to detect discontinuities in the laminate (delamination, porosity, inclusions, etc.) but underestimates damages that only cause deformation of the material, like BVID. If the defect is not a delamination or debonding, other techniques are required to be used together with the UT to ensure an adequate level of detectability [24].

Regarding the use of LUT, there are ideas for *smart facilities* able to inspect the aircraft in a fast and automated way [25]. The intelligent use of hangar roof built-in mirrors allows this facility to scan larger areas and parts of difficult accessibility.

6.3.2 Thermography

6.3.2.1 Overview

This NDI technique analyses the thermal response of a specimen to a thermal excitation, allowing to determine the location and size of defects within it [26]. After heating the surface with a pulse of light, an infrared camera takes pictures, which are analysed by a computer to provide temperature profile and distribution. Finally, these profiles permit to locate the defects and their types [27].

The main advantages and drawbacks of thermography are presented in Table 6.3 [26, 28–30].

6.3.2.2 Techniques

There are many different thermography techniques, as can be seen in Figure 6.12 Active ones are the most commonly used in aircraft inspection [31]. Its principle was explained before: a heat source provides excitation to the part and creates the thermal contrast needed to inspect the component [32].

Figure 6.13 depicts the concept of the active thermography and all the components required: a heat source, an IR camera and a PC for post-processing.

Moreover, inspection can be performed in a reflection (one side access required) or transmission configuration (both sides' access required) [34]. This project will be

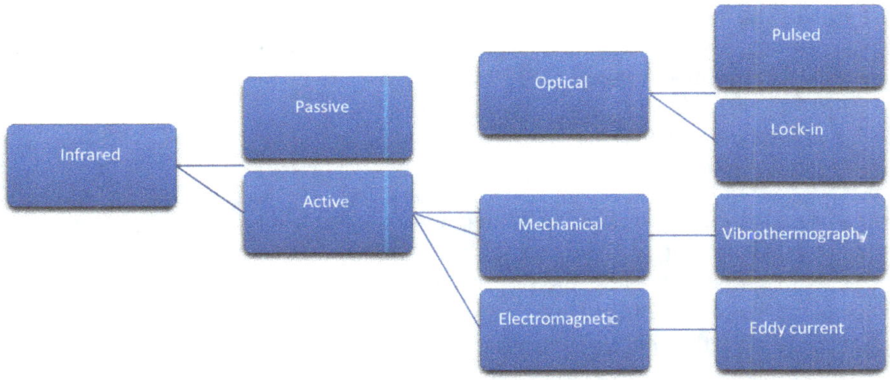

Figure 6.12 *Different types of thermography*

focused on the reflection configuration, as it is easier to perform in an *in situ* inspection [35].

Specific techniques work better with particular damages: when the depth of the damage is not known, pulsed thermography is agreed to be the most suitable technique [36].

6.3.2.3 Applicability

The detectability strongly depends on the capacity of the thermal excitation device as well as the sensitivity of the IR camera [29]. Thermography is considered to be very useful to detect subsurface damages and impact damage [28]. Besides, as the technique is based on the difference in thermal properties, it also permits to detect foreign material in the structure [29].

When comparing the different techniques, lock-in thermography is useful to detect surface cracks and fibre orientation [37]. Similarly, pulsed thermography enables revealing large defects, but not submillimetre ones. Both of them are good to detect delamination and disbonding [38].

Table 6.3 *Benefits and drawbacks of thermography*

Benefits	**Drawbacks**
High sensitivity	Difficult detection of very small and deep defects
Suitable for large areas inspection	Sensitive to thermographic data noise and non-uniform heating
Extremely fast	Quality of inspection decreases with thickness
Contactless	Cost depends on the equipment sensitivity but usually high
Possibilities of portability and automation	

Finally, as thermography is sensitive to the environmental noise, some methods have been developed to remove interferences and enhance both quality of inspection and the detectability of defects [28, 39, 40].

6.3.3 Shearography

6.3.3.1 Overview

Shearography, also known as speckle pattern shearing interferometry, is a holographic NDT technique based on the use of an interferometric system and the analysis of a speckle while the tested component is stressed. Indeed, it uses a laser beam (monochromatic and coherent light), and its reflection permits to get a speckle pattern [41]. Then, the material is stressed, modifying the speckle pattern depending on the loading: the stressed pattern is then compared to the unloaded state one. Differences between them are analysed by a computer, allowing to detect flaws [41]. Interpretation of interferometric images can be done with several methods [42]. The schematic principle of the technique is shown in Figure 6.14

The main advantages and limitations of shearography are shown in Table 6.4 [42, 44].

Different types of loading can be used: vacuum excitation, heating excitation, vibration excitation, etc. For the inspection of CFRP components, the most suitable technique seems to be the heating excitation [43].

6.3.3.2 Applicability

A shearography system developed by The Welding Institute is able to detect disbonds ('out-of-plane strain') and cracks ('in-plane strain') [41]. This technique also permits to detect delamination, wrinkles, kissing bonds, voids and BVID's [44].

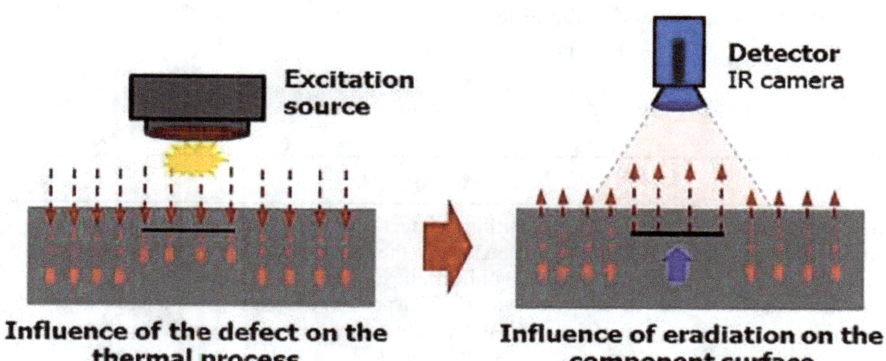

Figure 6.13 Active thermography [33]

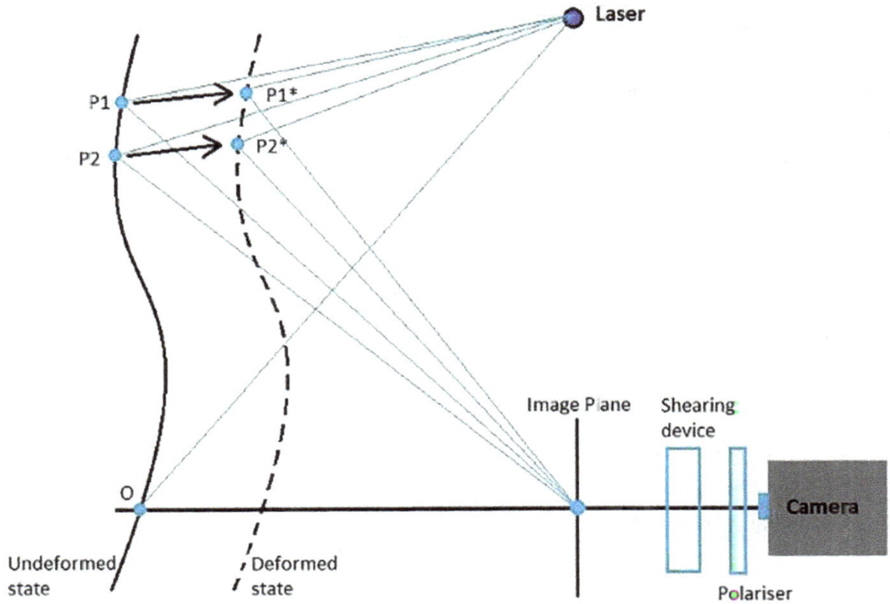

Figure 6.14 Principle of shearography (adapted from Reference 43)

6.3.4 Combination of thermography and shearography

The use of these two techniques together allows to enhance inspection quality by combining the advantages of both; indeed, each technique can detect some defects (types and/or size) that the other cannot. This combination permits a quick detection of most of the damages in large composite structures, providing more details about type, shape and criticality of defects than the use of just one technique alone [29]. Nevertheless, it is apparent that a combined technique will not be cost effective in short term. However, upon the reliability and accuracy it provides from the composite repair and damage regulatory perspective, it can become feasible for long-term use via mitigating repetitive unreliable inspections.

There are proposals for a combined method using the long-wave infrared wavelengths [39]. One example of its use is the European project 'Full-field advanced

Table 6.4 Benefits and drawbacks of shearography

Benefits	Drawbacks
Fast inspection	Difficulty to analyse results
Contactless	Lack of standardization
Inspection of large areas	Lack of operator experience
Not sensitive to environmental noise	Needs more development

non-destructive technique for on-line thermo-mechanical measurements on aeronautical structures' (FANTOM), aimed to the inspection of aircraft composite structures. Launched in 2008, this project was based on the combination of thermographic and holographic techniques in order to improve the quality of inspection and decouple temperature and deformation for results [45]. Its main benefits are 0% uncertainty in the correlation of the results of both techniques and a reduced sensitivity to the environmental noise [46].

6.3.5 Other NDI techniques

Finally, a brief overview of other promising, less popular techniques, is provided in Table 6.5.

Table 6.5 Other NDI techniques

Technique	Description	Detectable defects
Sonic infrared thermography [47]	Sonic actuators induce a vibration at the interface in order to heat the structure leading to stress concentrations	Multiple delamination BVID
X-ray computed tomography [48, 49]	Provides the cross-sectional view of the damaged area Associated health and safety issues	Voids, cracks, inclusions, dry fibres, defective fibre alignment Poor detection of thin delamination
Electromagnetic induction methods [50]	Induce a current in the material. Changes in its impedance lead to damage detection	BVID Disbonds delamination
Resonance methods [50]	Microwaves are emitted from a probe. Defects are detected by changes in resonance frequency	BVID Poor penetration
Vibration-based methods [51, 52]	Damage detection by observation of changes in vibrational response applicable for in-service monitoring	BVID
Electronic speckle pattern interferometry [53]	Contactless method capable of measuring displacements at various points simultaneously by laser illumination of surface	BVID
Acoustic emission [54]	Measurement of transient elastic waves generated by defects applicable for in-service monitoring	Fibre fracture, matrix cracking, delamination

6.3.6 Post repair inspection

Once the damage has been duly assessed and repaired in a controlled environment, it is essential to inspect the repaired area and confirm the degree of strength returned to the original part [55].

Therefore, repairs are evaluated through NDI techniques and need to be tested regarding pre-determined standards. For instance, visual inspection is not sufficient to guarantee the reliability of the repair in the aerospace industry. Thus, some different NDI techniques can be used to assess the repair such as the discussed thermography, ultrasonic and also shearography. Ultrasonic seems to be the most used to assess repaired areas [55].

Nevertheless, there is still a great limitation regarding the evaluation of the residual strength of a repair and the confirmation of the bond quality: currently, there is no non-destructive technique capable to do that. Thus, the best way to have a quantitative assessment is to prepare samples (test coupons) exactly in the same conditions (environmental conditions, material, monitoring, etc.) as the repaired part. Then, standardized mechanical tests such as climbing drum peel test, wedge test, lap shear test and 'T'-peel test are carried out on these samples, following specific conditions given by American Standards [56].

Currently, research on post-repair inspection is focused on the development of non-destructive techniques in order to confirm the adhesive's strength and comparative bondline strength prior to peeling [55].

6.3.6.1 Repair process

The damage location and characterization, achieved using the techniques previously discussed, is only the first step of the repair process. The actual operation commences now, and it is important to keep an eye on the following:

1. material removal
2. surface preparation
3. patch design and manufacture (including bonding/curing)
4. other relevant topics: bonding vs. bolting, quality control, structural health monitoring (SHM) and repair certification

All these topics are duly addressed in this section.

6.3.7 Material removal

Once the damage has been located and assessed, the next step in the sequence of the repair is the removal of the damaged region, generating a round or oval shape [57]. To achieve this, an accurate machining system is required.

There are two possible co-bonded repair methods: stepped and tapered/scarf, with different material removal processes. In the stepped approach, each layer is removed in steps to provide a landing surface for each replacement layer in the repair patch. In the tapered/scarf, however, a tapered-scarf angle (approx. 3° [57])

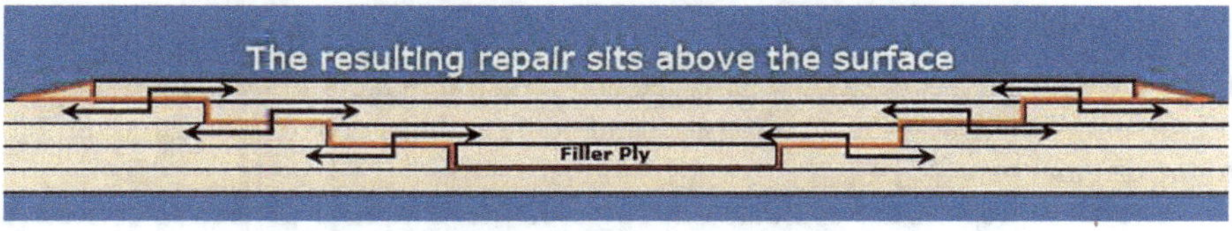

Figure 6.15 Load transmission to the repair patch in a stepped repair [58]

is machined through the composite structure, in order to expose each layer along a gently angled slope. Then, each repair ply lays over the corresponding exposed layer (along the tapered angle if required) [58].

The different geometries also cause different transmission paths, as can be seen in Figures 6.15 and 6.16.

In general practice, the second method is preferred: it is easier and faster with an automated system, keeps the overall thickness of the laminate and presents a lower risk of damaging underlying layers. Additionally, as can be seen in Appendix A in Figures 6.2 and 6.3, the shear stress distribution is better. Nevertheless, it may not be applicable to thick structures [58].

Composite machining is a relatively new practice, still in development stages [59]. Its machinability strongly depends on the fibre characteristics [18]. Matrix tolerance for high temperatures is a strongly limiting factor and the use of coolant fluids is problematic [59].

Some typical defects can appear after a machining process such as edge damage, inter-laminar delamination, matrix-cracking, fibre pull-out, etc. [18]. The quality standard in aviation does not allow delamination at all in primary structures [59].

6.3.7.1 Removal techniques

A brief explanation of the techniques is given, with a final table summarizing its main advantages and drawbacks.

6.3.7.2 Conventional machining

This includes milling, routing, abrasive machining and grinding. Tool wear, cutting forces and temperatures, surface quality and chip formation are the parameters taken into account to assess the machinability of the composite material and the suitability of the process [59].

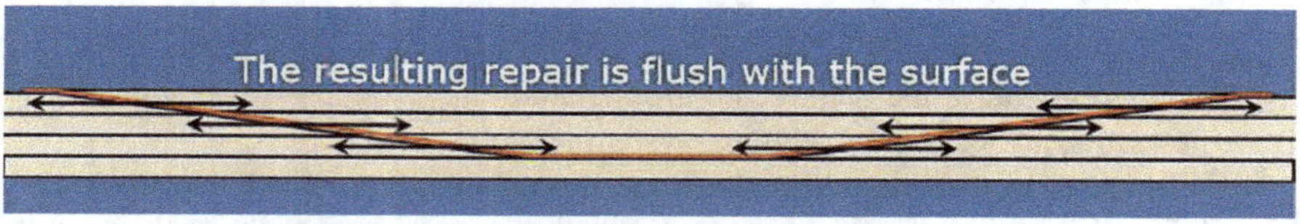

Figure 6.16 Load transmission to the repair patch in a tapered-scarf repair [58]

Conventional machining of composites is complicated: intermittent micro-fracture is a key characteristic, happening because of the different response of matrix and fibres to the applied mechanical forces [18].

There are possibilities of automation of these processes; however, complex systems are required to achieve automated scarfing of laminates for bonded composite repair [18].

6.3.7.3 Abrasive waterjet machining

A high-velocity waterjet is used together with abrasive particles, machining the composites through an accelerated micro-erosion process without causing heat damage or distortion [59]. Then, straight or stepped scarf geometries in composites can be machined.

Other, further developed, non-traditional systems are usually preferred because they are cheaper and more efficient [18].

6.3.7.4 Laser machining

The use of laser produces a local melt shearing and vaporization in the material [59]. A pulsed laser is preferred in CFRP machining than a laser working in continuous mode because it generates a smaller heat affection [18]. Complex laser–material interaction takes place, leading to a different removal mechanism for matrix and fibre [18].

It is a promising technology with research going on regarding its CFRP application [60, 61]. It could also be combined in a device capable of doing laser surface preparation, enormously simplifying the repair process.

6.3.7.5 Electrical discharge machining (EDM)

In this process, a spark discharge is used to thermally erode electrically conductive materials [59]. Hence, it is only suitable for carbon fibre composite materials, being impossible to use this process to cut glass fibre composites.

6.3.7.6 Comparison

Table 6.6 presents the main benefits and drawbacks of the techniques explained above.

6.3.8 Surface preparation

Adhesion is the attraction experimented between two substances (substrate and adhesive) caused by intermolecular forces. It is a key factor in the governance of the mechanical properties of composite materials [62]. Several theories have been proposed to understand the mechanisms governing adhesion; however, these are still not fully understood [18].

Therefore, the final characteristics (strength, stiffness, durability) of bonded joints are extremely dependent on the surface properties. However, when performing a repair, both the processes (material removal) and conditions (humidity, pollutants, chemicals) often contaminate the surface before bonding [18]. Then, a treatment is required prior to adhesive bonding, in order to facilitate one or more of the following [62]:

- surface de-contamination (prevent formation of weak boundary layers)
- creation of a high free energy surface (maximize molecular contact)
- increase surface roughness (increase contact area)
- passivation of the surface of the adherent

An appropriate surface treatment promotes adherent–adhesive interface bond strength, improving the structural performance of the joint [62]. Table 6.7 summarizes the most commonly used techniques.

6.3.9 Surface treatments

6.3.9.1 Peel-plyA

Sacrificial layer of woven fabric is added to the composite's surface during part fabrication, designed to form a weak bond with the composite part. This ply is just peeled off before the secondary bonding operation, in order to expose a clean, active

Table 6.6 Advantages and drawbacks of material removal techniques

Technique	Benefits	Drawbacks
Conventional machining	More mature processes Potential automation (although complex and expensive)	Heat transfer issues Use of coolant Potential material damage Debris generation Slow and complex process
Abrasive waterjet machining	No heat affected zone Absence of tool wear Fast and simple setup Environment-friendly process	Need a thorough drying process after machining Potential delamination Noise Lack of development High cost (investment and operational)
Laser machining	Contactless (no cutting forces) Independent of material strength and hardness Flexibility (of shapes) Ease of automation	Heat affection (although smaller than in conventional processes) Generation of toxic by-products
Electrical discharge machining	Fine surface finish High accuracy No cutting forces	Only applicable for electrical conductive materials Extremely low material removal rate

Table 6.7 Summary of surface preparation methods for composite materials [63]

Treatment type	Nature of treatment	Surface tension	Surface roughness	Surface chemistry	Durability
Abrasion and solvent wipe	Remove mould release		X		Good
Grit blasting	Remove mould release		X		Good
Acid etch	Etch	X		X	Poor
Peel-ply	Remove mould release				Good
Tear-ply	Remove mould release		X		Good
Corona discharge	Oxidizing	X		X	Good
Plasma treatment	Ablation and/or oxidation	X	X	X	Good
Flame treatment	Oxidizing	X			–
Laser treatment	Ablation and/or oxidation		X	X	

surface for bonding [62]. This method could be applied to the repair patch, but not to the base part being repaired.

6.3.9.2 Surface abrasion or grit blasting

This method permits to remove the outermost layer of matrix resin. However, it is labour-intensive, can lead to unconformity and may cause damage to the fibres [62].

6.3.9.3 Plasma treatment

Surfaces' exposition to ionized gases makes them suffer chemical modifications, enhancing surface free energy and wettability [64]. Low-pressure plasma treatments significantly improve adhesion in composite bonding, but they are difficult to use in large components. Atmospheric plasma treatments are easier to use, being good candidates for the treatment of big and complex surfaces.

This is a contactless, easy-to-automate process, able to provide surface uniformity and process repeatability. Nevertheless, it is expensive and surface properties can be lost over time when exposed to ambient conditions [18].

6.3.9.4 Laser treatment

This is a fast, precise and reproducible method that modifies the physical and chemical characteristics of the surface through photothermal and photochemical processes [18]. It is easy to scale up and automate and does not generate waste by-products [65].

However, laser-material interactions are complex and require a complete knowledge of the material properties and the laser parameters, as well as the optical and thermal characteristics [18]. It is also notably more expensive than conventional surface preparation systems.

Nevertheless, there may be a promising future for integrated laser systems, capable of damage removal and surface preparation, allowing for faster, cheaper and repeatable (and thus, certifiable) repair systems [66]. Additionally, Lufthansa Technik, a well-known MRO, is convinced of laser surface treatment and investing a lot in its research [67].

6.4 Material and conditions

Once the surface has been treated, a fitting repair patch must be applied in order to restore the part. Various parameters must be controlled to ensure a reliable repair: patch material, resin, adhesive and curing temperature and pressure. Properties of an ideal repair are described by Whittingham [56].

Aircraft downtime is critical in terms of cost for airlines; thus, all possibilities of speeding up the repair (e.g. use lower curing temperatures and shorter times) are of special interest. This may lead to rather new and unconventional methods, like out-of-autoclave (OOA) curing.

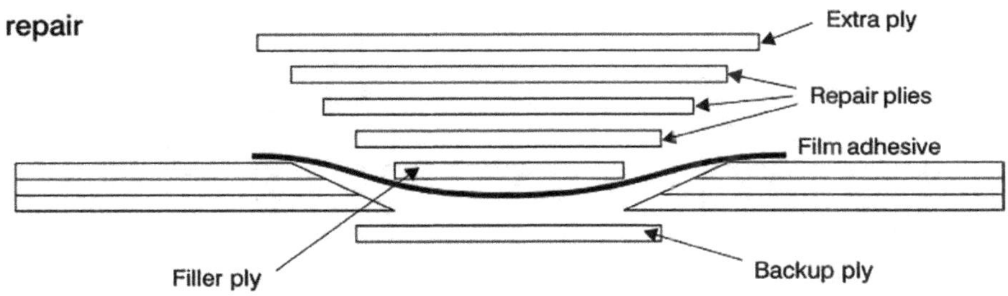

Figure 6.17 Taper sanded repair

6.4.1 Hard and soft patch

Repair patches may be of two types: pre-cured (hard patch) or produced *in situ* (soft patch). A hard patch is either manufactured with a mould that matches the scarf shape or machined from a larger patch. Experimental trials indicate that the machined patch comes closer to ideal than the moulded one [57]. The patch is then secondary bonded through applying and curing an adhesive film/paste.

On the other hand, the soft patch approach is based on tailoring a prepreg laminate and then co-curing it, often adding an extra adhesive layer. It is also possible to manufacture the patch from dry fabric and resin infusion. However, producing fitting properties with dry fabric is a big challenge [18]: prepreg is considered to be an easier approach for patches with varying thickness [68].

This technique is comparable with the 'taper sanded repair or scarf repair' described by Hexcel and shown in Figure 6.17: the overall thickness remains almost the same, and a more optimal load path is generated. Nevertheless, it is time consuming and requires experienced workman skills [69].

Some experiments indicate that there is no big difference in the failure load between hard and soft patches: for a similar 35-mm-diameter patch repair, the soft patches' failure load was just 6% higher [68]. But information by the EASA clarified that they would not accept a soft patch if the original structure was manufactured in an autoclave. A repair must always recover the original structure without any fundamental changes in its load path. The stiffness should match within 5–10% [70].

Table 6.8 shows the main advantages and drawbacks of both patch manufacturing techniques.

6.4.1.1 Cure conditions

When focussing on thermoset polymer composites, which are undergoing extensive chemical, refer to Figure 6.4 in Appendix A, and morphological changes during curing, it is obvious how sensitive this process is. To estimate the rate of change of degree of cure requires knowledge of the particular response for the individual resin and the temperature history for the resin. Once the resin has gelled, the glass transition temperature is often used as an index of the degree of cure [75]. The key parameters to control during the curing reaction are temperature and pressure, which gets supported by Figure 6.4 bonding report from LHT in Appendix A. Possible

Table 6.8 Comparison of recent repair system developments

System	Scanning	NDI	Material removal	Surface preparation	Portability	Benefits and comments
Lufthansa Technik: Composite Adaptable Inspection and Repair CAIRE [71]	Automation W+R GmbH: 3 types of optical scanning systems	RollerFORM ultrasound phased-array wheel probe (manual)	Robotic milling with software from iSAM AG that generate the scarf	LABOA project: laser surface activation to improve bonding strength and generate defined failure mechanisms [67]	Easily moved to the aircraft, hoisted using a forklift and mounted via suction cups	Improves scarfing efficiency by 60% vs. current manual processes; 1m^2 machinable area
DMG MORI & SAUER: ULTRASONIC mobile BLOCK [71]	Laser surface scanning	No	Five-axis ultrasonic milling unit	Atmospheric pressure plasma	90 kg; attachment to aircraft surfaces via 12	
DLR: milling machine [72]	Laser scanning after machining	No	Engineer gives CNC directions for robotic milling	No	45kg	Two to three times less time
American GFM: Inspection & Repair Preparation Cell (IRPC) [73]	3D digitalizing system via accordion fringe interferometry	Planned	Six-axis ultrasonic cutting and routing machine	Not stated	Not mobile	Labour time will be cut by at least 30%; preparation of repair materials; demonstration at C-130 radome
EADS IW: digital repair process [66]	GOM GmbH: ATOS high- resolution optical 3D scanner	Phased-array UT	Engineer gives CNC directions for robotic milling	Plasma treatment	Under development	Designed to address repair of Eurocopter rotor blades; usage of pre-cured patch, film adhesive and vacuum bag

(Continues)

Table 6.8 Continued

System	Scanning	NDI	Material removal	Surface preparation	Portability	Benefits and comments
WIWeB: precious mobile milling unit (PM3) and future mobile repair system (FMRS) [74]	Laser-line scanner (PM3)	UT (FMRS)	Five-axis milling unit (FMRS)	Atmospheric plasma jet treatment (FMRS)	Mobile; vacuum suction system (PM3),~60kg (FMRS)	Patch generator software (PM3), the FMRS elements are not one integrated system yet [135]

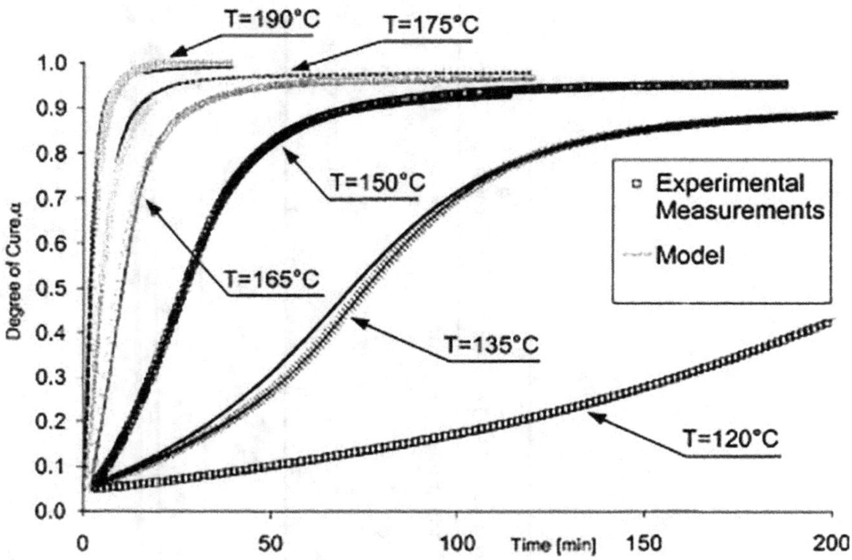

Figure 6.18 *Measured and predicted degree of cure for FM300-1k adhesive [77]*

difficulties arise from the low thermal conductivity of many composite materials, leading to thermal gradients, different degrees of cure and residual stresses [18].

The curing temperature is crucial, as it has a direct impact on the bonding quality, given that the degree of cure can be predicted with both temperature and time, as shown in Figure 6.17 [76]. Faster heating rates may be used, as it has been found that this has, up to a certain limit, no effect on the final degree of cure nor on the overall void content, which can be seen in Figure 6.18 [77].

Controlling the heating process is especially important when using soft patches: single-sided heat sources may lead to thermal gradients, resulting in a non-uniform cure [18]. Typically, heat blankets are used as thermal sources; however, unconventional heating methods are becoming increasingly promising, despite difficulties in sufficiently controlling the cure cycle [18]. Systems such as the 'TeQua Warm Air Box' and 'TeQua Cure Box', both from Marcotodo, are already used in structural repair of aerospace components [78]. Similar temperature boxes, or a traditional heat blanket, are used by the MRO Lufthansa Technik [79]. Another technique for adhesive curing is the electromagnetic induction, suitable for local and rapid heating: results show that both induction and oven-cured joints achieve similar levels of joint strength [80].

The pressure control during curing is essential to achieve an adequate bondline thickness, minimize bondline porosity and ensure adhesive wetting. The vacuum bag is the most popular method for pressure control: applicable to complex geometries and capable of applying uniform pressure while removing enclosed air and volatiles [18].

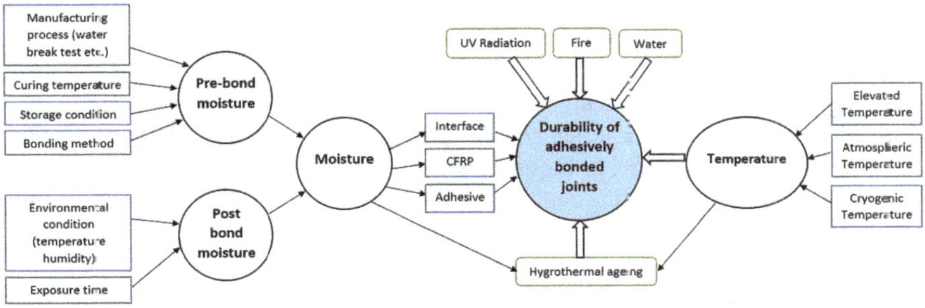

Figure 6.19 *Environmental parameters influencing the durability of adhesive bonded joints [86]*

Experimental data show an achievable repair efficiency of 75% with vacuum bag techniques, compared to 81% when using an autoclave procured patch [81].

Finally, the choice of adhesive strongly depends on the application requirements such as strength, durability, predicted service temperature, etc. Film adhesive is the most used, as it is easy to handle, does not require mixing prior usage, has a more even composition and produces a more consistent bondline. Nonetheless, it has a short shelf life and requires cold storage [82].

Ultimately, adhesive properties strongly depend on the operating temperature: Following FAA guidance, there must be at least a 28°C difference between the maximum operating temperature and the glass transition temperature [83].

6.4.1.2 Protective environment

Composite material usage requires careful control of the manufacturing environment, especially if prepregs are used. They require storage at −18°C and manipulation under certain conditions before curing to avoid moisture and contamination [84]. Experiments showing the environmental impact on prepreg composite manufacturing have been carried out [85]. Environmental threats for bonded joints are summarized in Figure 6.19

Therefore, a system capable of providing the required range of temperature, humidity and pollutant content is required. Many classes of controlled areas/rooms exist, differentiated by the number and size of particles admitted inside. A portable clean room has been developed by HAECO for aircraft structural repair [87].

6.4.2 To bond or to bolt?

The behaviour of a bonded patch is difficult to predict, although analysis of bonded joints has been carried out [86]. Ideally, the patch design would lead to a failure in the parent material because the bonded patch provides a higher strength. Additionally, its optimization reduces the amount of undamaged material removal and consequently the repair time. However, it is important to keep in mind that uncontrolled process conditions can lead to poor-performing bonds [18].

At present, bonded repairs are mainly used in thin skins. It is acceptable to repair minor damages in primary structure when the achieved strength is still above the design limit load, as can be seen in Figures 6.6 in Appendix A [88]. They may also be applied to thicker and highly loaded structures when it can be assured that the residual strength has not fallen below the design requirements. However, it is a major certification challenge, as these repairs are usually one-off occurrences [88].

There are pieces of evidence of strength increases (up to 21%) in bonded joints over bolted joints. Furthermore, the strength even decreases by 3.6% when using bolts on a bonded joint. This practice of adding bolts to a bonded joint is common among the MRO providers [83]. Breuer provides a list of failure modes for bolted joints and reasons to extend the application of bonded repairs [89].

The only clear statement found about bolting vs. bonding [90] is from 2007 and advises to

- bond if the laminate thickness is up to 1/8 inch
- do a repair design analysis when the laminate thickness is between 1/8 and 1/2 inch
- bolted doubler repair if the laminate is thicker than 1/2 inch

This statement got verified during one of the expert interviews [70]; the EASA contact confirmed that he was aware of this limitation but made clear that the values are not fixed. In fact, those size boundaries always depend on the original damage and the material.

The main issue of bonding repairs is that weak adhesive bonds (kissing bonds) cannot be reliably found by current NDI techniques [91]. Thus, requiring structural proof testing with systems such as laser stress shock may help to overcome that problem [88]. Another promising proof test, which reached TRL6 a year ago, is under development in Australia [71]: bonded repair coupons are adhered to the parent structure simultaneously with the repair patch, in order to validate the bond strength (detecting poor surface treatment and/or under-cured adhesive). This test is also an indicator of fatigue damage and gets described in Figures 6.7 in Appendix A [88]. Furthermore, this technique was discussed during an interview [70] and the EASA expert is well informed about it, describing it as being on the edge, but missing a serious application. This must be tested in active service, and he doubts if the correlation between what happens on the outside and inside the repair is reliable enough. The thermal interaction is different, and it is still complicated to draw conclusions about the actual bondline.

6.4.3 Quality control

For each process, standards must be respected: both international standards and internal procedures (company standards). Regarding the repair process, quality must be especially controlled during surface preparation, curing process and characterization of the surface after bonding [92]. Simon stated that the quality within each stage of the repair must be ensured [70].

6.4.4 Structural health monitoring

Structural health monitoring (SHM) integrates sensors into the aircraft structure to achieve a periodic inspection. It provides reliable data for the identification of damages, eliminating the need for manual inspection and allowing a base maintenance on the condition of the aircraft rather than in a fixed time period, with the consequent savings in both time and money.

SHM techniques can be broadly divided into two categories: local and global, with the latter being used to monitor simultaneously the condition of the whole aircraft [93]. In 2013, guidelines for the implementation of SHM techniques in fixed-wing aircraft were published by SAE International as Aerospace Recommended practice ARP-6461. The example of such an SHM technique employed in the Boeing's narrow-body aircraft is comparative vacuum monitoring (CVM) sensor capable of detecting the crack type damage and its propagation [93].

6.4.4.1 Techniques in service

CVM sensors are being flight tested in Boeing's single-aisle aircraft. Acoustic emission sensors are used to monitor fatigue tests in B777 HTPs, as well as eddy current foil sensors and crack wires in many different load tests: they are distributed all over the structure (wing bow, fuselage) and provide information about crack growth and propagation [94].

Airbus has also developed its own SHM sensors, such as the tail strike indication system, which detects contact between the tail and the ground by using crack wires, or the 'Life time monitoring system': strain gauges that measure and record structural loads in order to facilitate maintenance operations planning [94]. Another example is that the CFRP skin surrounding the passenger doors in the A350 is equipped with smart layer sensors.

A brief overview of the existing SHM techniques can be seen in Table 6.9

6.4.4.2 Techniques under development

The concept of global health monitoring for aircrafts by receiving in-service data is still under debate. Indeed, proving the detection capability of the technique in order to get certified and integrating the sensor systems across the length of the fuselage are big challenges.

Basically, the main challenge for SHM is the perception with which SHM is viewed by the aircraft industry, where the understanding of the technology is nascent and will be under scrutiny until proven with satisfactory real-time testing of full-scale components. The big hurdle of certification of integrated SHM sensors in repair patches is also a major issue: it will require exhaustive testing at both laboratory scale and in-service, leading to the requirement of big economic investments. Other main concerns are the development of reliable data processing and damage detection software to distinguish between environmental changes and those caused by damages. Another problem arises from the installation of sensors, because proving the durability of the bond is still a challenging task [11].

Table 6.9 Description of SHM techniques

Technique	Description	Damage assessment	Benefits	Challenges
Acoustic ultrasonic [95]	Smart layer piezoelectric sensors distributed in a polyimide film, sending and receiving ultrasonic surface waves	Comparison of patterns between the baseline and the in-use response of the structure	Passive nature (no external excitation required)	Transient event detection: only one chance to detect the damage Possibility to miss event
CVM	Measurement of the airflow created by the differential between the vacuum and the atmospheric galleries due to the presence of cracks	Very small cracks, disbonds, delamination	Sensitive to very small damage Can conform to various complex surface profiles	Miniaturization Difficulty to monitor internal damages as compared to surface damages
Fibre optic sensors [96]	Sensors of diameter 50–125 µm embedded into the CFRP structure during the manufacture	Measurement of temperature and stress fields during curing	Light weight Small size Flexibility High durability Can either be bonded or embedded	Durability of bond between sensor and structure Sensitivity to environmental changes
Vibration-based SHM	Damage will result in change in structure's dynamic properties by using accelerometers	Cracks, delamination	Wireless sensors Possibility of real-time monitoring. Global method for damage detection	Sensitivity to small damage Differentiation of changes due to normal operation and caused by damage

Some specific techniques look promising for its application in composite patch repairs, such as small-diameter fibre grating sensors to monitor debonding [97] or lamb waves observation to assess adhesively bonded. However, all these techniques still present several specific challenges to address before being adopted in the industry.

To sum up, SHM is considered to be a promising technique, capable of in-service monitoring of the structure, leading to consequent maintenance cost reductions. However, it is a technology still under development with many demanding tasks to solve before being adopted. The certification issues are the biggest concern. Therefore, it is decided not to include it in the proposed repair system, but it is strongly advisable to consider its future inclusion provided that the technology is mature enough and has been approved by the certification authorities.

Besides the difficulties to prove the robustness of the system, Simon also identified the sensitivity of the adhesive film layer to transmit the actual signal without being affected by noise or external sources as another key challenge [70]. Some of the problems related to the inclusion of sensors in the system are as follows.

- One of the most fundamental problems is to actually distinguish the source of the signal being transmitted by the system (e.g. is the change due to environmental/operational effect or due to damage).
- To provide a definitive evidence that the embedding of sensor does not affect the strength of the repair patch.
- It is a daunting task to travel through the test pyramid to get certified as the composite laminate is now considered as a new material.
- Ensuring reliability of the sensor is still a challenge itself [70].

6.4.5 Repair certification

In order to perform structural bonding, stringent process control is crucial. Hence, the material must be certified, the process must comply with the maintenance requirements and workforce must be trained as well as qualified.

Even though a lot of research has been carried out by numerous companies and institutes, the possibility of applying a bonded structural repair on commercial aircrafts is still very restricted. This is mainly due to regulations set by the airworthiness authorities, especially named in the 14 CFR §§ 23.573 'Damage Tolerance and Fatigue Evaluation of Structure', which states: The limit load capacity of each bonded joint, the failure of which would result in catastrophic loss of the airplane must be substantiated by one of the following methods.

1. The maximum disbond of each bonded joint consistent with the capability to withstand the loads must be determined by analysis, tests or both. Disbondings of each bonded joint greater than this must be prevented by design features.
2. Proof testing must be conducted on each production article that will apply the critical limit design load to each critical bonded joint.

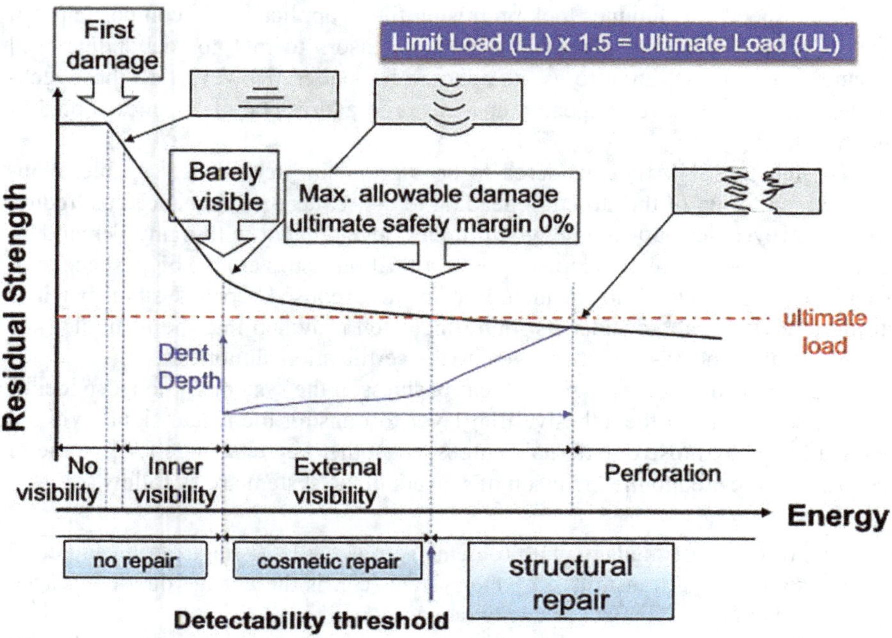

Figure 6.20 CFRP residual strength as a function of impact energy [90, p. 149]

3. Repeatable and reliable NDI techniques must be established which assure the strength of each joint.

Fulfilling requirement (1) requires either strictly limiting the repair size or the use of additional rivets (often referred to as 'chicken rivets'). A lot of effort would be needed to satisfy (2) because every single bonded joint would need to be loaded. Last but not least, no NDI technology complies with requirement (3) up to now. The insufficient NDI reliability is described in more detail in Reference 88.

Rather than using chicken rivets, which only help for mode one failure, Simon suggested multiple bondings or thickened-up parts [70].

Furthermore, there have been recent attempts by the EASA to decide on a size limit of bonded repairs: 'All critical structure will have a repair size limit no larger than a size that allows Limit Load strength to be achieved with the repair failed or failed within constraints of the arresting design features (in the repair or base structure)' [98]. Please refer to Figure 6.20 to visualize the important loads.

Guidance on the bonding environment is fragmentary [83].

- AC 20-107B: the production environment should match the environmental condition employed for the manufacture of items used qualification and proof of structure testing.

- AC 21–26: limits the humidity either at 75 °F below 46% or at 65 °F below 63%, which is also advised in the component production to reduce porosity.

CS 25.571 states '(a) General. An evaluation of the strength, detail design, and fabrication must show that catastrophic failure due to fatigue, corrosion, or accidental damage, will be avoided throughout the operational life of the aeroplane'. Historically, with metal aircrafts, fatigue and corrosion were critical, but nowadays that shifted more towards the accidental damage [99]. Hence, the EASA admits that further regulator-developed guidance is necessary. Simon added that the manufacturer might state other limits in his structural repair manual than the previous described ones. Finally, there are different levels of criticality with changing expectations from the regulator's point of view. He raised, for example, the awareness of principle structure elements [70].

Summing those issues up, there is a need for standardization of certification requirements regarding composite structure. Whatever repair process is about to be developed, it must be in discussion with the regulators as well as the manufacturers. The automation of the repair process (especially the material removal) will be an important step forward to support the certification of bonded repairs for primary aircraft structures.

6.5 Needs assessment

The following section is a brief overview of systems currently under development with some degree of integration/automation of the techniques and processes discussed previously. The objective of this analysis is to identify the gaps that need to be addressed by the proposed system. Additionally, the situation of the composite industry in the UK will be addressed.

6.5.1 Available systems

Several systems with similarities to the proposed one are presently under development or in limited commercial use, such as the Composite Adaptable Inspection and Repair (CAIRe) by Lufthansa Technik or the Inspection and Repair Preparation Cell (IRPC) by American GFM, among others. Table 6.8 compares the main characteristics of them.

The CAIRe system is the most comprehensive one, integrating diverse techniques in an automated and mobile prototype. The combination of three different scanning systems allows it not only to detect deviations from the digital design but also to take into account the fibre orientation as well as the surface roughness. However, patch design and manufacturing are not part of the system so far: the software rather sends the required information about the repair plies to an automated ply cutter, and these prepreg plies are manually placed [71].

Other recent research outcomes include the manually operated repair station by the German Aerospace Centre (DLR), which uses a laser cutting system and induction heating for the curing process [100], or the recently filed patent application by

Boeing for 'Automated production and installation of patches for reworking structures', including carbon nanotube additive [101].

However, all these systems are individual steps towards the final one: envisioned as a fairly mobile module that features a robotic arm with interchangeable heads performing digital scanning, NDI testing, milling/grinding, automated tape laying (ATL) and repair patch application, being attached with suction cups and still directly influenced by the technician via an onboard data link [66].

6.5.2 Gap analysis

Combining all the information provided in the previous sections (objectives from the introduction, state-of-the-art processes and techniques from the literature review and, finally, the brief market analysis summarized in Table 6.8), a gap analysis can be done.

There are new promising technologies ready to be incorporated into the repair of composites, such as 3D scanning, automated NDI techniques such as thermography or shearography or new milling procedures, such as laser material removal. The use of these could easily help to achieve the stated objectives of integration, automation and lower cost.

Regarding the analysed systems, there is only one (CAIRe) integrating all the required steps for the repair; the rest of them are still only partial solutions. Hence, there is a market need for a fully integrated system, capable of performing all the required steps in a controllable and repeatable way, in order to ease the problems of certification.

Besides, the CAIRe project does not take into consideration the manufacture of the patch nor its application: a system taking this into account is also desirable, integrating patch design, lay-up, curing and post-cure inspection.

As both CAIRe and IRPC systems demonstrate, huge savings in terms of labour utilization, time and materials can be achieved through automation and integration. Therefore, a system providing an even more integrated approach could increase these savings, as well as ensuring repeatability and reliability.

Summarizing, there is a clear need in the market for a portable, integrated system, capable of providing the following:

- fast scanning of the whole structure (damage detection) and point scanning around detected damages (damage assessment) with reliable NDI techniques
- automated geometry scanning, material removal and surface preparation
- patch design, manufacture, curing and post-cure inspection

6.5.3 Situation of the composite industry in the UK

The UK composites industry has, over the past three decades, experienced long-term growth based on global economic development and higher penetration into key markets, including the automotive and aerospace industries [102], both of which have been identified as two out of four key important pervasive and secondary technologies for future UK manufacturing activities according to the report by the Department for Business, Innovation and Skills [103].

According to the UK Composites Strategy [104], market barriers exist that may hinder further commercial investment in composites and technological development. The UK composites industry is fragmented and lacks a single voice to articulate industry requirements. This has resulted in coordination failures as few firms have the critical mass to invest in equipment to make composite structures at the speed and cost the industry requires. Consequently, technology and skill transfer between companies and sectors has been hindered by the industry's structure. The existing UK composites landscape has an important role to play in supporting the wider manufacturing sector in its transition to the use of advanced high-performance materials, which could transform products and their performance in order for the UK being able to deliver a £10bn growth in sales of advanced composites by 2030 [104].

The UK currently faces intense, and growing, competition from developed countries in the high-tech sector and from low-wage countries in the more mature traditional sectors. For example, Europe's market share of the current £70bn global composites market is only 24% (UK included), which is unsatisfactory compared to that of North America and Asia at 37 and 32%, respectively [102]. Hence, the UK needs to invest in composites research to increase both its innovation and its global market share and contributes to Europe's 2020 targets of creating 3.7 million new jobs and increasing annual GDP by €800bn [105].

6.6 Feasibility

During the feasibility study, key decisions are made to guarantee that the combined system is feasible and meets all the requirements of the project. In this section, the following topics will be covered to justify the main features of the proposed system: NDI system, machinable area, material removal system, surface treatment, 3D scanning, patch-related decisions and protective environment.

6.6.1 NDI system

NDI is the first step to be carried out by the system so it must be quick, effective and reliable to enable a fast repair. At least two NDI techniques are required to properly assess the damaged structure, as two independent techniques validate each other if they lead to the same findings.

As mentioned in the methodology section, the NDI techniques studied along the literature review are evaluated based on several factors, such as defects type, cost, automation capabilities, materials compatibility, repair stage applicability and size requirements. Table 6.10 presents the key features of the most promising techniques analysed in the literature review.

The aim of the NDI inspection is to assess the damage size and characteristics, as well as gather all the information required to design the repair patch and procedure.

Pulsed thermography together with phased-array ultrasonic testing (PAUT) seems to be a reasonable combination to guarantee that most of the defects are addressed with the required accuracy level. Shearography is a promising technology but still under development, which is the main reason why it is discarded for this project.

The system will first perform a fast scanning using an automated, large-scale method (pulsed thermography) in order to detect damage locations. Then, these damages are evaluated in detail by using the PAUT, allowing real-time imaging as well as quantification of damage size and depth [88]. This is done in order to increase the accuracy and reliability of the damage assessment.

However, it is important to highlight the main drawback of PA UT: it does not perform well under thin laminate conditions (e.g. under 2 mm). Nevertheless, the laminates to be inspected (in the fuselage structure) are within a range of 2–25 mm thickness, so this should not pose a major problem. Therefore, this technique looks promising to extract as much detail as possible about the damages in the structure, and it is already being extensively used in the aerospace field.

Experimental validation of the proposed combination of techniques, assessing the real capabilities of both pulsed thermography and PA ultrasonic, is discussed later in the Experimental Testing section.

6.6.2 Commercial options

6.6.2.1 Ultrasonic equipment

The commercial solutions cost within the range of 25–50 k £, depending on the features. Two proposed models are

1. Olympus OmniScan MX2 [106]

Table 6.10 NDI comparison table

Parameter	Thermography	Shearography	Ultrasonic
Contact	No	No	Yes/no
Scanning time	Fast	Fast	Fast/slow
Type of defects	Cracks Subsurface damage Inclusions Delaminations BVID	Cracks Delamination BVID Porosity Not good at bonded lap-joints	Cracks Delamination Debonding Porosity
Cost	Moderate High (if increase sensitivity)	High post-processing	Low cost
Size	Medium-large defects	Large area scanning	Local/large
Portability	Good	Good	Good
Automation	Good	Good	Depends
Main limitation	Size of defects	Lack of training Non-developed technology	Accessibility to the part

Table 6.11 Dimension and weight comparison of ultrasonic systems

Equipment	Size	Weight
Olympus OmniScan MX2	325 × 235 × 130 (mm) + 226 × 183 × 40 (mm) (phased-array module)	approx. 4 kg+1.6 kg (PA module)
Sonatest VEO+	8-inch screen (no more info available)	No information

2. SIUI Supor [107]

In case a bigger area is required to be inspected, a glider is available, so the scanning area can be increased for later processing. Each of them has its own software for data processing, and both of them are compared because they have the possibility of using a phased array, which increases the resolution of the system compared with a traditional one. Table 6.11 summarizes its main features.

The equipment was a Sonatest Veo for immersion ultrasonic testing. However, the proposed system would be a contact ultrasonic equipment because in-place measurement is required.

6.6.2.2 Pulsed thermography

There are three main parts that have to be considered for pulsed thermography: the first one is a light source (optical excitation), which produces a short light pulse to heat up the structure. The second part is an IR camera, which captures the IR radiation emitted by the analysed part. There are some commercial solutions for both excitation sources and IR cameras. The last part of the equipment is the software for image processing.

One of the main manufacturers of a complete portable kit of pulsed thermography equipment is LOT Quantum Design [108]. There are two options for portable systems: handheld devices and more complex systems, offering higher resolution and higher automation capabilities like the ThermoScope II used . This system consists of three different parts: IR Camera (FLIR SC7000), excitation source (flashes) and software to set up all the image parameters.

6.6.3 Machinable area

First of all, the size of the working area must be demined. To dimension this maximum area where the repair will take place, a correlation between the defect size and repair size must be found.

Two key parameters define the dimensions: the laminate thickness and the scarf angle. This relation can be seen in Figure 6.21, for scarf angles of 3° and 5°. The use of these numbers is justified later in section 5.5.1.

The fuselage of an Airbus A350 is usually 2–5 mm thick but can reach up to 25 mm around the doors [109]. Assuming a 1-m² circular patch as the maximum

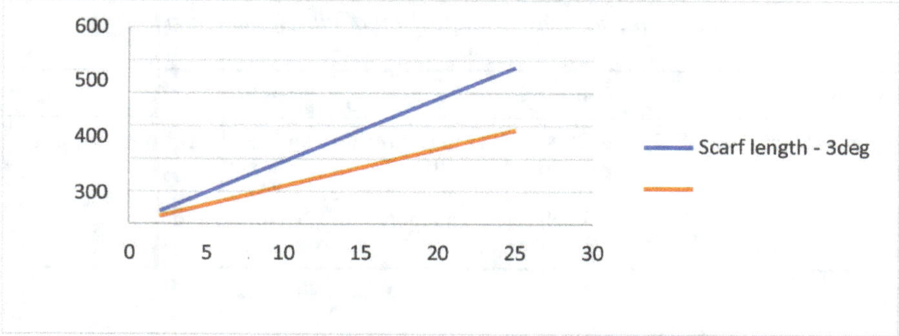

Figure 6.21 *Scarf length as a function of the laminate thickness*

repairable dimension, the system will be able to repair a round 809 mm damage in a 5-mm-thick laminate with a scarf angle of 3°. This maximum size is justified in light of Boeing's latest development of thick bonded repairs up to 1 m² for the B787 [66].

It is important to mention that auxiliary equipment, such as the protective environment box, will need to be slightly bigger than 1 m² in order to provide enough space for the system to be operated within it.

6.6.4 Material removal

As mentioned in the literature review, there are various techniques available for the removal of the damaged material. Table 6.12 compares conventional machining, abrasive waterjet, laser and EDM.

Three techniques get a comparable score. Hence, EDM is discarded at this point. Abrasive waterjet is discarded too, as water and composite do not go well with each other. Finally, a laser system is chosen for micro-machining and surface preparation, while conventional machining is used for rough machining.

Table 6.12 *Desirability of various machining processes (adapted from Reference 110)*

	Conventional	Abrasive waterjet	Laser	EDM
Capital cost	5	4	4	2
Running costs	4	5	5	2
Tool/Consumable cost	3	5	6	1
Processing time	5	6	3	1
Compactness and mobility	5	4	6	1
Quality/Damage to material	4	3	4	5
Health and safety	4	5	3	5
Technology readiness	6	2	5	2
Total desirability scale	36	34	36	19

1 represents least desirable and 6 most desirable.

Figure 6.22 *ProshiP II conventional machining system [113]*

Despite all the advantages of a laser system (like being contact-free and relatively compact, mobile – laser machining head can be up to 200 m [110] away from the power source – and easily automatable), the rate for material removal is considered to be not sufficient for the scope of this project. Experts talk about a 10-hour lead time for a 20-cm repair in a 4-mm-thick CFRP [111], while the conventional machining of a 3-mm laminate with a scarf ratio of 1:20 is predicted to take only about 40 minutes [112]. All this justifies the integration of conventional techniques for rough machining.

6.6.5 Conventional machining

As described in section 4.1, most of the recent developments of automated material removal systems for aerospace applications are based on conventional machining. An example of this is the research carried out by the Fraunhofer institute to automate the mobile machining of aircraft components with high precision [113]. A prototype of this system, especially designed for manufacturing CFRP, can be seen in Figure 6.22

The system chosen for our application consists of a robotic arm with multiple, interchangeable heads to perform the rough machining. Other required subsystems, such as a debris aspiration one, are also included.

6.6.6 Laser system

Many different laser types are available in the market. Table 6.13 summarizes their main characteristics in order to select the most suitable one.

A fibre laser system is chosen: it can work in pulsed mode, which is required to reduce the size of the heat-affected zone, and it provides an extremely small diffraction of the beam, which is critical for achieving a precise control and/ or surface quality. It

Table 6.13 Comparison of different industrial laser systems [114]

	CO_2	Fibre	Nd-YAG	Direct diode
Power	Up to 45 kW	Up to 50 kW	mW to kW range	kW range
Efficiency	10–15%	25–30%	3–10%	30–50%
Mode of operation	Continuous	Continuous and pulsed	Continuous and pulsed	Continuous and pulsed
Spectrum	UV and IR	IR	Near IR	Visible and IR
Wave length	10.6 μm	1.064 μm	1.064–1.55 μm	0.926 μm
Type	Gaseous state	Solid state	Diode pumped Solid state	Diode
Advantages	Good beam quality Widely used	Small focal diameter High intensity Reduced maintenance Compact	High DPI capability	Narrow wavelength
Drawbacks	Different focal lengths required for different applications	Slow cutting speed	Slow for thick materials	Low maturity level

is also better in economic terms: better efficiency (lower power consumption), virtually no maintenance (does not require servicing/consumables and no need for nozzle changes because of less reflection of the beam), and it has a longer lifetime, of up to 25,000 hours, in comparison to other systems capable of ~10,000 hours. In general, it presents a great combination of power, efficiency and quality.

6.6.6.1 Commercial options

Fibre laser commercial systems suitable for CRPF cutting application are:

- YLS-CUT Series, by IPG Photonics [115]
- redPOWER qube, by SPI lasers [116]

Fibre lasers have been tested for its use in carbon fibre and glass fibre reinforced plastics [117], working in continuous mode and using a lower power output (about 50 W). Therefore, the chosen system can be used for both micro-material removal and surface preparation [118].

6.6.7 3D scanning

Three-dimensional laser scanning technology has been used in the aerospace and automotive industry for reverse engineering and metrology application for many years, due to its high potential to generate rapid and accurate topographies of solid objects in a digital format. Laser scanning works on the principle of controlled steering of lasers and analysing the distance measurements originating from the target objects through a camera [119]. The generated digital file can be easily processed by a CAD software for additional measurements and the generation of a virtual surface [120].

Extensive research is being carried out to explore the applicability of this technology to aircraft maintenance, in order to accurately map the topography of damaged surfaces. This enables taking quick and reliable decisions when repairing damaged structures by avoiding manual, visual inspection that depends on the inspector's skills [121]. It is important to mention that this technology is used within the project to define the repair dimensions, and not for damage detection.

6.6.8 Commercial options

Table 6.14 summarizes the main features of 3D scanning systems from different manufacturers.

The proposed equipment is the CREAFORM HandySCAN 3D 700: a portable, handheld scanner, chosen for its high accuracy, good resolution, less weight and wide working environment. It uses a dynamic referencing system (TRUaccuracy), generating the digital file in real time, based on individually set parameters. An example of a topological image is presented in Figure 6.23

In this integrated repair system, 3D scanning will be carried out both before and after machining and surface preparation, in order to obtain data regarding the dimensions of the patch. This will enable an accurate manufacturing of the scarf angle during the machining of the hard patch.

The pre-material removal scanning is done to assure the curvature conformance with the fuselage diameter. However, it is a challenge to extrapolate from the damaged structure to the desired curvature. This can be achieved by comparison with a model, based on a free-form model, which is fitted to the real geometry. An example of such a software application is shown in Figure 6.24

Table 6.14 Comparison of 3D laser scanner manufacturers [122–124]

Model	Scan rate (million points/ second)	Scan range (m)	Resolution (Pixel)	Accuracy (μm)	Dimensions (mm)	Weight (kg)
Surphaser SR_100HSX	0.208–1.2	1–50	7,200 × 7,200	N/A	381 × 219 × 120	11
3D Digital corporation 500 L	N/A	N/A	2,950 × 1,920	125	575 × 100 × 150	4
Creaform 3D Handy SCAN 700	0.48	0.1–4	No limit	30	77 × 122 × 294	0.85
Creaform 3D Handy SCAN 300	0.205	0.1–4	No limit	40	77 × 122 × 294	0.85

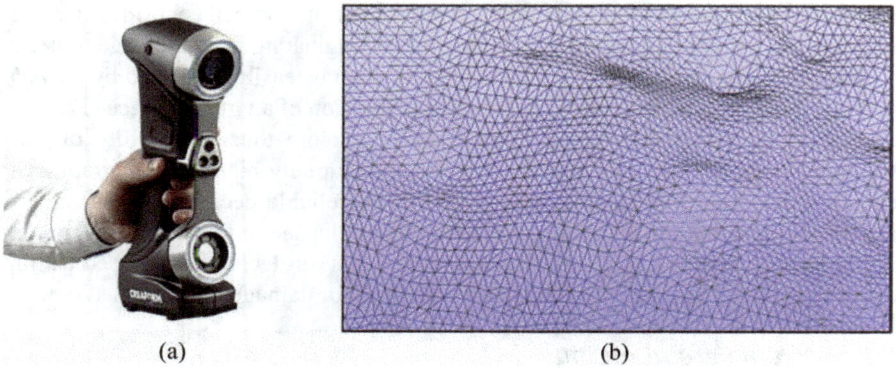

Figure 6.23 *HandySCAN 3D 700 from CREAFORM (left) and the generated topography of a fuselage structure with hailstorm damages (right) [121, 124]*

In conclusion, a 3D laser scanner can be used to generate the necessary data for manufacturing the hard patch (both dimensions and curvature). Its advantages are the portability of equipment, rapid data generation, level of accuracy and complete automation capability.

6.6.9 Patch

As discussed in the literature review and highlighted again in Table 6.15, there are two options for patch design and production: soft and hard patch. In the aerospace industry, the hard patch is the standard for primary structure as it guarantees that the curing stage of patch is done under controlled conditions. Whereas the soft patch production still presents some challenges from the manufacturing repeatability point of view. To avoid further difficulties when dealing with aerospace certification methodologies, the system will work under a hard patch production scheme.

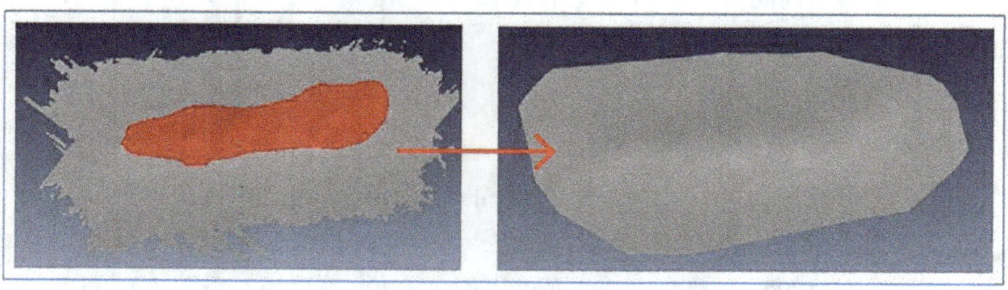

Figure 6.24 *Damage portion in red (left) and a reconstructed surface (right) [121]*

Table 6.15 *Comparing the production of hard and soft patch*

Parameter	Hard patch	Soft patch
Time	2 hours (7 bar, 120°)	Depends on the number of plies
Cost	1	5
Complexity	2	4
Automation	5	2
Industry acceptance	5	4

1 represents the most negative impact and 5 the best one.

6.6.10 Patch design

The main issue faced by tapered scarf and stepped-lap repairs is to avoid generating stress concentrations. Studies show that a scarf angle of 3° is optimal to ensure minimal stress concentration [125]. This can be seen in Figure 6.25, which is unfortunately partially confidential and thus, not totally recognizable.

Generally speaking, the smaller the taper angle is, the lower the adhesive bondline peak stresses are. Moreover, another paper highlights a method to optimize the shape of scarf repairs for different scenarios [126], presenting a correlation between scarf angle and joint efficiency, which can be seen in Figure 6.26

Based on Figure 6.26, the scarf angle might either be even smaller or up to 6°. This justifies selecting another scar angle of 5°, which was recommended in Reference 57 for adhesively bonded scarf repairs of the thick composite aircraft structure. This will enable the system to repair larger damages in the proximity of doors, where the thickness is maximum. It should be kept in mind that even a scarf angle of 1° might be necessary [69] and individual analysis must be carried out.

Additionally, the application of an overply is suggested by some papers to reduce the stress concentration in the patch borders [127], and Bendemra even defined a

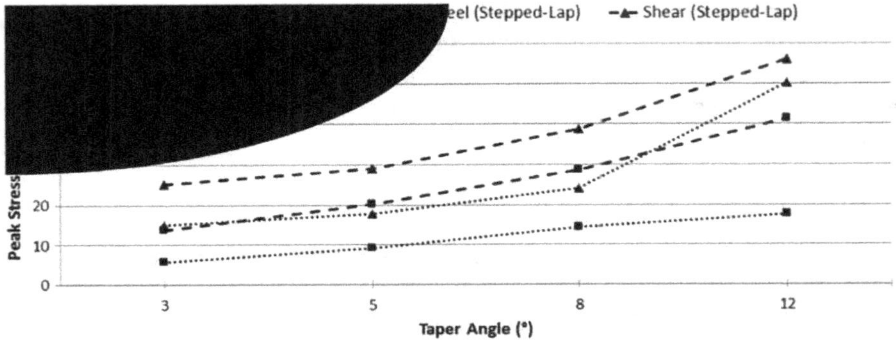

Figure 6.25 *Normalized peak peel and shear stress in the adhesive bondline for different taper angles [125]*

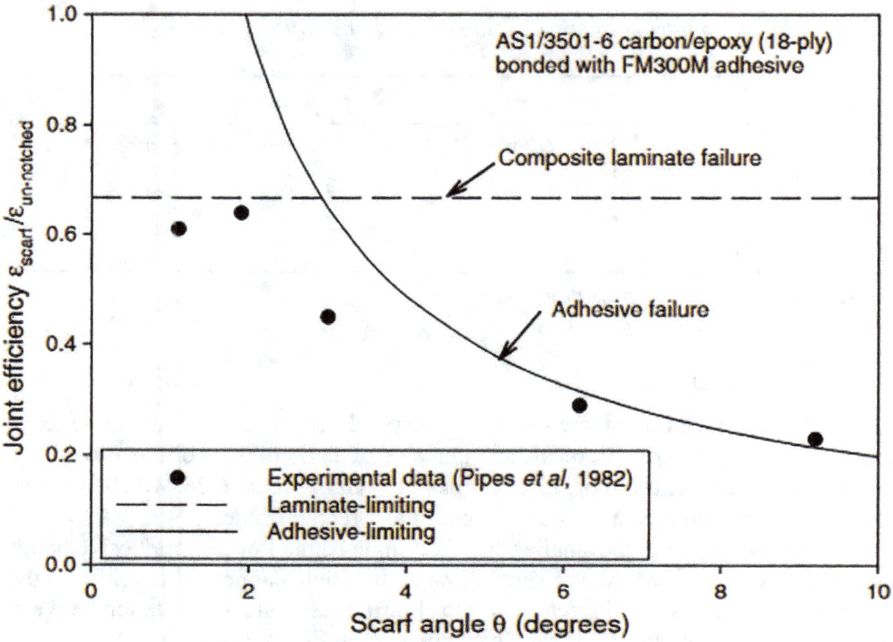

Figure 6.26 *Comparison between experimental data and predictions by the proposed methodology [126]*

critical overply length [125]. Nevertheless, this introduces a too complicated repair process for the system being developed because it would need manual placement of the layer and additional curing.

When talking about the patch design, it is not possible to totally neglect the repair via bonded doublers. For them, the overlap length is crucial, because if this is not sufficient, the joints are not creep resistant, and it may result in premature plastic behaviour of the whole joint. To determine the overlap, which achieves the maximum strength the load capacity method is recommended. Additionally, it should be kept in mind that a load attraction occurs at the end of the doubler because of an increased stiffness of the repair [83].

6.6.11 Patch production

When producing a hard patch, it is difficult to assure the tight tolerances of the scarf after curing. Hence, it is advisable to produce the patch in an individual over-size mould each time to ensure the 3D properties, and to machine the scarf afterwards. Finally, this is decided by the aircraft manufacturer.

The hard patch production implies special equipment requirements. The standard aerospace curing practise is to use autoclave processing and since every MRO is compelled to have one anyways [128], the system is based on using it for the patch

Table 6.16 Comparison of film and paste adhesive

Parameter	Film adhesive	Paste adhesive
Curing temperature	From 80°C	From room temperature
Pressure required	Vacuum	At least mechanical pressure
Automation capability	High	Medium
Temperature working range	Low to High temperatures	Moderate to low temperatures
Bondline thickness control	Good	Requires special additives

production. However, there have also been recent attempts to use microwave ovens for patch production, which could also be an option.

ATL is already implemented in the aerospace industry, and it may help to guarantee the layers' orientations are accurate enough when manufacturing the patch. Actually, Airbus is developing an automated fibre-patch preforming machine [129].

6.6.12 Adhesives

The adhesive selection plays a vital role in the system conceptualization because the system must be capable of applying it properly. The first required decision is whether to select a film or paste adhesive. In Table 6.16, the most important features for both of them are presented.

Since it is more difficult to control the bond thickness of paste adhesive and automating its application will be much more complex, film adhesive is selected to be implemented in the system. To make the final decision, Table 6.17 lists the major different manufacturers and their products, based on their individual selector guide information.

This table justifies the selection of FM 209–1 because it is the only film adhesive being especially described as AOO suitable. In the datasheet, its unique combination of toughness and shear strength is mentioned. Under vacuum pressure, the cure cycle is 121°C for 90 minutes.

6.6.13 On-site curing

Based on the process requirement of the picked adhesive, there are some different heating methods to ensure that the required curing temperature is achieved:

- heat blanket
- infrared lamp
- induction heating

It is advised to apply pressure while curing, which also affects the choice of heat source. To ensure the quality control during curing, thermocouples are attached to the structure. They are used to ensure that the structure is not overheated and to provide assurance of a full cure. Many MROs use a standard configuration for them:

90–120° apart around the repair, which is not sufficient [83]. However, the configuration would be decided on a case-by-case basis in our system.

6.6.13.1 Protective environment

HAECO has developed a portable clean room to 'replicate the environment conditions of a clean room facility that complies with aerospace standards at the location of damage on the aircraft'. These standards require a maximum of 14.1 million of particles per square meter (maximum of 400,000 particles of 0.5 µm and maximum of 3,000 particles of 5.0 µm), a relative humidity below 75% and a temperature range of 18–30°C. The environment needs to be dust free and clean without condensation.

As the typical damages are not very big, HAECO tried to reduce the size of its big portable tent (see Figure 6.27), developing the portable clean room [87]. However, the dimensions provided by this manufacturer (122 cm × 61 cm × 43 cm) are not big enough to cover the maximum machinable area expected in this project. Besides, it is designed to be used by an operator rather than by an automated system. Therefore, a further development, on the lines of this project, is required, in order to design a bigger portable clean room, attachable to the fuselage of the aircraft by using suction cups and with enough ports to allow the correct use of an automated system.

Table 6.17 Major aerospace adhesive manufacturers and some of their products

Manufacturer	Product name	Film or paste	Process
Cytec Solvay Group	DAPCO 3012	Paste	Service temperature <160°C, OOA
Cytec Solvay Group	FM 209–1	Film	Service temperature <160°C, OOA, cure temperature 71–160°C
Henkel	Loctite EA 9392 Aero	Paste	Cure temperature 25°C (5–7 days) or 82°C (1 hour)
Henkel	Loctite EA 9690 Aero	Film	Cure temperature 121°C (1 hour)
Hexcel	Redux 312	Film	Cure temperature 120°C (1 hour)
Huntsman	Epibond 1539-A/B	Paste	Service temperature 80°C, cure temperature 23°C
Magnolia Advanced Materials Inc.	6,368	Paste	

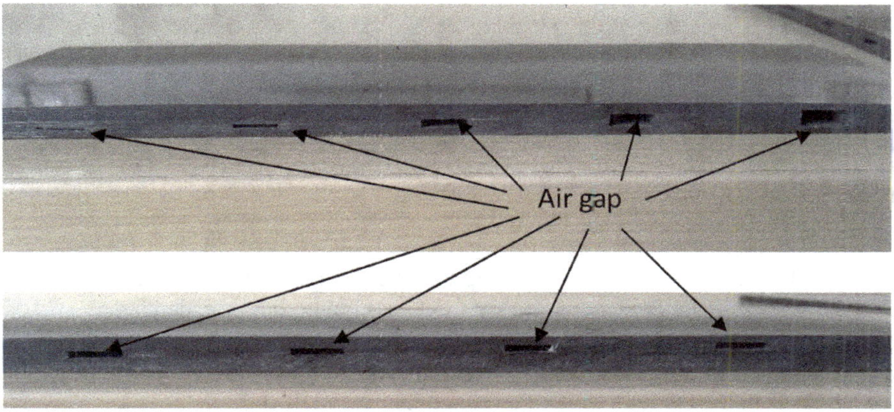

Figure 6.27 Air gap samples of 10 mm width through the laminate thickness

6.7 Implementation

Three main areas are covered in this section: first of all, the results of the experimental tests carried out to validate the NDI proposal. Following the system integration via 3D modelling, and finally, a brief discussion about the costs is presented.

6.7.1 Experimental testing

6.7.1.1 Experimental setup and methodology

As previously mentioned, the experimental validation of the NDI proposal consists of two parts: pulsed thermography and ultrasonic testing. Hence, similar but slightly different set-ups and methodologies are used for both tests.

6.7.1.1.1 Samples information

Two different kinds of samples were analysed: impact samples and samples with known air gaps (defects) introduced while manufacturing.

The impact samples correspond to a 3.7 mm thickness of 2 × 2 twill prepreg (TenCate 8020) laminate that was impacted under three different energy conditions: 8, 16 and 24 J, respectively.

The air gap samples (quasi-isotropic layup) were manufactured to simulate a debonding/delamination condition on the laminate that might occur in service. The samples can be observed in Figure 6.27

These induced defects are known in both size and position, so these samples are used to determine whether the NDI technique used can detect a range of damages in a different location through the laminate thickness dimension. There are two samples for air gap laminates:

Table 6.18 Impact samples' information

Sample	Size	Impact energy	Impactor velocity	Impactor mass	Name
Impact	150 × 100 mm	8J	4.6 m/s	2.28 kg	I_8J
Impact	150 × 100 mm	16J	4.6 m/s	2.28 kg	I_16J
Impact	150 × 100 mm	24J	4.6 m/s	2.28 kg	I_24J

1. Air gap laminate A: constant air gap thickness and at different depths
2. Air gap laminate B: varying air gap thickness but at the same depth

In Table 6.18, the most important parameters of the samples are collected, as well as their nomenclature for future referencing along the report.

6.7.1.1.2 Pulsed thermography testing order

For the impact samples, there are three different energy levels to be analysed; all of them lie within the low velocity and low energy impact limits. The experiments will be done from higher to lower energy values because of the easiness of damage detection.

There are two samples for each energy level (8, 16 and 24 J, respectively). Each sample will require two different thermal analyses (front and back face) to perform a deeper analysis of the damage. The procedure can be easily seen in Figure 6.28.

Once the information is collected, the post-processing stage gets underway. Pulsed-thermography post-processing is done by using ThermoStudio (in-house developed code by Dr. Yifan for pulsed thermography image processing). And the results obtained are later compared to the NDI of an air gap sample, so the accuracy of the technique can be demonstrated.

6.7.1.1.3 Ultrasonic testing order

For the ultrasonic testing, the samples are only analysed from one side, which reduces the number of experiments to the half. However, the order followed to

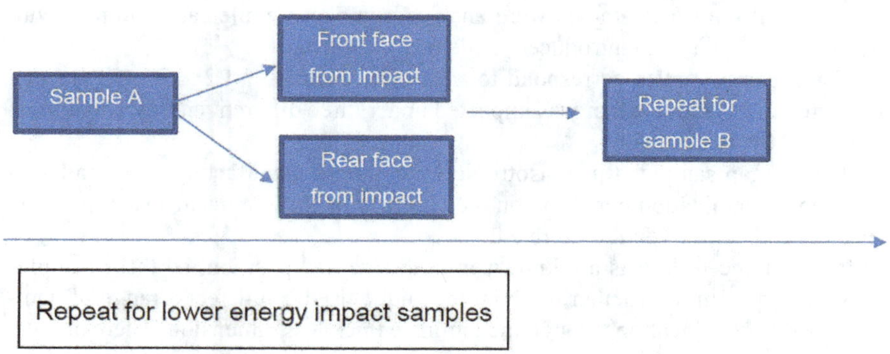

Figure 6.28 Sample testing order

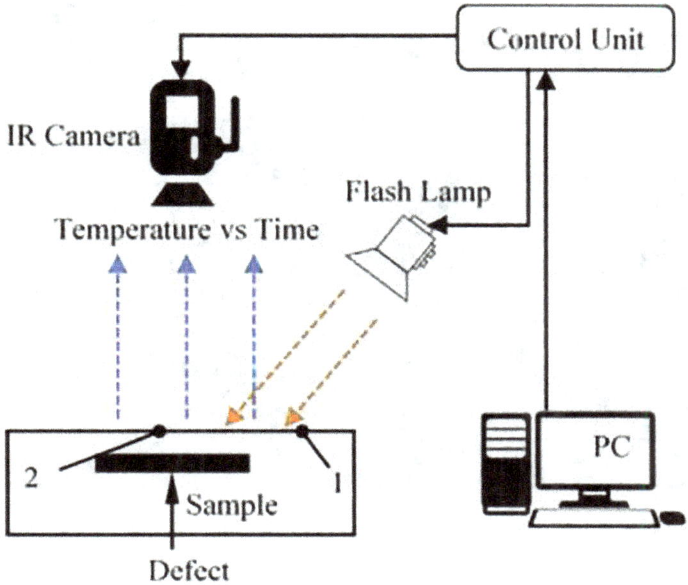

Figure 6.29 *Pulsed thermography system diagram [130]*

analyse the samples will be the same as for the pulsed thermography, from higher to lower impact energy values.

6.7.1.1.4 Experiment setup

Pulsed thermography

The experiment is carried out using the pulsed thermography system (active), which contains two halogen flash lamps (Xenon) powered by a capacitor bank, which is fixed inside a metallic hood capable of internally reflecting the light rays, and a personal computer is used to capture and store the recorded data. The infrared radiometer used is of FLIR SC7000, which consists of an Indium Antimonide sensor with a spatial resolution of 640 × 512 pixels. Figure 6.29 schematically shows how the system is mounted, and the actual setup can be observed in Figure 6.30

The composite laminate impacted sample is fixed in a holding fixture with its surface perpendicular to the line of sight of the infrared camera and the distance between the lens of the camera and the laminate is fixed to 300 mm. The sampling rate was decided to be 25 Hz considering the low thermal diffusivity of the twill carbon fibre laminate and its thickness (3.7 mm), respectively.

The software used to capture the thermal image is the Mosaiq, and the initial parameters (frequency and frames) are set using Cirrus software. Once the sample is placed in the fixture, the metallic hood is covered with the help of a black covering material (to reduce the interference of outside radiation sources). The scanning time of 30 seconds is chosen, and accordingly, the number of frames is chosen by the data

Figure 6.30 Pulsed thermography setup (1. IR camera; 2. excitation source; 3. sample holder)

setup. As soon as the irradiation takes place, the data capturing starts and the damage can be observed on the monitor screen. Thereafter, the region of interest (damaged area) can be chosen to carry out all the data post-processing, which will be explained later in this section.

Ultrasonic testing

The ultrasonic testing equipment used is Sonatest Veo (phased-array, pulse-echo-based and immersion system) (Figure 6.31) where the impacted sample to be tested is placed inside water and the ultrasonic probe is moved over the sample to obtain the information of the damaged area while it is in contact with the liquid medium. The probe is placed over a rail that has marked positions to make sure that all the feeds are stored in the right order guaranteeing the continuity of the data collected.

All the data are stored for later post-processing in the Sonatest UTStudio software. When opening the scanning data in this software, it directly offers the integrated image of the feeds due to the positioning system explained before.

6.7.1.2 Data analysis methodology

6.7.1.2.1 Pulsed thermography

Thermography provides a vast amount of data to process.

Data obtained in the test are stored as raw data at each pixel, and it requires a polynomial fitting to be applied according to the camera sensor used. This processing is known as thermographic reconstruction signal (TSR) and helps to reduce the noise as well as to get information from the derivatives of the indicated polynomial.

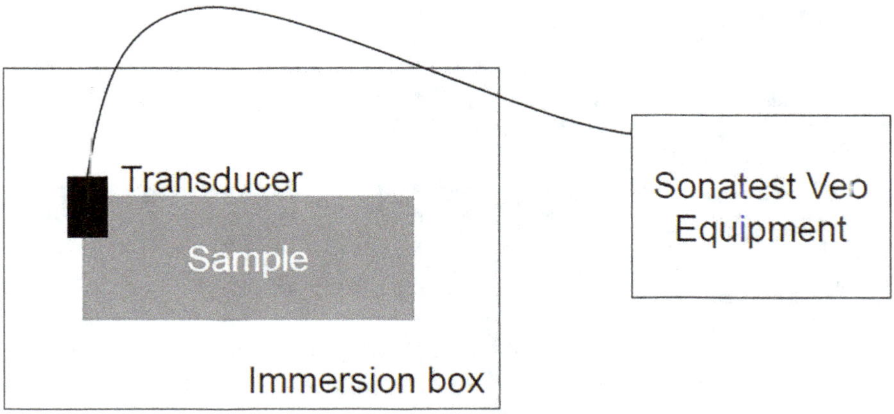

Figure 6.31 *Ultrasonic testing experiment layout diagram*

In this project, a polynomial of degree seven is used as well as an interpolation factor (m) of 2. Thus, it considers the value of the two pixels surrounding the pixel of interest to compute the value in the point of interest. The derivatives will give information about the temperature variation rate in the surface and this is directly affected by what is underneath this surface.

The main information to obtain is the logarithmic temperature–time curves (T–t), but they are point information based (pixel in this case). These curves will give us the exact time position where the contrast is supposed to be maximum, and then the defects will be detected easier. This point is the amplitude peak on the second derivative. At this point, the first derivative curve will reach its maximum slope value and the TSR curve will indicate the transition between transient cooling and steady-state of the temperature field in the sample. This point is shown in Figure 6.32.

6.7.1.3 Ultrasonic testing

Ultrasonic testing data are collected and then analysed with UTStudio Viewer, a software developed by the Sonatest Veo manufacturer (equipment used for the test).

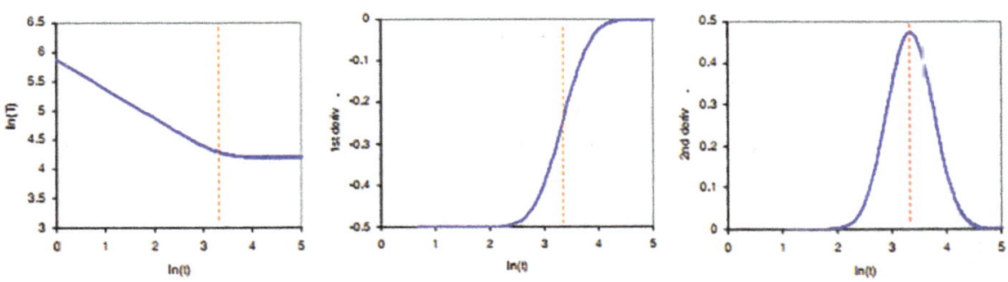

Figure 6.32 *Generic curves for TSR and its derivatives [131] to show the curve patterns*

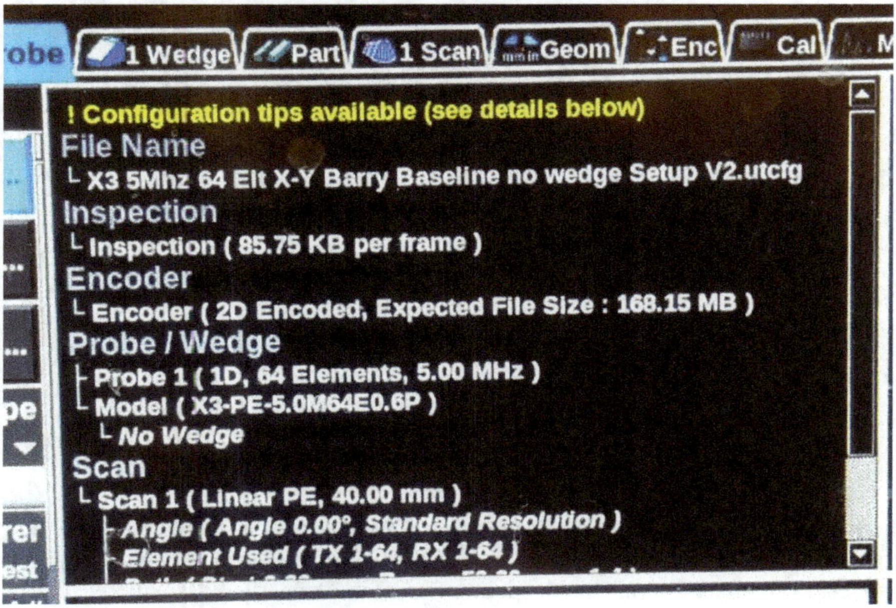

Figure 6.33 Ultrasonic testing equipment parameters for the inspection

Once the data are loaded into the software, the information about A-, B- and C-scan is shown. The test conditions and equipment setup are shown in Figure 6.33.

The results for the A-scan allow us to set the gauges to filter the data acquired and optimize the visualization of the results. This means the results only the signal received by the transducer that is over the set threshold value is shown in the C-scan and B-scan.

At this point, it is important to remember that the main goal of the UT is to compare the results with the ones obtained by pulsed thermography. This is the reason why only C-scan will be addressed in the results and discussion section. However, with the UT technique, information regarding in-depth measurements is possible by using the B-scan.

6.7.1.4 Pulsed thermography results

6.7.1.4.1 Impact samples: 24J

The results obtained for the 24Jimpacted samples can be observed in Figure 6.34

In Figure 6.34, the results for the TSR curve are shown at different points. These curves display the cooling of the sample (heat goes down) after the excitation flash. As it can be observed, there is a difference in the curves depending on the sample point or pixel picked. For the damage area, temperature evolution along time is slower because of some differences in the new material structure after the impact.

More information can be extracted from the derivatives of the TSR. For instance, in Figure 6.35, the TSR first derivative curves are shown. The initial part of this

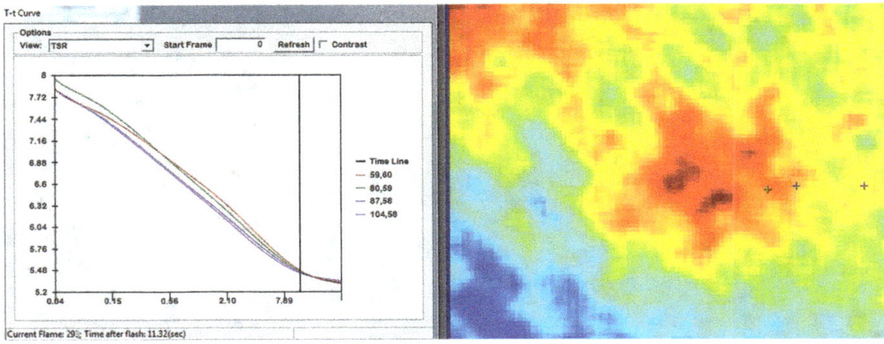

Figure 6.34 *24 J impacted sample – temperature decay (TSR) curves along time; 11.32 seconds after flash*

curve is not of interest, but what happens from the second 2.1 and beyond gives us important information about the impacted sample, hence the damage area. The first derivative contains information about the heating/cooling rate of the sample and how fast the heat penetrates through the material. As it can be observed, the sample retains the heat more in the damaged area than in the outside area (slower heat transfer). The impact generates, among many other things, a delamination, so some air is trapped within the damage area. Thermal diffusivity of the air (αair =19 mm^2/s) is lower than the CFRP (αcarbon composite =216.5 mm^2/s) of the surroundings; then the temperature change rate is slower. The slopes for different points in the time between 2.1 seconds and the end of the data show this pattern.

From the second derivative, the most relevant frame can be obtained because of the amplitude peak generated by the thermal excitation (Figure 6.36). At this point, the temperature change rate will be maximum too and the cooling curve of the material will stabilize at room temperature.

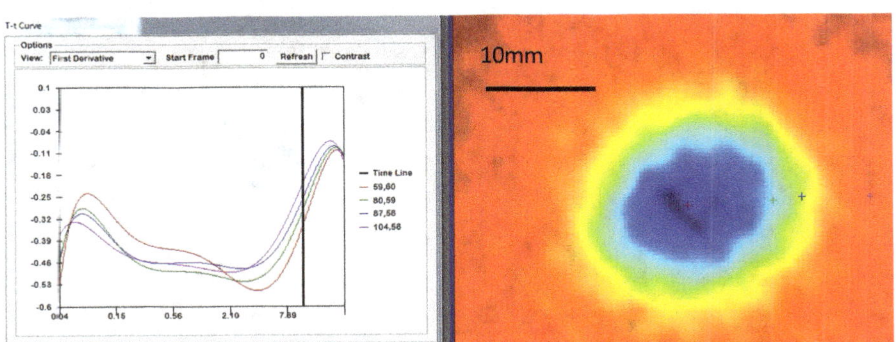

Figure 6.35 *24 J impacted sample – temperature first derivative curves along time; 11.32 seconds after flash*

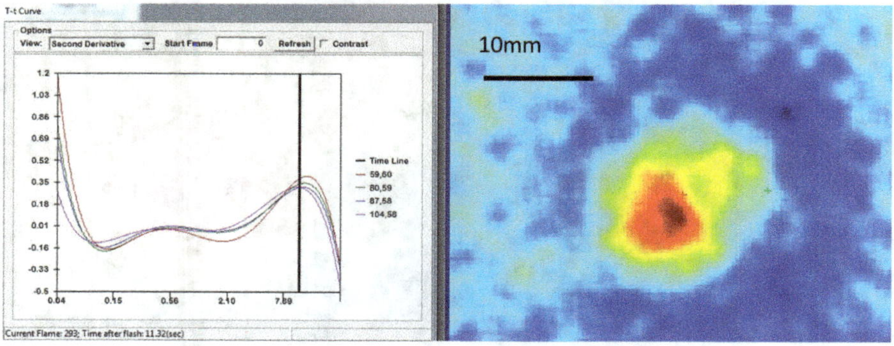

Figure 6.36 *24 J impacted sample – temperature second derivative curves along time; 11.32 seconds after flash*

The damaged area will be determined by using the first derivative image, and its value is calculated in pixels (Figure 6.37) based on a connected-component analysis (CCA). As well as one pixel corresponds to 0.36 mm approximately for the camera used, the total amount of average damaged area for the 24 J impacted samples will be.

As it can be seen, the measurements are done on both faces to compare the results (Table 6.19). However, for the scope of this project, only front face measurements will be of interest due to null access to interior face in actual in-service measurements on fuselages.

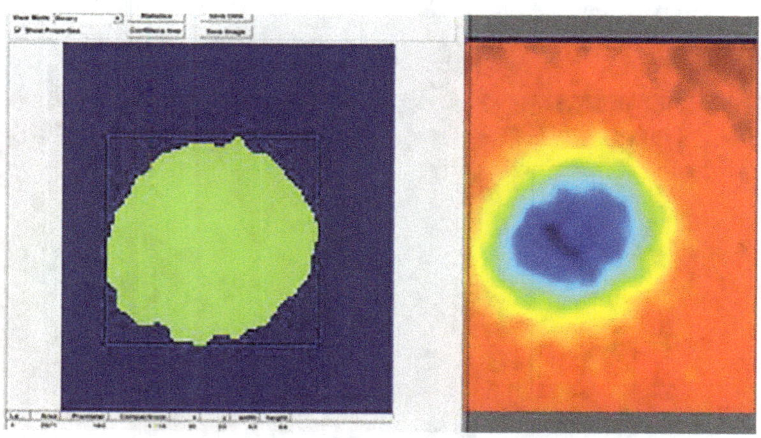

Figure 6.37 *24 J impacted sample – damage size prediction based on CCA method*

Table 6.19 *24 J impact samples results: pulsed thermography*

Face	Width average (mm)	Height average (mm)
Front	24.5 ± 1.8	23.4 ± 0.4
Back	28.8 ± 1.8	23.4 ± 0.7

6.7.1.4.2 Impact samples: 16 J

The same post-processes are shown for the 16 J impacted samples. In Figure 6.38, it can be observed how the offset of the damaged area curve with respect to the rest is reduced compared with the other measuring points.

The first derivative curves show the same pattern as TSR; the temperature change rate value is closer between the damaged and non-damaged areas (Figure 6.39).

For the second derivative, the curves are almost overlapped (Figure 6.40).

The damaged area size calculation is done, and the value of the affected area is also reduced as expected (Figure 6.41). It is important to highlight that for the damaged area calculation (all impact energies), two images will be overlapped. The important information is contained in the binary image; the colourmap is shown just to demonstrate that the calculation is done with the same sample as before.

Averaged results for all the samples of 16 J energy level are collected in Table 6.20.

6.7.1.4.3 Impact samples: 8 J

Finally, the same analysis is carried out for the 8 J samples, which correspond to the lowest energy level.

In Figure 6.42, the TSR curves in damaged and non-damaged areas are very similar with a little offset for those points corresponding to the damaged area. As can be seen in Figures 6.43 and 6.44, the curves for the first and second derivatives are almost overlapping. Hence, predicting the damage becomes more difficult.

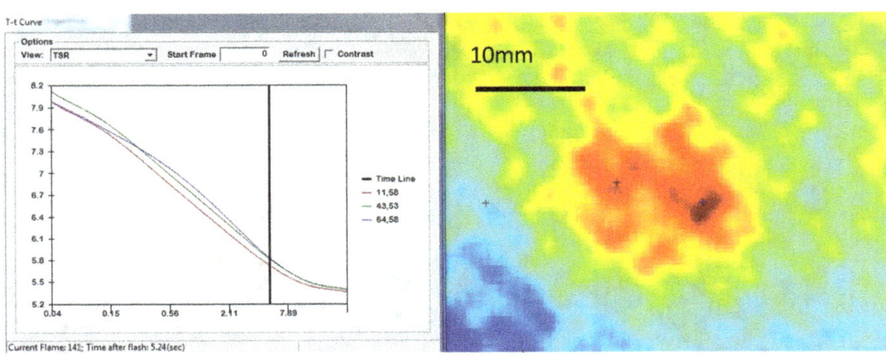

Figure 6.38 *16 J impacted sample – TSR curves along time; 5.24 seconds after flash*

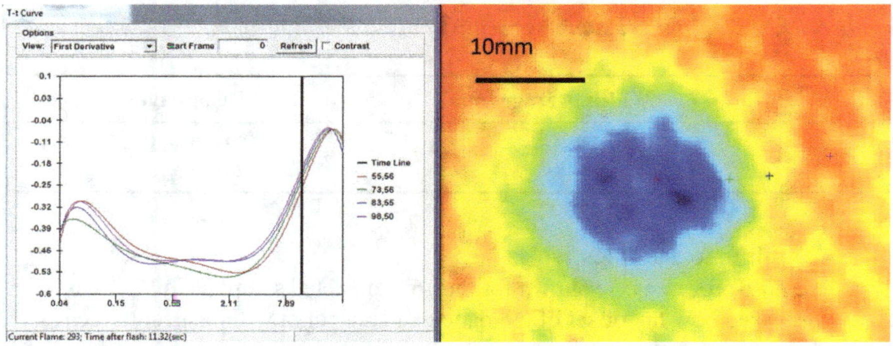

Figure 6.39 16 J impacted sample – temperature first derivative curves along time; 11.32 seconds after flash

The damaged area size is calculated following the same procedure as for previous samples, and the results can be seen in Figure 6.45.

Averaged results for all the samples of 16 J energy level are collected in Table 6.21

6.7.1.4.4 Pulsed thermography results' discussion

According to the results obtained, it is easy to justify that the damage area will be greater for the higher impact energy (Figure 6.46).

Regarding the damage detection, it has been demonstrated that this technique is helpful to determine whether there is a damage or not, but as the impact energy is reduced, the error incurred is higher. This means that the technique might not be useful to determine the accurate size of a damage by its own. This effect can be observed in Figure 6.47

Another interesting fact that can be extracted from Figure 6.46 is that the width and height are almost the same for all the damages, so the actual shape of the damage

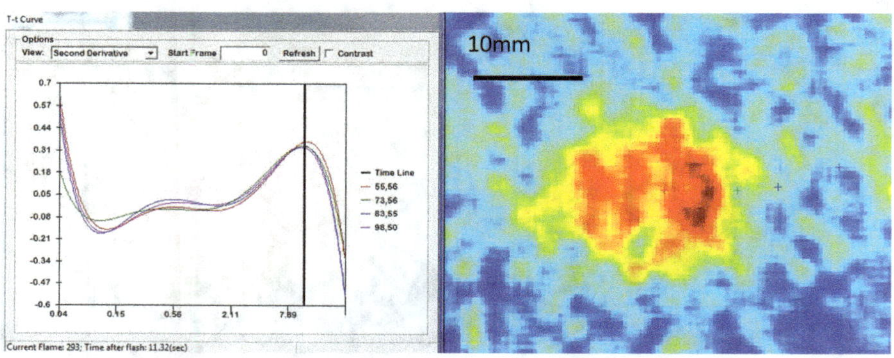

Figure 6.40 16 J impacted sample – temperature second derivative curves along time; 11.32 seconds after flash

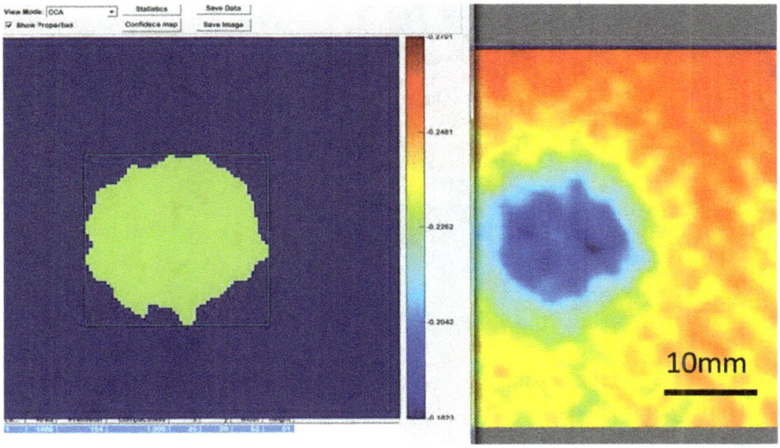

Figure 6.41 16 J impacted sample – damage size prediction based on CCA method

area can be approximated to a circumference. However, this can be caused by the material used. The twill fibres act as damage barriers so the damage gets trapped within the warp directions.

These results will be compared later with other NDI techniques to evaluate its validity and accuracy.

6.7.1.5 Pulsed thermography results: air gap samples

As previously mentioned, two samples with manufacturing-induced defects (air gaps) were analysed. A variety of information can be obtained by analysing these samples. However, the scope of this project is to check the dimensions obtained by performing a basic analysis and compare to the actual damage size, which is known for this sample. Other challenges will arise with these samples with the defect depth and the effect of the damage thickness (air gap thickness). It is important to highlight that the goal of this part is not to quantify the defects, but to determine (in a qualitative way) the damage assessment as well as list the main difficulties found along the measurements.

Table 6.20 16 J impact samples results: pulsed thermography

Face	Width average (mm)	Height average (mm)
Front	20.2 ± 1.1	20.0 ± 1.62
Back	20.3 ± 2.0	19.3 ± 3.4

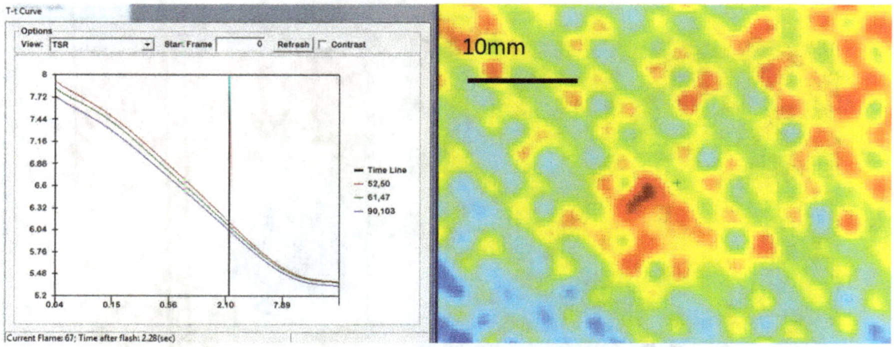

Figure 6.42 *8 J impacted sample – TSR curves along time; 2.28 seconds after flash*

6.7.1.5.1 Laminate with constant air gap thickness at different depths

The sample with constant air gap thickness (A_con) is expected to show a different pattern along the time for different air gaps because of different defects' in-depth location. To improve the visualization, the TSR second derivative has been used. In Figures 6.48 and 6.49 two different time instants are shown, together with the second derivative curves of the single pixel measurements. Colour points on the infrared image correlate the curves with the analysed pixel.

As it can be observed in Figure 6.48, 1.16 seconds after the flash, the air gap closer to the surface is detected and the second one can be spotted but not measured yet. However, in the instant $t = 2.28$ seconds, the following (and deeper) air gaps appear. The fourth air gap (in depth) is hardly measurable because the thermal gradient becomes blurry at that depth, so its effect in surface temperature variation is lower than the rest. The depth of the air gap could be calculated from the variation of the surface temperature, according to the equation [132]:

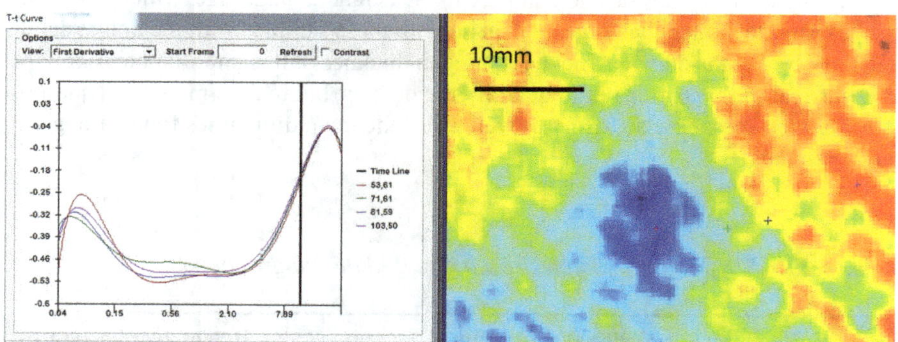

Figure 6.43 *8 J impacted sample – temperature first derivative curves along time; 11.32 seconds after flash*

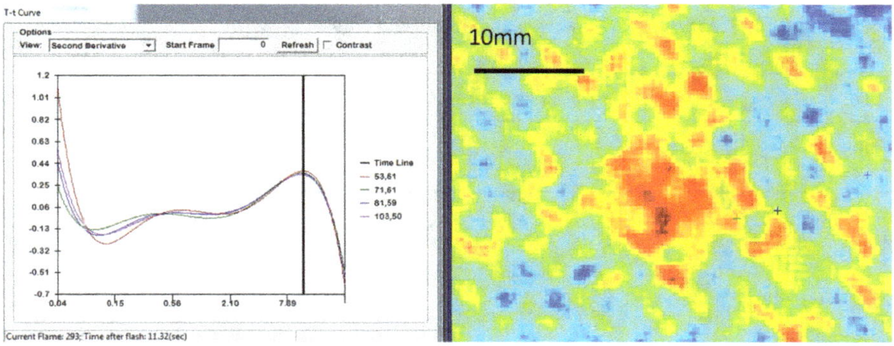

Figure 6.44 8 J impacted sample – temperature second derivative curves along time; 11.32 seconds after flash

$$\Delta T(t) = \frac{Q}{\sqrt{\pi \rho c k t}}\left[1 + 2\sum_{n=1}^{\infty} e^{\left(\frac{-n^2L^2}{\alpha t}\right)}\right] \tag{6.1}$$

where ΔT is the surface temperature variation as a time-dependent function;Q is the pulse energy (known from the experiment setup) in joules;ρ is the material density;c is the heat capacity;k is the thermal conductivity of the material;α is the thermal diffusivity of the material.

This expression was further developed by Zeng *et al.*. The idea was to derivate the expression to get the maximum value of the first derivative, which corresponds to the second derivative peak. Because it is a time-dependent function, this value corresponds to a certain time instant. The solution obtained was [130]:

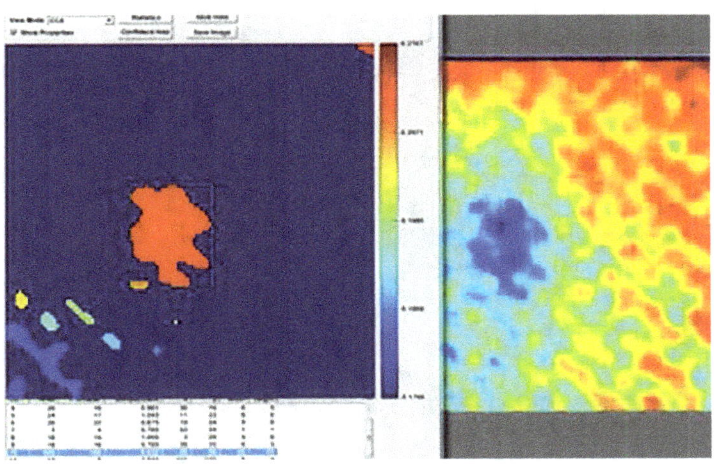

Figure 6.45 8 J impacted sample – damage size prediction based on CCA method

Table 6.21 8 J impact samples results: pulsed thermography

Face	Width average (mm)	Height average (mm)
Front	13.5 ± 4.1	15.8 ± 4.0
Back	12.0 ± 2.9	12.24 ± 1.4

$$t_{APST} = \frac{L^2}{1.93\alpha} \tag{6.2}$$

where t_{APST} corresponds to the time where the second derivative takes place, α stands for the thermal diffusivity of the material and L for the defect depth. Hence, it makes sense that the deeper the defect, the higher the time value, as can be observed in Figure 6.49

However, the information expected from the air gap samples for the Aerospace Composite Repair System project is simpler than that because only the size of the defect is required. For the determination of the size of the air gap (i.e. width), it can be seen that the width is 28 pixels (30 × 0.36=10.08 mm) in Figure 6.50 This is quite close to the actual value of the air gap, which is a width of 10 mm.

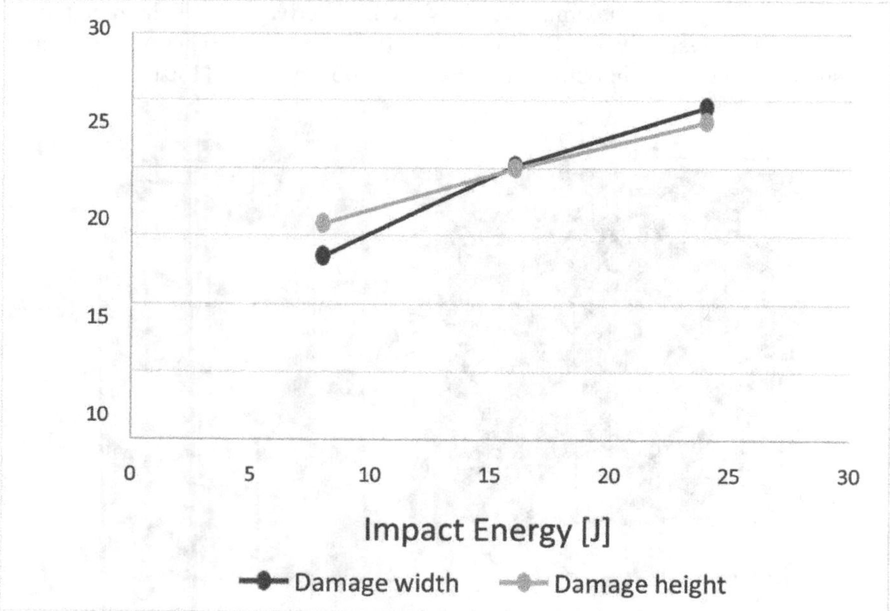

Figure 6.46 Damage dimensions based on orientation and impact energy

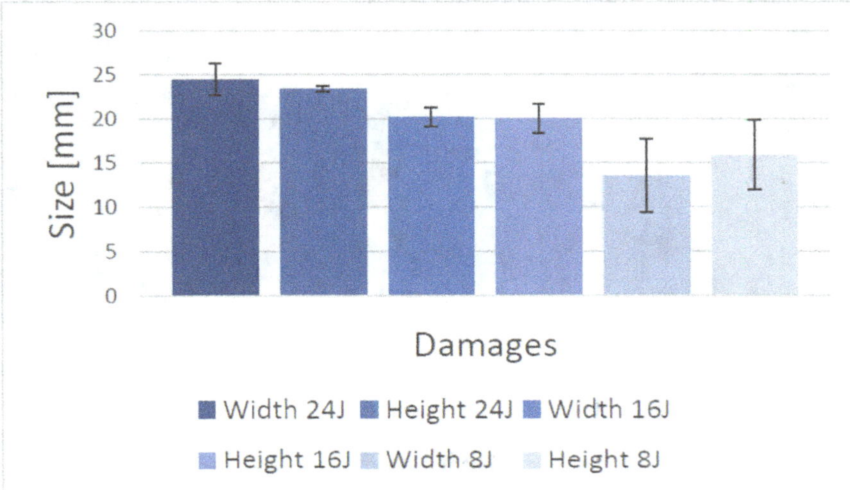

Figure 6.47 Averaged damage sizes obtained in pulsed thermography

6.7.1.5.2 Laminate with different air gap thicknesses at the same depth

The same analysis is carried out for sample A_diff, which contains five different defects of five different thickness levels, but all are located at the same depth with respect to the analysed sample face.

As it can be seen in Figure 6.51, the second derivative amplitude peak is located at almost the same time position for two different air gaps, so it can be demonstrated that the air gap depth is the same for both air gaps. This information is also visible in the colour map obtained for $t = 1.92$ seconds.

For the same sample, the temperature decay curves obtained by TSR fitting show how the four different points follow four different cooling trends (Figure 6.52). The

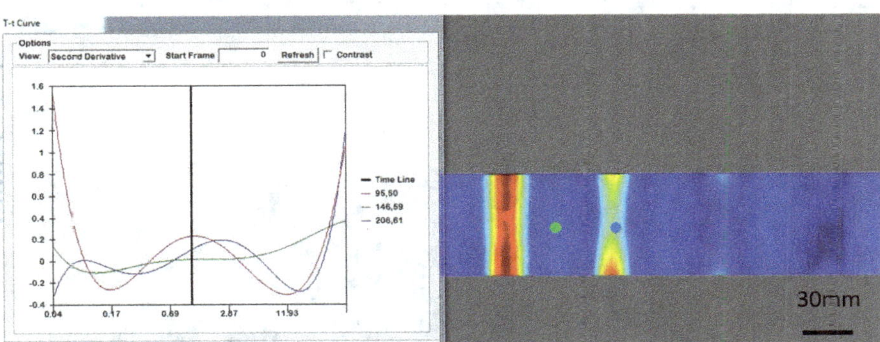

Figure 6.48 A_con sample: temperature second derivative instant 1

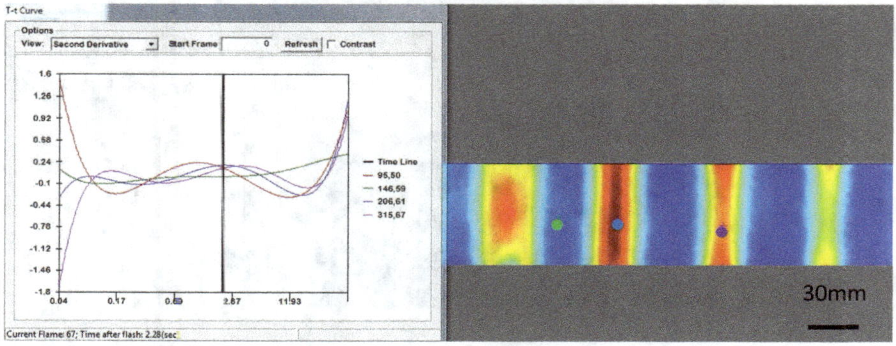

Figure 6.49 A_con sample: temperature second derivative instant 2

green and blue points follow the same pattern with a little offset (sound area), and the red and purple points follow different patterns (air gaps) with little differences at the beginning caused by the difference in thickness of the air gap, so the cooling rate changes because of material diffusivity, as explained before. The defect size is measured like in the previous sample obtaining the same results.

6.7.1.5.3 Air gap analysis conclusions

Some conclusions were obtained after the air gap analysis.

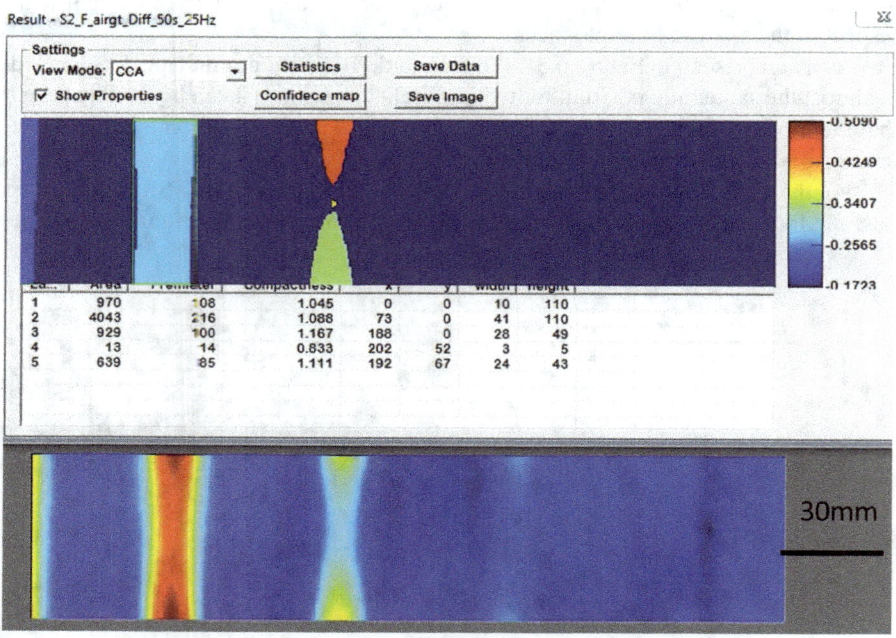

Figure 6.50 First air gap size for different depth samples (A_con)

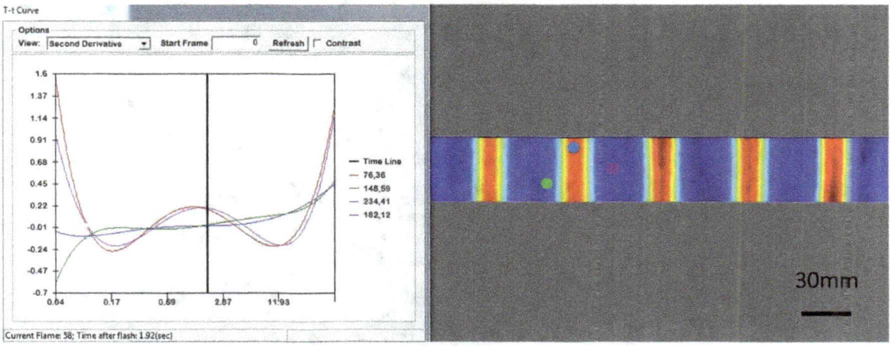

Figure 6.51 Second derivative of sample A_diff

- Pulsed thermography is limited by the thickness of the sample. The thicker the sample, the more difficult to get reliable information. This effect was appreciated when analysing samples with air gaps in a depth higher than 3.5 mm. For those depths, the surface temperature is affected in a weak or null way, so no information can be extracted from them.
- The air gaps in a depth greater than 3.5 mm from the analysed surface are very difficult to analyse because the data are not good enough.
- For constant air gap thickness at different depths, data can be extracted easily for the first three defects (1.5, 2 and 2.5 mm depth, respectively). Second derivative peaks are delayed in the time for three different air gaps, so different depths between them can be concluded.
- If the correct frame is selected, the damaged area calculation can be accurate, but it requires a deep analysis and data post-processing.

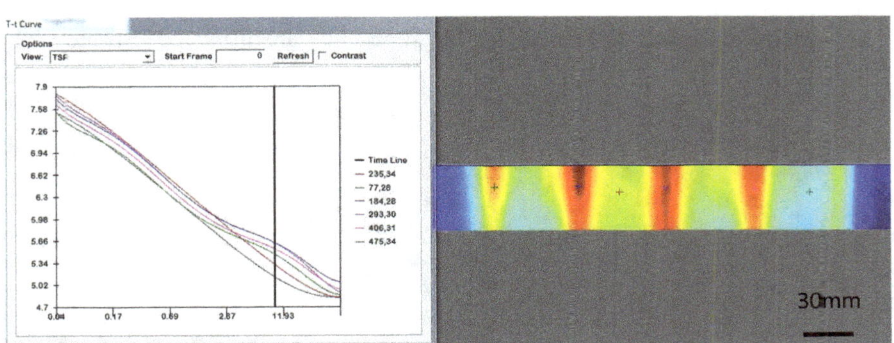

Figure 6.52 TSR image for different air gaps and locations

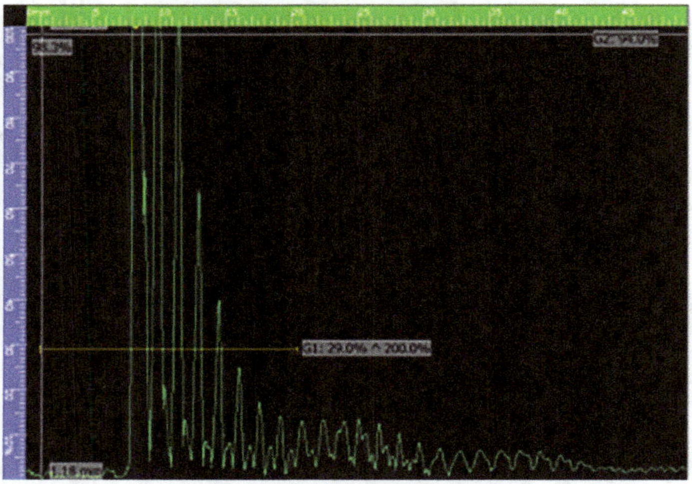

Figure 6.53 A-scan for 24 J impacted sample with gauge applied

6.7.1.6 Ultrasonic testing results

For the ultrasonic testing, only the impacted samples were considered. The obtained results are shown in the following subsections.

6.7.1.6.1 Impact samples: 24 J

In Figure 6.53, it can be observed how the A-scan results have been filtered. The gain gauge is set to 29%, and the width is 19.5 mm. Only the peaks within the gauge are considered for the analysis because they are the most important ones. As it can be observed, there is some signal attenuation along the time.

The result for the C-scan according to this A-scan analysis is shown in Figure 6.54 There are two spots surrounding the damage area that reflects the signal, following the same pattern as the damage area. However, these areas can be neglected because they correspond to the nuts used for supporting the sample in the immersion box.

The damaged area is near round, and the dimensions are displayed in Table 6.22.

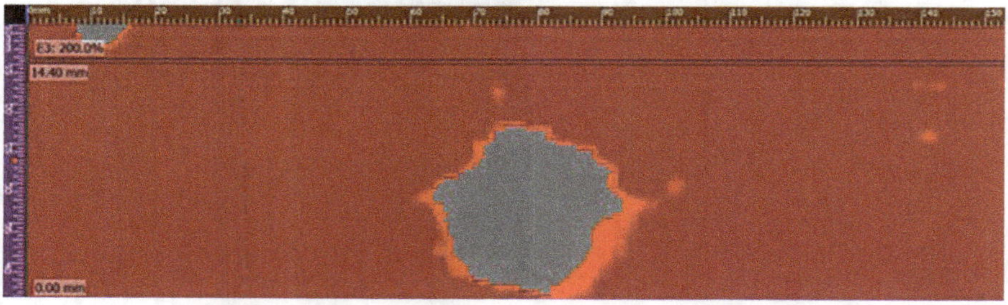

Figure 6.54 C-scan for 24 J impacted sample

Table 6.22 24 J Impact samples results: ultrasonic testing

Impact energy	Width (mm)	Height (mm)
24 J	30.3	21.8

No statistical data are available because the damage obtained by the UT for the second sample is not consistent with the other one and with the results obtained in the pulsed thermography. A problem with the data acquisition might have occurred.

6.7.1.6.2 Impact samples: 16 J

The A-scan gauges have been set to 30%, and the width is 14.93 mm. The result can be observed in Figure 6.55

The corresponding C-scan image obtained is shown in Figure 6.56 Non-reliable information can be extracted from this sample because of the shape of the damage area. The shape does not match with the results obtained in the pulsed thermography. Besides, it is not coherent with the rest of the results for other energy levels in which the damage area is always round shaped. Then the results for the 16 J are considered non-conclusive, and more tests would be needed to extract reliable information for this impact energy level.

6.7.1.6.3 Impact samples: 8 J

The A-scan gauges have been set to 24%, and the width is 16.25 mm. The result can be observed in Figure 6.57

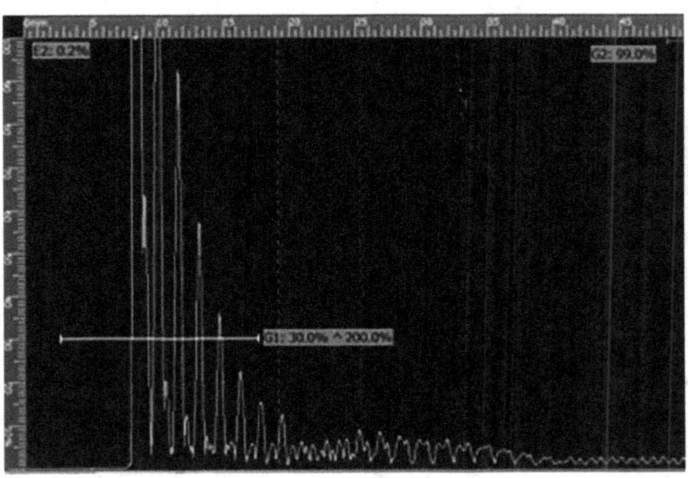

Figure 6.55 A-scan for 16 J impacted sample with gauge applied

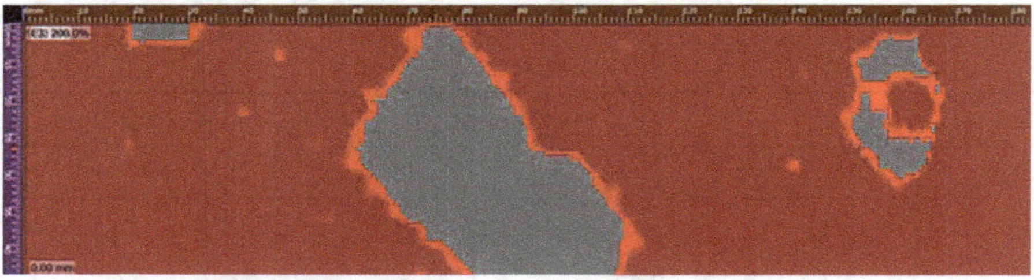

Figure 6.56 C-scan for 16 J impacted sample

The C-scan results are shown in Figure 6.58. Now it can be observed how the total size is reduced while keeping the rounded shape. For 8 J both samples, data were stored correctly and some averaged information could be extracted.

Grey areas on the sides are the nuts used to hold the sample, so no information must be extracted from them. The results for damage size measurements by using UT are collected in Table 6.23

6.7.1.6.4 Ultrasonic testing results' discussion

The most important information extracted from the ultrasonic testing is summarized as follows.

- Results are not reliable because of the lack of statistical data for measurement errors.
- The same trends as in pulsed thermography are observed.
- Because of the physics behind this technique, the results are more accurate than the pulsed thermography ones. The error incurred for 8 J is considerably lower than the one incurred for pulsed thermography.

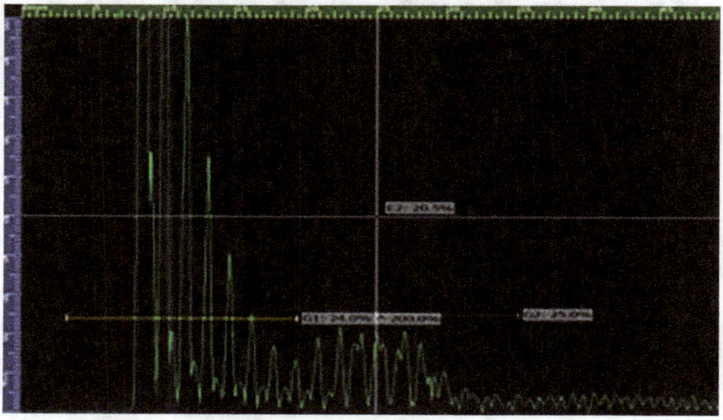

Figure 6.57 A-scan for 8 J impacted sample

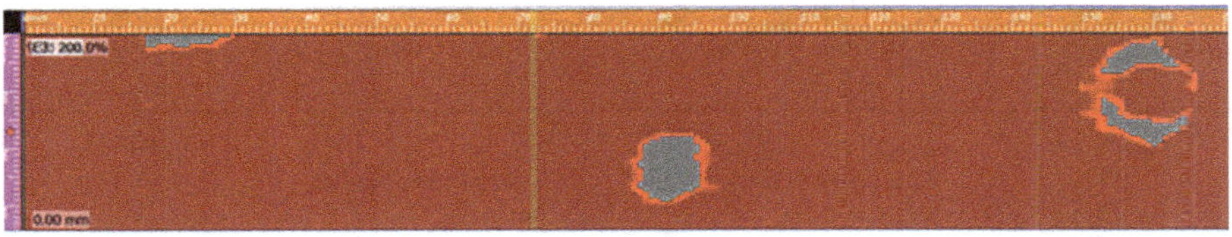

Figure 6.58 C-scan for 8 J impacted sample

- Small damage effects could be neglected if something bigger is happening because data filtering is required.
- Thickness is not an issue for this technique and allows us to take in-depth measurements easily.

6.7.1.7 Comparison of results of different techniques

After completing the measurements of the damage with both pulsed thermography and ultrasonic testing, the results can be compared between both techniques as well as the optical inspection of the sample.

First, a comparison of the damaged area estimation is done. The results for the damaged area calculations (approximated to a rectangle based on width and height dimensions obtained) are collected in Table 6.24

The results show how all the NDI techniques detect more damage than the optical inspection. This is because there is some damage underneath the visible surface, which is undetectable during an optical inspection. This damage can be fibre/matrix breakage as well as delamination effects caused by the impact in the surroundings.

Furthermore, it can be observed that pulsed thermography predicts a bigger damaged area than ultrasonic testing. This fact could be possible because of the thermal gradients generated in the surroundings of the damaged area. Heat dissipation phenomena take place, so a thermal gradient is present in the sample and this may lead to less accurate measurements.

The second parameter to consider is the biggest damaged dimension. The results are included in Table 6.25

The results for the damage dimensions are not as clear as the ones for the damaged area. It can be noted how some optical inspection measurements are bigger than the damage size measured by NDI techniques. This is possible because the NDI techniques are meant to focus on the most important damage. Thus, a surface micro-crack (the ones measured by the optical inspection) could be negligible compared

Table 6.23 8 J Impact samples results: ultrasonic testing

Impact energy	Width (mm)	Height (mm)
8 J	13.1 ± 0.9	13.5 ± 0.1

Table 6.24 Damaged area calculation. Techniques' comparison

Impact energy (J)	Pulsed thermography (mm^2)	Ultrasonic testing (mm^2)	Optical inspection (mm^2)
24	624.2	599.9	308.0
16	458.8	186.7	180.0
8	349.3	176.1	50.0

to a subsurface phenomenon that is happening. With the NDI techniques, damage assessment is performed and the technique will be useful to provide information about the biggest damage within the sample. Different post-processing stages would be required to analyse different damage scales.

6.8 Conclusion

In this section, a brief summary of the project is provided in the form of a hypothetical case study. After that, the main outcomes and future scope of the project are presented. To recapitulate all the steps considered within the Aerospace Composite Repair system, a fictitious case study is presented, explaining how a problem would be addressed using this system. The case study involves, of course, an aircraft with a primary structure made of composite material (Airbus A350, Boeing 787, Bombardier CSeries). Among the most typical causes of damage in aircraft, low-velocity impact is the most prevalent. In this case, a tool drop is considered, with the following characteristics:

- impact energy: 24 J
- equivalent mass: 1.6 kg
- equivalent height: 1.5 m

This impact has happened in the fuselage, the most commonly damaged aircraft structure: in particular, in an area made of a 5-mm-thick CFRP laminate. Almost no damage is appreciable by optical inspection. However, as explained

Table 6.25 Maximum damaged dimension: technique comparison

Impact energy (J)	Pulsed thermography (mm)	Ultrasonic testing (mm)	Optical inspection (mm)
24	26.3	30.3	22.0
16	21.6	15.4	18.0
8	19.8	13.5	10.0

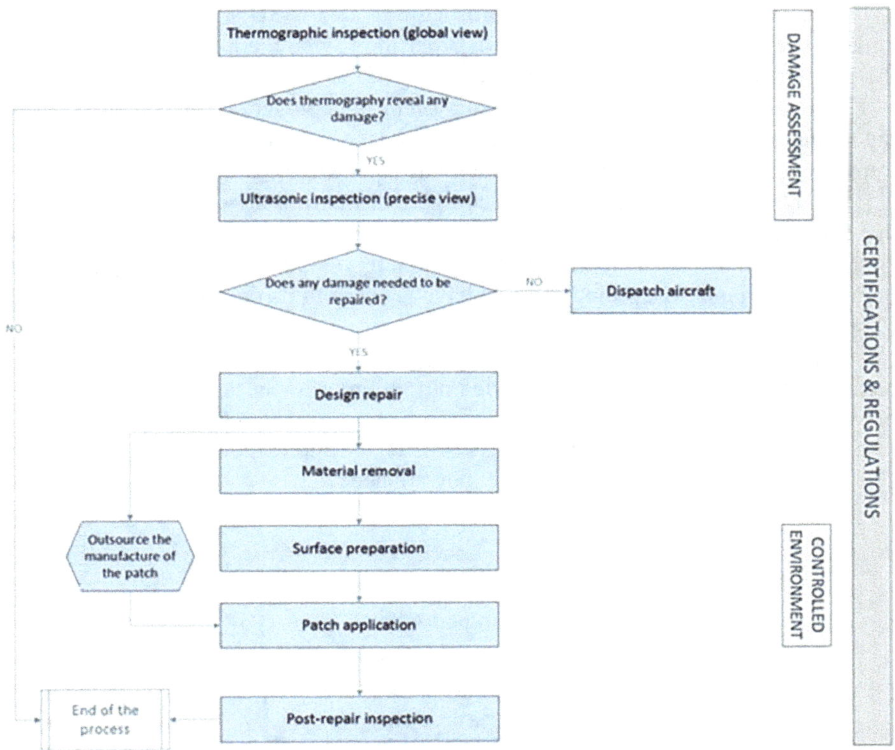

Figure 6.59 Process flow diagram of the system

during the literature review, major damage could be hidden in the underlying layers, usually in the form of delamination, which can pose a threat to the airworthiness of the aircraft.

The first step is to determine if there is a damage and, if it exists, assess its location and size. To do so, the NDI capability of the system is used. A quick pulsed thermography scan is carried out by the automated robotic arm; it shows a damaged area. Later, the same robotic arm uses an ultrasonic testing probe in the area where thermography spotted a damage: this allows assessment of the damage with a greater accuracy.

The damage is considered to be serious enough to require a repair. Hence, the robot uses the 3D scanning tool, in order to acquire the curvature of the damaged area: this information will be used to design the repair patch.

The material removal operation starts afterwards: a conventional machining head is mounted in the robot, performing a fast, rough machining. Once the material has been removed to near-net shape dimensions, the laser machining system is used: it removes the last bits of material in a more controlled and less damage-causing process.

When the machining process is finished, the area is 3D scanned again: this is done to acquire the final scarf shape, in order to provide the manufacturer with all the information required for the manufacturing of the hard patch.

The portable environmental control system is deployed at that point to cover the repair area: once it is working, the laser surface treatment can start, in order to improve the adhesion in the surface where the patch is going to be placed.

When the surface preparation is completed, the film adhesive is applied and the hard patch is carefully positioned by the robotic arm. Afterwards, thermocouples are installed (to monitor curing parameters), and the vacuum bag and heat blankets are manually applied: the adhesive is cured to its design temperature and time.

The portable clean room is then retired, as well as the curing equipment. A post-repair NDI inspection with UT is carried out, in order to check that everything has been done correctly.

Finally, in a case-by-case basis, it is studied if the application of bolts is required. They will be applied manually, in case they are needed.

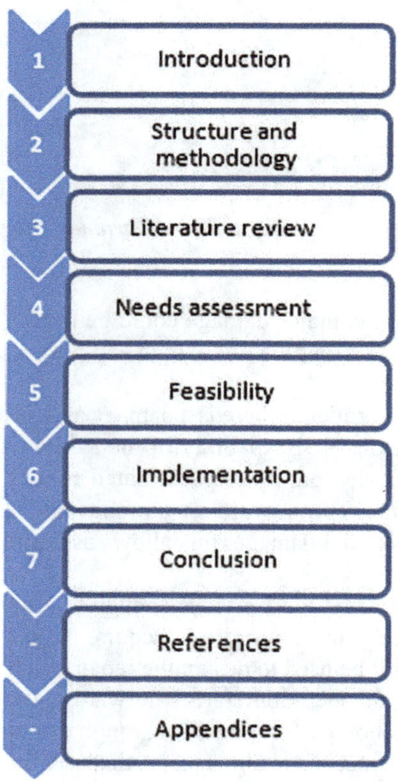

Figure 6.60

Of course, it is important to mention that certification requirements have been strictly followed throughout the process, with careful control of process variables and parameters to ensure quality in every step. This is required to be sure that the performed repair is in compliance with the existing regulation [70].

Figure 6.59 depicts the process followed by the Aerospace Composite Repair system, as explained in the previous paragraphs.

References

[1] ATAG *ATAG-facts & figures* [online]. 2016. Available from http://www.atag.org/facts-and-figures.html [Accessed Apr 2017].

[2] Civil Aviation Authority Commercial industry [online]. 2015. Available from https://www.caa.co.uk/Commercial-industry/ [Accessed Apr 2017].

[3] SKYbrary Continuing airworthiness [online]. 2016. Available from http://www.skybrary.aero/index.php/Continuing_Airworthiness [Accessed Apr 2017].

[4] British Airways *Aircraft maintenance [online]*. 2012. Available from http://www.ba- mro.com/baemro/aircraftMaintenance.shtml [Accessed Apr 2017].

[5] Spafford C., Hoyland T., Medland A. *Turning the tide: a wave of new aviation technology will soon hit the MRO industry turning the tide: a wave of new aviation technology will soon hit the MRO industry*. MRO Surv; 2015.

[6] Wikipedia Technology readiness levels [online]. 2017. Available from https://en.wikipedia.org/wiki/Technology_readiness_level#/media/File:NASA_TR L_Meter.jpg [Accessed Apr 2017].

[7] Modern Airliners Boeing 787 dreamliner SPECs [online]. 2015. Available from http://www.modernairliners.com/boeing-787-dreamliner/boeing-787-dreamliner- specs/ [Accessed Mar 2017].

[8] SAE International. *Composite materials handbook. Volume 3, Polymer matrix composites materials usage, design, and analysis*. SAE International; 2012.

[9] Baker A., Dutton S., Kelly D. *Composite materials for aircraft structures*. Reston, VA: American Institute of Aeronautics and Astronautics; 2004.

[10] 1001crash.com Worries about new composite made airplane [online]. 2011. Available from http://www.1001crash.com/index-page-composite-lg-2.html [Accessed Mar 2017].

[11] Giurgiutiu V. *Structural health monitoring of aerospace composites*. London: Elsevier; 2016.

[12] Roach D.P., Rice T.M. 'A quantitative assessment of advanced NDI techniques for detecting flaws in composite Laminate aircraft structures'. 2014.

[13] Nezhad H.Y., Merwick F., Frizzell R.M., McCarthy C.T. 'Numerical analysis of low-velocity rigid-body impact response of composite panels'. *International Journal of Crashworthiness*. 2015;20(1):27–43.

[14] Heimbs S., Ortiz R. 'Fe models and reports concerning impactor database, study on pre-strain effect and multi-scale modelling strategy Part 1: impactor database'. 2010.

[15] Dolbeer R.A., Wright S.E., Weller J., Begier M.J. *Wildlife strikes to civil aircraft in the united states 1990-2008*. Washington, DC: U.S. DEPARTMENTOF TRANSPORTATIONFEDERAL AVIATIONADMI NISTRATION; 2009.

[16] Paper C. 'Fibre-reinforced aerospace composite'. 2015.

[17] NDT Education Ultrasonic testing (UT)-NDT course material. Introduction to ultrasonic testing [online]. 2001. Available from https://www.nde- ed. org/EducationResources/CommunityCollege/Ultrasonics/cc_ut_index.htm [Accessed Mar 2017].

[18] Katnam K.B., Da Silva L.F.M., Young T.M. 'Bonded repair of composite aircraft structures: a review of scientific challenges and opportunities'. *Progress in Aerospace Sciences*. 2013;61(22):26–42.

[19] Campbell F.C. *Structural Composite Materials*. United States of America: ASM International; 2010.

[20] Fahr A. 'Ultrasonic C-scan inspection of composite materials'. *Engineering Journal of University of Qatar*. 1992;**5**:201–22.

[21] D'Orazio T., Leo M., Distante A., Guaragnella C., Pianese V., Cavaccini G. 'Automatic ultrasonic inspection for internal defect detection in composite materials'. *NDT & E International*. 2008;41(2):145–54.

[22] Hsu D.K. 'Nondestructive inspection of composite structures: methods and practice'. *17th World Conference on Nondestructive Testing*; 2008. pp. 1–14.

[23] Choquet M., Héon R., Padioleau C., Bouchard P., Néron C., Monchalin J.-P. 'Laser-ultrasonic inspection of the composite structure of an aircraft in a maintenance hangar'. *Review of Progress in Quantitative Nondestructive Evaluation*. 1995;**14**:545–52.

[24] Růžek R., Lohonka R., Jironč J. 'Ultrasonic C-scan and shearography NDI techniques evaluation of impact defects identification'. *NDT & E International*. 2006;**39**(2):132–42.

[25] Shin H.-J., Lee J.-R. 'Development of a long-range multi-area scanning ultrasonic propagation imaging system built into a hangar and its application on an actual aircraft'. *Structural Health Monitoring*. 2017;16(1):97–111.

[26] Hung M.Y.Y., Chen Y.S., Ng S.P., Shepard S.M., Hou Y., Lhota J.R. 'Review and comparison of shearography and pulsed thermography for adhesive bond evaluation'. *Optical Engineering*. 2007;46(5):051007.

[27] Usamentiaga R., Venegas P., Guerediaga J., Vega L., López I. 'Feature extraction and analysis for automatic characterization of impact damage in carbon fiber composites using active thermography'. *NDT & E International*. 2013;54(1):123–32.

[28] Usamentiaga R., Venegas P., Guerediaga J., Vega L., López I. 'Automatic detection of impact damage in carbon fiber composites using active thermography'. *Infrared Physics & Technology*. 2013;58(1):36–46.

[29] Burleigh D.D. 'A portable, combined thermography / shearography NDT system for inspecting large composite structures'. *Proceedings of SPIE - The International Society for Optical Engineering*. 2002;**4710**:578–87.

[30] GmbH A.T. 'Active thermography for nondestructive composites testing'. *Quality Magazine*. 2013.

[31] Avdelidis N.P., Almond D.P., Dobbinson A., Hawtin B.C., Ibarra-Castanedo C., Maldague X. 'Aircraft composites assessment by means of transient thermal NDT'. *Progress in Aerospace Sciences*. 2004;40(3):143–62.

[32] Maierhofer C., Myrach P., Reischel M., Steinfurth H., Röllig M., Kunert M. 'Characterizing damage in CFRP structures using flash thermography in reflection and transmission configurations'. *Composites Part B: Engineering*. 2014;57(1):35–46.

[33] NTC Active thermography and IRNDT [online]. 2015. Available from https://ttp.zcu.cz/en/laboratories/irndt/active-thermography [Accessed Apr 2017].

[34] Bates D., Smith G., Lu D., Hewitt J. 'Rapid thermal non-destructive testing of aircraft components'. *Composites Part B: Engineering*. 2000;31(3):175–85.

[35] He Y., Tian G., Pan M., Chen D. 'Impact evaluation in carbon fiber reinforced plastic (CFRP) laminates using eddy current pulsed thermography'. *Composite Structures*. 2014;109(1):1–7.

[36] Guillebert V., Nyingifa E., Frangolho D., *et al.* 'Development of a thermographic NDI system for service damage identification in inaccessible areas'. *Procedia CIRP*. 2013;11:124–9.

[37] Zhang H., Hassler U., Genest M., *et al.* 'Comparative study on submillimeter flaws in stitched T-joint carbon fiber reinforced polymer by infrared thermography, microcomputed tomography, ultrasonic c-scan and microscopic inspection'. *Optical Engineering*. 2015;54(10):104109.

[38] De Angelis G., Dati E., Bernabei M., Leccese F. 'Development on aerospace composite structures investigation using thermography and shearography in comparison to traditional NDT methods'. *2nd IEEE International Workshop on Metrology for Aerospace*. 2015;**1**:49–55.

[39] Georges M.P. 'Comparison between thermographic and holographic techniques for nondestructive testing of composites: similarities, differences and potential cross-fertilization'. *Proceedings of SPIE - The International Society for Optical Engineering*. 2015;**9660**:966002.

[40] Zheng K., Chang Y.-S., Wang K.-H., Yao Y. 'Improved non-destructive testing of carbon fiber reinforced polymer (CFRP) composites using pulsed thermograph'. *Polymer Testing*. 2015;46:26–32.

[41] The Welding Institute *Laser shearography [online]*. 2017. Available from http://www.twi-global.com/capabilities/integrity-management/non-destructive- testing/ndt-techniques/laser-shearography/ [Accessed Mar 2017].

[42] Santos F., Vaz M., Monteiro J. 'A new set-up for pulsed digital shearography applied to defect detection in composite structures'. *Optics and Lasers in Engineering*. 2004;42(2):131–40.

[43] De Angelis G., Meo M., Almond D.P., Pickering S.G., Angioni S.L. 'A new technique to detect defect size and depth in composite structures using digital shearography and unconstrained optimization'. *NDT & E International*. 2012;45(1):91–6.

[44] Dynamics D. Non-Destructive testing (NDT) -laser shearography [online]. 2017. Available from https://www.dantecdynamics.com/shearography-non-destructive-testing [Accessed Mar 2017].

[45] Georges M. *FANTOM project overview basic principles*; 2011.

[46] Seventh Framework Programme FANTOM (full-field advanced non destructive technique for on-line thermo-mechanical measurements on aeronautical structures) [online]. 2009. Available from http://www.fantom-ndt.eu/

[47] Gaudenzi P., Bernabei M., Dati E., De Angelis G., Marrone M., Lampani L. 'On the evaluation of impact damage on composite materials by comparing different NDI techniques'. *Composite Structures*. 2014;118(9):257–66.

[48] Amenabar I., Mendikute A., López-Arraiza A., Lizaranzu M., Aurrekoetxea J. 'Comparison and analysis of non-destructive testing techniques suitable for delamination inspection in wind turbine blades'. *Composites Part B: Engineering*. 2011;**16**(2):68–76.

[49] Li Z., Meng Z. 'A review of the radio frequency non-destructive testing for carbon-fibre composites'. *Measurement Science Review*. 2016;16(2):68–76.

[50] Pérez M.A., Oller S., Felippa C.A., Gil L. 'Micro-mechanical approach for the vibration analysis of CFRP laminates under impact-induced damage'. *Composites Part B: Engineering*. 2015;**83**:306–16.

[51] Katunin A. 'Nondestructive damage assessment of composite structures based on wavelet analysis of modal curvatures: state-of-the-art review and description of Wavelet-Based damage assessment benchmark'. *Shock and Vibration*. 2015;2015(4):1–19.

[52] Pagliarulo V., Rocco A., Langella A, *et al*. 'Impact damage investigation on composite laminates: comparison among different NDT methods and numerical simulation'. *Measurement Science and Technology*. 2015;**26**(8):85603.

[53] Scott I.G., Scala C.M. 'A review of non-destructive testing of composite materials'. *NDT Int*. 1982.

[54] Halliwell S. 'Repair of fibre reinforced polymer (FRP) structures'. *NetComposites*. 2012:1–33.

[55] Bevan K.B., L. G.C., Armstrong W.F. 'Post-repair inspection' in *Care and repair of advanced composites*. 2nd ed. Warrendale, PA: SAE International; 2005. pp. 330–33.

[56] Whittingham B., Baker A.A., Harman A., Bitton D. 'Micrographic studies on adhesively bonded scarf repairs to thick composite aircraft structure'. *Composites Part A: Applied Science and Manufacturing*. 2009;**40**(9):1419–32.

[57] Dorworth L.C., Gardiner G.L., Mellema G.M. *'Essentials of advanced composite fabrication and repair'*. . . *Aviation Supplies & Academics, Inc*. 2009.

[58] Ahmad J. *Machining of polymer composites*. New York: Springer; 2009.

[59] LZH Making lightweight construction suitable for series production [online]. 2017. Available from http://www.lzh.de/en/publications/pressreleases/2017/making-lightweight-construction-suitable-for-series-production [Accessed Apr 2017].
[60] LZH Repairing defects in fiber-reinforced plastics more efficiently [online]. 2017. Available from http://www.lzh.de/en/publications/pressreleases/2017/repairing-defects-in-fiber- reinforced-plastics-more-efficiently [Accessed Apr 2017].
[61] da Silva L.F.M., Öchsner A., Adams R.D. *Handbook of Adhesion Technology*. Heidelberg: Springer; 2011.
[62] Baldan A. 'Adhesively-bonded joints and repairs in metallic alloys, polymers and composite materials: adhesives, adhesion theories and surface pretreatment'. *Journal of Materials Science*. 2004;39(1):1–49.
[63] Wingfield J.R.J. 'Treatment of composite surfaces for adhesive bonding'. *International Journal of Adhesion and Adhesives*. 1993;13(3):151–6.
[64] Palmieri F., Belcher M., Wohl C., Blohowiak K., Connell J. 'Laser Ablation Surface Preparation of Carbon Fiber Reinforced Epoxy Composites for Adhesive Bonding'. *SAMPE*. 2013.
[65] Gardiner G. 'Primary structure repair: the quest for quality'. *CompositesWorld [Online]*. 2011.
[66] Adugna T., Schmutzler H. 'Automated machining and surface preparation (caire)'. *Lufthansa Tech*. 2016.
[67] Cheng P., Gong X.-J., Hearn D., Aivazzadeh S. 'Tensile behaviour of patch-repaired CFRP laminates'. *Composite Structures*. 2011;93(2):582–9.
[68] Hexcel Composites. 'Composite repair'. 1999.
[69] Waite S. *EASA expert interview*. Cologne: EASA; 2017.
[70] Kwak M., Robinson P., Bismarck A., Wise R. 'Microwave curing of carbon–epoxy composites: penetration depth and material characterisation'. *Composites Part A: Applied Science and Manufacturing*. 2015;75(3):18–27.
[71] Ehrhart B. 'Quality assessment of bonded primary CFRP structures by means of laser proof testing'. 2016.
[72] Black S. *Automated aircraft repair closer to reality*. CompositesWorld; 2006.
[73] Holtmannspötter J., Czarnecki J.V., Feucht F., *et al*. 'On the fabrication and automation of reliable bonded composite repairs'. *The Journal of Adhesion*. 2015;91(1–2):39–70.
[74] 'WIWeB präsentierte sich auf Der ILA 2014'. 2014.
[75] Djokic D., Johnston A., Rogers A., Lee-Sullivan P., Mrad N. 'Residual stress development during the composite patch bonding process: measurement and modeling'. *Composites Part A: Applied Science and Manufacturing*. 2002;33(2):277–88.
[76] Agius S.L., Magniez K.J.C., Fox B.L. 'Cure behaviour and void development within rapidly cured out-of-autoclave composites'. *Composites Part B: Engineering*. 2013;47(2):230–7.

[77] TeQua heat transfer tools Marcotodo waermetechnik [online]. Available from http://www.marcotodo.com/modulsystem - english.htm

[78] Adugna T. *MRO expert interview*. Hamburg: Lufthansa Technik; 2017.

[79] Kim H.-J., Gama B.A., Yarlagadda S., Gillespie J.W., Mahdi S. 'A comparison of oven-cured and induction-cured adhesively bonded composite joints'. *Journal of Composite Materials*. 2003;**37**(6):519–42.

[80] Elaldi F., Elaldi P. 'A study on curing processes and environmental effects for rapid composite repair'. *Journal of Reinforced Plastics and Composites*. 2011;30(9):749–55.

[81] Vlot A., Verhoeven S., Nijssen M. *Aerospace Materials11: Bonded Repairs for Aircraft Fuselages, Series 07*. Delft: Delft University Press; 1998.

[82] Davis M. 'Adhesive bond damage tolerance and failure assessment'. 2016.

[83] Corporation H. *HexPly Prepreg Technology*. 28. Hexcel Regist. Trademark; 2013.

[84] Anderson J.P., Altan M.C. 'Formation of voids in composite laminates: coupled effect of moisture content and processing pressure'. *Polymer Composites*. 2015;36(2):376–84.

[85] Budhe S., Banea M.D., de Barros S., da Silva L.F.M. 'An updated review of adhesively bonded joints in composite materials'. *International Journal of Adhesion and Adhesives*. 2017;72:30–42.

[86] HAECO. 'Composite Repair - Portable Clean Rooms Rev 02'. 2016.

[87] Baker A., Gunnion A.J., Wang J. 'On the certification of bonded repairs to primary composite aircraft components'. *The Journal of Adhesion*. 2015;91(1–2):4–38.

[88] Breuer U.P. *Commercial aircraft composite technology*. Heidelberg: Springer; 2016.

[89] Dorworth L. *Essentials of Advanced Composite Repairs*. Abaris Training; 2007.

[90] Ehrhart B., Valeske B., Muller C.-E., Bockenheimer C. *Methods for the quality assessment of adhesive bonded CFRP structures-a resumé*. NDT in Aerospace; 2010. pp. 1–9.

[91] Gardiner G. *Aircraft composites repair moves toward maturity. composites world [online]*. 2016. Available from http://www.compositesworld.com/articles/aircraft-composites-repair-moves- toward-maturity

[92] Staszewski W.J., Mahzan S., Traynor R. 'Health monitoring of aerospace composite structures – active and passive approach'. *Composites Science and Technology*. 2009;69(11–12):1678–85.

[93] GARDINER G. *Structural health monitoring: NDT-integrated aerostructures enter service*. Composites World; 2015.

[94] Van Dam J., Bond L.J. ' Acoustic emission monitoring of wind turbine blades'. *Smart Materials and Nondestructive Evaluation for Energy Systems*. 2015;**9439**:94390C.

[95] Brunner aJ, Barbezat M. 'Composites from piezoelectric fibers as sensors and emitters for acoustic applications'. *Journal of the Acoustical Society of America*. 2004;**22**:127–37.

[96] Takeda S., Yamamoto T., Okabe Y., Takeda N. 'Debonding monitoring of composite repair patches using embedded small-diameter FBG sensors'. *Smart Materials and Structures*. 2007;16(3):763–70.
[97] EASA. 'EASA proposed cm – S – 005 issue: 01 bonded repair size limits'. 2014.
[98] Waite S. 'Composite materials: developing continued airworthiness issues'. 2014.
[99] DLR *Patches for high-performance structures – DLR researchers develop new repair process for fibre-reinforced polymers [online]*. 2015. Available from http://www.dlr.de/dlr/en/desktopdefault.aspx/tabid-10081/151_read-13789/year- 2015/151_page-4/#/gallery/19609 [Accessed Apr 2017].
[100] Bertrand. *Automated Production and Installation of patches for reworking structures.pdf*. United States Patent Application Publication; 2015.
[101] Olympus OmniScan MX2 [online]. Available from http://www.olympus-ims.com/en/omniscan-mx2 [Accessed Apr 2017].
[102] JEC JEC Europe final report [online]. 2015. Available from http://reports.jeccomposites.com/jec-europe-2015/ [Accessed Apr 2017].
[103] The UK government Future of manufacturing: a new era of opportunity and challenge for the UK-summary report [online]. 2013. Available from https://www.gov.uk/government/publications/future-of-manufacturing/future-of- manufacturing-a-new-era-of-opportunity-and-challenge-for-the-uk-summary- report [Accessed Apr2017].
[104] Strategy C. Composites leadership lightening the load the 2016 UK composites strategy what is composite material. 2016. Available from http://nccuk.com/sites/default/files/news-attachments/CLF- %20The%202016%20UK%20Composites%20Strategy_1.pdf
[105] EC. Europe 2020 Flagship Initiative - Innovative Union. 2020.
[106] SIUI SUPOR phased array ultrasonic detector. 2017. Available from http://www.siui.com/cpzxLY/info_114.aspx?itemid=242&lcid=50 [Accessed Apr 2017].
[107] LOT Quantum Design Lot quantum design [online]. 2016. Available from http://www.lot-qd.co.uk/ [Accessed Apr 2017].
[108] Barman W. 'Airbus employee'. 2017.
[109] Negarestani R. *Laser Cutting of Carbon Fibre - Reinforced Polymer Composite Materials*. Thesis metd; 2010.
[110] Dittmar H. *Laser expert interview*. Hannover: LZH; 2017.
[111] Holzhüter D., Pototzky A., Hühne C. 'Automated Scarfing process for bonded composite repairs'. *Adaptive, tolerant and efficient composite structures*. Springer; 2012. pp. 297–307.
[112] Ohle M. *Robot on demand: mobile machining of aircraft components with high precision [online]*. 2016. Available from http://www.ifam.fraunhofer.de/en/Press_Releases/robot_on_demand.html [Accessed Apr 2017].
[113] Naeem M. 'Laser cutting of composites and ceramics'. *Laser Processing of Polymer, Metal and Ceramic Composites*. 2008.
[114] Power H., Fiber Y. 'YLS-CUT series YLS-CUT series'.

[115] Against B.P., Threats W., Benefits K. 'Key benefits and features'. 2010.
[116] French P.W., Naeem M., Clowes J., Sharp M.C. Fibre laser material processing of Aerospace Composites. *Proc. Fifth Int. WLT-Conference Lasers Manuf*; 2009. pp. 339–46.
[117] Suder W. *Laser expert interview*. Cranfield University; 2017.
[118] engineering.com *Quality basics: how does 3D laser scanning works [online]*. Available from http://www.engineering.com/AdvancedManufacturing/ArticleID/12390/Quality-Basics-How-Does-3D-Laser-Scanning-Work.aspx
[119] CAD/CAM SERVICES INC *3D scanning technology [online]*. Available from http://info.cadcam.org/blog/how-aerospace-uses-3d-scanning-technology
[120] Allard P., Lavoie J., Fraser J. 'Improvement of aircraft mechanical damage inspection with advanced 3D imaging technologies'. *5th International Symposium on NDT in Aerospace*. 2013:13–15.
[121] Surphaser. *3D laser scanners: 100HSX*; 2015.
[122] 3d Digital. '500 series'.
[123] Creaform. 'Portable 3D scanners: HandySCAN 3D'.
[124] Bendemra H., Compston P., Crothers P.J. 'Optimisation study of tapered Scarf and stepped-lap joints in composite repair patches'. *Composite Structures*. 2015;130(1):1–8.
[125] Wang C.H., Gunnion A.J. 'On the design methodology of Scarf repairs to composite laminates'. *Composites Science and Technology*. 2008;68(1):35–46.
[126] Breitzman T.D., Iarve E.V., Cook B.M., Schoeppner G.A., Lipton R.P. 'Optimization of a composite Scarf repair patch under tensile loading'. *Composites Part A: Applied Science and Manufacturing*. 2009;40(12):1921–30.
[127] EASA. *Foreign Part-145 approvals: composite repair workshop*. EASA; 2015.
[128] Airbus. *Automated fibre-patch preforming*. Airbus; 2016.
[129] Zhao Y., Mehnen J., Sirikham A., Roy R. 'A novel defect depth measurement method based on nonlinear system identification for pulsed thermographic inspection'. *Mechanical Systems and Signal Processing*. 2017;85(1):382–95.
[130] Shepard S. 'Flash thermography of aerospace composites'. *IV Conf. Panam. END Buenos Aires*; 2007. pp. 7.
[131] Lau S.K., Almond D.P., Milne J.M. 'A quantitative analysis of pulsed video thermography'. *NDT & E International*. 1991;24(4):195–202.
[132] Zeng Z., Zhou J., Tao N., Feng L., Zhang C. 'Absolute peak slope time based thickness measurement using pulsed thermography'. *Infrared Physics & Technology*. 2012;55(2–3):200–4.

Chapter 7

Augmented reality-equipped composites bonded repair

Xi Wang[1], Hamed Yazdani Nezhad[1,2], Samuel Court[3], Bikram Thapa[3], and John Erkoyuncu[3]

The prosperity of aircraft transportation together with revolutionary promotion of composite components employed in commercial aircrafts poses enormous challenges (uncertainties associated with polymer process control, zero-thickness defects and non-destructive inspection and testing capability to inspect process-induced defects, etc.) to the industry of aircraft composite repairing. An industry solution owning merits on efficiency, reliability and repeatability is in highly demand.

Augmented reality (AR), a human–computer interaction technology built up above the development of virtual reality, possesses its exclusive superiority on its capability of inflicting digital mock-up into physical environment. The above property of AR provides colossal opportunities to be utilised into the industry territory to contribute to the realisation of automatic, efficient, streamlined and reliable production line. The current project aimed at developing an AR system to explore its way of putting AR technology into aircraft composite repairing industry as a guidance tool to instruct technicians' repairing operation, mitigate human errors, and reduce the duration of repair and assembly. Stepped scarf repair embedded with soft composite patches was selected as the archetype to be brought into effect though hard patches were partially examined as well.

The developed AR system focuses on patch installation and vacuum bagging processes to address the predicament of miscellaneous steps and fibre directions to be followed. It contributes to provide a reliable, repeatable, reproductive and efficient composite assembly operation.

[1]Enhanced Composites and Structures Centre, School of Aerospace, Transport and Manufacturing, Cranfield University, Bedford, UK

[2]Aeronautics & Aerospace Research Centre Department of Mechanical Engineering & Aeronautics City, University of London, London, UK

[3]Through-life Engineering Services Centre, School of Aerospace, Transport and Manufacturing, Cranfield University, Bedford, UK

7.1 Introduction

7.1.1 Project background

The current individual research project is a continuation of previous group research project *Augmented Reality-Equipped Composite Repair* (CURES 7714/2019) carried on by the current author under the supervision of Academic Research Group at the School of Aerospace, Transport and Manufacturing at Cranfield University. The research has been carried out collaboratively between the Enhanced Composites and Structures Centre and the Through-life Engineering Services Centre.

The research aims at developing a set of digital toolkit equipped with augmented reality technology to facilitate an automatic, digital and reproductive composite repairing process to remit the overload of maintenance, repair and overhaul tasks of commercial aircraft brought about by the persistent incremental adoption of composite material.

7.1.2 Previous group project outcome

- Literature review on common defects and damages of commercial aircraft and their distributions.
- Literature review on typical composite repairing process and technology.
- Literature review on the state-of-the-art application of augmented reality technique into aircraft composite material manufacturing.
- Pristine digital mock-ups capable of indicating stepped scarfed patch laying up directions.

7.1.3 Gap analysis

The previous group research project succeeded in achieving sufficient literature research on composite defects and typical composite repairing technique as well as the frontage industrial application cases of augmented reality technology pitched into aircraft manufacturing. Following that, a pristine prototype capable of indicating patch laying up directions was built up. However, there are still distance and gaps to be covered considering the pre-set project objective as an entire scarfed patch laying up process instruction able to be pitched into industrial operating environment.

The distance and gaps between the existing prototype and the pre-designed objective functions were analysed as follows:

- Not entire patch assembly process but fragments of patch laying up direction indication.
- No available user interface to enable the technicians accessing and operating the applet.
- No interactive function between user and operating system was realised, which did not stick out the superiorities of augmented reality technique.

- No step-by-step representation of process information but a simultaneous information presentation of the entire procedures.

7.1.4 Aim and objectives

After comparing pre-designed objective functions with the existing prototype, the functions expected to be further exploited under the current individual research project are as follows:

- Build up user interface to enable the technicians to access and operate the system.
- Create user buttons to realise the interaction between system and user.
- Set up linkage between each button and each operating step to realise a stepwise process information presentation and illustration instead of an instantaneous information explosion.
- Extend the instruction from merely patch laying up direction indication into entire laying up process and vacuum bagging process.
- Enable the bilateral output terminal: tablet-Based and Microsoft HoloLens.
- Re-establish the space location of each digital mock-up.

7.2 Methodology

Several phases have to be gone through to guarantee the arrival of the pre-supposed destination. These are primarily gap analysis, objective setting, literature review, concept design and implementation. Their inner link and sequence are demonstrated in Figure 7.1.

Gap analysis between pre-designed functions and the existing prototype has to be triggered to obtain the cognition on the disparity to be covered up, which is the prerequisite of any following steps.

Objective setting is eligible to be proceeded directly after gap analysis. Objectives of the following project should be built up onto the consideration of covering existing research gaps and distance.

Literature review is a mandatory part to be gone through in every research project. Considering the sufficient background knowledge acquired during the group

Figure 7.1 Project methodology overview

research project on composite defects and repairing expertise, literature review of current individual research project would lean more emphasis on vacuum bagging and augmented reality aspects.

Concept design of the current individual project will basically focus on finalising the target process to be equipped into augmented reality applet, the layout of user interface and how to realise a stepwise information presentation.

Implementation under which software, platform and programming language are employed to proceed further development has to be chosen. Of the above, Applet developing ranged from importing image target and building up digital mock-ups until creating user interface would be the dominating parts of this section.

7.3 Literature review

To guarantee the integrity and readability of the current report, certain closely relative background knowledge on composite defects and repair technique acquired during group project has to be presented in this section, but in an even laconic manner.

Apart from the above, some enhancive knowledge on vacuum bagging and augmented reality technique is also indispensable in case of conducting further applet developing.

7.3.1 Defects and damages in composite materials and structures

Two categories of defects of composite parts can be defined based on the stage when they are generated, which are defects generated during manufacturing and damages given rise to during its service life [1].

7.3.2 Manufacturing defects

Porosity, contaminants and non-uniform fibre are the three primary sorts of composite defects generated during manufacturing process [2–9].

Porosity, one of the most typical composite fabrication defects, is always found as tiny spherical cavities emerged in the inner-laminate of composite panel, which is mainly introduced by improper curing parameters such as temperature, pressure and duration [10].

Contaminants mean any foreign objectives not meant to be existed in the composite matrixes. It could be wispy hairs, greasy dirt of fingers, and so forth [2].

Non-uniform fibre can be introduced from fibre breakage [11], fibre misalignments [11], ply misalignments [11], wrinkles [12] arisen during laying-up (Figure 7.2), ply cracking [11] and unconformity of thickness due to excess or inadequate resin [11].

7.3.3 Common service-life damages

Depending on the service ambient and the sensitivity of the material itself, a wide variety of defects such as delamination and diverse types of bond failure can be generated [11].

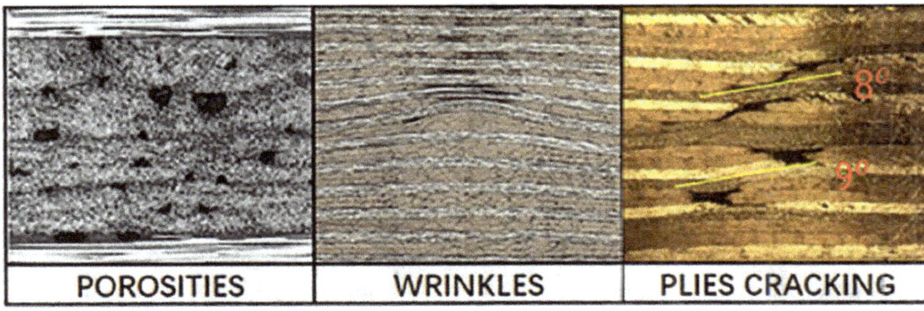

Figure 7.2 Typical composite defects generated during fabrication [12]

Delamination A most common damage generated from fatigue, impact or bearing damage. In most circumstances, delamination develops from cracks [11].

Bond failure also termed as disbond, commonly manifests in four modalities: cohesion failure, adhesion failure, mix-mode failure and ply separation failure. Figure 7.3 exhibits the physical appearance of these four types of disbond.

- **Cohesion failure** indicates a disbond within adhesive layer, which usually generates from overload against design limit.
- **Adhesion failure** is a disbond against interface between adhesive and adherend, which is always caused by improper processing like contamination and inadequate process.
- **Mix-mode failure** is a combination of disbond of adhesion failure and cohesion failure, where failure exists both within the adhesive itself and between the interface of adhesive and adherend. In most of the circumstances, mix-mode failure shares the same inducement as adhesion failure.
- **Ply separation failure**, as indicated in Figure 7.3, means the first ply of laminate separating from the composite matrix, which usually occurs when bond strength beats inter-laminar shear in thick composites.

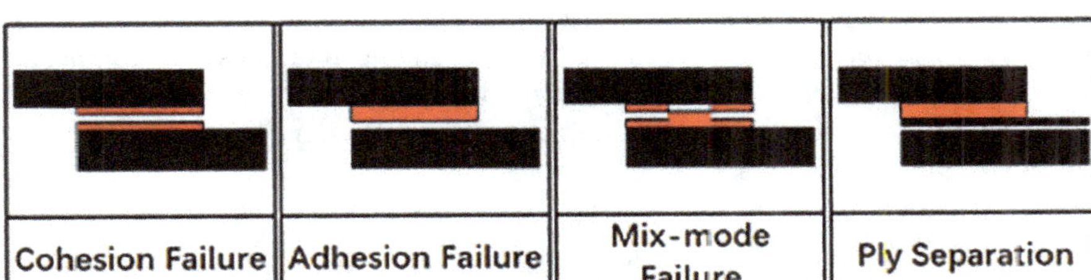

Figure 7.3 Four modalities of disbond(black area represents adherends and red one represents the adhesive)

7.3.4 Alternative classification of composite impact damage

An alternative type of classification concentrating on impact defects also exists, where high-energy impact defect, intermediate-energy impact defect and low-energy impact defect are nominated as three dominating types.

High-energy impact defect is a full penetration damage, the exit diameter of which is larger compared with the entry diameter in most circumstances, which is always treated by damage removal and scarf or doubler repair.

Intermediate-energy impact defect bears broader damage area compared with high-energy impact. Dent usually appears in the inflicted side; however, in some cases, fibre breakage is found at the other terminate face. The repairing method for the majority cases is damage removal and scarf or doubler repair.

Low-energy impact defect is the most insidious damage which is not even visible, no fibre fracture caused but delamination and resin fracture caused, can be generated from a terrific low-impact energy level such as tool drop, also termed as BVID (barely visible impact damage).

7.3.5 Composite material repairing technique

The most commonly used composite repairing methods will be introduced in this section, followed by a step-by-step interpretation of scarf-based repairing method which gains good graces among aeroplane industry considering its virtue in providing aerodynamic surface.

7.3.6 Classification of composite material repairing method

Resin injection repair and patches repair are pair of frequently used repairing methods in case of composite material [13].

Resin injection repair is a provisional means to stop damage from diffusing by perfusing resin into the defect region [14]. On consideration of the lower complexity of this technique itself as well as not being a permanent solution, a particular illustration will not be presented here.

Patches repair is the imparity on the option of removal damage block or not, one-side treatment or both-sides treatment, shape of removal temper differentiates diverse patch repair methods [14]. Figure 7.4 exhibits the branches of patch repair and its classification basis. Table 7.1 compares these patch-repairing techniques on the points of finishing surface, machining difficulty and operation accessibility.

Scarf-based repair gains increasing traction among the aerospace composite MRO (Maintenance, Repair and Overhaul) industry on consideration of providing aerodynamically smooth surface and effective stress transfer [15, 16]. A step-by-step interpretation will be made in section 7.3.7.

7.3.7 Scarf-based repair process

Material removal, surface preparation and repair fabrication are the three fundamental steps to be gone through in case of scarf-based repair [15].

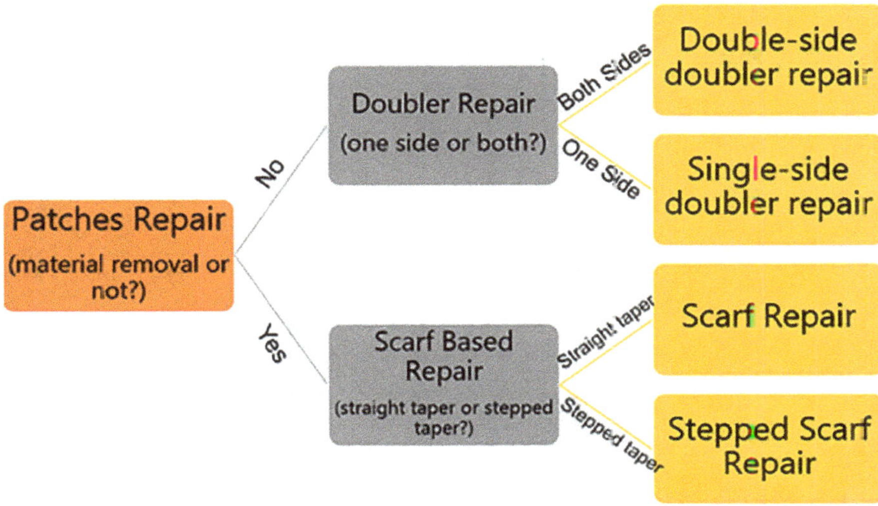

Figure 7.4 Patches repair classification

Material removal

It is the process of removing materials of damage area to create a hollow cavity for the embedded patches. Scarf repair and stepped scarf repair are partitioned based on the configuration of taper machined [14]. The discrepancy of the shapes of tapers between scarf repair and stepped scarf repair will be presented in Figure 7.5.

There are mainly three commonly used techniques being employed to proceed material removal for scarf-based repairing, which are conventional machining (cutting, grinding and sanding), laser machining and abrasive waterjet machining (AWJ) [14]. Table 7.2 will compare these three types of techniques in the aspects of accessibility, operability, economic efficiency and HSE (healthy, safety and environment).

Surface preparation

Surface preparation is an indispensable process to facilitate the surface activity of the interface further to guarantee the required adhesion force. As to composite repairing process in which brand new surface (scarf surface or stepped scarf surface) generated from previous machining process, surface preparation is set as a compulsory motion according to the regulation of European Aviation Safety Agency (EASA) [17]. This newly formed surface needs to be treated to reach surface uniformity and wipe off any debris, dusts or any type of contaminant caused by the upstream processes or its surroundings [18–21].

There is always misunderstanding on surface preparation which deems a clean surface as the exclusive requirement to attain satisfactory adhesive quality [22]. It has been proved that beyond a clean surface, close contact and flow, wet bonding surfaces are also elements contributing to a good adhesion [20]. Surface

Table 7.1 Patch-repairing methods comparison (information from Reference 14)

	Double-side doubler repair	**Single-side doubler repair**	**Scarf repair**	**Stepped scarf repair**
Surface	None aerodynamically smooth surface because of overlay patches		Aerodynamically smooth surface because of inlay patches	
Machining difficulty	No material removal process		Exist	
One or both sides operation	Both sides operation, may not accomplish due to restriction on accessibility (assembled parts)	One side operation	One side operation	One side operation

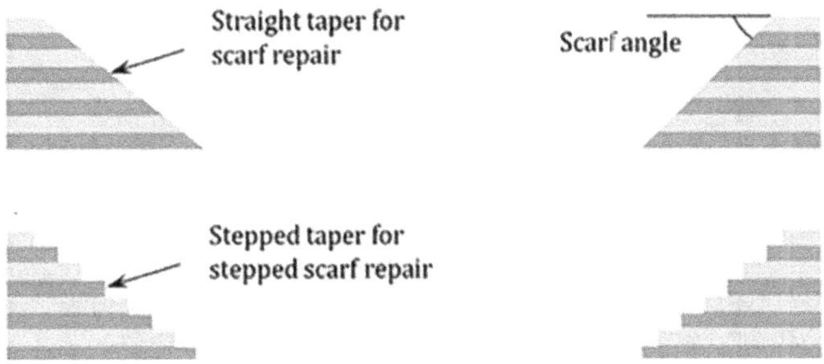

Figure 7.5 *Taper shapes of scarf repair and stepped scarf repair [14]*

wetting happens on the condition that the surface energy of the adhesive is lower than that of the adherend [23]. Abrasion/solvent cleaning, plasma treatment and laser treatment are some commonly used techniques capable of enhancing adhesive strength and persistence by promoting surface energy, aggrandising surface roughness and activating surface chemistry [24]. Table 7.3 compares the above-mentioned three types of treatment methods on the points of surface quality determinant, availability, surface uniformity achievability, surface persistence, process repeatability and cost.

Repair fabrication

Suitable material matrix and controlled curing conditions are duple critical systems contributing to a robust and reliable patch repair structure, where the material matrix refers to materials such as patch materials, resins and adhesives, as well as that temperature and pressure are elements to be paid attention to as to curing conditions [25, 26].

Material selection

Repairing time spent is a primary consideration to be taken into account in case of commercial aircraft. Based on the above, materials which can be stored in room temperature, capable of being cured under relative lower temperature and bearing shorter curing time are consummate. Nevertheless, an unneglectable factor to be put into the first priority is the compatibility between the additive material and their mother panel [14].

Curing conditions

An adequate curing reaction guarantees an excellent compaction of laminate piles when achieving adaptive temperature, pressure and vacuum conditions [14].

Curing pressure

An appropriate curing pressure helps enhancing consolidations and decreasing porosities by providing proper bondline thickness, enabling adhesive flowing, wetting the surface and squeezing out trapped air [25–27]. Vacuum bagging

Table 7.2 Comparison on different composite material removal technique (information from Reference (12)

Comparison between different composite material removal techniques			
	Conventional Machining	Laser Machining	Abrasive Waterjet Machining
Technical maturity	Industry experienced	Under developing	Under developing
Industrial penetration	High	Low	Low
Machining accuracy	Dependent on operating skills	High	High
Surface quality	Dependent on operating skills	Medium	Depending on process parameters and abrasive
Workability on complex curve	Difficult	Achievable	Achievable
Sensibility to processing parameters	High	High	High
Residual stresses	Yes	No	No
Setup complexity	Medium	Medium	High
Auxiliary system	Complex	Computer controlled	Complex
Tool wear	Yes	No	No
HAZs	No with coolant	Yes	No
Coolant needed or not	Yes	No	No
Moisture absorption	Yes	No	Yes
Cost	Low	High	High
HSE	Healthy hazard without dust extraction systems	Toxic by-products	Environmental-friendly process

Table 7.3 *Surface treatment methods comparison (information from Reference 14)*

Comparison between different surface treatment methods				
		Abrasion/ Solvent Cleaning	**Plasma Treatment**	**Laser Treatment**
Surface quality determinant	Material property	No	Yes	Yes
	Treatment parameters	No	Yes	Yes
Thin surface only		No	Yes	No
Pre-surface cleaning required		No	Yes	Yes
Contactless treatment		No	Yes	Yes
Surface uniformity		No	Yes	Yes
Surface persistence		Yes	No	Yes
Process repeatability		No	Yes	Yes
Cost		Low	High	High

[28] and mechanical fixtures [29] are alternative methods to inflict pressure on the repairing matrix. A comparison between these two methods on pressure uniformity, availability on high pressure and the need of special jigs are made in Table 7.4.

Curing temperature

To achieve a fine bond structure, it is critical to keep control on curing temperature. Eliminating thermal gradient is still a conundrum due to poor thermal conductivity in the through thickness direction of composite laminates which can lead to non-uniform stress transfer and ineffective bonding. Heat blanket is the most commonly used heating equipment under situ-repairing environment. Induction-curable adhesive and resin are under development; however, limitations are still posed on curing cycle control [30–33].

Table 7.4 *Pressure applying methods comparison (information from Reference 14)*

	Vacuum bagging	**Mechanically fixture**
Uniform pressure	Yes	No
Special designed fixtures	No	Yes
Accessibility on high pressure	No	Yes
Others	Removing volatiles holding heat blanket	

Table 7.5 Hard patch versus soft patch (information from Reference [14]

	Hard patch	Soft patch
Special mould required	Required	Not necessary
Surface preparation required in case of patches	Required	Not necessary
Coherence with panel in fibre-volume fraction and priorities	High	Low
Coherence with panel in shape	Relatively low	High
Additional process	Yes	Not necessary
Processing time	Long	Relatively short
Cost	High	Low
Others		· Accurate machining · Rigid storage conditions · Potential of low fibre-volume fraction and porosities · Potential of wrinkling defects · Accessibility depends on defects location

Hard patch and soft patch

Hard patches mean pre-cured patches manufactured either by moulding or machining [34].

Soft patches refer to patches fabricated on-site either utilising prepreg or wet laying-up [34].

Comparisons on occupation of special mould, surface preparation, shape coherence with mother panel, additional process required or not, efficiency and cost were made between hard and soft in Table 7.5.

Vacuum bagging technique

Vacuum bagging is a popular technique adopted among aircraft composite repairing industry to realise a uniform curing pressure. The uniform mechanical clamping force is obtained from pressure differential between normal ambient and sealed vacuum system. The sealed vacuum system can be achieved by applying vacuum pump to extract detained air and using sealant tape to create isolated inter-space [35].

It serves functions as follows [35]:

- Squeezing out the air trapped between laminate layers
- Reducing humidity
- Enabling efficient force transmission and disabling fibre orientation twisting by compacting fibres.
- Optimising fibre-to-resin ratio which is a key parameter to obtain strength-to-weight superiority of composite matrix.

Figure 7.6 exhibits a basic composition of vacuum bagging system employed by a majority of aircraft composite manufacturers. Vacuum, sealant and bagging are three fundamental elements to be illustrated in the following context.

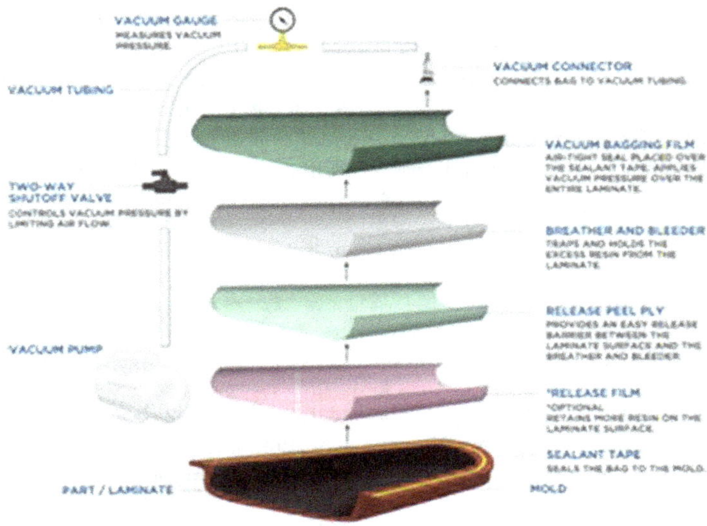

Figure 7.6 Composition of vacuum bagging [35]

Vacuum system constitutes vacuum pump acting as power provider, vacuum tubing as air communicator, vacuum gauge as pressure regulator and vacuum connector functioned as interface between vacuum system and bagging system [35].

Sealant is a crucial factor to achieve an air-tight bagging system. In the majority situations, the sealant of the bagging system is undertaken by sealant tape especially for aircraft industry. However, the sealant tapes employed have to be gotten through the rigorous validation procedure according to aircraft regulator [17, 35].

Bagging system is a combination of diverse layer of auxiliary materials whose functions ranged from easy peeling, absorbing until isolation. The composition of bagging layers is not universal but case-by-case depending on the factors of material, temperature and equipment employed [35]. Table 7.6 lists common types of auxiliary materials and its relevant functions.

Heating system is an optional element whose existence depends on the curing requirement of the resin adopted. However, aircraft composite manufacturing industry employs heating system very frequently ranging from autoclave, oven until

Table 7.6 Common auxiliary materials and its functions (information from Reference 35)

Materials	Functions
Adhesive film	Bonding patches and panels
Peel ply	Assisting easy splitting from cured compound
Release film	Isolating auxiliary material from laminate
Breather fabric	Absorbing excess resin
Bagging film	Sealing entire bagging system

heating blanket. In case of repair terrain, heating blanket is the minion considering its virtues on manoeuvrability and cost-effectiveness [35].

Another non-negligible unit in the heating system is the thermal couple which is employed by being distributed into different locations of the panel to be cured to monitor the curing temperature. Three thermal couples allocated respectively into part centre and tail end are mandatory according to most of the aircraft manufactures' specifications [35].

7.4 Augmented reality

7.4.1 Augmented reality technique

Augmented reality, compared with virtual reality which is a thorough virtual interface isolating from physical ambient, enhances human beings' perception of realistic surroundings by a combination of digital perceptual information with physical existence [36].

These computer-generated sensorial information can be overlaid in two formalisations, i.e. constructive and destructive, where constructive means additional digital data being inflicted into natural surroundings and destructive means the computer-generated sensorial information will over-ride the related physical subsistence [37, 38].

Four types of augmented reality can be classified based on the technique adopted for positioning, which are marker-based augmented reality, markerless augmented reality, projection-based augmented reality and superimposition-based augmented reality [37, 38].

Marker-based augmented reality: As its verbal meaning, the computer-generated information is positioned based on the pre-set relationship between marker and virtual mock-ups. The marker is sensed and captured by a reader which is normally a camera [37, 38].

Markerless augmented reality employs GPS, digital compass, velocity meter or accelerometer to locate the digital information into the desired place. The blossom of smartphones accelerates its pullulating, to some extent. It is commonly used to develop applets functioning as mapping directions such as assisting travellers to quickly locate the nearby business and service of an unacquainted street [37, 38].

Projection-based augmented reality projects artificial lights onto the physical world surface. Computer-human interaction is realised by sensing user's touch of that projected lights. Three-dimensional interactive hologram projected into the mid-air has been achieved currently [37, 38].

Superimposition-based augmented reality is the above-mentioned destructive technique which overlays virtual views into the original view. Object recognition is the crucial function to realise this technique. IKEA utilises this superimposition-based augmented reality to enable customers to grasp how their place will be after replenishing their furniture without the entities arrival but downloading an app [37, 38].

Table 7.7 MiRA technology

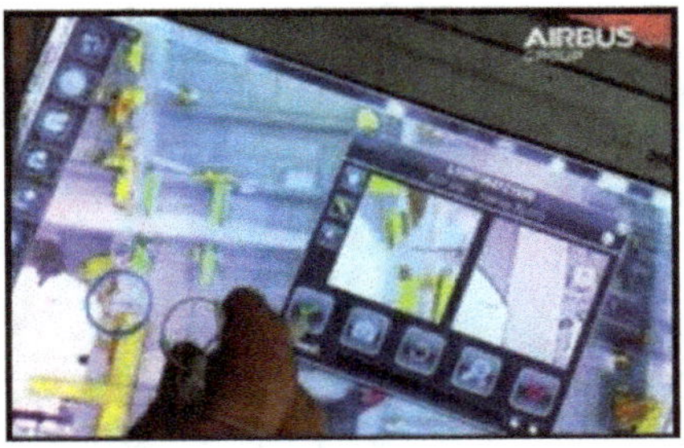

Application area	Aircraft assembly inspection
Data capture equipment	Digital camera
Output terminal	Tablet
Digital data transmission	Yes
Benefits	Reliable performance elevated efficiency descending labor cost

7.4.2 Application of augmented reality into aircraft industry

Aircraft manufacturing industry whose feathers of capital- and labour-intensive, stringent process control and high additional value vest potential and incentive to explore the application of augmented reality into its manufacturing and inspection fields.

Three typical representatives of augmented reality's application into aircraft manufacturing industry and a promising collaboration program will be addressed.

MiRA technology [39]

Airbus Group Innovation develops software applied for brackets inspection during assembly process by employing augmented reality technique (Table 7.7). A camera was equipped into a tablet computer to capture the practical position of the brackets. Inspectors identify losing, incorrect positioning and defects of brackets by comparing the built entities with the virtual models superimposed.

IRIS-integrated AR [40]

Safran Composite declared a precedent on utilising augmented reality technique to present composite defects on the exactly position of the parts where it occurs (Table 7.8).

A robot named as IRIS (Infra-Red Inspection System) adopts infrared thermography technique to implement non-destructive inspection to the parts. Digital data are captured and transmitted to the AR system. On the surface of the exact location of the defect generated, an image of inner defect will be presented synchronously. The area capable of being covered ranges from 3 square metres to 12 square metres.

PlyMatch in ACE [41]

The live video of composite parts was created with the assistance of digital camera when parts were positioned (Table 7.9). A coordinate measure machine was employed to obtain the digital information of the specific spatial position of parts. These dual

Table 7.8 IRIS-integrated augmented reality

Application area	Composite NDT
Data capture equipment	Infrared thermography
Output terminal	Digital image into physical panel
Digital data transmission	Yes
Benefits	Legible and intuitional presentation of composite defects and its location assisting correct judgment and decision

digital information were continuously fed into the servo system. Instructions on carbon wave plies and sequence of laying up were overlaid into the live video of the parts by the implementation of augmented reality technique.

CAIRe [42]

Lufthansa Technik persists in developing **C**omposite **A**daptable **I**nspection and **Re**pair project (CAIRe) objecting at developing an automatic and integrated composite repairing solution (Table 7.10). Potential industrial capabilities and adaptabilities on repair calculating, material subtracting, surface activating, automated scarfing and contamination inspection were exhibited from the current system. Considerable areas exist in this system to collaborate with AR technique to be transformed into an interactive, intelligent and integrated composite repairing system.

7.5 Concept design

7.5.1 Project scope selection

As the current individual research project is an extension of previous group research project, the pre-set project scope will be solidified as developing an interactive and

Table 7.9 Summary of PlyMatch

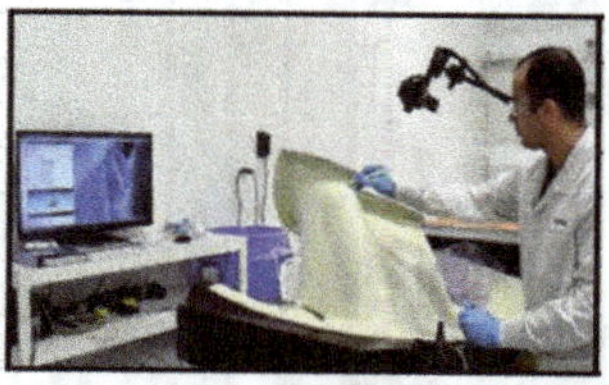

Application area	Lay-up instruction and record
Data capture equipment	Digital camera and CMM
Output terminal	Digital image generated onto computer
Digital data transmission	Yes
Benefits	Repeatable and reliable process efficiency promoting traceability achieved by data recording and storage

Table 7.10 *CAIRe*

	Integrated functionality	Repair calculation Material removal Surface activation Automated scarfing Contamination detection
	Benefits	Integrated system processing time reduction high accuracy new design possibilities

instructional applet equipped with augmented reality technique dedicating into composite repair territory.

7.5.2 Reparation method

Based on the discussion in Section 7.3.5, stepped scarf combined with soft patch was selected as the prototype method to be implemented, considering points as follows:

- Aerodynamically smooth surface provided.
- Monodirectional operation instead of bidirectional operation increases accessibility and manoeuvrability in case of hanger repair.
- Relatively lower machining difficulty in case of straight interface.
- Cost and time efficiency due to no additional process steps like machining patches together with no additional tooling called.
- Superiority on surface fit in case of airplane's complex curvature.

7.5.3 Target process

Figure 7.7. is the process flow generated based on the type of reparation method selected.

Laying up together with vacuum bagging process was selected as the target processes to be casted into program developing. The major consideration points are as follows:

- It is crucial to guarantee a uniform fibre orientation between repaired panel and scarfed patches to realise the continuity of mechanics.
- It will be an intricate process in case of dozens of plies to be applied together with the disparate orientation requirement.

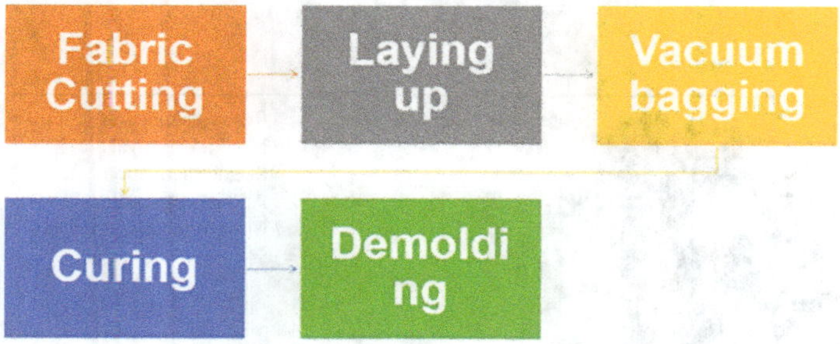

Figure 7.7 Soft patch scarf repair forming process

- It should be a case-by-case scenario in case of auxiliary material employed depending on the material specification of carbon and resin as well as the type and the location of defects.
- It is highly possible that technicians operate incorrectly or improperly during these laying up and vacuum bagging process.

7.5.4 Scenario design

To concentrate more on the design and developing of the applet itself, the panel and patches' dimensions of a previous Cranfield research scenario were inherited and utilised.

7.5.5 Panel

The panel was designed as a 300 mm by 300 mm flat square composite laminate.

The thickness was 6.8 mm.

The defect was designed to be found in the centre of the laminate with dimensions 68 mm by 72 mm.

7.5.6 Patches

The quantity of the patch steps was designed as three to avoid repeated programming.

Dimensions of each patch are shown in Table 7.11.

7.5.7 Bagging layers design

Apart from adhesive film and carbon fibres, a set of auxiliary materials functioning as absorbing excess resin, demoulding and sealing was mandatory to be applied to assist the entire curing process.

Figure 7.8. illustrates the combination of auxiliary materials to be applied in our scenario, among which (Figure 7.9.)

Dry peel ply was utilised for easily splitting all the consumable materials above it from the cured composite entities.

Table 7.11 Panel and patches dimensions

Unit: mm		Diameter/Sides length	Thickness	Sketch
Patches	Patch 1	76	2.1	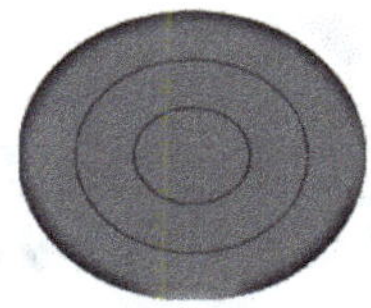
	Patch 2	138	2.35	
	Patch 3	200	2.35	
Panel		300	6.8	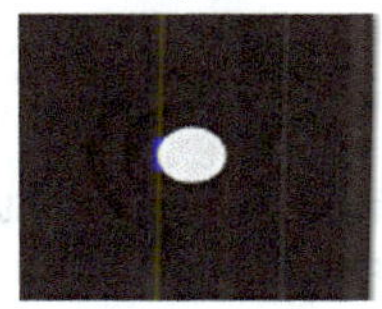

Perforated release film was employed to isolate other auxiliary materials from laminates, simultaneously, pores allocated enabling the releasing of trapped air below.

Absorbing/Breather fabric wrapped below and above the thermal blanket was applied for absorbing excess resin and protecting the thermal blanket from being contaminated by chemical liquid.

Thermal blanket should be attributed as equipment for heating source other than any auxiliary materials. Here it is introduced merely for explicit explanation.

Bagging film functioned as sealing the whole system to guarantee vacuum condition together with preventing any pollution from outside ambient.

7.5.8 Other assisting materials and equipment

Apart from what was mentioned formerly, some other materials and facilities are still indispensable to make up a vacuum bagging system, such as sealant tape to seal the

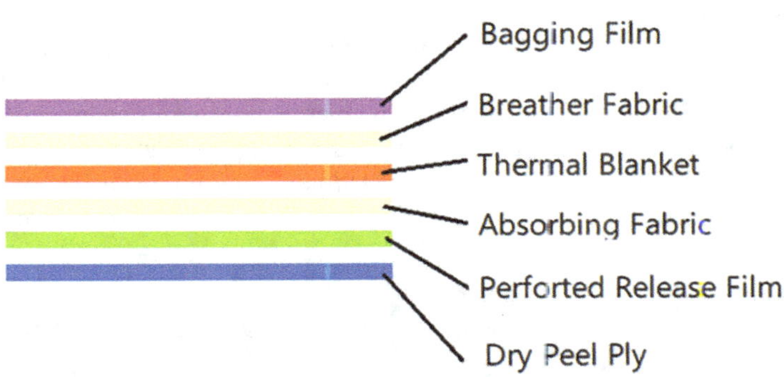

Figure 7.8 Auxiliary material lay-up sketch

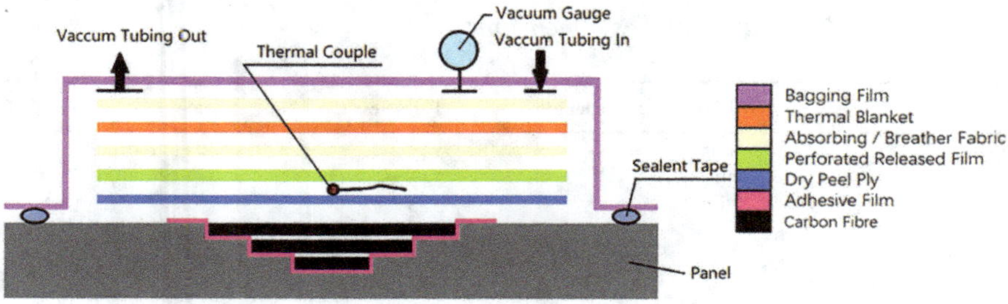

Figure 7.9 Sketch of Vacuum bagging system

periphery, heating system which normally is composed by the before-mentioned thermal blanket as heating source and thermal couples to monitor and control the curing temperature, together with vacuum networks consisting of vacuum pump (not relevant to lay-up however) vacuum tubing, connectors, vacuum gauge, etc.

7.5.9 Overall vacuum bagging system

Equipped with all the before-mentioned layers and facilities, the entire vacuum bagging system was tailored and set up as the following Figure 7.10. presented.

7.5.10 Functionality design

The primary functions expected to be realised in the current individual research project are user interface, user interaction with system and stepwise instruction.

User interface will be comprised by the step buttons.

The layout of the step buttons (Figure 7.11.) was designed to be standing in one horizontal line. These buttons will be built up from left to right following the operating sequence.

Each button will be named as the exact material or equipment it indicates, except for carbon layers in which additional information on laying-up direction will be presented on the buttons as well, due to its considerable importance.

Digital model of each entity to be employed during laying up and bagging process should be created as well to allow operators bearing direct perception on the dimension and position of each material or facility to be applied. Once the pre-set marker is recognised, the virtual matrix will emerge in front of the operator's view.

Interaction between user and system was built up through the trigger events set up above each button. In our case, the virtual mock-up of each layer will be colourised once its related button being triggered. The colour rendered to each layer

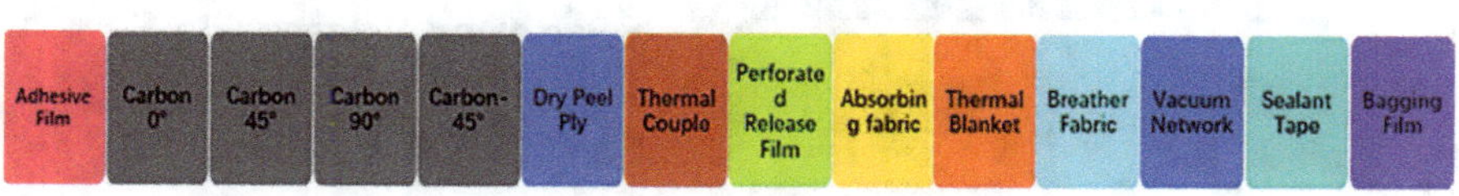

Figure 7.10 Button layout

Figure 7.11 *Microsoft HoloLens*

is not uniform but in accordance with the colour of its respective button to intensify identification function.

Stepwise instruction was realised through the massive bulk of process information being split and endowed into each button.

7.6 Implementation and result

7.6.1 Hardware configuration

The entire developing process was divided into two phases based on the different output equipment, which are tablet-based and HoloLens-based.

During the tablet-based phase, the only hardware employed is a digital camera-embedded tablet computer. The digital camera was exerted to capture the feather points of the image target to enable positioning the digital mock-ups. The tablet computer was utilised to operate the system and as the output termination to display the combination scene between virtual mock-ups generated from computer and physical operating circumstance captured by the embedded digital camera.

Further on, Microsoft HoloLens was introduced into the system as the output terminal to substitute the original tablet screen terminal. Thus, the image target will be captured by the digital camera embedded upon the HoloLens other than the previous computer-integrated camera, and the synthetic scene will present directly in front of operator's view via Microsoft HoloLens as medium.

7.6.2 Developing environment configuration

The primary developing platform was selected as Unity (Version 2018.3.14 f1 64-bit) equipped with Vuforia Engine considering the following points:

- Wide-range compatibility on operating systemguarantees extensive client base.
- Formidable and sophisticated computer graphic technique 43, 44 satisfies the current developing requirement.
- Broad user base and user-friendly developing interface secure user acceptability and adaptability of the applet.

- Compatibility to external resource can be realised by *drag and drop* functionality.
- Supportability on scripting Application Program Interface in C# by plugin.

Apart from Unity, a set of associative configuration should be equipped into computer to drive Microsoft HoloLens, which are

- Windows 10 PC operating system with developer mode enabled as the supportive system of running Unity developing.
- Visual Studio 2019 (Version 16.1 or Higher) to compile code, debug, test and deploy.
- Windows 10 Software Development Kit (Version 10.0.18362.0) to assist the building of HoloLens.
- Mixed Reality Toolkit (MRTK v2) for certain advanced and intermediate functions.

7.6.3 Coding language

C# currently is the exclusive coding language adaptable to Unity-developing environment. Boo and JavaScript had been adopted before, however, having been weeded out sequentially.

7.6.4 Developing process

Based on the pre-supposed function to be realised under the current system, there are three major stages to be undertaken.

Stage 1: Import image target to enable graphic recognition

Stage 2: Build up 3D models representing diverse materials and facilities to be applied under the image target as its child. Thus, once the image target was captured by the digital camera and recognised by the system, these child objectives would emerge at the pre-set position.

Figure 7.12. exhibits the pre-set image target and the matrix of digital mock-ups built upon the marker. A vertical view was deliberately designed to be presented in front of users' view to allow the operator possessing global view of the whole set of stuffs expected to be applied.

Stage 3: Create user interface (UI)

Step 1: Creating canvas to put user interface elements inside.

Step 2: Setting up render mode and layout. As what was supposed during the concept design phase, a horizontal layout was expected to be built up to indicate a left-to-right sequence of operating (Figure 7.13.). Here, *Horizontal Layout Group* tool was employed to realise a horizontal line layout automatically.

Step 3: Create interactive components.

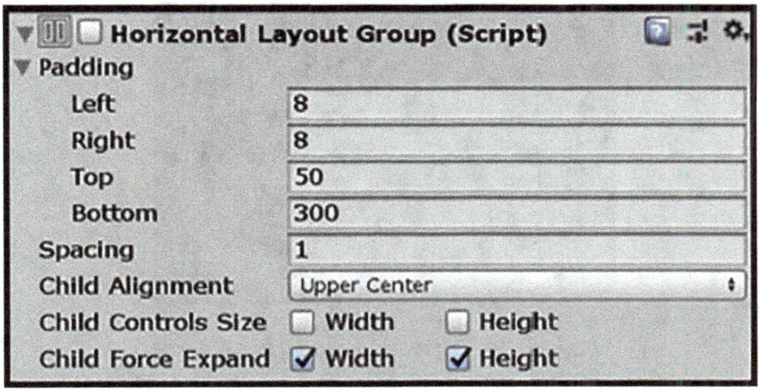

Figure 7.12 Auto layout tool

- Buttons representing each step were built up to realise the interaction between user and system.
- Relevant virtual mock-up will be colourised once its indicating button being clicked by setting up click event to each button. Especially, the rendering of digital mock-ups of carbon fibres (Figure 7.14.) was set to be the exact textures and orientations to be followed which delivers operator an instinct view of the exact orientation to be initiated.
- Diverse colours were endowed to indicate different stages of each button like normal colour (Figure 7.15.), highlighted colour, pressed colour and disabled colour.
- The normal colour of the button and the render colour of its related digital models were set as a uniform colour to build up another potential linkage between both (Figure 7.16.).

Step 4: Building up an indicator to trace the current steps. Fifteen buttons have to be established due to the tremendous steps of materials and equipment to be applied (Figure 7.17.). To avoid any labyrinth, a tracker was designed as an indicator to stand at the top of the current operating button. This function was realised by C# strips. For related coding, see Appendix B.

Stage 4: Converting the output terminal from tablet screen into HoloLens.

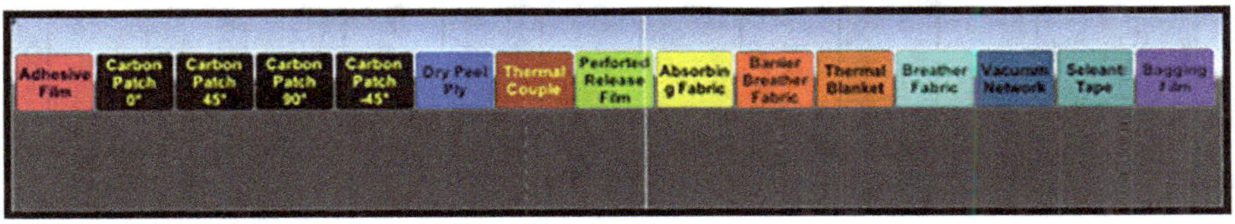

Figure 7.13 Buttons

Figure 7.14 Carbon rendering texture

Further attempt was made to transform the output platform into Microsoft HoloLens which bears superiorities on offering immersive ambient to user and hands liberation.

Some modifications had to be made to coincide with Microsoft HoloLens configuration:

- Image target was placed horizontally instead of vertically during tablet-based output.
- Stepwise process information were presented by activating the pre-set invalid mock-up to indicate disparate operating steps (Figure 7.18.), which means the monolithic mock-ups will be split and come up one after another based on the button being clicked instead of being colourised brick by brick.
- Materials were rendered as the texture of its physical entity to reinforce the indicating function of virtual mock-ups (Figure 7.19.), which caused the cancellation of the previous potential colour linkage between buttons and its related processing layer as mentioned in step 3 of stage 3.
- Step tracker was abandoned. Dynamic floating text (Figure 7.20.) was introduced to instruct operator the current step with its name broadcasting at the top of user interface.

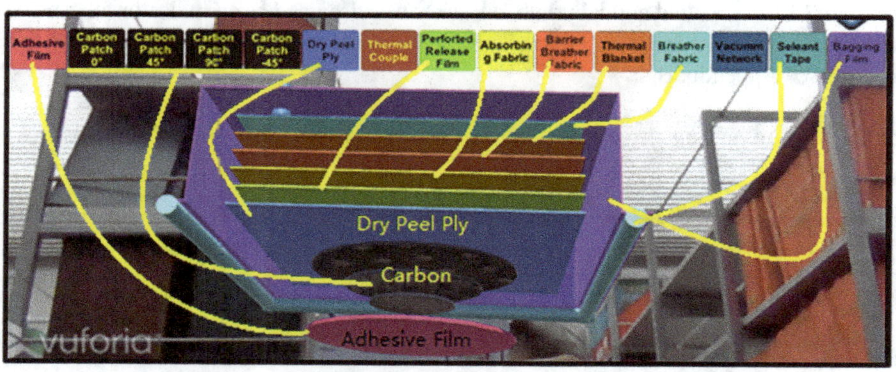

Figure 7.15 Button normal colour and mock-ups rendering colour

Figure 7.16 Image of steps tracker

- Four buttons functioned respectively as previous steps, next steps, reset and remote view to simplify the converting process from tablet program to HoloLens program (Figure 7.21.).

7.6.5 Program validation

There are available reports proved that the implementation of augmented reality technique into aerospace manufacturing instructing area facilitates promotion and upgrading on productivity and quality. A 90% increase in first-time quality and a 30% decrease in operating time were announced by Boeing by exerting augmented reality technique instructing to install electrical wiring of the airplane [45].

The current program can be validated through two separate directions, which are functionality validation and effect validation.

Functionality validation

Functionality validation mainly focuses on its sensitivity of image recognition and system's operability to composite technicians.

Image recognition sensitivity was tested in different environments like office building and composite laboratory. The average time taken to recognise and emerge relevant digital mock-ups was less than 1 s. The factor influencing the most was the distance and the angle to place the image target.

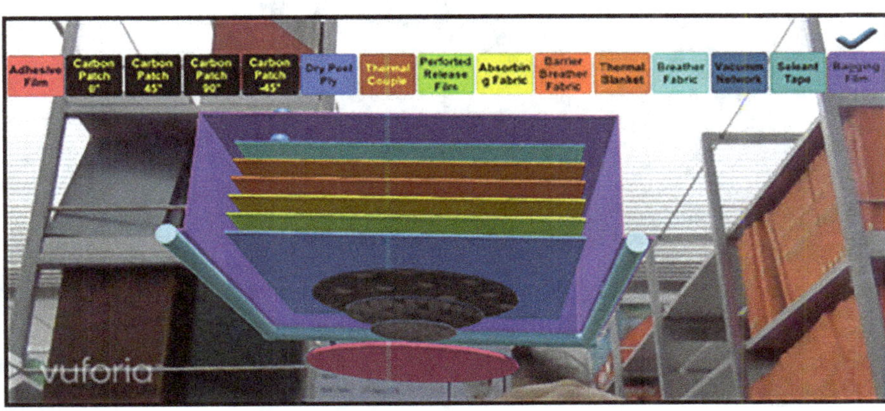

Figure 7.17 Real-time running screenshot of tablet-based program

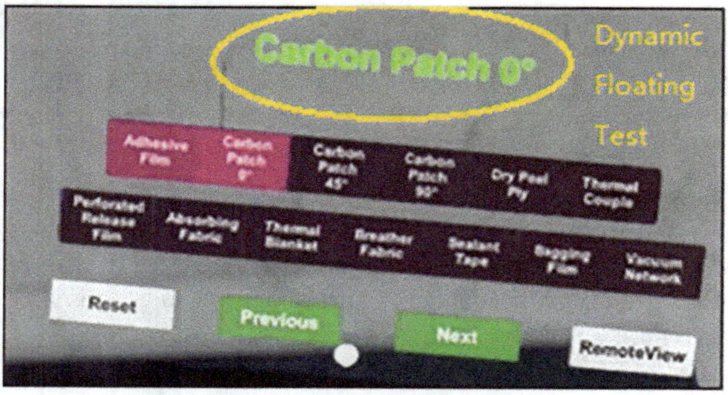

Figure 7.18 User interface of HoloLens terminal

System's operability was tested in composite laboratory through implementing it as a presentation which is a professional commercial aircraft composite repairing training course organised by the Cranfield University. During the presentation phase, technicians can operate the program from recognising the image target until the final bagging steps without stops and doubts, which demonstrates the readability of the user interface and the operability of the entire system.

Effect validation

Comparison experiment can be designed to validate some key performance indicators like percentage of processing time decreasing, rate of unconformity deduction, which can be carried out through allocating two operators bearing equivalent operating skill level into parallel processing environment performing same reparation on identical parts. The exclusive discrepancy would be the one referring traditional paper work instruction and the other utilising our developed augmented reality applet as guidance.

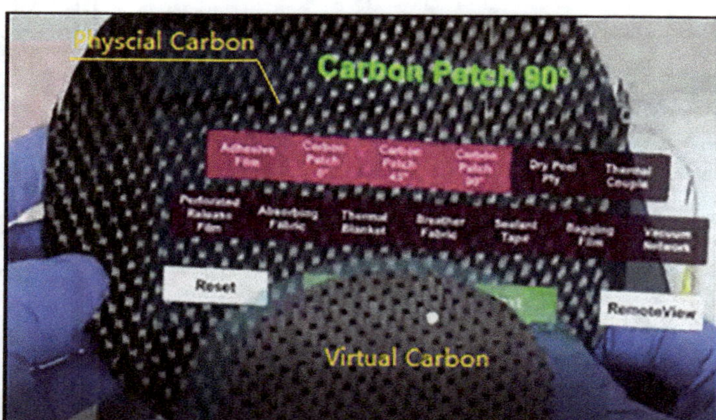

Figure 7.19 Contrast between physical carbon and virtual carbon

Figure 7.20 Superimposition-based augmented reality [37]

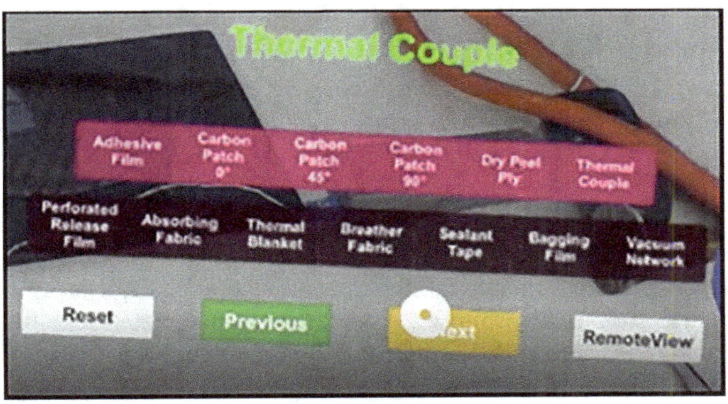

Figure 7.21 Screenshot of pressing button

Due to loads of software developing task taking up the majority of time, the effect validation was merely under designing phase and did not have any time to implement.

Further validation suggestion
A questionnaire can be designed to distribute to technicians who have used current applet to have some appraisal and suggestions on interface developing, loop design, user viewing and information presentation.

7.7 Discussion

7.7.1 Strengths

This current developed augmented reality toolkit provides a digital and automatic way of instructing composite repairing process. Compared with the conventional pattern in which process instruction is delivered through papery working instruction and 2D repairing sketch, it exhibited superiorities on the following aspects:

Processing time reduction

By implementing the developed augmented reality applet into aircraft composite repairing site, technicians click buttons displayed over tablet screen, alternatively gazing the options plus air-tab gesture when equipped with Microsoft HoloLens, instead of looking through papery specification and contrasting with repairing drawings page by page, which will decrease the processing time to a great extent.

Process reproducibility fortification

Process reproducibility enhanced through congelation operating sequence, material implemented and process parameters into augmented reality system, which prevents from any individual preference and human uncertainty factors.

Products quality promotion

Through design and experimenting different combination of sequence of procedures and parameters, a ultimate optimised procedure and parameters can be solidified into augmented reality applet to promote quality improvement.

Threshold of operators' skill descending

Reliance on operators' skills will minish considering the dominating expertise has already been embedded into the system. With the upgrade of the system, gradually, the sole requirement to operator will be evolved into being computer-literate.

7.7.2 Challenges

Limitations still exist in the current system which is still under development in the aspect of image recognition, position accuracy, system synchronism and scenario's compatibility.

Image recognition

Optical tracking is employed based on the current hardware and software configuration, which results in a strict restriction on distance between marker to camera, dimensions and pixel of target, digital camera's resolution, surrounding light conditions and even reflectivity of the object.

Accuracy of position

Challenge exists in guaranteeing the positioning accuracy of virtual mock-ups in case of further development on sensing, testing and judging patch's overlap.

System synchronism

A distinct delay was found when implementing Microsoft HoloLens as the displaying terminal. The live video presented over tablet was of 2 s hysteresis than what was acquired through HoloLens.

Scenario compatibility

No default module was tailored for industrial application. Virtual modules have to be designed and built from blank, which will aggravate the compiling task if pitched into industrial environment considering repairing scenarios are case-by-case depending on the material adopted and the defect type itself.

7.7.3 Future work

Some valuable points and functions could not be achieved within the scope of the current project constrained by limited time. However, suggestions were made in this section.

Integration with displacement sensor

It would be a tremendous academic value to combine displacement sensor into the current system to realise carbon patches' overlap distance detection to spread its function into positioning and judgement of overlap distance of carbon patches.

Correctness judgement and error warning

Moving further on the combination with displacement sensor, the system can achieve more interactive and guidance functionality on the judgement and error warning based on the technician's operation, such as cross symbol, red signals or even text broadcasting 'error operation'.

Data and process recording and transmission

The conventional method of operating control and process recording by papery manufacturing travellers can be terminated through the exploitation and implementation of data and information (Parts No., Serial No., Operating Steps, Material Batch No., Other Process Parameters, and so on) recording under the current system, the area of which will be of great business and industry values.

7.7.4 Prospect

The future composite repairing medium should develop towards a high technique assisted (augmented reality), integrated, automatic, intelligent and one-stop industry solution.

Lufthansa Technik's CAIRe project of an integrated aircraft composite repairing systems which synthesises calculation, material removal, surface activation and patch scarfing into one platform [43] and Boeing's patent in 'Automated production and installation of patches for reworking structures' are a couple of precursors within the composite repairing industry. These triumphant antecedents demonstrated the industry feasibility to develop an integration repairing system.

Nevertheless, it is infrequent to seek out a composite repairing platform which are both integrated and collaborated with augmented reality technique.

Thus, an augmented reality-based synthetical, functional, automatic system covering the complete process realisation and instruction from repair design and calculation, patch installation until curing fabrication owning feathers and function as follows will be exclusive, promising and of business interest.

Repair solution tailoring: The system is of intelligent enough to tailor reparation solutions based on different scenarios of defects, such as doubler or scarf, hard or soft, bond or bolt.

Material removal instruction, implementation and inspection: Robotic milling integrated with augmented reality visualising and interactive technique to realise the instruction to robot to position cutters as well as sensing and inspecting the correctness of the workpiece have been performed.

Surface treatment method selection and implementation: Intelligent selection surface treatment method based on different type of defects, materials, patch and repair solutions.

Patch installation and bagging instruction: This is what has been implemented in the current project.

Curing instruction, monitoring and control: Assisted with portable cleanroom, heating blanket, augmented reality visualisation technique, the instruction, monitoring and control on processing parameters as curing time, curing temperature and curing pressure can be realised.

7.8 Conclusion

The current individual research project accomplished an attempt on applying augmented reality technique to realise a real-time instruction of aircraft composite repairing process. Patch installation and bagging process were selected as the target process to be pitched into applet design. A previous Cranfield composite repair case was selected as a prototype to accommodate system's concept design and as the foothold to generate system's operating flow. Bagging scenarios were designed based on this specific composite repairing scenario. Digital mock-ups to indicate different implementing materials and facilities were created. User interface was designed and built up. User and system interaction were realised through creating interactive buttons and endowing trigger event to the related buttons. The blending between digital mock-ups and physical panels was realised through pre-set image target recognition. Dual output terminals (Tablet screen and Microsoft HoloLens) were realised. Functionality validation was executed through putting use into CIRJ training course. Effect validation route was designed and further validation suggestion was made.

The developed applet contributes on exploiting routes to employing augmented reality technique into commercial aircraft maintenance, repair and overhaul industry, and exhibiting the potential capability and space to be pitched into industrial environment to remit the increasing workloads brought about from continuous incremental adoption of composite material in airplane manufacturing through an efficiency, reliable and reproductive platform. The video of a demonstration of the technology is provided via https://streaming.cranfield.ac.uk/Watch/Xb62Awi5.

References

[1] Ghobadi A. 'Common type of damages in composites and their inspections. *World Journal of Mechanics*. 2017;07(7):24–33.

[2] Daniel I.M., Wooh S.C. *Defect and damage characterization in composite materials*. Evanston, IL: Northwestern University; 1987.

[3] Liu Y., Zhang X., Lemanski S., Nezhad H.Y., Ayre D. 'Experimental and numerical study of process-induced defects and their effect on fatigue debonding in composite joints'. *International Journal of Fatigue*. 2019;125(August):47–57.

[4] Nezhad H.Y., Stratakis D., Ayre D., Addepalli S., Zhao Y. 'Mechanical performance of composite bonded joints in the presence of localised process-induced zero-thickness defects'. *Procedia Manufacturing*. 2018;16(5):91–8.

[5] Liu Y., Lemanski S., Zhang X., Ayre D., Nezhad H.Y., Yazdani Nezhad H. 'A finite element study of fatigue crack propagation in single LAP bonded joint with process-induced disbond'. *International Journal of Adhesion and Adhesives*. 2018;87:164–72.

[6] Deane S., Avdelidis N.P., Ibarra-Castanedo C., *et al*. 'Application of NDT thermographic imaging of aerospace structures'. *Infrared Physics & Technology*. 2019;97(3):456–66.

[7] Sirikham A., Zhao Y., Nezhad H.Y., Du W., Roy R. 'Estimation of damage thickness in fiber-reinforced composites using pulsed thermography'. *IEEE Transactions on Industrial Informatics*. 2019;15(1):445–53.

[8] Yazdani Nezhad H., Zhao Y., Liddel P.D., Marchante V., Roy R. 'A novel process-linked assembly failure model for adhesively bonded composite structures'. *CIRP Annals*. 2017;66(1):29–32.

[9] Bhanushali R., Ayre D., Nezhad H.Y. 'Tensile response of adhesively bonded composite-to-composite single-lap joints in the presence of bond deficiency'. *Procedia CIRP*. 2017;59(2):139–43.

[10] Jeongguk K., Liaw P.Ketc. 'The NDE analysis of tension behavior in Nicalon/SiC ceramic matrix composites'. *Journal of the Minerals*. 2003;55:1–13.

[11] Smith R.A. 'Composite defects and their detection'. *Materials Science and Engineering*. 2020;**3**. Available from https://www.eolss.net/Sample-Chapters/C05/E6-36-04-03.pdf

[12] Hallett S. *Modelling of defects and failure in composites*. Bristol: University of Bristol; 2013.

[13] Soutis C., Hu F.Z. 'Design and performance of bonded patch repairs of composite structures'. *Proceedings of the Institution of Mechanical Engineers, Part G: Journal of Aerospace Engineering*. 1997;211(4):263–71.

[14] Katnam K.B., Da Silva L.F.M., Young T.M. 'Bonded repair of composite aircraft structures: a review of scientific challenges and opportunities'. *Progress in Aerospace Sciences*. 2013;61(2013):26–42.

[15] Armstrong K.B., Cole W, W., BevanG. *Care and Repair of Advanced Composites*. Vol. 2005. London: SAE International; 2005.

[16] Wang C.H., Gunnion A.J. 'Optimum shapes of Scarf repairs'. *Composites Part A: Applied Science and Manufacturing*. 2009;40(9):1407–18.

[17] Waite D.S. *Slides of bonded structure and repairs – a regulators perspective*; 2018.

[18] Da Silva L.F.M., Öchsner A., Adams R.D. *Handbook of adhesion technology*. Heidelberg: Springer-Verlag; 2011.

[19] Baldan A. 'Adhesively-bonded joints and repairs in metallic alloys, polymers and composite materials: adhesives, adhesion theories and surface pretreatment'. *Journal of Materials Science*. 2004;39(1):1–49.

[20] Fourche G. 'An overview of the basic aspects of polymer adhesion. Part II: application to surface treatments'. *Polymer Engineering and Science*. 1995;35(12):968–75.

[21] Wingfield J.R.J. 'Treatment of composite surfaces for adhesive bonding'. *International Journal of Adhesion and Adhesives*. 1993;13(3):151–6.

[22] Banea M.D., da Silva L.F.M. '2009 adhesively bonded joints in composite materials: an overview'. *Journal of Materials: Design and Applications*. 2009;223(1):1–18.

[23] Kendall K. 'The adhesion and surface energy of elastic solids'. *Journal of Physics D: Applied Physics*. 1971;4(8):1186–95.

[24] Da Silva L.F.M., Öchsner A. *Handbook of adhesion technology*. Heidelberg: Springer-Verlag; 2011.

[25] Chester R. Roberts JD. Void minimization in adhesive joints. *International Journal of Adhesion and Adhesives*. 1989;9(3):129–38.

[26] Tzetzis D., Hogg P.J. 'The influence of surface morphology on the interfacial adhesion and fracture behavior of vacuum infused carbon fiber reinforced polymeric repairs'. *Polymer Composites*. 2008;29(1):92–108.

[27] da Silva L.F.M., Adams R.D., Gibbs M. 'Manufacture of adhesive joints and bulk specimens with high-temperature adhesives'. *International Journal of Adhesion and Adhesives*. 2004;24(1):69–83.

[28] Newsom C.M. *Portable Vacuum Device and Method of Using Same*. US Patent 4554,036; 1985.

[29] Kociemba F., Soro D. *Equipment for On-Site Repair of a Composite Structure with a Damaged Zone and Corresponding Method*. US Patent 6206,067 B1; 2001.

[30] Guo Z., Du S., Zhang B. 'Temperature field of thick thermoset composite laminates during cure process'. *Composites Science and Technology*. 2005;65(3-4):517–23.

[31] Elaldi F., Elaldi P. 'A study on curing processes and environmental effects for rapid composite repair'. *Journal of Reinforced Plastics and Composites*. 2011;30(9):749–55.

[32] Penn L.S., Chou R.C.T., Wang A.S.D., Binienda W.K. 'The effect of matrix shrinkage on damage accumulation in composites'. *Journal of Composite Materials*. 1989;23(23):570–86.

[33] Djokic D., Johnston A., Rogers A., Lee-Sullivan P., Mrad N. *Residual Stress*; 2002.

[34] Whittingham B., Baker A.A., Harman A., Bitton D. 'Micrographic studies on adhesively bonded scarf repairs to thick composite aircraft structure'. *Composites Part A: Applied Science and Manufacturing*. 2009;40(9):1419–32.

[35] *Vacuum Bagging Equipment and Techniques for Room-Temp Applications*. Available from https://www.fibreglast.com/product/vacuum-bagging-equipment-and-techniques-for-room-temp-applications/Learning_Center.

[36] *Augmented reality* [online]. Available from https://en.wikipedia.org/wiki/Augmented_reality

[37] *RealityTechnologies.com* [online]. Available from www.realitytechnologies.com/augmented-reality

[38] *What is augmented reality (AR) and how does it work* [online]. Available from https://thinkmobiles.com/blog/what-is-augmented-reality/

[39] *overair 's butterfly evtol prototype to integrate composites-intensive design* [online]. Available from https://www.compositesworld.com/news/spirit-aerosystems-deploys-airbus-augmented-reality-tool-for-composite-fuselage-brackets

[40] Sauer F., AliK. *Augmented reality visualization of ultrasound images: system description, calibration, and features*. New York, NY: IEEE; 2001.

[41] *camera-based ply placement cuts time, cost on complex rotorcraft parts [online]*. Available from https://www.compositesworld.com/articles/camera-based-ply-placement-cuts-time-cost-on-complex-rotorcraft-parts

[42] Adugna T. 'Dr. Henrik Schmutzler slides of automated machining and surface preparation'.

[44] Using DirectX11 in Unity 4 *Unity technologies* [online]. 2013. Available from https://en.wikipedia.org/wiki/Unity_Technologies [Accessed Feb 2013].

[45] Ismail K. *8 augmented reality companies changing the digital workplace* [online]. 2018. Available from https://www.cmswire.com/digital-workplace/8-augmented-reality-companies-changing-the-digital-workplace/ [Accessed Apr 2019].

Chapter 8

3D printing of multi-material polymer composite systems

Feroz Khan[1] and Hamed Yazdani Nezhad[1]

This report overviews the 6-month address into building a multiphysics application to study the behaviour of multi-functional composites in 3D printing systems. Considering the huge number of 3D printing systems, this project lies spotlight on Fused Deposition Modelling (FDM), due to its accessibility and simplicity for embedding reinforcement filler. The study is built in COMSOL Multiphysics 5.6. It performs a non-Newtonian study on the flow of molten polymer through an E3D v6 HotEnd assembly. The study was built with Acrylonitrile butadiene styrene (ABS) as a validation case. The rest of the model being parametric and simple to simulate other isotropic, nano-particulate reinforcement in a polymer matrix. Another study allows the capability to estimate some thermo-physical properties to aid the development of examining select specimens for the flow study. Results include temperature distribution and phase shear plots to correctly identify best operating procedure for select specimens. Finally, details of possible future work have been noted.

8.1 Introduction

The aim of this project is to study existing AM techniques of multi-material polymer composite systems in order to aid in the development of a state-of-the-art laboratory scale 3D printing setup and procedure. Given the atmosphere regarding the COVID-19 pandemic during the summer of 2021, the project took a more computational approach utilising numerical analysis methods.

The project was underway over the course of 6 months, involving an intense literature review to identify and categorise the existing AM technologies and their capabilities. There are many types of 3D printing techniques all uniquely beneficial to a separate craft. Considering the scope of this project, as well as the short timeline, focussed intent lay on FDM style of printers. This technique is a type of

[1]Aeronautics and Aerospace Research Centre, Department of Mechanical Engineering and Aeronautics, City, University of London, London, UK

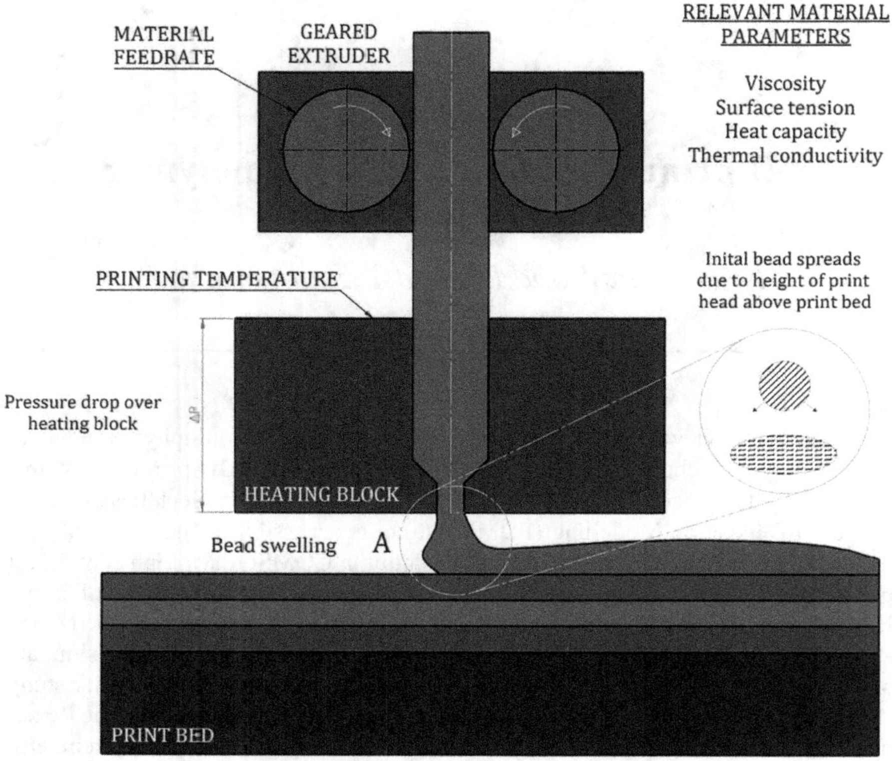

Figure 8.1 FDM process broken down. Overviewing material feeder, liquefier, layer building and consequent pressure drops [4].

material extrusion that deposits a 'wire' of polymer (hereafter referred to as filament) through a heated nozzle. FDM printing is one of the earliest types of AM and is heavily documented. Due to its depositing mechanism, it shows the most promise for multi-material composite printing without too much blue-sky thought or development of unique hardware.

Alongside this review, a COMSOL Multiphysics interactive application was developed. This COMSOL study models heat transfer from the heating block in the extruder to the material filament during the 3D printing procedure. The results display temperature distributions and the phase changes of the filament in the form of contour plots. This will be discussed thoroughly in the results below.

The model was developed under the material properties for the most common 3D printing filament, Acrylonitrile butadiene styrene (ABS). This was set as a validation case, as there is plenty of documentation and real-life experimentation on the expected outputs.

Building the COMSOL simulation application offered the most sophisticated way to parametrise this model, making it easy to pick up later with no set up\cite{comsol_how}.

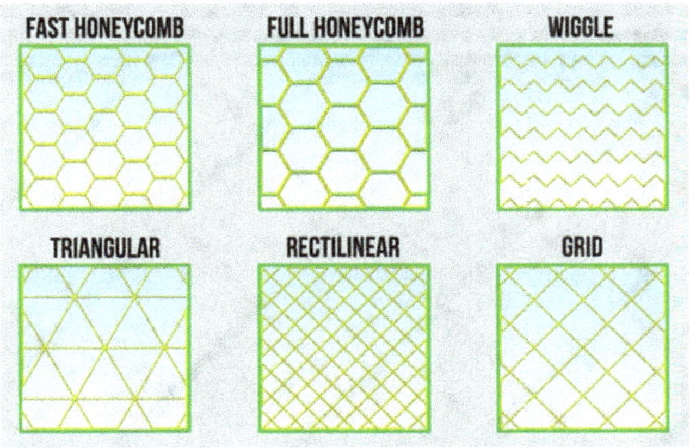

Figure 8.2 Various infill patterns to maintain a desired strength: weight

The input parameters can be edited to simulate a different material with new volumetric feed rates. These simulated materials make some assumptions that will be detailed fully in the following sections.

This model has the potential to show the relationship of temperature with feed rate of the polymer filament as well as develop an understanding for the relationship of viscosity and phase change of the molten polymer.

This report will overview a background into the current state-of-the-arts and how technologies are being developed in the realm of composite 3D printing. There will be a focus on the results developed in this study and an in-depth discussion of the approach taken to achieve these results.

8.2 Theoretical background

Additive manufacturing (AM) has brought a revolution in the world of manufacturing processes. Conventionally, components in real products utilise complex geometries. In manufacture, this results in the component undergoing multi-axis machining, a multi-part mould or several components whereas AM would be able to achieve this in one defined print. In short, AM removes hurdles of design complexity for manufacture, due to its inherent feature of building in layers. Some assemblies can even be 'printed in place', this refers to a mechanism printed in-situ that will work right away. In this sense, AM is ideal for bespoke manufacture and small batch production [1]. AM has also reduced production cost and increased the efficiency of the manufacturing process. Almost 90% of material waste is reduced by the AM process.

There are many types of 3D printing techniques all uniquely beneficial to a separate craft, e.g., Selective Laser Melting (SLM), Selective Laser Sintering (SLS), Laser Engineered Net Shaping (LENS), Electron Beam Melting (EBM),

Figure 8.3 *A rectilinear infill pattern, 20%*

Stereo-lithography (SLA), Digital Light Processing (DLP), Daylight Polymer Printing (DPP) and FDM.

SLM and SLS offer the capability to print in metal, ceramic and polymer. It utilises the *powder bed fusion* technique, whereby extremely fine powder is accurately levelled, whilst a precise laser sinter cross sections of the desired component. For each subsequent layer, fine powder is levelled over, repeating the sintering process. The outcome can build extremely intricate components, all possible due to the unsintered powder supporting the model throughout the printing process [2].

LENS introduces a unique concept where a high-power laser is used to melt metal powder that is deposited. The metal powder is injected with multiple converging powder streams fixed around the laser's focal point. The build platform moves in a raster pattern as the laser adjusts layer height. To aid in layer adhesion a substrate is applied for surface wetting.

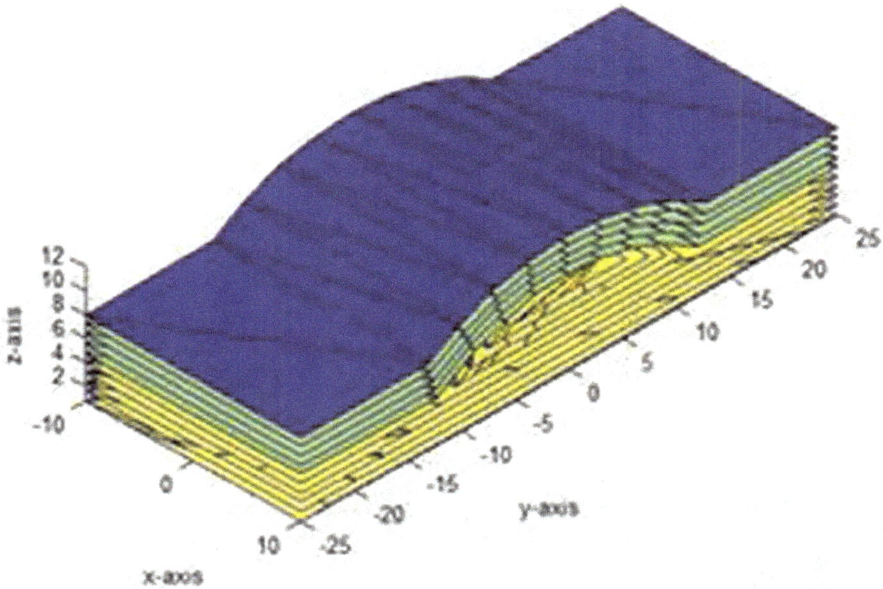

Figure 8.4 A cross section of an aerofoil undergoing smooth surface plane slicing [6]

SLA is a *vat photo-polymerisation* technique, where photo-polymer is cured using a precise ultraviolet laser. SLA (and other resin-style printers) is usually used in fine detail applications where the component will not undergo any stress. DLP and DPP also fall under this category of *vat photo-polymerisation*; however, they are not strictly limited to UV wavelengths. DPP, as the name suggests, can utilise visible light wavelengths, specifically, the blue light spectrum. This allows for the opportunity to use a simple LCD to cure photopolymer. Meaning XY resolution is resolved with the pixel density of the LCD. This is an example of a printing process that

Figure 8.5 Outlining the three main types of reinforcement that can be applied to a polymer matrix

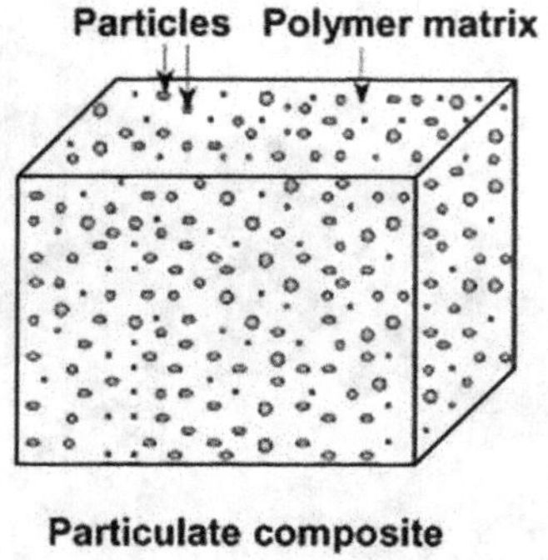

Figure 8.6 Particulate matrix + Polymer matrix = Effective matrix

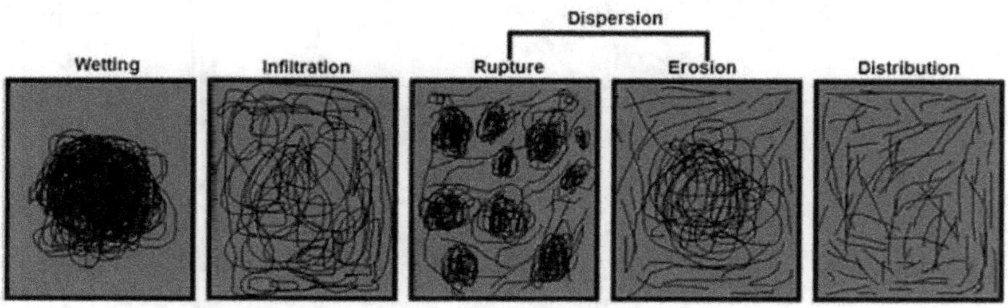

Figure 8.7 Mechanism of CNT dispersion [18]

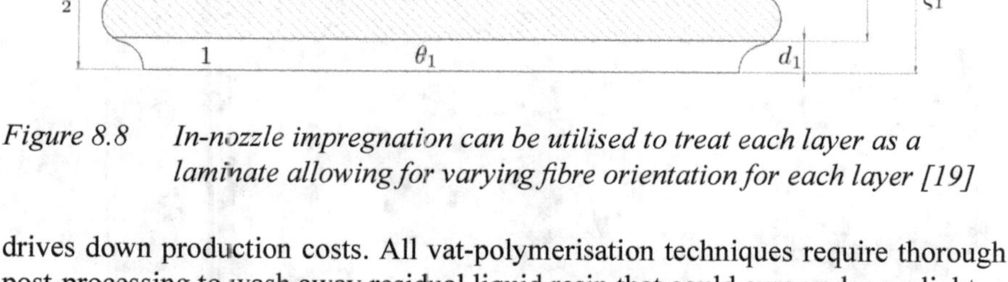

Figure 8.8 In-nozzle impregnation can be utilised to treat each layer as a laminate allowing for varying fibre orientation for each layer [19]

drives down production costs. All vat-polymerisation techniques require thorough post-processing to wash away residual liquid resin that could cure under sunlight.

FDM offers an opportunity for robust, strong components produced relatively quickly with little post processing. The availability of materials is advancing rapidly, and even more so, the availability of these systems. Most hobbyists already own an FDM 3D printer [3].

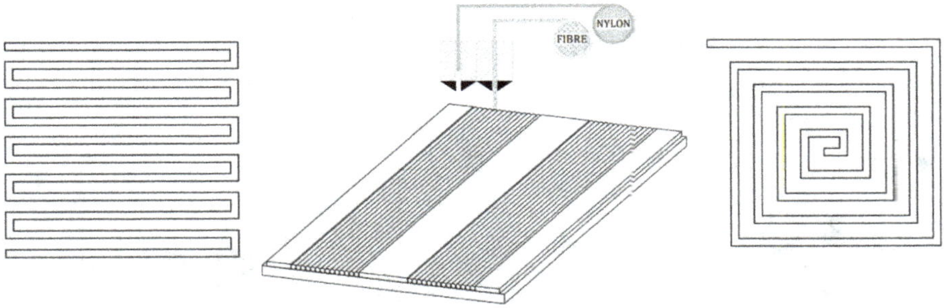

Figure 8.9 *Continuous fibre printing offers freedom to autonomously orientate fibrous lamina in optimum directions [20]*

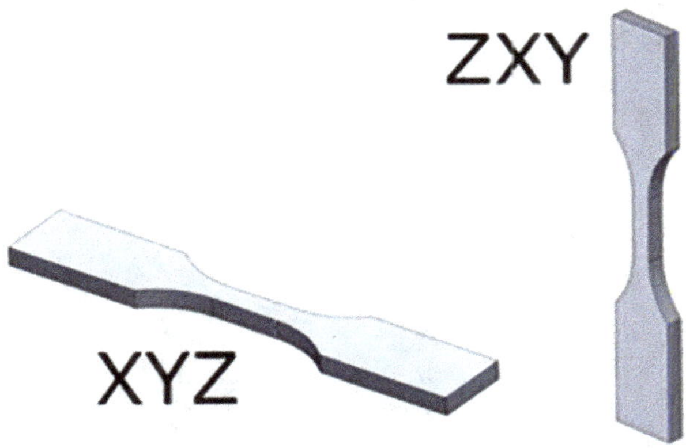

Figure 8.10 *Tensile test specimen printed in different orientations to achieve different anisotropic results [21]*

FDM is a form of material extrusion whereby solid polymer filament is passed through a heated nozzle (Figure 8.1). The molten filament is precisely deposited in a raster pattern on a build platform to form each layer of the component. The final product is produced by the successive layers. Resolution of this product is determined by the thickness of each layer. The most common layer thickness is ~100 microns. Researchers widely use *material extrusion*-based techniques due to their cost effectiveness, scalability and larger range of material processing ability [5].

From a prototyping perspective, 3D printing can be considered as tactile rendering. FDM systems are relatively simple machines. They consist of a gantry frame containing linear actuation for the 3-local axes. This system has just about reached a plateau in hardware advancement. Most can reach a sufficient resolution in XY coordinates and in Z layer height. Furthermore, most are closed loop systems.

Introduction of AM into the free open-source market has allowed consumers to develop software upgrades that improve printing capabilities and further reduce post

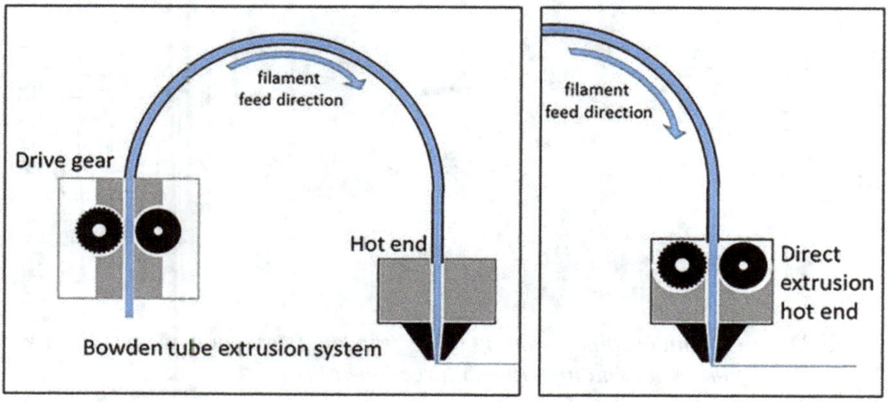

Figure 8.11 Bowden and direct drive schematic [26]

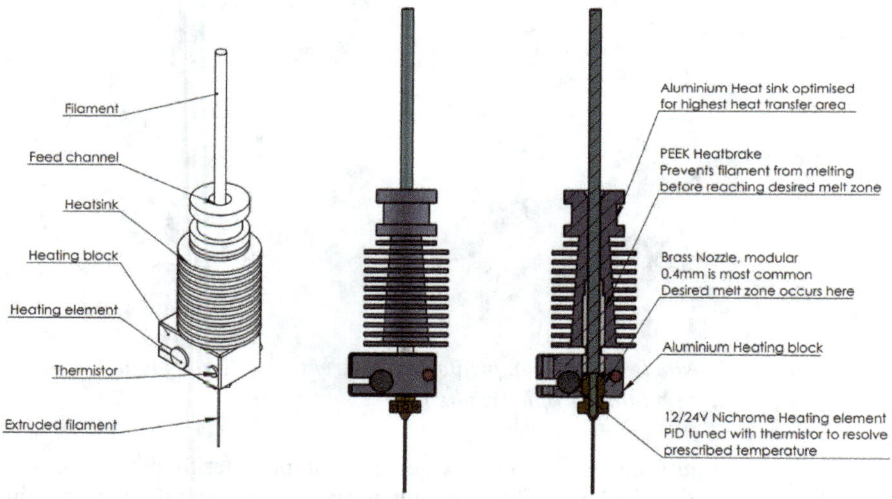

Figure 8.12 E3D v6 HotEnd schematic, made in SolidWorks

processing needs. To tackle warping, shrinkage and delamination software integrations can simplify the geometry.

Rafts aid in layer adhesion; this reduces the likeliness of corners warping or curling.

Infill patterns can reduce weight while providing the same strength, and less cooling polymer means less shrinkage, examples of infill patterns can be seen in Figure 8.2. Figure 8.3 illustrates the raster patterns for two separate layers, pink first, then black. This is laid up in a 20% infill, reducing the overall weight and cost of material, while still providing sufficient strength gains.

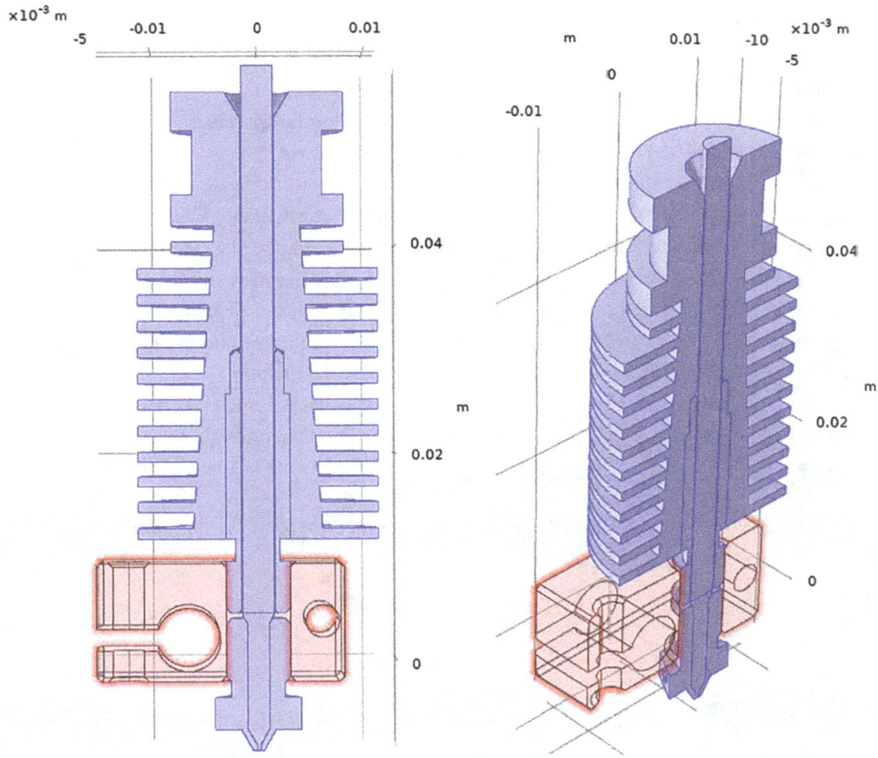

Figure 8.13 Heating block assumption

Non-planar printing (Figure 8.4) is a recent addition that allows for the opportunity for top surface smoothing whereby the first layers are printed normally, but the final layers are contoured to achieve a smooth surface, mitigating the staircase effect usually seen in 3D printing. This can subdue delamination issues.

These are just software implementations. The most significant lacking factor in 3D printing lies in the materials and the characteristics that can be amplified. Properties of the final product depend upon the properties of the material of the filament. The FDM process uses thermoplastics as the main polymer matrix. This means failed parts can be shredded and re-extruded into filament relatively easily. This also implies it is relatively simple to embed reinforcements into the polymer matrix in the form of fibres, whiskers and particles to achieve a multi-material composite. These reinforcements can also offer multi-functional capabilities. FDM is commonly used in 3D printing of fibre reinforced polymer composites [7].

▾ Parameters

Name	Expression	Value	Description
T_melt	105[degC]	378.15 K	Melting temperature
lat_heat	207[kJ/kg]	2.07E5 J/kg	Latent heat
dT_half	75[K]	75 K	Temperature transition zone half width
k_fil_solid	0.3[W/(m*K)]	0.3 W/(m·K)	Heat conductivity filament solid
Cp_fil_solid	1200[J/(kg*K)]	1200 J/(kg·K)	Heat capacity filament solid
rho_fil_solid	1050[kg/m^3]	1050 kg/m³	Density filament solid
k_fil_liquid	0.2[W/(m*K)]	0.2 W/(m·K)	Heat conductivity filament melt
Cp_fil_liq...	1797.6[J/(kg*K)]	1797.6 J/(kg·K)	Heat capacity filamentliquid
rho_fil_liq...	1050[kg/m^3]	1050 kg/m³	Density filament liquid
Q	13.5[mm^3/s]	1.35E-8 m³/s	Filament volume flow rate
T_amb	20[degC]	293.15 K	Ambient temperature
T_block	240[degC]	513.15 K	Temperature of the heating block

Figure 8.14 Filament parameters loaded with ABS properties

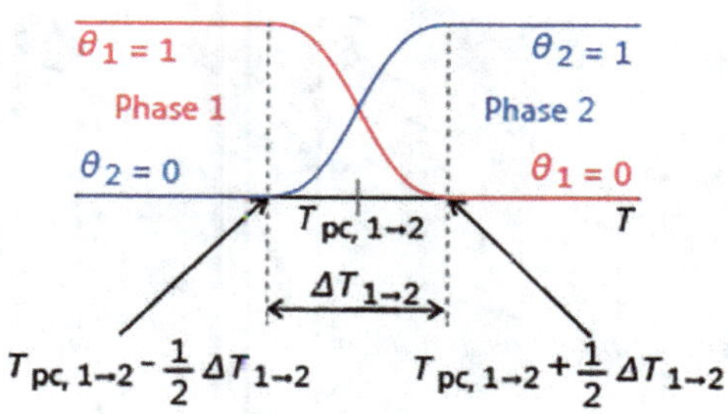

$$T_{pc,1\to 2} = T_{melt}$$

$$\Delta T_{1\to 2} = dT_{half} \times 2$$

$$L_{1\to 2} = h_{melt}$$

Figure 8.15 Phase change material

8.2.1 Material science

This section will address the advancement in material science in the realm of FDM printing. Starting with pure polymer, the fundamentals of polymer composites and approaching multi-functionality.

Some of the most common polymers used as a feedstock in FDM are ABS, polylactic acid (PLA), polycaprolactone (PCL) and polyamide – commonly known as nylon (PA).

ABS has been used extensively in all sorts of plastic manufacture; this results in plenty of documentation in the behaviour of this polymer. It is a very suitable material for 3D printing because of its high strength, moderate detail performance and functional dimensional integrity. However, it has limitations due to which good mechanical and thermal properties of 3D printed components cannot be obtained by using pure ABS. It has very poor resistance to heat and ozone. These can develop shrinkage, warping and deformation.

PLA is a naturally existing polymer that is primarily extracted from the starch of plants. It has excellent properties of biodegradability and biocompatibility. It also has a low viscosity and melting point. These characteristics of PLA make it a very suitable material for FDM but issues arise due to its high brittleness.

PCL is non-toxic material with very good bio-compatible and biodegradable properties, as such it is commonly used as a drug control release system and biodegradable products in medical sciences. It has the ability to store and restore deformation. These properties make it a suitable material for 3D printing, but its mechanical and thermal properties are lacking [8].

PA is an incredibly strong and robust material. Sadly, these properties do not get advertised when printing in PA due to their overcritical behaviour. It has a low coefficient of friction and is excellent for printing small makeshift bushings or thin flexible parts. However, it requires a heated enclosure to print large and prefers to be printed solid as opposed to wide spaced infill usually employed by 3D printers. This makes PA likely to warp and shrink when printing large volumes.

8.2.2 Polymer blends

The most fundamental polymer composite is a polymer blend. This as the name suggests, involves mixing various polymers, not unlike a metal alloy. Experimentation shows that these blends can be engineered to form superior flame retardant materials, and increase flexural and elastic properties. However, this is at the cost of longevity and mechanical strength [9].

Glassy ABS	**Value**	**Unit**
Thermal conductivity	0.3	W/(m.K)
Density	1050	kg/m^3
Ratio of specific heats	1	1
Rubbery ABS	**Value**	**Unit**
Thermal conductivity	0.2	W/(m.K)
Density	1050	kg/m^3
Ratio of specific heats	1	1

Figure 8.16 Thermal breakdown of filament

Description	Data Used
Processing temperature	378[K]
Temperature transition zone half width	75[K]
Heat of glass transition	207[kJ/kg]
Heat capacity at constant pressure, glassy state	1200 [J/(kg.K]
Heat capacity at constant pressure, rubbery state	1797.6
Ambient temperature	300 [K]
Melt inlet temperature	378 [K]
Casting speed	1.0 [mm/s]
Heat transfer coefficient, ABS	2000 [W/(m^2.K]
Heat transfer coefficient, air	10 [W/(m^2.K)]
Surface emissivity	0.95

Figure 8.17 Thermal breakdown of filament

8.2.3 *Fillers and reinforcements*

A plethora of polymers is available for FDM 3D printing. However, this does not necessarily imply functional printing and performance. The limiting factors here are the material characteristics. With the right reinforcement, desired mechanical and thermal properties can be stabilised. This allows for the possibility of multi-functional composites. Those within immediate reach are structural reinforcement,

$$\rho = \theta_1\rho_1 + \theta_2\rho_2$$

$$C_p = \frac{1}{\rho}(\theta_1\rho_1 C_{p,1} + \theta_2\rho_2 C_{p,2}) + L_{1\to 2}\frac{\partial\alpha_m}{\partial T}$$

$$\alpha_m = \frac{1}{2}\frac{\theta_2\rho_2 - \theta_1\rho_1}{\theta_1\rho_1 + \theta_2\rho_2}$$

$$k = \theta_1 k_1 + \theta_2 k_2$$

$$\theta_1 + \theta_2 = 1$$

Figure 8.18 Phase change equations

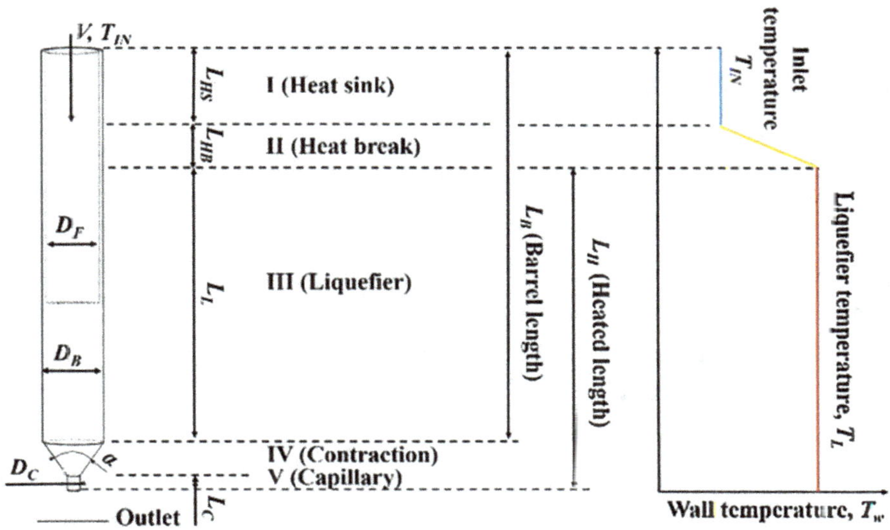

Figure 8.19 Thermal breakdown of filament [28]

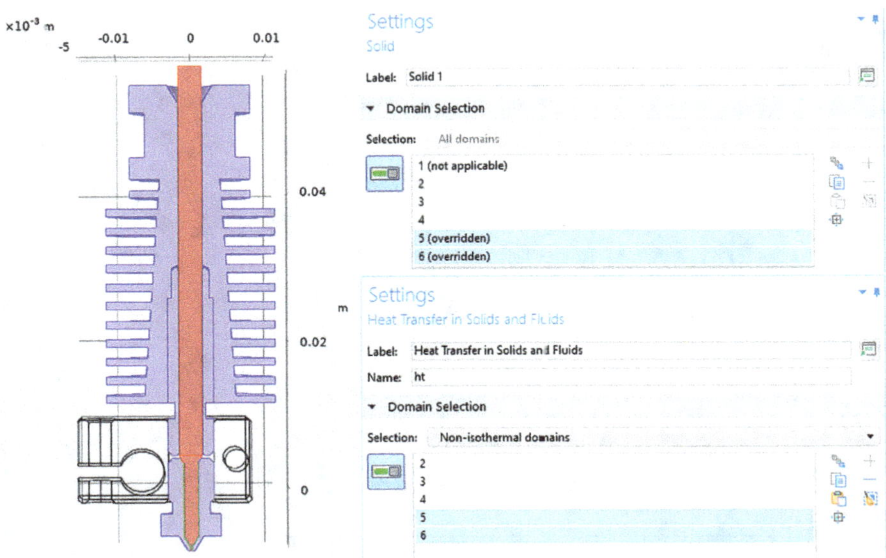

Figure 8.20 Settings for heat transfer

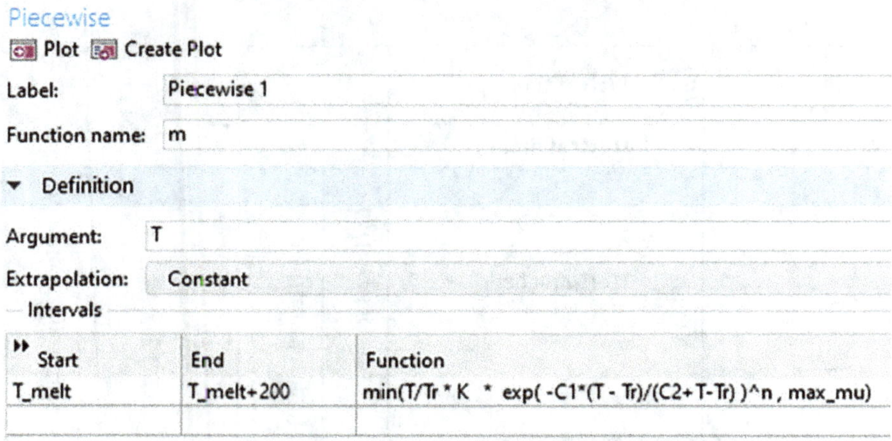

Figure 8.21 Piecewise function for m(T)

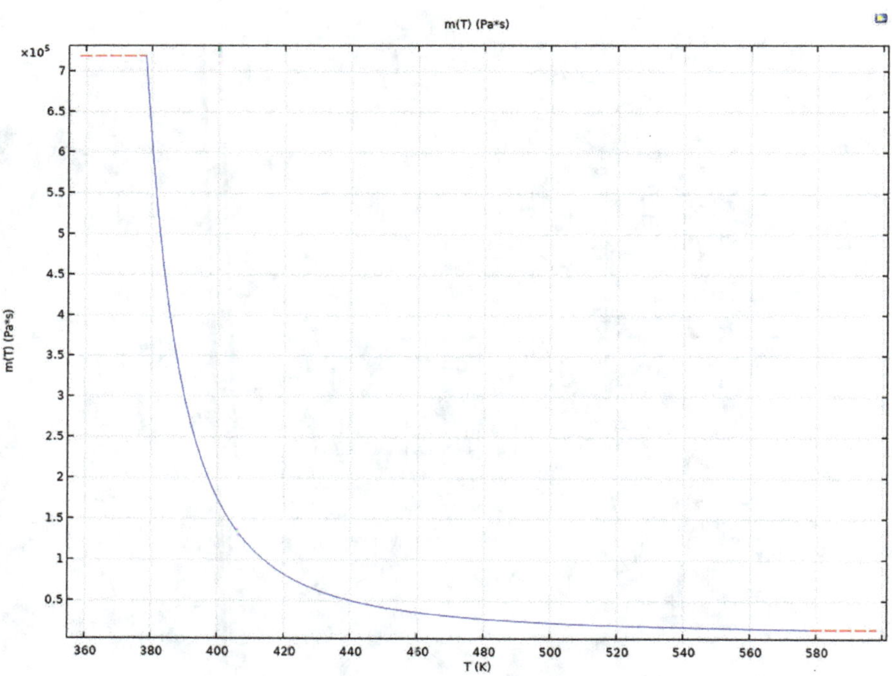

Figure 8.22 Computed values of fluid consistency coefficient

Figure 8.23 Nojac() operator

thermal conductivity and electric conductivity. Properties of newly developed composites depend upon filler material added in polymers, and the ratio of these matters greatly [10]. There are various sub-divisions of fillers and reinforcement types, as illustrated in Figure 8.5. These will be elaborated in the following sections.

8.2.4 Particulates

There are several types of polymer composites. Particulates refer to those that have a percentage of filler embedded in the polymer matrix (Figure 8.6). This can be done in several ways. For particle reinforced composites the filler is made up of micro/nano particles that are finely distributed in the polymer matrix. Experimentation shows that smaller particle size improves thermo-mechanical properties far more significantly. Particle reinforced composites have been shown to exhibit conductive properties and improved printability, often stabilising the polymer matrix that attempts to shrink [9].

On the microscopic scale, polymers are made up of cross links and interconnected networks, but they have low deformation resistance and modulus of elasticity. Fillers made up of different materials like ceramics, metals, natural and synthetic fibres have been used by researchers, depending upon the properties that are required based on their specification [11].

In the manufacturing of composites, it is very important to determine the interface between the polymer matrix and the reinforcement. When there is a strong interface, it will improve the strength of composite and aid against debonding and pulling. There are also some other filler parameters that have a significant effect on strength of composites such as size, shape, orientation and distribution [11].

Particle filler materials have become very common in polymer production even before the conception of 3D printing. But there are still some limitations. Sometimes desired quenching properties, flow properties, thickness of layer and voxelization size are not achieved by using particle filler materials. They can also clog up the heated nozzle. In some poor matchups, it increased the likelihood of delamination [9].

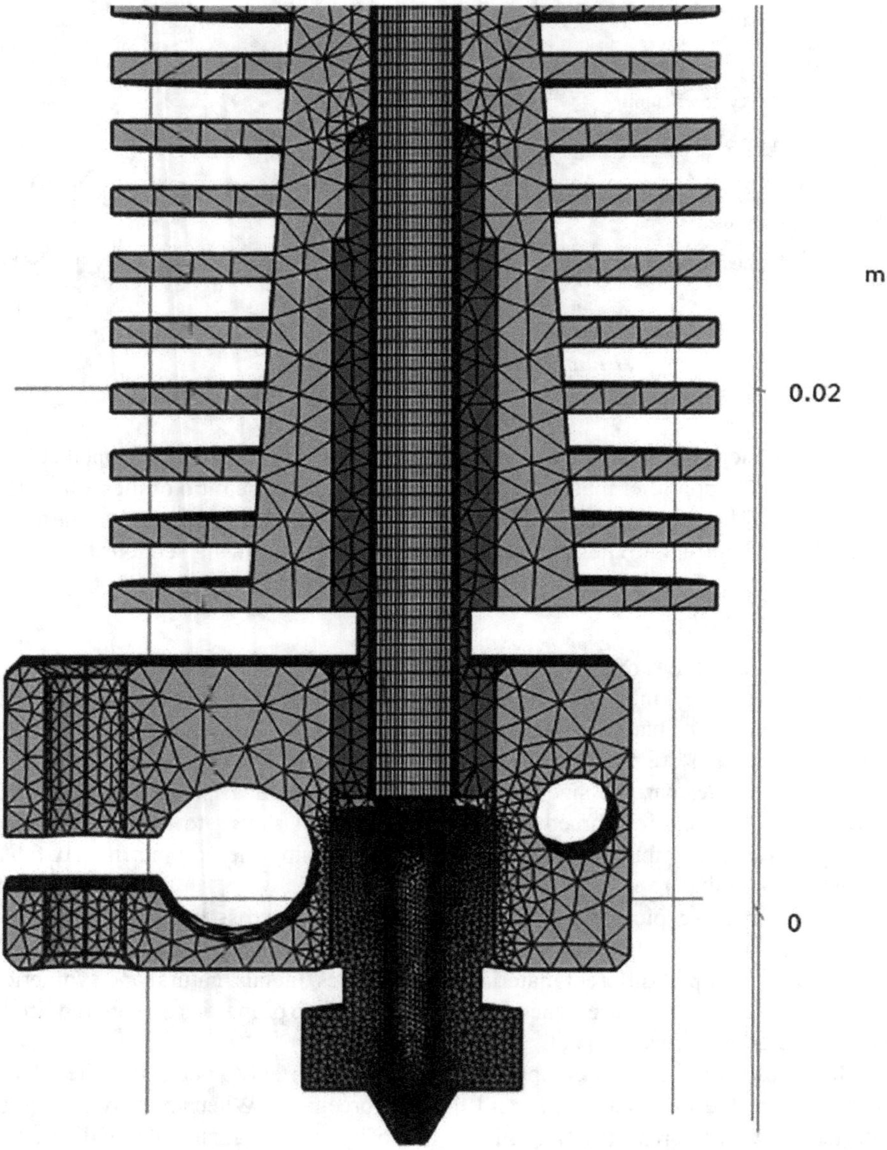

Figure 8.24 Graduating finer mesh

For this reason, it is very important to be mindful of the size of filler. The nozzle can also be clogged by the aggregation of particulates so the concentration of filler material should also be kept in mind. Hard metal will in fact wear away the soft brass nozzle over time, affecting tolerances. Furthermore, it is important to keep the balance; if the density and size of filler material are limited too greatly then desired

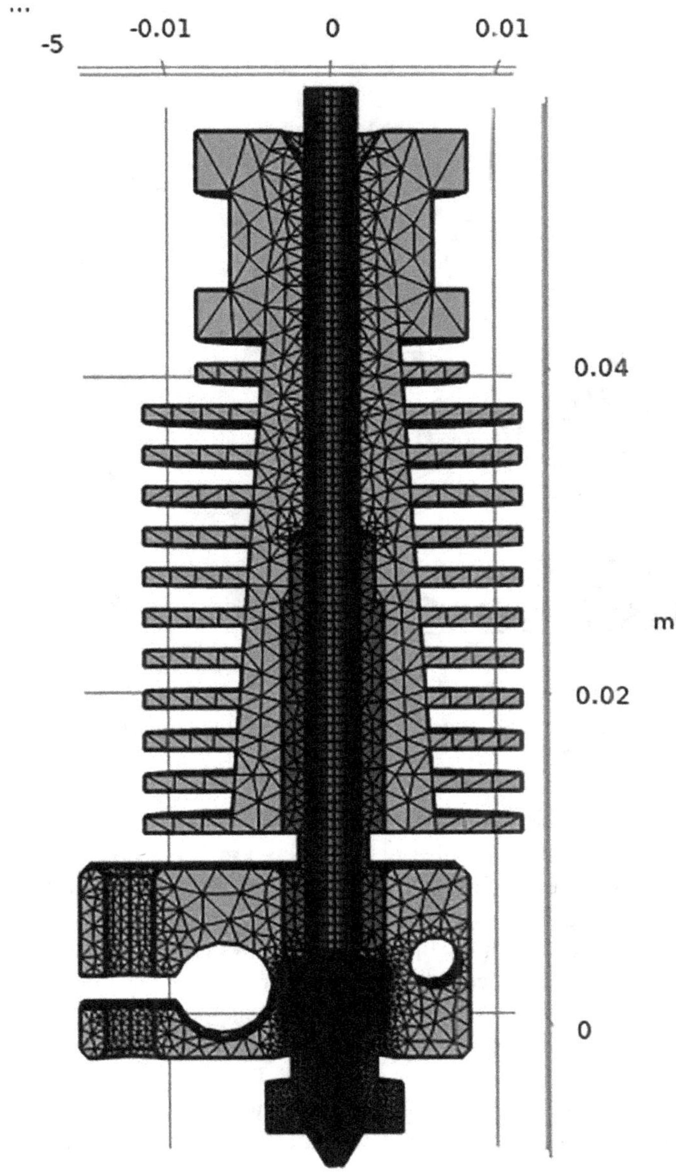

Figure 8.25 Free tetrahedral mesh

properties are not achieved. In order to deal with this problem, researchers are now shifting towards nanoparticles for example, Nano-silica, nano-clay and graphite nano-platelets [9].

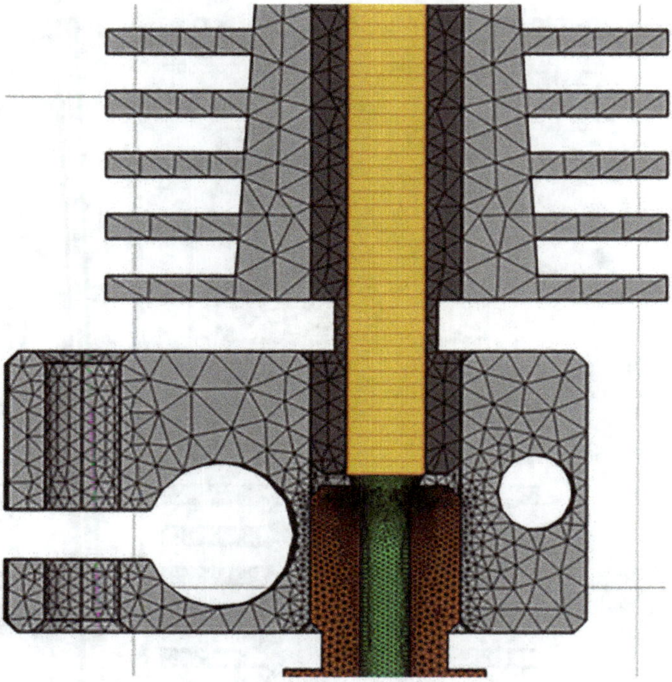

Figure 8.26 Swept mesh distribution for solid filament with corner refinement

Inorganic particles are the most common and widely used filler materials to enhance desired properties of polymer matrix used in 3D printing. They are cheap and can be easily mixed to form robust effective matrices. Different types of inorganic particles like metallic, carbon and ceramic particles can be used as filler material with excellent functionality and thermophysical properties. For example, reinforcement with iron and copper particles improves storage modulus and tensile strength; reinforcement with aluminium and alumina (Al_2O_3) improves wear resistance; and addition of tungsten and ceramics enhances the dielectric permittivity. All these advancements in inorganic particle reinforced polymer composites made it possible to produce fully functional, load-bearing, and high-performance parts. Moreover, the addition of inorganic particles also enhances the shape stability of polymers [1].

8.2.5 Metallic polymer composites

It is shown in research that entrenched metallic particles are very effective and promising additives [1]. The size of the particles has a remarkable effect on strength of the reinforced polymer. They exhibit the best mechanical properties when particle size is 40−100 μm. Consistency in the size of a metal particle is very important [9].

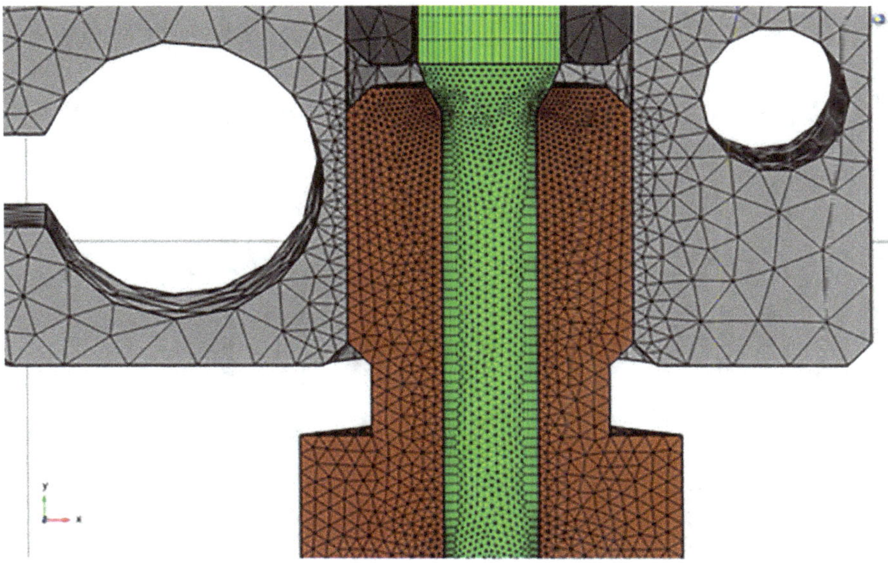

Figure 8.27 Extra fine mesh and non-slip condition at nozzle inlet

Iron/PA is a very common particle reinforced polymer composite. These composites have the highest value of tensile strength and tensile modulus when 30% vol of iron is added in PA. The highest values of tensile modulus and tensile strength of iron/nylon composites are 54.52 [MPa] and 3.87 [MPa], respectively [9].

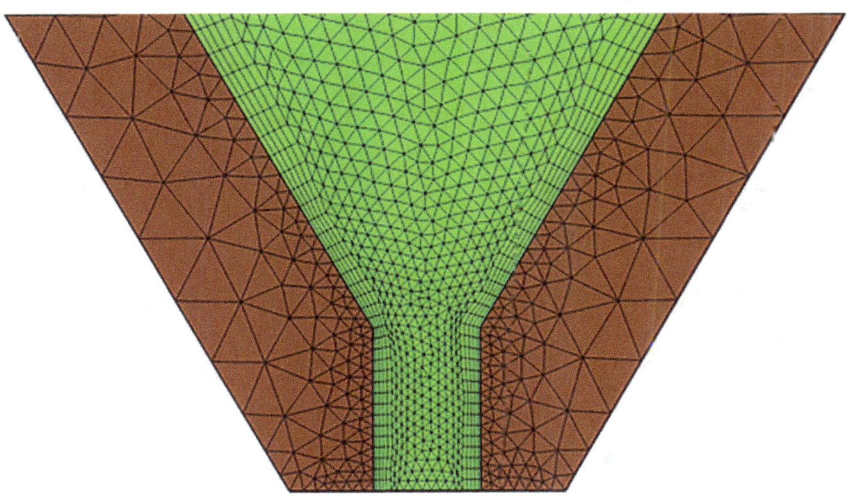

Figure 8.28 Extremely fine mesh approaching outlet

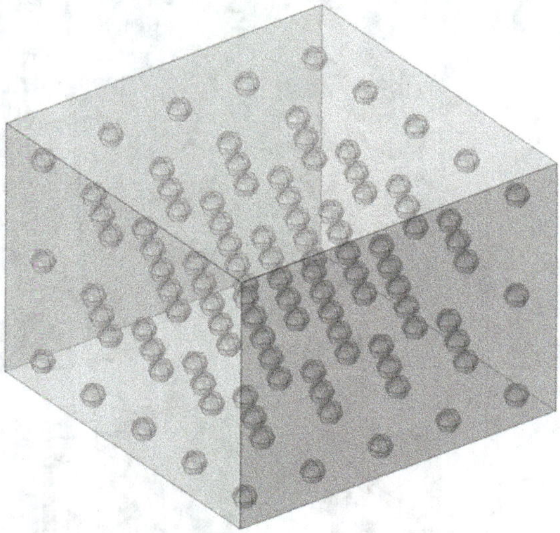

Figure 8.29 Estimating thermo-physical properties mesh

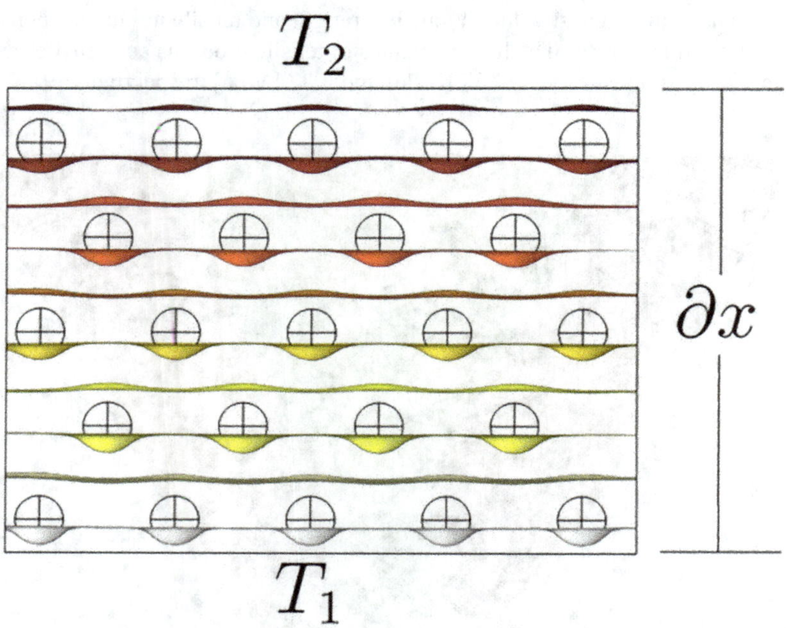

Figure 8.30 Thermal conductivity FEA, one boundary pair

	ρ	cp	k
Fe	7600	465	42.75
Cu	8945	385	394
Al	2690	900	201
TiO2	4010	690	7
Al2O3	3895	705	31.5

Figure 8.31 Chosen filler particulates, values provided from CES selector

Iron/ABS also produced satisfactory results. It is found that iron/ABS composites have maximum value of tensile strength of 12.8 [MPa] when iron particles are added by 24% vol [12]. The thermal conductivity of ABS is also found to be increased remarkably by the reinforcement of copper and iron particles [13].

▾ Parameters

▸▸ Name	Expression	Value	Description
k_ABS_solid	0.3[W/(m*K)]	0.3 W/(m·K)	Heat conductivity ABS solid
Cp_ABS_solid	1200[J/(kg*K)]	1200 J/(kg·K)	Heat capacity ABS solid
rho_ABS_solid	1050[kg/m^3]	1050 kg/m³	Density filament ABS solid
k_ABS_liquid	0.2[W/(m*K)]	0.2 W/(m·K)	Heat conductivity ABS liquid
Cp_ABS_liquid	1797.6[J/(kg*K)]	1797.6 J/(kg·K)	Heat capacity ABS liquid
rho_ABS_liquid	1050[kg/m^3]	1050 kg/m³	Density ABS liquid
T_melt	105[degC]	378.15 K	Melting temperature
lat_heat	207[kJ/kg]	2.07E5 J/kg	Latent heat
dT_half	75[K]	75 K	Temperature transition zone half width
k_filler	42.75[W/(m*K)]	42.75 W/(m·K)	Heat conductivity filler
Cp_filler	465[J/(kg*K)]	465 J/(kg·K)	Heat capacity filler
rho_filler	7600[kg/m^3]	7600 kg/m³	Density filler
d_filler	10[um]	1E-5 m	Mean particle diameter filler
perc_mass_filler	10[%]	0.1	Filler mass percentage

Figure 8.32 Applying filler parameters

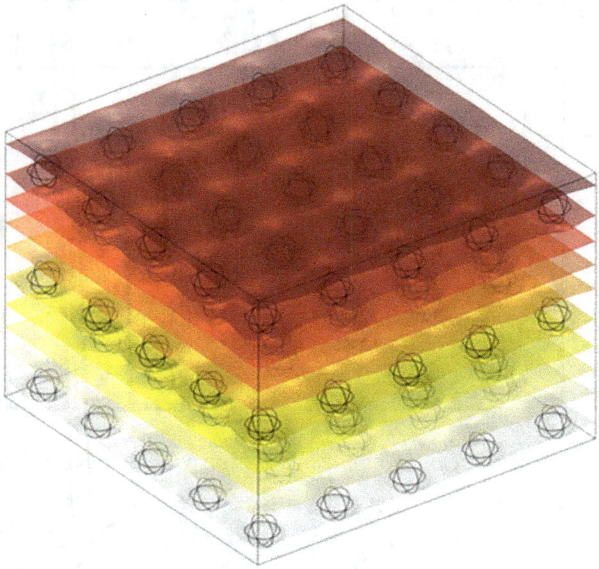

Figure 8.33 *This method provides ρ, cp and k for solid and liquid phases*

8.2.6 Ceramic polymer composites

Due to their excellent mechanical, thermal and biological properties, ceramic particles are also used as filler materials. Ceramic polymer composites are classified as one of the most important and influential types of 3D printable composites. A recent addition is PA12/ZrO_2/HAP. Here, PA12 polymer matrix is reinforced by zirconia (ZrO_2) and hydroxyapatite (HAP) particles. These reinforcements reduce flexibility and toughness of PA12 but increase the thermal stability, strength and stiffness. The degree of crystallinity of PA12 is also found to decrease. Due to these properties, PA12/ZrO_2/HAP is used to produce medical implants [14].

Reinforcement with Al_2O_3 yields poor mechanical performance while improving thermal conductivity vastly. The tensile strength of pure ABS is 33.96±1.74 [MPa], and the tensile strength of ABS/Al_2O_3 composites is found to be 28.8±2.62 [MPa]. This is because of voids present between lamina that presented as brittleness under stress. Further testing could be performed with lower filler concentrations or with the addition of a binding substrate [15].

A very useful reinforcement to consider is tungsten carbide (WC). The effective matrix of PA12/WC composites possesses enhanced mechanical and thermal properties. Young's modulus, tensile strength and elongation of PA12 polymer matrix are found to be increased from 700 to 1300 [MPa], 25–60 [MPa] and 260% to 700%, respectively [1].

Composite designation	Polymer matrix	Filler			Mixture SOLID			Mixture LIQUID		
		Type	Size (um)	Loading (%)	Thermal conductivity	Heat capacity	Density	Thermal conductivity	Heat capacity	Density
					(W/m K)	(J/kg K)	(kg/m3)	(W/m K)	(J/kg K)	(kg/m3)
O1	ABS	(-)	(-)	(-)	0.3	1200.000	1050.000	0.2	1797.600	1050.000
A1	ABS	ZnO	100	2	0.303339	1186.060	1067.352	0.202200	1771.708	1067.352
A2				5	0.308649	1165.150	1094.482	0.205828	1732.870	1094.482
A3				10	0.318376	1130.300	1142.899	0.212318	1668.140	1142.899
A4				20	0.341593	1060.600	1253.331	0.227870	1538.680	1253.831
B1	ABS	ZnO	45	2	0.303408	1186.060	1067.352	0.202285	1771.708	1067.352
B2				5	0.308695	1165.150	1094.482	0.205798	1732.870	1094.482
B3				10	0.318435	1130.300	1142.899	0.212326	1668.140	1142.899
B4				20	0.341595	1060.600	1253.831	0.227869	1538.680	1253.831
C1	ABS	Cu	10	5	0.305620	1159.250	1098.477	0.203754	1726.970	1098.477
C2				10	0.311845	1118.500	1151.646	0.207864	1656.340	1151.646
C3				20	0.326495	1037.000	1275.081	0.217654	1515.080	1275.081
C4				30	0.345516	955.500	1428.153	0.230336	1373.820	1428.153
C5				40	0.371550	874.000	1622.991	0.247716	1232.560	1622.991
D1	ABS	Cu	45	5	0.305671	1159.250	1098.477	0.203757	1726.970	1098.477
D2				10	0.311845	1118.500	1151.646	0.207866	1656.340	1151.646
D3				20	0.326506	1037.000	1275.081	0.217665	1515.080	1275.081
D4				30	0.345493	955.500	1428.153	0.230333	1373.820	1428.153
D5				40	0.371550	874.000	1622.991	0.247715	1232.560	1622.991
E1	ABS	Fe	45	5	0.306345	1163.250	1097.284	0.204262	1730.970	1097.284
E2				10	0.313429	1126.500	1149.028	0.209053	1664.340	1149.028
E3				20	0.330438	1053.000	1268.680	0.220443	1531.080	1268.680
E4				30	0.352452	979.500	1416.149	0.235235	1397.820	1416.149
E5				40	0.383416	906.000	1602.410	0.256065	1264.560	1602.410
F1	ABS	Al	25	5	0.318364	1185.000	1083.014	0.212311	1752.720	1083.014
F2				10	0.338938	1170.000	1118.171	0.226129	1707.840	1118.171
F3				20	0.391173	1140.000	1195.809	0.260928	1618.080	1195.809
F4				30	0.460916	1110.000	1285.032	0.307520	1528.320	1285.032
F5				40	0.541209	1080.000	1388.643	0.361166	1438.560	1388.643
G1	ABS	Alumina	30	5	0.312338	1175.250	1089.801	0.208299	1742.970	1089.801
G2				10	0.326147	1150.500	1132.738	0.217599	1688.340	1132.738
G3				20	0.359420	1101.000	1229.630	0.240020	1579.080	1229.630
G4				30	0.410560	1051.500	1344.649	0.274893	1469.820	1344.649
G5				40	0.466016	1002.000	1483.406	0.312164	1360.560	1483.406
H1	ABS	TiO2	45	5	0.310869	1174.500	1090.238	0.207553	1742.220	1090.238
H2				10	0.323016	1149.000	1133.683	0.216015	1686.840	1133.683
H3				20	0.351922	1098.000	1231.861	0.236190	1576.080	1231.861
H4				30	0.392990	1047.000	1348.655	0.265665	1465.320	1348.655
H5				40	0.441706	996.000	1489.915	0.299790	1354.560	1489.915

Figure 8.34 Parametric sweep of filler specimens under various percent loads and sizes

Another consideration is strontium titanate ($SrTiO_3$), commonly used as a ceramic filler to produce ABS/$SrTiO_3$ composite. The ultimate tensile strength of ABS decreases significantly due to the presence of voids [15].

More advanced composites include tricalcium phosphate, TCP ($Ca_3(PO4)_2$), embedded in a polypropylene (PP) matrix. The highest values of their compressive and tensile strength are 12.7 [MPa] and 19.8 [MPa], respectively. They are commonly used to produce bespoke medical parts thanks to 3D printing [16].

B3	ABS	ZnO	45	10
C3	ABS	Cu	10	20
E4	ABS	Fe	45	30
G5	ABS	Alumina	30	40
H2	ABS	TiO2	45	10

Figure 8.35 Specimens for examination

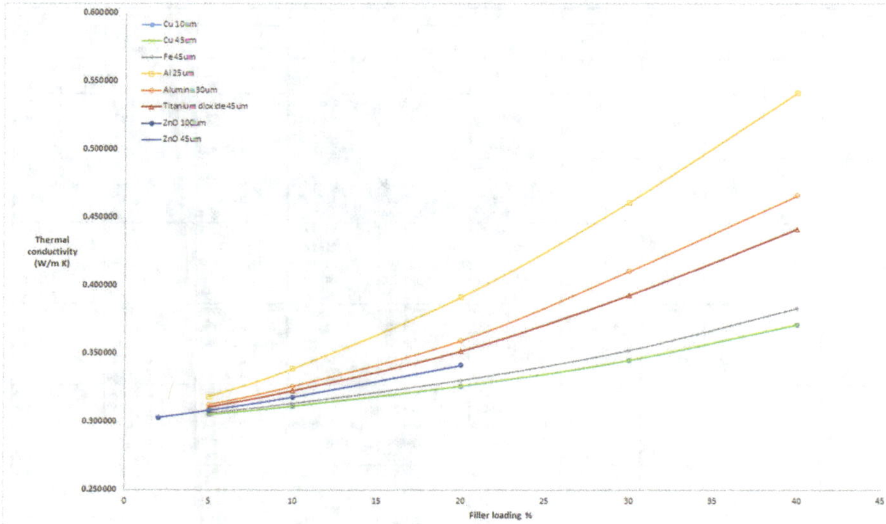

Figure 8.36 Thermal conductivity as a function of loading percentage

8.2.7 Carbon polymer composites

Various carbon particles (CP) like graphene, carbon black (CB), carbon nanotubes (CNT) and graphite usually show outstanding electrical, thermal and mechanical properties.

Warping is one of the major issues occurring while printing with pure polymer. Adding CP, 20% weight/centage ratio (wt.) in PA66 proved to provide much needed stability to nylon. Pure PA66 deviates around 0.1953% when printed pure. With the addition of CP, there is a reduction to 0.0928% [1].

Graphene has been used by many researchers as a reinforcing nanofiller for polymer-based composites and has become a topic of interest due to its unique properties. When 6% wt. of graphene was added to PA12 polymer, the modulus of elasticity increased by 7% in the direction of printing, and thermal conductivity increased by 51.4% [1].

CB reinforcement increases electrical conductivity and thermal stability. As such, they have potential application in the formation of sensors, electromagnetic shielding and 3D printed circuits. When CB particles are added to PP, it reduces its electrical resistivity. At 30% wt., CB/PP achieves an electrical resistivity of 2 Ω, making it a suitable option for 3D printable electrical circuits [17].

8.2.8 Fibres and whiskers

Fibres refer to long strands that are usually chemically dispersed amongst the matrices (Figure 8.7). In some cases, this can mean continuous fibres, but often the result is long discontinuous fibres entrenched within the polymer matrix. Whiskers allude to a halfway point between particulates and fibres. Here, the whiskers are in fact

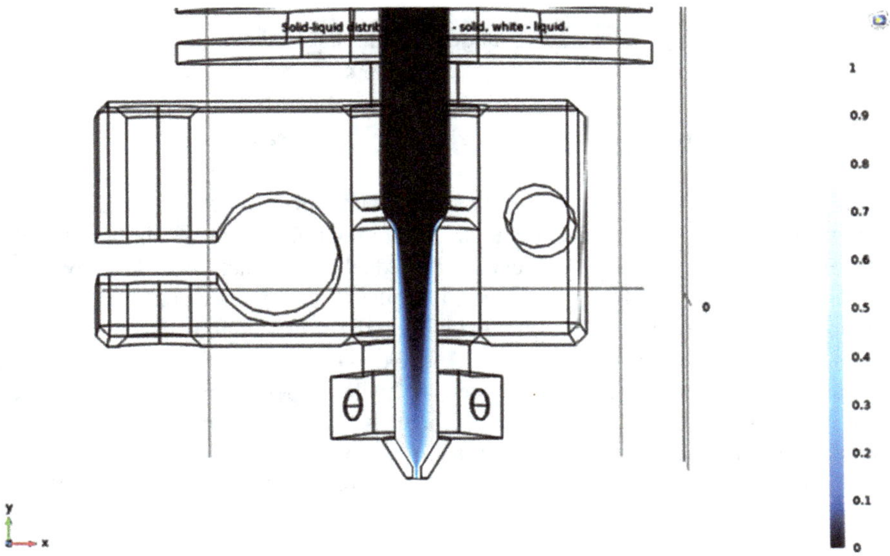

Figure 8.37 Phase share percentage

very short discontinuous fibres. Fibre reinforcement yields significant improvement in tensile strength and ductility. Many fibres reinforced polymers result in an overall reduction in thermal resistance, forming thermal bridges.

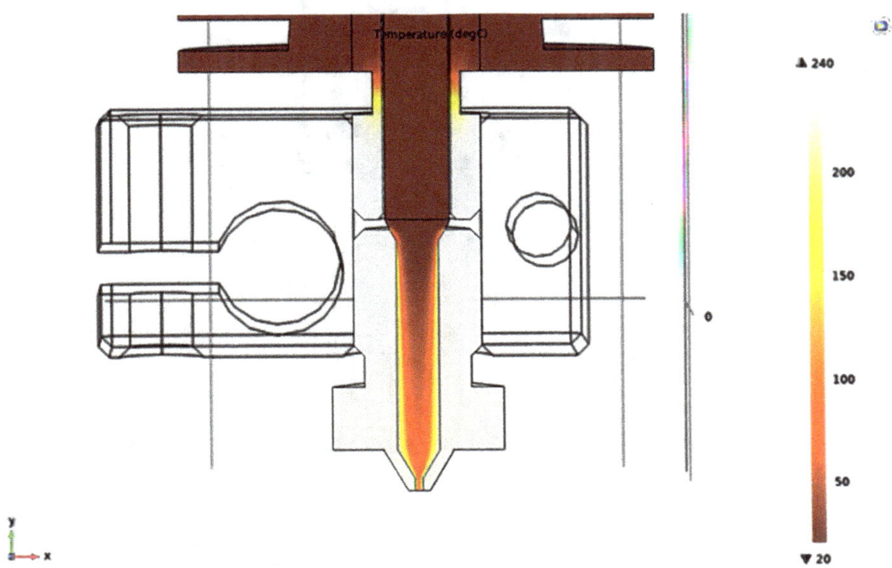

Figure 8.38 Temperature distribution

Outlet temperature (degC)	Outlet liquid share (1)
151.18	0.79445

Figure 8.39 Averaged surface outlet readouts

Fibre reinforcement can also refer to composites formed via in-nozzle impregnation. This is when fibres are not embedded into the polymer matrix but print with the polymer. This can also be achieved with a dual print head, i.e., a layer will be printed in your desired polymer, and the subsequent layers will be continuous fibres (Figure 8.8). The benefit here lies in deciding on the raster angle for each layer and generating a consistent cell periodicity. This is commonly seen with glass fibres or carbon fibres. The outcome exhibits superior tensile properties, but issues can arise from the misalignment of fibres. Another factor is that the final product results in a very porous outcome. That said, this is the most common avenue for fibre reinforced FDM printing, and many functional printed components have debuted from these systems.

Traditionally, fibre-reinforced polymer composites are considered to be lightweight and high-strength materials composed of strengthened, load-bearing fibres united together by an epoxy matrix. These fibres are usually carbon or glass fibres approximately 5–6 μm in diameter and having the length of tens of centimetres.

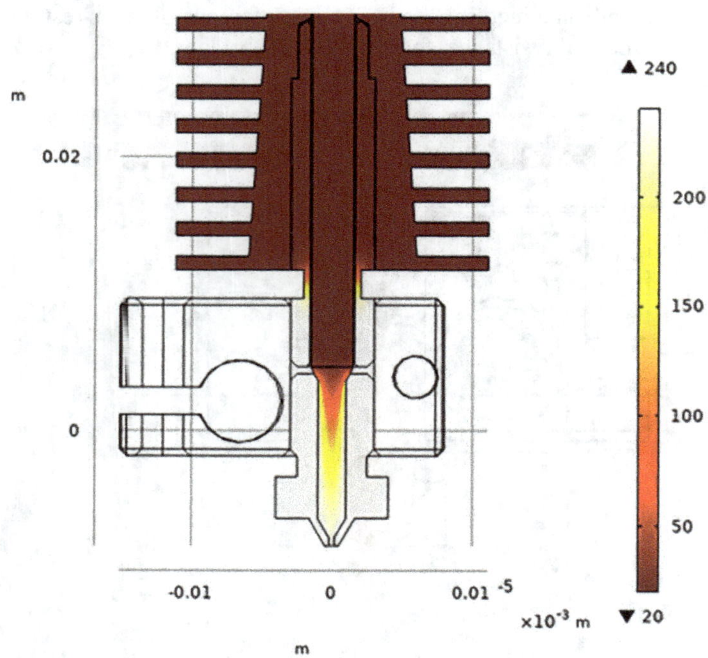

Figure 8.40 Temperature distribution for neat ABS at 2.4 mm³/s

Outlet temperature (degC)	Outlet liquid share (1)
235.44	1.0000

Figure 8.41 Phase share percentage for neat ABS at 2.4 mm³/s

However, in recent years, there is an emerging group of applications employing fibre-reinforced soft composites (FrSCs), which have proved to be very appealing towards new research and technological advancements. FrSCs are the composites that are composed of polymeric fibres implanted in some other soft-polymer matrix. The evolving applications of FrSCs involve 3D printing, 4D printing, biomimetic composites and entrenched actuation.

All these applications emphasise the imminent need to integrate 3D printing techniques with the direct deposition of polymeric fibre networks. Direct deposition of fibres opens the opportunity for varying orientations for each layer, as shown in Figure 8.9. Void fraction and orientation fibres play a vital role in the determination of the mechanical properties of FrSCs. Fibres can be oriented differently within printed parts depending upon the component's position on the building platform to produce anisotropic mechanical properties (Figure 8.10.). Best results of stiffness and strength FrSCs are obtained when there is fibre parallel to layer interface, but

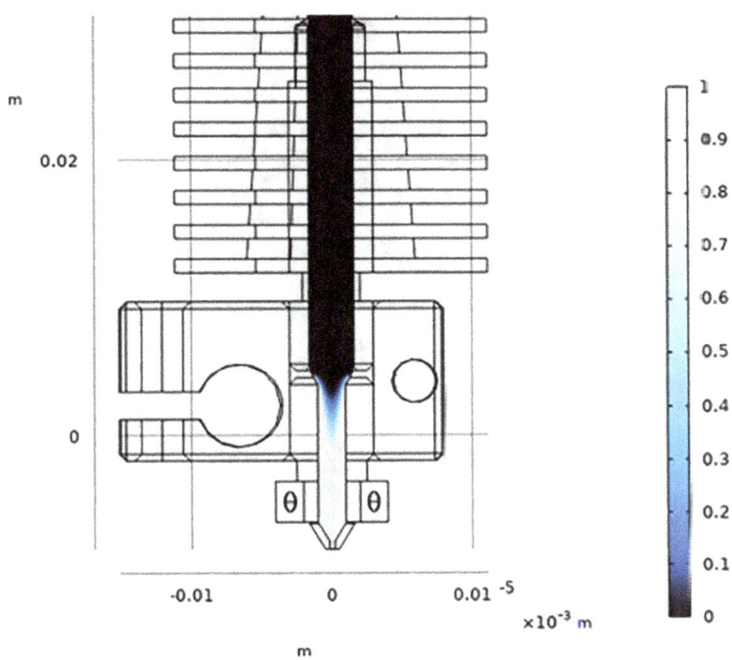

Figure 8.42 Averaged surface outlet readouts, neat ABS at 2.4 mm³/s

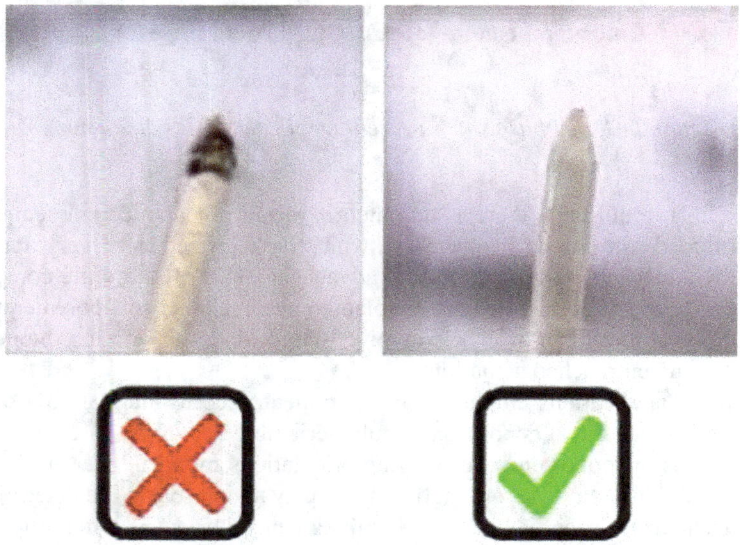

Figure 8.43 Atomic pull on the filament

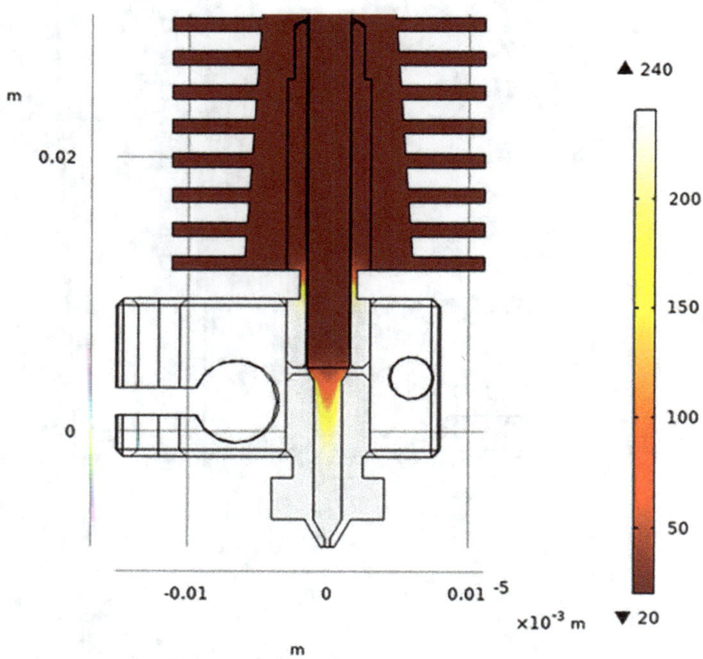

Figure 8.44 Temperature distribution for neat ABS at 1.2 mm³/s

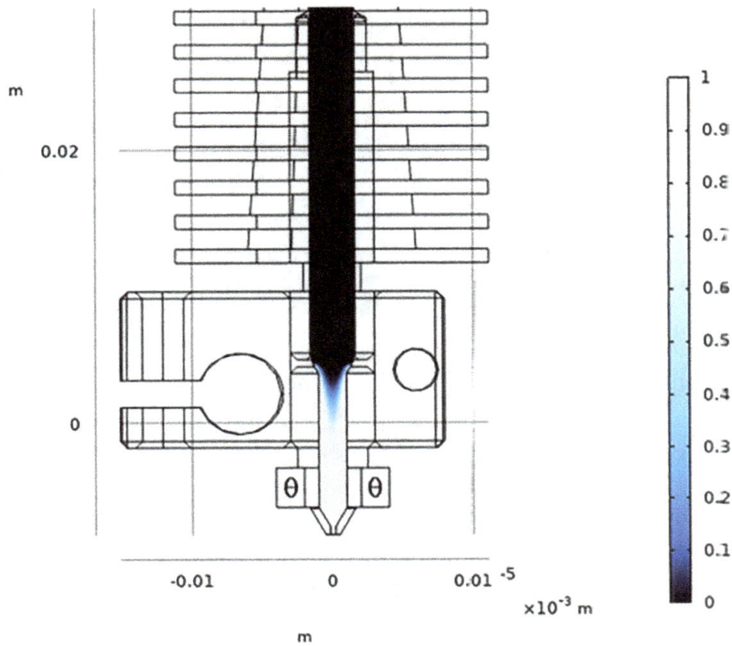

Figure 8.45 Phase share percentage for neat ABS at 1.2 mm³/s

these properties are increased by 60% when measured in a perpendicular direction to the fibres [22].

As expected, carbon fibres have been found to be a very useful filler material and an excellent additive to increase tensile strength, modulus values and flexural strength without compromising impact property. When 10% wt. of carbon fibre was added to the PA12 matrix, it was found that its tensile strength was increased by 102.2% and a remarkable increase in its flexural strength was observed, of 251.1%. It was also found that this effective matrix had a remarkable improvement in the thermal properties. Degradation temperature and the crystallisation peak temperature improved by 7.50% and 3.46%, respectively. Thermal conductivity was found to be increased from 0.221 [W/m·K] to 0.835 [W/m·K] [23].

It is found that reinforcement of continuous carbon, glass and Kevlar fibres increases the tensile strength of PA 6.3× [20]. The effective matrix of continuous carbon fibre (cCF) and PLA at 20% wt. was tested to show a flexural strength of 335 [MPa] and an elastic modulus of 30 [GPa] [9].

In the quest to find the ideal effective matrix for polymer composites, some researchers also tried natural fibre as a filler material. Satisfactory results were not produced due to their thermal degradation and the presence of large voids between lamina. When jute fibre was added to ABS, the ultimate tensile strength reduced from 17.73±2.52 [MPa] to 8.63±3.59 [MPa] [15].

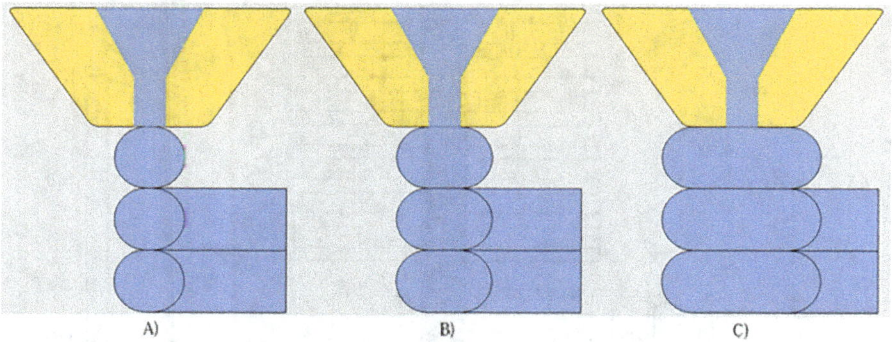

Figure 8.46 *Bead width: a) Too small; b) Minimum; c) Maximum [34]*

8.2.9 Final impressions

The current concern is to roll out novel lightweight polymeric multi-functional composites that can be utilised in biomedical, aerospace, automotive and energy sectors. There are promising candidates that are available in the immediate future.

With ceramic particulates offering great biocompatibility; various fillers providing tailored strength; and carbon reinforcements offering the potential for printed sensors and circuits, there is a clear benefit to the aforementioned industries. Most importantly, all of these are 3D printable, this means that bespoke manufacture and batch production is cost efficient.

Most research poses the issue of choosing the right filler size and concentration, then embedding it in a polymer matrix. Here are three individual choices to be made that can be streamlined. Numerical modelling offers the opportunity to predict aspects of the printing process, reducing the elements of failure when resorting to physical testing.

Finite element analysis (FEA) can be performed to calculate thermo-physical properties of the composite's effective matrix [24]. Furthermore, simulations can provide insight into the relationship between the materials' characteristics and environmental factors [25]. This ensures experimentation is not only more confident to be fruitful but also more cost effective.

This leaves more time to be left to focus on developing design guidelines not unlike the ones seen in other forms of plastic production. Ideally, this would be based on observing the behaviour of the composite undergoing the printing procedure. Prefacing this, empirical data on the performance of the printed component must be collated to make fair comparisons. This literature review is an attempt at summarising the current affairs and capabilities of multi-functional polymer composites.

8.3 Objectives

The time available for this project was short and issues related to the current pandemic subdued some original plans. That said, this only meant that greater resolution had to be applied to the spotlighted interactive multiphysics model.

As shown above, an in-depth identification of existing manufacturing methods has allowed for the categorisation of their capabilities for printing multi-functional polymer composites. Considering the scope of this project, attention was centred on FDM printing alone. FDM is very accessible, it is already used extensively in laboratory environments and the method of embedding reinforcement is rather simple.

The main objective was to build a sophisticated Multiphysics model for theoretical study and analysis. This simulation is intended to be used to better document the handling of material filament. This can be achieved by displaying temperature distributions and phase change shears with pertinent volumetric feed rates.

Polymers are not true solids; they do not have a fixed melting point and behave like a step function. They melt gradually in some span of temperatures. Polymers with fillers and reinforcements generally undergo abroad temperature transition zone, usually over several degrees in which both solid and molten polymer coexist in a ‘mushy’ zone. It is in this transition zone where experimentation on the relationship between temperature and feed rate can really be explored. Theoretically, phase shears could indicate temperature transition zones for each specimen.

As such it was imperative to decide on a select few polymeric composite specimens. Another cornerstone was deciding on which reinforcement method to simulate. While there is plenty of documentation on fibres and whiskers, namely CNT, the realm of building this computation was outside of capable learning scope considering the short-time frame. For simplicity purposes, the studies are implemented on particulates, and the material and size were chosen based on literature. That said, the benefits of particulates have already been outlined above. They are simple to embed, and the combinations are endless.

Finally, consideration should be taken on what kind of geometry can be presented to determine successful multi-material 3D printing. Perhaps a standardised model. This is to develop a best practice procedure. Printing compression, bending and tensile testing specimens are certainly useful; but for long-term use, the behaviour of the material during the printing process needs to be assessed. Using the resultant temperatures and feed rate from this simulation one should be able to launch a prescribed print that will help assess the functionality of the chosen specimen.

Outlet temperature (degC)	Outlet liquid share (1)
235.44	1.0000
239.80	1.0000

Figure 8.47 Averaged surface outlet readouts neat ABS at 1.2 mm³/s

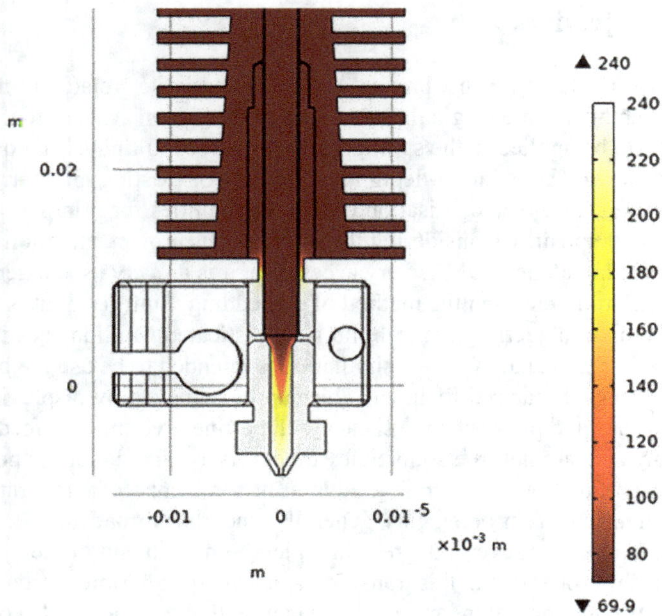

Figure 8.48 Temperature distribution for neat ABS at 2.4 mm³/s

Material and process need to be tested in conjunction. As such, part of the outcome is to propose a robust test for printing capabilities.

8.4 Numerical simulation

8.4.1 Schematic

The study was undertaken on COMSOL Multiphysics. To start this analysis some data must be addressed. FDM printers utilise an extruder to drive filament and heat up the polymer for extrusion. The extruder can be separated into two sub-assemblies. The drive system and HotEnd. In an arrangement where the drive system is right above the HotEnd, we refer to this as a direct drive system. In a system where the drive is set away from the printer, we refer to this as a Bowden tube system (Figure 8.11).

The schematic in Figure 8.12 was reverse engineered from an existing modular HotEnd. The HotEnd as the name suggests is the assembly that heats up in the extruder of an FDM system. It was decided to use this as the main geometry for study. This is due to its accessibility and modularity. The company E3D is known for making standardised all-metal HotEnd and extruders for handling many types of materials. This particular version is advertised as having the largest scope of materials. They also specialise in developing HotEnds that allow for colour mixing, water cooling and other mounting options for many printers.

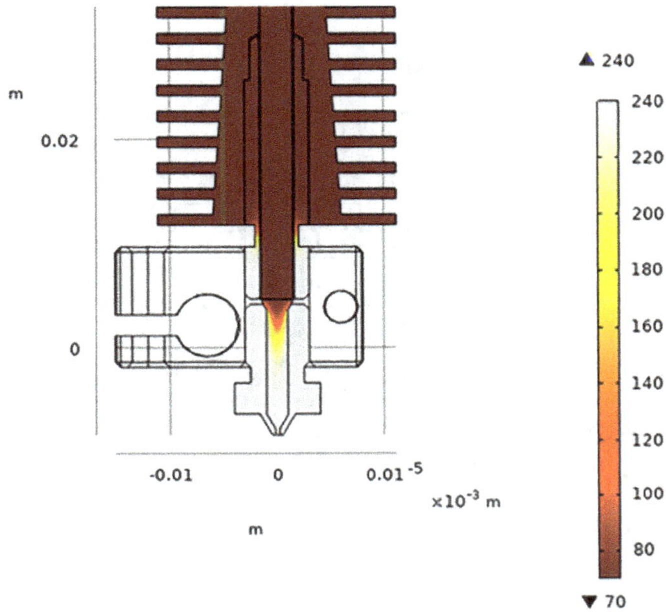

Figure 8.49 Temperature distribution for neat ABS at 1.2 mm^3/s

In this study, the boundary of this system encapsulates the HotEnd only. For the COM-SOL CAD import, the stated materials from Figure 8.12 were assigned. Nozzle diametre or materials are simple parameters and are easily edited, this allows for simulations of different fillers that would otherwise clog or damage the nozzle. For the filament domain, we need to consider that it will undergo a phase change; hence, Boolean partitions were performed to split it into two domains, one solid and one liquid.

During the printing process, the operator prescribes a temperature. The heating element is a 40 W cartridge-type heater that switches on. There is no voltage regulation. It heats up, approaching the prescribed temperature it will turn off. What you end up with is a relay control, a ripple of heating and cooling which, with PID tuning the range can be kept small. For the study, it is assumed that the PID tuning is optimised and that the whole heating block is at the prescribed temperature. There are a couple of reasons for this. First, it is unnecessary computation, a printer operator should optimise PID settings before performing experimentation. Second, COMSOL computes thermal equilibrium state which leads to vast increases in temperature and increased time for convergence.

Figure 8.13 shows a section plane cut. This is due to the plane of symmetry in geometry. It helps reduce computation time, and more importantly it provides a better viewing platform for the inner workings of this system.

We have already considered that process cannot be the same for each material, as such it was vital to build this model parametrically from conception. Figure 8.14

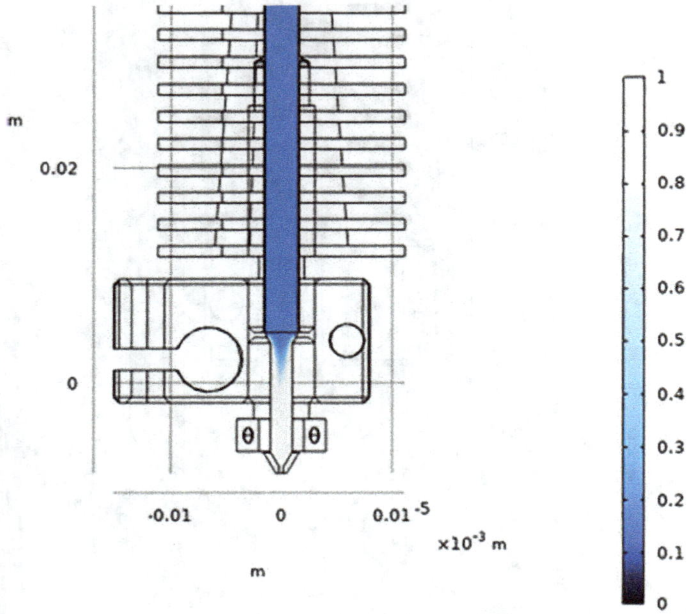

Figure 8.50 Phase share percentage for neat ABS at 2.4 mm³/s

displays the core parameters for this study. These parameters allude to the filament. By changing them, we are simulating a new isotropic filament.

8.4.2 Parameters

Melting temperature (T_{melt}) refers to the starting point of a polymer's melting range. As discussed before, polymers do not have a fixed melting point.

Latent heat (h_{melt}) is the required energy needed for the material to begin phase change.

Temperature transition zone, half width (dT_{half}): Figure 8.15 illustrates the phase shear diagram of a material. Phase (θ) is indicated as two states. In the middle, there is a transition zone where both solid and molten polymer coexists. The intersection of both shear lines indicates the melting point.

Furthermore, COMSOL documentation shows phase change is dependent on density (ρ), heat capacity (c_p) and thermal conductivity (k).

Further analysing Figure 8.14, *ambient temperature* (T_{amb}) allows for simulating environment control, i.e., if your printer was set in an enclosure, heated enclosures are preferable for some polymers, like polycarbonate (PC). It is assumed that the printer will be operated in a temperature-controlled enclosure, this means the air velocity around the nozzle can be assumed to be zero. This is not just to aid in convergence, it is also shown in the literature review that draughts can affect layer adhesion during printing. Finally, *heating block temperature* (T_{block}) refers to the

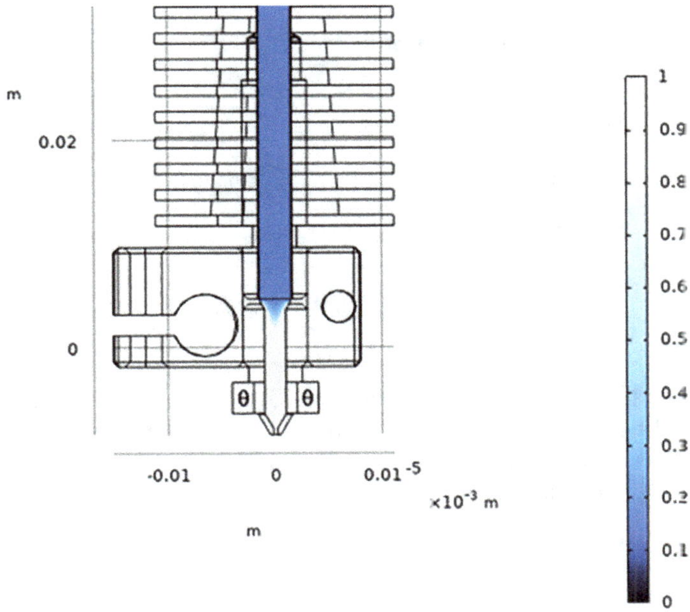

Figure 8.51 Phase share percentage for neat ABS at 1.2 mm³/s

temperature prescribed by the printer operator, and the drive system is simulated by a *volumetric feed rate(Q)* parameter.

Values for ρ, c_p and k are thermophysical properties of ABS. Figure 8.14 shows these values have a solid and liquid partition. These values were obtained from [27], an excerpt snapped in Figures 8.16 and 8.17.

These inputs allow the study to build a validation case. ABS is a highly documented material for all sorts of plastics manufacture and extrusion systems. Most importantly, I have first-hand experience in the behaviour of printing with ABS.

The filament is partitioned into two separate 'materials' where the phase share percentage can be calculated based on the equations provided by COMSOL

Outlet temperature (degC)	Outlet liquid share (1)
239.80	1.0000
239.84	1.0000
239.83	1.0000
239.84	1.0000
239.95	1.0000
239.85	1.0000

Figure 8.52 Outlet readouts for 1.2 mm³/s

Outlet temperature (degC)	Outlet liquid share (1)
235.44	1.0000
235.91	1.0000
235.73	1.0000
235.88	1.0000
238.60	1.0000
236.14	1.0000

Figure 8.53 Outlet readouts for 2.4 mm³/s

(Figure 8.18). Values of these partitions are overridden as phase change materials and need to be computed. The breakdown of the filament melting zones are detailed in Figure 8.19. The settings for establishing heat transfer within COMSOL are shown in Figure 8.20.

8.4.3 Boundary conditions

The physics couplings utilised in this study are heat transfer in solids and fluids and laminar flow. Let's consider the fluid flow.

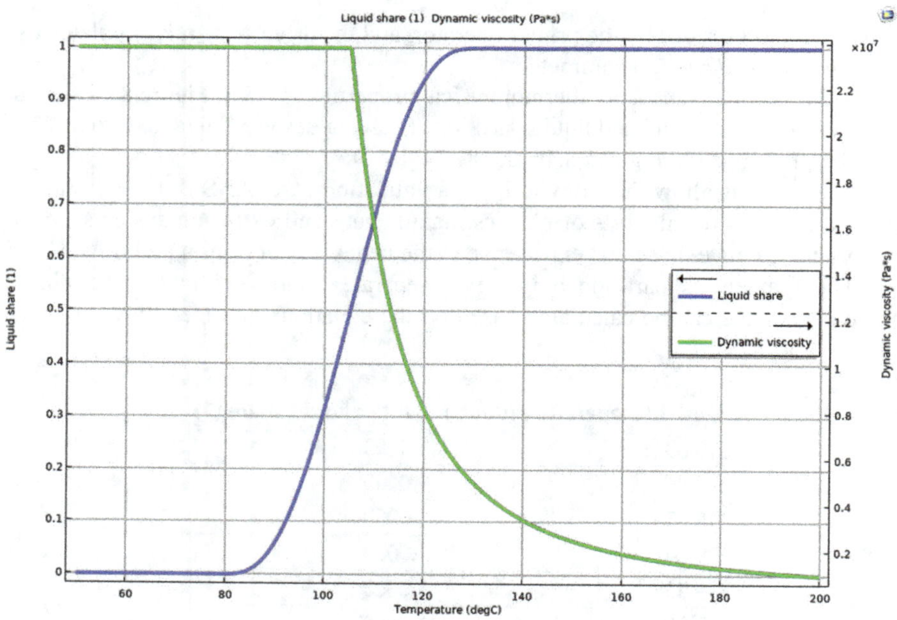

Figure 8.54 Influence of temperature on dynamic viscosity for some phase share percentage

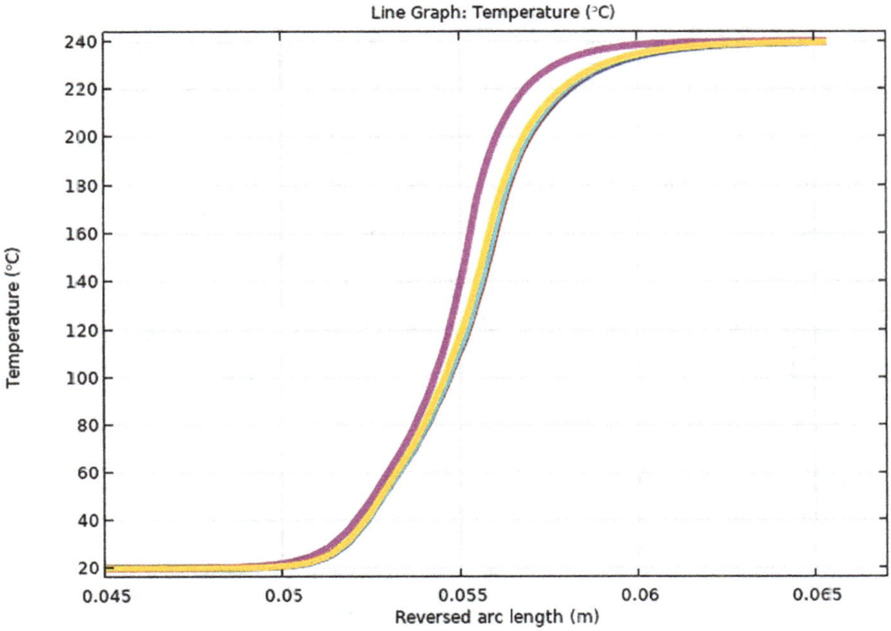

Figure 8.55 Temperature along centreline; 1.2 mm³/s Q

Here is the Navier-Stokes momentum equation. Displayed on the left are terms for mass and acceleration, and on the right lie the combination of forces that induce fluid flow.

$$\underbrace{\rho\frac{\partial\vec{u}}{\partial t}+\rho\vec{u}\cdot\nabla\vec{u}}_{\text{Inertial forces}}=\underbrace{-\vec{\nabla}p}_{\text{Pressure forces}}+\underbrace{\nabla\cdot\bar{\bar{\tau}}'}_{\text{Viscous forces}}+\underbrace{\vec{f}_{B\rho}}_{\text{External forces}} \tag{8.1}$$

Conservation of mass:

$$\frac{\partial p}{\partial t}+\nabla.(\rho\vec{u})=0 \tag{8.2}$$

In this problem, the flow is characterised as laminar, but also allows for the very low Reynolds numbers that molten polymer will exhibit. Under this regime, we develop creeping flow and so the inertial forces are very small in comparison to the viscous forces [29]. In COMSOL, we select to neglect the inertial term.

$$\underbrace{\rho\left(\frac{\partial\vec{u}}{\partial t}+\vec{u}\cdot\nabla\vec{u}\right)^{0}}_{\text{Inertial forces}}=\underbrace{-\vec{\nabla}p}_{\text{Pressure forces}}+\underbrace{\nabla\cdot\left[\mu\left(\nabla\vec{u}+(\nabla\vec{u})^{T}-\frac{2}{3}(\nabla\cdot\vec{u})\vec{I}\right)\right]}_{\text{Viscous forces}}+\underbrace{\vec{F}}_{\text{External forces}} \tag{8.3}$$

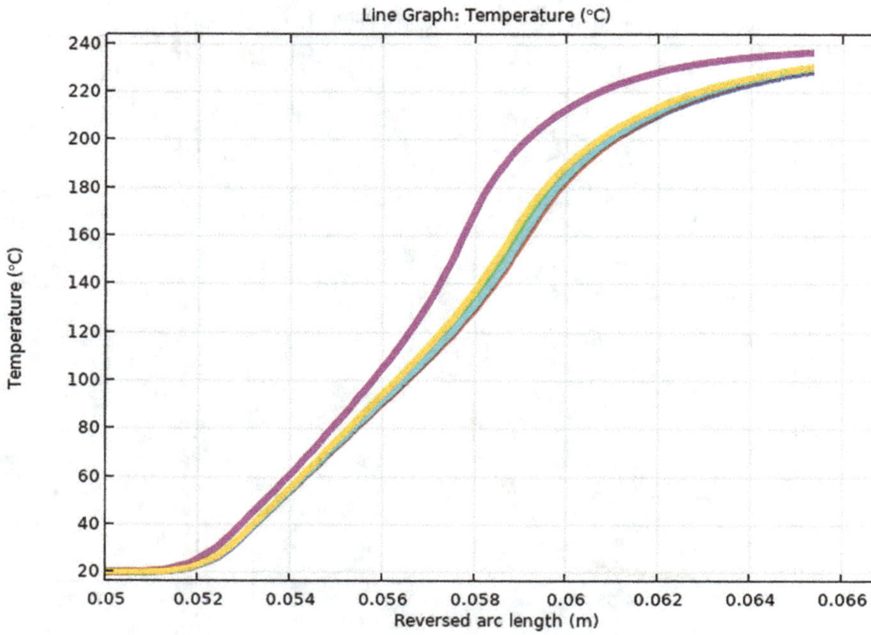

Figure 8.56 *Temperature along centreline; 2.4 mm³/s Q*

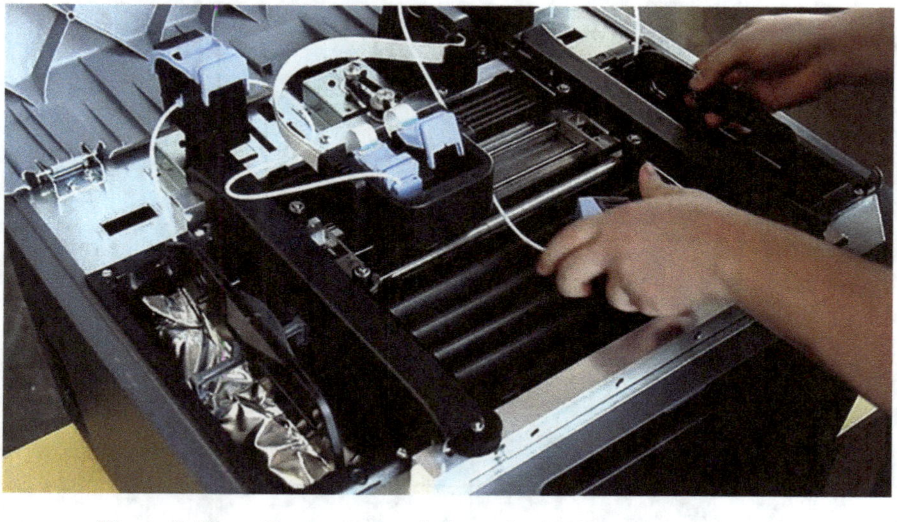

Figure 8.57 *Concertina-style heat shield: Stratasys FDM Mojo*

Outlet temperature (°C)	Outlet liquid share (1)
254.13	1.0000
255.20	1.0000
254.60	1.0000
255.00	1.0000
263.04	1.0000
255.79	1.0000

Figure 8.58 Outlet readouts for 3.2 mm³/s @ 270◦C nozzle and 70◦C chamber

The flow is considered incompressible, divergence velocity is zero:

$$\therefore \nabla \cdot \vec{u} = 0 \tag{8.4}$$

$$\Longrightarrow 0 = \underbrace{-\vec{\nabla} p}_{\text{Pressure forces}} + \underbrace{\nabla \cdot \left[\mu \left(\nabla \vec{u} + (\nabla \vec{u})^T - \frac{2}{3}(\nabla \cdot \vec{u})\vec{I} \right) \right]^0}_{\text{Viscous forces}} + \underbrace{\vec{F}}_{\text{External forces}} \tag{8.5}$$

$$\Longrightarrow 0 = \underbrace{-\vec{\nabla} p}_{\text{Pressure forces}} + \underbrace{\nabla \cdot \mu \left(\nabla \vec{u} + (\nabla \vec{u})^T \right)}_{\text{Viscous forces}} + \underbrace{\vec{F}}_{\text{External forces}} \tag{8.6}$$

Viscosity is a macroscopic property that dictates how molecular forces interact, it relates stress to rate of strain. Viscosity can behave in a few different ways.

Outlet temperature (°C)	Outlet liquid share (1)
265.19	1.0000
265.74	1.0000
265.61	1.0000
265.75	1.0000
268.59	1.0000
265.97	1.0000

Figure 8.59 Outlet readouts for 2.4 mm³/s @ 270◦C nozzle and 70◦C chamber

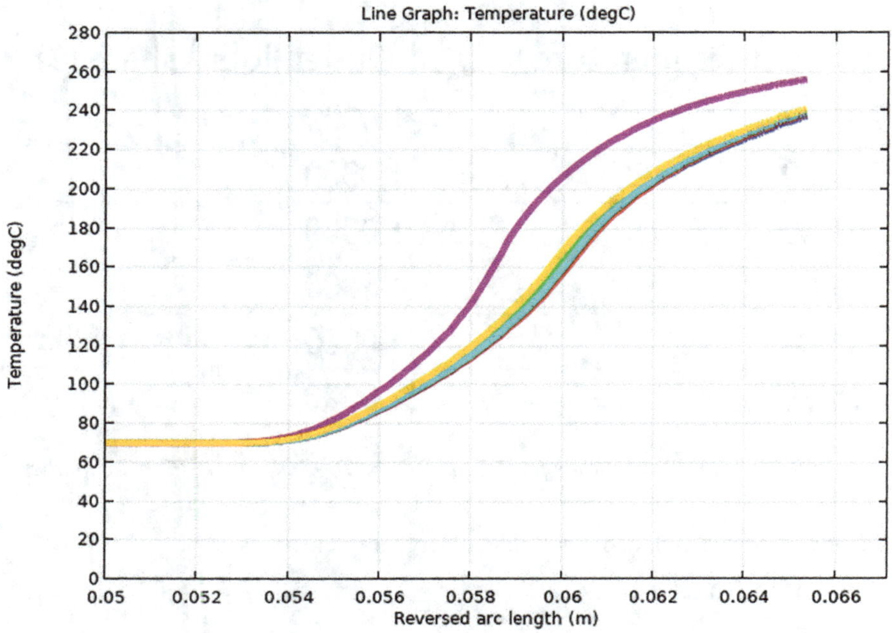

Figure 8.60 Temperature along centreline; 3.2 mm³/s Q @ 270◦C

Where dynamic viscosity (μ) is a function of:- temperature:

$$\mu = f(T)$$

shear rate magnitude ($\dot{\gamma}$):

$$\mu = f(\dot{\gamma})$$

pressure and time:

$$\mu = f(p, t)$$

If any of these apply, the observed fluid is characterised as non-Newtonian. In the case of molten polymer, we have shear thinning fluid. From COMSOL, the fluid is selected to be the inelastic non-Newtonian model. In this situation, the Power Law can be used.

$$\mu(\dot{\gamma}) = m\dot{\gamma}^{n-1}$$

Here, viscosity is a function of shear rate magnitude. Where n = flow behaviour index and m = *fluid consistency coefficient*. Molten polymer is a shear thinning fluid, therefore n<1.

Shear rate tensor $\bar{\bar{\bar{S}}}$:

$$\bar{\bar{\bar{S}}} = \nabla\vec{u} + (\nabla\vec{u})^T$$

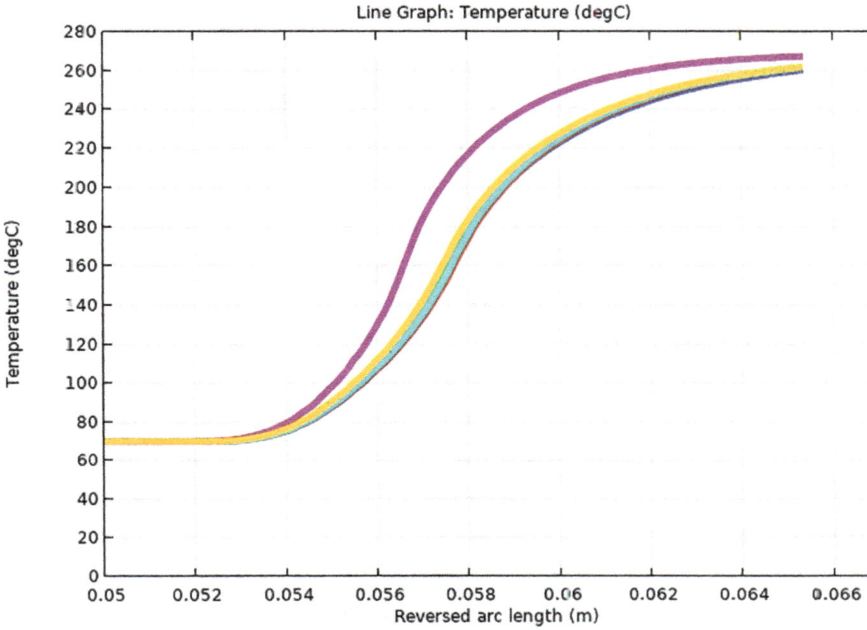

Figure 8.61 Temperature along centreline· 2.4 mm³/s Q @ 270°C

The magnitude of this is $\gamma^{\cdot}$.

If non-Newtonian dynamic viscosity (μ_{nN})

$\mu_{nN} = f(\dot{\gamma})$

Then we have a non-linear problem.

From literature, measured rheological flow behaviour of ABS was found. Furthermore, the governing Power Law equation can be expanded [28].

Equation (9) from [28]:

$$\mu_{nN}(\dot{\gamma}, T) = \frac{T}{T_r} \cdot \boldsymbol{a}^{n}_{T \cdot k \cdot \dot{\gamma}^{n-1}}$$

where;

reference temperature, $T_r = 473.15[K]$

flow behaviour index, $n = 0.24$

fluid consistency coefficient, $m = \frac{T}{T_r} \cdot \boldsymbol{a}^{n}_{T \cdot k}$

flow consistency at ref. temp., $k = 30104\ [Pa \cdot s^n]$

The Williams-Landel-Ferry (WLF) equation determines the temperature dependence; time-temperature shift factor, $a_T(T) = \exp \frac{-C_1(T-T_r)}{-C_2+(T-T_r)}$

WLF material constant 1, $C_1 = 8.97$

WLF material constant 2, $C_2 = 155.2$ [°C]

Using this excerpt, the fluid consistency coefficient, m can be computed as a function of temperature. For every temperature reading, a coefficient can be computed. ‘To improve the numerical convergence, the maximum viscosity was limited

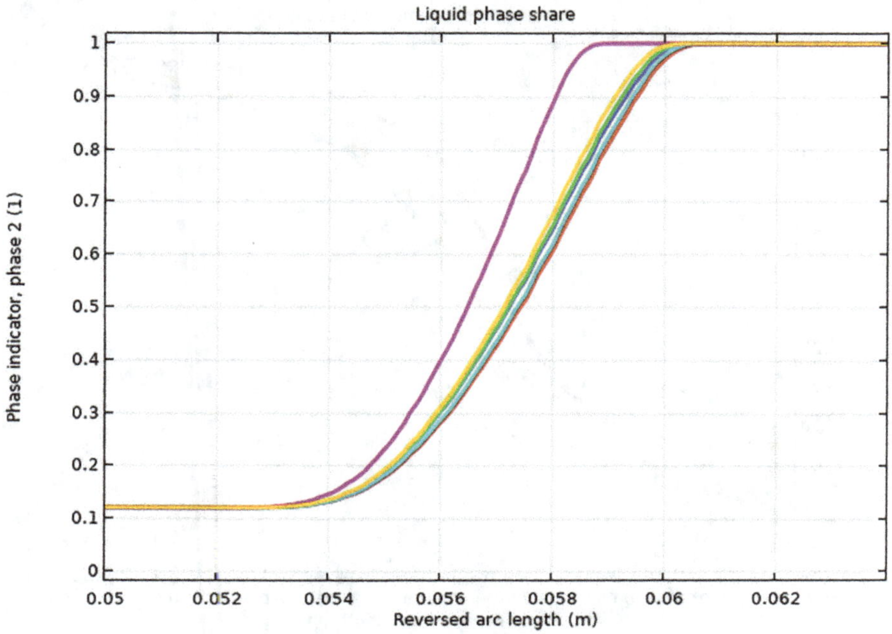

Figure 8.62 Phase along centreline; 3.2 mm³/s Q @ 270◦C

to *10⁶ P a·s*' [28]. Evidence of these computations are shown in the following figures. Figure 8.21 indicates the application of the piecewise function that was derived by Serdeczny *et al.*, [31]. Figure 8.22 shows the computed values of the fluid consistency coefficient, while limiting the max viscosity.

This piecewise function enables the capability to use either the computation of $m(T)$ or 10^6 P a · s, whichever is smaller.

Clearly, the fluid consistency coefficient is non-linear. This harms computation and may, in fact, never converge. Using the *nojac*() operator (Figure 8.23) helps reduce the array to accelerate convergence [30].

8.4.4 Mesh discretisation

To use FEA the geometry needed to be discretised to reduce the degrees of freedom (DOF) from infinite to finite. This makes the problem numerically solvable. In Figures 8.24 and 8.25, there is an overview of the mesh layout. A free tetrahedral mesh was used to correctly fit the complex geometry. The main point of interest lies in the heat transfer and flow of the polymer. As such approaching the nozzle the mesh elements refine gradually.

However, the filament is a simple cylinder and does not hold complex geometry. A swept mesh distribution opted for the solid partition of the filament (Figure 8.26). The swept mesh is more economic when addressing the number of elements in ratio to DOF. Furthermore, as the filament passes through the feed channel there is little

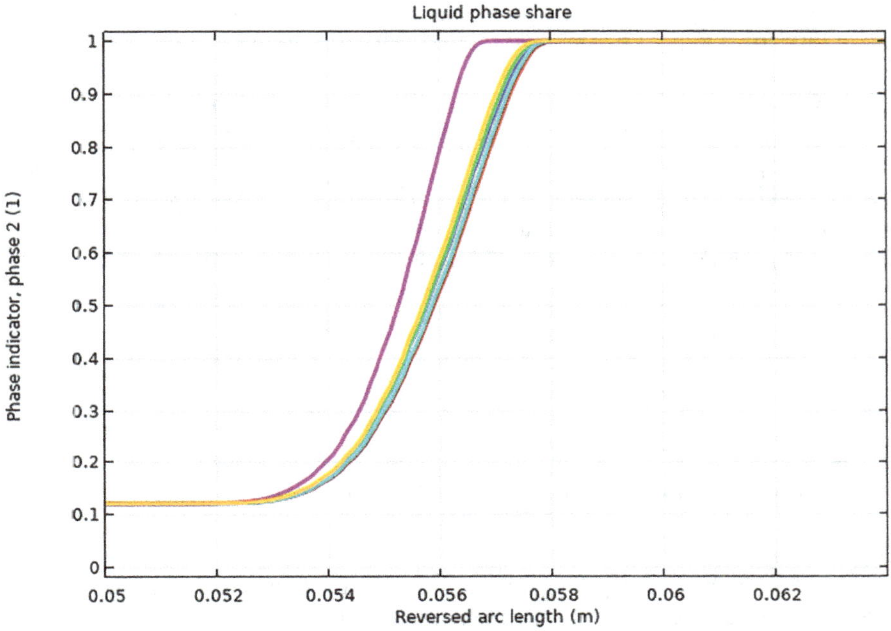

Figure 8.63 Phase along centreline; 2.4 mm³/s Q @ 270°C

friction. The combination of the heat-brake and heatsink prevents heat transfer from occurring further up the HotEnd. This means a slip condition can be applied to the walls to the swept mesh.

As the filament enters the nozzle inlet, it begins to melt. A non-slip condition is applied here to properly display the fluid flow (Figure 8.27). Here the elements have been reduced to five boundary layers. This nozzle boundary is the main interest in the flow simulation. An extremely fine mesh was characterised to properly compensate the computation (Figure 8.28).

Using a consistent mesh would unnecessarily increase computation time, hence it was vital to control the element distortion throughout the geometry.

8.4.5 Estimating thermo-physical properties

The described study in the subsections above utilises ABS as a validation case. For this study to properly assess composites, the thermo-physical properties need to be parameterised to simulate a new material. However, literature does not provide the necessary data. Viscosity is dependent on ρ, c_p and k. Furthermore, the study requires these parameters in solid and liquid phases.

ABS properties appeared to be the only polymer matrix that was documented heavily. Numerous papers had experimental data on ABS, examples including [27] and [28].

As discussed in section 8.3, focus was applied to particulate fillers. To properly move forward, another FEA method needed to be utilised to estimate thermophysical properties.

In COMSOL, a 3D unit block was generated. Assumptions are made to simplify the estimation. First, the particles are distributed uniformly and therefore the material is isotropic. As seen in Figure 8.29, the filler particles are hexagonally packed. Second, the filler particulates are indeed spherical. Finally, it is necessary to assume that no chemical reaction will occur within the effective matrix. This final assumption means that density (ρ) and heat capacity (c_p) can be calculated as a volume average of the filler and the polymer.

$$\rho_{mix} = \rho_{filler} \cdot V_{filler} + \rho_{matrix} \cdot V_{matrix} \tag{8.7}$$

$$c_{p,mix} = c_{p,filler} \cdot \%_{filler} + c_{p,matrix} \cdot (100\% - \%_{matrix}) \tag{8.8}$$

Considering the reputable ABS properties found earlier, the polymer matrix for all specimens will be ABS. This means that a filler material can be chosen from literature and their respective properties can be used to calculate the properties of the mixture.

Thermal conductivity depends on heat flux, this means the methods to calculate it will differ slightly. Using Fourier's Law of Conduction, a 1 degree Kelvin difference can be applied to each opposite boundary of the mesh in Figure 8.29. With three pairs of boundaries there are three solvers, so the final results are averaged [31]. A contour plot for one boundary pair is shown in Figure 8.30.

$$\dot{\boldsymbol{q}}[W/m^2] = -\boldsymbol{k}[W/m \cdot K] \cdot \nabla \boldsymbol{T}[K/m] \tag{8.9}$$

Considering 1D conduction:

$$\dot{q} = -k \cdot \frac{\partial T}{\partial x} \tag{8.10}$$

$$\frac{\dot{Q}}{A} = -k \cdot \frac{T_1 - T_2}{L} \tag{8.11}$$

Using *Granta EduPack: CES selector*, material properties for baseline materials were found and collated (Figure 8.31). Parameters for filler material, size and concentration were based on the literature review from section 8.2.1. These were applied to the COMSOL parameter setup (Figure 8.32). This finite element method provides a reasonable estimate for density, specifc heat and thermal conductivity for both solid and liquid phases the composite polymer (Figure 8.33).

Solving the composite's effective material properties (ρ, $_{cp}$ and k) allows for the flow study to properly model viscosity. But melting temperature, transition temperature and latent heat cannot be obtained from this estimator.

That said, it can be assumed these will not change significantly. Researchers added up to 40% fly ash to ABS. Measuring the melting temperature and glass transition temperature, it was clear that these values did not change significantly [32].

Further evidence was provided in Reference [15].

This solver was run as a parametric sweep to promptly gather properties for various materials, sizes and loading percentages.

8.5 Results and discussion

8.5.1 Estimated thermo-physical properties of specimens

The estimator study delivered a parametric sweep of 38 possible combinations of filler partnered with ABS. These were simply exported to an excel format for post-processing (Figure 8.34). Of this selection, five materials were chosen for further examination under the flow study. The chosen materials are shown in Figure 8.35.

The chosen specimens were selected based on literature, but also to gain some variance in concentration and size.

Figure 8.36 shows the relationship of thermal conductivity as a function of particulate loading percentage. This insight can help a user extrapolate ranges for specific operation.

8.5.2 Slicing

Figures 8.37 and 8.38 indicate a poor slicing profile. The melt cone is not fully formed. The extrusion has 20.56% of solid core (Figure 8.39). This melt profile would clog the nozzle over time, requiring frequent maintenance [1].

Usual printing temperature for ABS varies between 210◦C and 250◦C. In practice, this is approximated based on intention, in combination with printing speed and build platform temperature. However, there are some set profiles somewhat agreed upon by the community.

For E3D v6 HotEnds on a typical RepRap setup:

1. Printing temperature: 240◦C
2. First layer print speed: 30 [mm/s]
3. Infill speed: 60 [mm/s]
4. Inner shell speed: 60 [mm/s]
5. Outer shell speed: 40 [mm/s]
6. Typical layer height: 100 [µm]
7. Nozzle diameter: 0.4 [mm]

Volumetric feed rate (Q) can be calculated as nozzle diameter × layer height × printing speed = Q.

In this case, the study considers 2.4 mm^3/s and 1.2 mm^3/s. This covers the fastest and slowest printing speeds for this slicing profile at 100 µm. In theory, these flow rates also consider print speed of 60 mm/s at 50 µm or 30 mm/s at 200 µm.

Figures 8.40 and 8.41 follow the regime enumerated in the list above. Here, the melt cone is more pronounced and outlet phase share is 100% liquid (Figure 8.42). As a validation case, this follows according to expectations. Typically, if one were to perform an Atomic pull [33], if the nozzle is clean and maintained well, the filament exits as a smooth cone (Figure 8.43).

ABS has a habit to warp drastically when printing. As such, printing is slow.

Figures 8.44 and 8.45 allude to printing at 100 µm, 30 mm/s. This refers to the first layer of printing. The first layer typically is the thickest bead of plastic This is to hold adhesion to the print bed. But a thicker bead needs to be applied slowly. (Figure 8.46). The liquid share is evidence in Figure 8.47.

Professional FDM printers usually have further scope than the enumerated list above. To further aid in layer adhesion the printing chamber is heated, and the build platform is heated. Bed temperature: 110∘C. Enclosure temperature: 70∘C.

This simulation does not account for a build platform. However, editing the ambient temperature will simulate a heated build chamber.

For the chosen specimens, the temperature distribution contours did not vary significantly from Figures 8.48 and 8.49, similarly with the phase share percentage in Figures 8.50 and 8.51. In fact, similar results are displayed for all the contours regarding feed rates of 1.2 and 2.4 mm3/s. The simple fact is the thermo-physical values do not vary greatly enough to have a distinguishing remark. The most significant response lies in the alumina composite. Here the filler concentration is high at 40%. While the thermal conductivity of alumina as an individual material (Figure 8.31) is much lower than copper, the outcome of the effective polymer matrix is greater than copper.

$$
\begin{aligned}
k_{Cu} &= 394\ [W/m \cdot K] \\
k_{Al_2O_3} &= 31.5\ [W/m \cdot K] \\
k_{ABS} &= 0.3\ [W/m \cdot K] \\
k_{Cu,ABS} &= 0.3771550\ [W/m \cdot K](@40\%wt) \\
k_{Al_2O_3} &= 0.466016\ [W/m \cdot K](@40\%wt)
\end{aligned}
$$

This is even equivalent to filler size and percentage. The density of copper is much greater than alumina, therefore the particles in the alumina effective matrix are much closer together than that of copper.

Printing slower means the outlet temperature is more precise. However, alumina has an outlet temperature of 238.6∘C for 2.4 mm^3/s. This figure stands out from the others in Figure 8.52. And yet, in Figure 8.53, alumina's figure does not stand out. This suggests the temperature plateaus eventually and holds a relationship with phase shear, or perhaps, viscosity.

By building this relationship there are clear similarities to Figure 8.15, the phase shear diagram. The glass transition temperature of ABS is approximately, 110∘C, this is from empirical data. Here in Figure 8.54, the same intersection occurs at 110∘C. As theorised, the temperature does begin to plateau as the shears thin.

In Figure 8.55, the outstanding magenta line is, unsurprisingly, alumina. The rest are so closely aligned that a higher magnification would lose context. A typical 'S' curve is displayed for the temperature gradient over the length of the HotEnd.

Now, considering the faster flow, here the prescribed temperature just barely hits before the nozzle outlet has arrived (Figure 8.56).

Viewing the phase share plots, it is clear the slower flow has 8 mm of pure liquid flow to the outlet, while 2.4 mm^3/s resolves half of that.

Appendix B.a displays plots for a heated build chamber. The anticipation was that it would help the filament ascertain the liquid phase earlier. For the slower flow, it actually took longer to get to 100% liquid share. This is likely due to thermal equilibrium taking longer as the gradient is less steep with a heated chamber.

The bottom section of the 'S' curve has become squashed. This is due to the ambient temperature now being 70◦C instead of 20°C. This adds a small percentage of liquid share. This appeared to travel up past the heat sink. In the professional FDM systems that employ heated build chambers, they typically install a concertina-type shield (Figure 8.57) to prevent heat flux from reaching components. For example, the filament past desired boundary conditions.

So, to print at a faster rate, it is assumed that it must be done at a higher temperature. 3.2 mm^3/s is equivalent to printing with a 0.4 mm nozzle with a 200 μm layer height at 40 mm/s. Effectively, the printer is still running at print speeds profiled in the regime earlier but printing at a lower resolution. For the higher flow rate, the upper limit of ABS printing temperature is used, 270◦C (Figures 8.58 and 8.59).

It is clear in the phase shear below that the polymer has sufficient time to become molten and liquid before reaching the outlet. 270 is the upper limit of ABS printing but the actual printing temperature occurs closer to 260. The slower flow rate performs a more pronounced likeness of the 'S' curve that is desired, but this printing speed appears to be fine (Figures 8.60 and 8.61). Phase shares for theses simualtions are shown in Figures 8.62 and 8.63. In fact, it is apparent that the printing temperature can be bought lower. It is ideal not to needlessly burn away ABS, the goal is to smoothly melt it to a glassy state. Over melting unnecessarily will incur more maintenance time or replacement nozzles. In the case of filler, over melting the polymer matrix will leech the filler into the nozzle contributing to clogs and erosion of the outlet.

8.6 Conclusion

Assuming the printer operator has taken care to maintain their printer and the filament is moisture free it is fair to assume that ideal printing temperatures and coinciding print speed can be estimated accordingly using this Multiphysics study. This study would be best utilised as a partnering application for future lab researchers.

In terms of behaviour of particulates as filler, it is clear that concentrations of filler under 40% have little influence on printing temperatures and speed. So, printing profiles from the polymer matrix can be used as a keen guideline. Appendix contours outline the similarities between all the chosen specimens.

Building the COMSOL flow study was certainly successful. I met my milestones to build the desired study. However, I underestimated the time associated with building a model like this. The original outline plan stated a future outlook into thermal stresses and reconsolidation of the polymer. There was no time to properly consider this let alone laying the groundwork for it. Shrinking and warping is one of the most concerning yet intriguing aspects to study in this field. However, there is further learning to be done in this area to progress to that level of scope.

In hindsight, estimation of properties is highly theoretical work and needed to be developed from conception to building a more robust model. The estimator study built for this project was an after-thought built in the final weeks of the project timeline, out of desperation. Initially, the impression was made that thermo-physical properties would be able to be obtained eventually either through literature or experimentation. The pandemic atmosphere thwarted capabilities to perform any in-house laboratory work. Literature proved to be unfruitful for the necessary properties. This only became clear from performing the review. Nevertheless, plotting the pathway to develop a rudimentary estimator for the three main thermo-physical properties was an enjoyable experience. Furthermore, had work begun on property estimators from the beginning, there would have been focused intent on CNT. CNT as a filler appears infinitely valuable. More enticingly, it proposes an interesting problem to solve.

The initial months of this project were trapped in free fall. The necessary modules for COMSOL Multiphysics were not made available till much later. Nevertheless, this time was used effectively to read up on COMSOL's genuinely insightful documentation that provided many sympathetic examples.

3D printing parameters are convoluted. To make a true analysis one would need to observe the actual printing process. This numerical study is good groundwork but will be best suited to be used in conjunction with the printing operator and physical experimentation. Literature has made it clear that process and material need to be tested together.

A note for the reader

Files for the COMSOL studies are supplied here. This contains a readme file explaining how to operate the studies: https://drive.google.com/drive/folders/1e5YOhjzQZmyzqnun1ip8_qSn9BwLLst9?usp=sharing

References

[1] Zhang X., Fan W., Liu T. 'Fused deposition modeling 3D printing of polyamide-based composites and its applications'. *Composites Communications*. 2020;21(8):100413.

[2] Ligon S.C., Liska R., Stampfl J., Gurr M., Mülhaupt R. 'Polymers for 3D printing and customized additive manufacturing'. *Chemical Reviews*. 2017;117(15):10212–90.

[3] Bryll K., Piesowicz E., Szymanski P., Slaczka W., Pijanowski M. 'Polymer composite manufacturing by FDM 3D printing technology'. *MATEC*. 2020;**2**.

[4] Blok L.G., Longana M.L., Yu H., Woods B.K.S. 'An investigation into 3D printing of fibre reinforced thermoplastic composites'. *Additive Manufacturing*. 2018;22(403):176–86.

[5] Daminabo S.C., Goel S., Grammatikos S.A., Nezhad H.Y., Thakur V.K. 'Fused deposition modeling-based additive manufacturing (3D printing): techniques for polymer material systems'. *Materials Today Chemistry*. 2020;16(17):100248.

[6] Huang B., Singamneni S. ‘A mixed-layer approach combining both flat and curved layer slicing for fused deposition modelling’. *Proceedings of the Institution of Mechanical Engineers, Part B: Journal of Engineering Manufacture*. 2015;229(12):2238–49.

[7] Dikshit V., Goh G.D., Nagalingam A.P., Goh G.L., Yeong W.Y. ‘Chapter 17 - recent progress in 3d printing of fiber-reinforced composite and nanocomposites’; 2020. pp. 371–94. Available from https://www.sciencedirect.com/science/article/pii/B9780128199046000177.

[8] Zhu Z.H., Zhang N., Wang T., Hao M.Y. ‘Short review of polymer composites for 3D printing’. *IOP Conference Series: Materials Science and Engineering*. 2020;758(1):012046.

[9] Valino A.D., Dizon J.R.C., Espera A.H., Chen Q., Messman J., Advincula R.C. ‘Advances in 3D printing of thermoplastic polymer composites and nanocomposites’. *Progress in Polymer Science*. 2019;98(Suppl. 1):101162.

[10] Wang F., Xu Z. ‘Graphene and graphene oxide reinforced 3D and 4D printable composites’. *3D and 4D Printing of Polymer Nanocomposite Materials*. 2020;9:259–96.

[11] Velu R., Raspall F., Singamneni S. ‘3D printing technologies and composite materials for structural applications’. *Green Composites for Automotive Applications*. 2019;8:171–96.

[12] Saude N., Ibrahim M., Ibrahim M. *Mechanical properties of highly filled iron-abs composites in injection molding for fdm wire filament [online]*. 2013. Available from https://www.researchgate.net/publication/269345844_Mechanical_Properties_of_Highly_Filled_Iron-ABS_Composites_in_Injection_Molding_for_FDM_Wire_Filament

[13] Nikzad M., Masood S.H., Sbarski I., Groth A. ‘A study of melt flow analysis of an abs-iron composite in fused deposition modelling process’. *Tsinghua Science & Technology*. 2009;5(38):29–37.

[14] Rahim T.N.A.T., Abdullah A.M., Md Akil H., Mohamad D., Rajion Z.A. ‘The improvement of mechanical and thermal properties of polyamide 12 3D printed parts by fused deposition modelling’. *Express Polymer Letters*. 2017;11(12):963–82.

[15] Torrado A.R., Shemelya C.M., English J.D., Lin Y., Wicker R.B., Roberson D.A. ‘Characterizing the effect of additives to ABS on the mechanical property anisotropy of specimens fabricated by material extrusion 3D printing’. *Additive Manufacturing*. 2015;6(24):16–29.

[16] Kalita S.J., Bose S., Hosick H.L., Bandyopadhyay A. ‘Development of controlled porosity polymer-ceramic composite scaffolds via fused deposition modeling’. *Materials Science and Engineering: C*. 2003;23(5):611–20.

[17] Kwok S.W., Goh K.H.H., Tan Z.D., *et al*. ‘Electrically conductive filament for 3D-printed circuits and sensors’. *Applied Materials Today*. 2017;9(6):167–75.

[18] Cholleti E.R., Gibson I. ‘Abs nano composite materials in additive manufacturing’. *IOP Conference Series: Materials Science and Engineering*. 2018;455:012038.

[19] COMSOL *'introducing the composite materials module' [online]*. 2018. Available from https://www.comsol.com/blogs/introducing-the-composite-materials-module

[20] Dickson A.N., Barry J.N., McDonnell K.A., Dowling D.P. 'Fabrication of continuous carbon, glass and Kevlar fibre reinforced polymer composites using additive manufacturing'. *Additive Manufacturing*. 2017;16(1):146–52.

[21] Blanco I. 'The use of composite materials in 3D printing'. *Journal of Composites Science*. 2020;4(2):42.

[22] Badini C., Padovano E., De Camillis R., Lambertini V.G., Pietroluongo M. 'Preferred orientation of chopped fibers in polymer-based composites processed by selective laser sintering and fused deposition modeling: effects on mechanical properties'. *Journal of Applied Polymer Science*. 2020;137(38):49152.

[23] Liao G., Li Z., Cheng Y., *et al*. 'Properties of oriented carbon fiber/polyamide 12 composite parts fabricated by fused deposition modeling'. *Materials & Design*. 2018;139:283–92.

[24] Matine A., Boyard N., Cartraud P., Legrain G., Jarny Y. 'Thermal properties of composite materials: effective conductivity tensor and edge effects'. *Journal of Physics: Conference Series*. 2012;**395**:012014.

[25] Jerez-Mesa R., Travieso-Rodriguez J.A., Corbella X., Busqué R., Gomez-Gras G. 'Finite element analysis of the thermal behavior of a RepRap 3D printer liquefier'. *Mechatronics*. 2016;36(1):119–26.

[26] Ahlers D., Wasserfall F., Hendrich N. '3D printing of nonplanar layers for smooth surface generation'. *IEEE 15th International Conference on Automation Science and Engineering (CASE)*. 2019.

[27] Rahman M., Schott N., Sadhu L.K. *Glass transition of abs in 3d printing [online]*. 2016. Available from https://www.comsol.com/paper/download/361301/rahman_paper.pdf

[28] Serdeczny M.P., Comminal R., Mollah M.T., Pedersen D.B., Spangenberg J. 'Numerical modeling of the polymer flow through the hot-end in filament-based material extrusion additive manufacturing'. *Additive Manufacturing*. 2020;36:101454.

[29] COMSOL. *Navier-stokes equations [online]*. Available from https://uk.comsol.com/multiphysics/navierstokes-equations

[30] COMSOL. *Accelerating model convergence with symbolic differentiation [online]*. Available from https://www.comsol.com/blogs/accelerating-model-convergence-with-symbolicdifferentiation/

[31] Vdovichenko I.I., Yakovlev M.Y., Vershinin A.V., Levin V.A. 'Calculation of the effective thermal properties of the composites based on the finite element solutions of the boundary value problems'. *IOP Conference Series: Materials Science and Engineering*. 2016;158:012094.

[32] Bharne M., Bhosle S. 'Mechanical thermal properties of fly ash filled abs'. *International Journal of Engineering Research & Technology (IJERT)*. 2014;**03**(2).

[33] 3dverkstan *The atomic method [online]*. Available from https://support.3dverkstan.se/article/10- the

[34] Carrier P. *3D printer print speed calculation: how to find the optimal speed for reliable and constant print quality [online]*. Available from https://dyzedesign.com/2018/07/3d-print-speedcalculation-find-optimal-speed/

Chapter 9
3D printing of composites for space applications

Sung Wook Paek[1] *and Sivagaminathan Balasubramanian*[2]

9.1 Introduction

Additive manufacturing (AM) has recently been adopting a wide variety of materials ranging from plastics, metals as well as their composites and alloys. Food, fabric, concrete and cement are amongst the new printable materials that have been considered commercially. The application areas of AM have also ever been expanding toward the field of space engineering beyond terrestrial applications. Researchers and entrepreneurs are actively investigating the potential of three-dimensional (3D)-printed composites to achieve the short-term goal of affordable access to space and the long-term goal of mankind's prolonged existence beyond Earth, namely, space habitats on the Moon, Mars, etc. The short-term goal involves AM of composites and metals, whilst the long-term goal necessitates AM of new printable materials encompassing nearly all facets of our everyday lives. Materials types other than composites are left as future work in accordance with the advent of novel AM methods and space exploration campaigns [1–3]. The focus of this chapter is to introduce (1) applications where spacecraft components made of 3D-printed composites can be used and (2) how these manufactured components can be assembled into a space system or structure.

The process of 3D-printing parts for space systems may occur on the ground (Earth) beforehand or in situ on orbits. There are several examples in which rocket engines or satellite components were 3D printed and then assembled into a ready-to-launch system on the ground. SpaceX first used a 3D-printed component in 2014, which was a liquid oxygen valve in one of the nine engines inside the Falcon nine rocket – nowadays, an entire engine of a small rocket can be 3D manufactured using large-scale metal printers with robotic arms, which reduce the number of assembled engine parts by an order of magnitude [4, 5]. These 3D metal printers are heavy and power-consuming because of their laser melting process. Composite

[1]City, University of London, Northampton Square, London, UK
[2]Samsung R&D Institute India, Doddanekundi, Marathahalli, Karnataka, India

3D printers, on the other hand, are free from these constraints because composites have much lower melting points than metals. Composite 3D printers have been installed in the International Space Station (ISS) for astronauts' use. Different versions and brands of commercial 3D printers have proved their functionality under microgravity. They are not being used outside the ISS under harsh space conditions such as vacuum, radiation, micrometeorites, etc. The first automated on-orbit AM will be demonstrated in the near future to produce structural components with large dimensions and aspect ratios. In light of this background and context, the latest development of AM technologies for composite manufacturing and assembling will be discussed within mechanical, electrical, electrochemical and medical applications in space.

9.2 3D-printed structures

Mechanical components and structures are one of the space applications wherein AM can be applied almost immediately thanks to their high technology readiness level. Those 3D-printed components exhibit mechanical performance comparable to conventionally manufactured space structures; they can also be functionalised during the AM process to possess shielding capabilities against heat and radiation. After the ongoing technological demonstrations of AM onboard the ISS, the next envisioned step would be to utilise AM for servicing, assembling and manufacturing in space. One of such efforts is the 2022 NASA mission where a satellite will autonomously manufacture deployable composite booms on orbit using its onboard 3D printer.

9.2.1 Heat shields for suborbital flight

Reinforced carbon–carbon (RCC) is a composite material that was originally developed for the re-entry parts of intercontinental ballistic missiles. In space engineering, RCC is used for thermal protection system (TPS) of re-entry vehicles for astronauts' safe return to Earth The use of RCC dates back to space shuttles. Prior to that, fibreglass honeycomb matrix infused with epoxy resin derived from phenols and formaldehyde was used for the ablative blunt-body heat shield of Apollo capsules [6]. Space shuttles used RCC for their wings' leading edge, nose cap and lower fuselage as depicted in Figure 9.1 [7]. Later, phenolic-impregnated carbon ablator (PICA) was used in Stardust (1999), a mission to collect comet samples and bring them back to Earth. Curiosity mission (Mars Science Laboratory 2011) also adopted PICA for its rover-carrying spacecraft underwent entry, descent and landing manoeuvres.

The Columbia Shuttle disaster happened during its Earth atmospheric re-entry in February 2003 after the RCC panels on the orbiter's left wing were damaged during its launch and the compromised TPS went undetected throughout the mission. After the orbiter disintegration incident, the development of multi-phased composite systems followed. The constituents of a carbon–carbon composite system are matrices, fillers and reinforcements [6]:

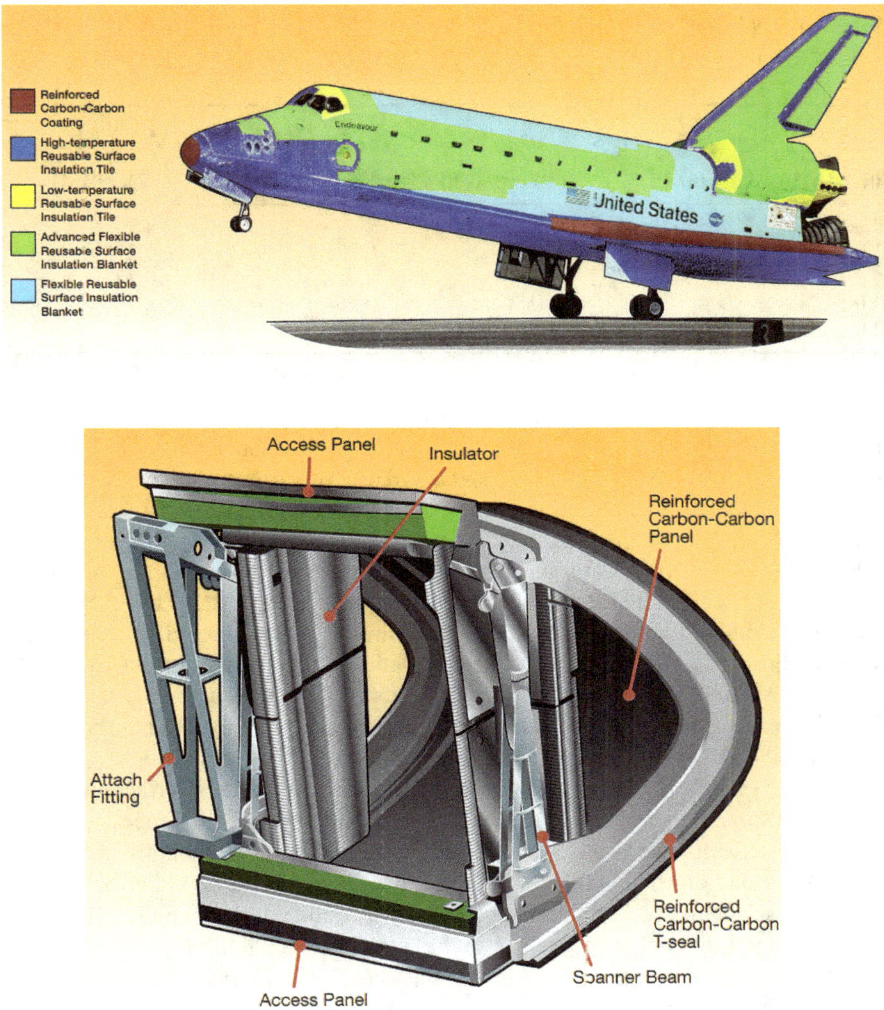

Figure 9.1 Thermal protection system of a space shuttle orbiter: tile placement (top) and wing panel assembly (bottom) [7]. Credit: NASA.

- Matrices: graphite, phenolic and polyamides
- Fillers: micro-filler (graphite and glass) and nanofiller (graphene nanoplatelet (GNP), carbon nanotube/fibre and silica)
- Reinforcements: glass fibre and continuous/chopped carbon fibre

Note that the presence of carbon nanotube (CNT) or graphene nanofillers, even with small proportions at a few weight (wt) %, increases thermal conductivity

Table 9.1 Thermal conductivity of polylactic acid composites with varying filler contents (W/mK) [7]

Filler (wt%)	Neat PLA	PLA + GNPs	PLA + MWCNTs	PLA + (MWCNTs + GNPs 1:1)
0	0.183	x	x	x
3.0	x	0.323	0.231	0.270
6.0	x	0.448	0.232	0.352
9.0	x	0.550	0.268	x
12.0	x	0.664	0.365	0.533

significantly. This feature would not be desirable for TPS and offset the advantage of improved mechanical properties by incorporating fillers into composites [7]. The linear increase in thermal conductivity in accordance of various filer contents is presented in Table 9.1, where thermoplastic polylactic acid (PLA, $[-CHCO-]_n$) in its pure composition was compared with mixtures containing variational filler wt% of multi-walled CNT (MWCNT) and GNPs. The result shows that adding any kind of fillers increased the thermal conductivity compared to neat PLA, but the degree of increases differed amongst the filler compositions. For a given filler proportion (wt%), addition of GNP into PLA resulted in relatively high thermal conductivity values; compared to GNP-doped PLA, MWCNT-doped PLA measured only about half of the thermal conduction coefficients. Both thermal conductivity and thermal resistivity have their own applications and could be utilised effectively once the physical phenomena behind these experiment results are understood. The interfacial thermal resistance (Kapitza thermal flow resistance) is affected by the difference in dimensional geometries between the 3D bulk polymer matrix and the nanofillers therein that could be one-dimensional (MWCNT) or two-dimensional (2D) (GNP). The 2D filler forms strong networks and homogeneous dispersion within the bulk composite, leading to better thermal conduction than the one-dimensional filler. It is possible to predict a composite's thermal characteristics affected by its matrix-filler morphology from molecular dynamics simulations that calculates the phonon density/modes in two phases and the phonon transport in between [10]. This computation-aided design of simulations could replace or complement time-consuming design of experiments (DOE), especially for the new types of additionally manufactured composites whose design space would be larger than that of traditionally manufactured composites.

Besides filler selection, the use of functionally graded materials (FGM) as reinforcements could be another design aspect to consider when it comes to the AM of TPS. The FGM layers or shells have gradually changed volume fractions along one or more dimensions of the solid. The idea of FGM was proposed in 1980s for thermal barrier applications [8]. Recently, in the context of AM, reinforcements are usually fibrous materials whose different heating/curing techniques have been suggested, including capillary-, ultrasound- or microwave-assisted approaches [9–11]. The FGM compositional gradient of SiC/C has a microstructure of sub-millimetre layers, from the pure SiC side to the pure C–C side with transitions, whose length scale is compatible with the AM resolutions. Both the SiC/C and the C/C composites reinforced with vapour-grown carbon fibre require high temperature (>1000 K) for oxidisation and carbonisation processes; as for pressure, these composites have a wide range of requirements ranging from several kPa to tens of MPa depending on fabrication steps. The AM settings for temperature and pressure will be much less demanding if TPS printing materials are available; it has been shown that functionally graded nanocomposites of phenolic resin and carbon nanofibre can be fabricated, and the AM implementation of FGM would be possible in the near future [12, 13]. Analytical methods and solutions have been developed for in-planar FGM problems to predict mechanical and heat conduction properties via finite element modelling, which is extendible to curved shapes theoretically and readily implementable in 3D printers [12, 14].

The first technical demonstration of TPS AM and assembly for space systems was made in 2019 when NASA Johnson Space Center and Oak Ridge National Laboratory investigated the feasibility of additively manufacturing heat shields with thermoplastics using a filament-based printer [15]. However, this method of fused deposition modelling (FDM), also known as fused filament fabrication (FFF), resulted in unstable char and swelling issues during a laboratory experiment mimicking atmospheric re-entry conditions. An alternative to this is thermoset resin mixtures and a customised commercial 3D printer that has an auger-based extruder. In this resin-based method, liquid-curable photopolymer is cured using a light source, whose optical-spot diameter determines the printing resolution [16]. As a result, the accuracy of this scheme is higher than that of FDM/FFF where the diameter and movement of a nozzle incur layer misalignment in x- and y-directions as well as layer squeezing and deformation in z-direction. The assembly of this ablative resin mixtures was also successfully demonstrated by depositing them onto a miniaturised mock-up representing a sphere-cone-shaped re-entry capsule. Scaling up this TPS-AM process without gaps may remain as a challenge, but the solution could be inspired from other disciplines such as seamless, continuous-fibre AM of wind turbine blades recently solicited by the US Department of Energy [17]. Above all, 3D printing of TPS has a potential for automating the present manufacturing approaches, which are human labour intensive and prone to errors. Both fabrication and quality inspection will take much less time and effort compared to the Apollo era when epoxy resin had to be filled into each honeycomb cell by hand. The same benefit would also apply to the maintenance of reusable launchers and suborbital airplanes.

9.2.2 Radiation shields in low-Earth orbits or deep space

Due to the absence of Earth's atmosphere and the surrounding magnetic fields, both robotic and crewed space vehicles in deep space are exposed to strong space radiation that is classified into ionising radiation and non-ionising radiation. The ionising radiation, because of its shorter wavelengths, has greater energy and higher possibility than non-ionising radiation to cause more damage to human DNA or microelectronics. Ionising

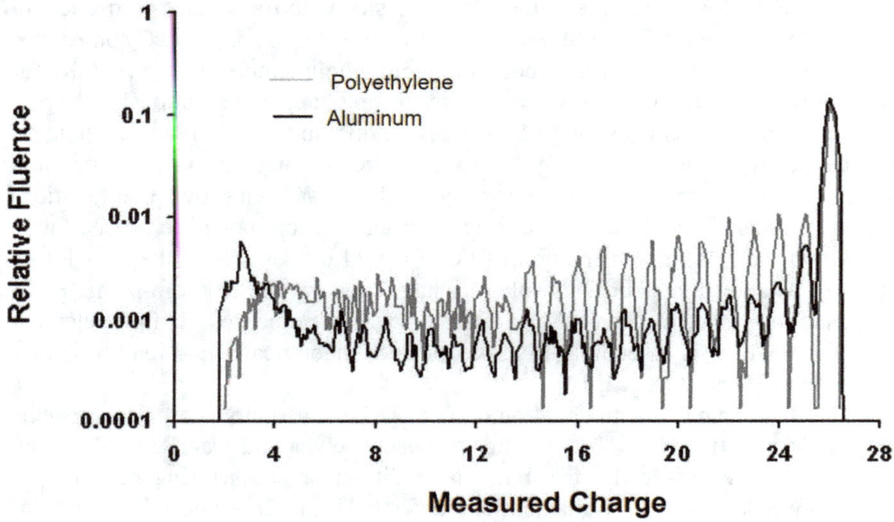

Figure 9.2 *Radiation-shielding properties of polyethylene and aluminium under accelerator measurements [19]. Reproduced with permission from Elsevier.*

radiation is subdivided into galactic cosmic ray (GCR) or solar particle event (SPE). Thought to originate from supernovae explosions, GCR contains high-energy nuclei (electron-free atoms) such as carbon, oxygen, magnesium, silicon or iron (Fe^{26+}). The heavy ions comprise only 2% of GCR, which are mostly protons and helium ions [18]. Also called HZE meaning high (H) atomic number (Z) and energy (E), these heavy ions are difficult to shield against with current technologies. Polyethylene (PE, $[-CH_2-]_n$) has been proposed as an alternative shielding material that absorbs more heavy iron ions than aluminium or lead. Figure 9.2 shows that PE absorbs more 1 GeV/amu (atomic mass unit) ^{56}Fe than aluminium over nearly all charges with a smaller areal density (4.5 g/cm^2 for PE and 5.0 g/cm^2 for Al) [19]. Here, the fluence is defined as the total number of particles passing through a sphere of unit cross section surrounding a point source of ionising radiation, meaning that relative fluence closer to nil provides better radiation shielding [20]. Customarily, PE is used with other materials in the sleeping quarters for astronauts aboard the ISS, and other recent applications involve the use of ultra-high-molecular-weight polyethylene (UHMWPE) composites [21, 22]. The shielding property of a material against GCR and secondary neutrons can further be enhanced as its hydrogen content exceeds that of PE (e.g., as a hydride, which is the anion of hydrogen) or is mixed with boron and nitrogen (e.g., borohydride or boronitride) [23, 24].

The process of SPE accelerating solar energetic particles via solar flares (solar atmosphere) or coronal mass ejection (interplanetary space) is relatively easier to observe and considered to be more predictable than GCRs of an interstellar origin. An historical SPE database and prototype nowcasting/forecasting tools provide the frequency (or probability) of solar flare and SPE occurrences and their intensity [25, 26].

Similar to GCR, protons from SPEs can be shielded in large space structures using water, food, clothing and other materials rich in hydrogen, but these options are not always available in smaller spacecraft where UHMWPE-like materials could be used instead [19]. Solar flares also emit X-rays (0.1–0.8 nm in wavelength) whose ionisation effect can damage or unexpectedly change the characteristics of spacecraft electronics. In a ground experiment, the total dose and rate of X-rays were shown to alter the voltage curve of metal-oxide-semiconductor field-effect transistors (MOSFET) that may lead to malfunctions in amplification or switching [27]. The deviation of threshold voltage (V_{th}) values was higher in p-type MOSFETs, whilst the change of voltage slope was more pronounced in n-type MOSFETs. Compared to X-rays, irradiation effects from heavy ion beams (^{35}Cl) were less marked, meaning that X-ray radiation shielding would be a higher priority for space electronics. Researchers demonstrated that polycarbonate (PC)–tungsten polymer matrix composite, 3D printed with tungsten-loaded PC filaments, resulted in X-ray radiation attenuation (~10%) compared to PC [28]. The volumetric loading and mass loading of tungsten microparticles were 0.3 and 5%, respectively, which did not affect the mechanical and electromagnetic properties of pure PC. Overall, these approaches may complement the existing, expensive radiation hardening techniques such as semiconductor chip shielding with borophosphosilicate glass, silicon on insulator, silicon on sapphire and so on [29].

Acrylonitrile butadiene styrene (ABS, $[-C_8H_8 \cdot C_4H_6 \cdot C_3H_3N-]_n$) is one of the first materials additively manufactured in orbit. Made In Space, a US company, used its Additive Manufacturing Facility (AMF) installed on the ISS to print and test radiation shields in mid-2017. In the tests, 3D-printed specimens with various thickness were attached to inner walls of the Bigelow Expandable Activity Module (BEAM), an inflatable annex to provide accommodation and workspace to astronauts (Figure 9.3 [30]). Each hemisphere-shaped specimen housed a USB-like Radiation

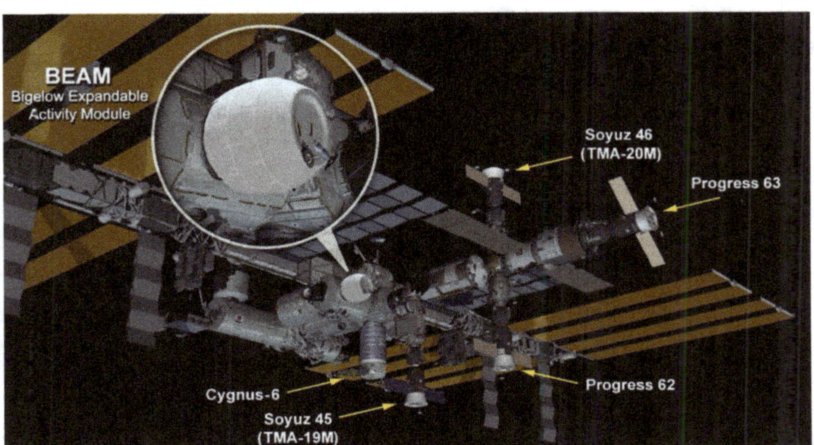

Figure 9.3 The BEAM module onboard the ISS [30]. Credit: NASA

Figure 9.4 *Radiation monitoring sensors and AM parts experimented inside the BEAM module [31]. Credit: NASA.*

Environment Monitor sensor, shown in Figure 9.4, to measure the radiation levels within the hemisphere, which was compared against those outside the hemisphere (still inside the BEAM) [31]. There were no noticeable changes in the radiation measurements for all plastic thicknesses (1.1, 3.3 and 10 mm) in this experiment, however. The ISS and other satellites in low-Earth orbits pass through the South Atlantic Anomaly (SAA) where the Earth's magnetic field has the local minimum, allowing X-ray and gamma-ray from solar flares to reach lower into the atmosphere and charged particles (protons) to be trapped therein [32, 33] (Figure 9.5). The influence of SAA ranges from laptop crashes in space shuttles to fatal software errors including one that ended the mission of Japan's X-ray astronomy satellite in 2016 [34]. Back in 2012 during its cargo mission to the ISS, SpaceX's Dragon spacecraft experienced a mild issue called single-event effect, which could be remedied by power cycling [35]. Astronauts passing through the SAA during their missions (Hubble, Mir and ISS) reported eye-related issues ranging from light flash (phosphene) to neuro-ocular syndromes [36]. Gamma-ray shielding might be possible with bismuth-ABS filaments according to a nuclear medicine research, which requires further verification in the space environment [37]. It is noteworthy that the efficacy of composite shielding will depend on the space weather conditions; solar maximum seasons would be desirable for deep-space travels because solar activity will deflect

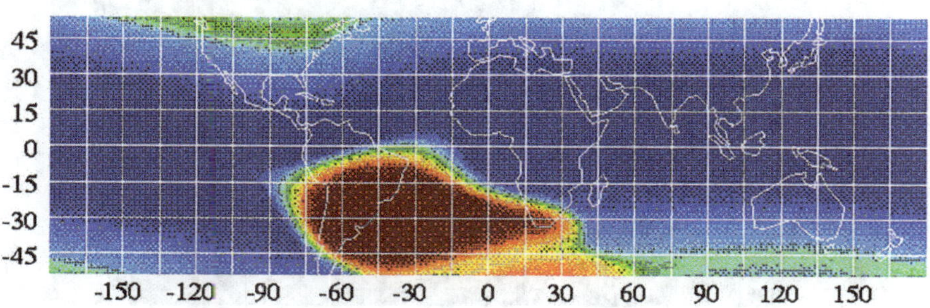

Figure 9.5 *South Atlantic Anomaly (SAA, red area) [32]. Credit: NASA.*

particles from distant galaxies (GCR) more efficiently [38]. A significant proportion of the GCR radiation consists of secondary radiation from neutrons and protons, whilst the SPE radiation is attributed to primary particles. Secondary radiation is produced from nuclear reactions between the primary particles and the spacecraft shielding, propagating toward the spacecraft interior in a multipath/pass fashion. Shielding secondary neutrons is especially important as they may comprise 20–50% of the equivalent dose where the upper band represents thicker shield materials and longer spaceflight duration, ultimately limiting human spaceflight. Shielded quarters or wearable shields might be 3D printed using silicon, for example [39]. Research facilities in the Lunar Gateway in lunar orbits, resembling the AMF in the ISS orbiting the Earth, would be useful to develop AM technologies for deep space travel.

The FDM/FFF method and the light-curing approach mentioned in section 2.1 required the heat source and filament materials. Direct ink writing (DIW) is another interesting approach that can broaden the material and geometry choices with a variety of curing methods: ultraviolet, elevated temperature and freeze drying to name a few [40]. Dimethylpolysiloxane (also called polydimethylsiloxane, PDMS, $[-Si(CH_3)_2O-]_n$), for example, exhibits neutron shielding properties after DIW using fillers such as boron, tungsten, tungsten (VI) oxide or/and gadolinium (III) oxide. Initial results from the neutron radiography proved the shielding capability in all test cases except when silica (silicon oxide) was filler. The ordering from the highest shielding specimen was: $B^{10}/Gd_2O_3 > B/Gd_2O_3 > B > Gd_2O_3$ ($>> SiO_2 \approx$ open beam without shielding) where B^{10} is a boron isotope. Other combinations reported in the medical sector for X-ray shielding, albeit not additively manufacturable yet, might take advantage of the DIW technique if the ink's rheological constraints are met. Composite materials are regarded as a potential alternative to toxic lead aprons and non-lead aprons whose alloy sheets could be still heavy. The alloy sheets contain non-lead heavy metals such as aluminium, tin, bismuth, tungsten and titanium; both lead and non-lead metals are used in combination with polymer-based composites in radiation protective garments [41, 42]. X-ray is a beam of high-energy photons induced from electron reactions and is shielded by lead-like materials dense in electron density, which is the main difference from high-energy-particle (GCR) shielding using hydrogen-rich materials. Similar to these radiation protective garments, polyethylene terephthalate (PET, $[-C_{10}H_8O_4-]_n$) fibres with barium sulphate ($BaSO_4$) was shown to provide lightweight, low-dose shielding for crews for aviation, which could be extended to suborbital or space flights [43]. Zirconium, gadolinium, tantalum, lanthanum and cerium are amongst other nanoparticle elements added as oxides to the polymer matrix for X-ray and gamma-ray shielding [44]. The reporting results differ in terms of irradiation energy levels, from a few keV to hundreds of keV, which needs adjust composite 3D printability is to be investigated.

9.2.3 *Issues in printing and assembling structural parts*

There are several applications for 3D-printing spacecraft structures with composites. After the printing of radiation shields, the AMF onboard the ISS has proven compatibility with 30 types of thermoplastics and polymers, producing over 200 parts and tools on orbit [45]. The AMF first used ABS and Green PE for manufacturing in zero gravity, the latter of which is a resin produced from sugarcanes [46]. Later, polyetherimide/PC (PEI/PC) was considered to produce tools used by astronauts during spacewalks. Unlike earlier 3D printable materials whose usage was limited to the interior of the ISS, PEI/PC has chemically stable properties adequate for exterior applications as well [47]; spare parts of the ISS were printed with PEI/PC due to its superior properties in harsh space environments including low outgassing in vacuum and durability against atomic oxygen or ultraviolet ray, to name a few.

One of the PEI product family is ULTEM™ series; for example, ULTEM 9085 is amenable to 3D printing because its flow property is enhanced by incorporating the PC copolymer, leading to the PEI/PC denotation [48]. This material was used to print an experiment apparatus (Slosh avionics box) delivered to the ISS when a fracture was found at the corner, as shown in Figure 9.6 [49]. Follow-on investigations have revealed that the fracture did not develop during or after the launch, ruling out the launcher's vibrations or crew handling from possible causes. One of the pictures taken during the pre-launch preparation indicated that a fracture did exist when packed for launch, which was not evident to naked eyes under the given lighting conditions (Figure 9.7, right; Figure 9.8). Traversal forces due to the misalignment of joined parts or radial stresses from the over-torqued screw were amongst the possible causes of fracture. Instead of pan-headed (cylinder shaped) or round (hemispherical) screws whose screwheads extrude from the surface, flat-head (inverted conical) screws were used to assemble the 3D-printed parts. Although flat-head screws save volume and prevent damage or injuries, they are susceptible to assembly errors in terms of both alignment and torque. This is because the conical head, sunk after fastening, cannot be checked easily whether it is under-torqued or over-torqued. The over-torqued scenario could be re-enacted in a laboratory experiment in 2015, but direct comparisons were still not possible because the experimented specimen differed with the avionics box in geometries.

On top of demonstrating the over-torqued scenario, the study investigated tensile/flexural behaviours with a DOE. Different settings for printing and post-processing (epoxy impregnation and surface sealing) were used. The extrusion paths were either solid (0″ gap) or gapped (0.004″ gap); five applicants in Table 9.2 were selected based on literature, amongst which Loctite 5110 is the least viscous and Hysol E-20HP is the tackiest [49–52]. Overall, gapped samples tended to have more gentle break characteristics than solid samples, as shown in Table 9.2 and Figure 9.9 with the following rubrics.

- Stayed together: the sample broke but was held together by a thin strand of material.
- Very gentle break: the sample failed and fractured but did not separate into two pieces.
- Very energetic break: the sample broke violently, impacting the walls of the test volume with considerable force.
- Unpredictable break: the type of break varied greatly from sample to sample.
- Solid samples broke under a higher load, but deflected.

Figure 9.6 A broken 3D-printed part around the screw (top, onboard the ISS) and the same part during packaging for launch (bottom) [48]. Credit: NASA.

Quantitatively speaking, gapped samples yielded higher strain and lower stress than solid samples for both control and experiments with most of the applicants. Extra spacing between adjacent extruding paths allowed for more deflection whilst limiting stress, with only few exceptions of BJB TC-1614 for load and Probuild Marine for displacement. Compared to control, Arathane 5750 A/B and Hysol E-20HP yielded higher deflection with similar loading limits, achieving higher flexural moduli. These applicants, however, resulted in a mass increase from 2% (solid) to 11% (gapped) relative to control. Not only the average increase in mass, higher standard deviations in gapped samples posed issues in processability

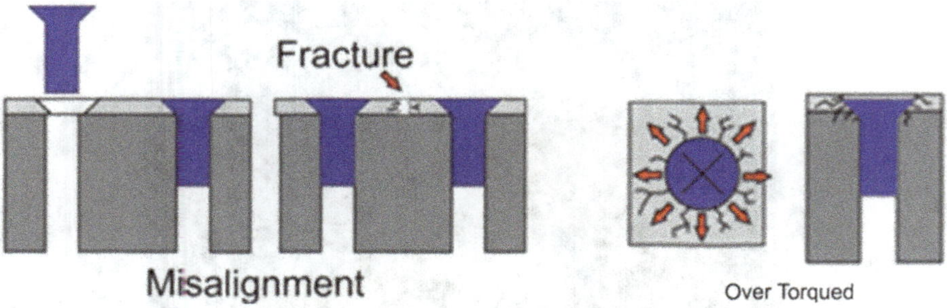

Figure 9.7 *Possible failure modes of countersunk flat-head screws: misalignment (left) and excessive torque (right) [48]. Credit: NASA.*

and reproducibility. Researchers have also been focusing on optimising several other printing parameters in addition to air gap; having a raster angle helps to reduce anisotropy because filaments can resist tensile loads in the length direction and the width direction of a sample. Similar rationale would apply in flexural modulus, as already shown by another NASA research [53]. More recent results, however, rather highlighted the role of other printing parameters; the most important were raster angle and raster width, followed by contour number and contour width, and the least important was air gap amongst five variables [54]. The role of air gap was less important for flexural strength contrary to its importance for tensile strength. The opposite phenomenon was observed in the other processing parameters, more influential to flexural strength but less influential to tensile strength, which would require further cross-verification.

As discussed so far, the mechanical properties of AM parts are often unsatisfactory compared to the conventionally manufactured (i.e., injection moulding) counterparts

Figure 9.8 *The threaded insert (left, avionics box) and radial cracks around it (right, specimen) [48]. Credit: NASA.*

Table 9.2 Break characteristics of samples according to applicant types and airgap configuration (B: brittle, D: ductile) [48–52]

	Applicant		**Airgap**	
	Name	**Viscosity (cPS)**	**No (solid)**	**Yes (gapped)**
Control	N/A	N/A	Energetic (B)	Stayed together (D)
Experiment	Arathane 5750 A/B	100–250	Energetic (B)	Very gentle (D)
	Hysol E-20HP (hardener/ resin)	5,500–8,000 40,000–90,000	Energetic (B)	Unpredictable (-)
	Loctite 5110	36–66	Very energetic (B)	Energetic (B)
	Probuild Marine	900–1,100	Very gentle (B)	Very gentle (B)
	BJB TC-1614	600	Very energetic (B)	Very energetic (B)

due to their anisotropic nature. Even after in-planar isotropy is achieved, inter-layer adhesions (layer stacking/building direction) are often weak. Instead of applicant post-processing, pre-deposition heating with a laser was proposed by researchers who studied ULTEM™ 1010 in Table 9.3. Along with ULTEM™ 1000, ULTEM™ 1010 is the one of the strongest FDM/FFF printable materials possessing superior tensile strength, flexural strength and modulus compared to ULTEM™ 9085 [54–57]. The pre-laser heating scheme increased the build-direction tensile strength of ULTEM™ 1010 by 178%; the corresponding isotropy value was up to 82.8%, which was defined as the strength ratio between the build direction and the build plane [58]. The maximum average tensile stress of 82.0 MPa is much closer to the manufacturer's injection-moulded rating of 105 MPa after the 178% increase. The optimal laser power level that led to this result was found to be 1.6 W; the scanning electron microscopy imagery of a fractured sample corroborates this isotropic-like behaviour, exhibiting fracture trajectories beyond the boundary of adjacent layers. The inter-layer interface is not distinguishable, which suggests the healing phenomena at the microstructure near interfaces as well as

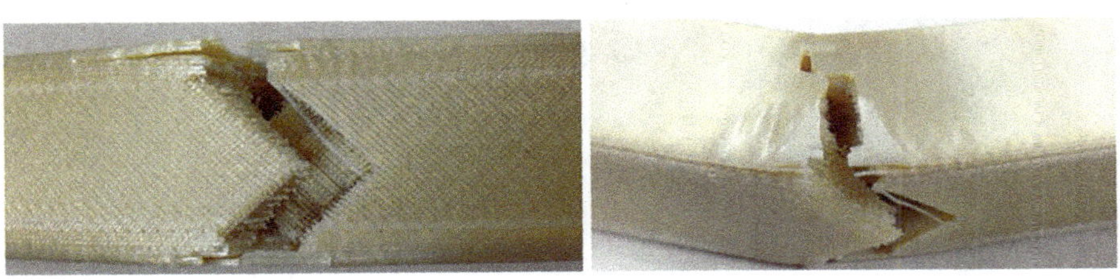

Figure 9.9 Gapped control sample that stayed together after the three-point beam test, viewed from beneath (left, rastered bottom/top) and sideways (right, contoured sides) [47]. Credit: NASA.

Table 9.3 *Mechanical properties of ULTEM series (units in MPa) [55–58]*

	ULTEM 9085				ULTEM 1000		ULTEM 1010	
	Injection moulded (Sabic)	FDM printed (Stratasys)	FDM (GRC)	FDM (UiS)	Injection moulded (Sabic)	FDM + C (GRC)	Injection moulded (Sabic)	FDM printed (UL)
Raster angle		0°	±45°	0°		0°/±45°		
Tensile strength	83	72	62	–	110	50/44	105	82
Tensile modulus	3,432	2,200	2,230	–	3,579	2,901 /2,248	3,200	–
Flexural strength	137	115	92	127	165	–	160	–
Flexural modulus	2,913	2,500	1,901	2,400	3,511	–	3,300	–

GRC, NASA Glenn Research Center; UL, University of Louisville; UiS, University of Stavanger.

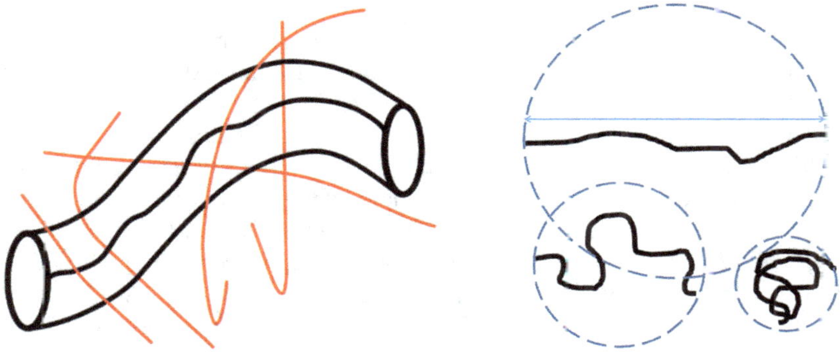

Figure 9.10 *Polymer kinetics: virtual tube of a polymer of interest (black) constrained by others (red) [60] (left); radii of gyration of a polymer chain in different conditions (concentration, extrusion, etc.) (right)*

the inter-layer interface. The power sweep between 0.3 and 2 W was performed, and the tensile strength decreased at 2 W power level with burning and smoke. Because the degradation temperature is 510°C, power levels equal to this or higher seem to generate local defects through decomposition during printing. When the power level is below 1.6 W, fewer entanglements occur between the two adjacent layers.

The effect of laser or ultrasound treatments have been experimentally demonstrated, but there have not been many attempts for theoretical explanations, predictions or verifications thereof [59]. In reptation theory, the relaxation time of an entangled chain system is proportional to the cube of molecular mass [60]. In Figure 9.10 (left), for enhanced mechanical properties of AM parts, a polymer chain must make snake-like (or reptile, from which the word reptation came from) wiggling motions out of a virtual tube it was occupying to proceed and eventually cross the interlayer boundary. It must then entangle with another chain on the other side of the interlayer boundary. The ratio of the strength of the interface (σ_t) to the maximum tensile strength of the material (σ_{max}) is proportional to the ratio of the interface healing time (t_{weld}) and the reptation time (τ_{rep}) to the power of one fourth, i.e. $(t_{weld}/\tau_{rep})^{0.25}$ [61, 62]. The healing time, which is the duration of temperature retention above the glass transition or melting threshold, can be extended in FFF via laser heating or ultrasonic pressure waves. The reptation time measures how long a molecule takes to diffuse a distance equal to the radius of gyration. As $\tau_{rep} = 2R_g^2\pi^{-2}D^{-1}$, the reptation time will increase if the centre of mass of a polymer chain does not displace much (small self-diffusion coefficient D); a similar result is obtained if a polymer coil is occupying more space with a large end-to-end distance (increased radiation of gyration R_g), as depicted in Figure 9.10 (right). Theoretical and experimental analysis of polymer film formation would shed light on optimising for 3D-printing space parts on Earth or in orbits.

Polyetheretherketone (PEEK, $C_{19}H_{14}F_2O$) is the final composite material to be introduced in this section. Both belonging to the polyaryletherketone (PAEK) family, PEEK and polyetherketoneketone (PEEK) have high stiffness and thermal resistance that distinguish them from PEI/PC despite their higher cost. PEEK is commonly found as either semi-crystalline or amorphous, whilst PEEK may have a crystallinity level up to 40% [63]. The amorphous nature of PEEK would facilitate inter-layer diffusion of polymer chains and layer adhesion, but PEEK has a higher continuous use temperature of PEEK (250°C) than that of PEEK (150°C, untreated) [64]. Because PEEK requires extra post-treatment to enhance its thermomechanical properties, it has been used for 3D-printing parts on the ground; PEEK is used in Boeing's CST Starliner, a space taxi for transporting crews and cargo to and from the ISS, and is also speculated to have been used for astronaut helmets manufactured by SpaceX [65]. On the other hand, PEEK is compatible with the AMF on the ISS and is being considered by industries for in-space or off-Earth manufacturing in the future.

9.3 3D-printed electronics

Printed electronics is a new research area for which electrically conductive or insulative thermoplastic filaments are needed. AM of reliable, high-performance circuitry is quintessential for internet of things (IoT) and electric-drive mobility for space/planetary exploration.

9.3.1 Traces and substrates

Earlier 3D-printed electronics date back to the printing and patterning of flexible microelectrodes that carry signals between microelectronic components [66]. Along with the nominal characteristics of conducting silver-ink traces, their failure modes due to excessive currents have been investigated [67], interconnects for solar cells or light-emitting diodes, whose compactness is comparable to satellite design. As shown in Figure 9.11, wiring harness must be distributed under the tight constraints in terms of both volume and geometries, routed through the leftover space inside a satellite structure. Wiring harness distribution is nontrivial but still has a low priority because accommodation of the payload footprints (main functions) is considered first, and then the bus volume (supporting functions) is minimised before wiring is done [68, 70, 71]. The nature of small satellite constraints often complicates the wiring process (Figure 9.11, left) but is not any more facile in large satellites, either (Figure 9.11, right) [69]. Embedded electronics with effective radiation shielding and thermal management would be useful for small satellites, robots and astronaut tools. However, their all-in-one nature would make it difficult to repair anomalies or recycle after the end of life.

Because the holistic realisation of smart structures of a satellite has the aforementioned issues, research activities of 3D-printed wiring are more focused on component applications including circuits, sensors and antennas. They consist of electrically conductive paths and components mounted on the nonconductive

Table 9.4 Classification of additively manufactured electronics

Class*	I	II	III	IV
Substrate structure	TM	AM	AM (post-print)	AM (dual-print)
Embedded device	TM	TM	AM (pre-print)	AM (dual-print)

*Class nomenclature is arbitrary.

substrate, either or both of which can be additively manufactured. Table 9.4 shows various approaches of additively manufacturing embedded electronics, with arbitrary class numbering used here for the sake of convenience. For most electronics in either ground-based systems or space-based systems, both substrates and devices are commercially available via traditional manufacturing (TM), as both marked "TM" in Class I of Table 9.4. As for the printed circuit board (PCB) substrates, flame resistant – four composite, made of woven fibreglass and epoxy resin, or glass fibre 10% has commonly been used in the National Electrical Manufacturers Association designation. Replacing these planar substrates with unusually or intricately shaped AM substrates provides extra degrees of freedom in functionalised design.

Copper traces on traditional PCBs conduct electricity with minimal losses because copper has an electrical resistivity as low as 1.6×10^{-8} Ωm (or 6×10^7 S/m); this resistivity value is slightly higher than silver, but copper has a major advantage in cost effectiveness. Compared to etched copper patterns, copper-based inks are 10–100 times more resistive at a laboratory scale and ≥ 1000 times at commercial production levels [72, 73]. Silver nanoparticle conductive inks have low viscosity (~10 cP) and thus are more adequate than copper conductive inks to be used in ink-jet printers [74, 75]. On the other hand, copper conductive inks are economical and mass producible, which has led to a wide variety of substrate applications including

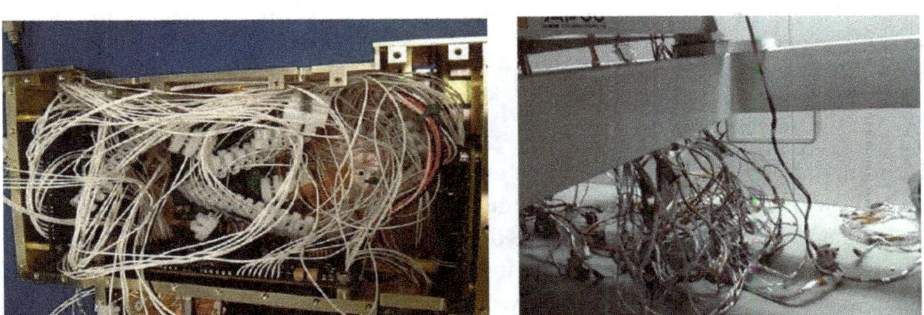

Figure 9.11 Wiring harness of Trailblazer CubeSat (left, 10 × 10 × 20 cm) and Galileo (right, 1.1 × 1.2 × 2.7 m) test facilities [68, 69]. Credit: University of Texas El Paso, NASA, ESA.

paper, fabrics, PET, PI and HJT PV cells for high precision additive screen printing [76]. The combination of these non-PCB substrates or traditional PCBs and copper-printed paths may be considered as a hybrid-manufactured substrate, which is an intermediate stage between Classes I and II [77]. Fabrication of more complicated circuit structures, double-sided or multi-layered, necessitates completely 3D-printed substrates (Class II). A study that tested several copper-mixed composites for soldering revealed that the tensile strength of substrate-component joints was the strongest for ABS, closely followed by polystyrene ($[-C_8H_8-]_n$), and the weakest for PLA [78]. From the industry side, it is now possible to print a ten-layer PCB with densely populated components [79]. Stacking multiple layers of a PCB increases manufacturing costs of the traditional electronics industry but is still a valid choice if the user application necessitates a high number of circuit components per unit area whilst keeping the PCB's thickness and volume largely unchanged. A circuit designer can use interconnects through these layers or allocate multiple power and ground planes in the interior layers for enhancing electrical isolation. These advantages of multiple layers already exist in traditional PCBs, but 3D-printed PCBs might match or outsmart the traditionally manufactured counterparts in terms of electromagnetic compliance, power quality and signal integrity by precisely controlling impedance of dielectric polymer ink [80]. Those extra layers could also be used for thermal management or other purposes, as briefly mentioned in the holistic approach. As for other ink materials, carbon black has higher electric resistivity than copper inks but could be used in harsh environments where metallic inks are prone to oxidation [81]. PEEK filaments doped with CNT or/and graphite nanoplates were investigated in an ESA research initiative, where the conductive PEEK had conductivity of 10 S/m, a hundredfold increase from pure PEEK [82].

Conformal components for IT devices are immediate terrestrial applications of printed flexible PCBs [83]. Electronics printer will be launched to be used onboard the ISS, and radio frequency components are being manufactured on the ground to be used on the ISS [84]. Even though the most conductive PEEK is too resistive to act as effective current paths, its thermal stability and thermoelectric property might be useful for (1) energy harvesting whereby waste heat is converted to electricity or (2) nuclear electric propulsion where heat is converted to electricity for ion propulsion.

9.3.2 Passive components

Passive circuit components consume energy (resistor) and store electric energy in an electric (capacitor) or magnetic (inductor) field. In circuit design, resistors lower the voltage, in the form of a voltage divider, to match the required input voltage of other components; capacitors and inductors allow alternating current and direct current to flow through, respectively, whose characteristics are utilised in high-pass filters and low-pass filters in analogue circuits. Transducers are also classified as passive components because they convert physical signals to electrical signals without amplifying functions [85]. Table 9.5 summarises the electrical characteristics of copper foil, which is conventional material choice for conducting paths on

Table 9.5 Electrical properties of copper foil (rectangular) and passive components (assumed circular) [75, 89, 90]

	Resistance	**Inductance**	**Capacitance**
Copper foil	$\frac{\rho l}{tw}$	$0.2l\left(ln\frac{2l}{w} + 0.2235\frac{w}{l} + 0.5\right)$ [μH] (l in mm, t negligible)	N/A
Circuit components	$\frac{\rho l}{A}$	$\frac{r^2 n^2}{8r+11w}$ [μH], Wheeler's approximation (r and w in inches)	$n\frac{k\varepsilon_0 A}{d}$

PCBs, in the first row and those of passive circuit components in the second row [72, 86, 87].

These formulas are derived from physics and are the same for traditionally manufactured components and 3D-printed components. The cross section of copper foil is usually rectangular, and circuit components have either circular (lumped) or rectangular (surface mounted) cross sections. In Table 9.5, ρ is the resistivity, l is the length, r is the mean radius, w is the width of a foil or coil and n is the number of turn (inductor coil) or stacks (capacitor); A stands for the cross-sectional area in a resistor and the overlapping area of stacks in a capacitor. The interconnects favour using copper because it has very low resistivity, resulting in negligible voltage drops despite micrometre-level thickness; for 3D-printed materials with a high resistivity value to be viable, the interconnect should have a large cross-sectional area or/and short length which will limit their applications, as depicted in Figure 9.12. For passive components, on the other hand, AM may diversify existing 2D geometries by printing unusual shapes or conformal patterns onto any curvilinear surfaces [88, 89]. Researchers were able to manufacture all basic circuit components including

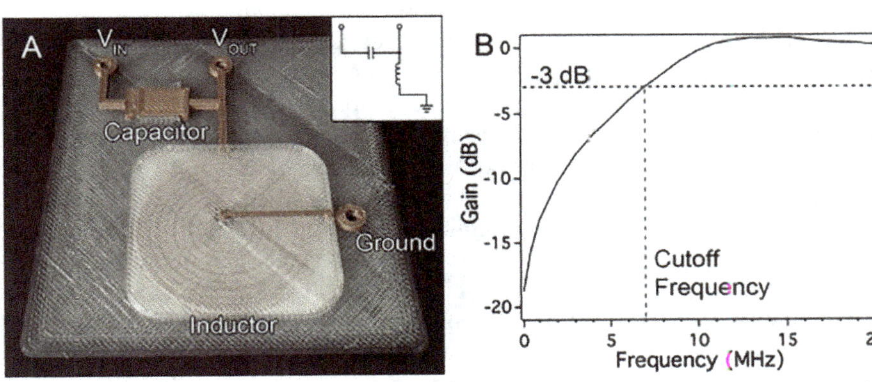

Figure 9.12 A fully 3D-printed high-pass filter: (a) circuit layout; (b) frequency response. Reproduced with permission from Elsevier.

resistors (R), inductors (L) and capacitors (C) that can be combined to function as filters or antennas (LC tanks); these components along with conducting traces are mounted on the surface of a circuit board. Another conceivable design is buried channels or components, which is filled with the liquid metal paste or suspension; strictly speaking, the metal is not 3D printed, but the substrate bearing internal channels is. One downside is that the increased injection/flow resistance for channel widths less than 400 μm places a limit in manufacturability [90]. The influence of microgravity in space or decreased gravity on the planetary surfaces must also be assessed because the packing of liquid metal pastes into subsurface channels would be affected. The ground experiment has been shown to reduce the paste's conductivity by a factor of ten compared to the ideal paste, and the influence of microgravity has not been researched yet. It would be worth investigating whether or not the reduced gravity and more uniform suspension of liquid metal is favourable to the electrical properties of buried channels or components. Another topic for future research would be fast and simple methods for improving the performance of AM circuit parts with conventional surface mounted, extruding geometries; one-step electrodeposition has proved to enhance the conductivity and other performance metrics (e.g., inductance for inductors) of 3D-printed electronic components [91].

Electrical machines or motors also benefit from composite 3D printing such as machine frame, shaft and cooling structures that could be used for traction systems for next generation rovers [92, 93]. An electric motor consists of a rotor and a stator that are attached to a shaft and a housing case, respectively. The stator, a hollow-cylindrical steel frame, is wound upright with copper wire for inductive properties [94]. Owing to the existence of copper winding, coolant layers are usually located around the motor housing and also referred to as a "water cooling jacket." Similar housing-embedded channels have been prototyped for the 3D-printed motor of an electric motorcycle [95]. For this prototype to be realized, motor housing, traditionally made of metal, needs to be 3D printed where thermosetting resins commonly used for injection moulding can be applied. It was claimed that composite housing for electric motors can achieve higher mechanical damping, corrosion resistance and electrical performance than metallic housing (e.g., aluminium) [96].

Researchers at the Fraunhofer Institute for Chemical Technology and the Karlsruhe Institute of Technology (KIT) injection moulded glass-fibre-reinforced thermoset (Vyncolit X7700) in their electric motor for traction applications [97]. Although they did not resort to AM, their choice of materials is easily transferable to AM. The above phenolic thermoset was chosen for housing due to its higher chemical stability than thermoplastics; it does not swell when in contact with water-glycol coolants or other automotive chemicals. Thermal conductivity of the material was not given a high priority because the inside stator, not housing, can pump out heat. This novelty of their stator water-circuit design can achieve reduced weight and increased power density. Table 9.6 compares indirect (conventional) and direct (proposed) cooling approaches. Their new direct-cooling scheme uses flat-section wires instead of round-section ones for conventional indirect cooling. The radial arrangement of flat wires leaves room for cooling channels within the stator that is closer to the heat sources, i.e., copper winding of the stator itself or the inside rotor,

Table 9.6 Cooling techniques for copper winding inside electric motors (injection moulding)

	Inner		**Outer**
	Wire	**Inter-wire space**	
Indirect cooling	Round	Irregular and left empty	Embedded cooling sleeve
Direct cooling	Flat	Cooling channel with triangular section	Polymer housing without cooling sleeve

than the housing where water cooling jacket used to be located. For stator fabrication, the iron frame and copper winding were overmoulded using epoxy resin, EME-A730E, which contains aluminium oxide fillers to be thermally conductive with a conductivity value of 3 $Wm^{-1}K^{-1}$. Mould cores are used to prevent the moulding process from filling the empty space where water channels should be. Lastly, the stator is assembled into the housing and fixed by connectors and union nuts.

Once AM is more widely used, the need for injection mould cores would be eliminated, enabling more challenging channel geometries, such as internally finned tubes, for stators or housings. It has been reported that moderately thermally conductive carbon-fibre reinforced nylon reduced the copper winding temperature rise by 44% [98]. Integrative AM of mechanical parts and electric parts (PCB circuit) of a motor could be envisioned, as demonstrated using injection moulding during the KIT study. Other components of a tractive system, such as brake rotors, may also be 3D printed [99].

9.3.3 Active components

In electronics, active components are defined as components that inject a net power into a circuit to perform various functions including signal manipulations, logical operations, optoelectronic displays and so forth. The active components are usually semiconductor based as seen in diodes, transistors and integrated circuits (IC). Chip bonding and assembly thereof have mainly been investigated in the context of stretchable printed electronics. For ultra-thin wearable/flexible circuits, electrically conductive adhesives should be used for chip bonding because soldering is inappropriate. Ink stretchability and bending/pulling failure modes were analysed for a variety of combinations of silver flake inks (siloxane, urethane and polyester based) and anisotropic conductive adhesives (epoxy, acrylic and silicone based) [100].

Their most delicate variants such as programmable devices require microfabrication steps of photolithography, deposition and etching whose nanometre-level resolution cannot be accomplished with AM; for example, a field programmable gate array is an IC consisting of hundreds of thousands of gates built on 28 or 14 nm process technologies [101, 102]. Although the most advanced AM technology providing near-micron resolution would not achieve this level of complexity, a relatively simple microcontroller system-on-chip (SoC) prototype was created in 2017 by a group of researchers from Air Force Research Laboratory using silver-infused thermoplastic polyurethane for conductive traces [103, 104]. This kind of flexible and stretchable SoC is to be adopted in a variety of IoT devices that personnel can wear whose target applications may include space activities. Since the SoC incorporates onboard memory and computing units, applications for soft robotics are also to be envisioned.

Drop-on-demand inkjet printing has been considered as one of the most promising AM technologies to manufacture transistors, diodes and displays that are active components commonly used in IC [105]. Small thickness of thin-film transistors (TFT) has made inkjet printing adequate for printing their vertically stacked structure. Printed logic gates, display driving circuits and switching relays consist of an array of TFTs whose ink and source/drain electrode films employ metal oxide, organic material, CNT or graphene [106]. To address the failure concerns of TFT, restoration of interconnects thereof was considered by triggering a self-engineering process [107]. In this scheme, an inherently nonconductive liquid dispersion around the interconnect becomes electrically conductive as metallic nanoparticles therein respond to an electric field strength between 0.7 and 1.3 V/μm. Open faults of 20 μm lengths could be healed, which corresponded to healing voltage values between 14 and 26 V across them. The healed TFT interconnects withstood repeated bending of 100 cycles, and the self-healing technology was implemented in a TFT-integrated voltage amplifier. The IoT and industrial IOT for space applications will make avail of other base technologies such as developing nontoxic inks and controlling the printing drop sequences [108, 109].

9.4 3D-printed batteries

Electrochemical energy storage devices in terrestrial applications store (charge) and release (discharge) energy for variegated purposes in our life. For space applications, batteries have been an important component in satellite operations and electric propulsion [110–112]. Planetary aircraft or rotorcraft resembling air taxis and NASA's Mars helicopter (Ingenuity) could be battery electric [113]. As for human spaceflight, the Environmental Control and Life Support System (ECLSS) requires constant and backup power sources. Specific properties considered whilst choosing batteries for space applications are energy density, cycle life, ease of maintenance, safety, reliability, peak power, sustainable power density (charging and discharging rates), range of operating temperature, self-discharge rates, cost, ease of manufacturing, etc. [114]. As different missions require different ranges of specifications, many different types of batteries, such as sealed lead–acid batteries, hydrogen batteries, Ni–Cd batteries, etc., have been studied and used for space applications in the past [115]. In recent days, lithium-ion battery (LIB) technology is finding prominent use and replacing conventional battery technologies in several of the space applications, because of its superior characteristics in many of the aforementioned properties.

This section mainly reflects on common AM technologies for making LIB that can address the demands of space applications.

Additive manufacturing technologies (AMT) are required for batteries, when conventional manufacturing technologies (CMT) do not meet the specifications (geometric shape, energy density, power density, etc.) demanded by the applications or where CMT is not feasible as in the beyond-Earth environment. A lithium-ion cell is a sandwich of roll coated anode and cathode with separator in between and pores filled with ion conducting electrolyte. The traditional approach of 2D coating limits the energy density and power density. The main advantage of AMT is its ability to leverage design and control of complex 3D geometries that can maximise the performance of materials through improved ion and electron transport, which CMT using 2D coating technology cannot achieve beyond a certain thickness. Also, AMT allows the flexibility to tailor the battery cell construction to meet the application or mission-specific demands, such as shape matching the available volumetric space, energy and power density, etc. LIB cells with 3D-printed electrodes have been shown to retain capacity of 85% even after operating at >10 C (charging/discharging the full capacity in 1/10 hour) for thousands of cycles [116]. Conventionally made cells can exhibit such a capacity retention only at an operation rate of <2 C (charging/discharging the full capacity in 1/2 hour).

Amongst the many different AMT, Pei *et al.* have identified four commonly used techniques – stereolithography (SLA), fused deposition modelling (FDM), DIW and inkjet printing – as most commonly suited for making LIB cells (refer to Figure 9.13) [117]. In LIB cell making, AMT is used for making the two electrodes and electrolyte with controlled geometry, e.g., interdigitated electrode for film, cube or gyroid-type electrodes and electrolyte composite to achieve better power density and energy density characterisations.

In the case of lithium metal anode cells, DIW method is used for making a lithophilic host on which lithium metal is deposited. The complex shape can withstand large volume change associated with lithium anodes in lithium-metal batteries [118]. In other electrodes, the electrode is made with a hierarchically porous structure using FDM or SLA [119]. FDM is the most commonly used technique to generate cells with complex 3D pore structures. The hierarchically porous structure decreases the charge transfer distance and enables cells to achieve power density as high as 10–20 C [116]. There are few studies that have concentrated on patterning the solid electrolyte interface (SEI), a protective layer created between anode and electrolyte. Film electrodes are usually made with inkjet printing technique [120]. In case of electrolyte, solid polymer electrolyte or garnet type ($Li_7La_3Zr_2O_{12}$) solid electrolyte is commonly used [121, 122]. Stencil printing, inkjet printing and SLA are commonly used [123]. The main objective of SEI printing or electrolyte printing is to decrease the impedance between electrode–electrolyte interfaces.

To maximise the desired outcomes, method of making complex geometry must match the rheological properties of materials. So active material for battery making is blended with cellulosic nano fibres, polymer network precursors, solvents, etc. Following printing or patterning, the structure undergoes heat treatment or photocuring to achieve desired characteristics, solvent removal, etc. Whilst a significant improvement in power and energy density characteristics has been achieved with AMT, further improvement in host material choice is needed for maintaining structural integrity. Even more improvement in precision in material deposition could be achieved with finer needles. However, the limitations arising from the clogging of nozzles used for material extrusion or inkjet printing limits the geometry choices. Advances in material choices could overcome these limitations.

Another notable class of energy storage devices is supercapacitors for self-powered electronics such as miniaturised sensors, biomedical implants and portable RFID tags. Compared to batteries, supercapacitors have limited energy density but provide compelling power density. Nickel (II) oxide (NiO) micro-supercapacitors were inkjet printed with nanoparticle-based thin film NiO for highly conductive electrodes and magnesium perchlorate aqueous solution for wide-voltage-window

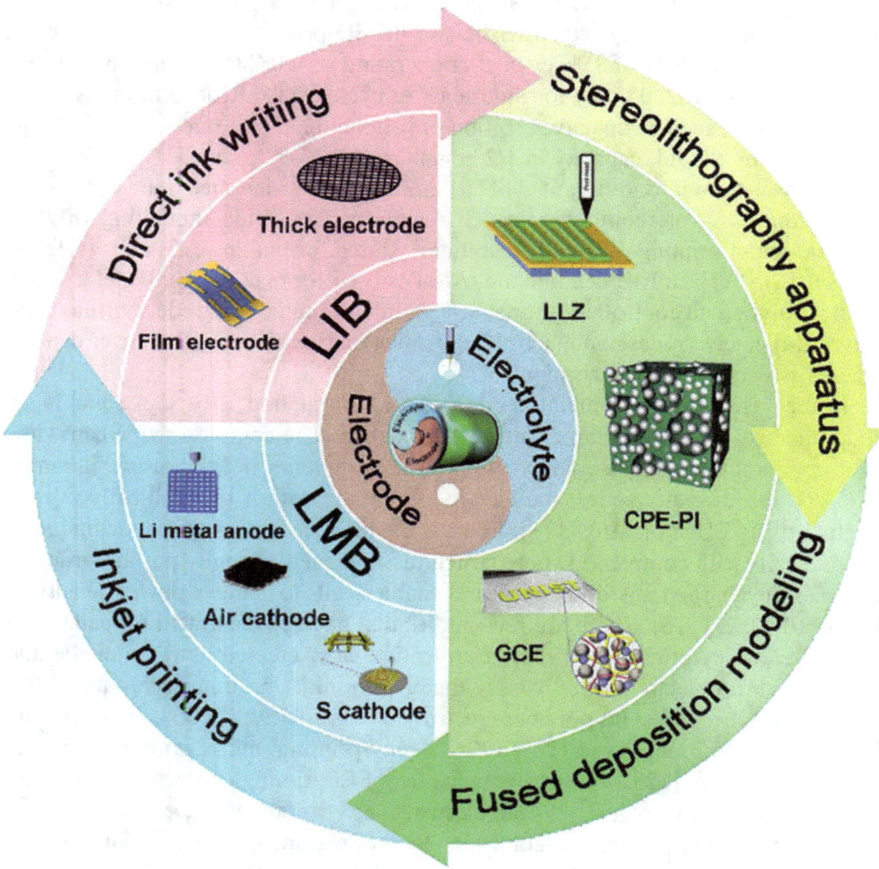

Figure 9.13 Possible role of different 3D-printing technologies in lithium-ion battery manufacturing [117]. Reproduced with permission from Royal Society of Chemistry.

electrolyte [124, 125]. A “3D-printed multiscale porous carbon aerogel (3D-MCA)” achieved 6.5 times higher capacitance than the non-3D-MCA, with a capacitance-per-mass density of 71.4 F/g at a high scan rate of 200 m/V [126]. Fabricated via direct inkjet writing and freeze-drying, the primary (500 μm) and secondary (nm sized) cavities increase the specific surface area up to 1750 m^2/g [79, 126]. The capacitance amounts to 148.6 F/g at −70°C if the scan rate is maintained low (5 m/V), which could endure nights on Mars without heating or nights on the Moon with heating.

Lastly, albeit non-power-storing, photovoltaic panels could be 3D printed using PET. The 200-micron-thick panel can be installed on nonplanar surfaces made of plastic or steel. Whilst the PV panels could be fully recycled, their efficiency of generating electricity from sunlight is an order of magnitude lower than conventional ones. Unlike this example, most 3D-printed energy storage devices suffer difficulties in recycling. Both volume manufacture methods (FDM) and film manufacture methods (inkjet printing and DIW) shown in Figure 9.13 should consider assembly/disassembly aspects to address this issue.

9.5 3D-printed devices for medical and life support purposes

Because medical assistance is not readily available in most space operations, it is essential to equip a spacesuit, space vehicle or outpost with health monitoring capabilities for its wearer or crew. On top of passive health monitoring for prevention or diagnosis of disease and injury, telemedicine would be desirable for crew medical care, providing necessary treatment and cure via remote or robotic means. However, medical and other life-supporting aspects complicate the logistics of crewed, long-duration missions as the mission planning should maintain a balance between supplies shortage and overstocking. If these life-critical supplies unexpectedly run out in the course of a deep space mission, restocking them from Earth is practically impossible. Carrying an excessive number of spares throughout the entire mission, however, will seriously degrade the feasibility and viability of the mission given the resource, space and weight limits [127]. Printing medical and life-supporting supplies on demand will reduce mission costs and contribute to crews’ health and well-being. From a broader perspective, the holistically optimised logistics via 3D printing will enhance the long-term sustainability of human space activities by minimising environmental impacts of rocket launch such as carbon emission or debris generation.

AM for medical usage is an emerging area even for terrestrial applications. For example, electric conductive paths, discussed in Section 3, are also necessary in medical devices for data connection and transmission. Electrically conductive paths, USB cables and USB-RS232 adapters made of PEEK have been shown to achieve a data transfer of 9600 bit/s (bps or baud) between two computers [128]. These data transfer interfaces support both wired and wireless communications; wired communications use adapters and wires between computers; wireless communication requires the same devices between the computer and an optical transmitter or receiver designed to transmit data to/from medical equipment [129]. For soft electronics whose conducting paths constitute sensors on human skins, multi-layer inkjet printing was applied to PDMS in conjunction with silver ink, the former of which mentioned in section two for gamma-ray shielding [130]. Incorporating 5G and IoT technologies will enable internet of medical things for prolonged human space activities.

The family of thermoplastics which PEEK belongs to is called PAEK. Other plastics that belong to the PAEK family are: polyetherketone (PEK), PEEK, polyetherketoneketone (PEEK), polyetheretherketoneketone (PEEKK) and polyetherketoneetherketoneketone (PEKEKK or Ultrapek®). In the above acronyms, "E" and "K" represent ether (R–O–R') and ketone (R2–C=O) groups, respectively, and R is the phenyl group (or phenyl ring) which is the simplest aryl. The wear performance of thermoplastic against itself has been investigated for PEK, PEEK and PEKEKK for spinal arthroplasty or small joints for the extremities in Reference 131. This biomedical application is relatively new and dominated by PEEK in terms of the market share although PEKEKK offers slightly higher melting temperature and tensile strength [132]. There are several commercial variants according to molecular weight and reinforcement fillers. Increased molecular weight leads to higher tensile elongation in unfilled PEEK. Filling glass and carbon fibres increases flexural strength by 67% and 100%, respectively, and tensile/flexural modulus by 200% and 500%, respectively, compared to unfilled PEEK [133]. In a laboratory setting, PEKEKNK was synthesised by inserting a naphthyl group and was shown to have higher rigidity and glass transition temperature then PEKEKK [134]. ESA and its industrial partners are developing a PEEK 3D printer operable in space and the prior ISS project (3D printing in Zero G) has demonstrated that microgravity did not generate significant engineering defects compared to the parts printed on the ground [127, 135].

Whilst PAEKs' medical usage is centred on surgical implants, medical/surgical instruments have been printed using less costly materials. Scalpel handles, sponge sticks, haemostats, forceps and clamps are amongst the common handheld instruments used in various surgical procedures, which have been printed with ABS and evaluated by surgeons in practice [136]. As mentioned earlier, ABS has been used on the ISS for zero-gravity 3D printing to hundreds of spare parts. However, ABS printers should be operated in a sealed environment because of its ultrafine particle emission whose rates could be as large as 200 billion per minute [137]. Surgeons participating in the study also did not prefer polymer-based cutting tools, making them one of the cases where metal-polymer composite hybrid structures shall be used [138]. Possibly related research activities from the metallic printing side include the Aiming toward Zero Waste & Efficient Production of High-Tech Metal Products (AMAZE) initiative by ESA and electron beam freeform fabrication technology (EBF3) by NASA Langley Research Center. Polyarylether sulphones (PAES) are manufactured as membranes for haemodialysis/haemofiltration applications [139]. The PAES family consists of polysulfone, polyethersulfone and polyphenylene sulfone, the last of which has better toughness than the first two in terms of elongation at break and impact strength [140]. Compared to their symmetric, homogeneously dense counterparts such as polyacrylonitrile and polymethylmethacrylate, PAES membranes have asymmetric pores with directionality. In the PAES membrane installation, the larger pores facing upstream act as a prefilter, which stops larger particles at the surface and allows smaller particles to enter [141]. Because of the PAESs are steam sterilisable with high-temperature structural stability, they are used for food/beverage-catering containers [142]; on account of the same advantage, they were also considered once for wastewater treatment apparatus in space-based ECLSS [143]. The application of PAES membranes was studied for different types of fuel cells (alkaline, proton exchange, etc.) which is another important part of the ECLSS[144]. The commercialisation of PAES filaments and compatible printers will expedite their use in space where applicable.

Lastly, environment sensors, alongside with wearable sensors discussed in Section 4, can be used for ECLSS purposes and would benefit from AM. Ink materials such as palladium–silver, aluminium (-tin), carbon–carbon–polymer composite are considered for the monitoring of NH_3, CO_2, CO, CH_4, H_2, humidity, temperature and pressure [145]. Some of these environment sensors or medical instruments might be expendable, and the polymer recycler/refabricator is being developed for ULTEM™ and other materials.

9.6 Conclusion

The assembly of 3D-printed composite structures has ever-growing applications in space as well as on Earth. Their immediate novel applications could be solar sail or drag sail by reason of the flexural properties of composites that allow spacecraft to roll and store composite-made appendages [146]. Extended-scale structures with high stiffness and strength will be manufactured on orbit autonomously aiming at fully blown, "in-space manufacturing (ISM)" [147]. The term ISM has a nuance that manufacturing processes are taking place on space orbital platforms, and "off-Earth manufacturing" may have a broader connotation of anywhere outside Earth, including the surface of other celestial objects [148].

Regardless of terminologies, advanced composite structures will play a vital role to extend human space activities from a single sortie to a series of campaigns, eventually leading to perpetual off-Earth existence [149]. Multidisciplinary efforts are indispensable in achieving these steps, however, because diverse applications will need different materials and methods for manufacturing. At microscopic levels, morphological optimisation for AM will enhance mechanical, shielding, electromagnetic or electrochemical properties where biomimicry or inspiration from other devices could be useful. Enhancement of a supercapacitor's power/energy delivering performance beyond that of planar printed counterpart illustrates a good example where the porous structure, common in conventionally manufactured batteries, increases the contact area between the electrodes and the electrolyte substantially. Multimaterial printing or assembling would require development of efficient joining methods and inspection thereof [138]. Autonomy or the internet of things will be crucial in the assembling stage as well as the manufacturing stage because human interventions will not be readily available at all times [145]. In a broader logistics perspective, more research and technological demonstrations are needed in the fields of 3D-printed composite recycling for sustainability or in-situ resource utilisation for collecting and processing raw materials in space [150, 151].

Acknowledgements

S.W. Paek thanks Dr. Akin Orunmuyi for useful discussions on the usage of composite materials in medical radiography.

References

[1] Lockney D. *3D-Printed composites for high temperature uses* [online]. 2021. Available from https://technology.nasa.gov/patent/LEW-TOPS-145 [Accessed 2 Oct 2021].

[2] *European space agency* [online]. 2017. Available from https://www.esa.int/Enabling_Support/Space_Engineering_Technology/3D_printing_CubeSat_bodies_for_cheaper_faster_missions [Accessed 2 Oct 2021].

[3] O'Neill P. *New study on the space durability of 3D-printed nanocomposites* [online]. 2020. Available from https://www.issnationallab.org/iss360/uiuc-durability-3d-printed-nanocomposites-misse/ [Accessed 3 Oct 2021].

[4] Salmi B. *The world 's largest 3D metal printer is churning out rockets* [online]. 2019. Available from https://spectrum.ieee.org/the-worlds-largest-3d-metal-printer-is-churning-out-rockets [Accessed 31 Dec 2021].

[5] Thryft A.R. *SpaceX reveals 3D-printed rocket engine parts* [online]. 2014. Available from https://www.designnews.com/design-hardware-software/spacex-reveals-3d-printed-rocket- engine-parts [Accessed 2 Oct 2021].

[6] AVCOAT-wikipedia. Available from https://en.wikipedia.org/wiki/AVCOAT [Accessed 3 Jan 2022].

[7] Chapline G., Rodriguea A., Snapp C., *et al*. *Thermal Protection Systems* [online]. Available from. Available from https://www.nasa.gov/centers/johnson/pdf/584728main_Wings-ch4b-pgs182-199.pdf [Accessed Accessed 2 Oct 2021].

[8] Sanoj P., Kandasubramanian B. 'Hybrid carbon-carbon ablative composites for thermal protection in aerospace'. *Journal of Composites*. 2014;**2014**(1):1–15.

[9] Spinelli G., Lamberti P., Tucci V, *et al*. 'Nanocarbon/poly (lactic) acid for 3D printing: effect of fillers content on electromagnetic and thermal properties'. *Materials*. 2019;**12**(15):2369.

[10] Matsubara H., Ohara T. 'Effect of the in-plane aspect ratio of a graphene filler on anisotropic heat conduction in paraffin/graphene composites'. *Physical Chemistry Chemical Physics*. 2021;**23**(21):12082–92.

[11] Koizumi M. 'FGM activities in Japan'. *Composites Part B: Engineering*. 1997;**28**(1–2):1–4.

[12] Yeong W.Y., Goh G.D. '3D printing of carbon fiber composite: the future of composite industry?' *Matter*. 2020;2(6):1361–3.

[13] Qiao J., Li Y., Li L. 'Ultrasound-assisted 3D printing of continuous fiber-reinforced thermoplastic (FRTP) composites'. *Additive Manufacturing*. 2019;**30**(3):100926.

[14] Shi B., Shang Y., Zhang P., *et al.* 'Dynamic capillary-driven additive manufacturing of continuous carbon fiber composite'. *Matter*. 2020;2(6):1594–604.

[15] Dhakate S.R., Mathur R.B., Dhami T.L. 'Development of vapor grown carbon fibers (VGCF) reinforced carbon/carbon composites'. *Journal of Materials Science*. 2006;**41**(13):4123–31.

[16] Bafekrpour E., Yang C., Natali M., Fox B. 'Functionally graded carbon nanofiber/phenolic nanocomposites and their mechanical properties'. *Composites Part A: Applied Science and Manufacturing*. 2013;**54**(11).

[17] C.C M., Chen Y.T. 'Theoretical analysis of heat conduction problems of nonhomogeneous functionally graded materials for a layer sandwiched between two half-planes'. *Acta mechanica*. 2011;221(3):223–37.

[18] NASA *3D printing heat shields* [online]. 2021. Available from https://techport.nasa.gov/view/93193 [Accessed 7 Oct 2021].

[19] Grames E. *Resin vs filament 3D printer (SLA vs FDM): The differences* [online]. 2021. Available from https://all3dp.com/1/sla-vs-fdm-resin-vs-filament-3d-printer-the-differences/ [Accessed 12 Oct 2021].

[20] Gardiner G. *Orbital composites to demonstrate containerized 3D printing robots for am wind blade manufacture* [online]. 2021. Available from https://www.compositesworld.com/news/orbital-composites-to-demonstrate-containerized-3d-printing-robots-for-am-wind-blade-manufacture

[21] Barcellos-Hoff M.H., Mao J.-H. 'HZE radiation non-targeted effects on the microenvironment that mediate mammary carcinogenesis'. *Frontiers in Oncology*. 2016;**6**:57.

[22] Guetersloh S., Zeitlin C., Heilbronn L., *et al.* 'Polyethylene as a radiation shielding standard in simulated cosmic-ray environments'. *Nuclear Instruments and Methods in Physics Research Section B: Beam Interactions with Materials and Atoms*. 2006;252(2):319–32.

[23] *Ionactive, fluence rate* [online]. 2018. Available from https://ionactive.co.uk/resource-hub/glossary/fluence-rate [Accessed 12 Oct 2021].

[24] Elmore B., Alyaqoub Y., Sussman M., Law T., Griffin C. *Solar particle event shielding* [online]. Available from https://events.engineering.oregonstate.edu/sites/expo.engr.oregonstate.edu/files/expo_poster_0_0.pdf [Accessed 12 Oct 2021].

[25] Zhong W.H., Sui G., Jana S., Miller J. 'Cosmic radiation shielding tests for UHMWPE fiber/nano-epoxy composites'. *Composites Science and Technology*. 2009;**69**(13):2093–97.

[26] Thibeault S.A., Fay C.C., Lowther S.E, *et al. Radiation shielding materials containing hydrogen, boron, and nitrogen: systematic computational and experimental Study—Phase I*. Washington, DC: NIAC Final Report; 2012.

[27] Bennington S., Lovell A., Headen T, *et al*. *Spacecraft shield. Patent application us 2014/0299794 A1*; 2014.

[28] *U.S. department of commerce, NOAA, space weather prediction center, solar proton events affecting the earth environment (January 1976 – March 2011)* [online]. 2011. Available from ftp://ftp.swpc.noaa.gov/pub/indices/SPE.txt [Accessed 15 Oct 2021].

[29] Papaioannou A. *Web-based Tools for Forecasting Solar Particle Events and Flares* [online]. 2019. Available from https://blogs.egu.eu/divisions/st/2019/01/14/web-based-tools-for-forecasting-solar-particle-events-and-flares/ [Accessed 15 Oct 2021].

[30] da Silveira M.A., Santos R.B., Leite F, *et al*. 'Radiation effect mechanisms in electronic devices'. Proceedings of Science. X Latin American Symposium on Nuclear Physics and Applications; Montevideo, 2013.

[31] Shemelya C.M., Rivera A., Perez A.T, *et al*. 'Mechanical, electromagnetic, and X-ray shielding characterization of a 3D printable tungsten–polycarbonate polymer matrix composite for space-based applications'. *Journal of Electronic Materials*. 2015;**44**(8):2598–607.

[32] Vázquez-Luque A., Marín J., Terrón J.A, *et al*. 'Neutron induced single event upset dependence on bias voltage for CMOs SRAM with BPSG'. *IEEE Transactions on Nuclear Science*. 2013;**60**(6):4692–96.

[33] Garcia M. *BEAM Fully Expanded and Pressurized* [online]. 2016. Available from https://blogs.nasa.gov/spacestation/2016/05/28/beam-fully-expanded-and-pressurized/[Accessed 9 Oct 2021].

[34] Wells N. 'Bigelow Expandable Activity Module (BEAM) ISS Year-Three Technology Demonstration, Utilization, and Potential Future Applications. *ISS R&D Conference*; Altanta, GA, USA; 30 July 2019.

[35] Snowden S.L. *South Atlantic Anomaly* [online]. 2018. Available from https://heasarc.gsfc.nasa.gov/docs/rosat/gallery/misc_saad.html [Accessed 12 Oct 2021].

[36] Hajdas W., Eggel C., Wigger C., Sanctuary H., Zehnder A., Smith D. *Spacecraft activation and south atlantic anomalyanomaly profiles measured with RHESSI satellite*. Vol. 536. Proceedings of the 7th European Conference on Radiation and Its Effects on Components and Systems: ESA Special Publication; 2004. p. 607.

[37] Witze A. 'Software error doomed Japanese Hitomi spacecraft'. *Nature*. 2016;533(7601):18–19.

[38] Bergin C. *Dragon enjoying ISS stay*, despite minor issues – Falcon 9 investigation begins [online]. 2012. Available from https://www.nasaspaceflight.com/2012/10/dragon-iss-stay-minor-issues-falcon-9-investigation/ [Accessed 13 Oct 2021].

[39] Macias B.R., Ferguson C.R., Patel N, *et al*. 'Changes in the optic nerve head and choroid over 1 year of spaceflight'. *JAMA Ophthalmology*. 2021;**139**(6):663.

[40] Ceh J., Youd T., Mastrovich Z., *et al*. 'Bismuth infusion of ABS enables additive manufacturing of complex radiological phantoms and shielding equipment'. *Sensors*. 2017;17(3):459.

[41] Dobynde M.I., Shprits Y.Y., Drozdov A.Y., Hoffman J., Li J. 'Beating 1 sievert: optimal radiation shielding of astronauts on a mission to Mars'. *Space Weather*. 2021;**19**(9).

[42] Talley S.J., Robison T., Long A.M, *et al*. 'Flexible 3D printed silicones for gamma and neutron radiation shielding'. *Radiation Physics and Chemistry*. 2021;**188**(10):109616.

[43] Brounstein Z., Zhao J., Wheat J., Labouriau A. 'Tuning the 3D printability and thermomechanical properties of radiation shields'. *Polymers*. 2021;**13**(19):3284.

[44] Jaquith K. *Determining which X-ray apron material is right for you* [online]. 2016. Available from https://blog.universalmedicalinc.com/determine-x-ray-apron-material-right/ [Accessed 18 Oct 2021].

[45] *Apc cardiovascular, xenolite garments [online]*. 2021. Available from https://www.apccardiovascular.co.uk/portfolio-item/xenolite-garments/ [Accessed 18 Oct 2021].

[46] Kim S.-C., Son J.S. 'Double-layered fiber for lightweight flexible clothing providing shielding from low-dose natural radiation'. *Scientific Reports*. 2021;**11**(1):1–9.

[47] Brounstein Z., Zhao J., Wheat J., Labouriau A. 'Tuning the 3D printability and thermomechanical properties of radiation shields'. *Polymers*. 2021;**13**(19):3284.

[48] *Redwire space, additive manufacturing facility (AMF)* [online]. 2021. Available from https://redwirespace.com/products/amf/ [Accessed 20 Oct 2021].

[49] Caliendo H. *Braskem's green PE used to 3D print parts in space* [online]. 2016 [Accessed 22 Oct 2021].

[50] Koslow T. *New plastic allows astronauts to 3D print spacewalk tools* [online]. 2017. Available from https://all3dp.com/new-plastic-allows-astronauts-to-print-spacewalk-tools/ [Accessed 30 Oct 2021].

[51] Zaldivar R.J., Witkin D.B., McLouth T., Patel D.N., Schmitt K., Nokes J.P. 'Influence of processing and orientation print effects on the mechanical and thermal behavior of 3D-printed ULTEM® 9085 material'. *Additive Manufacturing*. 2017;**13**(3):71–80.

[52] NASA Ames Research Center,. *3-D Printed Ultem 9085 Testing and Analysis* [online]. 2017. Available from https://ntrs.nasa.gov/api/citations/20150017060/downloads/20150017060.pdf [Accessed 7 Oct 2021].

[53] Krayden Inc., Arathane 5750-A/B (LV) [online]. Available from http://phxcubesat.asu.edu/sites/default/files/general/arathane_5750_conformal_coating_tds_huntsman.pdf [Accessed 15 Nov 2021].

[54] *Henkel adhesive technologies, loctite EA E-20HP* [online]. 2021. Available from https://www.henkel-adhesives.com/cn/en/product/structural-adhesives/loctite_ea_e-20hp.html [Accessed 20 Oct 2021].

[55] *BJB materials, TC-1614* [online]. 2021. Available from https://bjbmaterials.com/epoxy/3d-printed-part-sealer/tc-1614-a-b/ [Accessed 29 Dec 2021].

[56] *Cass polymers, probuild marine epoxy systems* [online]. 2010. Available from https://www.michiganfiberglass.com/specs/data%20sheets/ProBuild%20Tech.pdf [Accessed 29 Dec 2021].

[57] Gebisa A.W., Lemu H.G. 'Influence of 3D printing FDM process parameters on tensile property of ULTEM 9085'. *Procedia Manuf.* 2019;**30**(3):331–38.
[58] Gebisa A.W., Lemu H.G. 'Investigating effects of fused-deposition modeling (FDM) processing parameters on flexural properties of ULTEM 9085 using designed experiment'. *Materials*. 2018;**11**(4):500.
[59] *ProtoLabs, ULTEM [*online]. 2021. Available from https://www.protolabs.com/media/1014801/ultem-1000-im.pdf [Accessed 28 Nov 2021].
[60] *E-polymer, ULTEM* [online]. 2017. Available from http://www.e-polymer.com/pdfs/1010.pdf [Accessed 8 Nov 2021].
[61] Han P., Tofangchi A., Deshpande A., Zhang S., Hsu K. 'An approach to improve interface healing in FFF-3D printed ultem 1010 using laser predeposition heating'. *Procedia Manuf.* 2019;**34**:672–77.
[62] Shivokhin M.E., Read D.J., Kouloumasis D., *et al.* 'Understanding effect of constraint release environment on end-to-end vector relaxation of linear polymer chains'. *Macromolecules*. 2017;50(11):4501–23.
[63] *Wikipedia. reptation* [online]. 2021. Available from https://en.wikipedia.org/wiki/Reptation [Accessed 28 Dec 2021].
[64] Tofangchi A., Han P., Izquierdo J., Iyengar A., Hsu K. 'Effect of ultrasonic vibration on interlayer adhesion in fused filament fabrication 3D printed ABS'. *Polymers*. 2019;11(2):31513 Feb 2019.
[65] Kim K.D., Sperling L.H., Klein A., Hammouda B. 'Reptation time, temperature, and cosurfactant effects on the molecular interdiffusion rate during polystyrene latex film formation'. *Macromolecules*. 1994;27(23):6841–50.
[66] Toscano A. *Peek vs PEEK, what is the difference?* [online]. 2021. Available from https://www.roboze.com/en/resources/peek-vs-pekk-what-is-the- difference.html [Accessed 18 Nov 2021].
[67] Hendrixon S. *Ultem, peek or PEEK? choosing between high-temperature plastics* [online]. 2019. Available from https://www.additivemanufacturing.media/articles/ultem-peek-or-pekk-choosing-between-high-temperature-plastics [Accessed 18 Nov 2021].
[68] Scott A. Exclusive: Boeing's space taxis to use more than 600 3D-printed parts [online]. 2017. Available from https://www.reuters.com/article/us-boeing-space-exclusive-idUSKBN15I1HW [Accessed 19 Nov 2021].
[69] Kwas A., MacDonald E., Muse D, *et al.* 'Enabling technologies for entrepreneurial opportunities in 3D printing of smallsats'. *Small Satellite Conference Website*. 2014.
[70] Ahn B.Y., Duoss E.B., Motala M.J., *et al.* 'Omnidirectional printing of flexible, stretchable, and spanning silver microelectrodes'. *Science*. 2009;323(5921):1590–3.
[71] Roberson D.A., Wicker R.B., MacDonald E. 'Microstructural characterization of electrically failed conductive traces printed from Ag nanoparticle inks'. *Materials Letters*. 2012;**76**:51–54.
[72] Becedas J., Caparrós A. 'Additive manufacturing applied to the design of small satellite structure for space debris reduction'. *Applications of Design for Manufacturing and Assembly*. 2018.

[73] Paek S.W., Kim S., de Weck O. 'Optimization of reconfigurable satellite constellations using simulated annealing and genetic algorithm'. *Sensors*. 2019;19(4):765.

[74] *European space agency, satellite harness* [online]. 2020. Available from https://www.esa.int/ESA_Multimedia/Images/2020/03/Satellite_harness [Accessed 25 Nov 2021].

[75] Ma W.-Y., Cheng Y.-Y., Chen J.-K., *et al.* 'Synthesis of antioxidative conductive copper Inks with superior adhesion'. *Journal of nanoscience and nanotechnology*. 2018;18(1):318–22.

[76] Flowers P.F., Reyes C., Ye S., Kim M.J., Wiley B.J. '3D printing electronic components and circuits with conductive thermoplastic filament'. *Additive Manufacturing*. 2017;**18**:156–63.

[77] Fernandes I.J., Aroche A.F., Schuck A, *et al.* 'Silver nanoparticle conductive inks: synthesis, characterization, and fabrication of inkjet-printed flexible electrodes'. *Scientific Reports*. 2020;**10**(1):1–1.

[78] Espera A.H., Dizon J.R.C., Chen Q., Advincula R.C. '3D-printing and advanced manufacturing for electronics'. *Progress in Additive Manufacturing*. 2019;**4**(3):245–67.

[79] *Copprint, print ready, nano copper conductive inks* [online]. 2021. Available from https://www.copprint.com/products/ [Accessed 6 Dec 2021].

[80] Salam B., Lai W.L., Albert L.C., Keng L.B. 'Low temperature processing of copper conductive ink for printed electronics applications'. *2011 IEEE 13th Electronics Packaging Technology Conference*; IEEE; 2011. pp. 251–5.

[81] Podsiadły B., Skalski A., Słoma M. 'Soldering of electronics components on 3D-printed conductive substrates'. *Materials*. 2021;**14**(14):3850.

[82] Belz L. *HENSOLDT and nano dimension achieve breakthrough in electronics 3D printing.* [online]. 2021. Available from https://www.hensoldt.net/news/hensoldt-and-nano-dimension-achieve-breakthrough-in-electronics-3d-printing/ [Accessed 12 Dec 2021].

[83] Shepard J. *What are the applications of 3D and 4D printed electronics?* [online]. 2021. Available from https://www.analogictips.com/what-are-the-applications-of-3d-and-4d-printed-electronics-faq/ [Accessed 10 Dec 2021].

[84] Leigh S.J., Bradley R.J., Purssell C.P., Billson D.R., Hutchins D.A. 'A simple, low-cost conductive composite material for 3D printing of electronic sensors'. *PLoS ONE*. 2012;**7**(11):e49365.

[85] Gonçalves J., Lima P., Krause B, *et al.* 'Electrically conductive polyetheretherketone nanocomposite filaments: from production to fused deposition modeling'. *Polymers*. 2018;**10**(**8**):925.

[86] *Octomec, 3D printed electronics-aerosol jet technology* [online]. 2021. Available from http://optomec.com/printed-electronics/aerosol-jet-technology/ [Accessed 15 Nov 2021].

[87] Thryft A. *3D printing for more circuits* [online]. Available from https://semiengineering.com/3d-printing-for-more-circuits/ [Accessed 15 Nov 2021].

[88] Richert F. *Introduction to electronic compontents: active vs. passive components* [online]. 2020. Available from https://www.power-and-beyond.com/introduction-to-electronic-compontents-active-vs-passive-components-a-893768/ [Accessed 10 Dec 2021].

[89] *Tech web, resistance and inductance of copper foil* [online]. 2018. Available from https://techweb.rohm.com/knowledge/dcdc/dcdc_pwm/dcdc_pwm03/5048 [Accessed 17 Nov 2021].

[90] Wheeler H.A. 'Simple inductance formulas for radio coils'. *Proceedings of the Institute of Radio Engineers*; 1928. pp. 1398–400.

[91] Huang J., Ware H.O.T., Hai R., Shao G., Sun C. 'Conformal geometry and multimaterial additive manufacturing through freeform transformation of building layers'. *Advanced Materials*. 2021;**33**(11):2005672.

[92] Peels J. *How conformal will change design and 3D printing* [online]. 2021. Available from https://3dprint.com/281436/how-conformal-will-change-design-and-3d-printing [Accessed 19 Dec 2021].

[93] S.Y W., Yang C., Hsu W., Lin L. '3D-printed microelectronics for integrated circuitry and passive wireless sensors'. *Microsystems & Nanoengineering*. 2015;**1**(1):1–9.

[94] Kim M.J., Cruz M.A., Ye S, *et al*. 'One-step electrodeposition of copper on conductive 3D printed objects'. *Additive Manufacturing*. 2019;**27**:318–26.

[95] Naseer M.U., Kallaste A., Asad B., Vaimann T., Rassõlkin A. 'A review on additive manufacturing possibilities for electrical machines'. *Energies*. 2021;**14**(7):1940.

[96] Lockheed Martin. Lockheed Martin and General Motors Team Up on Lunar Rover [online]. 2021. Available from https://www.lockheedmartin.com/en-us/news/features/2021/lunar-terrain-vehicle.html [Accessed 10 Dec 2021].

[97] Composites World, *Reducing-heat-in-ev-motors-via-fiber-reinforced-thermoset-housing-and-stator* [online]. 2019. Available from https://www.compositesworld.com/news/reducing-heat-in-ev-motors-via-fiber-reinforced-thermoset-housing-and-stator[Accessed 10 Dec 2021].

[98] *University of Nottingham, PHD student wins additive world design challenge Award – campus news* [online]. 2016. Available from https://exchange.nottingham.ac.uk/blog/phd-student-wins-additive-world-design-challenge-award/ [Accessed 10 Dec 2021].

[99] Van Dine P., Odessky V., Spencer B.E., Smith J.S. *Inventors; Electric Boat Corp, Assignee. Method for Making a Composite Electric Motor Housing*. 2000.

[100] Langheck A., Reuter S., Saburow O, *et al*. 'Evaluation of an integral injection molded housing for high power density synchronous machines with concentrated single-tooth winding'. *8th International Electric Drives Production Conference (EDPC)*; IEEE, 2018. pp. 1.

[101] Sixel W., Liu M., Nellis G., Sarlioglu B. 'Cooling of windings in electric machines via 3-D printed heat exchanger'. *IEEE Transactions on Industry Applications*. 2020;**56**(5):4718–26.

[102] Zelinski M. *Lighter, better-performing brake rotor from 3D printing* [online]. 2021. Available from https://www.additivemanufacturing.media/articles/lighter-better-performing-brake-rotor-from-3d-printing-the-cool-parts-show-27 [Accessed 31 Dec 2021].

[103] Behfar M.H., Khorramdel B., Korhonen A, *et al*. 'Failure mechanisms in flip-chip bonding on stretchable printed electronics'. *Advanced Engineering Materials*. 2021;**17**:2100264.

[104] Xilinx 7 series FPGAS data sheet: overview (DS180) [online]. 2021. Available from https://www.xilinx.com/support/documentation/data_sheets/ds180_7Series_Overview.pdf [Accessed 26 Dec 2021].

[105] Davidson A. *A new FPGA architecture and leading-edge finfet process technology promise to meet next-generation system requirements* [online]. 2020. Available from https://www.intel.com/content/dam/www/programmable/us/en/pdfs/literature/wp/wp-01220-hyperflex-architecture-fpga-socs.pdf [Accessed 26 Dec 2021].

[106] Alia-Novobilski M. *AFRL, American semiconductor create flexible system-on-chip for 'internet-of-things'* [online]. 2017. Available from https://www.wpafb.af.mil/News/Article-Display/Article/1402631/afrl-american-semiconductor-create-flexible-system-on-chip-for-internet-of-thin/ [Accessed 9 Dec 2021].

[107] Alia-Novobilski M. *AFRL, Harvard researchers invent new hybrid 3-D printing method for flexible electronics* [online]. 2017. Available from https://www.af.mil/News/Article-Display/Article/1345387/afrl-harvard-researchers-invent-new-hybrid-3-d-printing-method-for-flexible-ele/ [Accessed 27 Dec 2021].

[108] Chung S., Cho K., Lee T. 'Recent progress in inkjet-printed thin-film transistors'. *Advanced Science*. 2019;**6**(6):1801445.

[109] Shao S., Liang K., Li X, *et al*. 'Large-area (64 × 64 array) inkjet-printed high-performance metal oxide bilayer heterojunction thin film transistors and n-metal-oxide-semiconductor (NMOS) inverters'. *Journal of Materials Science & Technology*. 2021;**81**:26–35.

[110] Ding L., Joshi P., Macdonald J., Parab V., Sambandan S. 'Self-healing thin-film transistor circuits on flexible substrates'. *Advanced Electronic Materials*. 2021;**7**(3):2001023.

[111] Nayak L., Mohanty S., Ramadoss A. 'A green approach to water-based graphene ink with reverse coffee ring effect'. *Journal of Materials Science: Materials in Electronics*. 2021;32(6):7431–42.

[112] Pervin Brishty F., Grau G. 'Machine vision methodology for inkjet printing drop sequence generation and validation'. *Flexible and Printed Electronics*. 2021;**6**(3):035009.

[113] Paek S.W., Balasubramanian S., Kim S., de Weck O. 'Small-satellite synthetic aperture radar for continuous global biospheric monitoring: a review'. *Remote Sensing*. 2020;**12**(16):2546.

[114] Paek S.W. 'Concurrent design optimization of earth observation satellites and reconfigurable constellations'. *Journal of the British Interplanetary Society*. 2017;**70**:19–35.

[115] Paek S.W., Kronig L.G., Ivanov A.B., de Weck O.L. 'Satellite constellation design algorithm for remote sensing of diurnal cycles phenomena'. *Advances in Space Research*. 2018;**62**(9):2529–50.

[116] Paek S.W., Kim S., Raj R.V. 'optimal endurance and range of electric aircraft with battery degradation'. *Transactions of the Japan Society for Aeronautical and Space Sciences*. 2020;**63**(2):62–65.

[117] Jha A.R. *Next-Generation batteries and fuel cells for commercial, military, and space applications*. Boca Raton: FL,CRC Press; 2012.

[118] Halpert G., Frank H., Surampudi S. 'Batteries and fuel cells in space'. *The Electrochemical Society Interface*. 1999;**8**(3):25–30.

[119] Hu J., Jiang Y., Cui S, *et al*. '3D-printed cathodes of limn1-x fe x po$_{-4}$ nanocrystals achieve both ultrahigh rate and high capacity for advanced lithium-ion battery'. *Advanced Energy Materials*. 2016;**6**(18):1600856.

[120] Pei M., Shi H., Yao F, *et al*. '3D printing of advanced lithium batteries: a designing strategy of electrode/electrolyte architectures'. *Journal of Materials Chemistry A*. 2021;**9**(45):25237–57.

[121] Chen C., Li S., Notten P.H.L, *et al*. '3D printed lithium-metal full batteries based on a high-performance three-dimensional anode current collector'. *ACS Applied Materials & Interfaces*. 2021;**13**(21):24785–94.

[122] Saleh M.S., Li J., Park J., Panat R. '3D printed hierarchically-porous microlattice electrode materials for exceptionally high specific capacity and areal capacity lithium ion batteries'. *Additive Manufacturing*. 2018;**23**(10):70–78.

[123] Lawes S., Sun Q., Lushington A., Xiao B., Liu Y., Sun X. 'Inkjet-printed silicon as high performance anodes for li-ion batteries'. *Nano Energy*. 2017;**36**:313–21.

[124] Yousefpour P., McDaniel J.R., Prasad V, *et al*. 'Genetically encoding albumin binding into chemotherapeutic-loaded polypeptide nanoparticles enhances their antitumor efficacy'. *Nano Letters*. 2018;**18**(12):7784–93.

[125] Cheng M., Jiang Y., Yao W, *et al*. 'Elevated-temperature 3D printing of hybrid solid-state electrolyte for li-ion batteries'. *Advanced Materials*. 2018;**30**(39):1800615.

[126] Kim S.-H., Choi K.-H., Cho S.-J., Yoo J., Lee S.-S., Lee S.-Y. 'Flexible/shape-versatile, bipolar all-solid-state lithium-ion batteries prepared by multistage printing'. *Energy & Environmental Science*. 2018;**11**(2):321–30.

[127] Giannakou P., Masteghin M.G., Slade R.C.T., Hinder S.J., Shkunov M. 'Energy storage on demand: ultra-high-rate and high-energy-density inkjet-printed nio micro-supercapacitors'. *Journal of Materials Chemistry A*. 2019;7(37):21496–506.

[128] Tomiyasu H., Shikata H., Takao K., Asanuma N., Taruta S., Park Y.-Y. 'An aqueous electrolyte of the widest potential window and its superior capability for capacitors'. *Scientific Reports*. 2017;**7**(1):1–2.

[129] Yao B., Peng H., Zhang H, *et al*. 'Printing porous carbon aerogels for low temperature supercapacitors'. *Nano Letters*. 2021;**21**(9):3731–37.

[130] Rajput S. 'A review of space surgery-what have we achieved, current challenges, and future prospects'. *Acta Astronautica*. 2021;**188**(1):18–24.

[131] Saunders S. *Esa and zortrax 3D print peek composites for built-in electronics* [online]. 2020. Available from https://3dprint.com/274117/supported-by-esa-zortrax-3d-prints-composite-parts-using-two-peek-polymer-blends/ [Accessed 23 Oct 2021].

[132] Lin J.C., editor *Advances in electromagnetic fields in living systems. London: Plenum Press*; 1994. p. 152.

[133] Mikkonen R., Puistola P., Jönkkäri I., Mäntysalo M. 'Inkjet printable polydimethylsiloxane for all-inkjet-printed multilayered soft electrical applications'. *ACS Applied Materials & Interfaces*. 2020;**12**(10):11990–97.

[134] Scholes S.C., Unsworth A. 'The wear performance of PEEK-OPTIMA based self-mating couples'. *Wear*. 2010;**268**(3–4):380–87.

[135] Urakawa O. *Encyclopedia of polymeric nanomaterials – polyaryletherketone* [online]. 2021. Available from https://link.springer.com/referenceworkentry/10.1007%2F978-3-642-36199-9_410-1 [Accessed 29 Dec 2021].

[136] *Victrex, material properties guide* [online]. 2021. Available from https://www.victrex.com/-/media/downloads/literature/en/material-properties-guide_us-4-20.pdf?rev=6e0e04abaf9f49ee971517316e6baa4c [Accessed 31 Dec 2021].

[137] Dosière M., Villers D., Zolotukhin M.G., Koch M.H.J. 'Comparison of the structure and thermal properties of a poly (aryl ether ketone ether ketone naphthyl ketone) with those of poly (aryl ether ketone ether ketone ketone)'. *E-Polymers*. 2007;**7**(1).

[138] *BeeVeryCreative, development of a 3D printer for the internation space Station* [online]. 2021. Available from https://beeverycreative.com/proj-imperial.php?lang=en [Accessed 11 Nov 2021].

[139] Wong J.Y., Pfahnl A.C. '3D printing of surgical instruments for long-duration space missions'. *Aviation, Space, and Environmental Medicine*. 2014;85(7):758–63.

[140] Stephens B., Azimi P., El Orch Z., Ramos T. 'Ultrafine particle emissions from desktop 3D printers'. *Atmospheric Environment*. 2013;**79**(Suppl. 2):334–39.

[141] Feistauer E.E., Santos J.F., Amancio-Filho S.T. 'A review on direct assembly of through-the-thickness reinforced metal–polymer composite hybrid structures'. *Polymer Engineering & Science*. 2019;**59**(4):661–74.

[142] Yumoto M., Nishida O., Moriyama K, *et al*. 'In vitro evaluation of high mobility group box 1 protein removal with various membranes for continuous hemofiltration'. *Therapeutic Apheresis and Dialysis*. 2011;**15**(4):385–93.

[143] Sastri V.R. *Plastics in medical devices: properties, requirements, and applications*. Elsevier; 2021.

[144] Sterlitech Advantages of asymmetric membranes [online]. 2018. Available from https://www.sterlitech.com/blog/post/advantages-of-asymmetric-membranes

[145] BASF U *(Psu, PESU, PPSU) (basf.com)* [online]. 2021. Available from https://plastics-rubber.basf.com/global/en/performance_polymers/products/ultrason.html [Accessed 2 Jan 2022].

[146] Slavin T.J., Oleson M.W. 'Technology tradeoffs related to advanced mission waste processing'. *Waste Management & Research: The Journal for a Sustainable Circular Economy*. 1991;**9**(5):401–14.

[147] NASA Marshall Space Flight Center. *International Space Station Environmental Control and Life Support System* [online]. 2004. Available from https://www.nasa.gov/centers/marshall/pdf/174687main_eclss_facts.pdf [Accessed 3 Jan 2022].

[148] Clinton R.J. Jr.*am in space: ISM and IRMA NASA initiatives* [online]. 2019. Available from https://ntrs.nasa.gov/api/citations/20190005001/downloads/20190005001.pdf [Accessed 23 Nov 2021].

[149] Serfontein Z., Kingston J., Hobbs S., Holbrough I.E., Beck J.C. 'Drag augmentation systems for space debris mitigation'. *Acta Astronautica*. 2021;**188**(A6):288-88.

[150] Paek S.W., de Weck O., Hoffman J., Binzel R., Miller D. 'Optimization and decision-making framework for multi-staged asteroid deflection campaigns under epistemic uncertainties'. *Acta Astronautica*. 2020;**167**(6533):23–41.

[151] European Space Agency *Off-earth manufacturing symposium: how to build a new home in space* [online]. 2021. Available from https://www.esa.int/Enabling_Support/Preparing_for_the_Future/Discovery_and_Preparation/Off-Earth_manufacturing_symposium_how_to_build_a_new_home_in_space [Accessed 24 Nov 2021].

Chapter 10

Development and manufacturing of thermoplastic composite booms for drag augmentation system of a small satellite

Chun Ho Yau[1], Zaria Serfontein[2], and Hamed Yazdani Nezhad[1]

In recent years, the number of small satellite launches has increased significantly due to technological advancement allowing space to be more accessible and to be commercialised. As of the end 2020, market research shows that the number of small satellite launches has reached a record number high bringing it up to a total of 1,163 small satellites, which is more than triple the amount to previous 2019 record reported by market research. The increase in launches has added to the already substantial amount of object into the low Earth orbit. Space debris is widely recognised as a critical threat by major space agencies as it creates a hazardous environment for future missions. Therefore, there is a need for a cost-effective drag augmented system to de-orbit satellite at the end-of-life missions as required by Inter-Agency Debris Committee (IADC). A drag augmented system called the de-orbit mechanism (DOM) was subsequently developed at Cranfield University, which uses a drag sail to de-orbit satellites in low Earth orbit, which enables satellites to be compliant with the debris mitigation requirement set upon by the IADC. The current DOM uses a copper beryllium (CuBe) boom and a lightweight alternative was needed. This project looks at different designs for lightweight composite booms via the use of finite element analysis. Lastly, the project explores the cost-effective way of additive manufacturing (AM), and using Markforged onyx series FDM printer, a set of collapsible tubular mast (CTM) booms were created.

[1]Aeronautics and Aerospace Research Centre, Department of Mechanical Engineering and Aeronautics, City, University of London, London, UK
[2]School of Aerospace, Transport and Manufacturing, Cranfield University, Bedford, UK

10.1 Introduction

10.1.1 Background

The number of satellite launches has significantly increased in recent years due to advancement in technology allowing spaceflight to be more accessible for commercialisation. Market research shows as of the end of 2020, the number of small satellite (<500 kg) launches has dramatically increased to a record high, bringing the number up to 1,163. shows that this is triple the number of launches recorded, by market research, in comparison to the year 2019's record. The number of future satellite launches is estimated to only increase as low-cost satellites attract commercialisation [1].

However, the number of launches invites imminent risks as the amount of objects in lower Earth orbit (LEO) would increase. Space debris is widely recognised by major space agencies, such as the European space agencies (ESA) and National Aeronautics and Space Administration (NASA), as a major critical threat as debris creates a hazardous environment for future missions. ESA publish reports annually to track the amount of debris in orbit. These reports highlighted the urgent need for a subset of internationally recognised Space Debris Mitigation (SDM) measures as debris increases exponentially as shown in graph in from the report conducted [2].

Subsequently, major space agencies from around the world such as ESA and NASA placed a high-priority on the end-of-life operations of a spacecraft. The IADC was formed to standardise mitigation measures and to create guidelines for the long-term sustainable use of outer space. IADC is an inter-governmental effort and since it was formed, major steps have been taken to safeguard future missions. It is now required for satellites to be removed from LEO within 25 years of end-of-life (ISO 24113). The failure to comply with IADC's guideline can result in the denial of launch licences from authorities. Therefore, there is a need to provide a low-cost solution to enable small satellites, considering the increasing number of satellites occupying LEO would be <500 kg and especially ones without propulsion system, to de-orbit at end-of-life in-order to meet IADC's SDM requirements [3].

10.1.2 Cranfield University's de-orbit mechanism

The drag augmented system called the DOM was subsequently developed at Cranfield University which consists of a drag sail to de-orbit the satellite in LEO, which enables the satellite to comply with the debris mitigation requirement set upon by IADC. The mechanism developed by Cranfield was designed to be low mass and is simple in design. The design of DOM is self-contained and allows installation to take place at a very late stage of the satellite development. The DOM operates by the deployment of four individual aluminised Kapton sail quadrants using deployable boom to enlarge the effective mass of the spacecraft. The increase in drag causes the orbital decay of the vehicle to increase, thus allowing it to re-enter and burn up in the Earth's atmosphere [3].

The DOM consists of several different components as shown in the exploded drawing of the assembled DOM below (Figure 10.1). The CuBe booms and the

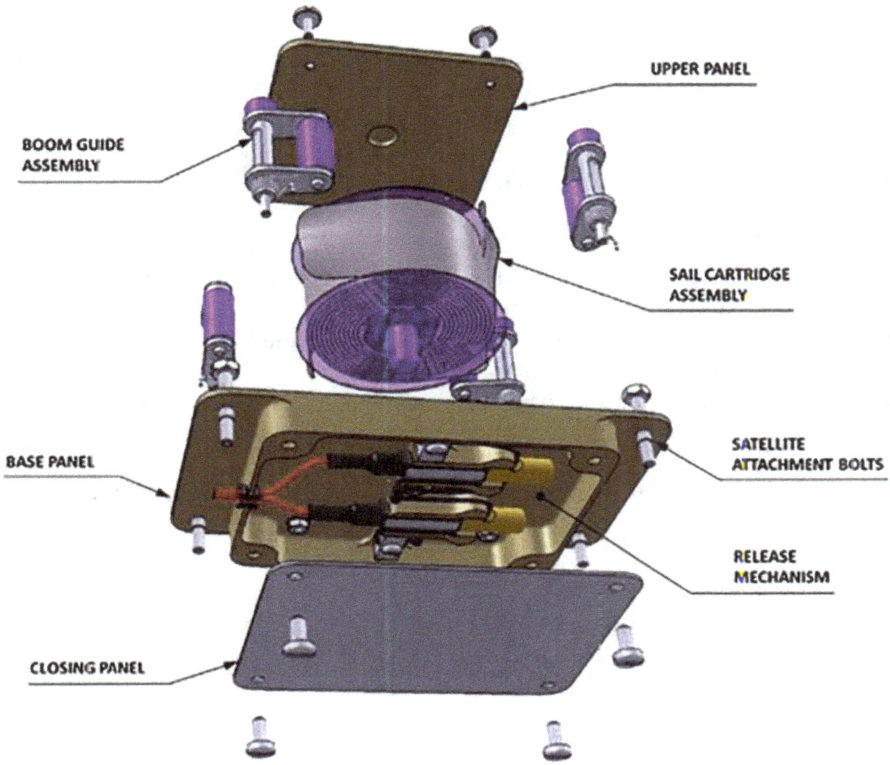

Figure 10.1 Exploded view of the DOM

aluminised Kapton sail quadrants are coiled around a central spool. This is held in place by Kevlar cords and the deployment of the boom is achieved through a series of commands which activates the two *CY PRES™* cord cutters. The cord cutter snaps the Kevlar cords to allow the booms and sails to be released by converting the stored strain energy to be converted into kinetic energy. The size of the device is restricted only by the length of the booms rather than the size of the host satellite. Investigation by Cranfield University suggests that the only components that will experience a large increase in mass disproportionately are the CuBe booms. Therefore, different materials should be considered to allow scaling up and for the reduction of mass. This can be achieved by using composite booms as suggested by Cranfield University's research, which will reduce mass significantly and lower the costs of launches. The benefit of composite is its high strength/stiffness to weight ratio and its tailoring characteristics which allows the properties of the laminate to be tailored to suit the current requirements, and there are recent advancement in AM that will be able to produce a part with carbon fibre (CF) reinforcement with agreeable mechanical properties and its feasibility to be used in the aerospace industry be explored [3].

10.1.3 Project's aims and objectives

The main objective for this project is to develop and manufacture a set of composite booms which can be used to be integrated into the DOM developed by the research team at Cranfield University. There are several commercial requirements set upon by Cranfield to be considered during the development of the booms, and the relevant requirements are adapted to current project's aims [3]. These are as follows:

- Scalability: The boom shall be able to scale up to a minimum length of 1 m.
- Mass: The DOM shall have a limit between 0.5 and 1 kg. The mass of the boom shall not cause an increase in the overall mass of the device.
- Lifetime: The boom must be able to deploy reliably after the host satellite operational period of 10 years in LEO.
- Lifetime: The composite boom shall have the ability to be stored on ground without the need for re-acceptance testing at the end of the 10-year storage period.
- States: The booms will have three states: stowed, deploying and deployed.
- Cost: The cost of the manufacturing process shall be economical.

10.1.3.1 Research aims and objectives

For this particular research, there are also a number of primary objectives to be met which can be split into two phases:

Design and analysis phase:

- Perform research on different designs of deployable composite booms and identify the superior configuration.
- Conduct research on AM and conventional manufacturing techniques.
- Investigation into Markforge's 3D printer [4].
- Research Markforge's flagship material Onyx and investigates its mechanical properties.
- Finite element analysis (FEA) for the chosen design of the boom, in which the flexural strength is identified through the performance of a three-point bending test on Abaqus CAE.

Manufacturing and testing phase:

- Manufacture a set of composite booms to the length of at least 1 m utilising the integrated industrial AM platform provided by Markforge.
- Multiple samples will be created before the final design is made to observe the part physically. Different printing orientation will also be tested and the most efficient orientation will be identified. Shortcomings and cons will also be identified through experimentation of different printing methods.
- Three-point bending test of the boom to identify the flexural strength and comparisons will be made with theoretical data that was recorded from FEA.

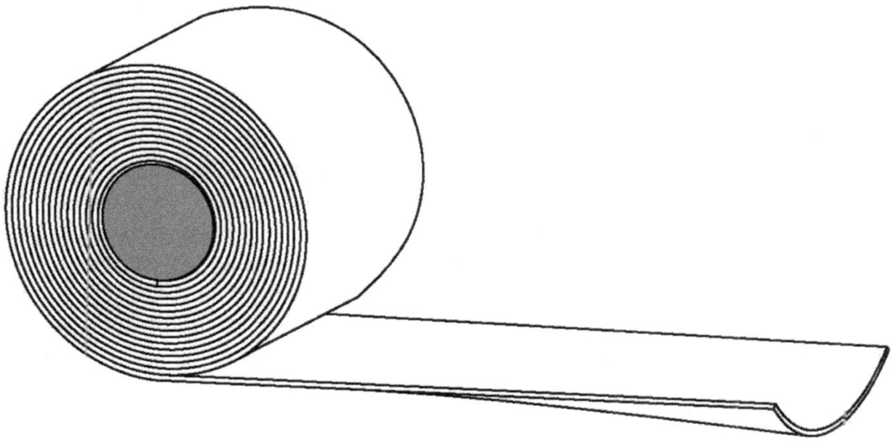

Figure 10.2 *CuBe boom design [6]*

10.2 Designs of booms

Deployable boom is often used in spacecraft architecture to provide a method to package and deploy structure such as a solar sail efficiently, since any planar structure would be too large to be stored within a small host satellite structure [5]. In the DOM, aluminised sail quadrants are stored and as mentioned in the previous chapter that the drag sail is to be deployed within 25 years of end-of-life mission to meet IADC's requirement. Therefore, a reliable design is required to ensure successful deployment, and a number of research has been carried out to identify the advantages and shortcomings of current available designs.

The CuBe boom currently used in the DOM is a simple single c-shaped design, as shown in Figure 10.2. An experiment was conducted by researchers at Cranfield University to identify the maximum length of the c-shaped boom design until it becomes unstable. The result shows that the maximum unsupported length until snap-fail occurs is roughly 1 m under 1 g condition. The current c-shaped design will, therefore, not be suitable for scaling up above 1 m and a different design is recommended.

Tubular extendible booms have shown popularity since its first introduction: triangular rollable and collapsible (TRAC), storable tubular extendible member (STEM) and CTM [6]. Similar to the current CuBe boom design on the DOM, these structures could also be flattened and coiled around a cylindrical spool. This section presents different types of designs in detail and discusses the manufacturability of each boom type [7].

10.2.1 TRAC boom

TRAC was developed by the United States Air Force Research Lab by (Murphy and Banik), and the design has been considered in various missions, for example, the

CubeSat Near Earth Asteroid (NEA) Scout project by NASA. The popularity stems from its superior packing efficiency, its ability to reliably deploy during missions and its flight heritage. TRAC consists of two tape springs which are bonded together along one edge, forming a triangular cross section. The simple geometry also allows the process of manufacturing to be relatively simple in comparison to other designs with higher geometric complexity. The design of the boom can be seen in the technical drawing adapted from previous research (). TRAC was designed to provide higher cross-sectional inertia or second moment of inertia for the same packaged height and stowed strain than other designs such as the STEM and CTM, which will be discussed further in sections below. However, the downside of the design is that it is still an open-section design, and the torsional stiffness is significantly lower. Packaging of the design also poses a challenge as energy balance between the stored energy from pinching the free end together and strain energy from rolling around a spool can be difficult to achieve. Failure to achieve this may result in the open-ended flange to bifurcate during the rolled configuration causing the packaging efficiency to drastically be lowered [3].

10.2.2 STEM boom

STEM booms, similar to TRAC booms, are stowed flat around a spool. STEM booms store a large amount of strain energy during stowage, and once uncoiled it will extend outwards and revert back into an unstrained tube configuration following a single kinematic path. A particular advantage of the design is its ability to pack efficiently around a spool due to the thin materials. STEM booms are often manufactured using highly elastic material, and deformation of the thin material remains elastic during its lifetime of stowage and in missions. The cross section of a composite STEM boom have a distinctive curvature, and bistability can be introduced during the manufacturing process by creating a stress distribution in the extendible member beyond the ploy region. The boom is considered bistable when it is both stable in stowed and deployed states. Bistability is often desired as it also ensures high packing efficiency, allowing the tape to tightly and stably wrapped around the coil. STEM booms can be identified as unstable, bistable and naturally stable through the strain energy stored in both the deployed and stored states. A mechanism with motors and guides controls the manner of the boom during deployment through commands given to the motor optimising its rate of deployment. The design is highly reliable, and therefore, it is often used in the space industry for antennas and instrument deployment. The reliability of the design stems from its ability to retain full stiffness characteristics throughout the entire length of the boom during deployment. This makes deployment highly predictable and it is very much desired in the space industry. This characteristics allow the motor of the mechanism to be designed to allow the length adjustment even after deployment. However, the open section design provides the weakest torsional stiffness out of the other deployable boom structure design, and to overcome this issue, the edges of the spring tapes are often designed to overlap to create a closed cross section, since a closed section geometry provides a more desirable torsional

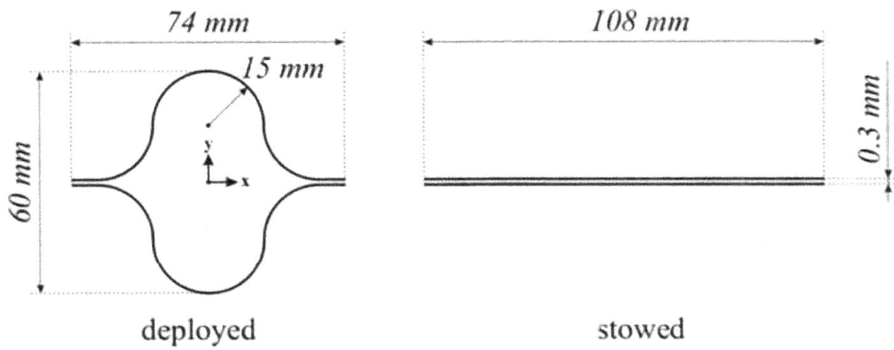

Figure 10.3 *Cross section of CTM boom in stowed and deployed states*

stiffness and avoid the risk of buckling under non-axial loading. presents the design of the STEM boom [3, 7].

Bi-STEM is a derivative of STEM which is gained by forming the boom with at least two strips of spring tapes. The two strips extend simultaneously, and each strip forms a partial section of the tube. The concave surface of the strips is placed to face each other, and the edges overlap to create a closed cross section. Some iterations also include slits and taps to interlock two tapes together to achieve a closed section to provide an even higher torsional stiffness. Bi-STEMs have overall higher bending and torsional stiffness with a narrower cross-sectional structure in comparison to conventional STEM booms, and with the only disadvantage of slight increase in mass since more materials are needed. shows a Bi-STEM boom design configuration used in past missions adapted from NASA [7].

10.2.3 CTM booms

CTM booms are designed to have a fully closed cross section, which gives these booms the higher torsional stiffness out of the other two designs discussed previously in this section. CTM boom consists of two thin omega-shaped shells bonded together using adhesives or is stitched together at the edge called the web. The cross section can be described in more detail and can be defined by joining three circles subtended at an angle α with equal size to create an omega shape as shown in Figure 10.3. The material used should be highly elastic to allow the boom to store in a similar way to previously mentioned designs. The elasticity of the material should allow the structure to collapse flat and rolled around a spool without any plastic deformation occurring allowing the boom to be stored for a long period of time and in an efficient and compact way. However, viscoelastic effects become a more prominent risk as creep, and stress relaxation effects become significant overtime during long-term storage for composite materials. Long-term storage in the flattened state will result in a smaller subtended angle, α. Figure 10.3 represents the geometry of a CTM boom. A smaller subtended angle α would result in a weaker in-plane moment of area and reduces the ability to withstand buckling load. Maximising moment of

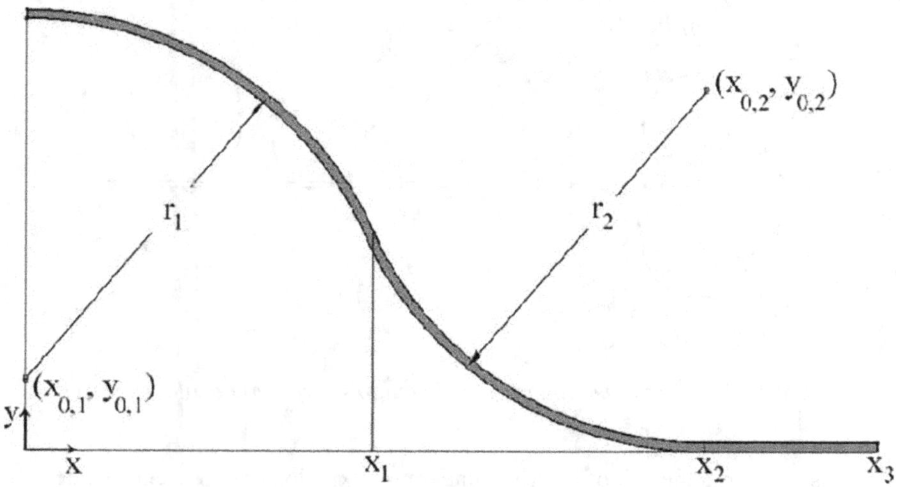

Figure 10.4 *Geometry of CTM boom with coordinate system for preliminary design calculation*

area also requires the largest subtended angle with the smallest web width ω; therefore, the stress that causes the deformation to flatten the boom must not be over the limit of the material to retain the original desired strength. Research conducted at the LaRC by Fernandez *et al.* [8] have shown that subtended angle greater than 80° for a web width of 3 mm will result in an unacceptable amount of flattening strain. Fernandez *et al.* also recommended that the boom should have a low axial Young's modulus during storage and deployment since a less energetic system is desired during both states. However, on the other hand, a high axial modulus is desired to maximise the overall stiffness (*EI*) of the boom to reduce global column buckling, the most common failure modes, in the deployed state [7].

10.2.3.1 Calculation of preliminary design parameters for CTM booms

In order for the bending stiffness (*EI*) of a CTM boom to be calculated, the moment of the cross section has to be estimated. *E* is the Young's Modulus, and *I* is the moment of area. The complexity of the geometry means that it can be very difficult to estimate the moment of area. Hakkak and Khoddam introduced the use of conventional moment of area equations for the estimation of these values. This assumption was validated through FEA data and accurate results were yielded. The calculation is carried out by first dividing the geometry of the CTM boom into quarters to simplify the calculation; this is permitted because the cross section has two perpendicular axes of symmetry. Figure 10.4 shows the divided quarter of the CTM boom within a coordinate system where the origin is set at the centre of the boom's cross section [9].

The following expressions are adapted from Hakkak and Khoddam's paper for the calculation of the moment of area about the x- and y-axis of the CTM boom (I_{xx} and I_{yy}) [9].

$$\begin{aligned} I_{xx} &= 2tr_1\left[x_1\sqrt{r_1^2-x_1^2}+(r_1^2+2y_{01}^2)\sin^{-1}\frac{x_1}{r_1}\right] \\ &+2tr_2(x_2-x_1)\left[\sqrt{r_2^2-(x_2-x_1)^2}-4r_2\right] \\ &+6tr_2^3\sin^{-1}\frac{x_2-x_1}{r_2}+8tr_1x_1y_{01} \end{aligned} \tag{10.1}$$

$$\begin{aligned} I_{yy} &= 2tr_1\left(r_1^2\sin^{-1}\frac{x_1}{r_1}-x_1\sqrt{r_1^2-x_1^2}\right)+\frac{\beta t}{3}(x_3^3-x_2^3) \\ &+2tr_2\left[(3x_2-x_1)\sqrt{r_2^2-(x_2-x_1)^2}-4x_2r_2\right] \\ &+2tr_2(x_2^2-r_2^2)\sin^{-1}\frac{x_2-x_1}{r_2} \end{aligned} \tag{10.2}$$

The equations follow the same coordinate system and the required input parameters are the same as in Figure 10.4. I_{xx} and I_{yy} are derived from the expression below, which is the ordinary formulation for moment of inertia. Note that the results are multiplied by four to account for the full section of the boom.

$$I_{xx} = 4\int y^2 A \tag{10.3}$$

$$I_{yy} = 4\int x^2 A \tag{10.4}$$

β, the edge thickness coefficient, from the above I_{yy} equation was introduced by Hakkak and Khoddam to replace the value "4" to represent the different thickness along the length of the edge. This is to accurately represent the geometry since the edges are either bonded with adhesives or stitched, the different thickness may instead be achieved during the manufacturing process. β can be determined using the relationship below [9]:

$$\beta = \frac{2t_{web}}{t} \tag{10.5}$$

where t_{web} is the total thickness of the web achieved after joining the process. Since the CTM boom as mentioned before is symmetrical in both the x- and y-axis. the calculated I_{xx} and I_{yy} from the above equations are the principal moments of area. The smaller value will therefore be the minimum moment of area of the section which should be greater than the specified moment from design requirement. The torsional moments of area of the boom can be determined using the equation below since the boom is a thin-walled hollow tube, where A is the area enclosed by the perimeter of the boom wall and the x- and y-axis, as the cross section is divided into quarters as previously mentioned. P is the perimeter of this enclosed area. Figure 10.5 represents the full area enclosed by the wall of the boom. Note that the web is omitted [9].

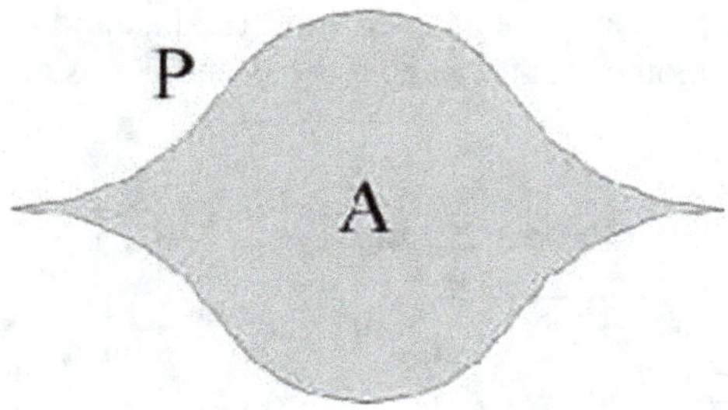

Figure 10.5 *The area and perimeter of the CTM boom excluding the web*

$$J = \frac{4A^2t}{P} \tag{10.6}$$

10.2.3.2 Optimal CTM geometry

Using the above equations a trade analysis was carried out on the [45PW/0] two-ply lay-up with the HTA40/MR60H spread-tow carbon by Fernandez from LaRC. As a result of this research, the optimal cross-section geometry of the boom was determined to maximises the principal moment of area and the torsional moment of area while ensuring that the flattening strains and rolling strains are kept at an acceptable level. A coordinate axis that was adopted by Fernandez to show the designed

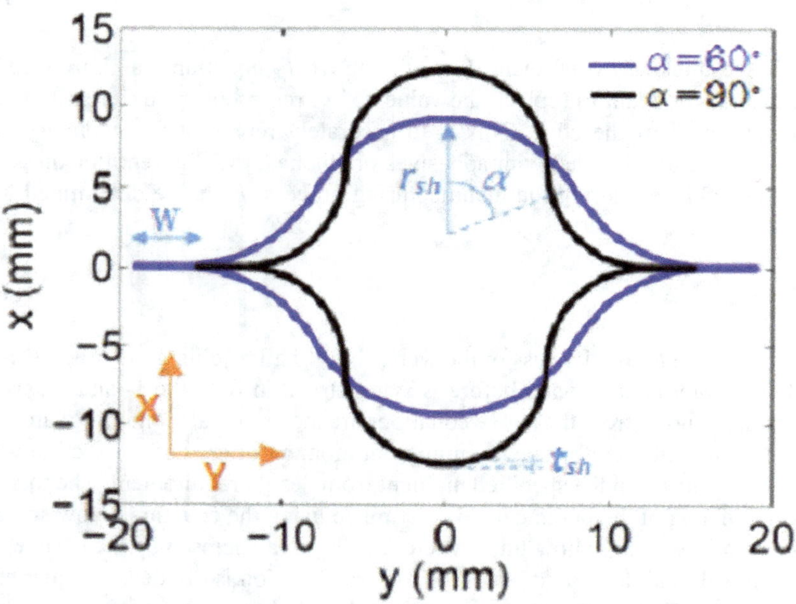

Figure 10.6 *CTM boom geometry with coordinate system*

cross-section geometry with their design parameters is shown in Figure 10.6.. A few design parameters were suggested after carrying out trade analysis and should be taken into consideration for future designs [8]:

- The subtended angle, α, shall be between 50° and 90°. The graph in shows the trade analysis results, and it can be observed that I_{yy} becomes more desirable as subtended angle increases. However, subtended angle above 80° with web width of 3 mm also results in unacceptable flattening strains after long stowage, as shown in . Since viscoelastic effects such as creep become severe during long-term stowage, design should try to counter this through having the largest possible subtended angle. Moreover, it should be noted that smaller flattened height will result in weaker I_{xx} [8].
- The width of the web should be: 3 mm $\leq \omega \leq$ 5 mm. A smaller web width will result in large shear stresses at the bonded edge and a large web would be perversely counter-intuitive as an optimal design is desired [8].
- Fernandez used thin-ply laminates with silicon-based materials and so different lay-ups were considered. The only similarity between the lay-ups was the outer surfaces, which were made from plain-weave laminates to prevent delamination. Note that axial unidirectional outer surface plies for this type of structure tends to promote premature delamination [8].

10.2.3.3 Manufacturing processes

The manufacturing of the CTM booms can be complicated due to the complexity of the cross-sectional geometry. The two omega shells were often manufactured separately using a mould and then bonded at the edge using adhesives to form the lenticular shape. Several factors must be taken into consideration during the fabrication process to ensure that the edge is up to a high standard of quality [8]. These are as follows:

- The surface of the bonding area must be thoroughly cleaned, ensuring that there is minimal dirt, before the applications of adhesives. This ensures a high bond strength and a longer lifespan of the joint. However, it should be noted that the surface preparation method is highly dependent on the material itself.
- The alignment of the boom's edge for bonding is an important factor to ensure that the geometry is as designed, and misalignment of the boom's edge can be disastrous for bond strength.
- The thickness of the adhesives should also be controlled appropriately with appropriate pressure ensuring no excess glue seeps through the edge affecting the strength of the bond.

Concerns on the reliability of bonding at the web were raised by previous researches. It was suggested that a different manufacturing process is needed. A single step manufacturing process was introduced by Deutsches Zentrum für Luft- und

Figure 10.7 Manufacturing process of CTM boom from NASA

Raumfahrt (DLR) during the DeOrbitSail Project which involves curing the two-omega shell simultaneously by placing an inner vacuum bag within the cavity of the boom and an outer vacuum bag which surrounds the whole tool. Excess materials will then be removed from the side to create the flat edge. However, the amount of material wasted can be as much as the actual boom itself, which will affect the cost of manufacturing. Despite this, the single-step process still provides a more cost-effective method in comparison to the conventional method.

Fernandez later suggested a different single step process with even better cost-effectiveness. This process requires a silicone plug to be manufactured using the two negative omega molds. Silicone was chosen for its flexibility to allow the plug to be removed from the cavity with ease. This is important as the plug acts as a male mould inside the cavity. This process uses a single female mould instead of the two suggested by DLR, and this makes the process highly cost-effective in comparison. Extra effort was made to minimise the difference of coefficient of thermal expansion (CTE) between the boom and the female mould. This is because a CTE difference will result in deflection errors and the boom may not end-up as straight as initially desired. This problem was addressed through the use of a carbon foam mould instead of the cheaper polyurethane foam since the boom in Fernandez paper is made from CF. The carbon foam yielded near perfect straight booms, justifying the high cost of such mould [7].

Figure 10.7. below shows the steps of the process introduced by Fernandez as mentioned above. The first step of the process involves laying and pressing down the material of the bottom half on top of the female mould, with the omega-shaped plug on top forming the shape. Adhesives are then applied on top with appropriate bond thickness. The top of the boom is then laid on top and pressure is then applied to form the whole lenticular cross section. Top release film, breather ply and vacuum bag are then placed on top in order as listed. Vacuum is then applied while ensuring no leaks. The mould will then be transferred on to an oven for the curing process, and once curing is completed the boom is removed from all the mould. Finally, the boom is then removed from the mould and thoroughly cleaned. The initial cost of using mould to manufacture can have high initial cost due to the fact that primary tool has to be designed and manufactured [8]. This project suggests using AM and thorough research carried out is discussed in the next section.

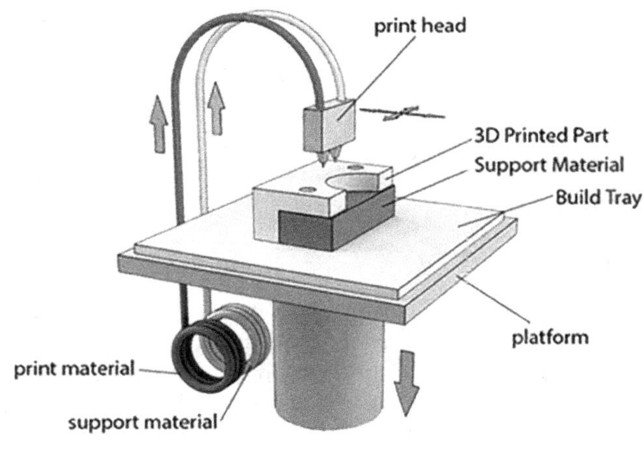

Figure 10.8 *FDM process*

10.3 Additive manufacturing

Additive manufacturing (AM) has been successful in various applications. AM is defined as the process of adding layers upon layers of material to create a 3D object as opposed to the conventional subtractive manufacturing where material is removed instead to form the designed shape. Fused deposition modelling (FDM) is an AM method that involves melting plastic filaments to form a 3D object using 3D model data created using computer aided design (CAD) software such as Solidworks. FDM is favoured in many industries due to its relatively low cost of an FDM printer and that a large variety of plastic filament could be obtained at very low cost. FDM is currently being employed for rapid prototyping in industries such as the automotive, aerospace, construction and medicine. The FDM process involves the extrusion of melted plastic fibre through a heated nozzle. The material is fused with each additional layer and after cooling the material hardens. Figure 10.8. shows a presentation of this process [10].

Despite FDM's popularity in industries, the mechanical properties of the printed structure are significantly weak. The low mechanical properties would mean that structure would not be acceptable for actual engineering applications and is only appropriate for prototyping. Since parts created in FDM process will have to be recreated in conventional manufacturing process that will hinder creativity and complexity of design, recent developments demonstrate that the merging of polymer and CF increases the mechanical properties and revolutionises the AM industry, allowing printed structure to be used in real engineering applications instead of only prototyping. Despite the increase in material's strength and stiffness, the layer-to-layer adhesion still remains low. L.J. Love discovered that with additional CF the CTE and thermal conductive differs. This is important commercially as most AM is often carried out inside an oven to reduce thermal gradients in part. This causes residual stress to be significant causing the part to warp and distort from plastic deformation. Moreover, L.J. Love also found that FDM with CF also enables the process to be done in room temperature without causing severe distortions, with the increase in thermal conductivity and a reduction of the CTE. FDM can tackle complexity of composite geometry which is often quite difficult to achieve when manufacturing of composite is done in a conventional way [11].

10.3.1 Markforged's FDM printer and materials

Markforged (MF) Mark two is an award-winning composite FDM printer, which is built around CF reinforcement process to create parts reliably. It utilises a dual nozzle system to allow embedment of different fibres within a part. This includes CF, Kevlar, high strength, high temperature glass fibres and plain glass fibres. The Mark two provides hardened extrusion to enable printing of MF's flagship material Onyx, which is a micro-CF filled filament [12]. The printer uses a browser-based software to aid the FDM process and provides a visual representation of the toolpath layer by layer, the estimated print time, volume of materials and the volume of reinforcement. A series of mechanical testing was carried out by researchers at Los Alamos National Laboratory (LANL) in the United States to characterise the mechanical properties of the materials supplied by MF for their printers. Tensile, compression three-point, Charpy impact, CTE and calorimetry tests were carried out on fibre reinforced samples and was carried out following standards. Figure 10.9. is a table adapted from LANL which shows the different standards applied [13].

The results provided by the labs confirm the ability of MF's continuous fibres printing process. It can be seen that with the addition of fibresleads to the increase the strength. However, it leads to parts being more brittle with little elongation at break. Kevlar improves the strength whilst maintaining ductility. CF reinforcement

Code or Standard No.	Date	Title
ASTM D695	2015	Standard Test Method for Compressive Properties of Rigid Plastics
ASTM D638	2014	Standard Test Method for Tensile Properties of Plastics
ASTM D790	2017	Standard Test Methods for Flexural Properties of Unreinforced and Reinforced Plastics and Electrical Insulating Materials
ASTM D7426	2013	Standard Test Method for Assignment of the DSC Procedure for Determining Tg of a Polymer or an Elastomeric Compound
ASTM E831	2014	Standard Test Method for Linear Thermal Expansion of Solid Materials by Thermomechanical Analysis
ASTM D6110	2018	Standard Test Method for Determining the Charpy Impact Resistance of Notched Specimens of Plastics

Figure 10.9 Standards for tests carried out by LANL

was proven by the lab that flexural modulus has improvement but can be affected greatly by print quality as results were different.

Onyx as mentioned is a nylon reinforced with short fibres and MF claims that parts printed with onyx are 30 percent stronger and stiffer than other materials. Not only that the parts could be used straight after the printing process. The materials provide the toughness of nylon plus the stiffness of fibres with heat resistance of up to 145 °C. The finishing product is smooth and has a matte surface finish. According to MF, the fibre path is roughly 1 mm and with two material shells on either side the minimum thickness of a part is 2.9 mm [12]. This is unfortunately thick and therefore reinforcing thin structure such as the CTM boom will not be possible. Therefore, onyx is an ideal material for 3D printing thin structure with the benefits of having short fibre reinforcement. Max James Sauer at the South Dakota State University carried out digital microscopy on Onyx to understand the fibres within a printed specimen with 100 percent infill. CF length, diameter, volume fraction and orientations were studied using a VHX-6000 from Keyence Corporation. The samples created were cut, cross-sectioned and polished with a seven-step process before observations were made under the microscope. Images were taken and using image processing software, the average length of the fibres was determined to be 108.2 μm and an average diameter of 7.36 μm. Further imaging post processing carried out to find that 9.179 percent fibre volume was in the specimen. show the microscope image taken using the Keyence Digital microscope by Max James Sauer. This is an important study as the modulus of elasticity and yield strength increase with increase in fibre volume fraction and fibre length. Moreover, the elongation at break also decreases with increasing fibre length and volume fraction, meaning it will be brittle [14].

10.4 Design of the CTM boom

The booms currently in used for the Dom are made from CuBe and as mentioned previously that the mass of the boom increases disproportionately with increasing length of the boom; this means that scaling up would not be ideal. The main factor of concern when scaling up is the mass, and more importantly increasing the DOM as a whole, the only components that will experience a large increase in mass are the booms. Therefore, a lighter composite material was suggested by researcher at Cranfield University. Composite materials are lighter in weight and has high strength/stiffness to weight ratio and tailoring characteristics allowing the fibre orientation to be tailored to the stiffness and strength required for specific applications. The composite booms created should abide by the same requirement previously set upon at Cranfield University. The main requirements relevant are as listed below [7]:

- The manufacturing process shall remain low.
- The boom shall have the ability to be stored for an extended period of time on ground without having to perform a reacceptance test.

Table 10.1 Onyx and CuBe properties

Properties	Onyx	CuBe
Young's modulus (GPa)	2.4	200
Tensile yield strength (MPa)	40	490
Density (g/cm^3)	1.2	7.85

- The boom's deployment shall be reliable after 10 years of storage on the host satellite in LEO.
- Boom shall be able to scale up to at least 1 m in length.
- The deployment of the booms shall also be controlled and predictable.

However, knowing that the booms will be manufactured in MF's printer using FDM process, the boom will instead be reinforced with short CF as opposed to continuous. This has to be the case since, as mentioned previously, that the minimum thickness of a part that could be reinforced is 2.9 mm. MF supplies a wide variety of filament for its FDM printers, and out of all the materials, Onyx provides the most desirable mechanical properties since it is their only material reinforced with short CF. It provides the benefit of Nylon's toughness with the stiffness of CF and can produced parts 30 percent stronger as claimed by MF. The cost of parts made in MF's FDM is significantly low, despite the high cost of buying a machine. The Onyx series from MF uses a superior hardened extrusion system to extrude Onyx and produce strong parts reliably [15].

10.4.1 Onyx and CuBe

A comparison of the mechanical properties of Onyx and CuBe is made in Table 10.1. Strength and stiffness are directly related to Young's modulus, which is a factor that measures how easily a material could bend or stretch. Tensile strength or yield strength is the stress that a material can withstand until plastic deformation and will no longer return to the original shape. From the table below, Onyx is significantly less stiff in comparison to CuBe. The lower Young's modulus suggests that onyx is highly elastic. The high elastic nature of onyx, despite having lower yield strength which is expected when comparing to metallic material, suits perfectly to our current need as the boom is stored by coiling around a spool where it will experience high strain and will be difficult to achieve with stiffer materials. Not only that, Onyx also has a significantly lower density in comparison to CuBe, using onyx will provide a huge reduction in mass [16].

10.4.2 Cross-section geometry

Extensive research has been carried out on CTM booms by Fernandez at LARC to determine the optimal cross-section geometry to maximise the two principal moments

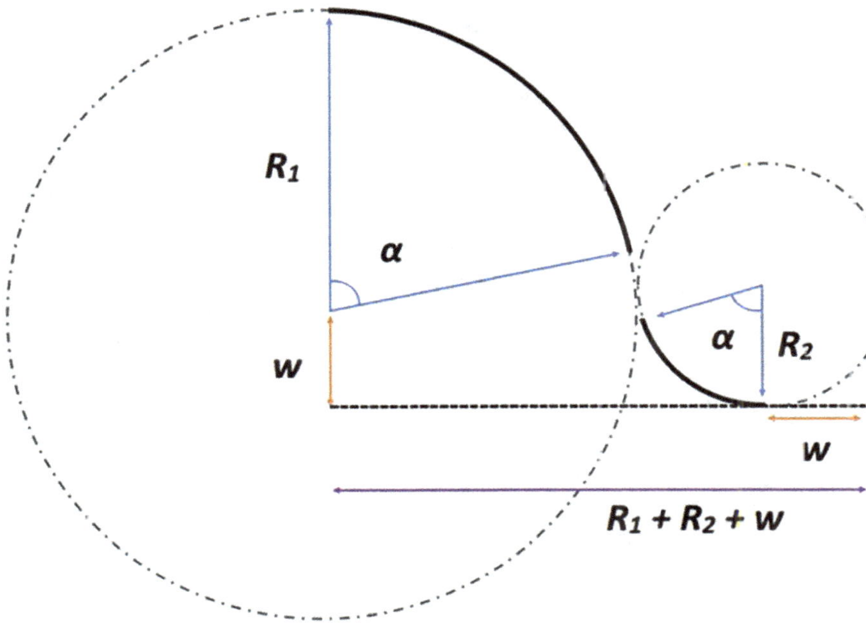

Figure 10.10 *The geometry relations of the CTM boom*

of area whilst maintaining the rolling and flattening strain at acceptable levels. For the design of the boom, several factors must be taken into considerations [8]:

- To maximise the moment of area around the *y*-axis, Iyy, the largest possible subtended angle is required with the smallest possible web width.
- Composite booms stored for an extended period of time have a high risk, where viscoelastic effects such as creep and stress relaxation effects will result in a smaller subtended angle than previously intended. This must be addressed by having the largest possible subtended angle to counter this effect.
- Note that a large, subtended angle will result in high strain during flattening which must be considered during the design phase.

As previously demonstrated, the optimal geometry is to have a subtended angle of 80 degrees and a web width of 3 mm to provide the maximum moment of area around the principal axes. The cross-section geometry of the CTM can simply be calculated using equation (10.7). Figure 10.10 also shows the relationships of these terms in relation to the CTM's geometry.

$$h = 2\omega + 2\alpha\left(R1 + R2\right) \tag{10.7}$$

h is the height of the boom at its flattened states, ω is the width of the web, α is the subtended angle, and finally the R is the radius of the circle segments which join to create an omega shape. The design was decided to have an equal circle segment with

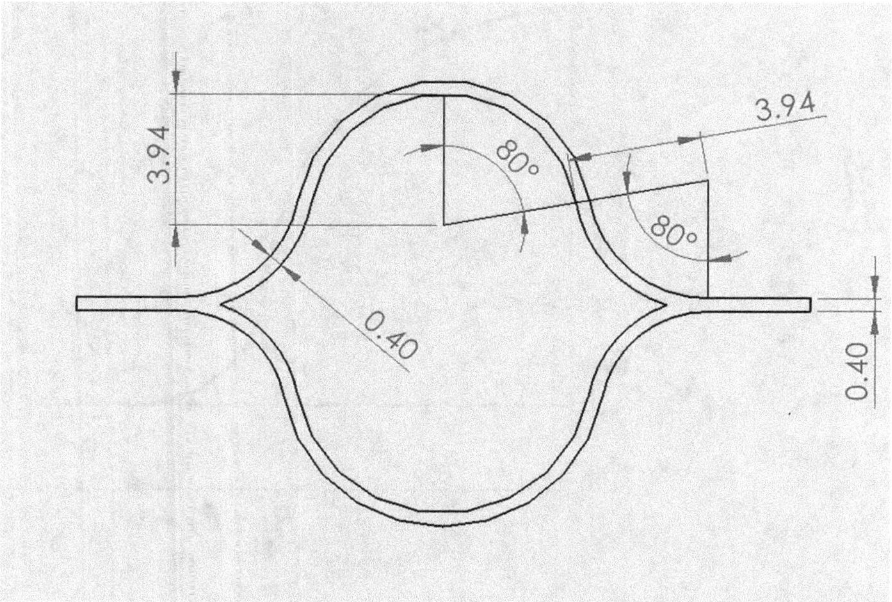

Figure 10.11 *The cross section of the designed CTM boom's geometry with 0.4 mm wall thickness*

a subtended angle between 50 and 90 degrees. This is to prevent having a small, flattened height. The equation can therefore be adapted into Reference 7.

$$h = 2\omega + 4\alpha R \tag{10.8}$$

where R represents the radius of the circles since it is equal. The flattened height h is restricted by the size of the DOM, which is 28 mm. Since the web width ω of 3 mm and subtended angle of 80 degrees was recommended to provide the maximum moment of area, the only remaining unknown term is the radius of the circle R. The radius of the three circle segments was calculated to be 3.94 mm. The height of the deployed boom can also be estimated using the equation below [7].

$$h = 2\left(R1 + R2 + \omega\right) \tag{10.9}$$

where $R1$ is equal to $R2$. The deployed height was estimated to be 21.76 mm. Figure 10.11 shows the resulting cross-section of the CTM created with aforementioned parameters. 2.1This was adapted from F Hakkak and Khoddam's paper to estimate the bending moment of area about the x- and y-axis, I_{xx} and I_{yy}, and the torsional moment of area, J, to identify the stronger orientation (Table 10.2), which will be discussed later. Table 10.3 is the geometric parameters of the CTM's boom cross-section following the coordinate system from [9].

Table 10.2 Calculated I_{yy}, I_{xx} *and* J *of the CTM boom*

Geometric parameters	CTM boom
I_{xx} (mm^4)	292.36
I_{yy} (mm^4)	517.38
J	326.17

10.4.3 Finite element analysis of CTM boom

A series of FEA has been carried out to the cross-section geometry designed in the previous section to simulate the behaviour of the boom when load is applied and allow experimental data to be compared with. Using FEA provides a cost-efficient way to conduct the experiment without having to provide and manufacture an extra set of prototype booms. Since the composite boom takes time to manufacture, it will also effectively save time.

10.4.3.1 Maximum length analysis methodology

Experiments were conducted by researchers at Cranfield University to determine the maximum unsupported length of the CuBe boom until snap-fails occur. The CuBe becomes unstable and fails at 995 mm [7]. The experiment was replicated on the cross-section geometry in both orientations, from the previous section, in FEA software Abaqus to identify whether it will provide an improvement in maximum length in 1 g condition, thus determining the design's ability to be scaled up. The first set of these analyses were carried in static, general. The brief description of the methodology is as follows and is the same for both orientations.

Table 10.3 Geometric parameters of the CTM boom for calculation of I_{yy}, I_{xx} *and* J

Geometric parameters	CTM boom
r_1 (mm)	3.94
r_2 (mm)	3.94
$y_{0,1}$ (mm)	3
x_1 (mm)	3.88
$x_{0,2}$ (mm)	7.76
$y_{0,2}$ (mm)	3.94
t (mm)	0.4
A (mm^2)	101.06
P (mm)	50.01

1. The structure of the boom is created in Abaqus by creating a shell with extrusion.
2. Material properties were then created according to MF Onyx data sheet, and a section was then created to assign the material properties on to the structure. The density, Young's modulus and Poisson's ratio were given for this set analysis.
3. An instance is created in the assembly and a reference point is added to the structure to apply coupling constraints to one end of the boom.
4. The boom is cantilevered by setting Encastre boundary condition on the opposite end of the reference point.
5. Structured meshing of the model is then carried out choosing appropriate size, mesh control and mesh type. Since high performance computing will be used for the analysis, the mesh density would be high to ensure accuracy without having to carry out mesh refinement.
6. Gravitational load of 1 g is then applied on to the whole model depending on the orientation.
7. When the analysis is completed, the length will then be increased by 200 mm and will be remeshed for additional analysis. Analysis is repeated until it reaches to a point when the boom becomes unacceptably bended.

10.4.3.1.1 Maximum length in horizontal direction

The effect of self-weight of the boom in 1 g can be seen effectively in the figures below captured from FEA. As expected, as the length of the boom increases, the unsupported end begins to bend downwards due to gravity. A comparison of each length in 1 g can be seen in Figures 10.12–10.20 . The figures also show that the stress concentration because of self-weight is near to the supported edge as expected, which is similar to when load is applied on one end. The results of this analysis show that the boom's performance was not as high which is expected since the analysis was carried out on the boom's weaker orientation as shown in with I_{xx}

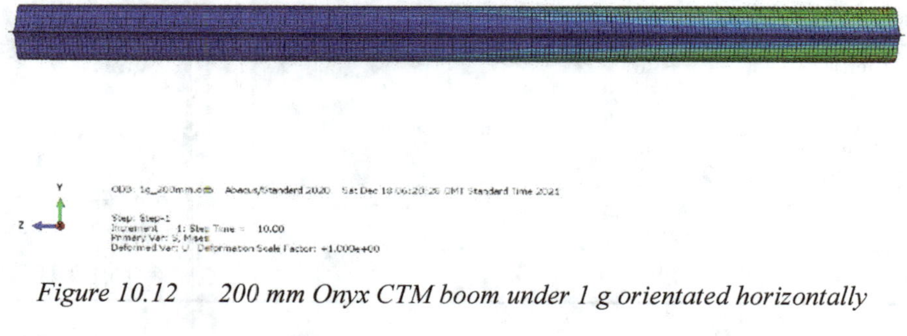

Figure 10.12 200 mm Onyx CTM boom under 1 g orientated horizontally

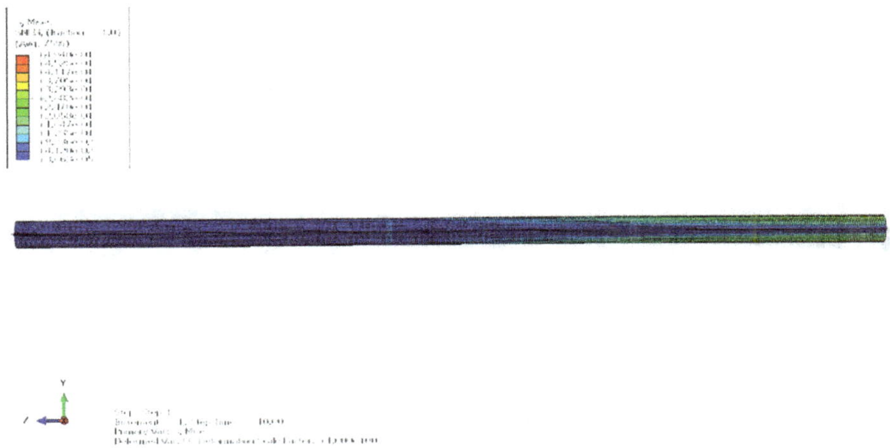

Figure 10.13 400 mm Onyx CTM boom under 1 g orientated horizontally

being smaller, as shown in Table 10.2. However, this has already provided a slight improvement as initial observation of the result shows that despite being orientated on the weaker side, the length of the boom seems to be able to hold its self-weight at length 1000 mm. Onyx is significantly lighter in comparison, and with the CTM geometry it provides even better maximum length performance with a huge reduction in weight.

10.4.3.1.2 Maximum unsupported length in vertical orientation

The experiment was then repeated, and gravitational load was applied on the boom in the vertical orientation, which is the stronger side as shown in Figure 10.21., to

Figure 10.14 600 mm Onyx CTM boom under 1 g orientated horizontally

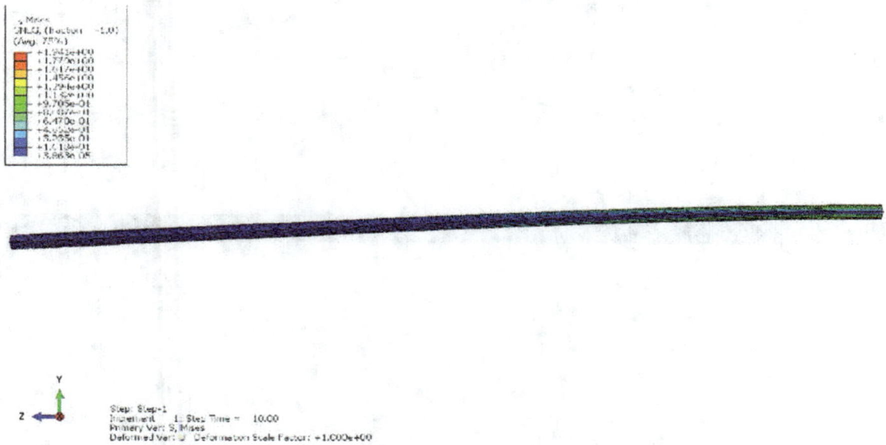

Figure 10.15 *800 mm Onyx CTM boom under 1 g orientated horizontally*

identify the maximum unsupported length. Since the maximum unsupported length of the CuBe boom is roughly 1 m, the analysis will be carried out from 1 m, and an increment of 0.2 m will be added after each analysis using the same methodology as presented in the previous section. As expected, the CTM boom out-performs the CuBe boom in both orientations. Figures 10.21–10.25 and TheFigures 10.34, 10.35 and 10.38 show the result of the analysis.

Despite the reliability of FEA, it is always better to compare with a set of results from an experimental test. Therefore, it is recommended that a maximum unsupported length experiment, similar to the one carried out by researchers at Cranfield University, should be carried out on a set of Onyx boom. Since the Onyx boom is

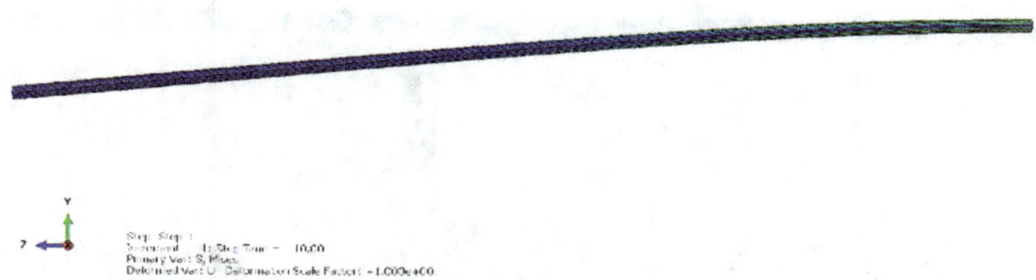

Figure 10.16 *1000 mm Onyx CTM boom under 1 g orientated horizontally*

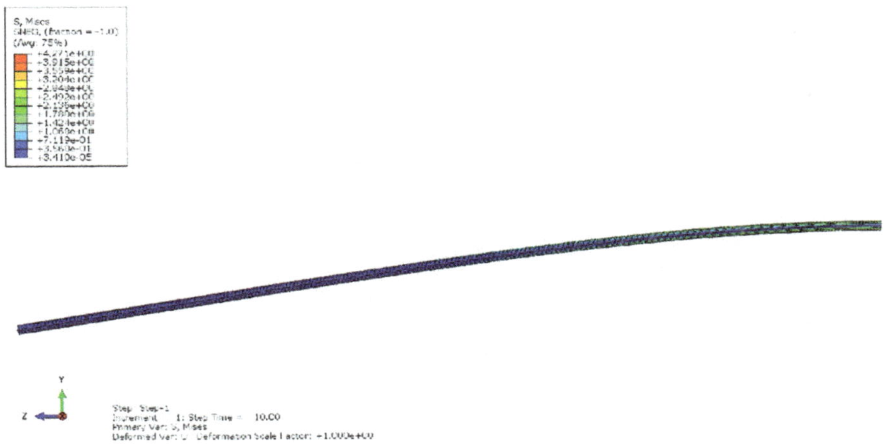

Figure 10.17 *1200 mm Onyx CTM boom under 1 g orientated horizontally*

manufactured with FDM, the experiment will be relatively cheap, but one limitation is that the length of the boom is dictated by MF printer size.

10.4.4 Three-point bending test methodology

Three-point bending test is performed to provide a measurement of flexural strength. Flexural strength is defined as the material's ability to resist deformation, and the value represents the highest stress within the material at the moment of rupture. A testing machine is normally used with a three-point bend fixture. The test is carried out with displacement being applied in the middle of the specimen using a top roller

Figure 10.18 *1400 mm Onyx CTM boom under 1 g orientated horizontally*

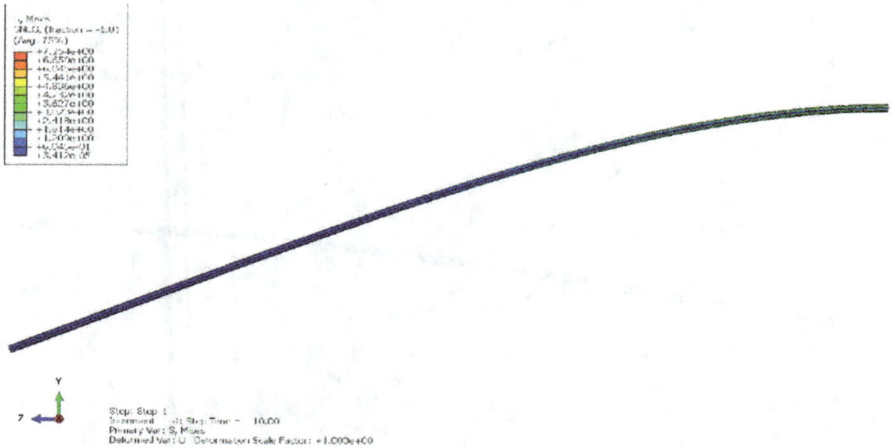

Figure 10.19 1600 mm Onyx CTM boom under 1 g orientated horizontally

with two pinned rollers under the specimen as demonstrated in . The supporting pin at the bottom of the specimen is set-up so that it can be rolled reducing friction [17].

MF carried out three-point bending test on their flagship material Onyx, and a flexural strength of 81 MPa was achieved on a specimen with 4.5 in (*L*) × 0.4 in (*W*) × 0.12 in (*H*). The values of which could be found in the datasheet provided by MF. The preparation of the test is relatively simple but is highly dependent on the specimen geometry, loading geometry and strain rate. The structure of the CTM boom being designed in Onyx, the flexural strength will obviously be different as geometry is different. A three-point bending test is recommended to provide a picture of the structure's ability to resist deformation under high stress. Three-point bending is

Figure 10.20 1800 mm Onyx CTM boom under 1 g orientated horizontally

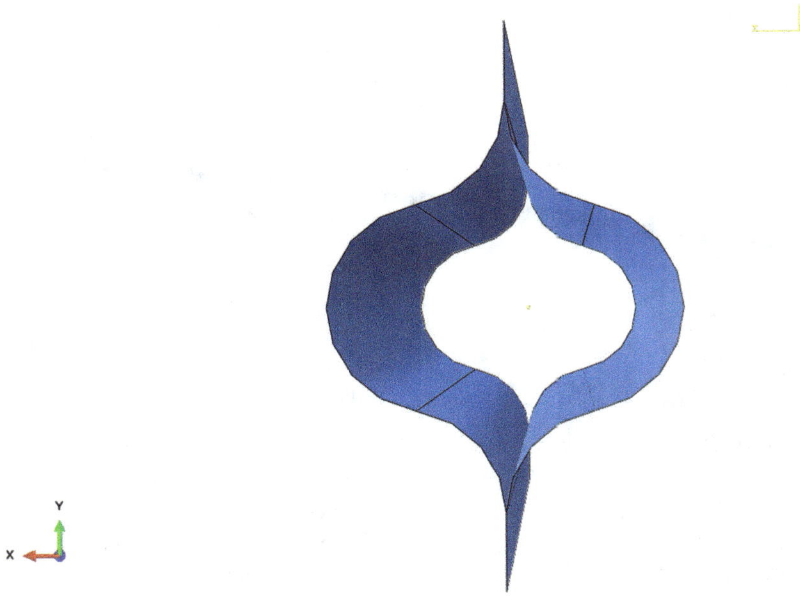

Figure 10.21 Onyx CTM boom orientated vertically

performed on a universal testing machine as mentioned previously, and it can also be performed in FEA. Obtaining both practical and theoretical results provide a clearer picture of what is happening to the structure. The FEA was carried out, and the methodology is described as

1. The structure of the boom is created in Abaqus by creating a shell with extrusion and the pin is created as a solid using extrusion.

Figure 10.22 1200 mm Onyx CTM boom under 1 g orientated vertically

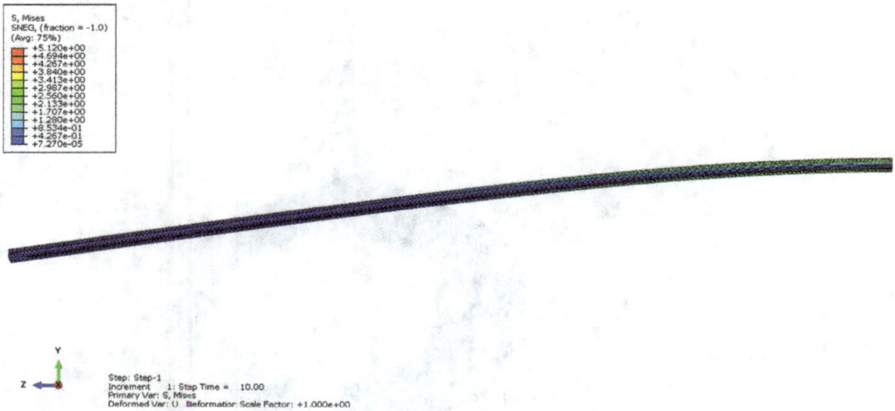

Figure 10.23 1400 mm Onyx CTM boom under 1 g orientated vertically

2. Material properties were then created according to experimental data adapted from literature to model Onyx's hyper-elastic behaviour, and a section was then created to assign the material properties onto the structure. Steel's material properties were also created and assigned on to the structure of the roller pins.
3. An instance is created in the assembly and the three roller pins are assembled with the CTM boom. Reference points were added on to rollers and rigid body constraint is applied Figure 10.26 .
4. Interactions were assigned where a fiction penalty was applied between the surface of the roller and the CTM boom.
5. Structured meshing of the model is then carried out choosing appropriate size, mesh control and mesh type. Since the analysis was carried out in explicit, dynamic, mesh refinement is not required. Same as before the analysis is carried

Figure 10.24 1600 mm Onyx CTM boom under 1 g orientated vertically

Figure 10.25 1800 mm Onyx CTM boom under 1 g orientated vertically

out using high-performance computing, it would be beneficial to run simulation with a high mesh density which would be high to ensure accuracy.

6. Encastre boundary conditions were given to the two-support roller pin below the specimen and a displacement of –10 mm was given to the top roller pin over the 10-s period ensuring a quasi-static analysis.
7. History output is created on the reference point at the top pin to record the displacement and reaction force. This will allow the flexural strength to be calculated (Figure 10.27).

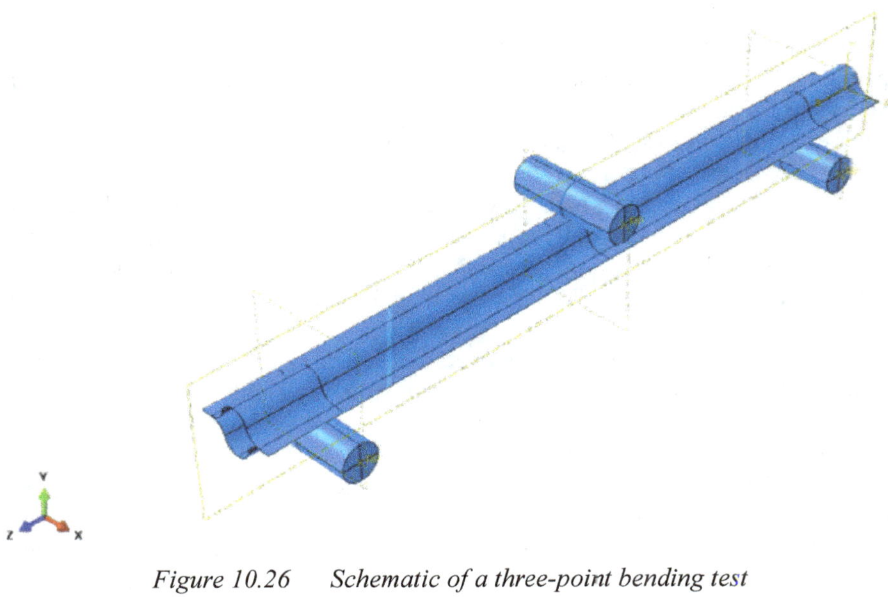

Figure 10.26 Schematic of a three-point bending test

Figure 10.27 Result of three-point bending test

10.4.4.1 Three-point bending results

Typically, hyperelastic materials are modelled with experimental test data as input. Providing the stress–strain data points, a continuous stress–strain curve to be generated in Abaqus by curve fitting the inputted test data, in order for the non-linearity to be simulated. Abaqus recognises uniaxial test, bi-axial, planar and volumetric test, as shown in [18, 19].

Abaqus uses strain energy potential U to relate stresses to strains. Different strain energy potentials are currently available in Abaqus. The choice of this model is highly dependent on the test data available. For example, for Ogden model, extensive data are needed, and at least uniaxial and biaxial data have to be included for the model to be stable. If data are limited Abaqus recommend using more simpler model such as the Arruda-Boyce model. Hyperelastic models are also categorised as either physically motivated or phenomenologically motivated. Physically motivated considers responses from a microstructure level such as the Arruda Boyce model and Van der Waals model which provides a good material behaviour under different loading. Phenomenologically motivated models on the other hand consider the "point of view of continuum mechanics" and are normally used for its simplicity, and it works reasonably well. The mathematical form of the strain energy is as given in [18].

The deviatoric term in the strain energy potential models the change in shape of material acting like rubber and volumetric term models the change in volume change behaviour in rubber-like materials. Depending on the analysis, different test data should be provided as input which is typically a nominal stress versus nominal strain data. The built-in material evaluation will generate the curve for all test types which include uniaxial, biaxial and planar test data. However, there are limited test data and research available for the onyx, so the result shown in Figure 10.14. of the model is inaccurate. Therefore, it is recommended that tests should be conducted in order for the model to be simulated accurately on Abaqus, especially the uniaxial and biaxial test as demonstrated in . This allows physically motivated model to be used to provide an accurate solution [18, 19]. However, the shape of the deformation is as expected and is close to what happens in reality as observed later after the booms were manufactured despite the model's inaccuracy.

10.5 Manufacturing of the CTM booms

The manufacturing of the boom using FDM is relatively simple on MF's Onyx series printer, as MF provides a browser-based software to aid with the process. Eiger is designed to read CAD files and print part seamlessly, in a 'secure, fast and intuitive' way. The software allows the part to be prepared for printing and store multiple parts in the system. It also allows monitoring during printing and accurately estimates the time and cost of the whole process. The tool path can also clearly be displayed at each layer and allows control of support materials to be rendered. However, one disadvantage that was discovered during this manufacturing phase was that it does not allow the tool path to be altered, despite that there are still many different customisation available in the system such as infill density and print orientation. These factors will, however, affect the mechanical properties of the structure so appropriate choices must be made [20].

The preparation before printing is simple and only requires the application of glue with a glue stick on to the print bed to allow ease of printing after the part is completed. The process is automated and only requires a press of a button to start the printing process. One other limitation is the size of the machine, which only provides enough space to manufacture a boom with a maximum length of 200 mm. Testing of different printing orientation and method was carried out. Figure 10.28 shows the CTM boom being printed in action in MF's onyx series printer.

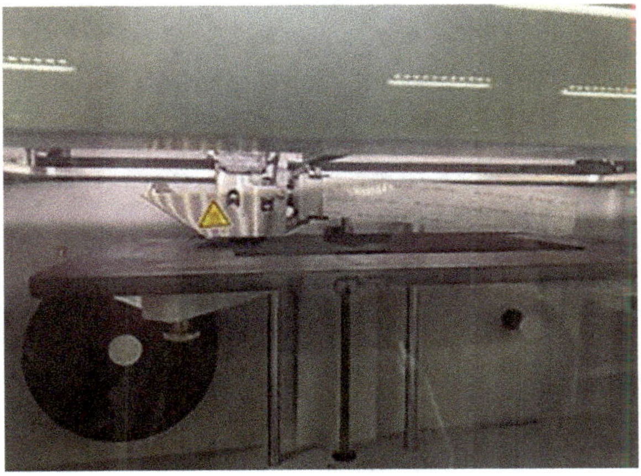

Figure 10.28 CTM boom being printed

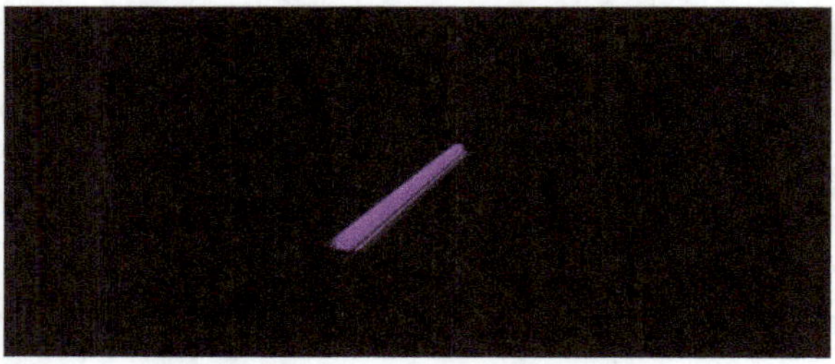

Figure 10.29 Screenshot of Eiger for 200 mm horizontal, half profile boom

10.5.1 Horizontal, half profile

This method of printing requires a lesser amount of materials as less supports are required. Figure 10.29 shows the screenshot from Eiger. The calculated cost of material is 1.49 USD with an estimated print time of 1 hour 39 m. However, the two halves manufactured have to be glued together with MF's recommended Loctite 401. Unfortunately, the glues are too rigid and as the boom is rolled together the shear causes the glue to break in a brittle manner by the outer boom being tensile and

Figure 10.30 The two halves of the boom are placed together

the inner boom being compressed causing a difference in length. Observing the half alone, the quality is exceptionally high with a nice matte surface finish. Figure 10.30 shows the two-halves placed together.

10.5.2 Vertical, full profile

This involves printing the booms vertical in full profile form layer-by-layer. This way requires no support, but the limitation is also with size. The maximum length of the boom is 50 mm which is not ideal. The smallest wall thickness achieved is only 1 mm, any thinner would result in the structure leaning on to one side during the printing process as observed. The result, as shown in Figure 10.31 is of high quality with a nice matte surface finish. Despite the high quality, the 1 mm thickness causes the structure to be too stiff and will not allow flattening and coiling around a spool. is a screenshot of Eiger and the cost of material calculated is 0.02 USD with a print time of 2 hours 49 m.

10.5.3 Horizontal, full profile

This method can be achieved by introducing extra supports on both ends of the web and inside the cross section of the tube. Building layer-by-layer the full profile can

Figure 10.31 50 mm vertical full profile boom

Figure 10.32 200 mm horizontal, full profile boom

be printed. This provides the best outcome compared to the two previously discussed printing orientations since the edge does not require any adhesives and it allows the full 200 mm to be printed. This, however, requires extra materials as support and it can be observed that the amount of materials required can be nearly as much as the boom itself. The support can also be very difficult to remove, and the zig-zag configuration of the support which already allows the ease of removal can still cause damage to the boom's structure itself. Extra care must be taken, and it is recommended that the removal should be done with tools such as a drill to remove the support inside the structure instead of pulling by hand. shows the damage that was caused by removal of support. Figure 10.32 show the manufactured boom and screenshot from Eiger, respectively. The cost of material calculated is 3.57 USD with a print time of 3 hours 25 m. As mentioned previously, this printing orientation provided the best outcome and thus will be used to manufacture the full set of booms.

10.6 Conclusion and future work

In conclusion, different designs of deployable booms were studied in this project, and based on gathered knowledge, a CTM boom is designed with an optimal cross section. Multiple FEA were conducted to and at the end a set of 200 mm CTM booms that adds up to 1 m were manufactured using MF's FDM printer.

The project was split into two main phases: the design and analysis phase and the manufacturing and testing phase. The objectives of each phase are largely met, and different topics were introduced in this project. Due to the lack of resources or time, some topics were not fully explored. In the analysis stage, FEA three-point bending was introduced, but there was lack of information available in research on the hyperelastic characteristics of MF's flagship material the Onyx, which is a nylon-based material reinforced with short CF. It was suggested in the previous sections that mechanical testing such as the uniaxial and volumetric test should be carried out to better characterise the material behaviour under certain loading conditions in FEA. It was suggested that a physical three-point bending test was to be carried out to determine the flexural strength of the CTM boom and this was not determined experimentally. A test in the future should be beneficial. The viscoelastic effect of the boom during long-term stowage inside the DOM should also be determined since creep and relaxation effects will lead to a smaller subtended angle which causes the boom to have weaker performance. Therefore, the effect should be studied further with relaxation test and creep test. Since the booms were manufactured as 200 mm each, a method to join them together has to be explored in order for the AM CTM boom to be scalable. Finally, the full length of the CTM boom should be tested against the CuBe boom to measure the improvement of this geometry compared to the c-shaped design.

References

[1] Satellite Applications Catapult *2020 Q4 small satellite market intelligence report-satellite applications catapult* [online]. 2021. Available from https://sa.catapult.org.uk/news-comment/market- reports/small-satellite-market-intelligence-report/small-satellite-report-registered/ [Accessed 15 Jun 2021].

[2] Sdo.esoc.esa.int *ESA's annual space environment report 2020 [online]*. 2021. Available from https://www.sdo.esoc.esa.int/environment_report/Space_Environment_Report_latest.pdf [Accessed 15 Jun 2021].

[3] Serfontein Z., Kingston J., Hobbs S., Holbrough I.E., Beck J.C. 'Drag augmentation systems for space debris mitigation'. *Acta Astronautica.* 2021;188(A6):278–88.

[4] Markforged *Threading the needle with ease: minimum fiber feature sizes [online]*. 2021

[5] Hu Y., Zhao Y., Tuo Z., Wang J. 'Parametric modeling and optimal design of space tubular extendable booms via a one-dimensional unified formulation'. *Advances in Structural and Multidisciplinary Optimization*. 2017::597–609.

[6] Serfontein Z. *Scalability and design refinement of Cranfield University's drag augmented system*. Cranfield University; 2018.

[7] Fernandez J. *Advanced deployable shell-based composite booms for small satellite structural applications including solar sails*; 2017.

[8] Fernandez J., Rose G., Stohlman O, *et al*. 'An advanced composites- based solar sail system for interplanetary small satellite missions'. *Aerospace Research Central*. 2018.

[9] Hakkak F., Khoddam S. 'On calculation of preliminary design parameters for lenticular booms'. *Proceedings of the Institution of Mechanical Engineers, Part G: Journal of Aerospace Engineering*. 2007;221(3):377–84.

[10] Ning F., Cong W., Qiu J., Wei J., Wang S. 'Additive manufacturing of carbon fiber reinforced thermoplastic composites using fused deposition modeling'. *Composites Part B: Engineering*. 2015;80(1):369–78.

[11] Love L.J., Kunc V., Rios O, *et al*. 'The importance of carbon fiber to polymer additive manufacturing'. *Journal of Materials Research*. 2014;*29*(17):1893–98.

[12] Sauer M. *Evaluation of the Mechanical Properties of 3D Printed Carbon Fiber Composites*. 2018.

[13] Mavrodontis N. *Modelling hyperelastic behavior using test data in abaqus. [online]*. 2021

[14] Markforged *Onyx one desktop 3D printer | markforged [online]*. 2021

[15] Support.markforged.com *Markforged support portal [online]*. 2021

[16] Ćurković L., Bakić A. 'Flexural strength of alumina ceramics: weibull analysis'. *Transactions of FAMENA*. 2010;34(1):13–19.

[17] Obbink-Huizer C. *Hyperelastic material models in abaqus. when to choose what?. [online]*. 2021

[18] Morgan R., McReynolds B., Husmann K, *et al*. *Markforged Continuous Fiber Composite Material Testing*. 2020.

[19] Markforged *eiger 3D printing software | markforged. [online]*. 2021. Available from https://markforged.com/software

Chapter 11

Adhesively bonded polymer composite joints

E. A. S. Marques[1], R. J. C. Carbas[1], A. Akhavan-Safar[1], and L. F. M. da Silva[2]

Adhesive bonding is a key technology for many different technological sectors, representing an effective, practical and versatile joining technology. In fact, adhesive bonding is especially important for applications where composite components must be assembled, since it provides a relatively uniform stress distribution, does not require drilling holes or other modifications and can have excellent adhesion to a wide range of substrates. There are, however, some important issues associated to composite bonding that must be fully understood, such as the susceptibility to delamination, which occurs due to the peeling loads generated by the adhesive layer on the low transverse strength substrates. In this chapter, the main considerations for bonding composite substrates are initially discussed, and a set of specific and recent research topics related to composite joints and adhesives are discussed in more detail. These include the use of modifications in the composite adherends, the use of functionally graded substrates, the mechanics and uses of bi-adhesive joints and the growing use of composite adhesives, especially those reinforced with natural materials.

11.1 Introduction

Lightweight structures are nowadays a key element of the products manufactured by a wide range of industrial sectors, especially those related to the transportation industry. For example, the automotive, shipbuilding and aerospace industries make use of advanced structural materials such as aluminium, titanium and composite materials in order to reduce the weight of their products. Structures such as aircraft fuselages and wings, high-speed train bodies and wind turbine blades are composed of bonded composite parts, which serve to reduce their weight and increase structural rigidity.

[1]Institute of Science and Innovation in Mechanical and Industrial Engineering (INEGI), Rua Dr. Roberto Frias, Porto, Portugal
[2]Departamento de Engenharia Mecânica, Faculdade de Engenharia, Universidade do Porto, Rua Dr. Roberto Frias, Porto, Portugal

Shipbuilding industries also extensively use bonded fibre-reinforced polymer (FRP) structures in ship construction. This is especially important for the construction of the taller components, such as the superstructures, since it lowers the centre of mass of the ship, increasing its seaworthiness [1]. In the specific case of the modern automotive industry, where emission regulations are becoming increasingly tighter, the adoption of lightweight vehicle structures is a key step for achieving better performance, enhanced fuel economy and reducing pollutant emissions. This is also true for the case of the most recent electric vehicles, since the large weight of the battery pack can be offset by a lighter composite structure, extending the vehicle range.

While composite materials are extensively used in these structures, they are rarely used alone. In fact, a combination of different materials is often used, mainly merging FRPs and metal. This is true, e.g., in aircraft fuselages, vehicle body panels and wind turbine blades. In these applications, aluminium- or titanium-based light weight alloys are commonly used in combination with composite materials. Furthermore, since structural components are of great complexity and/or size, it is often impractical to manufacture these products as a single piece in most applications. Thus, the sub-components, which compose the structure, must be assembled through an appropriate joining technique, which should be able to ensure the stiffness and integrity of the assembled structure and support the expected service loads.

Experience shows that the use of fasteners, welding, riveting, adhesive bonding and hybrid connections (combining different connection techniques in a single joint) is a joining technique, which is suitable for many types of high-performance structures using composite components. Still, some of these techniques are not ideal for joining composite materials, especially those which necessitate the creation of holes or require high temperatures. This is because most of the composite materials are highly sensitive to damage and defects (caused by holes, excessive temperature, etc.) and have low transverse strength, which makes them highly susceptible to failure under large peeling stresses, as we will see in greater detail next.

11.2 Bonding of composite materials

Composite joining techniques have been the target of extensive experimental and numerical research, with most of these research works being dedicated to improving joint design and manufacturing procedures, thus ensuring that these joints can withstand service loads without any type of failure. While fasteners and rivets are often used in practice to join composite components, this type of connection can increase the weight of the structure [2] and brings some other particular issues such as an irregular joint surface, critical for the construction of aerodynamic profiles where a smooth surface is desirable. In addition, these mechanical joining methods require the use of an additional sealant to ensure that watertightness can be achieved. Furthermore, the hole that must be created in the composite material creates discontinuities in the reinforcement fibres and creates stress concentration points around the drilled holes. In all, this will dramatically decrease the strength of the composite substrate and will greatly increase the susceptibility to failure by delamination.

Delamination failure is a failure mode where the plies of a laminated composite material lose cohesion (delaminate) because of peel stresses (stresses which act perpendicularly to the plies of the laminate). The combination of all these effects dramatically reduces the performance of a bonded joint, since neither the composite nor the adhesive is being stressed to their ultimate capabilities. Alternative joining solutions have been developed to overcome some of these issues, and include, e.g., fusion or ultrasonic welding of thermoplastic composites [3], although the suitability of these techniques for structural applications is quite limited still.

Considering these limitations, adhesive bonding has grown to be industrially accepted as the most practical joining technique for joining composites. Using modern adhesives, a properly designed bonded joint with composite substrates can be relatively simple to manufacture and allows for significant structural weight reductions while retaining the overall performance of the complete structure. With the use a continuous adhesive layer, the bonded joints experience a generally more uniform stress distribution. Additionally, since composite structures often include other types of material, the capability for bonding a wide range of materials exhibited by modern adhesives, provided adequate surface preparations are made, is of great value. Furthermore, very thin substrates can also be easily bonded since no modifications to the substrates are necessary.

Adhesives have an important visco-elastic response, which is reflected in a higher fatigue life when compared with mechanical joining techniques. This behaviour also has a significant influence on the damping capacity of the bonded structures, allowing these bonded composite structures to dampen vibrations and exhibit sufficient strength under highly dynamic loadings, such as impacts. Adhesive bonding of composites is also a process, which requires a very limited set of manufacturing operations, which leads to lower productions costs and is also more flexible regarding geometrical configuration of the adherends and the joint, lending itself to joining complex geometries without the need for highly specific producers.

In several critical applications, there is the option to increase operational safety margins by combining two or more different joining techniques in a single overlap. These are known as hybrid joints and are especially useful for the aeronautical sector, which greatly values safety and durability. The most typical of these consists in the combination of riveting and adhesive bonding, connecting aluminium–composite joints in components such as wings [4]. Still, it is quite common to find applications where adhesive bonding is used almost exclusively, such as some applications in the automotive industry [2].

As referred before, although the use of adhesive bonding is advantageous for composite bonding, one must always consider the low transverse strength of the composite substrates and the associated danger of delamination failure. This failure mode occurs if the peel stresses generated at the edges of the bond line exceed the interlaminar strength of the composites. Composite materials are known for their very high strength in the direction of the fibres, but their strength in transverse direction pales in comparison, being orders of magnitude lower. In the transverse direction, the fibres do not effectively reinforce the matrix, and thus the strength of the composite in this direction is in fact closer to that of the much weaker matrix. For

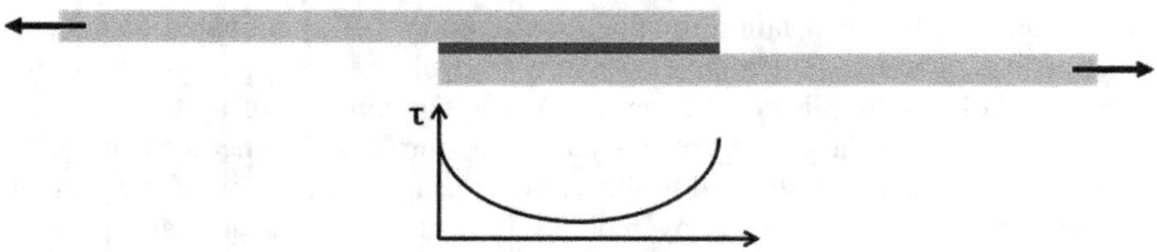

Figure 11.1 SLJ configuration and shear stress distribution along the overlap length

example, the single-lap joint (SLJ), the most widely used adhesive joint configuration, has a non-uniform stress distribution (shear and peel), generated along the overlap length. Important stress concentrations can be found at the ends of the overlap (Figure 11.1).

The peel stress concentration located at the ends of the bonded area can easily induce composite failure, loading the composite almost exclusively in the low strength, transverse direction. In this case, the composite adherend fractures under the peel stresses, resulting in a greatly diminished shear transfer capacity between both adherends [1].

Avoiding this issue has been the target of intense academic research, which has explored diverse methodologies, such as novel composite configurations and materials and new joint geometries. The nature of the adhesive and the adherends, joint types and dimensions, surface preparation, environmental conditions, etc. can all alter the susceptibility for delamination of a composite bonded joint. Most academic research in this field focuses on the reduction of the stress concentration at the edges of the bonded area [5] or in strengthening or reinforcing the composite materials in the transverse direction. In some cases, the adhesive itself can be modified with the particles to ensure that the failure mode transitions from delamination of the adherends to fully cohesive failure within the adhesive layer. This approach works mainly by reducing the strength of the adhesive, but it does not necessarily lead to joint strength reduction, since it is compensated by a more benign stress distribution and positive changes in the joint failure mode. For example, the work of Hsiao *et al.* [6] showed an improvement in joint strength by 45% by adding 5% multi-walled carbon nanotubes to the adhesive and changed the failure mode from interfacial to the cohesive failure within the adhesive layer.

Selecting a less stiff adhesive and using larger bonded areas are also effective in transferring the load through the composite joints, without the generation of excessive peel stresses and the risk of fibre delamination. The work of Shang *et al.* [7] evaluated techniques for improving the performance of bonded composite joints and concludes that the joint geometrical configuration has a large influence on the joint strength. Still, the authors also point out that in practice it is often difficult to implement the suggested geometrical changes, since these processes can be costly, or the adherend geometry is often fixed by design. A novel design procedure for bonded composite joints is also described in the work of Chamis and Murthy [8]. This was achieved by analytically studying the stress distribution in different types of joints, based on the free body diagrams. The authors arrived at a butt joint configuration that was experimentally subjected to different loading and environmental conditions. The authors suggest avoiding asymmetric composite joints, given their inefficiency with regard to the load carrying capability. Other works present in the literature studied diverse changes to the geometry of SLJs with composite substrates, searching from improvements in joint strength. Some of these studies found success by altering the geometry of the complete bonded joint [9, 10] while others only rely on modifications to the spew fillet [11, 12]. The use of bi-adhesive (or mixed adhesive joints), where a ductile adhesive is used at the overlap ends to reduce the peel stresses, and a stiffer adhesive is applied at the middle of the overlap [13, 14] has also been reported as useful. More information regarding this method is given in a subsection of this chapter.

11.3 Composite adherend modifications

As discussed in the introductory section, one of the key principles for increasing the strength of composite strengths is the avoidance of delamination via the reduction of the peel stresses acting on the composite substrates or instead by strengthening and toughening the composite adherends [7]. This section is devoted to the latter of these approaches, describing a set of effective methodologies for increasing the performance of composite substrates.

11.3.1 Transverse toughened and reinforced adherends

Toughened composites can be obtained by the lamination of two or more different laminate materials, creating a material which has novel, synergistic properties. Composites toughened with metals are known as fibre metal laminates (FMLs) and consist of metallic layers with FRPs such as carbon and glass fibre. They can be used to enhance specific properties of the laminate. For example, the use of FML reinforced with aluminium alloys can be used to greatly improve the impact strength of carbon fibre composites used in aircraft components [15].

Hybrid joints using reinforced adherends, e.g., those using an FML type of construction, have been shown to be able to fully avoid or even mitigate delamination failure of the joint [7, 16, 17]. This is because the substrates present an increase in the transversal strength induced by the presence of the metallic (or composite) reinforcement layers [7, 15]. Researchers at the Delft University of Technology were the creators of the first FMLs in 1978, first using these materials to improve the behaviour of aluminium alloys under fatigue loads [17, 18]. Subsequent research also identified an increase in strength when aramid fibres were used to reinforce aluminium sheets, and this resulted on the creation of a special type of FML, known as aramid reinforced aluminium laminate (ARALL), which was eventually followed by additional FML configurations using other fibre materials as reinforcement, such as carbon ARALL and basalt ARALL. Nowadays, FMLs are mainly used in the aerospace industry, permitting the construction of lightweight, yet tough, aircraft components. The literature shows many other FML configurations, which explore

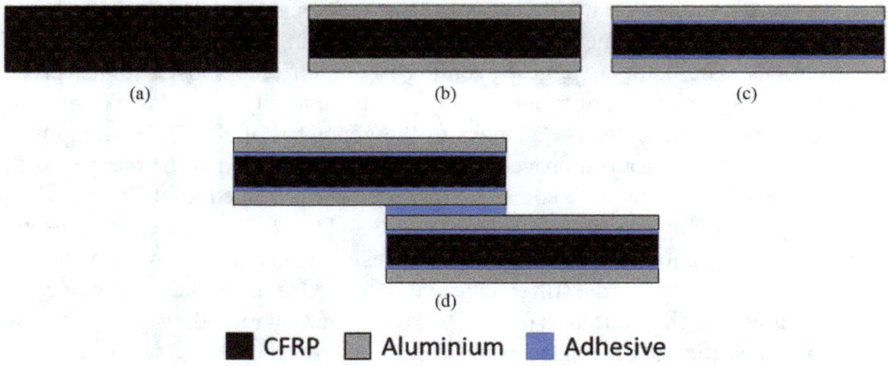

Figure 11.2 *(a) Fibre only, (b) fibre + metal, (c) fibre + metal with additional adhesive layer and (d) adhesive joint with FML + additional adhesive layer (adapted from [17])*

the use of other materials, such as titanium and the addition of other layers besides the fibre and the metal layer. For example, additional layers of a film adhesive can be integrated in the laminate, furthering the adhesion between the composite resin and reinforcement of the metal. This effect is shown schematically in Figure 11.2 [18].

Santos *et al*. studied the performance of SLJs bonded with three different FML configurations, using only carbon FRP (CFRP), CFRP and titanium and CFRP with titanium joined by an intermediate adhesive layer [17]. An improvement of the joint strength over the reference CFRP-only joint was found for the two modified designs under study. In fact, the substrates using additional adhesive layer led to the optimal joint performance, with a failure load improvement of approximately 30% (Figure 11.3). Also importantly, both modified designs led to cohesive failure,

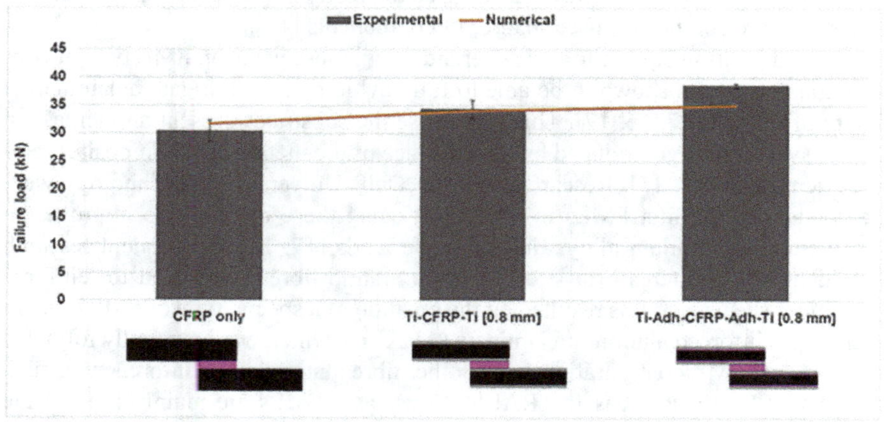

Figure 11.3 *Numerical and experimental failure loads [17]*

while the reference CFRP substrates failed by delamination [17]. Numerical simulations using cohesive zone modelling were found to be able to precisely replicate the experimentally obtained failure mode and predict joint performance.

Morgado *et al.* [16] further assessed the performance of adhesive joints with FML substrates. In this work, CFRP substrates were modified with aluminium layers, bonded and tested under quasi-static and impact conditions. Under quasi-static loads, the failure mode transitioned from complete delamination in non-FML specimens to a cohesive type of failure, increasing the failure load by 30%. Under impact loads, delamination occurred in both cases, although the failure load increased by 22.5%, indicating that the FML substrates can also be beneficial under impact loads.

Polymeric layers can also be used to enhance the toughness of composite layups, reducing delamination failures under static loading [7]. Although few works have been published on this subject, the literature describes three techniques for reinforcement of CFRP using polymers. These are local toughening, global toughening by composite layup and global toughening by the introduction of additional adhesive layers.

The local toughening technique was proposed by Schollerer *et al.* achieved with the introduction of a Polyvinylidene fluoride (PVDF) thermoplastic layer, 0.1 mm thick, at the overlap ends [19]. This layer is co-cured with the prepreg. The purpose of this PVDF layer is to locally increase the toughness of the adherends in the areas of the joint where the higher peel stresses are generated. PVDF is a highly ductile material, which is able to absorb a large component of the strain acting on the composite substrate, thus, reducing stress concentrations [19]. Bonded joints using these substrates exhibited a 38.2% reduction of peel stress and a 33.7% reduction of the shear stresses. Overall, this resulted in an increase of 84% of the joint strength, compared with a reference, non-reinforced joint. An example of a locally toughened layout is shown schematically in Figure 11.4.

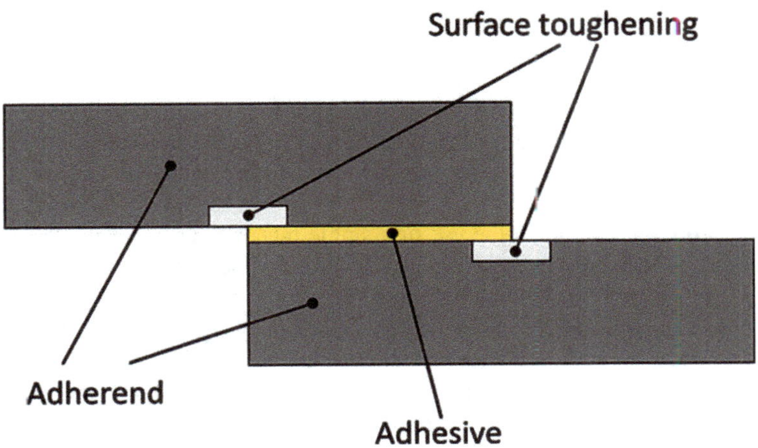

Figure 11.4 Locally toughened SLJ (adapted from [19])

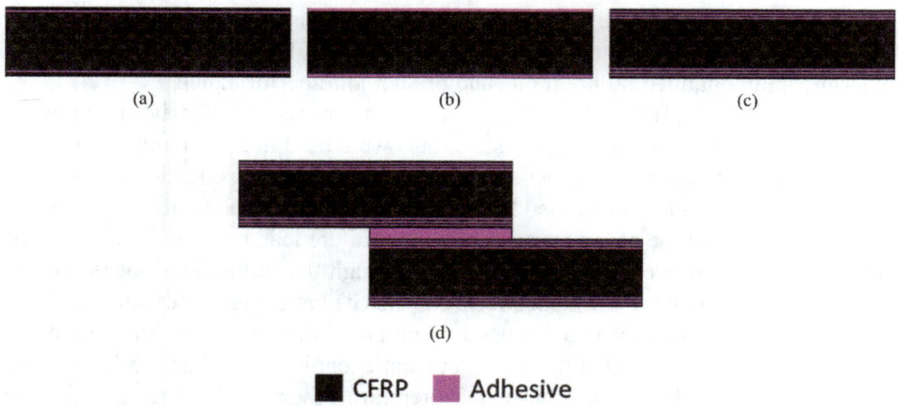

Figure 11.5 *(a) CFRP with one interlaminar adhesive layer (AF 163–2K), (b) one superficial adhesive layer (ALR), (c) three interlaminar adhesive layers and (d) adhesive joint with three interlaminar adhesive layers (adapted from [21])*

The alternative technique of global toughening by composite layup is achieved via the use of a fibre reinforced, high toughness resin, co-cured with a CFRP plate. The result is a composite plate, which has a very stiff core clad by a tougher surface layer on both sides. First proposed by Shang *et al.* [20], this method has been shown to significantly reduce delamination on composite joints. SLJs using these global toughened layers demonstrated a strength increase of 22% and no delamination failure. This especially significant, since the joint was bonded using a very brittle and stiff adhesive, which generates large peel loads [20].

Last, toughening composite materials can also be achieved with the introduction of fully polymeric layers in the composite layup. In fact, this technique is quite close to the FML concept and was developed to minimize the delamination of the composite, by reducing and spreading the peel stresses acting on the composite substrate, using additional plies of resin/adhesive with different configurations [20, 21]. Figure 11.5 shows a scheme with different configurations of composite material toughened with polymer layers. The case, where an adhesive is used to reinforce the external layer (d), is known as the adhesive layer reinforced (ALR) configuration [21].

Morgado *et al.* evaluated the use of a single interlaminar adhesive layer, three interlaminar adhesive layers and a single external adhesive layer (the aforementioned ALR configuration) in bonded joints. The optimized designs presented better results than the reference CFRP-only joint, loaded under quasi-static loads, with an increase of approximately 23% of the failure load achieved by the laminates with superficial/interface reinforcement. This was closely followed by the specimens with three interlaminar adhesive layers and the single interlaminar adhesive layer [21]. In both these cases, cohesive failure on the adhesive layer was encountered, not on the reinforcement. Moreover, the ALR joint presented a strength increase of

17% in comparison with the CFRP-only joint when tested under impact conditions (3 m/s) [21].

11.3.2 Functionally graded substrates

Functionally graded materials (FGMs) are valuable for a diverse range of applications, where the possibility of having a gradual variation in the mechanical and physical properties of the material is advantageous. Adhesive joints can also benefit from these materials, as they can be used to solve one of the major issues associated with adhesive joints, which is the damage generated by the stress concentration at the overlap ends. Ideally, FGMs can be employed to create adherends with highly optimized properties, suited not only to withstand the loadings to which the structure is submitted but also to provide any additional and desirable physical properties in specific locations (e.g., variable thermal and electrical conductivity).

Ganesh *et al.* [22] was the first to introduce the use of this technology in bonded joint, in 2002 and, since then, only a limited number of works present experimental results for adhesive joints with FGM adherends. This is related to the most important challenge associated with this technique, which is the difficulty in creating a practical process, able to produce FGMs effectively. Some of the first graded adherends described in the literature employ graded perforations, as proposed by Melograna *et al.* [23]. The authors used stainless steel specimens and glass fibre composite sheets, perforated with circle- and triangle-shaped holes of several different sizes and orientations, allowing to obtain an optimized distribution of graded stiffness [23]. The holes can be created using a waterjet cutter (for composites and metals) or laser cutter (for metals only). The steel and composite plates were then bonded in joints using a vinyl ester adhesive, as shown in Figure 11.6. Experimental testing identified the optimal graded distribution to be seven to nine rows of either circular- or triangular-graded holes. For both glass fibre materials under study, the increase in joint strength was always above 20%, although failure was still reported to occur in the glass fibre substrate.

The performance of other adherend shaping techniques has also been experimentally validated [24]. Some works have studied fully homogeneous, continuous adherends but that exhibit a variable thickness ratio between a free face and a restrained face. When used in a joint, this tapering of the substrate effectively results in a graded stiffness variation through the overlap length. Graded adherends can be

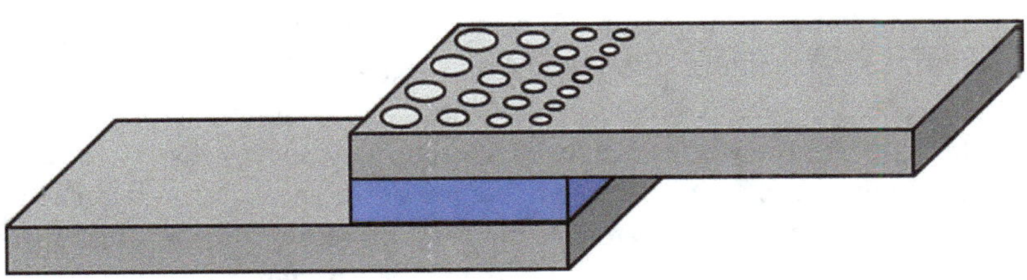

Figure 11.6 Three-dimensional (3D) scheme of an adhesive joint with graded perforations

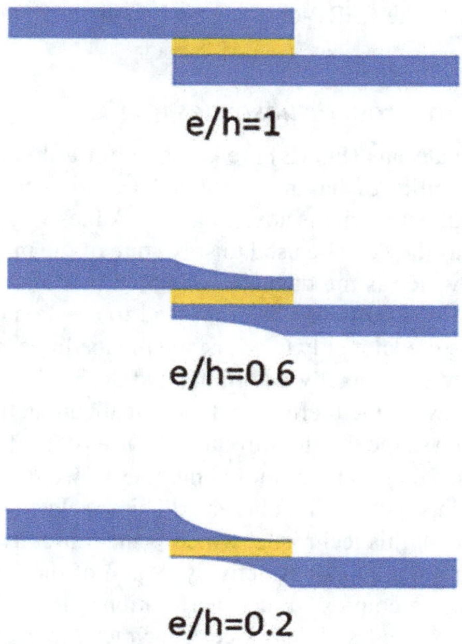

Figure 11.7 *Specimens with different values of e/h tapering ratio (adapted from [24])*

manufactured using diverse methodologies, such as lamination in a mould, machining operations or the use of multi-material 3D printer, able to create the whole sample at once [24]. In fact, the use of additive manufacturing greatly increases the possibilities associated to FGMs, allowing to obtain complex geometries and combining polymers with different properties in a single component, something which has already been done for adhesive layers but not for adherends [25, 26]. Figure 11.7 shows examples of e/h ratios that were experimentally evaluated, where e represents the thickness on the free face and h represents the thickness of the restrained face.

The results shown in Figure 11.8 clearly demonstrate that geometrically graded SLJs lead to a large increase in failure load when compared with a reference, non-modified, joint geometry. The best performing joints are those with more aggressive tapering (e/h = 0.2), attaining a very significant increase in 80% of the failure load. This level of strength improvement matches that found for much more complex joints, which combine optimized reverse tapering with spew fillets [7, 9, 27–29]. This increase can be explained with the reduction of the shear stress associated with the tapering configuration, generated by the low stiffness of the very thin material in the edge of the substrate [24]. The authors further reported that failure initiates at the interface and then transitions to the middle of the bondline as a fully cohesive failure. This indicates that the stress concentration peaks on the free faces, typically

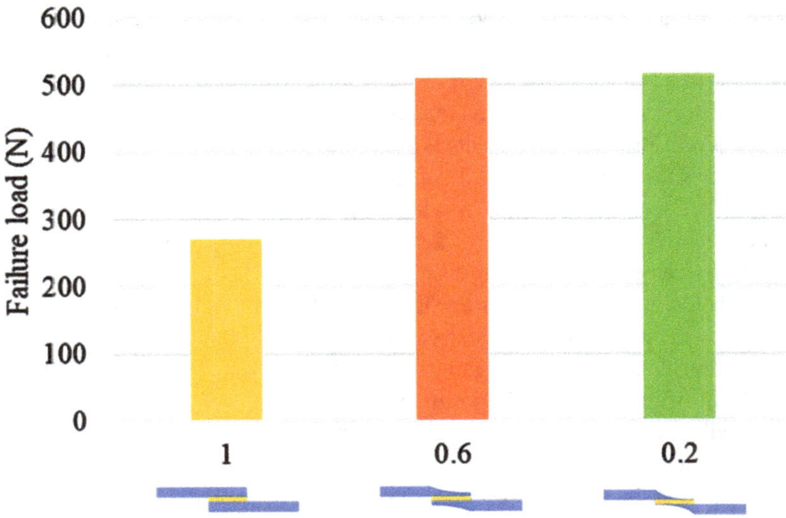

Figure 11.8 Failure load for printed specimens with different graded tapering ratios (e/h) (adapted from [24])

exhibited by SLJs, have been successfully mitigated by the grading technique. Such geometry can enhance the energy absorption and greatly increase joint strength.

Another experimental work employing the use of a graded adherend in bonded joints explores a material with honeycomb structures of variable sizes and has been specifically designed for impact attenuation in helicopter seats [30], aiming to reduce the load transmitted to the pilot and passengers in a crash. The design concept is inspired by the cross section of bones, presenting anisotropic properties, optimized to a specific loading direction. The honeycombs were created using a film adhesive to join thin metal sheets, shaped to form half of the honeycomb geometry [30]. The graded design, with variable sheet thickness per layer, generates a structure with variable stiffness and energy absorption per layer, something which is ideal for supporting impact loads, since the deceleration of the impact mass is achieved in a gradual manner. Under impact conditions, the lower honeycomb cells are responsible for deforming and absorbing the maximum amount of energy, reducing the load transmitted, while the upper honeycombs stay relatively undeformed, providing the necessary stiffness to the structure. The structure was optimized using genetic and sequential quadratic programming algorithms, verified using numerical simulations that were found to accurately represent the experimental results obtained.

11.4 Bi-adhesive joints

Overlap ends, especially in dissimilar composite/metal bonded joints, experience high stress levels. Different optimization approaches have been proposed by researchers to reduce the stress level at the overlap ends and improve joint strength by creating a more uniform stress distribution along the bondline. The better known improvement methods include tapering the adherends and using fillets at bonding edges [9]. However, despite the advantages of these methods they suffer from some drawbacks. Forming controlled fillets is practically complicated especially for low-viscosity adhesives. On the other hand, tapering the substrates in practice is not always possible, especially for joints with composite adherends.

Results show that load-bearing adhesives with high shear performance can increase joint strength. But the joint strength is restricted by peel stresses generated at overlap ends. On the other hand, due to the higher flexibility, ductile adhesives show a better stress distribution but less strength. Considering these points, bi-adhesive joints were introduced and investigated by some researchers. Bi-adhesive joints are a simple form of graded adhesive joints in which two different adhesives, one stiff and one flexible, are used along the overlap. The stiffness of the joint is controlled by the stiff adhesive, which is used in the middle of the overlap while the flexible adhesive placed at overlap ends controls the stress distribution. Using flexible adhesives at bonding ends, especially for joints with longer overlaps, can significantly improve the stress distribution by reducing the maximum peel stresses. However, due to the application of ductile adhesives, the strain state at the ends of the overlap will increase compared with the joint with a single stiff adhesive. Although, generally, the bi-adhesive technique can significantly improve the mechanical performance of the bonded structure, it should be noted that an improvement is achieved when an appropriate combination of the ductile and stiff adhesive is used, and the part of the overlap covered by each adhesive is correctly adjusted. Several studies have dealt with these types of joints, and different aspects of bi-adhesives have been investigated. Most of the works are related to single-lap adhesive joints but there is research in which other types of joints were analysed. In this section, bi-adhesive joints as a simple form of graded adhesive joints are briefly reviewed in terms of joint type, joint geometry and adhesive properties. However, before that, the manufacturing procedure of bi-adhesive joint is discussed.

11.4.1 Manufacturing techniques

The two ductile and brittle adhesives used in bi-adhesive joints are sometimes very different in terms of their physical properties. Different viscosities, different curing temperatures and times and different pot-life lengths make the manufacturing of bi-adhesive joints challenging. Alternative approaches have been introduced and used by diverse authors. For example, using silicon rubber or nylon tape at the boundary of the two stiff and ductile adhesive was considered in some studies [29, 31], while other authors placed a wire at the overlap and along the width of the joint to define the boundary of the two adhesives [32, 33] (see Figure 11.9). Machado *et al.* [34] used nylon tape and showed that using this technique it is possible to minimize the mixing of the two adhesives and reduce the bubbles formation at the boundary of the adhesive layers.

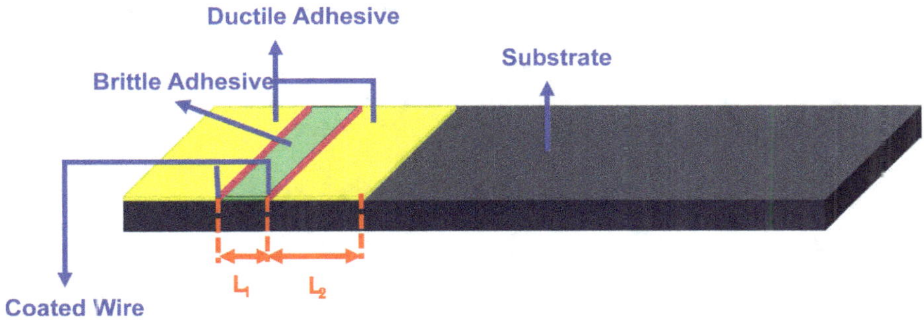

Figure 11.9 *Scheme of a bi-adhesive joint utilizing wire to prevent adhesives mixing*

For some specific ductile and stiff adhesive combinations, the large difference between the viscosities can also help to avoid adhesive mixing. Placing a gap between the two adhesives [35] and also applying the adhesives in two time steps [36] are other techniques considered in some studies. Although this method is used almost in all the mentioned works, authors believe that they could successfully apply the two ductile and stiff adhesives on a single overlap. To avoid overcuring and the degradation of the adhesive properties due to overheating, the lowest possible curing temperature should be considered. It should be noted that curing at different temperatures can change the stiffness of the adhesives as shown by Carbas *et al.* [37–39].

11.4.2 Bi-adhesive SLJs

One of the most common bi-adhesive joint geometries is based on the SLJ. Due to the eccentricity of the load acting on this joint, a bending moment is induced to the joint, which generates large peel stresses at the ends of the overlap. By using more ductile adhesives in these locations and applying the stiff adhesive at the middle of the overlap, the induced peel stresses at bonding ends will decrease.

The stress distribution in bi-adhesive joints is different from the SLJs bonded with a single adhesive. In bi-adhesives, there are two stress peaks along the overlap. The first peak is close to the overlap ends, and the second peak is at the boundary of the two flexible and stiff adhesives. Figure 11.10 schematically shows the stress distribution of the shear stresses along the overlap in a single-lap adhesive joint bonded with single- and bi-adhesives.

In bi-adhesive SLJs, the peel stress at overlap ends is positive but the stiff adhesive experiences compressive normal stress. A significant drop in peel stress due to the presence of the ductile adhesive takes place at the bonding end of the bi-adhesive joints compared with the joints with a single stiff adhesive. However, due to the use of a ductile adhesive at the end of the overlap, the strain levels are higher in bi-adhesive joints.

As mentioned above, to obtain the maximum strength improvement in bi-adhesive joints, the flexible and ductile adhesives should be selected properly, and the length of each adhesive on the overlap should be correctly set. Several authors

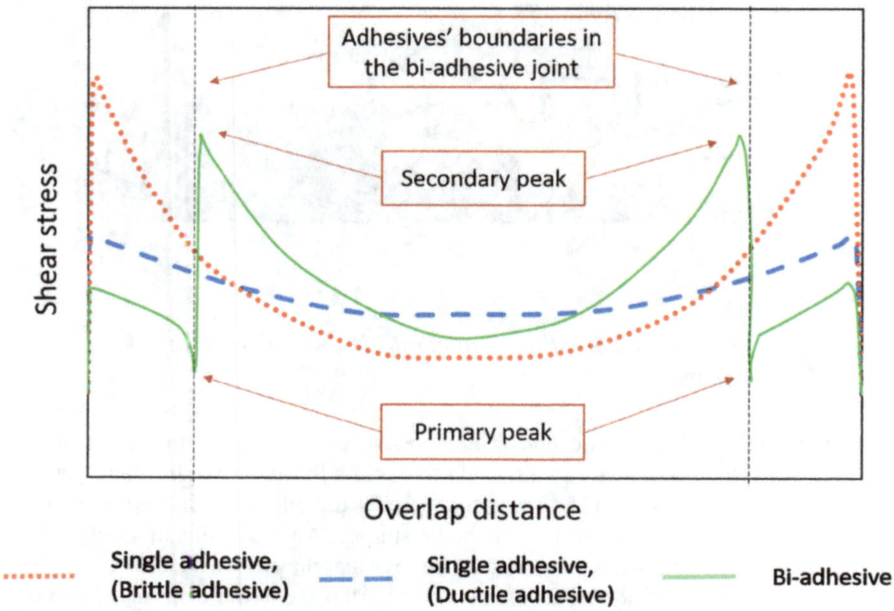

Figure 11.10 Effect of the bi-adhesive joint configuration on the shear and peel stress distributions

have investigated the role of adhesive properties and joint geometry on the strength of bi-adhesive SLJs. In the following sections, the effects of each parameter are discussed.

11.4.2.1 Length ratio

One of the most influencing parameters in the bi-adhesive joining method is the length ratio (LR). LR is defined as the length of the overlap bonded by the ductile adhesive (L_1) to the overlap length where the more brittle adhesive is used (L_2), as shown in Figure 11.9. Accordingly, the LR is defined as LR = L_1/L_2 where the total overlap length is $l = 2L_1 + L_2$. LR is zero whenever the total overlap length is covered by the brittle adhesive. Based on the results [2], there is always an optimum LR where the joint reaches its maximum strength. The optimum LR is a function of adhesive thickness and is influenced by the mechanical properties of the two brittle and ductile adhesives. Accordingly, different optimum LRs have been proposed by authors. For example, Özer and Öz [40] suggested an optimum LR of 1.3, Piresa *et al.* [36] found that the joint reaches its optimum strength when LR = 0.7 and Kumar *et al.* [41] reached the optimum LR of 1.25. Ramezani *et al.* [42] experimentally analysed peel strain distribution in bi-adhesive SLJs. They found that the increase in normal compressive strains at the boundaries of the two ductile and brittle adhesives improves the strength of the bonded bi-adhesive SLJs. However, as the adhesive's LR increases, these compressive strains decrease which leads to a decrease in bond strength. Accordingly, there is an optimum LR where the strength of the bi-adhesive joints is maximum.

As mentioned above, one of the main reasons behind the different optimum LRs is the different mechanical properties of the considered brittle and ductile adhesives. For example, E_{ratio}, which is defined as the ratio of Young's modulus of the ductile adhesive; E_D, to Young's modulus of the brittle adhesive; E_B, affects the optimum LR. The effect of E_{ratio} on the mechanical performance of bi-adhesive joints is discussed in section 11.4.2.2.

11.4.2.2 Effects of the E_{ratio}

It was discussed in section 11.4.2.1 that the optimum LR is a function of E_{ratio}. In terms of the stress distribution, the lower the Young's modulus of the ductile adhesive, the lower the stress level at bonding ends. But on the other hand, the higher the stiffness of the adhesive, the higher the strength of the joints. As studied by authors [43], by increasing the E_{ratio} the peel stress close to the overlap ends increases, which reduces the joint strength. But, on the other side, for very low E_{ratio} where the stiffness of the ductile adhesive is very low, joint strength will be even less than that of the joint with a single brittle adhesive. Carbas *et al.* [38] showed that the more uniform the variation of Young's modulus of the adhesives along the bondline, the higher the failure load of the joints. They could increase the mechanical performance of the joints by increasing the failure loads and displacements. Not only the E_{ratio} but also the bondline thickness affects the optimum LR. Section 11.4.2.3 deals with the effects of adhesive thickness on the strength of the bi-adhesive joints and on the optimum LR.

11.4.2.3 Adhesive thickness effects

Although the numerical results propose a thicker bondline where the load is distributed over a bigger area and within a higher amount of the adhesive and the numerical stress is lower, experimental evidence is in contrast with this analysis. According to the experimental data, there is an optimum bondline thickness where the joint reaches its maximum mechanical performance. The optimum adhesive thickness is a function of the stiffness of the adhesive. Generally, for more brittle adhesives the optimum thickness is lower than that of ductile adhesive. However, for bi-adhesive joints, the situation is more complicated as both the ductile and brittle adhesives are used on a single overlap. On the other hand, not only the optimum thickness is a function of the properties of each of the ductile and brittle adhesives but also it is a function of the LR as well. Nonetheless, similar to joints with a single adhesive, there is an optimum adhesive thickness where the optimum strength of the joint is obtained [42]. Results show that the role of adhesive thickness is more significant for bi-adhesive joints with lower values of LR and E_{ratio} [38–42, 44].

11.4.2.4 Effect of the overlap length

Overlap length has similar effects in bi-adhesive joints as the joint with single adhesive [45]. The higher the overlap length, the higher the peel stress applied at overlap ends. Accordingly, in bi-adhesive joints where a ductile adhesive is used at bonding ends, for longer overlaps, the peel stresses are more uniformly distributed compared with the joint with a single brittle adhesive. Consequently, the performance of bi-adhesive SLJs with longer overlaps is more pronounced. However, to obtain the maximum strength, other parameters such as the LR should be correctly adjusted. Fitton and Broughton [46] showed that by increasing the overlap length, the optimum LR increases.

11.4.2.5 Temperature effects

Although the bi-adhesive joints have been mainly analysed at room temperature, in some industrial applications, joints may be exposed to very high or very low temperatures. Adhesive manufacturers usually suggest specific high- or low-temperature adhesives for joints used at high or low temperatures, respectively. But in some applications, such as in aeronautical industries, adhesive joints may experience very low-, room- and high-temperature conditions in service. Bi-adhesive joints can be a solution for such conditions as suggested by Hart-Smith [47]. In this condition, an adhesive compatible with high-temperature applications is used in the middle of the joint while a low-temperature adhesive is applied at overlap ends. In this case, the E_{ratio} changes in service with the ambient temperature (E_{ratio} increases by decreasing the temperature). At low temperatures, the adhesive at the overlap ends controls the strength of the joints while the joint stiffness is mainly supported by the high-temperature adhesive when it is subjected to high-temperature conditions [43]. However, it has been shown that the performance of bi-adhesive joints is more pronounced at room temperature as the high and low temperatures can decrease the strength of the bi-adhesive joints [34] but still, bi-adhesive joints are recommended as they can be employed for a higher temperature ranges comparing with the joints with a single ductile or brittle adhesive.

11.4.3 Other types of joints

As mentioned before, SLJ is the most common joint geometry considered by researchers on bi-adhesive joints. However, other types of joints such as double-lap [35], T [48], plate [49] and double-strap [50] bi-adhesive joints have been also analysed.

11.4.3.1 Double-lap joints

Figure 11.11 shows a typical double-lap bi-adhesive joint.

Similar to the SLJ bi-adhesive joints, LR plays a key role in the mechanical performance of bi-adhesive double-lap joints (DLJs). Figure 11.12 shows the shear stress distribution along the bondline in a bi-adhesive DLJ. As shown in Figure 11.12, using two adhesives along the overlap can significantly decrease the shear stress at

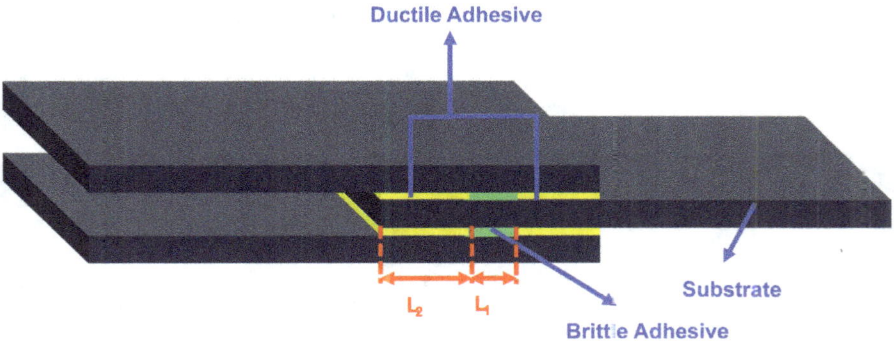

Figure 11.11 Schematic configuration of a bi-adhesive DLJ

overlap ends while the shear stress in the middle of the joint increases. Similar to the shear stress, peel stress in bi-adhesive DLJs decreases at the ends of the overlap, compared with the joints bonded with a single brittle adhesive. In contrast with the SLJs where the stress distribution is symmetric along the overlap, the stress along the bondline in bi-adhesive DLJs is not symmetric due to the asymmetric geometry. In another study, da Silva and Adams [51] predicted the failure load of bi-adhesive DLJs where titanium was used as the adherend, and the brittle and ductile adhesives were Redux 326 and Supreme 10HT, respectively. They studied two different LRs as 0.16 and 0.5. They considered either the shear yielding of the adhesive or tensile

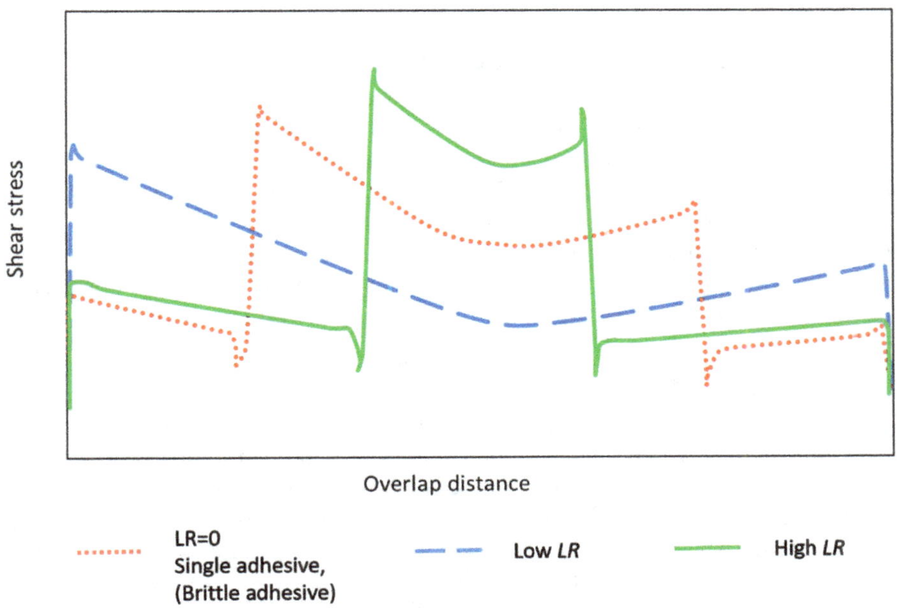

Figure 11.12 Effect of the adhesives LR on the shear and peel stress distributions

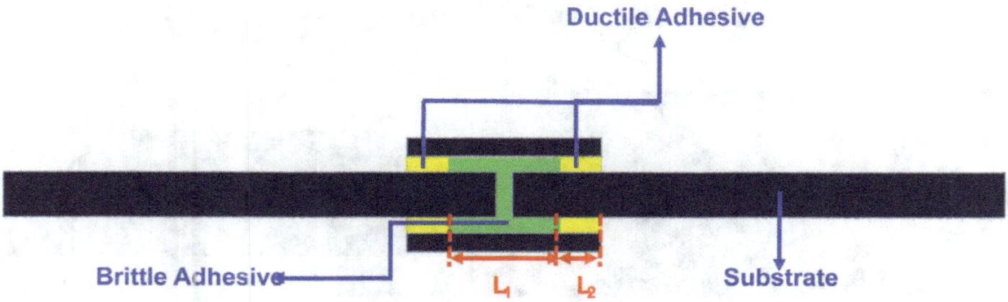

Figure 11.13 Schematic configuration of the double-strap adhesive joint

yielding of the adherends as failure criterion. Results showed that at high temperatures, failure load is higher for joints with the LR of 0.16 where the joint experiences a lower shear stress level in the brittle adhesive.

Effects of E_{ratio} on double-lap bi-adhesive joints have been also analysed by authors [43, 51]. Results show that increasing the E_{ratio} raises the shear stress level in the brittle adhesive and declines the stress level in the ductile adhesive placed at overlap ends. Several authors [51, 52] have investigated the effects of the temperature on bi-adhesive DLJs. Effects of temperature on DLJs are similar to that of bi-adhesive SLJs. At high temperatures, the load is mainly supported by the high-temperature (brittle) adhesive located in the middle of the joint while at low temperatures the low-temperature (ductile) adhesive used at the ends of the joints bears the applied loads.

11.4.3.2 Bi-adhesive double-strap joint

Double-strap bi-adhesive joints have been considered in some studies [29, 50]. A bi-adhesive double-strap joint (DSJ) is schematically shown in Figure 11.13. Temiz [50] showed that peel stress increases by LR. Three-point bending tests were conducted by some researchers [53] on bi-adhesive DSJs where FM73 and SBT9244 adhesives were used as the brittle and ductile adhesives, respectively. They found that for the tested bi-adhesive DSJs, increasing the substrate thickness improves the failure load at all the tested LRs [50, 53]. In terms of the overlap length, it was also found that the optimum LR decreases by increasing the overlap length.

11.4.3.3 Plate joints

Plate joints are used in many real applications such as in aeronautical structures where two parts of the fuselage are bonded together through a long overlap [54]. Using bi-adhesive technique can improve the performance of such joints as these structures are usually exposed to a big range of temperatures. Marques *et al.* [55] experimentally analysed bi-adhesive plate joints. They found an increase in failure load of bi-adhesive plate joints at high temperatures compared with the joints with a single brittle adhesive. But based on the room temperature results [56], the strength of the joint with a single adhesive can be higher than bi-adhesive plate

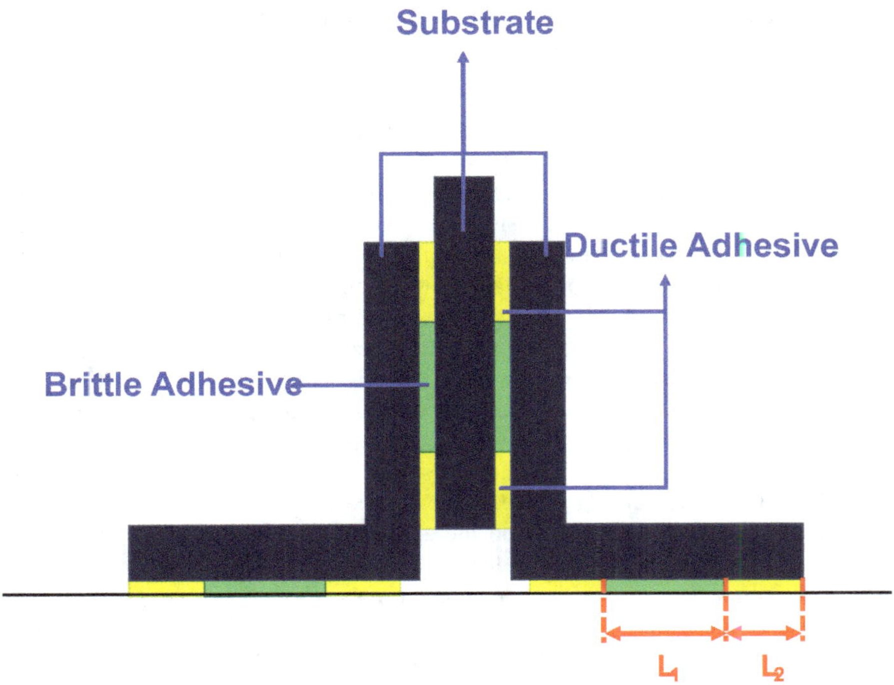

Figure 11.14 Schematic configuration of the bi-adhesive T-joint

joints. However, it should be noted that changing the temperature changes the E_{ratio}, which can significantly affect the joint strength in bi-adhesive joints.

11.4.3.4 T-joints

Despite the extensive application of T-joints in industrial structures, a few studies have been conducted on the use of the bi-adhesive techniques in T-joints. A typical bi-adhesive T-joint is shown in Figure 11.14. According to the results published by Akpinar *et al.* [48], using bi-adhesive method the stress reduced at overlap ends and while the stress level in the middle of the joint where the brittle adhesive is applied increases, which create a more uniform stress distribution along the overlap. Akpinar *et al.* [48] could increase the failure load of T-joints by 20% by using bi-adhesive technique.

To summarize, it can be concluded that using bi-adhesive joints as a simple form of graded adhesive joints can significantly improve the mechanical response of the joints. An increase in failure load due to the use of bi-adhesive technique will lead to a weight reduction and will also decrease maintenance costs. The application of two adhesives along a single overlap has some challenges such as the difference in pot life and the issues related to the curing procedures of the two ductile and brittle adhesives at the same time. However, using the bi-adhesive technique is advantageous, due to the significant improvement that is obtained in the mechanical performance of bi-adhesive joints.

11.5 Composite adhesives with natural fibres

Along with all the advantages of the adhesive joints, structural adhesives usually suffer from the lack of sufficient resistance against crack growth. In these types of adhesives, usually a crack initiation can easily lead to a joint failure since the crack propagates suddenly. On the other hand, a main disadvantage of the more flexible adhesives is their low strength. To overcome these drawbacks, several methods have been proposed by researchers to improve the mechanical response of the adhesive materials. One of the most common techniques is using additives. Additives can significantly improve the fracture energy as well as the strength of the bonded joint. Although synthetic fibres/particles have been widely considered by authors as a second phase in adhesives to improve the joints' performance in terms of toughness or strength natural fibres/particles are less investigated.

However, recently the studies on the application of natural fibres as an eco-friendly and biodegradable material have increased in response to global environmental concerns. Except for these inherent properties, natural fibres have more advantages such as low cost, good/high mechanical performance and availability and abundant sources. Several types of natural fibres have been considered by researchers such as banana, jute, sisal, hemp, flax and coir [57–59]. Natural fibres are used not only for research purposes at laboratory levels but also for real applications for the strengthening of components [60, 61].

Besides the advantages, natural fibres suffer from some drawbacks such as the moisture absorption, dependency of the mechanical, physical and even the chemical composition of the natural fibres to the area and climate at which the fibres have grown. One of the main concerns about natural fibres is their hydrophilic properties, which lead to a weak bond between the adhesive and the fibre. However, as discussed in section 11.5.1 there are some techniques to improve the adhesion properties of the natural fibres.

11.5.1 Surface preparation

Fibre treatment is an essential part of manufacturing composite adhesives with natural fibres. Using an adequate surface treatment, the lignin and natural oils covering the fibres will be removed. A surface preparation commonly used by researchers is alkaline treatment, which is mainly used for natural fibres. Some authors have analysed different concentrations of NaOH for preparing the surfaces of the fibres. Adel Salih *et al.* [62] and Boumaaza *et al.* [63] analysed the effects of NaOH on the properties of bamboo, sisal, flax and jute fibres. Khakpour *et al.* [64] conducted an extensive analysis on the effects of NaOH treatment on the strength of the date palm tree fibres and the tensile strength enhanced with these fibres. They analysed samples with fibres without surface treatment and with treated conditions where two NaOH solutions with 1% and 6% were used. According to their results, they found that alkaline treatment can significantly increase the strength of the fibres, and that increasing the concentration of the alkaline solution from 1% to 6% will improve the strength of the date palm tree fibres. Delzendehrooy *et al.* [65] used date palm tree fibres to enhance the strength of SLJs. They manufactured joints with different treatment methods by considering 1% and 6% NaOH solution. Their results showed that increasing the concentration of the alkaline solution from 1% to 6% improves the mechanical response of the bonded SLJs. According to these results, it is concluded that surface treatment not only improves the adhesion between the adhesive and the fibres but also can improve the tensile strength of the fibre itself.

11.5.2 Cork particles

Cork particles extracted from oak trees have been used for 5,000 years for different applications, from bottle stoppers to thermal insulators in aeronautics. Cork particles are mainly known for their properties in acoustic [66] and thermal insulation [67]. However, the effect of particles on the strength and fracture toughness of bonded joints is less studied. Effects of size and the density of cork particles on the toughness of an epoxy adhesive were studied by Barbosa *et al.* [68]. Their results showed that the impact energy absorption of the joint can be improved by adding 1 wt% of cork particles. As mentioned before, one of the main drawbacks of the natural particles is the water absorption. However, results experimentally obtained by authors [69] showed that the water uptake and drying process in adhesives are not significantly influenced by the presence of this specific natural particle. Some authors [70] investigated the role of cork particles in the glass transition temperature (T_g) of an adhesive. It was found that cork can decrease the T_g of the adhesive, which should be considered in joint design. In terms of stress–strain behaviour, Barbosa *et al.* [71] showed that the presence of cork particles can improve the ductility of the adhesive.

Although cork particles have been used widely in industrial contexts, the high-temperature tests conducted on SLJs showed that the strength of the joints decreases by the amount of cork particles [72]. The role of cork particles is to increase the mechanical performance of the joints by increasing the ductility of the adhesive. However, at high temperatures, the adhesive itself is ductile and cork particles act as a defect and cause a significant weakening in the bondline. However, the results of the joints enhanced with microparticles of corks tested at –20°C showed that the particles do not significantly change the failure load.

The effects of cork particles on the fatigue life of SLJs have been experimentally analysed by Akhavan-Safar *et al.* [72]. A significant increase in the total fatigue life of the joints enhanced with microcork particles was obtained. They could improve the fatigue life by 70% by adding cork particles to an epoxy-based adhesive. Cork particles can improve the mechanical performance of the bonded joints by increasing the ductility of the adhesives, stabilizing the crack propagation and preventing unstable crack growth specially in fatigue loading conditions. They also act as a crack stopper. However, the higher the amount of the microcork particles, decrease the joints strength by increasing the particle aggregation which can lead to crack bridging between two adjacent particles and between particles and voids. On the

Figure 11.15 *Different parts of a date palm tree: (a) mesh, (b) bunch, (c) petiole and (d) rachis*

other hand, the possibility of cork-matrix deboning increases with higher amount of particles. Although these phenomena, which cause weakening of the joint, are not desired, however, some researchers [73] showed that weakening the adhesive layer can change the failure mechanism from the interfacial failure to the cohesive failure, which can be a desired mechanism in some applications. They experimentally and numerically showed that by increasing the size and amount of the particles in the adhesive, the crack tends to propagate within the adhesive layer, not the interface. Based on their results, a small reduction in failure load was observed.

11.5.3 Date palm tree fibres

Date palm trees are a family of palm trees, which can be found over the world. Low cost and availability are two main advantages of date palm tree fibres compared with the synthetic fibres. Palm trees usually have a long life and can sustain a wide climate range. Figure 11.15 shows the different parts of a date palm tree used for extracting the fibres. Several authors have worked on composite materials enhanced with date palm fibres [74, 75] but a few works have been conducted on the application of date palm tree fibres on the mechanical response of adhesives and adhesive joints. By extracting fibres from different parts of the date palm tree, Khakpour *et al.* [64] analysed the effects of fibre type, fibre size and fibre density on the strength of an epoxy-based adhesive. They found that date palm fibres can increase the tensile strength of the adhesives by 60%. They also showed that there is an optimum fibre density in which the adhesive reaches its maximum strength. In terms of the fibre length, it was found by Khakpour *et al.* [64] that higher improvement is achieved by fibres with longer lengths. Delzendehrooy *et al.* [65] analysed the static strength of single-lap adhesive joints bonded with an epoxy adhesive enhanced with date palm tree fibres.

Different fibre types including bunch, mesh, rachis and petiole extracted from different parts of a date palm tree with two different sizes (short and long) were considered by Delzendehrooy *et al.* [65]. They also investigated the effects of fibre density on the strength of SLJs. Results showed that fibres can significantly improve the joints' strength. According to their results, rachis fibre with 10% density in short form can improve the strength of the joint by 118% when the fibres are treated with 6% alkaline solution. They also found that adding bunch and rachis fibres the area under the average shear stress–strain curves was surprisingly increased by 12.63 and 8.55 times, respectively. Higher fibre densities were considered by Kaddami *et al.* [76] where the tested adhesive was enhanced with 17 wt% of date palm fibres. Mahdavi *et al.* [77] improved the properties of the adhesives by considering 20 wt% and 40 wt% of leaflet, trunk and brunch. Tripathy *et al.* [78] reported that 10 wt% is the optimum amount of the considered fibre for strengthening the studied adhesive. Accordingly, the optimum density of the date palm fibres is highly influenced by the adhesive properties but based on the reports the optimum weight ratio for the tested adhesives is above 10 wt%. Studies on mode I fracture energy of the adhesives enhanced with date palm fibres also showed that palm tree fibres not only can increase the strength of the joints but also can significantly improve the fracture energy of the adhesives. Considering 10 wt% of rachis and 5% of bunch fibres improved the fracture energy of an epoxy adhesive by factors of 7.6 and 6.9. Crack arresting, fibre pullout, fibre bridging and fibre delamination are different failure mechanisms that improve the toughness of the adhesives.

References

[1] Noury P., Hayman B., Mcgeorge D., Weitzenböck J. 'Lightweight construction for advanced shipbuilding – recent development'. Presented at Proceedings of the 37th WEGEMT Summer School; Madrid, Spain, 2002.

[2] Jeevi G., Nayak S.K., Abdul Kader M. 'Review on adhesive joints and their application in hybrid composite structures'. *Journal of Adhesion Science and Technology*. 2019;33(14):1497–520.

[3] Davies P., Cantwell W.J., Jar P.-Y., Bourban P.-E., Zysman V., Kausch H.H. 'Joining and repair of a carbon fibre-reinforced thermoplastic'. *Composites*. 1991;22(6):425–31.

[4] Möller F., Thomy C., Vollertsen F., Schiebel P., Hoffmeister C., Herrmann A.S. 'Novel method for joining CFRP to aluminium'. *Physics Procedia*. 2010;5(1):37–45.

[5] Hart-Smith L. *Delamination and Debonding of Materials*. ASTM International; 1985.

[6] Hsiao K.-T., Alms J., Advani S.G. 'Use of epoxy/multiwalled carbon nanotubes as adhesives to join graphite fibre reinforced polymer composites'. *Nanotechnology*. 2003;14(7):791–.

[7] Shang X., Marques E.A.S., Machado J.J.M., Carbas R.J.C., Jiang D., da Silva L.F.M. 'Review on techniques to improve the strength of adhesive joints with composite adherends'. *Composites Part B: Engineering*. 2019;177(4):107363.

[8] Chamis C.C., Murthy P.L.N. 'Simplified procedures for designing adhesively bonded composite joints'. *Journal of Reinforced Plastics and Composites*. 1991;**10**(1):29–41.

[9] da Silva L.F.M, Adams R.D. 'Techniques to reduce the peel stresses in adhesive joints with composites'. *International Journal of Adhesion and Adhesives*. 2007;**27**(3):227–35.

[10] Kaye R.H., Heller M. 'Through-thickness shape optimisation of bonded repairs and lap-joints'. *International Journal of Adhesion and Adhesives*. 2002;22(1):7–21.

[11] Lang T.P., Mallick P.K. 'Effect of spew geometry on stresses in single LAP adhesive joints'. *International Journal of Adhesion and Adhesives*. 1998;18(3):167–77.

[12] Rispler A.R., Tong L., P. Steven G., Wisnom M.R. 'Shape optimisation of adhesive fillets'. *International Journal of Adhesion and Adhesives*. 2000;20(3):221–31.

[13] da Silva L.F.M., Adams R.D. 'Adhesive joints at high and low temperatures using similar and dissimilar adherends and dual adhesives'. *International Journal of Adhesion and Adhesives*. 2007;27(3):216–26.

[14] Ramezani F., Ayatollahi M.R., Akhavan-Safar A., da Silva L.F.M. 'A comprehensive experimental study on bi-adhesive single LAP joints using DIC technique'. *International Journal of Adhesion and Adhesives*. 2020;102(24):102674.

[15] Vogelesang L.B., Vlot A. 'Development of fibre metal laminates for advanced aerospace structures'. *Journal of Materials Processing Technology*. 2000;103(1):1–5.

[16] Morgado M.A., Carbas R.J.C., Marques E.A.S., da Silva L.F.M. 'Reinforcement of CFRP single LAP joints using metal laminates'. *Composite Structures*. 2019;230(12):111492.

[17] Dos santos D.G., Carbas R.J.C., Marques E.A.S., da Silva L.F.M. 'Reinforcement of CFRP joints with fibre metal laminates and additional adhesive layers'. *Composites Part B: Engineering*. 2019;**165**:386–96.

[18] Schijve J V.L., Marissen R. 'Laminate of aluminum sheet material and aramidfibers'. 1983:United States Patent 4500589.

[19] Schollerer M.J., Kosmann J., Völkerink O., Holzhüter D., Hühne C. 'Surface toughening – a concept to decrease stress peaks in bonded joints'. *The Journal of Adhesion*. 2019;95(5-7):495–514.

[20] Shang X., Marques E.A.S., Machado J.J.M., Carbas R.J.C., Jiang D., da Silva L.F.M. 'A strategy to reduce delamination of adhesive joints with composite substrates'. *Proceedings of the Institution of Mechanical Engineers, Part L: Journal of Materials: Design and Applications*. 2018;37(3):146442071880571.

[21] Liu H., Wang Y., Morgado M.A., Carbas R.J.C., Santos D.G.D., da Silva L.F.M. ' tertiary amine and tooth mineral hydroxyapatite facilely trigger self-cure of 10-MDP based adhesives '. *International Journal of Adhesion & Adhesives*. 2019;**92**:73–79.

[22] Ganesh V.K., Choo T.S. 'Modulus graded composite adherends for single-lap bonded joints'. *Journal of Composite Materials*. 2002;**36**(14):1757–67.

[23] Melograna J.D., Grenestedt J.L. 'Improving joints between composites and steel using perforations'. *Composites Part A: Applied Science and Manufacturing*. 2002;33(9):1253–61.

[24] Kumar S., de Tejada Alvarez A. 'Modeling and experimental evaluation of geometrically graded multi-material single-lap joints'. *Presented at the 56th AIAA/ASCE/AHS/ASC Structures, Structural Dynamics, and Materials Conference*; Kissimmee, Florida, 2015.

[25] Kumar S., Wardle B.L., Arif M.F. 'Strength and performance enhancement of bonded joints by spatial tailoring of adhesive compliance via 3D printing'. *ACS Applied Materials & Interfaces*. 2017;9(1):884–91.

[26] Khan M.A., Kumar S., Cantwell W.J. 'Additively manufactured cylindrical systems with stiffness-tailored interface: modeling and experiments'. *International Journal of Solids and Structures*. 2018;152-153(1):71–84.

[27] Hildebrand M. 'Non-linear analysis and optimization of adhesively bonded single lap joints between fibre-reinforced plastics and metals'. *International Journal of Adhesion and Adhesives*. 1994;14(4):261–7.

[28] Belingardi G., Goglio L., Tarditi A. 'Investigating the effect of spew and chamfer size on the stresses in metal/plastics adhesive joints'. *International Journal of Adhesion and Adhesives*. 2002;22(4):273–82.

[29] Marques E.A.S., da Silva L.F.M. 'Joint strength optimization of adhesively bonded patches'. *The Journal of Adhesion*. 2008;84(11):915–34.

[30] Galehdari S.A., Khodarahmi H. 'Design and analysis of a graded honeycomb shock absorber for a helicopter seat during a crash condition'. *International Journal of Crashworthiness*. 2016;21(3):231–41.

[31] Silva M.R.G., Marques E.A.S., da Silva L.F.M. 'Behaviour under impact of mixed adhesive joints for the automotive industry'. *Latin American Journal of Solids and Structures*. 2016;**13**(5):835–53.

[32] Fitton M. *Multi-modulus adhesive bonding of advanced composite materials*. Ph.D. Thesis. Oxford: Brookes University; 2004.

[33] Broughton J.G., Fitton M.D. 'Science of mixed-adhesive joints' in *Hybrid adhesive joints*. Springer. P; 2011. pp. 257–81.
[34] Machado J.J.M., Marques E.A.S., da Silva L.F.M. 'Influence of low and high temperature on mixed adhesive joints under quasi-static and impact conditions'. *Compos Struct*. 2018;**194**(2):68–79.
[35] da Silva L.F.M. 'Analytical models of adhesively bonded joints-Part I: literature survey'. *International Journal of Adhesion and Adhesives*. 2009;29(3):319–30.
[36] Pires I., Quintino L., Durodola J.F., Beevers A. 'Performance of bi-adhesive bonded aluminium LAP joints'. *International Journal of Adhesion and Adhesives*. 2003;23(3):215–23.
[37] Carbas R.J.C., Silva L. da, Critchlow G.W. 'Effect of post-cure on adhesively bonded functionally graded joints by induction heating'. *Proceedings of the Institution of Mechanical Engineers, Part L*. 2015;**229**(5):419–30.
[38] Carbas R.J.C., da Silva L.F.M., Critchlow G.W. 'Adhesively bonded functionally graded joints by induction heating'. *International Journal of Adhesion and Adhesives*. 2014;48(2009):110–8.
[39] Carbas R.J.C., Marques E.A.S., da Silva L.F.M., Lopes A.M. 'Effect of cure temperature on the glass transition temperature and mechanical properties of epoxy adhesives'. *The Journal of Adhesion*. 2014;90(1):104–19.
[40] Özer H., Öz Ö. 'A comparative evaluation of numerical and analytical solutions to the bi-adhesive single-lap joint'. *Mathematical Problems in Engineering.*. 2014;**2014**(2):1–16.
[41] Kumar S., Pandey P.C. 'Behaviour of bi-adhesive joints'. *Journal of Adhesion Science and Technology*. 2010;**24**(7):1251–81.
[42] Ramezani F., Ayatollahi M.R., Akhavan-Safar A., da Silva L.F.M. 'A comprehensive experimental study on bi-adhesive single LAP joints using DIC technique'. *International Journal of Adhesion and Adhesives*. 2020;102(24):102674.
[43] das Neves P.J.C., da Silva L.F.M., Adams R.D. 'Analysis of mixed adhesive bonded joints Part II: parametric study'. *Journal of Adhesion Science and Technology*. 2009;23(1):35–61.
[44] Akhavan-Safar A., Ayatollahi M.R., da Silva L.F.M. 'Strength prediction of adhesively bonded single LAP joints with different bondline thicknesses: a critical longitudinal strain approach'. *International Journal of Solids and Structures*. 2017;109(3):189–98.
[45] Machado J.J.M., Marques E.A.S., Silva M.R.G., da Silva L.F.M. 'Numerical study of impact behaviour of mixed adhesive single LAP joints for the automotive industry'. *International Journal of Adhesion and Adhesives*. 2018;84(3):92–100.
[46] Fitton M.D., Broughton J.G. 'Variable modulus adhesives: an approach to optimised joint performance'. *International Journal of Adhesion and Adhesives*. 2005;**25**(4):329–36.
[47] Hart-Smith L.J. *Adhesive bonded single lap joints*. NASA Contractual Report, NASA CR-112235; 1973.

[48] Akpinar S., Aydin M.D., Özel A. 'A study on 3-D stress distributions in the bi-adhesively bonded T-joints'. *Applied Mathematical Modelling*. 2013;**37**(24):10220–30.

[49] Marques E.A.S., da Silva L.F.M., Flaviani M. 'Testing and simulation of mixed adhesive joints for aerospace applications'. *Composites Part B: Engineering*. 2015;74(5):123–30.

[50] Temiz Ş. 'Application of bi-adhesive in double-strap joints subjected to bending moment'. *Journal of Adhesion Science and Technology*. 2006;**20**.

[51] da Silva L.F.M., Adams R.D. 'Joint strength predictions for adhesive joints to be used over a wide temperature range'. *International Journal of Adhesion and Adhesives*. 2007;27(5):362–79.

[52] da Silva L.F.M., Adams R.D. 'Adhesive joints at high and low temperatures using similar and dissimilar adherends and dual adhesives'. *International Journal of Adhesion and Adhesives*. 2007;**27**(3).

[53] Temiz Ş., Adin H., Sülü I.Y. 'Behaviour of bi-adhesive in double-strap joint with embedded patch subjected to bending'. *Journal of Theoretical and Applied Mechanics*. 2015;45(3):83–96.

[54] da Silva L.F.M., Adams R.D. 'Stress-free temperature in a mixed-adhesive joint'. *Journal of Adhesion Science and Technology*. 2006;20(15):1705–26.

[55] Marques E.A.S., Magalhães D.N.M., da Silva L.F.M. 'Experimental study of silicone-epoxy dual adhesive joints for high temperature aerospace applications'. *Materialwissenschaft und Werkstofftechnik*. 2011;42(5):471–7.

[56] Marques E.A.S., Campilho R.D.S.G., da Silva L.F.M. 'Geometrical study of mixed adhesive joints for high-temperature applications'. *Journal of Adhesion Science and Technology*. 2016;30(7):691–707.

[57] Devireddy S.B.R., Biswas S. 'Thermo-physical properties of short banana–jute fiber-reinforced epoxy-based hybrid composites'. *Proceedings of the Institution of Mechanical Engineers, Part L: Journal of Materials: Design and Applications*. 2018;232(11):939–51.

[58] Gupta M.K. 'Thermal and dynamic mechanical analysis of hybrid jute/sisal fibre reinforced epoxy composite'. *Proceedings of the Institution of Mechanical Engineers, Part L: Journal of Materials: Design and Applications*. 2018;232(9):743–8.

[59] Oliveira L. Á., Santos J.C., Panzera T.H., Freire R.T., Vieira L.M., Scarpa F. 'Evaluation of hybrid-short-coir-fibre-reinforced composites via full factorial design'. *Composite structures*. 2018;202:313–23.

[60] Park G., Park H. 'Structural design and test of automobile bonnet with natural flax composite through impact damage analysis'. *Composite Structures*. 2018;184:800–6.

[61] Beber V., de Barros S., Banea M., *et al.* 'Effect of Babassu natural filler on PBAT/PHB biodegradable blends: an investigation of thermal, mechanical, and morphological behavior'. *Materials*. 2018;11(5):820.

[62] Adel Salih A., Zulkifli R., Azhari C.H. 'Tensile properties and microstructure of single-cellulosic bamboo fiber strips after alkali treatment'. *Fibers*. 2020;8(5):26.

[63] Boumaaza M., Belaadi A., Bourchak M. 'The effect of alkaline treatment on mechanical performance of natural fibers-reinforced plaster: optimization using RSM, Journal of natural fibers'. *Journal of Natural Fibers*. 2020:1–21.

[64] Khakpour H., Ayatollahi M.R., Akhavan-Safar A., da Silva L.F.M. 'Mechanical properties of structural adhesives enhanced with natural date palm tree fibers: effects of length, density and fiber type'. *Composite Structures*. 2020;237(4):111950.

[65] Delzendehrooy F., Ayatollahi M.R., Akhavan-Safar A., da Silva L.F.M. 'Strength improvement of adhesively bonded single LAP joints with date palm fibers: effect of type, size, treatment method and density of fibers'. *Composites Part B: Engineering*. 2020;188(4):107874.

[66] Hernández-Olivares F., Bollati M.R., del Rio M., Parga-Landa B. 'Development of cork–gypsum composites for building applications'. *Construction and Building Materials*. 1999;13(4):179–86.

[67] Sierra-Pérez J., Boschmonart-Rives J., Dias A.C., Gabarrell X. 'Environmental implications of the use of agglomerated cork as thermal insulation in buildings'. *Journal of Cleaner Production*. 2016;126(Part 3):97–107.

[68] Barbosa A.Q., da Silva L.F.M., Öchsner A., Abenojar J., del Real J.C. 'Influence of the size and amount of cork particles on the impact Toughness of a structural adhesive'. *The Journal of Adhesion*. 2012;88(4-6):452–70.

[69] Barbosa A.Q., da Silva L.F.M., Öchsner A. 'Hygrothermal aging of an adhesive reinforced with microparticles of cork'. *Journal of Adhesion Science and Technology*. 2015;29(16):1714–32.

[70] Barbosa A.Q., da Silva L.F.M., Abenojar J., del Real J.C., Paiva R.M.M., Öchsner A. 'Kinetic analysis and characterization of an epoxy/cork adhesive'. *Thermochimica Acta*. 2015;604:52–60.

[71] Barbosa A.Q., da Silva L.F.M., Öchsner A. 'Effect of the amount of cork particles on the strength and glass transition temperature of a structural adhesive'. *Proceedings of the Institution of Mechanical Engineers, Part L: Journal of Materials: Design and Applications*. 2014;228(4):323–33.

[72] Akhavan-Safar A., Barbosa A.Q., da Silva L.F.M., Ayatollahi M.R. 'Micro failure analysis of adhesively bonded joints enhanced with natural cork particles: impact of overlap length and particles volume fraction'. *Frattura ed Integrità Strutturale*. 2018;12(46):266–74.

[73] da Silva C.I., Barbosa A.Q., Carbas R.J.C., Marques E.A.S., Akhavan-Safar A., da Silva L.F.M. 'Influence of cork microparticles on the fracture type in single LAP joints'. *Proceedings of the Institution of Mechanical Engineers, Part C: Journal of Mechanical Engineering Science*. 2021;235(3):497–507.

[74] Shalwan A., Yousif B.F. 'Investigation on interfacial adhesion of date palm/epoxy using fragmentation technique'. *Materials & Design*. 2014;53(4):928–37.

[75] Alsaeed T., Yousif B.F., Ku H. 'The potential of using date palm fibres as reinforcement for polymeric composites'. *Materials & Design*. 2013;43(6):177–84.

[76] Kaddami H., Dufresne A., Khelifi B., *et al.* 'Short palm tree fibers – Thermoset matrices composites'. *Composites Part A: Applied Science and Manufacturing*. 2006;37(9):1413–22.
[77] Mahdavi S., Kermanian H., Varshoei A. 'Comparison of mechanical properties of date palm fiber-polyethylene composite'. *Bioresources*. 2010;5:2391–403.
[78] Tripathy S., Dehury J., Mishra D. 'A study on the effect of surface treatment on the physical and mechanical properties of date-palm stem fiber embedded epoxy composites'. Presented at the 5th National Conference on Processing and Characterization of Materials; Rourkela, India, 2016.

Chapter 12

Design principles and recent developments in adhesively bonded joints of fibre-reinforced plastic composite structures

Roya Akrami[1], Sakineh Fotouhi[2], Ali Tabatabaeian[2], Saeid Lotfian[1], Mahdi Bodaghi[3], and Mohammad Fotouhi[4]

Fibre-reinforced plastic (FRP) composite components must be joined such that the overall structure maintains its structural integrity while it is under mechanical and environmental loads. This chapter introduces adhesively bonded joints (ABJs) as a main joining technique of FRP composite structures, their design principles, characterisation methods and recent developments. Advantages and disadvantages of ABJs are discussed and their performance is compared with mechanical joining techniques. Design considerations and effective parameters such as adhesive properties, surface preparation, joint configuration and environmental factors on the joint behaviour are described. Different techniques for characterising ABJs' static and in-service performances such as fatigue and ageing are briefly discussed. A summary of the analytical and numerical methods developed for stress analysis of ABJ is also provided. Novel trends on ABJs, including bio-inspired designs to reduce stress concentration, self-healing and mechanical interlocking, are briefed, given new directions for research and design of ABJ of FRP composite structures.

12.1 Introduction

Adhesive bonding and mechanical fastening are two commonly used joining techniques for FRPs. Figure 12.1 shows a schematic of an ABJ, which is a material joining process that results from physical and chemical interactions between the

[1]Faculty of Engineering, University of Strathclyde, Glasgow, UK
[2]School of Engineering, University of Glasgow, Glasgow, UK
[3]Department of Engineering, School of Science and Technology, Nottingham Trent University, Nottingham, UK
[4]Microlab, Faculty of Civil Engineering and Geosciences, Delft University of Technology, Delft, the Netherlands

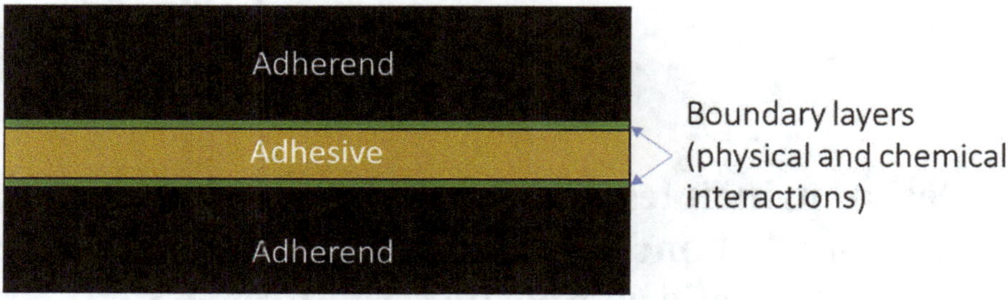

Figure 12.1 Schematic of an ABJ

adhesive and the adherend surfaces in the boundary layers. ABJs are increasingly replacing mechanical joints in engineering applications in different industrial sectors such as aerospace, automobile, oil and gas, sports, marine and civil.

Nowadays, there are increased demands on the use of adhesive bonding for the repair and assembly of composite structures. The application of ABJs in structural components made of fibre-reinforced composites has increased significantly in recent years. The traditional fasteners usually result in the cutting of fibres, and hence the introduction of stress concentrations, both of which reduce structural integrity. Table 12.1 summarises advantages and disadvantages of ABJs [1, 2]. Among these

Table 12.1 Advantage and disadvantages of ABJs in FRP composites

Advantages	• Continuous and uniform stress distributions • Good strength to weight ratio • Reduced weight • Design flexibility • Ease of manufacture • Improved production speed • Low capital and production costs • Dissimilar materials joining • Better production sequences • Good fatigue performance • Good shock absorption • Reduced galvanic corrosion • One-side access to the adherend is enough to make a joint • Improved sealing • Ability to join heat sensitive materials
Disadvantages	• Careful surface preparation • Considerable cure times • High temperature and pressure to cure some adhesives • Specific jigs and fixtures to locate components during curing • Rigid process control to get consistent results • Difficult to detect defects • Special storage condition for some adhesives • Poor resistance of adhesives to tensile stresses • Degradation of adhesives above an upper temperature

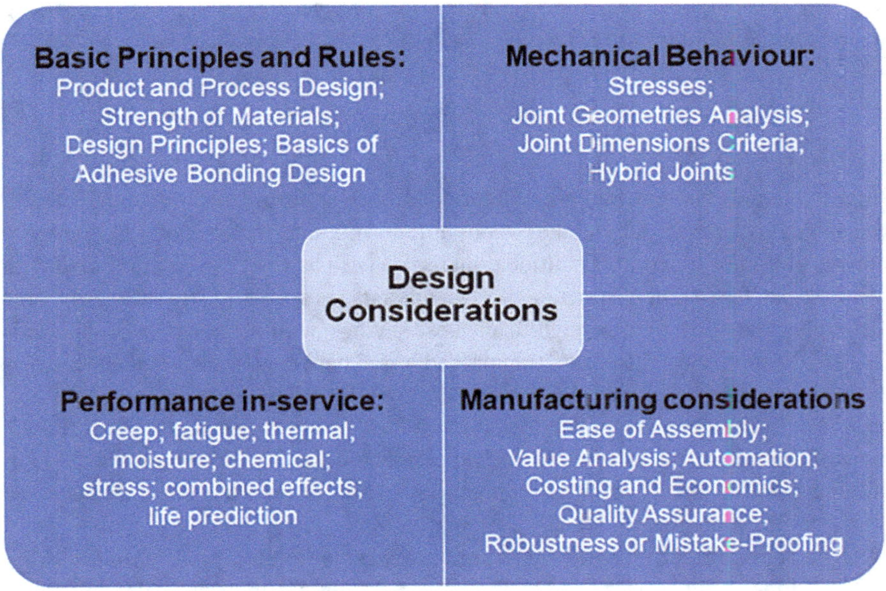

Figure 12.2 *Summarises the most important parameters that are necessary to consider during the design of ABJs*

advantages is the continuous connection and avoiding stress concentrations at fastener hole, and ABJs better performance such as high strength to weight ratio, design flexibility, better damage tolerance, lower fabrication cost and higher fatigue resistance compared with conventional joining techniques such as mechanical joints.

A substantial amount of research and development has been done on the ABJs in FRP composites. Some review papers and several handbooks were published [3–17] on different aspects such as adhesive materials, fatigue strength, analytical and finite element methods and environmental durability. This chapter presents an overall overview of different studies on ABJs for FRP composite structures. The topics covered in the following sections are on: (i) design considerations and effective parameters, (ii) characterisation methods, (iii) analytical and finite element models and (iv) recent developments such as bio-inspired designs.

12.2 Design considerations and effective parameters in ABJs in FRP composites

Figure 12.2 summarises the most important parameters that are necessary to consider during the design of ABJs. As mentioned before, in comparison with other manufacturing methods, the adhesively bonded structure's fabrication is essentially fast during the assembly process. Even if during the assembly process, curing is required, this could be accommodated 'off-line' or combined with other processing stages such as paint curing [1].

12.2.1 Basics of adhesive bonding design and manufacturing considerations

Adhesive bonds demonstrate broad freedom of choice during material specification. Using this joining strategy makes it possible to mix and match material combinations to suit joint function. Traditional materials such as steel can be combined with new composites to give distinctive product advantages as hybrid structure. Adhesive assembly strategies provide significant cost savings with a reduction in materials costs and simplifying the production procedures. There is abundant evidence showing cost-effectiveness when the design and manufacturing functions co-operate to design or redesign the product. Applying adhesive bonding usually involves the manual procedure, but even when the process is mechanised or automated for high volume manufacturing, the final component is lighter than the one fabricated by other joining strategies [6, 18, 19].

The series of activities to fabricate an adhesive-bonded joint include design and manufacturing processes. Design activities contain material selection, joint designing and adhesive selection steps, while the manufacturing procedures comprise surface pre-treatment, assembly and curing processes and final inspection [20, 21].

It is necessary to consider the specific topics related to the design and assembly of adhesive joints design in addition to basic principles and rules. It is essential to evaluate the mechanical behaviour of materials and joints structure, stresses on the joint, geometries and dimensions criteria. Regarding the adhesive joint's performance during the in-service, it is vital to investigate the joints' structural integrity by analysing the failure modes such as creep, fatigue and moisture effect. Other important parameters related to adhesive bonded are the adhesive's curing mechanism and the assembly process's time factor. The automated robotic system, bulk dispensing system, portable hand-held dispensing cartridge, using a spatula, are the usual methods to apply the adhesive. The procedure is chosen based on the volume of adhesive being used. It is worth mentioning that in heat-curing of adhesives, plates, autoclaves and vacuum presses may be required [22].

The series of activities to fabricate an adhesive-bonded joint include design and manufacturing processes. Design activities contain material selection, joint designing and adhesive selection steps, while the manufacturing procedures comprise surface pre-treatment, assembly and curing processes and final inspection.

Some necessary activities related to the acceptable bonding process are included as follows: (a) sourcing and storing adhesive system; (b) adhesive application (metering, mixing, dispensing and control); (c) adherend preparation; (d) assembly (order of assembly, environment and tooling); (e) cure (equipment and control); (f) inspection and (g) remanufacturing.

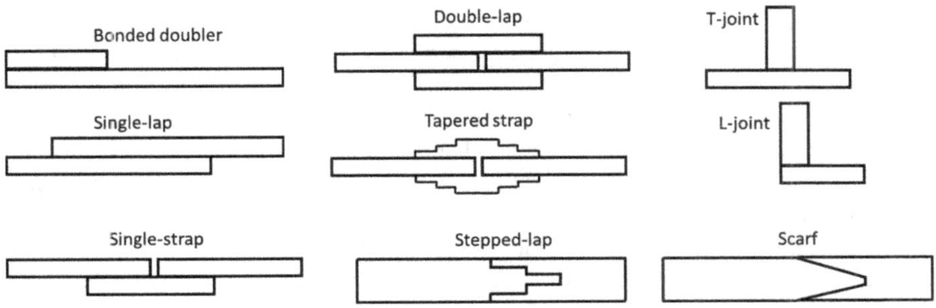

Figure 12.3 Examples of adhesive-bonded joint configurations

12.2.2 Joint configuration: geometries and dimensions

The main challenges in the design of structures are related to the joints. The design of composite structures, in particular, needs extra consideration as the geometry of the system is discontinuities, materials properties are different in the joint area, and as results, high local stress concentrations exist in the joint area [20, 23].

During the design activities usually, the decision is made about the joint configuration. As shown in Figure 12.3, a wide variety of joints are available to the designer, e.g., single-lap joints, double-lap joints, scarf joints and stepped-lap joints. In adhesive joints, the load may be taken in shear, compression, tension, peel and/or cleavage. Load in peel and cleavage should be minimised by applying the principles of good joint design. The strength of a designed joint in the absence of environmental factors depends on the given load, the fact that the load is static or cyclic and the joint's stress distribution, which depends on adhesive/adherend geometry and mechanical properties. The common practice is to design the joint to minimise stress concentrations [24, 25].

The single-lap joint is the most common joint used due to its simplicity and efficiency. However, the main problems associated with this joint are that the stress distribution (shear and peel) is concentrated at the overlap's ends. Researchers have adopted different strategies to improve the efficiency of single-lap joints, including altering the adherend geometry, adhesive geometry and spew geometry. Regarding the adherend geometry, the focus was on adherend tapering and stepping. Taper angle and taper edge thickness were the most considered variable parameters [6, 18, 19].

12.2.3 Stress distribution and failure modes

It is mentioned that the joint strength depends on adhesive strength. The ductility and stiffness are two critical parameters that indicate the strength. The adhesives with high ductility and flexibility usually show low strength, but if they are used in a joint, their ability to distribute the stress uniformly along the overlap (lower stiffness) and to deform plastically could give a joint strength much higher than adhesives with higher strength and less ductility. A low modulus adhesive gives a more uniform stress distribution compared to a stiff adhesive where there is a high-stress concentration at the ends of the overlap. For the overlap in part, a ductile adhesive can redistribute the load and use less stressed in the joint part. In contrast, a brittle adhesive concentrates the load at the ends of the overlap, giving low average shear stress [1, 26, 27].

As individual materials and components, the adhesives could be strong, brittle and stiff, or weak, ductile and flexible. The ideal system is to achieve a strong, ductile and flexible adhesive, but reaching such a system is a huge challenge. This is very difficult to achieve, although the properties are independent. The guideline recommendation is to use ductile adhesives, but this also depends on the overlap structure. During crack propagation, the ductile adhesives show more resistance compared with brittle adhesives by giving a much higher toughness. As a result, the fatigue strength is generally lower for brittle adhesives, meaning that the fatigue life of joints with ductile adhesives is considerably higher than that of joints with brittle adhesives. This is because of the more uniform stress distribution and the higher damping energy associated with ductile adhesives. When the loading is non-uniform such as peel, cleavage or internal (residual) stresses, a joint with a ductile adhesive will also demonstrate a better response. It is worth mentioning that the adherend properties also have an impact on joint strength. The important parameters that affect the joint strength are the adherend modulus and its strength. The higher the adherend modulus, the lower its deformation at the ends of the overlap, where the load transfer occurs [6, 18, 28].

It is essential to notice that the bond's quality determines the failure modes at each interface, specimen geometry and loading. Thus, it is necessary to characterise them and investigate the adhesive and joint properties. Different standards are in place to give guidance based on adherend and adhesive materials. In FRP composite adhesive joints, according to the standard ASTM D5573, there are seven typical characterised modes of failure: (1) adhesive failure, (2) cohesive failure, (3) thin-layer cohesive failure, (4) fibre-tear failure, (5) light-fibre-tear failure, (6) stock-break failure and (7) mixed failure (Figure 12.4). Many researchers experimentally investigated the influences of various parameters on the failure behaviours on composite-bonded joints. The typical bonding parameters are surface conditions such as contamination, abrasion, plasma treatment, fillet, bond line thickness, surface ply angle, stacking sequence, environmental conditions, etc. Several studies also cover the failure predictions for composite-bonded joints performed. The composite-bonded joints' failure prediction is a difficult task considering all mentioned parameters, various bonding methods and the complexity of stress state and its effect on failure strength. Primary failure modes for composite sandwich structures are buckling, local delamination and fatigue/fracture. Buckling is one of the most important failure modes for composite structures, which have a low modulus of elasticity [29].

12.2.4 Adhesive type

Several methods can be used to classify adhesives; none of them is universal. The classification could be based on the curing methods (transform from a wetting liquid to a load-bearing solid), their usage method during assembly, or chemical type [18, 19, 29].

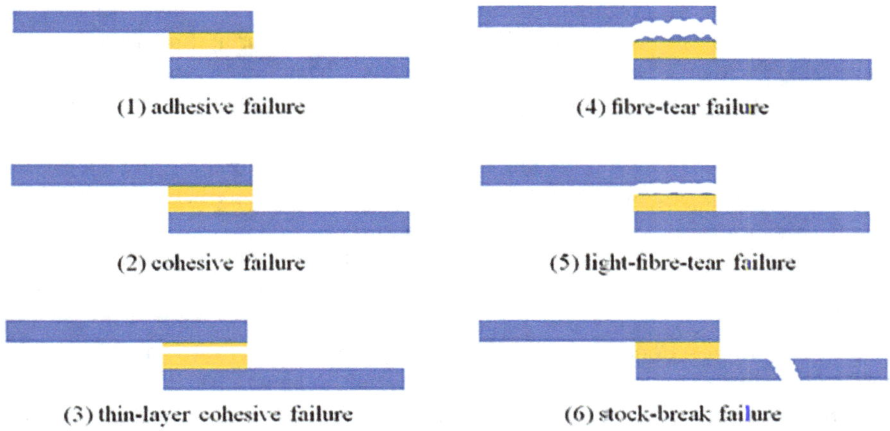

Figure 12.4 *Possible failure modes in bonded joints between FRP composite adherends based on the standard ASTM D5573*

Based on the curing reaction, the strong adhesives are set by a chemical reaction, whereas less strong types are harden by physical change such as evaporation of a solvent. The major classes are described below:

Anaerobics adhesives designed for fastening and sealing applications in which a tight seal is required without light, heat or oxygen. They are also known as 'locking compounds' because they are usually used to secure, seal and retain turned, threaded or similarly close-fitting parts. Their service temperature is −55 to +150°C, and they are cured in the absence of air or oxygen at room temperature as they are based on synthetic acrylic resins.

Cyanoacrylates adhesives are used for fast bonding in plastic and rubber. It can be cured through reaction with moisture held on the surface to be bonded at room temperature. They have service temperature between −30 and +80°C and need close-fitting joints, and usually, solidify very fast (in seconds or minute). It worth noting that they are a particular type of acrylic resin.

Toughened acrylics are categorised as versatile adhesives with fast curing capabilities. They can tolerate dirtier and less prepared surfaces, and after curing, they offer high strength and toughness. Their service temperature is between −40 and +120°C, and they bond well to a wide range of materials as their curing mechanism is based on free radical.

Epoxies have high strength and show temperature resistance. Epoxies adhesives consist of an epoxy resin plus a hardener. They allow great versatility in formulation since there are many resins and many different hardeners. They demonstrate relatively low cure temperature, low shrinkage, ease to use and low cost. Epoxy adhesives can be used to join most materials. However, they can have low peel strength and flexibility and are brittle. The epoxies service temperature is between −40 and +100°C.

Polyurethane shows good flexibility at low temperatures and resistant to fatigue, impact resistance and durability. They have a fast curing time as they have chemically reactive formulations. They provide strong impact-resistant joints and have better low-temperature strength than any other adhesive. Polyurethanes are useful for bonding glass FRP and have better low-temperature strength than any other adhesive. The fast cure makes them possible choice to apply the adhesives by machine. The polyurethane adhesive service temperature is between −200 and +80°C.

Silicones are excellent sealant for low-stress applications. Silicones as adhesives are known for their high degree of flexibility, high-temperature resistance and excellent durability. They are often used as bath and shower sealants, but they are capable of sealing or bond materials of various nature. Their curing time is long, but they are used in a vast temperature range between −60 and +300°C.

Phenolics were the first adhesives for metals and had a long history of successfully joining metal to metal and metal to wood. Despite having good strength retention for short periods of time, their resistance to thermal shocks is limited. Their service temperature is between −40 and +175°C.

Polyimides are mainly used in applications that exploit their ability to withstand temperature up to 350°C but this thermal stability, dependent on several factors. They are available as liquids or films, and although they have good strength retention at high temperature, they have the disadvantage of being expensive and show difficult processability.

Bismaleimides are very rigid and show low peel properties. Their service temperature is between −50 and +200°C.

12.2.5 Surface preparation

Diverse industrial sectors used synthetic adhesives for several years. It is accepted that a pre-treatment is required to obtain the optimum performance from an adhesive joint. It is noteworthy that in a few applications, there is no need to surface preparation. Preparation can cover the treatments between the best surface preparation, economics of component manufacture and health and safety issues. The benefits of using appropriate surface preparation are: (a) enhanced mechanical performance of joint, (b) improved joint durability in harsh environments, (c) increased lifespan of component and (d) ability to bond difficult adherends, e.g., polyolefins and polytetrafluoroethylene [6, 19, 30–32].

The main targets of surface pre-treatment are achieving one or more of the following:

- To remove or prevent the formation of weak boundary layers, e.g., dust, dirt, grease, oils and even finger marks.
- Intimate molecular contact between adhesive and adherend is required to form an adequate bond.
- To protect the surfaces before bonding. This is usually required in metals that, after surface preparation, have a highly reactive surface towards adhesives and atmospheric contaminants.

- To produce a specific adherend surface topography to increase the bondable surface area.

There is a wide range of surface treatments available through the suppliers. The best practice is to collect the appropriate surface preparation procedure through the standards, e.g., ISO 13895 (for plastics). Techniques can be classified into five groups, according to the nature of treatment:

a. **Cleaning/degreasing**: removal of loose solids can be done using a clean brush or blast of clean, dry air. An organic solvent or alkaline aqueous solution removes organic materials such as grease, oil and wax from adherend surfaces. Also, it can be accomplished by wiping, dipping or spraying.
b. **Surface roughening**: with these techniques, abrasive materials are employed to remove unwanted layers and generate a roughened surface texture.
c. **Chemical treatments**: the surface can be partially dissolved or chemically activated by immersing the adherend in an active solution. An electrochemical reaction can also be included where the current is transferred through an electrolyte between an auxiliary electrode and the adherend surface, e.g., anodising. Usually, such a technique introduces several additional factors to be controlled.
d. **Physical treatments**: techniques where the surface is cleaned and chemically modified by exposure to exciting charges or species, e.g., corona discharge, plasma, flame or exposure to ultraviolet/ozone.
e. **Primers**: it is an alternative surface treatment, often simpler than chemical or physical methods, applied by dipping, brush or spray. They can chemically alter the surface (e.g., silane coupling agents and chromate conversion coatings) or protect the preferred surface already generated by another method (e.g. aerospace primers after anodising).

In general, the above-mentioned types of surface preparation can be divided into three major categories: mechanical, chemical and energetic (see Table 12.2). These can be further subdivided into a given technique or method of surface preparation [19, 28].

The selection of preparation techniques should be based on cost, production, durability, health and safety. Pre-treatment facilities can include equipment,

Table 12.2 Surface pre-treatments for adherends

Mechanical	Chemical	Energetic
Alumina grit blast	Solvent cleaning	Plasma
Cryoblast	Detergent wash	Corona discharge
Soda blast	Acid etch	Flame
Peel ply	Anodising	Excimer laser
Silicon carbide abrasion	Primer	

chemicals and consumables. Key surface features are wettability, roughness, soundness, stability, contamination, uniformity and adhesive compatibility. Surface preparations techniques have to be appropriate to performance requirements and be compatible with manufacturing procedures and costs [18, 28, 33].

12.2.6 Environmental parameters

ABJs may be exposed to various environmental conditions during their service life. The performance of adhesive systems can considerably deteriorate when exposed to harsh environments. The environmental factors that can affect an ABJ's properties can affect the mechanical performance of the joint. These factors should be thoroughly taken into account in determining the long-term durability of ABJs and must be carefully identified and related to the type of service the material will perceive. The main environmental factors in climatic exposure are temperature and humidity. The prolonged exposure or even short-term exposure to elevated temperatures will often produce irreversible chemical and physical changes within adhesives. As the temperature increases, the bond strength decreases. The moisture absorbed in a polymeric material can also lead to a wide range of reversible and irreversible effects, including swelling and degradation.

The presence of moisture in adhesive joints may weaken the adhesive's physical and chemical properties and the interface between the adhesive and the substrate. In composite joints exposed to humid environments, the degradation mechanisms are quite different compared to adhesively bonded metal joints. Unlike metals, adhesion of composite to epoxy joints does not negatively affected by the presence of water, thus decreases the likelihood of interfacial failure on ageing. Also, the composite adherend will absorb water, affecting the kinetics of water absorption into the adhesive. Temperature and moisture can also influence the mechanical properties of the composite matrix material, and the interface between fibres and matrix may be weakened in the presence of moisture. Researchers suggested a wide range of degradation mechanisms, including plasticisation, hydration, microcracking of the polymer and fibre–matrix weakening in composite joints. The degradation can be assessed by constitutive and fracture tests [6, 34, 35].

12.2.7 Ease of assembly and costs

Joint design must allow cost-effective assembly. It is usually better to design joints such that the parts are pressed down on the adhesive rather than slid into position. Placement on rather than sliding over ensures that the adhesive is not pushed out of the joint detail. However, placement can result in trapped air, and appropriate precautions must be taken. The design must also allow the adhesive to be positioned in an appropriate pattern. Stripes and crosses are better than closed-loop patterns because they help avoid air entrapment when the components are brought together. Large-bonded areas require careful and even dispensing adhesive; otherwise, considerable pressure is required to squeeze out and spread the adhesive to a uniform thickness.

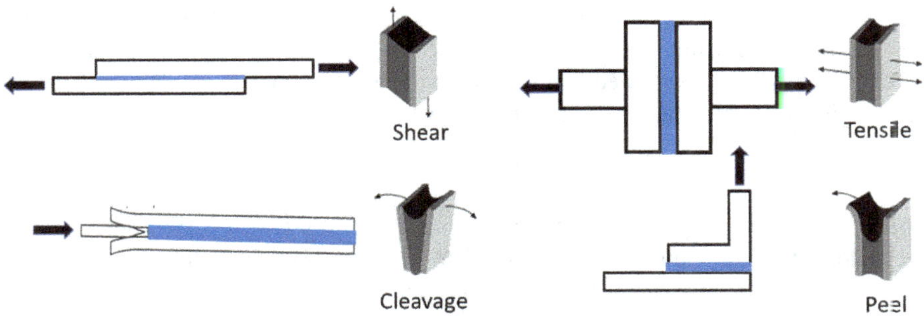

Figure 12.5 *Basic loading modes experienced by adhesive joints*

12.3 Characterisation methods for adhesively bonded composite-to-composite joints

Different types of joints need to be characterised. As the joints can be complex, their characterisation can be very expensive; therefore, more basic and mostly standardised characterisation tests are usually defined to characterise the adhesive and the interface properties to make the characterisation process less expensive and easier to compare. Before manufacturing the actual joints or prototypes, a designer should make a good prediction based on the joint's adhesive and interfacial properties. Thus, adhesion tests can be divided into methods that provide data on the bulk adhesive's mechanical properties that help select the adhesive and methods that determine the quality of the adhesive joints. As illustrated in Figure 12.5, the bonded joints have four main loading mode including: (1) peel loads produced by out-of-plane loads acting on thin adherends, (2) shear stresses produced by tensile, torsional or pure shear loads subjected on adherends, (3) tensile stresses produced by out-of-plane tensile loads and (4) cleavage loads produced by out-of-plane tensile loads acting on stiff and thick adherends at the ends of the joints also show the basic loading modes experienced by adhesive joints. There are currently many ASTM and ISO standards that have been written to analyse and practical evaluation of adhesives and joints properties [36–38]. These standards provide a basis for testing. Typically, test methods that have been developed and used to obtain the properties of the adhesives and joints include: tensile tests, shear tests, peel test and wedge cleavage tests, as shown in Figure 12.5.

12.3.1 Tensile test

Tensile tests are among the most common tests used for evaluating adhesives. Tensile butt joints test and pull-off test are the two most commonly tensile test methods.

12.3.2 Shear test

There are several shear test geometries including lap joints (single, double and modified), V-notched beam (Iosipescu), Arcan, plate-twist, torsion rod, torsion of cylindrical butt joint, thick adherend, napkin ring, skin doubler), each intended to create a completely homogeneous shear state within the bulk adhesive or adhesive layer. Among these mentioned methods the lap-shear tests are the most widely used method for producing in-situ shear strength data of an ABJs.

12.3.3 Peel test

Peel resistance is defined as the average force per unit test specimen width, measured along the bond line that is required to separate progressively two adherend members of a bonded joint. There are three popular peel tests including: T-peel, climbing drum peel and floating roller method.

12.3.4 Wedge cleavage test

Cleavage stresses (essentially concentrated tensile stresses applied perpendicular to the bond line) represent the most severe form of loading that a bonded joint can experience. The wedge cleavage test is widely used to characterise the cleavage strength.

12.3.5 Double cantilever beam (DCB), end-notched flexure (ENF) and mixed-mode tests

A major consideration in the design of adhesively bonded structures is the possibility of crack growth, either within the adhesive or at the adhesive-adherend interface. Crack propagation can be catastrophic when the fracture toughness G of the adhesive-adherend system has been exceeded. DCB, ENF and mixed-mode bending (MMB) (see Figure 12.6) are standard test methods used to characterise Modes I and II and mixed-mode fracture toughness in ABJs.

12.4 Analytical and finite element models for analysis of adhesively bonded composite-to-composite joints

For analysing structural joints in real, applications, finding stresses, strains and possible failure points are necessary. Two main mathematical approaches have been used in order to analysis ABJs consisting of analytical analyses (closed form) and finite element analysis (FEA). Analytical analysis is based on the formulation of the physical principles and FEA simplifies complicated structures by dividing into small and manageable parts [39]. The basic and most common analytical analysis and FEA of composite materials are briefly discussed in the following.

12.4.1 Analytical analysis

The simplest type of joining in composite materials is a single joint, being a typical standard test specimen to calibrate adhesive properties [40]. The closed-form solutions for adhesive stresses and edge bending moment for balanced single-lap joints are presented in this chapter.

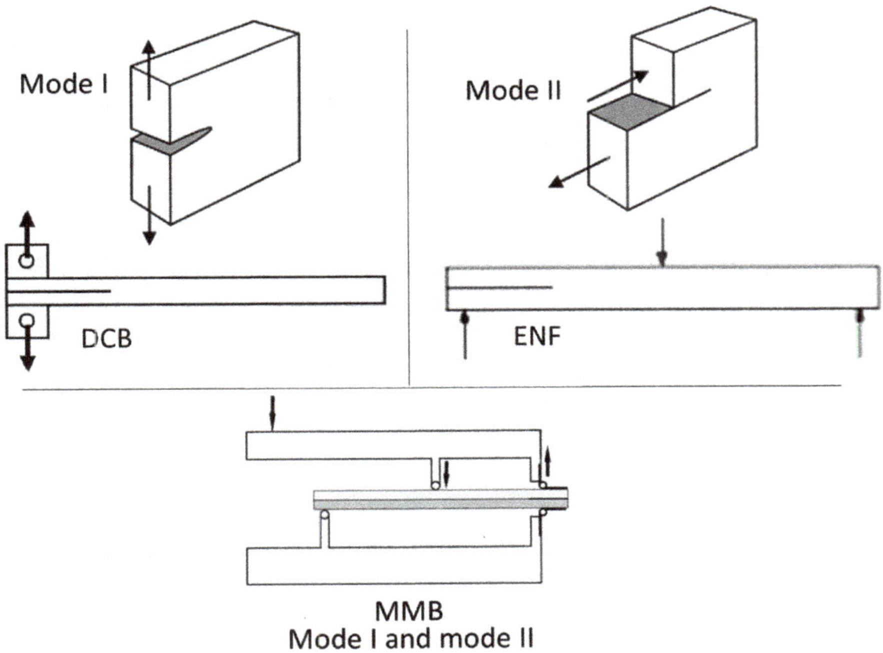

Figure 12.6 *DCB, ENF and MMB tests [12]*

12.4.1.1 Average shear stress model

In this model, the adherends are assumed to be un-deformable rigid bars and the adhesive is a deformable soft part (see Figure 12.7a). The adhesive shear stress τ is equal to $F(2c)$ (F shows the tensile force per unit width and $2c$ is the overlap-jointed length).

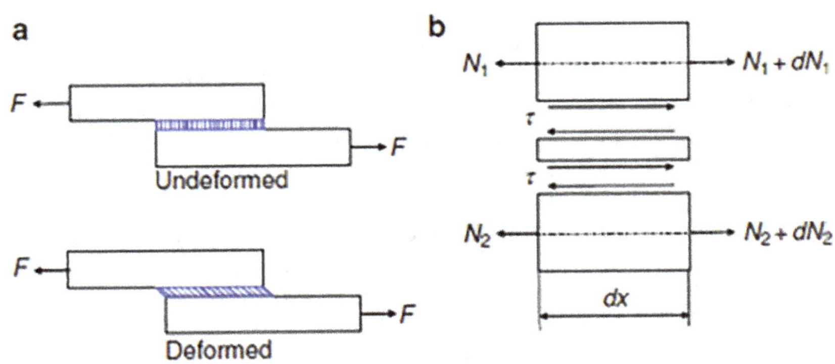

Figure 12.7 *(a) The average shear stress model; (b) free body diagrams of the shear stress and Volkersen model [40]*

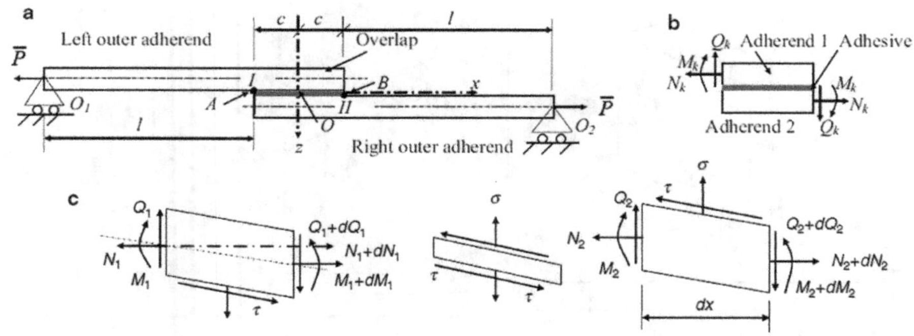

Figure 12.8 *(a) Configuration, (b) forces and (c) free body diagrams of Goland and Reissner model [40]*

12.4.1.2 Shear lag model (Volkersen)

Volkersen [41] proposed a model that is also called shear lag model. Volkersen analysed the axial or longitudinal deformed adherend. The equilibrium and constitutive equations (see Figure 12.8) for adherends are

$$\frac{dN_1}{dx} + \tau = 0; \frac{dN_2}{dx} - \tau = 0 \tag{12.1}$$

$$N_i = A_{di}\frac{du_i}{dx} \; (i = 1, 2) \tag{12.2}$$

A_d is the extensional stiffness and i denotes the adherend's number.

The adhesive shear stress is defined as

$$\tau = \frac{G_a}{t_a}\left(u_2 - u_1\right) \tag{12.3}$$

where G_a and t_a are the shear modulus and thickness of the adhesive. Solving (12.1), (12.2) and (12.3) gives the shear stress [40, 42].

12.4.1.3 Adhesive-beam model (Goland and Reissner)

Adhesive-beam model for the single-lap joint is shown in Figure 12.8. This model was presented by Goland and Reissner [43] and they considered the rotation of the adherends, using elastic medium and Euler beams assumptions for modelling the adhesive and adherends, respectively.

The equilibrium equations of the adherends are

$$\begin{aligned} &\frac{dN_1}{dx} + \tau = 0; \frac{dQ_1}{dx} + \sigma = 0; \frac{dM_1}{dx} + \frac{t_1}{2}\tau - Q_1 = 0 \\ &\frac{dN_2}{dx} - \tau = 0; \frac{dQ_2}{dx} - \sigma = 0; \frac{dM_2}{dx} + \frac{t_2}{2}\tau - Q_2 = 0 \end{aligned} \tag{12.4}$$

t_1 and t_2 show the thicknesses of adherends, and τ and σ are the adhesive shear and peel stresses, defined by da Silva *et al.* [40]:

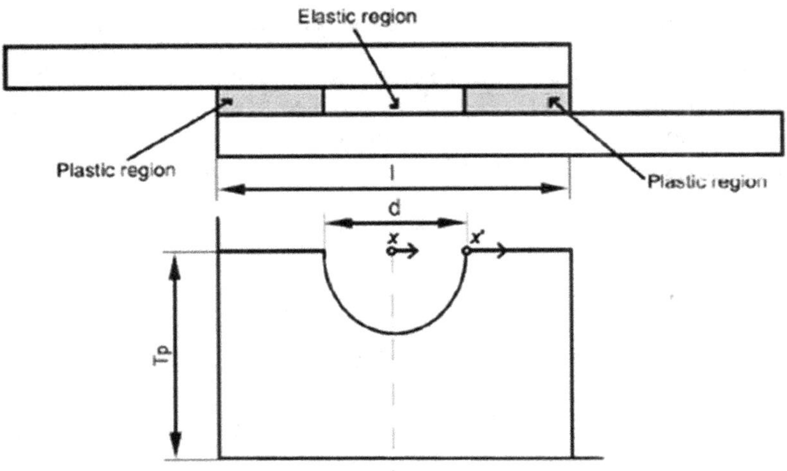

Figure 12.9 *Regions considered by Hart-Smith [44]*

$$\tau = \frac{G_a}{t_a}\left[(u_2 - u_1) + \left(\frac{t_1}{2}\frac{dw_1}{dx} + \frac{t_2}{2}\frac{dw_2}{dx}\right)\right]; \sigma = \frac{E_a}{t_a}(w_2 - w_1) \qquad (12.5)$$

The adhesive shear stress distribution with t is found by solving (4) and (5) [42].

12.4.1.4 Hart-Smith

Hart-Smith [44] developed Goland and Reissner's model to consider the inelastic behaviour (adhesive plasticity) of the adhesive and investigated the coupled formulation [40, 42]. A bi-linear elastic-perfectly plastic approximation was used for modelling shear stress. Three areas are assumed in the overlap: an elastic region in the centre (with the length d in Figure 12.9) and two plastic regions in the surrounding. Coordinates x and x' are defined, as shown in Figure 12.9. When τ_p is equal to the shear yield strength, the elastic zone starts to plastificate. This leads to failure initiation of the bonded joint [42].

The shear stress is found in the elastic area using (6):

$$\tau_{(x)} = A_2 \cosh\left(2\lambda' x\right) + \tau_p (1 - K) \qquad (12.6)$$

where

$$\lambda' = \sqrt{\left[\frac{1 + 3\left(1 - \upsilon^2\right)}{4}\right]\frac{2G_a}{t_a Et}}; A_2 = \frac{2G_a}{t_a Et}\left[\bar{P} + \frac{6\left(1 - \upsilon^2\right)M}{t}\right]\frac{1}{2\lambda' \sinh(2\lambda' c)}$$

$$M = \bar{P}\left(\frac{t + t_a}{2}\right)\frac{1}{1 + \xi c + \left(\frac{\xi^2 c^2}{6}\right)}; \xi^2 = \frac{\bar{P}}{D}$$

D shows the bending stiffness for adherend and is equal to: $\frac{Et^3}{12(1-\upsilon^2)}$. E, G_a, υ and t_a are the adherend's modulus, adhesive shear modulus, Poisson's ratio and the thickness of the adhesive, respectively. $\bar{P}$ is the applied tensile load per width's unit. Also, c and t show half length of the overlap and thickness of the adherend.

For the plastic area, shear strain is

$$\gamma_{(x')} = \gamma_e \left\{1 + 2K\left[(\lambda' x')^2 + \lambda' x' \tanh(\lambda' d)\right]\right\} \tag{12.7}$$

τ_p is the plastic adhesive shear stress and $A_2 = \frac{K\tau_p}{\cosh(\lambda' d)}$.

K and d are solved by iterative approach using the following equations:

$$\frac{\bar{P}}{l\tau_p}(\lambda' l) = 2\lambda'\left(\frac{l-d}{2}\right) + (1-K)(\lambda' d) + K\tanh(\lambda' d) \tag{12.8}$$

$$\left[1 + 3k(1-\upsilon^2)\left(1 + \frac{t_a}{t}\right)\right]\frac{\bar{P}}{\tau_p}\lambda^2\left(\frac{l-d}{2}\right) = 2\left(\frac{\gamma_p}{\gamma_e}\right) + K\left[2\lambda'\left(\frac{1-d}{2}\right)\right]^2 \tag{12.9}$$

$$2\left(\frac{\gamma_p}{\gamma_e}\right) = K\left\{\left[2\lambda'\left(\frac{1-d}{2}\right) + \tanh(\lambda' d)\right]^2 - \tanh^2(\lambda' d)\right\} \tag{12.10}$$

where $\gamma_e \wedge \gamma_p$ are the elastic and plastic adhesive shear strain.

Other coupled formulation and non-linear effects were also approached in the literature. For example, Ojalvo and Eidinoff [45] modified some coefficients in the shear stress equations of Goland and Reissner model and considered new boundary conditions for bond peel stress calculation. Oplinger [46] considered sheer but no peel stress effect. Luo and Tong [47] extended Timoshenko beam theory and dealt with fully coupled formulation. In most cases, it is noteworthy that FEA is used to ease the solution of the formulations.

12.4.2 *Finite element analysis*

FEA provides a numerical approximate solution to problems using different nodes and elements. The most common formulation in FEA involves a symbolism matrix of $K\delta = F$, where δ is the vector of the field quantity at the nodes, F is a vector of known loads and K shows the elements' material property [39]. As the matrix K contains complicated integrals, a numerical integration scheme (e.g., Gauss quadrature) is used to estimate the K matrix. Then the δ values are achieved through the numerical calculation, allowing the calculation of strains and stresses through the entire structure [39].

Modelling of composite materials requires the definition of local coordinates for ensuring the coincidence of layers' orientation with Model 1 (see Figure 12.10) [39].

For modelling an ABJ, continuum solid elements with 2D analysis with in-plane strain condition are commonly used. This is applicable for most of linear and non-linear analyses. ABJs mostly fail from inside. Because the thickness of the adhesive

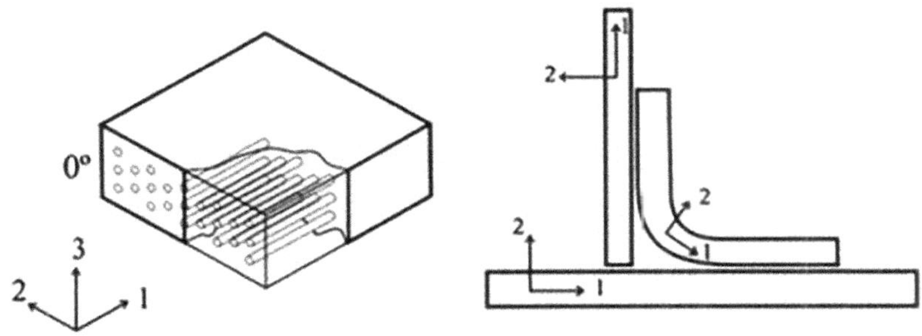

Figure 12.10 *Local coordinates in the case of laminated composites [39]*

is much smaller than the thickness of the adherends, so the finite element mesh sizes for the adhesive should be much small than that of the rest of the model. Besides, a smooth transition between the adherends and adhesive should be applied [48]. Both implicit and explicit mathematical codes are used for a finite element problem of ABJs, while explicit coding is more beneficial for higher failure deformations [39].

The strains and stresses of FEA are usually used in different failure criterion in the literature. For brittle adhesives, Adams *et al.* [49] suggested maximum principal stresses for brittle joints. Yield criterion (Von Mises) was also used by Ikegami *et al.* [50] to study the strength of scarf joints between glass fibre composites and metals. It should be noted that this criterion is more applicable to material yielding than strength. In addition, other studies used maximum shear stress for predicting the strength of different types of joining [43, 51, 52]. This criterion was proven to be valid for brittle adhesives and short overlaps; however, it lacks the effect of normal stresses [17, 53]. When it comes to ductile adhesives, maximum principal strain, maximum shear strain, both shear and peel contributions, strain energy, plastic energy density failure criteria have been used [5, 44, 54–58]. However, all of them are applicable to continuous structures only. Defects or two system materials are not consistent with such an assumption. Stresses calculated using continuum mechanics are infinite at the crack tip, while it is physically finite. In addition, because of material discontinuity in composite material, the stress distribution is not uniform so continuum mechanics lack the exact prediction of all kind of materials. Fracture mechanics could deal with the problems that could not be dealt with continuum mechanics.

Linear elastic fracture mechanic (LEFM) or modified version of LEFM usually help to offset continuum mechanics' lacking assumptions [59–62]. When materials deform plastically, the LEFM concepts have to be extended into elasto-plastic fracture mechanics. In this regard, the J-integral approach has been used by many researchers to predict strength of pre-cracked adhesive joints with reasonable results [63–66]. During damage propagation, different damage mechanisms are initiated, making the failure simulating a complicated process using the J-integral method.

Meanwhile, damage mechanics allow the gradual simulation of damage and fracture at a pre-cracked path until the complete failure [67]. Damage modelling

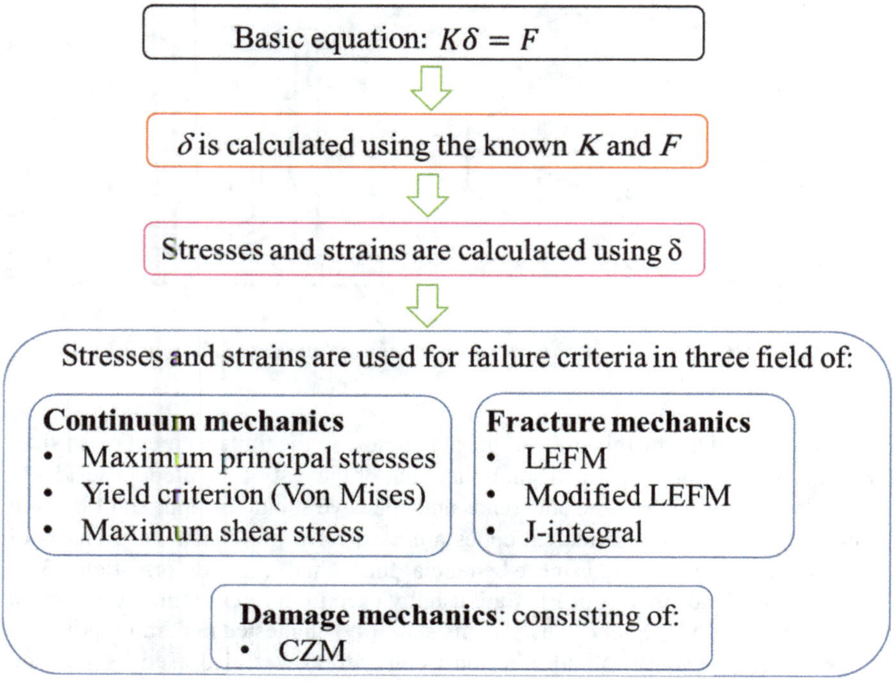

Figure 12.11 *The overall steps of FEA of ABJs. The dots show the most common examples of each method*

is recently used in commercial finite element software such as Abaqus using local technique and continuum technique. The local technique is used for modelling the failure between different materials such as 'adhesive and adherend' or 'composite adherends' layers' [68–70]. The continuum technique is used for the failure simulation of the adhesive bond [71, 72]. Simulation of static or fatigue damage is commonly done using cohesive zone models, which are based on the concepts of stress and fracture mechanics, which suit to both local or continuum techniques. Figure 12.11 gives an overview of how FEA contributes to the analysis of ABJs.

12.5 Design principles and recent developments in high-performance composite-to-composite ABJs

Great effort has been dedicated to introduce novel bio-inspired design strategies based on the optimised topology to improve the strength-to-weight ratio, damage tolerance and mechanical performance of composite joints [73]. Recent work by Burns *et al.* [74, 75] studied the mechanical properties of bio-inspired composite T-joints. Biological structural features of tree joint designs were incorporated in two case studies. The first study focused on embedded design [74] that mimics the tree

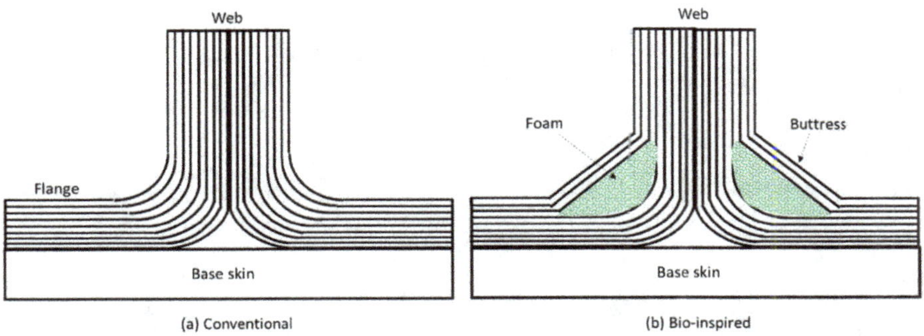

Figure 12.12 Conventional T-joint design versus the bio-inspired design [77]

branch embedded within the trunk, and the second one [75] examined bio-inspired optimisation by tailoring the ply-stacking sequence of the stiffener to reduce interlaminar stresses in composite joints. They found that considering the selected design methods separately led to the improvement of one mechanical property at the expense of another. The joint toughness defined as strain energy to failure was improved by embedding the carbon fibre-epoxy plies of the stiffener flange into the skin laminate [74]. On the other hand, the disadvantage of the embedded design method was a reduction to the damage initiation load and consequently design load by up to 50%. This was because of an increase in the interlaminar tensile and shear stresses in the radius/delta-fillet area. It was shown that the optimisation of the ply pattern of the stiffener lay-up without embedded plies enhanced the failure initiation load by 40% under bending loads; however, there was not seen any improvement in the joint toughness [75]. In another study [76], they reported that hierarchical joints resulted in a significantly improved tensile strength in comparison with traditional T-joints. The novel design also showed higher strain energy absorption to failure load under tensile and bending loadings. It was furthermore found that the hierarchical T-joints had a highly reduced critical joint cross-sectional area because of the embedded design. Akrami *et al.* [77] introduced a novel bio-inspired design strategy based on the optimised topology of bird bone's joint (see Figure 12.12) to improve the strength-to-weight ratio and damage tolerance of composite T-joints. A wet layup method was applied for manufacturing of the T-joints, as summarised in Figure 12.13. They carried out quasi-static results showing that the bio-inspired T-joint design had significant improvements in comparison with traditional T-joints in the elastic stiffness (~60%), peak load (~40%) and mechanical energy absorption (~130%). These enhancements were achieved by only 3% wt increase in the bio-inspired T-joint compared to the conventional one. The fatigue results also revealed a considerable improvement for the bioinspired design verifying the efficiency of the new bio-inspired design for quasi-static and cyclic loadings.

Recent progress in additive manufacturing often referred to as 3D printing, particularly the capability to 3D print different materials at meso-scale, has provided researchers a robust tool to design and fabricate bio-inspired composite materials and

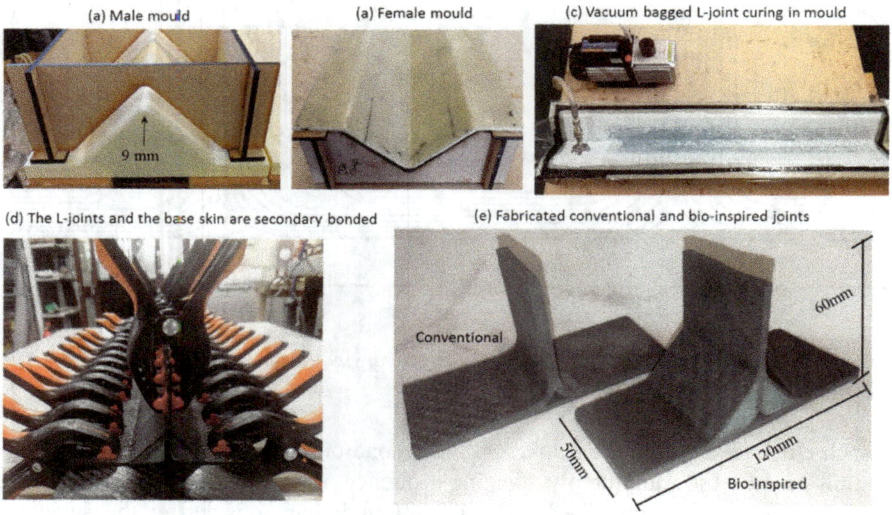

Figure 12.13 The overall manufacturing procedures of the T-joints [77]

joints [78]. The capabilities of advanced 3D printers combined with computational modelling have given researchers more prospects for design, optimisation and test of composite materials and joints to achieve high mechanical performance and reliability [79, 80]. Khan and Kumar [81] demonstrated enhanced performance of tubular-bonded joints with compliance-tailored bondlayer via experiments utilising multi-material 3D printing and modelling. An edge-compliance was imparted to end areas of the bondlayer of tubular-bonded joints mimicking the cortical bone microstructure via 3D printing. Experimental tensile tests on 3D-printed tubular-bonded joints revealed a 100 and 70% increase in load carrying capacity and toughness, respectively, compared to traditional joints with homogeneous bondlayer. Finite element modelling of 3D-printed joints showed that the peak shear and peel stresses in the bondlayer were lower because of increased edge-compliance. Dahmen [82] investigated the suitability of dual cure epoxy 3D printed by digital light synthesis method as bonding material for composite T-joints structures. The joints co-cured with the dual cure epoxy were compared to joints fabricated by conventional bonding materials. The bonding strength of the 3D-printed epoxy was first studied by single lap-shear joints and was then implemented in the 3D printing of composite T-joints, see Figure 12.14. The T-joints by means of 3D-printed epoxy as a bonding material exhibited sufficient joint strength in tensile tests in comparison with other traditional bonding approaches. Moreover, the 3D-printed joints provided a highly reproducible, defect-free bond with improved geometric accuracy.

Koga and Todoroki [83] investigated the performance of 3D-printed snap-in single-lap specimens using standard shearing tests. Snap-in single-lap specimens were 3D printed with male/female snap-in joints, carbon FRP structures, tabs for the clamping to join the male/female parts and holes to enable pining of the jointed

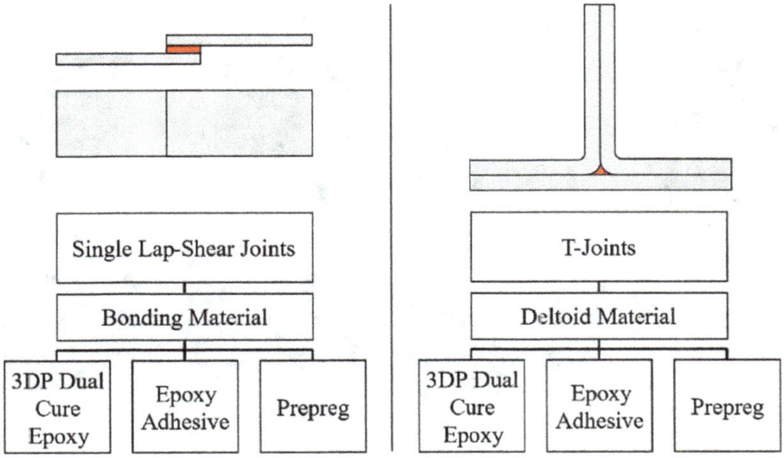

Figure 12.14 *Joint geometries and design of experiments to assess the mechanical properties of different structures and bonding materials [82]*

sections with Ajax-pins, see Figure 12.15. The joint system offered comparable strength to that of adhesive and metal bolt joints, while adding negligible weight to the parts.

Interface design is critical for bonded joints in carbon fibre-reinforced composites. A smart designing method is always desired to minimise the damage and failure incurred by weak bond interfaces. Garcia and Prabhakar [84] implemented

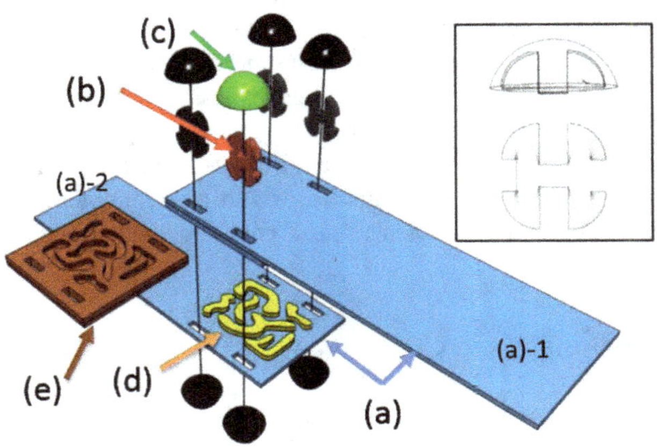

Figure 12.15 *Joint system on snap-in joint and Ajax-pin/cap [83]. (a) The specimen. (b) Ajax-pin. (c) Ajax-cap. (d) Male side of snap-in joint. (e) Female side of snap-in joint. Schematic shows cross-sectional side view of separated Ajax-cap/pin.*

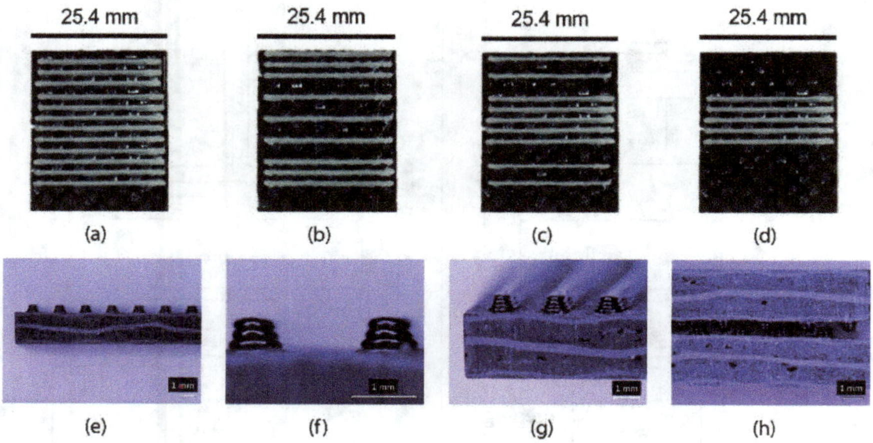

Figure 12.16 *Images of the substrate surface with printed reinforcements for Models 1–4, respectively; (E)–(H) digital microscopy of 3D-printed substrates with visible gaps in side view of lap joint [84]*

3D-printing technology for imparting texture to bond regions in ABJs. An improvement in adhesively bonded single lap joints' apparent shear strength values was accomplished by fusing structural reinforcements to the adherents through fused deposition modelling technology, see Figure 12.16. Pure adhesive joints were manufactured followed by the fabrication of 3D-printed adhesive joints. Pure adhesive joints failed mostly adhesively leading to low peak loads and shear strength, whereas 3D-printed adhesive joints showed higher average peak loads and shear strengths (up to ~832 %) with predominantly cohesive failure.

Kumar *et al.* [85] explored compliance-tailored multi-material adhesive interface designs via 3D printing and performed experimental strength tests. While retaining joint stiffness, a more than 100% increase in strength and a 150% increase in toughness were observed for both centre-stiffness- and edge-compliance-tailored adhesive designs of the joints in comparison with their nontailored and constant moduli counterparts due to reduction in peak peel and shear strains at the ends of the adhesive.

12.6 Conclusions

There has been a substantial demand in the use ABJs in FRP composite components due to their excellent performance including high strength to weight ratio, lower cost, high fatigue life, etc. This chapter provides an overview on the design principles and effective parameters, characterisation methods and recent developments of ABJs. In addition, recent bio-inspired design concepts and additively manufactured ABJs are summarised showing the way forward for further research and development. It was found that despite significant research and development, there is still significant need to improve the adhesive materials, introduce novel design solutions to reduce stress concentration and to use additive manufacturing techniques to one-shot print ABJs.

References

[1] Pizzi A., Mittal K.L. *Handbook of adhesive technology*. 3rd ed. Boca Raton: CRC press; 2017.

[2] Budhe S., Banea M.D., de Barros S., da Silva L.F.M. 'An updated review of adhesively bonded joints in composite materials' *International Journal of Adhesion and Adhesives*. 2017;72:30–42.

[3] Katnam K.B., Da Silva L.F.M., Young T.M. 'Bonded repair of composite aircraft structures: a review of scientific challenges and opportunities'. *Progress in Aerospace Sciences*. 2013;61(22):26–42.

[4] Vallée T., Tannert T., Fecht S. 'Adhesively bonded connections in the context of timber engineering – a review'. *The Journal of Adhesion*. 2017;93(4):257–87.

[5] da Silva L.F.M., das Neves P.J.C., Adams R.D., Wang A., Spelt J.K. 'Analytical models of adhesively bonded joints—Part II: comparative study'. *International Journal of Adhesion and Adhesives*. 2009;29(3):331–41.

[6] Banea M.D., da Silva L.F.M. 'Adhesively bonded joints in composite materials: an overview'. *Proceedings of the Institution of Mechanical Engineers, Part L: Journal of Materials: Design and Applications*. 2009;223(1):1–18.

[7] da Silva L.F.M., Öchsner A., Adams R.D. *Handbook of adhesion technology*. 2nd edn. Cham: Springer; 2018.

[8] da Silva L.F.M., Dillard D.A., Blackman B., Adams R.D. *Testing adhesive joints*. Hoboken, NJ: Wiley; 2012 Sep 19.

[9] da Silva L.F.M., Sato C. 'Design of adhesive joints under humid conditions' in *Design of adhesive joints under humid conditions*. Berlin, Heidelberg: Springer; 2013.

[10] Banea M.D., da Silva L.F.M., Campilho R.D.S.G., Sato C. 'Smart adhesive joints: an overview of recent developments'. *The Journal of Adhesion*. 2014;90(1):16–40.

[11] Abdel Wahab M.M. 'Fatigue in adhesively bonded joints: a review'. *ISRN Materials Science*. 2012;2012(3):1–25.

[12] Chaves F.J.P., da Silva L.F.M., de Moura M.F.S.F., Dillard D.A., Esteves V.H.C. 'Fracture mechanics tests in adhesively bonded joints: a literature review'. *The Journal of Adhesion*. 2014;90(12):955–92.

[13] Heshmati M., Haghani R., Al-Emrani M. 'Environmental durability of adhesively bonded FRP/steel joints in civil engineering applications: state of the art'. *Composites Part B*. 2015;81:259–75.

[14] Marques E.A.S., da Silva L.F.M., Banea M.D., Carbas R.J.C. 'Adhesive joints for low- and high-temperature use: an overview'. *The Journal of Adhesion*. 2015;91(7):556–85.

[15] Costa M., Viana G., da Silva L.F.M., Campilho R.D.S.G. 'Environmental effect on the fatigue degradation of adhesive joints: a review'. *The Journal of Adhesion*. 2017;93(1-2):127–46.

[16] Pethrick R.A. 'Design and ageing of adhesives for structural adhesive bonding – a review'. *Proceedings of the Institution of Mechanical Engineers, Part L: Journal of Materials: Design and Applications*. 2015;229(5):349–79.

[17] da Silva L.F.M., das Neves P.J.C., Adams R.D., Spelt J.K. 'Analytical models of adhesively bonded joints—Part I: literature survey'. *International Journal of Adhesion and Adhesives*. 2009;29(3):319–30.

[18] Vassilopoulos A. *Fatigue and fracture of adhesively-bonded composite joints*. Amsterdam: Elsevier; 2015.

[19] TWI *Adhesive bonding of composites*. TWI; 2007.

[20] Adams R.D. *Adhesive bonding: science, technology and applications*. Cambridge: Woodhead Publishing; 2021.

[21] Maheri M.R., Adams R.D. 'Determination of dynamic shear modulus of structural adhesives in thick adherend shear test specimens'. *International Journal of Adhesion and Adhesives*. 2002;22(2):119–27.

[22] Matthews F.L., Kilty P.F., Godwin E.W. 'A review of the strength of joints in fibre-reinforced plastics. Part 2. Adhesively bonded joints'. *Composites*. 1982;13(1):29–37.

[23] Adams R.D., Wake W.C., Adams R.D., Wake W.C. 'The nature and magnitude of stresses in adhesive joints' in *Structural adhesive joints in engineering*. Dordrecht: Springer; 1984.

[24] Chamis C C., Murthy P.L.N. 'Simplified procedures for designing Adhesively bonded composite joints'. *Journal of Reinforced Plastics and Composites*. 1991;10(1):29–41.

[25] Tong L. 'An assessment of failure criteria to predict the strength of adhesively bonded composite double LAP joints'. *Journal of Reinforced Plastics and Composites*. 1997;16(8):698–713.

[26] Kim J.K., Kim H.S., Lee D.G. 'Investigation of optimal surface treatments for carbon/epoxy composite adhesive joints'. *Journal of Adhesion Science and Technology*. 2003;17(3):329–52.

[27] Kim K.-S., Yoo J.-S., Yi Y.-M., Kim C.-G. 'Failure mode and strength of uni-directional composite single LAP bonded joints with different bonding methods'. *Composite Structures*. 2006;72(4):477–85.

[28] Jeevi G., Nayak S.K., Abdul Kader M. 'Review on adhesive joints and their application in hybrid composite structures'. *Journal of adhesion science and technology*. 2019;33(14):1497–520.

[29] ASTM International 'D5573-99 standard practice for classifying failure mode in fiber-reinforced-plastic (FRP) joints'. *ASTM*. 2019.

[30] Baldan A. 'Adhesively-bonded joints and repairs in metallic alloys, polymers and composite materials: adhesives, adhesion theories and surface pretreatment'. *Journal of Materials Science*. 2004;39(1):1–49.

[31] Davis M., Bond D. 'Principles and practices of adhesive bonded structural joints and repairs'. *International Journal of Adhesion and Adhesives*. 1999;19(2-3):91–105.

[32] Allen K.W. 'Adhesion and adhesives'. *Chemistry in Britain*. 1986;22(5):451–55.

[33] Molitor P., Barron V., Young T. 'Surface treatment of titanium for adhesive bonding to polymer composites: a review'. *International Journal of Adhesion and Adhesives*. 2001;21(2):129–36.

[34] Ashcroft I.A., Hughes D.J., Shaw S.J. 'Adhesive bonding of fibre reinforced polymer composite materials'. *Assembly Automation*. 2000;20(2):150–61.

[35] Ashcroft I.A., Wahab M.M.A., Crocombe A.D., Hughes D.J., Shaw S.J. 'The effect of environment on the fatigue of bonded composite joints. Part 1: testing and fractography'. *Composites Part A: Applied Science and Manufacturing*. 2001;32(1):45–58.

[36] Lee H.K., Pyo S.H., Kim B.R. 'On joint strengths, peel stresses and failure modes in adhesively bonded double-strap and supported single-lap GFRP joints'. *Composite Structures*. 2009;87(1):44–54.

[37] Stamoulis G., Carrere N., Cognard J.Y., Davies P., Badulescu C. 'On the experimental mixed-mode failure of adhesively bonded metallic joints'. *International Journal of Adhesion and Adhesives*. 2014;51(3):148–58.

[38] Melcher R.J., Johnson W.S. 'Mode I fracture toughness of an adhesively bonded composite–composite joint in a cryogenic environment'. *Composites Science and Technology*. 2007;67(3-4):501–6.

[39] da Silva L.F.M., Campilho R.D.S.G. 'Advances in numerical modeling of adhesive joints' in *Advances in numerical modelling of adhesive joints*. Berlin, Heidelberg: Springer; 2012.

[40] da Silva L.F.M., Öchsner A., Adams R.D. *Handbook of adhesion technology*. 2nd edn. Cham: Springer; 2018.

[41] Volkersen D.O. 'Niektraftvertleiling in zugheanspruchtenmitkonstantenlaschenquerschritten'. *Luftfartfoeshing*. 1938;15:41–47.

[42] Rodríguez R., Paiva W., Sollero P., Albuquerque E.L., Rodrigues M.R.B. 'Analytical and numerical tools for bonded joint analysis'. *Mecánica Computacional*. 2010;**2**(76):7557–69.

[43] Goland M., Reissner E. 'The stresses in Cemented joints'. *Journal of Applied Mechanics*. 1944;11(1):A17–27.

[44] Hart-Smith L.J. *Adhesive-Bonded Single-Lap Joints*. NASA-CR-112236. 1973.

[45] Ojalvo I.U., Eidinoff H.L. 'Bond thickness effects upon stresses in single-lap adhesive joints'. *AIAA Journal*. 1978;16(3):204–11.

[46] Oplinger D.W. 'Effects of adherend deflections in single LAP joints'. *International Journal of Solids and Structures*. 1994;31(18):2565–87.

[47] Luo Q., Tong L. 'Analytical solutions for adhesive composite joints considering large deflection and transverse shear deformation in adherends'. *International Journal of Solids and Structures*. 2008;45(22–23):5914–35.

[48] He X. 'A review of finite element analysis of adhesively bonded joints'. *International Journal of Adhesion and Adhesives*. 2011;31(4):248–64.

[49] Adams R.D., Wake W.C. *Structural Adhesive Joints in Engineering*. Netherlands: Springer; 1984.

[50] Ikegami K., Takeshita T., Matsuo K., Sugibayashi T. 'Strength of adhesively bonded scarf joints between glass fibre-reinforced plastics and metals'. *International Journal of Adhesion and Adhesives*. 1990;10(3):199–206.

[51] John S.J., Kinloch A.J., Matthews F.L. 'Measuring and predicting the durability of bonded carbon fibre/epoxy composite joints'. *Composites*. 1991;22(2):121–7.

[52] Lee S.J., Lee D.G. 'Development of a failure model for the Adhesively bonded tubular single LAP joint'. *The Journal of Adhesion*. 1992;40(1):1–14.

[53] da Silva L.F.M., Carbas R.J.C., Critchlow G.W., Figueiredo M.A.V., Brown K. 'Effect of material, geometry, surface treatment and environment on the shear strength of single LAP joints'. *International Journal of Adhesion and Adhesives*. 2009;29(6):621–32.

[54] Harris J.A., Adams R.A. 'Strength prediction of bonded single LAP joints by non-linear finite element methods'. *International Journal of Adhesion and Adhesives*. 1984;4(2):65–78.

[55] Adams R.D., Mallick V. 'A method for the stress analysis of LAP joints'. *The Journal of Adhesion*. 1992;38(3–4):199–217.

[56] Clark J.D., McGregor I.J. 'Ultimate tensile stress over a zone: a new failure criterion for adhesive joints'. *The Journal of Adhesion*. 1993;42(4):227–45.

[57] Adams R.D., Harris J.A. 'The influence of local geometry on the strength of adhesive joints'. *International Journal of Adhesion and Adhesives*. 1987;7(2):69–80.

[58] Zhao X., Adams R.D., da Silva L.F.M. 'Single LAP joints with Rounded Adherend corners: experimental results and strength prediction'. *Journal of Adhesion Science and Technology*. 2011;25(8):837–56.

[59] Dillard D.A., Singh H.K., Pohlit D.J., Starbuck J.M. 'Observations of decreased fracture toughness for mixed mode fracture testing of adhesively bonded joints'. *Journal of Adhesion Science and Technology*. 2009;23(10–11):1515–30.

[60] Kinloch A.J. *Adhesion and adhesives*. Vol. 8(2). Dordrecht: Springer Science & Business Media; 1987.

[61] Hashemi S., Kinloch A.J., Williams J.G. 'Corrections needed in double-cantilever beam tests for assessing the interlaminar failure of fibre-composites'. *Journal of Materials Science Letters*. 1989;8(2):125–9.

[62] Charalambides M., Kinloch A.J., Wang Y., Williams J.G. 'On the analysis of mixed-mode failure'. *International Journal of Fracture*. 1992;54(3):269–91.

[63] Fernlund G., Papini M., McCammond D., Spelt J.K. 'Fracture load predictions for adhesive joints'. *Composites Science and Technology*. 1994;51(4):587–600.

[64] Choupani N. 'Interfacial mixed-mode fracture characterization of adhesively bonded joints'. *International Journal of Adhesion and Adhesives*. 2008;28(6):267–82.

[65] Sørensen B.F., Jacobsen T.K. 'Determination of cohesive laws by the J integral approach'. *Engineering Fracture Mechanics*. 2003;70(14):1841–58.

[66] Banea M.D., da Silva L.F.M., Campilho R.D.S.G. . 'Temperature dependence of the fracture Toughness of Adhesively bonded joints'. *Journal of Adhesion Science and Technology*. 2010;24(11–12):2011–26.

[67] Duan K., Hu X., Mai Y.-W. 'Substrate constraint and adhesive thickness effects on fracture toughness of adhesive joints'. *Journal of Adhesion Science and Technology*. 2004;18(1):39–53.

[68] Yang Q.D., Thouless M.D. 'Mixed-mode fracture analyses of plastically-deforming adhesive joints'. *International Journal of Fracture*. 2001;110(2):175–87.

[69] Yang Q.D., Thouless M.D., Ward S.M. 'Elastic–plastic mode-II fracture of adhesive joints'. *International Journal of Solids and Structures*. 2001;38(18):3251–62.

[70] Turon A., Costa J., Camanho P.P., Dávila C.G. 'Simulation of delamination in composites under high-cycle fatigue'. *Composites Part A: Applied Science and Manufacturing*. 2007;38(1):2270–82.

[71] Song S.H., Paulino G.H., Buttlar W.G. 'A bilinear cohesive zone model tailored for fracture of asphalt concrete considering viscoelastic bulk material'. *Engineering Fracture Mechanics*. 2006;73(18):2829–48.

[72] Kafkalidis M.S., Thouless M.D. 'The effects of geometry and material properties on the fracture of single lap-shear joints'. *International Journal of Solids and Structures*. 2002;39(17):4367–83.

[73] Boaretto J., Fotouhi M., Tende E., *et al.* 'Biomimetics and composite materials toward efficient mobility: a review'. *Journal of Composites Science*. 2021;5(1):22.

[74] Burns L.A., Mouritz A.P., Pook D., Feih S. 'Bio-inspired design of aerospace composite joints for improved damage tolerance'. *Composite Structures*. 2012;94(3):995–1004.

[75] Burns L.A., Mouritz A.P., Pook D., Feih S. 'Strength improvement to composite T-joints under bending through bio-inspired design'. *Composites Part A: Applied Science and Manufacturing*. 2012;43(11):1971–80.

[76] Burns L., Mouritz A.P., Pook D., Feih S. 'Bio-inspired hierarchical design of composite T-joints with improved structural properties'. *Composites Part B*. 2015;69(6):222–31.

[77] Akrami R., Fotouhi S., Fotouhi M., Bodaghi M., Clamp J., Bolouri A. 'High-Performance bio-inspired composite T-joints'. *Composites Science and Technology*. 2019;184(5):107840.

[78] Xiang Gu G., Su I., Sharma S., Voros J.L., Qin Z., Buehler M.J. 'Three-dimensional-printing of bio-inspired composites'. *Journal of Biomechanical Engineering*. 2016;138(2):021006.

[79] Dugbenoo E., Arif M.F., Wardle B.L., Kumar S. 'Enhanced bonding via additive Manufacturing-Enabled surface tailoring of 3D printed Continuous-Fiber composites'. *Advanced Engineering Materials*. 2018;20(12):1800691.

[80] Harris C.G., Jursik N.J.S., Rochefort W.E., Walker T.W. 'Additive manufacturing with soft TPU – adhesion strength in multimaterial flexible joints'. *Frontiers in Mechanical Engineering*. 2019;5:37.

[81] Khan M.A., Kumar S. 'Performance enhancement of tubular multilayers via compliance-tailoring: 3D printing, testing and modeling'. *International Journal of Mechanical Sciences*. 2018;140:93–108.

[82] Dahmen V., Redmann A.J., Austermann J., Quintanilla A.L., Mecham S.J., Osswald T.A. 'Fabrication of hybrid composite T-joints by co-curing with 3D printed dual cure epoxy'. *Composites Part B: Engineering*. 2020;183:107728.

[83] Koga Y., Todoroki A. 'Three-dimensionally printed designable joint for carbon fibre reinforced plastics'. *Advanced Composite Materials*. 2019;28(2):147–61.

[84] Garcia R., Prabhakar P. 'Bond interface design for single LAP joints using polymeric additive manufacturing'. *Composite Structures*. 2017;176(1):547–55.

[85] Kumar S., Wardle B.L., Arif M.F. 'Strength and performance enhancement of bonded joints by spatial tailoring of adhesive compliance via 3D printing'. *ACS Applied Materials & Interfaces*. 2017;9(1):884–91.

Chapter 13

Mechanical degradation of composite bonded joints subjected to environmental effects

Sonia Ruth Rincon Urbina[1], *David Ayre*[2], *and Hamed Yazdani Nezhad*[2,3]

The purpose of the current study was to evaluate the joint strength of a composite single lap joint (SLJ) adhesively bonded, exposed to a hostile environment. The aged joints under hygrothermal cycles were tested under static and fatigue load. A combined experimental-numerical cohesive zone model (CZM) predicted the joint strength degradation and damage propagation. The composite SLJ of T800/M21 bonded with FM94 was subjected to hygrothermal cycles in an environmental chamber (maximum 70°C and minimum 20°C), at maximum 85% RH.

The mechanical test results showed that the strength degraded with the increase in the number of cycles. The strength was reduced by 42% under static load after 714 cycles when compared to unaged joints. The degradation was accelerated in the initial 84 cycles but fell away slightly after this point. The fatigue life was evaluated at 30, 40 and 45% (ultimate static load) to a maximum of 1,000,000 cycles, resulting in a continuous fatigue life reduction with the increase in the number of ageing cycles. A characterization of the moisture diffusion parameters was performed on the adhesive (FM94) and composite laminate (T800/M21) subjected to hygrothermal cycles. Finding that the adhesive reached its moisture saturation level of 1.54% wt, while composite laminate was 0.68% wt, in both cases, the moisture diffusion obeyed Fick's second law. A displacement-diffusion analysis was conducted to determine the effect of moisture on the elasticity of the adhesive, predicting an elasticity modulus reduction of 20%.

The displacement-diffusion model results and shear lap test results were employed to establish the degradation parameters of the CZM, thus predicting the degradation of the joint with an accuracy of 13% at 714 hygrothermal cycles.

[1]Aerospace Technology Research Center, Colombian Air Force, Valle del Cauca,, Colombia
[2]Enhanced Composites and Structures Centre, Cranfield University, Bedford, UK
[3]Aeronautics and Aerospace Research Centre, Department of Mechanical Engineering and Aeronautics, City, University of London, London, UK

13.1 Introduction

The main goal of this research is to evaluate the structural strength of an adhesive joint in a composite laminar system of carbon fibre reinforced polymer (CFRP), which is exposed to a harsh environmental plan. The degradation plan includes mechanical and chemical effects including temperature, moisture and stress (fatigue) to identify the behaviour of the joints.

The aircraft industry continuously faces repair issues of composite structures; engineers have to decide what type of joint is able to maintain the integrity for a long time, encouraging the necessity to know about bonded joint characteristics over the long term. Adhesive joints have shown a significant advantage in comparison to traditional mechanical fastening [1] because of the lower weight, reduced stress concentration, high specific stiffness and strength that improve the structural performance [2].

However, the use of bonded joints in primary structures and repairs has been restricted because of some Airworthiness Authority's concerns about aviation-safety issues related to lack of knowledge about long-term durability, difficulty with quality assurance, limited standardization of the manufacturing process and repair techniques [3]. For that reason, a study for joint strength prediction in terms of life/strength will provide confidence when using adhesively bonded joints instead of mechanically fastened joints because they induce weakening in two properties: tension, 40–60% and compression, 15% [4].

The development of new failure criteria and propagation model is complicated by the difficulty in obtaining repetitive experimental data, even more so because composite joint analysis started from models adapted from the typical experiments of metals taken as the first approach. There was evidence that failure mechanisms and damage propagation of composites are more complex and have different behaviour from metals This is why there is a lack of reproduction and adaptation of laboratory results considered equivalent [5].

The aim of this research is to evaluate the structural strength of an adhesive joint in a CFRP system under a hostile environment (temperature, humidity, load), searching to predict joint behaviour by means of a model.

The scope of this investigation is limited by the time and the mechanical behaviour of the laminar system CFRP in the degradation environment. It means that durability issues such as fatigue, temperature and moisture resistance are covered by the scope. However, optimization of the materials, execution and production, as well as economic issues, is outside the scope of this research.

A research question was raised: 'Can joint behaviour, with respect to degradation, be predicted?', and as such, the aim was set to demonstrate the experimental degradation behaviour trends and mechanical properties of aerospace grade adhesively bonded composite joints (joint strength and failure mechanisms). The data were used to validate an analytical model designed to predict joint behaviour.

13.1.1 Objectives

- To measure the performance of adhesive under mechanical and chemical effects including temperature and moisture cycles and stress (fatigue).

- To identify an analytical model (joint degradation) to predict the behaviour of the joint for long-term use.
- To evaluate the adhesive joints with non-destructive inspection techniques to verify the structural performance and characterize the damage.

13.2 Literature review

This literature review provides the knowledge of the more relevant factors that affect the strength and durability in bonded joints, and they should be identified and controlled during design, manufacture and service life. This review has been focused mainly on: (1) design aspects, (2) environmental impacts, (3) studies of the durability of adhesive joints and (4) damage predictions.

13.2.1 Adhesive joints

The bonding of two materials is due to adhesion by the molecular attraction between two parties to transfer the loads. These materials can be epoxies, polyesters, polyurethanes, methacrylate, etc. [4]. The most common types employed for are epoxies [6]. Adhesive bonding is structurally efficient compared to conventional joining due to its weight reduction, elevated stiffness and strength and better stress distribution under cyclic loading [2, 7]. However, disadvantages of bonded joints are (a) the absence of a non-destructive test (NDT) technique that can assure the quality of the manufacturing process [8] and (b) adhesive joints cannot easily be disassembled to replace damaged parts [9]. Additionally, joining of thick sections can be susceptible to high peeling stresses [8].

The durability of bonded joints can be affected by many parameters as listed by Bardis and Kedward [10], who also have described the variables of each one (see Table 13.1). These factors can be classified into two main groups: the first is related to materials and manufacturing procedures, such as materials substrate, adherend lay-up, type of adhesive, surface preparation, manufacturing process and type of joint, which can be controlled in the design stage, and a second group dependent on the service life, such as load and environmental conditions (i.e. temperature, moisture, chemicals) [11].

13.2.2 Design aspects

13.2.2.1 Type of joint

The SLJ is the most common type of joint used to measure the behaviour of adhesive joints; as it is economic and easily repeatable. However, the single lap is the weakest joint as a result of the eccentricity of loading. The misalignment causes bending moments (through-thickness direction), resulting in high peeling stress of adherend and non-uniform shear stress in the adhesive layer [12–14]. shows the shear and peel stress distribution along the overlap.

Table 13.1 Factors affect the bond's durability modified [10]

Factor			**Variables**
Controlled in the design: Materials and manufacturing procedures	Materials	Adherend Material	Fibre, composite, matrix, metal
		Adherend Lay-Up	0°[n], quasi-isotropic, other lay-up ; orientation of ply on bonding surface
		Adhesive type	Film adhesive, type of carrier cloths, Type of filler, percentage of filler
		Adhesive	Hand-mixed, machine-mixed, apply vacuum to remove trapped air
	Manufacturing	Type joint	Single lap, double lap, tapered lap, scarf, butt and double strap, stepped lap.
		Bonding process	Secondary bonding, co-cured, co-bonding
		Bondline Thickness Control	Glass microbeads/silane treatment, wires tabs/ tape, applied pressure
		Compressed "Shop Air" Blowing	Pressure, exposure time
	Surface preparation	Grit Blast	Pressure, grit size, number of passes, speed of passes
		Hand Sanding	Grit size, number of passes, applied pressure
		Peel Ply or Release Film	Nylon, polyester, none
		Solvent wiping	Acetone, isopropyl alcohol, number of wipes, applicator type
		Temperature Exposure	Temperature, exposure time, prebond postbond, under load
		Humidity Exposure	Humidity %, exposure time, prebond, postbond.

(Continues)

Table 13.1 Continued

Factor		Variables
Controlled by the service life	Environmental Degradation	Humidity, temperature, pressure, hygrothermal cycles, solar radiation, water
	Corrosion	Acids, hydraulic fluid, jet ful,anti-iceing additives, pollution
	Load	Fatigue, bending, flexure

The linear-elastic Vokersen's analysis described the shear stress (τ) distribution with (13.1). This analysis assumes that the adhesive is deforming only in shear and the adherends deform elastically only in tension. *A*1 and *A*2 are arbitrary constants, defined by the boundary conditions [15]:

$$\tau x = A_1 \cos hwx + A_2 \sin hwx \tag{13.1}$$

Then the maximum shear stress (τ_{max}) is presented in (13.2), where *G* is the adhesive shear modulus, *E* the adherend modulus, *l* overland length and *ts* and *ta* the adherend and adhesive thickness:

$$\tau_{max} = \overline{\frac{Gl^2}{\sqrt{(2E_t t)sa}}} \tag{13.2}$$

Hence, based on (13.2) it is understood that the strength of the joint depends on both properties of the adherend and the bondline interface, as well the geometry parameters [16].

- Effects of increasing either adherend thickness or adherend stiffness

The load-bearing capacity is increased because the peak stress levels are reduced and the stress distribution in the bondline becomes more uniform [12]. However, the adherend thickness increment is limited to an appropriate range in relation with the bondline thickness [17].

Reference [2] found that the peak load and shear strength increased with the increment of the adherend thickness. As can see in , with the increment of the adherend thickness, the ultimate failure load and the shear strength increase, but its relation is not directly proportional to an efficiency factor of approximately 0.25.

- Increment the adhesive thickness or reduction of shear modulus of the adhesive

The peak adhesive stress is reduced and the shear strains are distributed along the whole dimension, resulting in lower strains per unit length. The authors of References 12 and 18 studied the effect of adhesive thickness on the strength, and they found a significant increment of the failure load and displacement with the reduction of the adhesive layer thickness.

- Effects of overlap length

The overlap length must be enough to reduce the stress concentration caused by the eccentricity path load and to ensure low stresses in the middle of the overlap. A low-stress region in the middle can avoid creep failure. It is recommended an overlap length approximately ten times the minimum adherend thickness. An overlap excessively longer does not increase substantially the static or fatigue strength, and it can introduce a weight penalty [12].

Increasing the overlap length increases the failure load and reduces the shear stress. The load increment is not linearly proportional to the overlap increase because the load transfer is governed by the edge of the bond area [12].

A comparison of strength for different overlap lengths of SLJs is shown in . It shows the reduction of the strength versus the overlap length increment. Song *et al.* [19] considered 25.4 mm overlap length as 100% joint strength because this length represents the medium value of strength (failure load 11.7 kN). And a length of 38.1 mm registered 24% lower strength (failure load 12.4 kN) than the baseline (25.4 mm), as shown in . Similar results were reported by Li *et al.* [2], who tested CFRP SLJ at different overlap lengths and reported that at longer overlaps the peel stress increased causing delamination failure.

However, Volkersen's theory does not take into account two important factors such as the eccentricity of the load path and the adherend bending. Hence, additionally to the adhesive shear stress, the design analysis has to evaluate the effect of the peel stress and the secondary bending stress.

Particularly in composite adherends, the bending stress leads to inter-laminar failure mode. The composite adherend is weaker under inter-laminar tensile stress than is the adhesive in peel [14]. As shown in , the peel stress near the end of the adherends leads to delamination. For that reason, fillets and end tabs are used as measures to minimize the eccentricity in SLJs and avoid the adherends failure [12, 20].

13.2.2.2 Laminate lay-up and stacking sequence

Laminate stacking sequence (LSS) affects properties such as strength, stiffness, stability and damage tolerance. According to the recommendations of the Military Handbook [17], to obtain an optimum strength, it is recommended homogeneous, balanced and symmetric LSS, and to have a minimum of four different ply angles to get quasi-isotropic material. In the NPL Manual, it is affirmed that the distribution of plies should be preferably 25% plies in 0°, 25% plies in 90° and 50% plies in ±45° [12].

Table 13.2 Examples of stacking sequence used for bonded and bolted joints

Joint	Author/year	Laminate stacking sequence
Adhesive	[1]	[0/–45/45/0]2S
	[21]	[0]16
	[22]	[0/90/0/90/0/90/0/90/0]
	[23]	[0]12T
	[2]	**[45/0/–45/90]2S**
Mechanical	[24, 25], (Atas and Soutis, 2014)	
	[26]	[45/0/–45/90]4S
	[25]	[0/90/45/–45]2S
	[27]	[90/45/0/–45]2S

In the case of bolted joints, the Europe Space Agency Structural Material Handbook states that the stress concentration is reduced with an addition of ±45° layers [12]. Additionally, this NPL Manual suggests that for bolted joints, the quasi-isotropic lay-up (0°/90°/±45°) shows a better performance. Moreover, if the laminate has a majority per cent of 0° layers and few transverse fibre layers, the failure will result in shear out.

Matthew and Tester [13] used bonded joints of CFRP to study the impact on the strength for the stacking sequence for 6, 8, 10 and 12 plies. They found that for a configuration of 8 ply laminates, the best performance is obtained using 0° layers in the outside of each substrate.

Researchers commonly use a balance and symmetric quasi-isotropic lay-up to develop their studies of both mechanical and bonded joints, as shown in Table 13.2. The authors used different orientation plies distributed in order to reduce the stress concentration and cleavage. Some studies employed only 0° layers and other researchers such as References [2, 24, 27] and [25] selected a stacking sequence used by the aircraft industry. Overall, there is not a unified stacking sequence to evaluate the strength.

13.2.3 Manufacture process

There are five conditions for manufacturing a durable adhesive joint: correct surface preparation, properly mixing or applying the adhesive, controlling bond line thickness, uniform pressure during the cure and proper time and temperature of curing [6].

13.2.3.1 Surface preparation

The adhesive bond quality is significantly affected by the surface preparation; ideally, the stress should be transmitted through the adherend/adhesive interface. Mainly an adhesive joint has to satisfy three characteristics: (a) it must avoid wet surface, (b) it allows bonding between adherend and adhesive and (c) it should be durable.

The durability of an adhesive bond depends on the chemical strength, and it is influenced by physical and chemical interfaces [28]. In order to improve durability, the surface roughness, surface chemical treatment and contaminants must be controlled.

There are several treatments to prepare the surface of composite adherends, mainly grit blasting, peel ply, atmospheric plasma and atmospheric laser and various chemical treatments.

shows the surface states caused by some of them.

Peel ply treatment consists of a layer of open-weave material (nylon, fibreglass) added on the surface of the prepreg lay-up, and it is removed after the curing immediately before bonding [29]. The surface result is clean and rough, but with low activated surface and high risk or contamination [30]. This technique is used in the aerospace industry for CFRP surface preparation; however, the increment of components with manufacturing complexity has caused this technique to be unsuitable for CFRP repair areas [31].

Mechanical treatment by abrasive paper or blasting (aluminium or sand grit) causes the first resin layer to be removed with the potential contaminants. The grit blasting process is cheaper, but it can damage the carbon fibre [32, 33].

Atmospheric pressure plasma (APP) is a physico-chemical process that uses an ionized gas (argon, oxygen, nitrogen and ammonia). The ionized gas molecules penetrate and activate the ions, electrons and radicals of the surface; as a result, the surface is modified without affecting the bulk properties [34]. Plasma treatment offers the best surface energy and joint strength [33]. However, it is not used in the aerospace industry [35].

Laser treatment can ablate a thin resin layer without damaging the fibre but removes all the contaminants [30]. It needs minimal previous preparation; it is carried out dry. The advantage of this process is that it is suitable for large and complex areas [31]. But, the final state of the surface has to be closely evaluated because of changes of the thickness of the resin layer.

Hart-Smith *et al.* [36] studied the use of nylon peel ply and identified that a mechanical abrasion is required to remove nylon residue as the nylon incompatibility with epoxy materials [36]. However, later Knight *et al.* (2012) affirmed that using polyester peel ply in carbon surface preparation may offer surface roughness and minimize the contamination because the polyester ply does not require the use of an agent to remove it [21].

Madge *et al.* [37] and Kumar Singh [32] evaluated the strength of bonded composite joints. They found that the peel ply surface preparation offers less strength behaviour than mechanical treatments, plasma and laser [32, 37].

Gomatam and Sancaktar [38] studied the effect of different surface treatments for bonded joints subjected to fatigue behaviour. Substrates of aluminium and steel were treated with grit blasting, silane coupling and plasma, subject to environmental conditions. The authors found that the grit blasting retarded the absorption of moisture, preventing the joint degradation [38].

Gude *et al.* [33] studied the influence of different surface treatments on the strength of joints They tested polyester ply, grit blasting and APP under environmental ageing, hygrothermal 55°C and 95% RH. They found that surface preparation with plasma showed a better adhesive joint strength before the environmental attack. However, after hygrothermal ageing, the joint treated with peel ply showed the best strength [33].

They concluded that the ageing can be harsher on high surface energy materials, as is the case of plasma treatment. On the contrary, peel ply treatment caused a relaxation of stresses in the interface. Encinas *et al.* [39] agree with the disadvantaged response of plasma treatment surface under humidity conditions [39].

13.2.4 Durability

13.2.4.1 Environmental effects

Despite the inherent advantages of the polymeric structures, there are concerns regarding performance under harsh and changing environmental conditions – the durability. For that reason, researchers have been evaluating the complex degradation mechanisms and testing configuration to identify; the durability of bonded joints; the effects of contamination of bonded surfaces and non-destructive techniques to predict the failure and fatigue [40, 41].

Several researchers analyse resistance to environmental degradation of polymers and composites through the physical, chemical and mechanical variables by monitoring changes in area and weight and analysing resistance by their loss and/or gain as a function of time.

Particularly with the moisture, some authors have shown changes of the properties of adhesives and polymer composite materials such as variation of plasticization, swelling, micro-cracking and crazing, hydrolysis, strength loss or an increase of ductility, as a result of the effect of humidity absorption, some being reversible and some not [42, 43].

All thermal, mechanical and chemical changes are related with the moisture transport in the polymer, which are linked with the molecular-sized free volume in the polymer and the water affinity of the polymer. The amount of free holes is influenced by the curing degree, chain stiffness, stoichiometry and cohesive energy density of the polymer. The water affinity depends on the polymer network's sensitivity to create hydrogen bonds with the water molecules, usually by a chemical reaction known as hydrolysis [44, 45].

The first mechanism, the moisture diffusion, occurs through the free volume, without interaction with the polymer molecules nor affecting the dimension. Second, at the hydrolysis reaction phase, the molecular structure is changed, and the moisture interrupts the Van Der Waals bonds between the polymer chains [44]. In epoxies, it has been observed that there is a change of the glass transition temperature (*Tg*), swelling and plasticization because of the interrupted interchange of hydrogen bonding of the polymer [45]. shows the effect of the moisture content on the s transition *Tg* of some structural adhesives.

Changes in temperature may also affect the behaviour of polymers.

At a high temperature above *Tg*, the properties suffer a continuous degradation, resulting in oxidation, chemical attack and mechanical creep [42, 46]. Under a humid environment, an increment of the temperature causes a reduction of the solubility of adhesive, increasing the diffusivity and the moisture contents [44].

At sub-zero temperatures, there is a change in water state, causing swelling, hygrothermal stresses and polymer interlocking breaks. Those local fractures are known as cracking/crazing. In this case, the molecular structure can be affected thus modifying the Tg [43].

Several efforts have been made to study the environmental effects on durability joints. Adhesive bonding of aluminium–aluminium, aluminium–composite, steel–composite and composite–composite has been studied. Banea and da Silva [42] affirm that for the specific case of composite–composite bonded joint, the kinetics of water sorption is different compared to metal-bonded joints [42]. The water is absorbed into the joint and causes degradation of the adhesive, the fibre reinforced polymer (FRP) composite and the interface, as illustrated in .

The adhesive when affected by the mechanism of plasticization causes swelling, cracking and hydrolysis [47]. Three damage mechanisms caused by moisture have been identified for composite substrate: resin matrix modification, interface fibre/matrix debond and fibre degradation. illustrates the different diffusion mechanisms discussed previously and the damages suffered by a fibre-reinforced polymer composite.

In bonded joints, the interface stability under humidity conditions plays an important role in the strength. In the interface, the diffusion is faster than the bulk adhesive. Initially, the moisture ingress by the bulk adhesive travels speedily through the interface and is absorbed into the composite adherend by diffusion and capillary phenomenon [47]. Those mechanisms can reduce the strength bond adhesion between the adherend and the adhesive. However, according to Banea and da Silva [42] in the epoxy/FRP composite interface, the thermodynamic work of adhesion remains positive after ageing in water, which can reduce the probability of interfacial failure.

Knight *et al.* [21] investigated the impact of the ageing of an SLJ under environmental conditions of 82°C and 85% relative humidity for 772 days. They used CFRP adherend of Toray T800H/3900–2 and secondary bonding of AF555M adhesive. They found a reduction in the shear strength and modifications of failure modes. Although they did not find indications of interfacial failure in the adhesive after a drying process, there was evidence of a reduction in the strength, which shows that some irreversible changes happened. They attributed the effect on the strength, after the water desorption, to new bonds between water molecules and adhesive networks. Parker [48] studied the strength of CFRP-epoxy bonded joints; the joints were ageing at 50°C/96% RH (oversaturated K2SO4) for 375 days and tested at 20, 60 and 80°C. They found evidence of interfacial failure after a long exposure time in hot humid conditions and joint strength reduction caused by moisture content and temperature increment.

13.2.4.2 Hygrothermal cycle effects

The moisture and temperature effects have been extensively studied simulating static environmental conditions, as it was discussed in the previous section. However, the strength degradation of adhesively bonded joints is heightened when the joints experience cyclic humidity and temperature [49, 50].

Composites have to resist stress while being exposed to an aggressive environment in aerospace applications. These extreme conditions include humidity, pressure, high and low temperatures, hygrothermal stockings and harsh solutions such as solvents, paint strippers, aviation fuel or lubricants, salt water, hydraulic and deicing fluids [33, 51, 52].

The operating temperatures in an aircraft have several changes due to the conditions of use; normally, low temperatures occur during high-altitude flight or when the plane flies in cold climate, it can decrease to around –54°C. High temperatures can be present according to the area in the structure and type of aircraft. On flights, the components near the engines can reach 130°C; also during the take-off, some parts are exposed up to 92°C. On the ground, the skin of the frame in tropical and hot climates can reach 71°C [52].

Jedidi *et al.* [53] designed an accelerated hygrothermal cycle to evaluate polymer matrix composites used under conditions of supersonic aircraft. The cycles modelled the time on the ground, take off, flight at subsonic and supersonic mode and landing. The study was focused on the drying effect at high temperatures (130°C) in the durability of the composites used in primary structures. The study showed a novel procedure to design an accelerated hygrothermal cycle to represent supersonic environmental conditions. They performed the flight-cycle conditions and the maintenance time. shows the moisture and temperature realistic changes during the flight operations. They used Fick's law to characterize the moisture state in conditions of absorption (during maintenance operation in ground) and desorption after flying 300 flight cycles [53].

Hu *et al.* [22] performed a mixed experimental–analytical study to characterize the effect of cycle temperature on an adhesive bonded joint of CRFP steel with adhesive Araldite AV138. They used a cohesive model to predict the strength degradation and validated with the experimental testing. SLJs were conditioned under cycles of 2 hours of heating up +80°C and cooling to –30°C. Samples were taken every 84 hours until 672 hours. Additionally, samples without degradation were tested in order to compare the level of degradation. They found that the tensile strength was reduced in relation to the number of cycles and that the mode of failure changed from cohesive failure to adhesive failure after the cycles [22].

Korta *et al.* [50] developed an experimental and numerical study to determine the effect of the humidity-temperature cycle of ageing on shear and tensile strength of multi-material bonded joints. They employed two different adhesives to bond single lap samples of CFRP to steel and CFRP to aluminium. The testing was performed according to SAE standard of the automotive industry, considering two different cycle types, first was at accelerated heating up 85°C/95% RH and holding for 5 hours, followed by a heating up 180°C and holding 1.40 hours and cooling to –40°C and the second cycle was at heating up 85°C/95% RH and holding by 8 hours to finalize with cooling to –40°C. To predict the behaviour, some finite-element analysis was carried out using Altair Radios solver. They found that, under this ageing, an important number of samples showed premature failure. The cycle of temperature led to the debonding of joints. The reduction of strength properties was more severe in the case of samples subjected to the first type of ageing. Also, samples of cylindric joints of CFPR–CFRP were tested under tensile load, after the ageing, the tensile performance of CFRP–CFRP joints was increased [50].

Agarwal *et al*. [52]performed freeze-thaw cycles to steel–CRFP lap joints; they found strength reductions of 28% after 40 cycles and change in the type of failure mode. The reduction can be attributed to microcracks caused by the expansion of the water molecules at low temperatures [54].

Additionally, some researchers have exposed that cooling an epoxy resin at sub-zero after saturating at high temperature (below *Tg*) drives more water absorption [55].

13.2.4.3 Fatigue durability

To predict fatigue is a difficult task due to complex modes of crack initiation, propagation, the variable behaviour of adhesive, substrate and interface front of moisture and temperature environments and, moreover, the geometry of joint and complex performance of composite under loading cycles. However, fatigue durability can be evaluated from the loss of strength and fatigue resistance when the element is subjected to diverse environmental conditions [56, 57].

The influence of humidity and temperature on fatigue has been extensively studied and especially joints of metal adherents.

Broughton and Mera [57] studied the fatigue behaviour of a double lap joint subject to environmental conditions of 70°C and 95% RH for 750 hours and testing at –55, 20 and 130°C. They used two different adhesives F6552/T650-35 and HP655 and substrates of CFRP. Fatigue life was reduced for the after ageing condition. Although, the elevated temperature test (130°C) impacted more significantly the cycles of life than the hygrothermal ageing.

Ashcroft and Shaw [8] studied the effect of temperature on fatigue crack propagation (FCP), they tested double lap and lap strap joints of CFRP at –50, 22 and 90°C and found that temperature has an important effect on the fatigue crack growth rate. Also, the induced failure shifted from the adherend layer to the adhesive layer as the temperature increased. Furthermore, they used a mechanical failure model in order to predict the threshold failure with a good approximation.

Despite the fact that humidity and temperature diminish the joint strength, the temperature has a high toll on the joint durability [58].

Abdo and Aglan [59] worked in determining the influence of aerospace cyclic thermal ageing on static and fatigue behaviour of the adhesive SLJ of aluminium 7075-T6 and adhesive 3M AF-163–2K. The thermal cycle was accorded with ASTM D1183 test procedure C, consisted in increasing the temperature up to 74.5°C/<10% RH for 48 hours, then a decrease of temperature to 23.5°C/100% RH for 48 hours, followed by a reduction in temperature to –58°C/100% RH for 8 hours and finally increasing the temperature up to 39°C/100% RH for 64 hours, as shown in Figure 13.1. Samples were tested at 2, 4 and 6 weeks thermal cyclic duration. Under a static condition, the joint strain increased with the number of cycles, and the displacement increased to 2.5 mm for the six thermal cycles. But the load strength

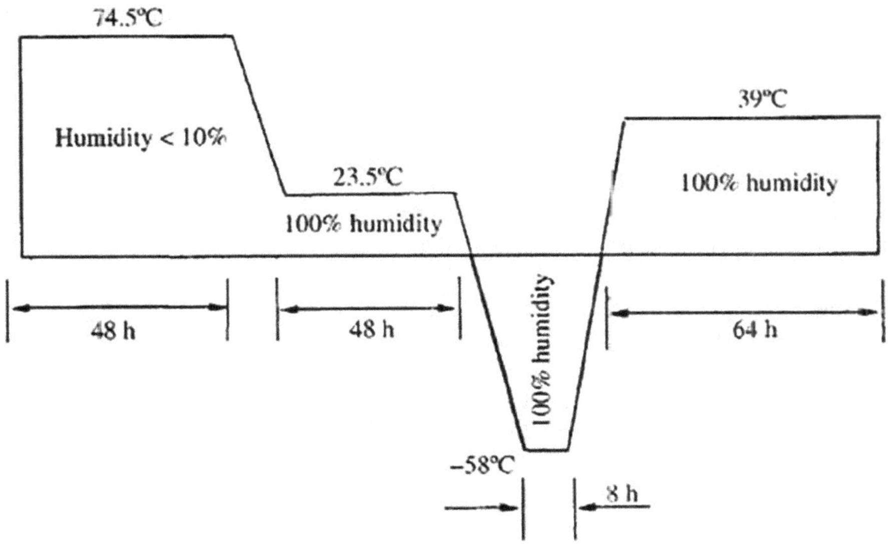

Figure 13.1 Thermal ageing cycle [59]

joint was not significantly affected by the cycling. For fatigue testing, the fatigue strength was reduced by 47% at six thermal cycles. The failure mode was a cohesive failure. The degradation by moisture was presented at the first cycles and caused an increase in the FCP. Moreover, a modified crack layer model was employed to predict the failure with successful results.

13.2.5 Non-destructive testing

According to Federal Aviation Administration (FAA), the continued airworthiness of both bonded structures and composite repairs needs to be assured with an inspection to detect and to quantitatively estimate any defect size [60]. Also, the assurance of the bond is an important part of the qualification of the joint in civil and automotive industries.

There is a lack of means to non-destructively measure the bond strength in primary structures of aircraft. Structural engineers do not have actual bond strength values, so they cannot predict the life. In manufacturing, the bonding process is more controlled than in repair. It is because the variability process can lead to degradation of strength. For this reason, repairing the development non-destructive techniques is a priority to determine the joint strength [6].

The composite inspections have been performed with techniques such as ultrasonic, X-ray radiography, visual inspection, leak test, bond test, acoustic emission, low-frequency vibration, thermography, shearography and holography [61]. They have successfully detected defects; however, they are not able to detect weak bonds (kissing bonds). So, the actual NDT is not effective to assure the quality in terms of bond strength, physical and chemical properties of adhesion or prediction of crack growth [62].

13.2.5.1 Types of defects

The NDI studies have been focusing on the characteristic of failure of the adherend/adherent interface. The main designers' concern focuses on the joint structures working at nearly full strength because the degradation causes reduction or loss of bond strength [6].

Examples of effects found in the adhesively bonded joint are delamination, debond, porosity, voids (high volume porosity), incorrect matrix cure and cracks, weak and kissing bonds [62]. Disbonds are related to moisture content, lack of surface preparation, impact damage or local over-heating . These disbonds can be detected by ultrasonic or acoustic inspection.

Weak bonds are associated with a design issue. The adherends are in intimate contact but the interface doesn't ensure a mechanical load transfer capacity, and this kind of defect is not detectable actually by non-destructive inspection [6, 63].

13.2.5.2 Non-destructive inspection methods

The testing methodology used to define the properties of materials is known as non-destructive evaluation (NDE). The NDE studies to measure the bond strength have been addressed to determine the chemical bonding in the interface [6].

The developments to inspect bonds have been using physical waves for the evaluation of the bond strength. The material response of front wave is seen as a mechanical property. The parameters such as frequency, signal shape and amplitude of the wave are used as criterion for the qualification of the adhesive joint [62]. Ultrasonic and laser are the most common methods used in the industry for bonded joint inspections Recently, studies also have tested shearography, tomography computational and holographic techniques to evaluate bonded structures.

Ultrasonic applies to composite using a frequency of 5 MHz; this technique includes methods such as pulse-echo, through-transmission, ultrasonic spectroscopy, harmonic imaging, oblique incidence, guided waves, ultrasonic acoustic and shear wave resonance. Although for the NDE technology, the ultrasonic methods have not allowed detecting changes of bond strength, they can provide information about the morphological and elastic characteristics of the interface of bonded joints [62].

Active thermography, this technique induces an external temperature gradient; the thermal flow in the structure is monitored to detect temperature resulting over the surface. It is unable to detect weak bonds due to the anisotropy of the composite materials and the small heat diffusion. However, thermography combined with the proper tool of excitation is a potential NDE method to measure the bond strength [62].

The laser shock adhesion developed by CNRS and Laser Bond Inspection created by Boeing and LSP Technologies are the most efficient tools. Laser shock wave generated by plasma propagates through the joint structure in compression with a reflection signal on the back surface that induces a tensile load that can rupture a weak bond, attenuating the signal read [62].

Although some limitations of thickness arise because of the attenuation, laser adhesion tests have been used to describe different levels of adhesion of weak composite bonds and correlated with the mechanical strength values and validated with post-failure tests such as interferometric confocal microscopy, ultrasonic and visual inspection [6].

13.2.6 Damage prediction

13.2.6.1 Models to predict durability

In order to predict the service life of adhesive joints exposed to a hostile environment, the moisture diffusion behaviour in the adhesive joints has to be known. Diffusion models have been developed to understand the phenomenon; those can be grouped based on the mechanism of diffusion moisture in the adhesive.

The first diffusion assumes that the water molecules diffuse into the epoxy and reside in free volume of the material. The second model assumes a chemical interaction between water molecules and the polymer matrix. A third mechanism, known as anomalous diffusion, combines physical and chemical interactions. Furthermore, a phenomenon of capillary diffusion affects some variables such as temperature, surface effects and swelling [11, 64].

There are three main approaches to predict the durability of adhesive joints; one is a stress-life approach, which is based on the stress and strain of the adhesive joint. The main advantage of this approach is that it is simpler than others, but it needs care with the refinement and converging of the mesh. The second approach is fatigue crack initiation/propagation approach. This method supposes that materials have defects and the cracks and crack growth follow a fracture criterion. The fatigue crack initiation method has been less studied because of the difficulty to link a model of nucleation of crack and to find the initiation phase. Additionally, a fracture mechanics method can be limited to the pre-existence of crack to establish the strain energy release rate.

Finally, the CZM based on a traction-separation constitutive law can model multiple cracks, without a need to know the directions of propagation, in order to predict the initiation and failure [56, 64].

13.2.6.2 Modelling moisture diffusion

To determine the moisture concentration, there are two methods of testing: first, there is no uniform moisture distribution, and the specimens are exposed to moisture for a fixed time. It is used as a quality control method and to compare batches of materials with similar geometry and manufacture conditions; in the second method, the specimens are subjected to moisture until they reach the equilibrium condition, and it is used to determine the properties of the material associated with diffusion process, and its condition depends on the thickness of the bond layer [65, 66].

The moisture diffusion in polymers is governed by mechanisms and physical phenomena, which are taken as factors in the mathematical model. Fick's law has been presented as the simplest mathematic equation to describe the diffusion in polymers. Fick's law assumes that molar flux caused by diffusion of a substance through a unit area is proportional to the concentration gradient as shown in (13.3). Fick's law assumes that the water molecules do not interact with polymer molecules [11]:

$$F = -D\frac{\partial C}{\partial x} \tag{13.3}$$

However, some polymers have not shown Fick's moisture diffusion behaviour [11, 64, 67]. Hence, based on different findings, the concentration models have been grouped as follows [44].

Concentration Independent of Fick's law

- Water molecules move freely.
- Mobility is smaller than the stress relaxation rate of the polymer structure.
- No interaction between molecules.
- The diffusion becomes zero at saturation level.

Concentration Dependent of Fick's law

- At temperatures above the glass transition temperature (*Tg*).
- Free volume, void size and mobility are increased.
- Internal stress relaxation is not instantaneous.
- Diffusion is dependent on moisture (history), boundary conditions and time.

Non-Fickian or Anomalous

- At temperatures below *Tg*.
- Hydrolysis of polymer molecules.
- Mobility and stress relaxation rate are similar.
- A dual-mode sorption, Fick diffusion and polymer relaxation phenomena.

The diffusion of moisture in polymers is usually studied with Fick's second law, which describes the concentration in non-permanent state for three-dimensional spaces.

Liljedahl [68], who studied the diffusion through FM73 adhesive, got a good agreement with the Fickian diffusion model. Gude *et al.* [33] employed the solution for Fick's law proposed by Crack [69] to find the concentration of moisture in a carbon fibre/epoxy bonded joint.

On the other hand, Reference 64 reported non-Fick diffusion behaviour of FM73 adhesive; it was immersed in water until saturation at 50 and 70°C and subjected to cycles of desorption. The moisture diffusion for absorption was not predicted by the Fickian model, while for desorption, Fick's model presented a good correlation with the experimental data.

Wahab *et al.* [67] worked to develop an accurate transient diffusion model based on the diffusion stress analysis model. They studied experimentally, numerically and analytically the diffusion of moisture in adhesive joints of CFRP; a sample of composite and adhesive was conditioned at 45°C/85% RH and 90°C/97% RH. The moisture absorption of adhesive showed Fickian behaviour and the sorption in composite at 90°C/97% RH presented non-Fickian behaviour.

Models have been employed to predict non-Fickian diffusion behaviour. One is a modification to Fick's law named dual Fickian model, which considers two mechanisms in parallel with different diffusion properties. Ameli *et al.* [70] proposed that a sequential dual Fickian (SDF) model may be used to predict the moisture concentration in adhesive joints subjected to fluctuating temperature and humidity.

Second uptake stage has been noted in epoxy adhesives, at low temperatures [71]. In these cases of the second stage of rapid diffusion, Amelia *et al.* [11] have affirmed that an SDF model can predict the diffusion behaviour and maximum moisture content.

13.2.6.3 Cohesive zone modelling

The Cohesive Zone Modelling (CZM) has been used widely to predict the failure load of bonded joints and recently employed to predict the degradation caused by moisture and temperature cycles (Katnam et al. (2011) [72]).

The mechanical fracture is described by a phenomenological model, which uses a material model represented by a traction-separation law or cohesive law. The cohesive model is used to predict the damage initiation, propagation and ultimate failure. Before the initiation, the material remains bonded by traction as cohesive zone elements. Then, two new surfaces are initiated while the crack is growing, the crack propagates and finally the material fails and separates, originating two new surfaces. The traction changes in relation to the displacement of the surfaces [73].

To model composite failure, micro mechanical concepts or cohesive laws can be used. As shown in Figure 13.2 , the interface between two materials is modelled as a cohesive layer of elements [75].

In numerical applications, there are three methods to formulate the cohesive layer.

1. Modelling as an interface between two bodies bonded; the thickness of the layer is overlooked. Therefore, the law constitutive material model considers that the fracture energy is intrinsic because of traction.

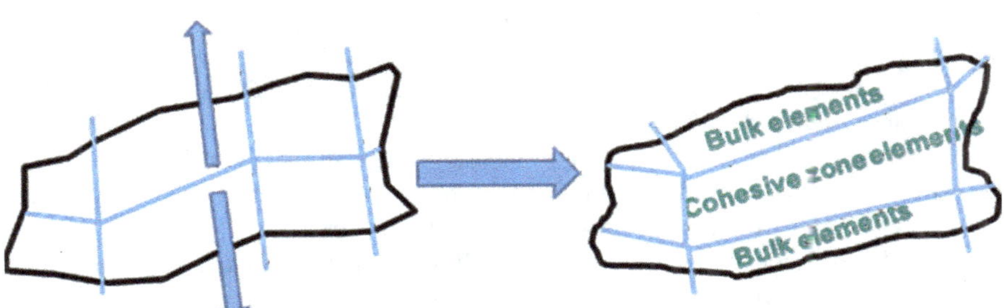

Figure 13.2 Crack growth in cohesive zone [74]

2. Modelling as an elastoplastic continuum with fracture parameters. The adhesive layer is considered by volume. The material properties are degraded progressively by the evolving damage parameters using traction separation relation.
3. Modelling as a bi-material interface consistent with layer thickness. The macroscopic response of such an interface is described under the cohesive law.

The first approach is commonly used to model interface in composite laminates, interface in bonded joints or interface between solid phases of materials [76]. The second method uses the continuum mechanical method and presents the highest degree of accuracy; but, it requires elastic properties as well as fracture adhesive parameters. It is also numerically more complex to implement. The continuum approach is used to model the cohesive fracture of the adhesive bond [76]. The balance is reached using modelling method 3, achieving greater accuracy than modelling method 1 and greater simplicity than modelling method 2 [77].

The cohesive law constitutes a relation between the traction acting in the interface and the interfacial separation between two bonded materials. There are mainly three parameters in cohesive law: cohesive energy (*Gc*), cohesive traction (*T*) and displacement (δ).

There is displacement in the direction normal (δn) to the surface interface and tangential (δt) to the surface interface, which depends on the traction acting on the normal direction (*Tn*) and tangential direction (*Tt*), respectively. The area under the curve is the energy (G_C) needed for separation.

Cohesive damage occurs when the maximum stress (*Tn*0,*Tt*0) is exceeded. After the peak traction, a linear softening stage continues, until the separation reaches the maximum value ($\delta^0 n$) ($\delta^0 t$). At this critical point, the stress is zero. There are three failure modes when separation occurs as a result of normal loading to the crack plane, the failure mode is known as mode I or opening. When failure occurs as a result of traction tangential to the crack plane, they are known as mode II or mode III. In mode II, the load is in-plane shear and mode III is out of plane shear, as shown in The combination of any separation mode is known as mixed failure mode [75, 78].

There are several cohesive zone law configurations described in the literature. They can mainly be categorized into the following groups [79].

- Polynomial cohesive zone laws [80]. It was used to model ductile materials failing at mode I and [81] for mixed failure mode.
- Piece-wise linear cohesive zone laws, multilinear, trapezoidal [82]. For modelling adhesively bonded sandwich structure and polymer matrix composites. Widely applied by its application for different materials.
- Exponential cohesive zone laws [83]. It was used for brittle materials, and Scheider and Brocks [84] used that for modelling ductile materials.
- Rigid-linear cohesive zone laws, either called bilinear or linear decreasing: [85] applied this law for brittle materials.

The CZM laws configurations can be adapted to the behaviour of the thin material or interface to be simulated. Actually, an accurate CZM can be selected through advanced techniques. However, the experimental tests are still challenging. Hence, the CZM is simplified and defined based on the knowledge of the materials to be modelled [76].

The CZM law effect depends on the geometry and mechanical behaviour of material and interface. Pinto *et al.* [86] found for SLJ with stiffer adherends that the CZM shape used has a significant effect on the accuracy of the results [86]. In this way, bonded joints of ductile adhesives have often been simulated with the trapezoidal law, and to simulate brittle materials, a bi-linear CZM (known as triangular shape) is frequently used [76].

Sugiman, Crocombe and Aschroft, (2013) [7] studied the effects of ageing in fatigue of an adhesive SLJ; joints were immersed in deionized water at 50°C for 2 years. They found that the fatigue strength is reduced with the increase of moisture, although there is saturation level that the trend of fatigue reduction gets to be stabilized. In order to determine the fatigue, they used a CZM, and they suggested non-linear fatigue degradation parameters to predict failure load and the failure process of the SLJ.

13.2.6.4 Environmental degradation models

Several degradation models have been developed to predict the damage of adhesive joints ageing under environmental chemicals.

Ashcroft *et al.* [87] used a model that combined with a diffusion moisture analysis and mechanical stress analysis to predict the fatigue resistance of a CFRP joint subjected to environmental ageing. The lap-strap joints were tested under fatigue load at –50, 25 and 90°C under humidity conditions and under high humidity testing at 25°C (95% RH) and 90°C (97% RH) [87].

First, the moisture concentration distribution was achieved through a mass diffusion modelling; the analysis needed the diffusion coefficient of the materials. Then, the diffusion elements are changed to solid elements for the stress analysis. The stress–strain of each adhesive element is determined, based on the moisture concentration. The analysis assumed for the CFRP that the properties weren't affected by the moisture. Finally, a failure criterion was employed to determine the damage in the joint; elastic fracture and nonlinear elastoplastic fracture conditions were used to predict the trend fatigue failure. This coupled model was validated with the experimental results. shows a diagram summary of the model steps.

Katnam *et al.* [88] used a couple mechanical diffusion and cohesive zone approach to predict moisture effect in the static and fatigue failure of adhesive bond of 2024-T3 with FM73 adhesive. Single lap and double lap joints were immersed in de-ionized water at 50°C during 360 and 720 days and tested under static and fatigue load. The frame of the mixed model used. Their assumptions for the numerical model were as follows.

- A bilinear traction separation response represents the failure in the adhesive bond line.
- Moisture degradation can be represented by the moisture-dependent traction-separation response.
- The bondline fatigue damage has degraded the bilinear traction-separation response
- Moisture degradation and stress damage can be accumulative.
- The moisture diffusion (D) doesn't depend on moisture concentration.
- The moisture gained by the adhesive obeys Fick's law.

Using an FE model in Abaqus and the analogy between heat transfer analysis and diffusion analysis, the moisture concentration in the adhesive bondline was calculated for 320 and 720 days. The concentrations were exported into the static and fatigue failure step to degrade the cohesive parameters. The parameters were degraded as shown in (13.2) of the static model:

$$\sigma_m = \sigma_m{}^d(1 - c) + c\sigma^s{}_m \tag{13.4}$$

$$\Gamma_m = \Gamma_m{}^d(1 - c) + c\Gamma_m{}^s \tag{13.5}$$

$$E = E^d(1-c) + cE^s \tag{13.6}$$

where (c) is assumed as the normalized moisture concentration and ($\sigma_m{}^d$, $\Gamma_m{}^d$, E^d) the dry conditions and (σ_s, $\Gamma_m{}^s$, E_s) moisture condition.

For the fatigue conditions, a strain-based model was assumed:

$$\frac{dD}{dN} = \alpha\left(\varepsilon_p - \varepsilon_t\right)^\beta \tag{13.7}$$

$$\hat{\sigma}_m = \sigma_m\left(1 - D\right) + b^2 = c^2 \tag{13.8}$$

$$\hat{\Gamma}_m = \Gamma_m\left(1 - D\right) \tag{13.9}$$

where D is the damage variable, N is the number of cycles, ε_p is the maximum principal strain in the adhesive bondline, ε_t is the threshold strain and α and β are constants.

The static strength reduction and the fatigue life were successfully predicted by the CZM and strain-based fatigue damage model, respectively.

Han *et al.* (2014) proposed an experimental work followed by a numerical simulation study to investigate the residual strength of degraded adhesively bonded joints. Their study focused on aluminium joints bonded with FM73 subjected to a combined mechanical-hygrothermal ageing (immersion in water at 50°C). Fick's model was used to calculate the diffusion coefficient and moisture concentration. The fully coupled approach consisted of two steps:

(1) to model the long-term ageing process of the joint under combined thermal-hygro-mechanical loading and (2) to simulate the quasi-static tensile testing of the adhesive joints, aged in step 1, using CZM. The schematic description of the general work is shown in .

The first step, a fully coupled degradation process was used to simulate the joint under creep and hot-wet conditions. The model contains four field variables (moisture diffusion, elasticity, creep and thermal and swelling expansion). The adhesive layer was modelled with 3D coupled thermal-displacement elements. The experimental results of bulk adhesive tests were the material input of model and the aged single lap test results of the validated numerical study. Finally, the results of the first step were transferred to step 2 (Han *et al.* 2014 [89]).

In the second step, a bilinear CZM was implemented to simulate the quasi-static tensile test. A quadratic stress criterion was introduced as initiation damage criteria. The adhesive layer was modelled with 3D cohesive elements. The unaged calibration parameters were taken from the experimental data, on the adhesive bulk tensile test and double cantilever beam test. The aged bulk adhesive tests were performed to determine the degraded Young's modulus(E) and degraded Shear modulus (G). The degraded normal stiffness and shear stiffness were calculated by dividing the E and G modulus by the adhesive layer thickness. Also, a linear relation was undertaken between the drop of the tripping traction and fracture energies and the reduction of the strength of the aged bulk adhesive.

The simulation of moisture distribution, stress and damage prediction showed a good correlation with the experimental results. The author proposed that this method can be used to model interfacial failure for bonded joints.

A similar approach was implemented by Liu *et al.* (2016) [90] to simulate the hygrothermal effects on composite double lap shear joints (immersed in water at 90°C for 60 hours). The moisture diffusion was simulated according to Fickian's second law, and two diffusivities were calculated for composite laminated. The diffusion perpendicular to the fibre direction was used in (13.10). The diffusion along the fibre direction was assumed as 1.85 times the diffusivity perpendicular to the fibre direction.

$$D = 0.57^{-\frac{4993}{T}} \tag{13.10}$$

T: Temperature (k).

The moisture diffusion of the joint was modelled in Abaqus through the mass diffusion model. With results of the diffusion analysis, the moisture concentration is transferred as a field variable and output for the hygrothermal stress analysis.

The hygrothermal analysis was performed to determine the strain induced by the thermal expansion and swelling using (13.11):

$$\{\varepsilon\} = \lfloor S \rfloor \{\sigma\} + \{\beta\} \Delta C + \{\alpha\} \Delta T \tag{13.11}$$

$$\{\sigma\} = [Q] \{\varepsilon\} - \{\beta\} \Delta C - \{\alpha\} \Delta T \tag{13.12}$$

A CZM was implemented to evaluate the interface degradation. The unaged and aged adhesive properties were taken from the tensile adhesive experimental data, and the degraded parameters for the cohesive layer were estimated by multiplying a factor:

$$T^h = T_0 \sqrt{\frac{\sigma^h}{\sigma} f} \tag{13.13}$$

T^h: Critical traction with moisture

T_0 : Critical traction without degradation

σ^h: Ultimate strength of adhesive with moisture

σ: Ultimate strength of adhesive without degradation

f: Degrading factor

The numerical model was validated with the experimental data, and either the failure load or the failure mode was predicted in good agreement with the experimental results.

13.2.7 Overview

The durability of adhesive bonded joints depends on factors of design and service conditions, such as materials, dimensions, bonding process and environmental conditions and type of load.

SLJs are frequently implemented due to the simplicity and repeatability. Elastic properties of adherend and adhesive, length overlap and surface preparation, are crucial factors when designing them under shear and bending adhesive stresses. Hence, testing allows one to obtain these properties under environmental effects, comparing materials and dimensions with ease.

Particularly with the moisture, some authors have shown changes of the properties of epoxy adhesive such as variation of plasticity, strength loss or increase of ductility as a result of the effect of humidity absorption. In the same way, the composite adherents absorb moisture, and this can affect the kinetics of water absorption in the adhesive [42]. Furthermore, at temperatures above glass transition temperature *Tg*, the composite matrix properties suffer a continuous degradation [42]. Additionally, the moisture diffusion is affected by the temperature in the degradation process [7]; at high temperature, the diffusion increases, and at sub-zero temperature, the diffusion decreases and the swelling causes polymer interlocking breaks.

Environmental effects on the bonded joints' durability are a concern for the aviation authorities and industry. Joint behaviour over time is still unknown, and without reliable NDT methods to determine joint strength, it is difficult to predict failure; the materials based on epoxies are affected by temperature and humidity due to the process of moisture diffusion (a phenomenon generally described by Fick's law). The moisture causes changes of the properties, plasticization, swelling, microcracks, hydrolysis, debonding and loss of strength. Moisture transport mechanism first occurs through the free volume, and then there are reactions between the polymer chain and the water molecules. In summary, those new bonds induce a reduction in the glass transition temperature by interrupting interpolymer chain Van der Waals bonds.

During the diffusion process, water molecules may form some strong bonds with the molecules of the adhesive. Then, despite a desorption process in the joint, the degradation and reduction of strength may be irreversible.

In a bonded joint, it is important to know and to evaluate the behaviour of adhesive, adherents and interphase. They work as a unique system with respect to any environmental degradation, each individual (laminate, adhesive, bondline interface) behaviour can affect the performance and the degradation mechanism of the joint.

In composite bonded joints, the bondline interface is the area most sensitive to the transportation of moisture. The diffusion through the interface can cause degradation of the bond adhesion between the adherend and adhesive. In contrast with the metal bonded joints, the thermodynamic work of adhesion in composite joints remains positive under humid conditions.

Aeronautical structures encounter hygrothermal (change humidity and temperature) cycles during their operating conditions. In normal conditions on the ground, the moisture is high, and in tropical areas, the temperature can rise to 70°C.

The current NDT such as ultrasonic, thermography, shearography, laser, radiography, acoustic emission, etc., are able to detect defects such as delamination, debond, porosity, void, and crack. But, due to anisotropic material performance, there still are not non-destructive techniques able to detect all kinds of defects in composite bonded joints. Studies have shown the application of ultrasonic testing to the evaluation of degradation of adhesively bonded joints with consistent results.

The environmental degradation models have shown a similar approach methodology, consistent on to predict the moisture effect, thermal effects and mechanical analysis, and using CZM predict the damage based on the degradation of the parameters of the traction-separation law.

Fick's second law is used to describe the moisture diffusion in composites laminates and adhesives. Even in absorption and desorption cycles, using this law is still applied due to the simple solutions and acceptable accuracy. However, this kind of analysis needs to know the moisture diffusion characteristics (diffusivity, coefficient thermal expansion and moisture content in saturation). Usually, the diffusion and moisture content are taken from the experimental test.

The cohesive zone law is a phenomenological model that describes the mechanical fractures with a traction-separation law. Using the CZM, the failure initiation is predictable. Bondline interface of adhesively bonded joints is frequently modelled with CZM.

13.3 Methodology, materials and methods

In this chapter, are described the methodology, materials, equipment and procedures used to develop the research. The research methodology designed is shown as a schematic diagram, which are highlighted in four phases linked between them to meet the objectives proposed.

13.3.1 Research methodology

To investigate the strength and durability of an adhesive joint ageing under a hygrothermal cycle; the methodology has consisted of four phases as shown in Figure 13.3.

Phase 1: The descriptive methodology was used in the preliminary part. The scope of the research is designed based on the literature review; previous studies and results were used to determine parameters for the experimental phase and to identify a numerical model. A campaign of preliminary testing and a familiarization were developed to understand the methodology, get training in the manufacturing

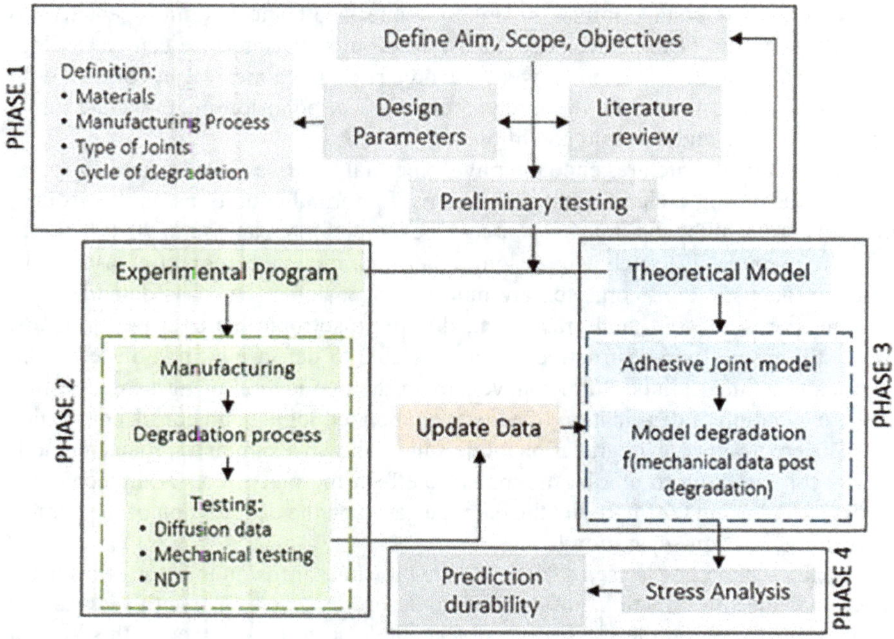

Figure 13.3 Methodology flow chart

procedures, tools and equipment of testing and develop the experimental program (phase 2) based on learned lessons.

Phase 2: An experimental campaign was performed to get data, and these data have been used to validate the model and to study the behaviour of the joint under static and cyclic load after ageing. Absorption characterization of the CFRP and bulk adhesive, mechanical tests and visual and ultrasonic inspection were used to evaluate and explain the joints' behaviour after the environmental attack.

The experimental program has been divided into three main parts: first, manufacture of the SLJs and the bulk adhesive samples; second, consisted of an environmental attack of the joints and specimens and third, a testing campaign was carried out to characterize the sorption materials, the joint strength, material properties, non-destructive inspection of the joints and failure analysis.

Phase 3: Development of an analytical model was undertaken as a parallel activity with phase 2. A numerical model was updated with experimental results from mechanical testing and the cohesive parameters.

Phase 4: The validated model was used to predict the damage and failure of the SLJ subjected to environmental cycles and, finally, the discussion and analysis of the results.

Table 13.3 Autoclave curing process [93]

ITEM	VALUE
Vacuum (bar)	0.94
Autoclave Pressure	Apply 7 bar gauge Reduce vacuum to a safety value of –0.2 bar when the autoclave pressure reaches 1 bar gauge
Rate heating (°C/min)	1–3.0
Holding (min)	120
Temperature T (°C)	180
Cooling (°C/min)	2–5
Venting pressure (°C)	60

13.3.2 Adhered manufacturing

The material prepreg Hexcel T800/M21 was cut in layers of 600 mm × 600 mm, to manufacture two panels of eight plies. The stacking sequence selected was a quasi-isotropic of [0°/45°/90°/–45°]S; according to the literature, a better performance is reached with a laminate quasi-isotropic lay-up, and the optimal distribution contains a 25% plies of 0° layers, a minimum of 10% of 90° and a 50% of plies in ±45 [19, 25, 91, 92].

Before starting the vacuum bagging process, the laminate surface was covered with a nylon peel ply. Peel ply surface preparation was done to improve the mechanical and chemical adherence, and to avoid substrate damages or contamination.

Additionally, a study of different surface preparation treatment of carbon bonded joints, ageing at 55°C and 95% RH for about 33 days, has shown that the peel ply surface treatment provided the strongest bond for the joint; it was attributed to the lower surface energy of the substrates and the thicker polymer layer caused by peel ply compared with grit blasting and plasma treatments [33].

The curing cycle was in an autoclave according to the manufacturer's instructions [93]. An initial vacuum pressure of 1 bar was used to remove air trapped. Then the curing in the autoclave was performed at 180°C for 2 hours. It is explained in Table 13.3.

Finally, the panels, once cured, were cut into sheets of 300 mm × 100 mm as shown in Figure 13.4.

13.3.3 SJL manufacturing

To manufacture the SLJs, a secondary bonding procedure was used. A bonding fixture was used to ensure overlap dimensions and accurate alignment of the joint (see Figure 13.5). Two pins in each border of the substrate were used to fix the overlap area as shown in Figure 13.5(c).

Before bonding, the panels were dried in the oven to remove moisture trapped in the surface. The peel ply was removed at the same time as the bonding process. A debonding strip of polytetrafluoroethylene (PTFE) of 0.1 mm thickness and 2 mm

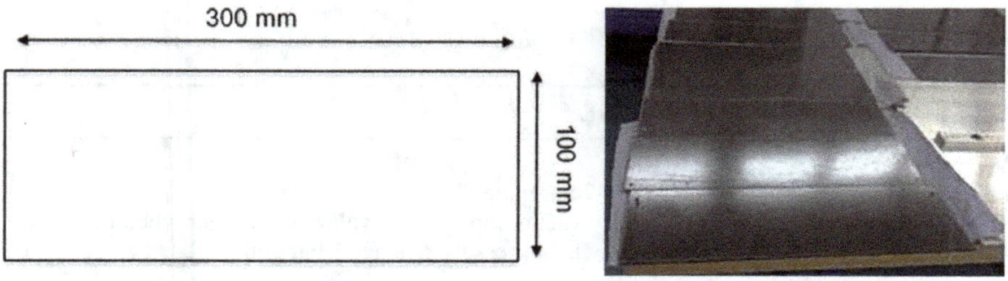

Figure 13.4 T800/M21 UD CFRP sheets before bonding

width was laid in the free ends of bonded joints (see Figure 13.5 (a)). They were used to avoid a fibre Tear failure (FT) [56], and hence to try and provide consistent failure results.

The secondary bonding was made according to datasheet Cytec FM 94 as described in Table 13.4. Althouth Cytec recommends to apply 0.28 MPa of pressure (which condition would require autoclave), the vacuum bag curing was chosen to simulate a repair scenario, as shown in Figure 13.6.

Balancing end tabs were bonded on opposite faces of the joint to minimize the eccentricity of load (NPL National Physical Laboratory Materials Centre, 2007), as shown in Figure 13.7 . Those CFRP tabs had dimensions of 50 mm × 150 mm. They were bonded with Huntsman Araldite 420 A/B, mixing ten parts by weight resin with four parts hardener. Clips were used to get the pressure and fixed the tabs. The cure was made in the oven at 60°C for 4 hours.

Finally, the bonded plates were cut in SLJs of 25 mm width, 175 lengths and 25 mm overlap. Figure 13.8 shows the geometry, and the PTFE crack initiators can be identified. The SLJ geometry is according to ASTM D5868-01 (reapproved 2014) with a variation of adhesive thickness and substrate length.

Furthermore, aluminium tabs were attached to the gripping zone, to prevent tear stress in the specimens in the grip area, during the mechanical testing.

All tabs were bonded with an aerospace adhesive, Araldite 420 A/B. The surface was mechanically abraded with sand paper and cleaned with acetone; to clamp the end tabs, clips were used as shown in Figure 13.9. The liquid adhesive was cured at 50°C for 4 hours in the oven.

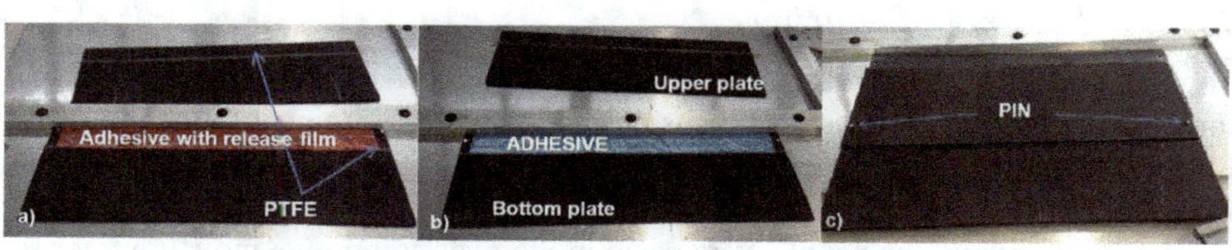

Figure 13.5 Joint bonded process: (a) PTFE tips and film adhesive setup, (b) before bonding and (c) final assembly bonded joint

Table 13.4 Curing adhesive process [96]

Properties	UNIT	Value
Pressure	MPa	0.1
Rate heating	°C/min	1.7–2.8
Holding	min	60
Temperature	T (°C)	121

13.3.4 Hygrothermal cycles

Flight operation conditions cause composite joint structures to be subjected to different environments, such as the combined effect of temperatures, dry and wet atmosphere [41]. Low temperatures occur at high-altitude flight operation or ground storage during cold weather. According to MIL-Handbook 17, the majority of commercial aircraft components are designed to the maximum temperature of 71°C and 85% RH, without taking into account components close to engines or landing gear [17].

An environmental chamber was used to simulate the conditions of temperature and humidity. Multiple cycles of heating to 70°C, with 85% RH holding for 4 hours, then cooling to –20°C holding for 2 hours and then heating again was carried out continually. The cycles are presented in Figure 13.10.

Joint samples and bulk adhesive samples were conditioned until a specific number of cycles, when they were removed to be tested under static and fatigue load. Following the approaching by [59], the time intervals defined to evaluate the behaviour were 42, 84, 168, 252 and 336 cycles.

13.3.5 Bulk adhesive specimens manufacturing

In order to perform tensile tests and moisture absorption tests, there was a need to manufacture sheets of bulk adhesive (1 mm thickness). Hence, a lay-up of four layers of FM94, each one of 0.25 mm thickness, were stacked together and cured.

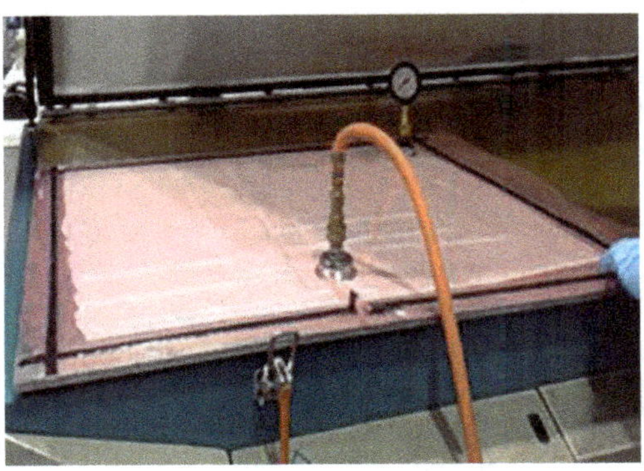

Figure 13.6 Vacuum bag for joints bonding

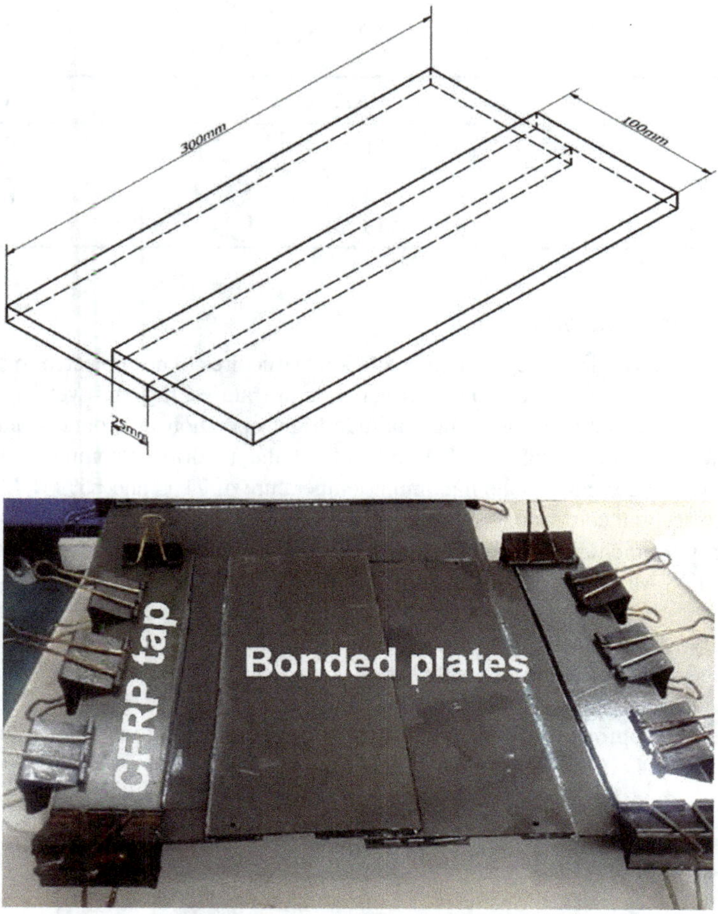

Figure 13.7 *CFRP end tabs bonded: (a) bonded plates'schematic and (b) bonded plates*

13.3.5.1 Preliminary test

Studies with FM94 adhesive were not found in the literature; however, there is widely research about FM73 film adhesive; it is produced by the same manufacturer CYTEC, hence others researchers' experiences were used to identify the optimal process to achieve the bulk adhesive samples.

- **Closed mould technique:** To manufacture FM73 bulk adhesive specimens, Mubashar [64] used a metallic close mould and hot pressure represented in Figure 13.11. For that reason, a closed mould technique was implemented with FM94 adhesive. Layers of 150 mm × 150 mm were cut, stacked and pressed using a steel roller to reduce the voids between layers. The rectangular metal

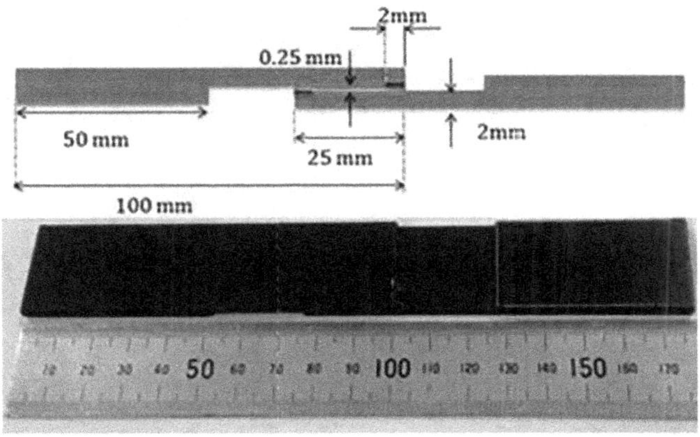

Figure 13.8 *Geometry of the SLJ specimen and final specimen*

mould was used to keep the flow during the curing. To close the mould, two metallic plates were fixed in the bottom and top. Films of PTFE were used as release to avoid adhesive sticking to the mould, as seen in Figure 13.11. Mould was made to cure the adhesive [64].

The adhesive cured at 121°C for 60 minutes, and the cured process was made in a GE more cure hot pressure [94]. Finally, the bulk adhesive plate cured presented a thickness controlled but a high voids content.

- **Degassed prior to close mould:** Liljedahl [68] employed a degassing process to minimize the amount of voids of FM73 bulk adhesive specimen. This procedure was hence used to manufacture the FM94 bulk adhesive. The layers stacked were placed in a vacuum oven, the adhesive was heated and degassed at 80–90°C per 10 minutes and the procedure was repeated three times. Figure 13.12 shows the process of degassing and how the air trapped comes up.

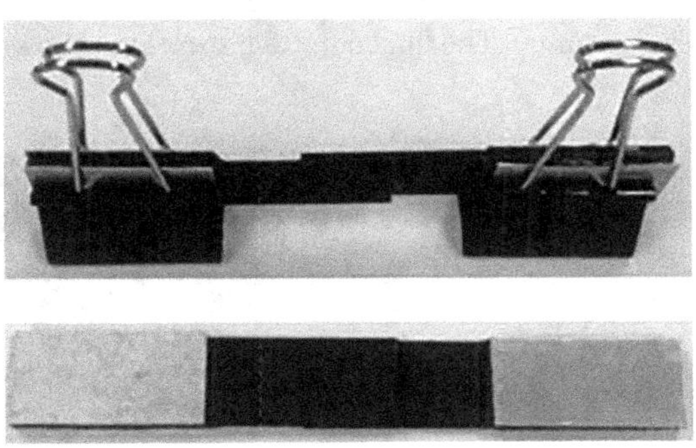

Figure 13.9 *The upper photo shows clamping before curing. The bottom photo shows the final specimen.*

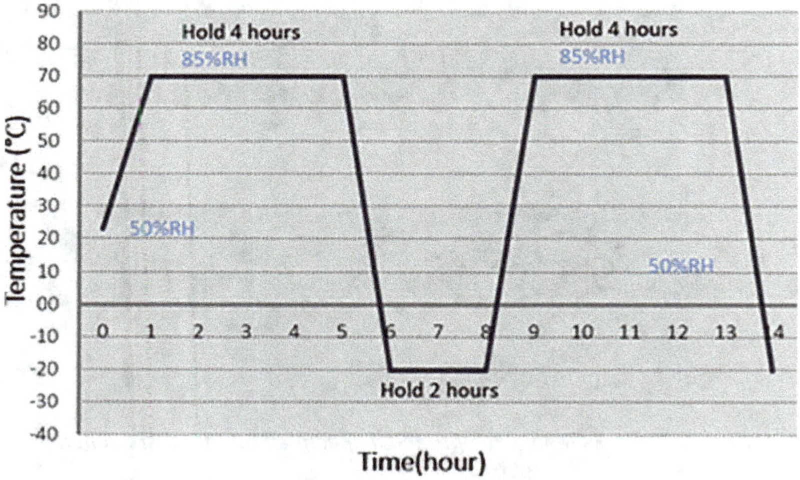

Figure 13.10 Cycle of temperature and humidity

During the degassing process, the adhesive under vacuum looks like a foam. Once the vacuum was removed, the air bubbles disappeared and the adhesive shape was recovered.

The degassed adhesive was set in the metallic mould. The mould was closed with the top plate. The adhesive was cured under hot pressure as described above. However, the specimen contained bigger voids than the technique without degassing.

- **Opened mould cured:** Four adhesive layers of 50 mm × 160 mm were stacked (to get one thickness bulk adhesive) and placed on a glass plate. Spacers were used to control the flow during the curing. The adhesive was cured at 121 °C for 60 minutes. As a result, the amount of voids was reduced and the sheet remained at a thickness of 1 mm. Also, the process resulted in a flat surface across the entire specimen. The final adhesive sheets are shown in Figure 13.13.

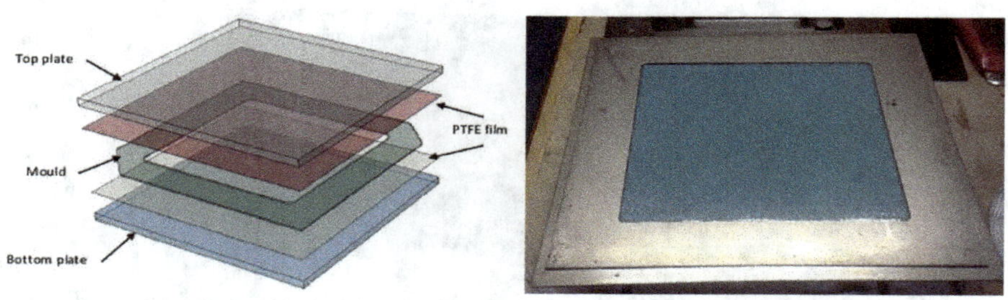

Figure 13.11 Mould made to cure the adhesive [64]

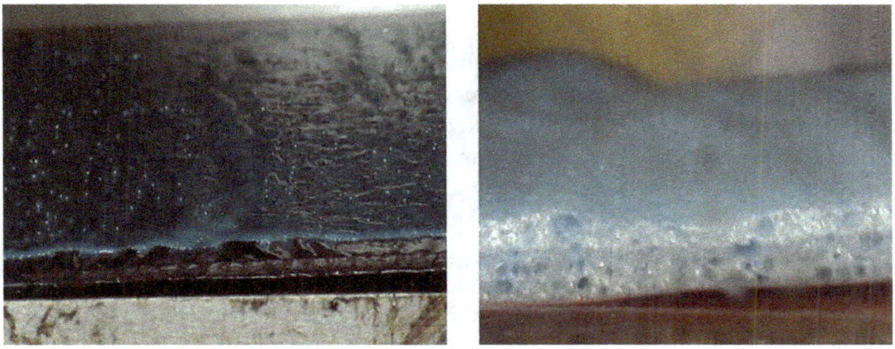

Figure 13.12 *Left: adhesive starting to degas. Right: adhesive during the degassing process.*

- **Degassed prior to cure in an opened mould:** Furthermore, a degassed adhesive and opened mould curing process was tested. The adhesive was degassed in the vacuum oven for 10 minutes at 80–90 °C. Then, the adhesive was placed on a glass plate and cured at 121 °C for 60 minutes. The sheet showed several voids.

13.3.5.2 Bulk adhesive sheet manufacturing

Once defined the opened mould cured was the method to produce the bulk adhesive sheet, six bulk adhesive sheets (50 mm × 160 mm dimensions) were manufactured. As was mentioned above, four adhesive layers were stacked over a glass (cover with PTFE tape to avoid sticking). The adhesive sheets were cured, and samples were cut with dimensions 150 mm × 10 mm.

Lastly, specimens after cutting were polished to avoid sharp changes in the borders. However, it is important to highlight that they presented a notable amount of voids.

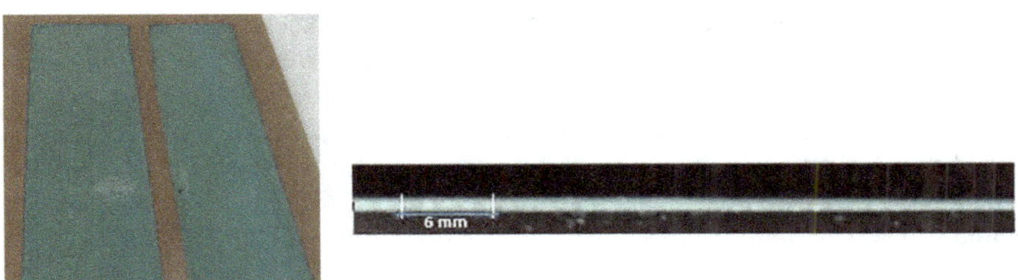

Figure 13.13 *Left: adhesive cured in opened mould. Right: adhesive cured in open mould.*

13.3.6 Materials' characterization

In order to characterize the moisture diffusion and mechanical properties of materials, specimens of the composite adherend and bulk adhesive were prepared. The specimen dimensions were defined based on standards for plastics: determination of water absorption BS EN ISO 62:2008 and determination of tensile properties from the adhesive BS EN ISO 527:2012.

The composite laminate mechanical properties were taken from the literature as described later in this section.

To make the adhesive bulk sheets free of voids was a significant challenge during the specimen's manufacturer. Several techniques were investigated to obtain a specimen with a number of voids negligible.

Several Fick's behaviours have been reported for both epoxies and carbon fibre composite with epoxy matrix at a temperature below the *Tg* [43, 67, 68]. Hence, a Fick's law solution has been implemented to determine the diffusion coefficient and to predict the moisture concentration in the bulk materials.

13.3.6.1 Gravimetric test

In order to evaluate the moisture diffusion in the adhesive and CFRP substrates, gravimetric testing was done according to ASTMD5229/D5229M-14 [95]. The moisture uptake during the cycles was determined to calculate the diffusion coefficient 'D' and the moisture concentration '*c*' at saturation. Hence, bulk adhesive and CFRP T800/M21 samples, with dimensions of 50 mm × 50 mm and 1 mm thickness, were conditioned in the environmental chamber with the same hygrothermal cycles used with the joints.

Under dynamic conditions of moisture and temperature, as was exposed in section 13.2.4.1, the moisture diffusion is directly influenced by the temperature. For that reason, it was considered to carry out the moisture characterization under two different conditions for the adhesive bulk. One series of specimens were removed from the environmental chamber at a dry low temperature. Another group of bulk adhesive samples were removed at hot-wet conditions. In the case of CFRP samples, they were removed only during hot-wet conditions. Procedures made for drying and weighing were the same for all of the samples (Table 13.5).

Initially, to determine the initial weight, all the specimens were cleaned with acetone and stored in a desiccator for 2 hours before they were weighed. Following that, the samples were dried in an oven at 50°C for 24 hours, and then specimens were removed and cooled in a desiccator before being weighed again. After weighing, they were returned to the oven, and this procedure was repeated until the mass reached the constant value. This final constant value will thus be considered the initial weight condition [65, 96]

To measure the moisture gained, all pre-dried samples were conditioned in an environmental chamber under the hygrothermal cycle. Heating until 70 °C-85% RH and holding for 4 hours, following the temperature is reduced to about –25°C for 2 hours. Each cycle takes 8 hours and runs continuously.

Table 13.5 Characterization moisture samples' description

Material	Dimensions	#Sample	Conditioning: Environment	Conditioning: Remotion for testing	Samples name
Adhesive FM94	50 mm x 50 mm, 1 mm thickness	7	Heating up 70°C/85% RH holding for 4 hours. Cooling until −20°C holding for 2 hours. Cycles 8 hours each.	About 70°C/85% RH	A1-A4
				About 10°C	A5-A8
T800/M21 UD CFRP	50 mm x 50 mm, 2 mm thickness	4		About 70°C/85% RH	C1-C4

The samples were periodically removed from the environmental chamber. They were kept in a bag sealed while the samples are moving to the balance area. Samples were wiped with a cloth to remove the moisture from the surface, and then they were weighed immediately on a balance with an accuracy of 0.01 g. Ending the weighing, the samples were closed in the bag again and moved to the environmental chamber.

The frequency of the gravimetric testing decreased gradually as the rate of gain of moisture diminished. Hence, for the first day, the samples were weighed three times. Then for the first week, two readings were taken for each day, and in the second week, one reading per day, followed by a reduction of the frequency as the rate of weight gain is smaller [65].

The per cent of mass change was calculated using (13.14) for each time interval. The measures were taken over a period until the moisture per cent achieved the equilibrium. The equilibrium or saturation is reached when the average moisture content changes by less 0.02% over at least three consecutive time intervals [95]:

$$w_t(\%) = \frac{w_t - w_0}{w_0} \times 100 \tag{13.14}$$

where w_0 is the mass of the test specimen (mg) after initial drying and before moisture exposition. w_t is the mass of the test specimen (mg), after moisture exposition.

The moisture content per cent wt(%) of each sample is recorded and plotted versus $\sqrt{(t/l)}$, t being the time of exposure and l the thickness of the sample.

The diffusion coefficient 'D' was calculated from the slope of the curve. The methodology to determine the diffusion coefficient is described later in section 13.3.4.2.

In addition, after weighing the samples removed periodically from the environmental chamber, the length and thickness were measured with a calliper and micrometre, respectively.

Lastly, descrption testing was performed; the samples preconditioned and saturated were weighed prior to start, heating to 75°C in a controlled temperature oven. The samples were removed and cooled in a desiccator, wiped and weighed. Then, they were returned to the oven for over 24 hours. The percent mass loss was calculated with (13.14), and the procedure was repeated until the per cent of change was less than 0.1%. The moisture loss as a function of √time drying is plotted, and the results are compared with the mass moisture gained.

13.3.6.2 Diffusion coefficient

Fick's law has been widely used to describe the moisture diffusion in composite and polymeric materials, as mentioned in section 13.1.6. Fick's second law of diffusion has been developed to predict a non-steady-state diffusion and for diffusion in three dimensions:

$$\frac{\partial C}{\partial t} = D_z \left(\frac{\partial^2 C}{\partial z} \right) \tag{13.15}$$

According to ASTM *D5229/D5229M-14,* the moisture content (percentage) in the material, as a function of time and temperature, can be predicted with (13.16) [95]:

$$M(T,\ t) = M_b + G(T,\ t)\left(M_m - M_b\right) \tag{13.16}$$

where T = temperature °C

M_b = moisture at start of experiment (sample after oven-dry), % (taken as zero)

M_m = moisture equilibrium content (%)

h = average specimen thickness (mm)

$G(T,t)$ = moisture absorption function (13.17):

$$G(T,t) = 1 - \exp\left(-7.3 \left(\frac{D_h(T) \times t}{h^2} \right)^{0.75} \right) \tag{13.17}$$

where, as all terms identified above except t = time

h = effective moisture content.

$D_h(T)$ = diffusion coefficient at temperature specific

The method to characterize the moisture absorption is described in section 13.3.6.1. The weight of every sample was recorded at different predetermined times.

illustrates a typical curve of moisture content fitted to the Fick diffusion law. Initially, the curve shows a linear acceleration increase of moisture content. This initial moisture gained is highlighted with a discontinued red line in . Gradually, the absorption rate decreases as long as the moisture uptake approaches a maximum value. The maximum moisture content is known as the saturation level (*Mm*). In , this level is highlighted as *Mm* horizontal line according to Reference 97.

The moisture diffusivity constant D_h is calculated by the slope of the initial straight line using (13.18)[95]:

$$D_h = \pi \left(\frac{h}{4M_m} \right)^2 \left(\frac{M_2 - M_1}{\sqrt{t_2} - \sqrt{t_1}} \right)^2 \tag{13.18}$$

where:

D_h = diffusion coefficient in the thickness direction.

h = average specimen thickness (mm), and the fraction is the slope of linear portion of moisture absorption versus time 1/2.

M_m = effective moisture equilibrium content.

M_1 = moisture value 1 at time 1 (%).

M_2 = moisture value 2 at time 2 (%).

t_1 = time value 1 (hours).

t_2 = time value 2 ($t2 > t1$).

For determining the linear slope, two points were selected. For this study;

*M*1 is assumed as baseline moisture content (*M*1 = 0%) at *t*1 equal zero. *M*2 is considered to be equal to *Mm*, with the corresponding time (*t*2). *M*2 value is the point of interception between the straight line and the horizontal *Mm* line.

The equilibrium moisture content (*Mm*) was obtained calculating the average between the maximum and minimum moisture content value after reaching saturation.

Finally, Wong [44] states that a Fickian diffusion behaviour can be recognized, by analysing predicted moisture uptake *M*(*T*, *t*) against $t^{1/2}$. The graph should fulfil the next three rules:

- There is an initial linear stage of absorption; it has to achieve at least up to 60% of *Mm*.
- After the straight line (initial linear slope), the absorption curve shows a concave zone; it extends to reach the equilibrium (*Mm*).
- Plots of moisture uptake against $t^{1/2}$/thickness are superimposable at different thicknesses for the same material.

13.3.7 Mechanical properties

13.3.7.1 Composite CFRP

The T800/M21 UD ply properties have been taken from the literature review [98] (Table 13.6).

Table 13.6 268 gsm^{-1} T800/M21 UD properties

	E1	**134,700**
	E2	**7,700**
Elastic modulus (MPa)	**E3**	**7,700**
Poisson's ratio	v23	0.5
	v12 – v13	0.37
Shear modulus (MPa)	G12 = G13	4.200
	G23	2.500

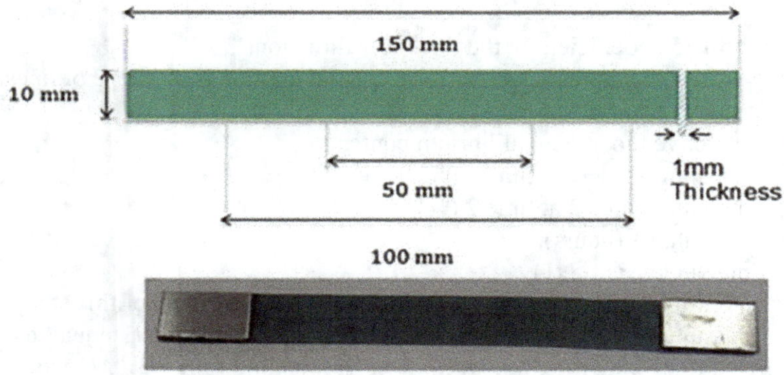

Figure 13.14 Specimen type II geometry [100], above the specimen

13.3.7.2 Adhesive FM94

Adhesive samples type II (dimension of 10 mm × 150 mm) were cut from the sheet of cured bulk adhesive. The bulk tensile test followed the standards EN ISO 527-1:2012 and EN ISO527-3:1996 [99, 100]. The test was carried out to determine the materials' tensile strength of unconditioned and conditioned bulk adhesive specimens. The tensile strength results were used to identify the effect of diffusion moisture on adhesive mechanical properties. Also, using stress–strain curves, the elastic modulus (*E*) values were determined for the specimens and the effect of hygrothermal cyclic on this modulus.

Before conditioning the samples to environmental attack, aluminium end tabs were bonded using Araldite 420 A/B liquid adhesive. The cure was carried out at room temperature for 7 days. The geometry description and a sample can be seen in Figure 13.14.

Four tensile samples of each set were tested. Unaged specimens were tested to identify the initial condition. Aged specimens were conditioned at different periods in cycles (weeks): nought, 84 (four) and 168 (eight). Specimens were not tested at 42 and 252 cycles because of the low rate of changes presented during the experiments.

Immediately before the tensile test, the specimens were removed from the environmental chamber and dried with a wipe. The testing was done at once to avoid change of moisture in the specimen. The tensile test was carried out in the universal machine under a maximum load of 10 kN with a rate of 0.5 mm/minute. The displacement was measured using a laser extensometer.

13.3.7.3 Adhesively bonded SLJ tests

13.3.7.3.1 Lap shear test

Shear lap testing was conducted to determine the effect of degradation process in the strength of the joints. The tests were done according to the standard test method ASTM D5868-01, 2001. The Electro-Mechanical Universal Machine, Instron 5500R with 30-kN load cell, was used for testing. A laser extensometer was employed to

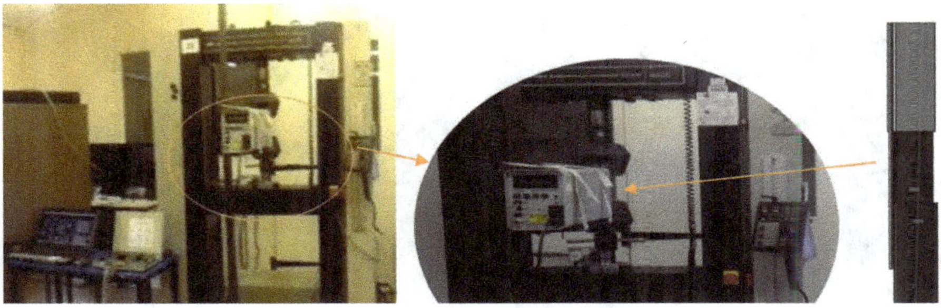

Figure 13.15 *Single lap shear test configuration: (a) general view, (b) laser extensometer and (c) SLJ schematic position in front of the laser extensometer*

measure the deformation of the joint, as shown in Figure 13.15. The test ran out until failure with a displacement rate of 1 mm/minute. At least three samples were tested by each cycle determined. The maximum failure load was used to define the load levels of the fatigue test

13.3.7.3.2 Fatigue test

The fatigue testing was performed according to the NPL Guide Fatigue and Creep testing of adhesives and thermoplastic joined systems [101], using a Denison Mayes-Group servo-hydraulic machine. Unaged and aged specimens were cycled up to 10^6 cycles under a sinusoidal wave with constant amplitude, stress ratio (R) equal to 0.1 and a frequency of 5 Hz.

Each set of specimens was removed from the environmental chamber once reaching the time to be tested. Then, the joints were stored in a hermetic plastic box under moisture condition (50% RH) and room temperature. The moisture condition was monitored with a thermos hygrometer DC 102.

13.3.7.4 Non destructive inspection –NDI

In order to characterize the adhesive joint quality (pre- and post-ageing), two methods were performed: first, visual inspection using an optical microscope at low and high magnifications, and ultrasonic C-scan testing.

13.3.7.5 Microscopy evaluation

To observe voids and changes caused by ageing, one control sample was observed pre- and post-condition. The control sample was removed every 2 weeks and observed by optical microscopy using a NIKON Optiphot microscope, under reflected light mode at different magnification lenses (2.5X, 5X 10X).

The bonding edges were polished with an abrasive paper, and they were inspected under microscope NIKON Eclipse and ME600 at different magnification

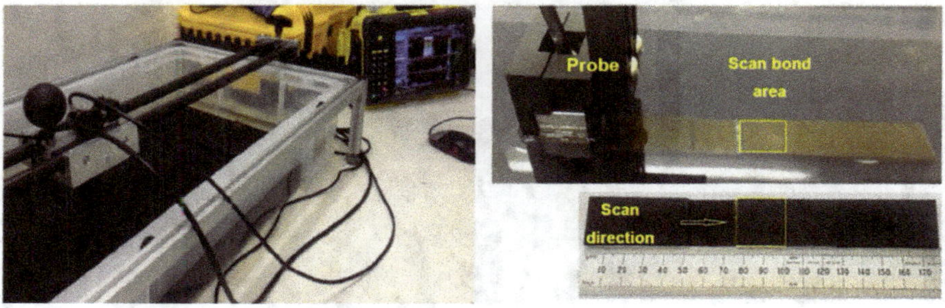

Figure 13.16 *Ultrasonic C-scan through immersion (a) scanning setup, (b) probe and joint arrangement and (c) scheme of scan direction and area*

lens. Furthermore, a fracture mechanical analysis was performed to evaluate and identify the failure mode of the joints.

13.3.7.6 Ultrasonic inspection

As mentioned in the literature, ultrasonic technique (UT) is used to evaluate bonded joint and to detect bond defects such as debond, cracks or porosity [62, 102, 103]. Also recent studies have undertaken UT to evaluate degradation of adhesively bonded joints [102].

Based on these studies, an immersion technique was implemented with the Sonatest Ultrasonic C-Scan model VEO 16:64, and an arrangement consisting of a probe type phased array of 5 MHz, a square pulse width of 100 ns, 64 elements and a gain of 31 dB and gain reference of 23.9 dB.

A control sample was used as a representative specimen, which was periodically removed from the environmental chamber and inspected to characterize the ageing effect on the bond quality.

Ultrasonic C-scans were taken of unaged joint and aged joints (at 84,168 and 252 cycles). The joint was removed from the environmental chamber each time interval, inspected by C-scan and returned to the chamber until the following evaluation; the arrangement of the ultrasonic through immersion of the SLJ is shown in Figure 13.16.

The ultrasonic signal penetrates the adherend and reflects the first interface adherend/adhesive and reaches the second interface and the back of the second adherend.

The C-scanned images from the bond region (yellow square, Figure 13.16) were reached by setting the gate on the reflections from the two interfaces and the back face of the second adherend. The C-scan image therefore is the mapping of the amplitude signal; the attenuation of amplitude signal is interpreted as the degree of bonding.

13.4 Experimental results

13.4.1 Introduction

The experimental programme to evaluate the behaviour of an adhesive composite joint exposed to environmental attack was described in Chapter 3. Moisture characterization and mechanical tests have been performed on bulk materials and the composite SLJs. The results of these tests are presented in this chapter.

13.4.2 Moisture characterization

In this work, three to four samples of each material, T800/M21 laminate and FM94 bulk adhesive were tested to determine the moisture diffusion under hygrothermal cyclic conditions (temperature and moisture). Diffusion coefficient results and Fick's model analysis are described in this section, according to the procedures of section 13.3.6.1.

13.4.2.1 FM94

In order to know how the temperature and humidity cycles influence the moisture diffusion, the gravimetric test was performed at two different conditions. The procedure was performed only for FM94 samples.

In the first series, four specimens were removed from the environmental chamber at hot-wet condition stage (about 70°C/85% RH). In the second series, samples were removed from the environmental chamber at cold-dry condition stage (about −10°C).

First series: The gravimetric test results of FM94 samples removed from the environmental chamber at hot-wet conditions (about 70°C and 85% RH) are illustrated in Figure 13.17. The graph shows the moisture uptake (wt %) vs the square root of time for 1 mm thickness; the moisture absorption of all specimens showed a similar trend. Initially, the moisture content rises linearly with time. The rate of absorption is faster during the first period, 24 hours (4.90 $h^{0.5}$), and then the increment gradually goes down to reach a stable moisture content. The maximum moisture content was obtained after 120 hours (10.95 $h^{0.5}$) exposure time.

After reaching the equilibrium moisture content, there were some peaks in the value moisture absorption. The peaks were presented at 18.97 h0.5, 22.45 h0.5 and 25.92 h0.5. The changes can be attributed to variants during the gravimetric measurement because although the samples were extracted from the chamber at about 70°C and 85% RH, it is possible that they were taken at different times of the cycles (beginning of the hot-wet stage or after some hours under hot-wet condition). These variations could affect the amount of water adsorbed by the surface, slightly affecting the measurement of mass gained.

The gravimetric test was performed over a much longer period than that to achieve the saturation; the last measurement was greater than 40 h0.5 (1,600 hours). The moisture uptake after this long period does not show sharp changes.

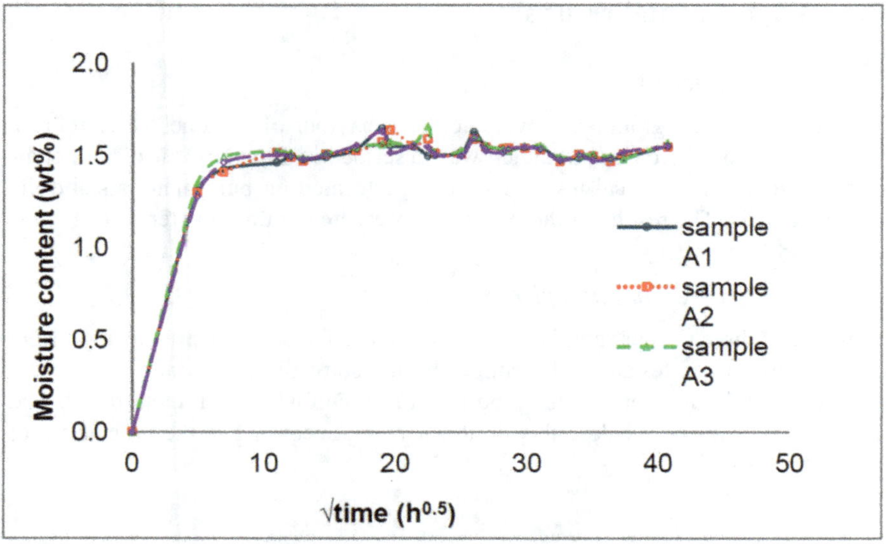

Figure 13.17 Moisture absorption for FM94 bulk adhesive (1 mm thick), samples removed at hot-wet condition

To calculate the diffusion coefficient (D_h) and moisture concentration at saturation (*Mm*), the procedure of section 13.3.4.2 was followed. Figure 13.18 shows the moisture uptake average against the √time, for 1 mm thick. The standard deviation for each data point was calculated and represented by the error bars; the average value for the standard deviation was 0.02, which means a low disparity between the data.

The saturation average value (*Mm*) was determined between the minimum and the maximum moisture content. These limits (minimum and maximum) are illustrated with brown and green dashed lines, respectively. The saturation content (*Mm*) obtained was 1.54%, and it is represented as an orange horizontal line. The diffusion coefficient (D_h) calculated with the (13.18) was 5.91 × 10–3 mm^2/hour, and it is represented as the straight line slope (red line).

Fick's model was performed using the values D_h and *Mm* in 4.2 [95]. The M(*T*, *t*) values were graphed vs √time, as shown in Figure 13.19. The *M*(*T*, *t*) curve obtained was compared to the experimental data, showing a good correlation of the experimental results. The prediction of moisture content was satisfactory.

Analysing *M*(*T*, *t*) based on [44], it can be stated that: first, the moisture absorption at the initial straight line achieved up to 1.31% mass uptake equivalent to 85% of *Mm* (moisture equilibrium). It was measured at 24 hours (4.9 $h^{0.5}$).

Second, the concave zone can be seen after the main increment of moisture is absorbed. This portion of the curve has slow moisture content increment after 24 hours and equilibrium at 120 hours (10.95 $h^{0.5}$). Third, the moisture absorption

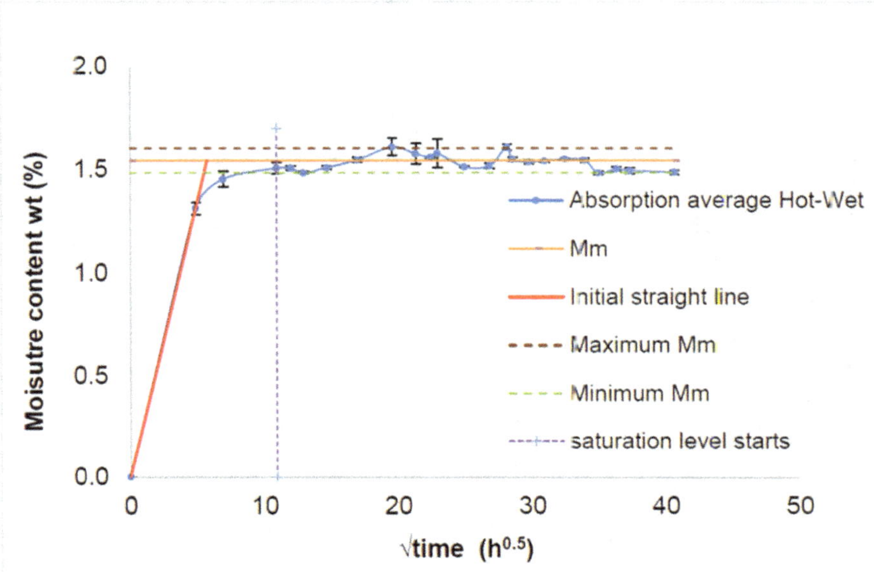

Figure 13.18 *Moisture absorption average FM94 bulk (1 mm thick), at hot-wet condition*

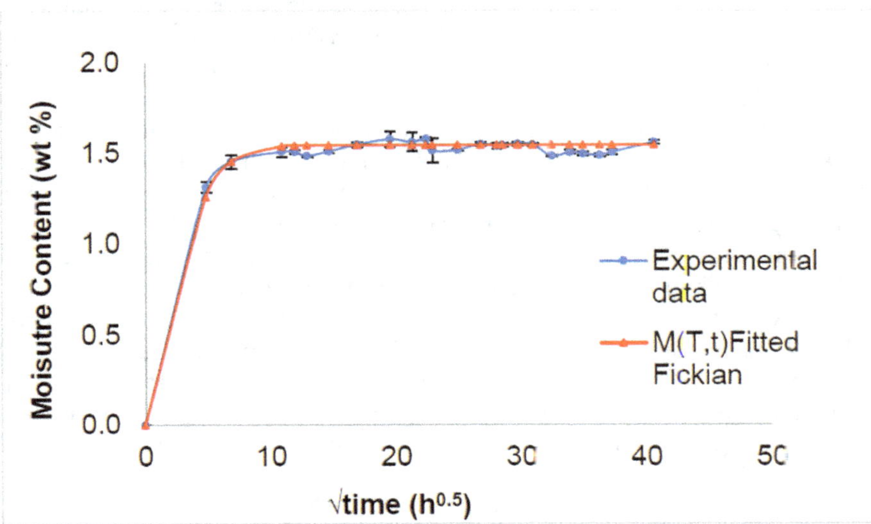

Figure 13.19 *Predicted moisture absorption (1 mm thick) at hot-wet condition compared with experimental moisture uptake*

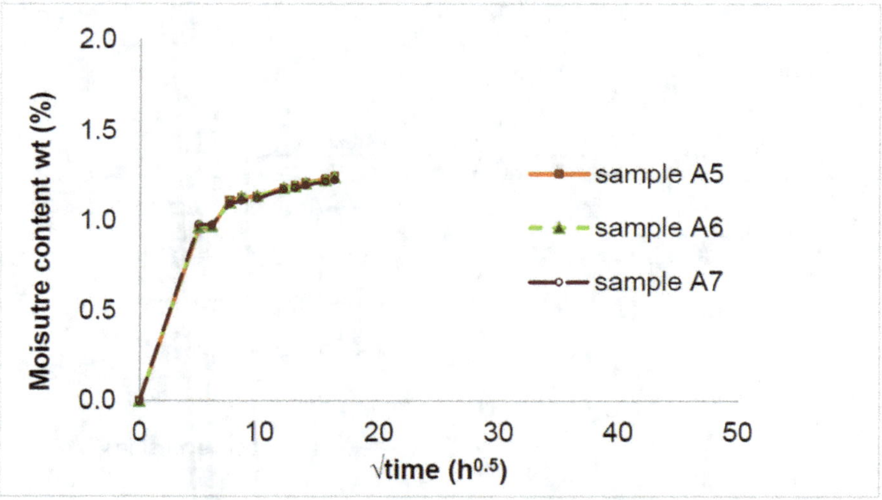

Figure 13.20 Moisture absorption for FM94 bulk adhesive (1 mm thick), at colddry condition

content was determined only at one thickness (1 mm). For that reason, a superimposable curve for different thicknesses cannot be confirmed.

In brief, two of the three requirements were accomplished. It can be concluded that the absorption was governed by Fick's diffusion law under hot-wet conditions.

Second series: Three bulk adhesive samples were gravimetric tested to get the moisture diffusion at cold-dry. The samples were removed from the environmental chamber at about –10°C. This test was carried out in less time than specimens removed at hot-wet conditions.

Figure 13.20 shows the moisture uptake over the time exposure for bulk adhesive samples (1 mm thick). The three samples showed similar diffusion behaviour to each other and to samples removed at hot-wet conditions (first series). The major moisture absorption occurred in the first stage before 24 hours (4.9 $h^{0.5}$).

In contrast to the hot-wet condition, the concave zone was not clearly observed; after 36 hours (6 $h^{0.5}$) of exposure, the absorption exhibited a very small increment with respect to the initial absorption and following a sharp increase at 60 hours (7.48 $h^{0.5}$). Those changes in the rate of absorption can be attributed to the short time interval between the first and the second reading. The moisture content tended to level off after 56 hours (7.48 $h^{0.5}$) – change of mass gained about 0.01%. According to ASTM D5229, the equilibrium state is reached when the moisture content changes less than 0.02% [104].

Figure 13.21 illustrates the moisture average content for the samples removed at the cold-dry condition; the error bars refer to the standard deviation at each data point. The standard deviation was 0.02, meaning a good repeatability of the test with a good control of the factor controlled by the researcher. The maximum moisture

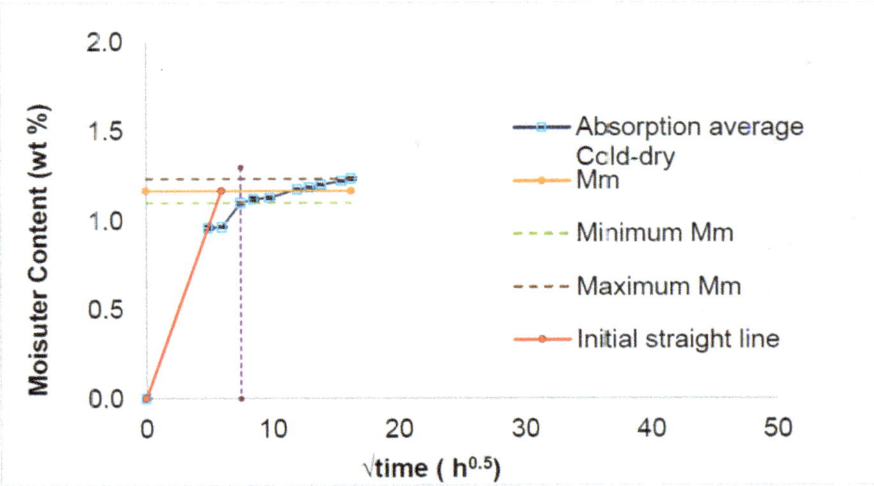

Figure 13.21 *Moisture absorption average FM94 bulk (1 mm thick), at cold-dry condition*

concentration defined was 1.17%, and the diffusion coefficient (D_h) calculated was 5.60×10^{-3} mm²/hour.

Figure 13.22 shows the predicted moisture absorption (1 mm thick) in cold-dry condition compared with experimental data, the Fick's fitted curve showed a good

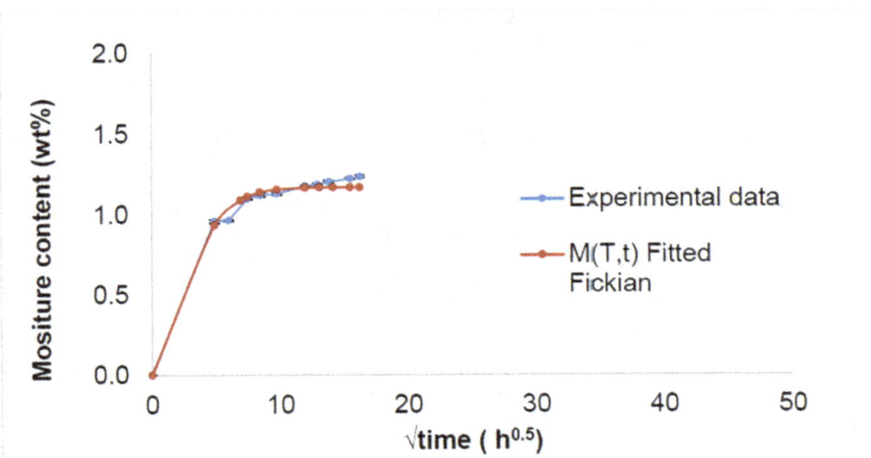

Figure 13.22 *Predicted moisture absorption (1 mm thick) at cold-dry condition compared with experimental data*

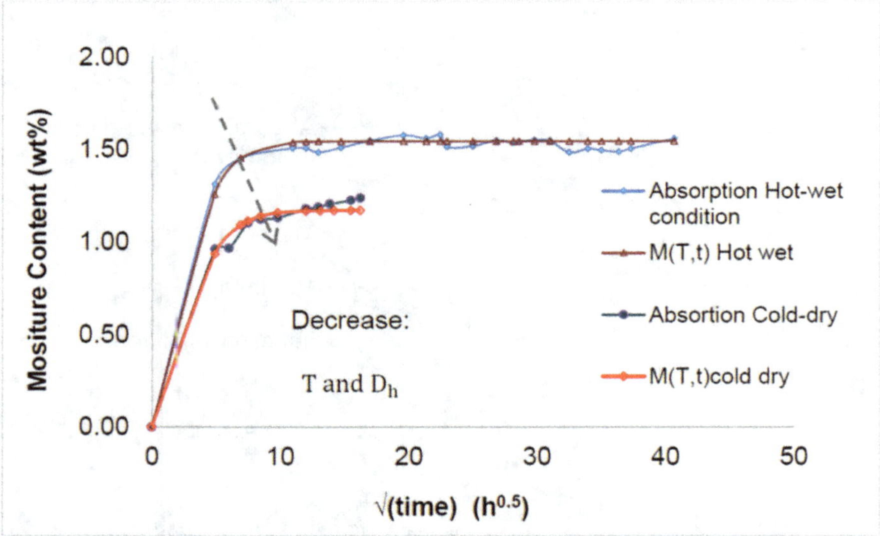

Figure 13.23 Moisture diffusion FM94 bulk, samples were taken out at hot-wet and cold-dry conditions

correlation with the experimental results (red line). However, an underestimated prediction of moisture concentration can be seen after 144 hours (12 $h^{0.5}$).

Analysing Fick's characteristics: the initial absorption (linear stage) climbed up 0.97% at 4.9 $h^{0.5}$ (24 hours); this value equals 83% of the maximum concentration M_m. Additionally, there is a concave portion before the saturation level. There are no experimental data for other thicknesses; hence, it cannot confirm the superimposable curve. In conclusion, similar to the diffusion under hot-wet conditions, moisture diffusion followed the second Fick law.

In fact, moisture diffusion of FM94 subjected to the hygrothermal cycles obeyed Fick diffusion behaviour. Here, changes in the diffusion rate can be observed between hot-wet and cold-dry conditions; these changes have evidenced diffusion dependence on the temperature. Temperature reductions caused a drop of diffusivity and in the maximum moisture content. It can be seen in Figure 13.23 where a comparison of moisture uptake at hot-wet and cold-dry is illustrated.

Table 13.7 summarizes the results for bulk adhesive FM94, 1 mm thickness. The diffusion rate at cold-dry was 5.3% slower than the diffusion determined for samples removed from the hygrothermal cycle at the hot-wet condition. This may be attributed to the increase of free volume, due to both molecular movements and thermal vibrations.

The maximum concentration achieved at cold-dry was 5% smaller than at hot-wet condition. That can indicate that the maximum moisture content was driven by the relative humidity and the temperature.

Table 13.7 FM94 moisture absorption comparison between hot-wet and cold-dry

Condition removed from environmental chamber	Mm %	Dh mm^2/h
70°C/85 %RH	1.54	5.91E-03
About −10°C	1.17	5.60E-03

Indeed, the diffusion regarding the hygrothermal cycle studied shows a dynamic behaviour. In other words, moisture diffusion content oscillates between an upper and lower value. A schematic description of the hygrothermal ageing cycles and the diffusion is shown in Figure 13.24. It can be seen during the 4 hours at high temperature and humidity (70°C/85% RH), the coefficient diffusion and the moisture uptake are higher, while at freezing conditions (−10°/0% RH), the diffusion and moisture content decrease.

13.4.2.2 Composite laminate T800/M21

Four T800/M21 UD CFRP samples were gravimetrically tested to determine the moisture diffusion parameters. Tests were performed following the same procedure described in section 13.3.4.1. In the case of composite specimens, the samples were only removed from the chamber at the hot-wet stage (about 70°C/85% RH).

The mass uptake, for all samples, as a function of the √time is illustrated in Figure 13.25. The moisture absorption profile was similar for the four specimens.

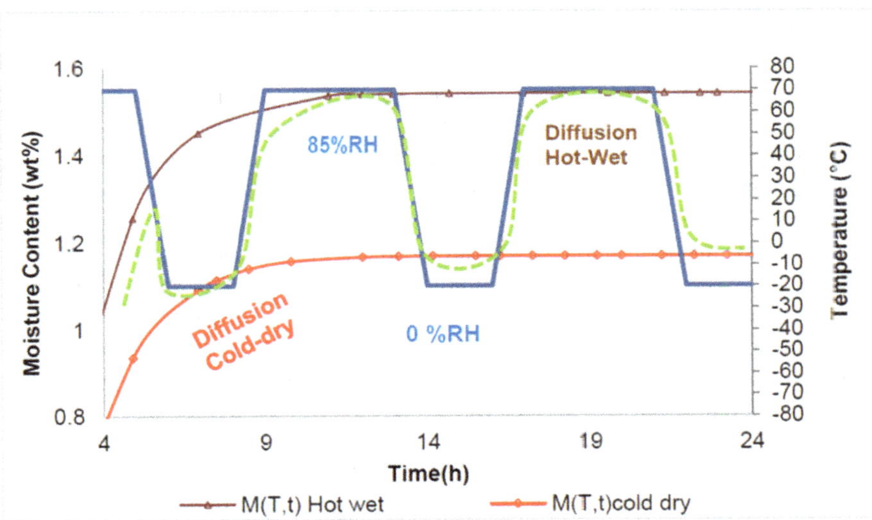

Figure 13.24 Schematic cyclic diffusion FM94

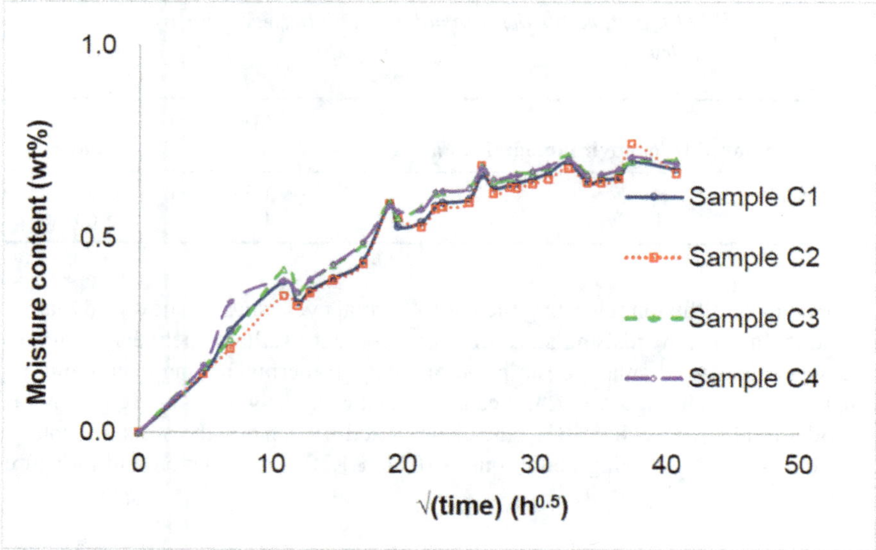

Figure 13.25 CFRP moisture absorption samples (2 mm thick)

Initially, moisture linearly rises with the root square of time; this rapid increment reached 10.95 h0.5 (120 hours). Later, a steep drop occurred, and the moisture gained decreased up to 10% at this point. The absorption continued to go up slowly until it reached a stabilization. There were drastic changes and peaks of moisture uptake as long the time. Reminding that samples were exposed to temperature and moisture cycles, the changes in the moisture gained are not unexpected as the conditions were changing.

The moisture average is shown in Figure 13.26; error bars indicate the standard deviation for each data point. The standard deviation was 0.01% that indicates low dispersion of data between samples. The spread of moisture absorption data under long time exposure can be attributed to another mode of interaction; the adsorption on the surface of CFRP.

In general, the absorption was steady until it reached the apparent moisture equilibrium at 25.92 $h^{0.5}$ (672 hours); after that point, the moisture uptake fluctuated between a maximum and minimum value. The T800/M21 saturation level average (*Mm*) was 0.68% (samples removed from the chamber at hot-wet conditions). The maximum and minimum values are represented in Figure 13.26 with purple and green discontinuous lines, respectively.

Due to the drastic changes during the initial stage of absorption and to avoid overestimate diffusion, two initial straight lines were drawn before defining the straight line diffusion. Two discontinuous lines were plotted; one – D upper – following the initial absorption until 10.95 $h^{0.5}$ and second –D lower – following the moisture uptake data after that time. Subsequently, the 'initial straight line' was

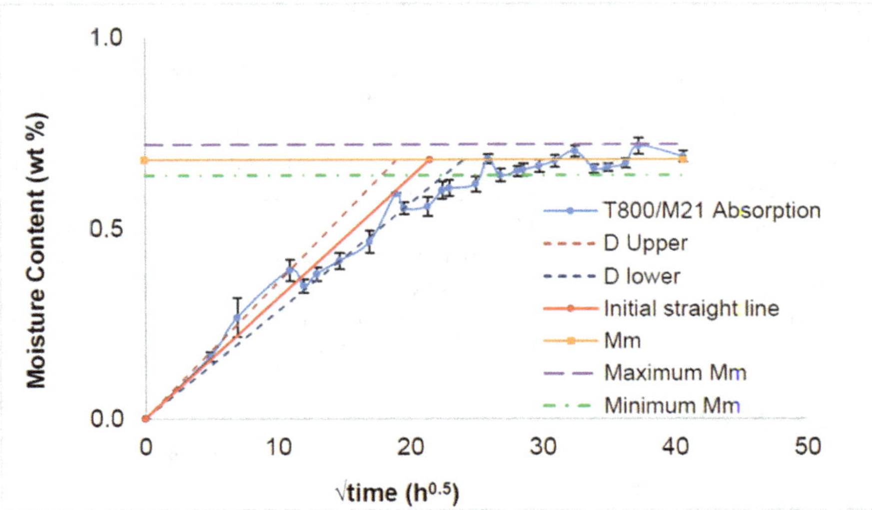

Figure 13.26 *CRFP moisture absorption average (2 mm thick)*

drawn as an average between both discontinuous lines, red line in . The diffusion coefficient (D_h) calculation for the trend slope was 1.7 E-3 mm²/hour.

Figure 13.27 shows the curve of the predicted moisture $M(T, T)$ as a function of root square of time. It is compared with the experimental data results. The model proposed based on ASTM showed good agreement [95]. Although the absorption

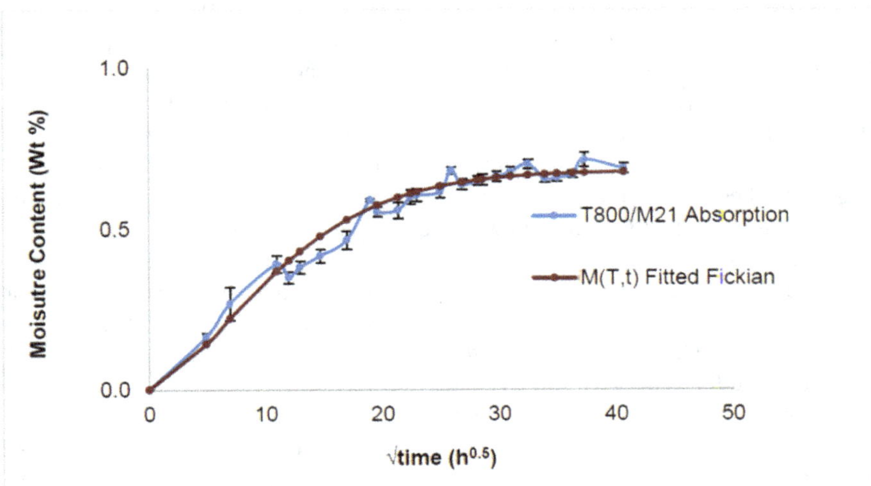

Figure 13.27 *CFRP predicted moisture absorption (2 mm thick) at hot-wet condition compared with experimental moisture uptake*

after 10 $h^{0.5}$ was overestimated, the moisture equilibrium content showed a good correlation with the experiment results.

Evaluating the predicted *M*(*T*, *t*) curve, the initial moisture absorption (straight line) achieved up to 0.43% absorption at 12.96 $h^{0.5}$. This absorption equals 63% of maximum moisture content (*Mm*). The concave zone can be identified in ; the slow increment can be seen between 15 $h^{0.5}$ and 31 $h^{0.5}$ time. The moisture absorption was evaluated only for T800/M21 laminate 2 mm thickness; hence, the superimposable curve for other thicknesses cannot be defined.

In conclusion, the moisture absorption of the T800/M21 laminate (2 mm thickness) exposed to the hygrothermal cycles obeys Fick's law behaviour.

Other authors have reported Fick's behaviour for composite fibre reinforcement, as in the case of carbon laminate IM7/977-2 exposed at 80C/80% RH [53].

13.4.2.3 Swelling

All samples after weighing were measured to identify dimensional changes. The tolerance on measurements was ±0.01 mm. Neither swelling nor length modifications were detected in samples of adhesive FM94 and composite T800/M21.

13.4.2.4 Desorption following absorption FM94 and CFRP T800/M21 UD

Moisture desorption after the saturation test was performed to identify the materials' desorption characteristics following the procedure described in section 13.3.4.1. Figure 13.28 illustrates the change of mass during absorption and desorption as a function of time.

Both FM94 and T800/M21 showed an accelerated moisture loss in the first stage followed by a gradual reduction in mass change until achieving the constant moisture content. Comparing the moisture desorption rate with the absorption rate, both materials presented a desorption rate faster than the moisture absorption.

After the desorption test, any moisture absorption was found to be reversible indicating little or no chemical reaction (e.g. hydrolysis of the polymer chain).

The rate of diffusion for desorption was higher than the rate of absorption diffusion that has been associated with the changes in the epoxy caused during the water absorption and interactions between water molecules and adhesive [55, 64].

The results from the work presented here show that the diffusion rate in the composite laminate T800/M21 is 30% slower than in the adhesive. This can suggest that the amount of moisture uptake by the joint is driven by the film adhesive. And therefore the joint strength will be affected by the adhesive (changes due to moisture absorption) and the stability of the interface.

Although the diffusion and moisture content under the hygrothermal cycles showed dynamic behaviour, in fact, the difference between both conditions was about 0.9%. It can be interpreted that the moisture diffusion profile was driven by the hot-wet conditions. Agreeing with this approach, commercial aircraft industries have affirmed that the moisture content depends mainly on the ground relative humidity during non-flight operations [105].

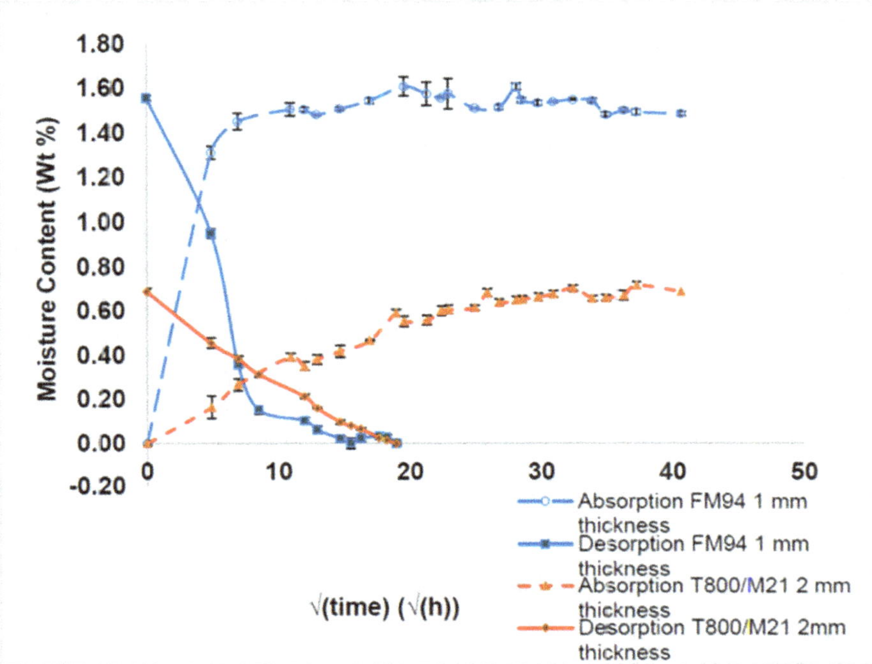

Figure 13.28 Desorption FM94 bulk and T800/M21 laminate

For that reason, the moisture properties used in the degradation model implemented in this study used the diffusion coefficient and moisture concentration measured at hot-wet conditions, see Figure 13.29.

13.4.2.5 Mechanical properties of adhesive FM94

To obtain the mechanical properties dependence of the hygrothermal cycles (i.e. tensile strength and elastic modulus), bulk adhesive specimens were tested under static load, as described in section 13.3.5.

Figure 13.30 shows the tensile stress against strain curves of unaged and aged bulk adhesive samples. Three specimens are plotted for each ageing condition (0, 84 and 168 cycles). It can be seen that neither tensile strength nor the strain showed changes by increasing the hygrothermal cycles.

The tensile strength average of all samples (unaged and aged) was 36 MPa with a standard deviation of ±1.56 MPa; the results agree with Roh and Sun, who estimated a failure stress of 5,000 psi failure stress that is equivalent to 35 MPa [106]. But the estimate was lower than the 50 MPa reported by Zavatta [107].

Figure 13.31 shows the variation of the tensile stress average versus moisture content. The moisture content values of 0, 1.51 and 1.54% were taken from the experiments (as seen in section 4.2.1) and corresponded to 0, 84 and 168 cycles,

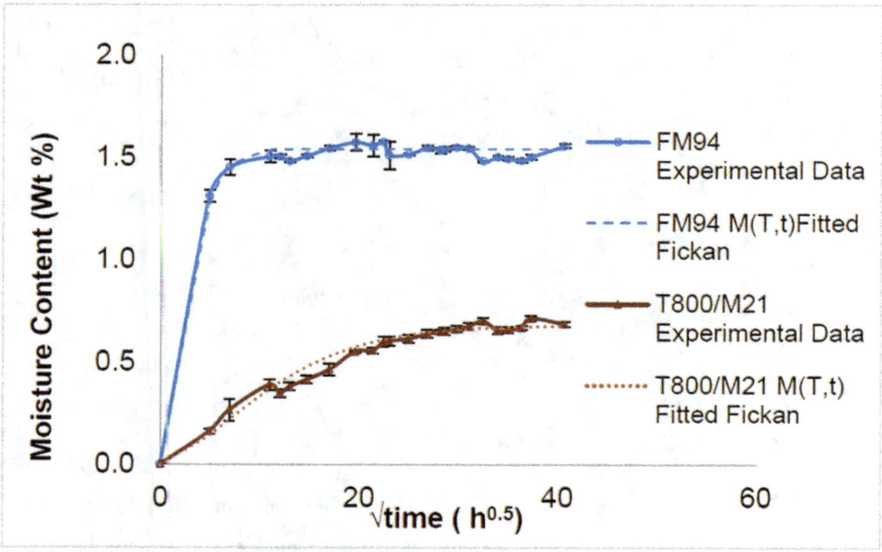

Figure 13.29 Moisture absorption FM94 and T800/M21

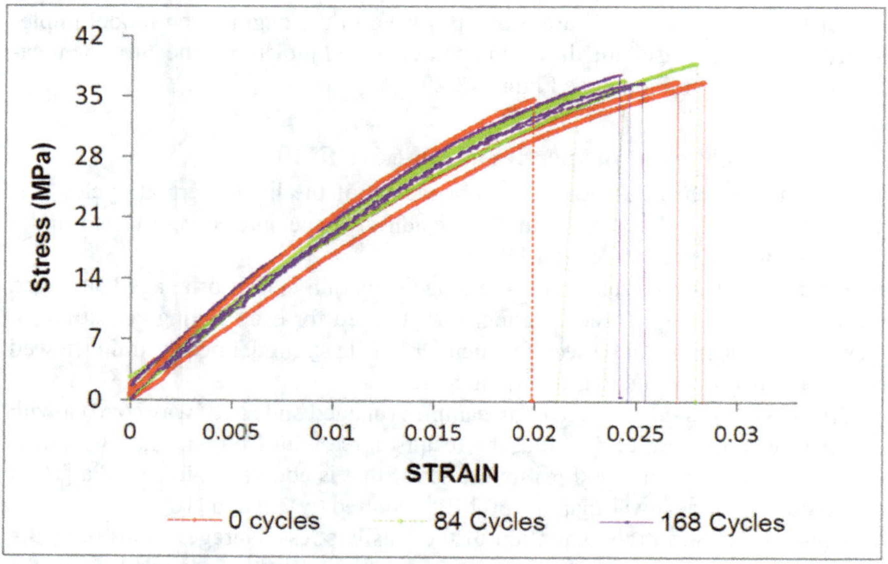

Figure 13.30 Effect of hygrothermal cycles on mechanical properties FM94

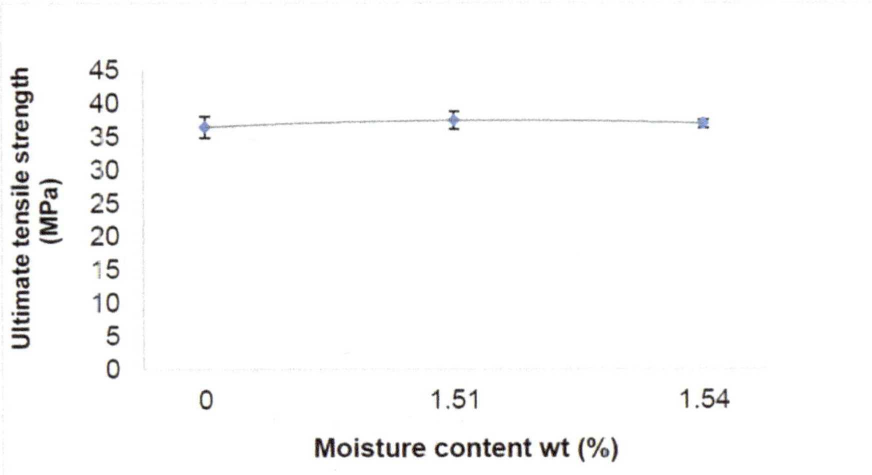

Figure 13.31 *Variation ultimate tensile stress versus moisture content wt (%)*

respectively. The error bars refer to the standard deviation for each set of samples tested.

Analysing Figure 13.31, 1.51% moisture content was a small increment of the tensile strength compared with 0% moisture (unaged state), and at 1.54%, the stress showed a reduction compared to the unaged state. However, all of these changes are between the error value, confirming that the tensile stress remained constant stress after the moisture absorption.

The elastic modulus (*E*) moisture dependence is shown in Figure 13.32. The elastic modulus average obtained for unaged specimens was about 1,750 MPa with a high standard deviation equal to 117 MPa. After 84 hygrothermal cycles at 1.51% moisture, the *E* modulus increased up to 1,900 MPa. Then, at 1.54% moisture content (168 cycles), the *E* average went down by 4%.

The *E* modulus of bulk adhesive didn't show a significant change after taking the maximum moisture content of 1.54%. The experimental elastic modulus is noticeably smaller than 3,000 MPa reported in Reference [107], but it agrees with 1,750 MPa reported in Reference [106].

According to these results, moisture uptake during the hygrothermal cycles apparently does not affect the mechanical properties of the FM94. However, as mentioned in Chapter 2, the moisture ingress in adhesives can cause changes of the strength, stiffness and strain [108], which will be discussed in Chapter 6.

13.4.3 Adhesive bonded SLJ test

SLJs were manufactured using laminates of CFRP Hexcel T800/M21 and film adhesive Cytec FM94K. The surface preparation used was peel ply and joining used secondary bond following the instruction of the Cytec's datasheet. Two tip initiation

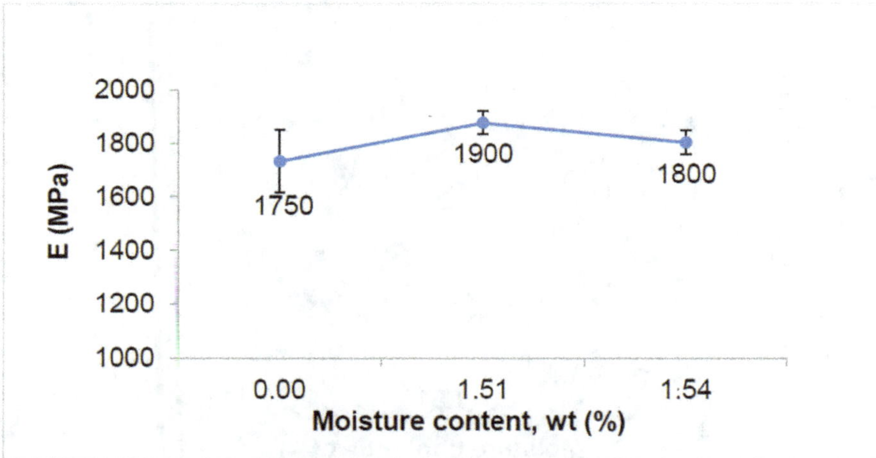

Figure 13.32 Elastic modulus as a function moisture content

cracks were included in the joints by the use of PTFE, 0.1 mm thickness and 2 mm width, laid in the free ends of bonded joint, as described in section 13.3.1.

13.4.3.1 Mechanical testing

The results of tensile test and fatigue test are presented in this section. Also, the fracture failure analysis is discussed. As it was mentioned in section 3.7.1 three to five samples were tested for each category.

13.4.3.1.1 Lap shear test

To investigate the influence of ageing on hygrothermal cycles on the joint strength, unaged and aged specimens were tested under shear lap load. The procedure is described in section 13.3.5.3.1.

The expected maximum load for the SLJ was 13.7 kN, a value calculated using the FM94 tensile shear of 23.8 MPa for secondary bonding of epoxy/graphite substrates for 575 mm^2 area [94].

Load–displacement curves of unaged and aged specimens are shown in Figure 13.33; one representative curve has been chosen for each set. Load–displacement measurements do not start from zero as closure grips introduce a preload.

The load–displacement curves have shown brittle behaviour and similar loading slope. However, the unaged sample showed a roughly linear trend, while the aged samples showed a moderate nonlinear behaviour before failure. Linear behaviour has been reported for CFRP bonded joints for 2 mm adherend thickness [2]. Park *et al.* [109] reported a variation from linear to nonlinear response after ageing (elevated temperature and wet conditions) [109].

An apparent reduction of the stiffness with the hygrothermal cycles increase is observed. Although a variation of the stiffness behaviour can be seen as load and

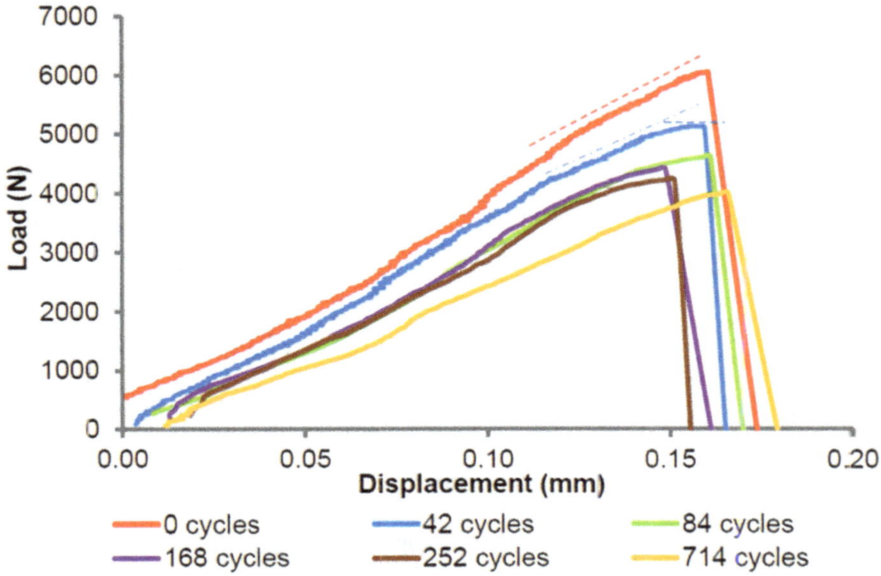

Figure 13.33 Load–displacement curve unaged and aged SLJ

displacement increase, this effect on the apparent stiffness is caused by the eccentricity of the SLJs. This rotation affects the displacement measurement recorded by the laser extensometer, causing low confidence in the data recorded by the laser extensometer.

The maximum failure load average was 6.58 kN for the unaged samples. It was 52% lower than the expected maximum load, as shown in Table 13.8. The lower ultimate load can be attributed to the influence of multiple factors, such as entrapment moisture on the adhesive bond, preparation surface and the pre-crack tip in the joint.

Table 13.8 Summary of test data unaged and aged SLJ

	Failure load (kN)		Failure displacement (mm)		
Hygrothermal cycles	**Average**	**Standard deviation**	**Average**	**Standard deviation**	**Initial stiffness (N/mm^3)**
0	6.58	0.43	0.18	0.03	54.40
42	5.09	0.42	0.17	0.02	54.20
84	4.41	0.48	0.15	0.02	51.23
168	4.60	0.18	0.16	0.02	57.22
252	4.19	0.63	0.13	0.02	52.29
714	3.79	0.19	0.13	0.03	52.15

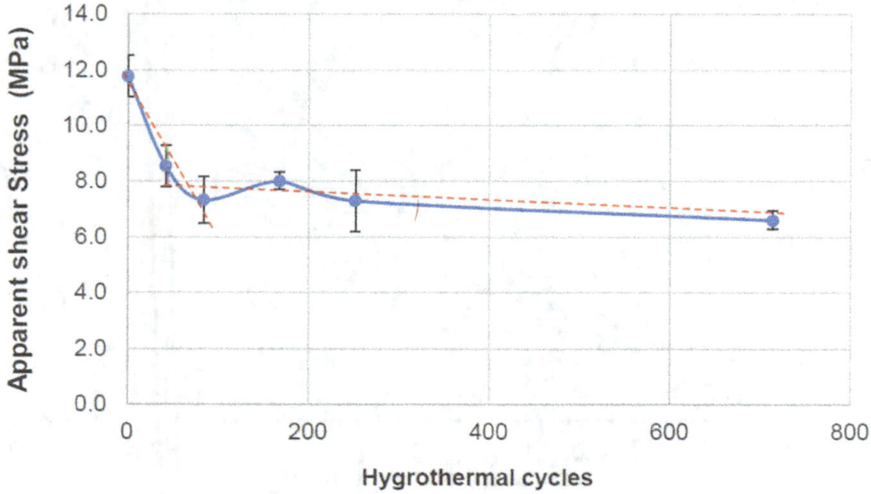

Figure 13.34 Shear strength as a function of hygrothermal cycles

The failure load decreased gradually with an increasing number of hygrothermal cycles. The main reduction occurred at 42 cycles, decreasing by 23% joint strength compared with unaged specimens. The decreasing trend continued until 84 cycles; after that, the strength reduction levelled off. After 714 cyles, the loss of joint strength achieved a figure of 42%.

Figure 13.34 begins with a dramatic decrease in strength as the number of cycles increases. This was followed by a small strength increase to 168 cycles and from that point onwards fell away to a figure of 6.6 MPa at 714 cycles. It must be noted that this fall is within the standard deviation.

A similar trend was reported by Knight *et al.* [21], who studied the ageing of CFRP-CFRP SLJ under hot wet conditioning over 2 years. The shear strength reduction after 500 days was 35% less than the control sample, and the trend to level off was attributed to the saturation.

Figure 13.35 shows a bar graph of the maximum displacement average against the hygrothermal cycles from SLJ. Due to an apparent elastic relaxation after failure (seen in Figure 13.33), the maximum displacement value was taken at the maximum failure load.

The displacement at failure fell with hygrothermal ageing, decreasing from 0.18 to 0.13 mm. The reduction was interrupted at 168 cycles with a displacement of 0.16 mm, when an apparent increment occurred. From this point, the fall continued to 0.13 mm at 252 cycles and levelled off at this figure.

Post-failure analysis

Microscopy was used to evaluate fracture surfaces in detail. Figure 13.36 shows the typical failure surface of joints (unconditioned and conditioned). In the photography, can be seen the fracture occurred along the adherend/adhesive interface.

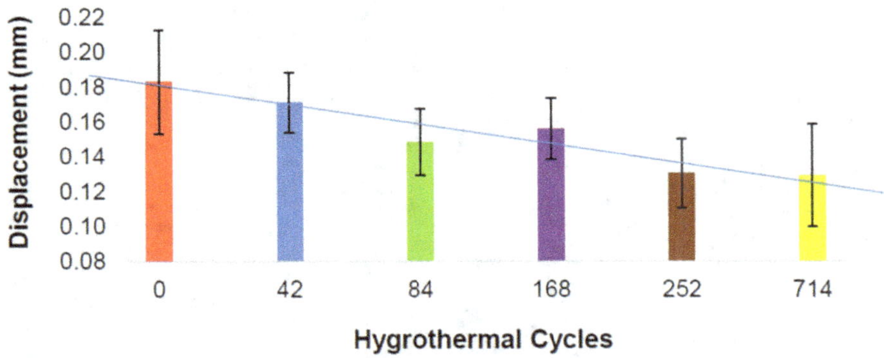

Figure 13.35 Displacement at failure of unaged and aged SLJs

The adhesive was fractured at both substrate sides, showing two failure modes: adhesion failure and thin cohesive layer (higher proportion).

The morphology can indicate that the cracks have initiated from both edges (PTFE areas) and propagated through the interface toward the opposite edge. Additional, a noticeable amount of porosity can be seen along the adhesive layer in .

The thin-layer cohesive failure (interface failure) can be seen as a light dusting of adhesive Figure 13.37a. The laminate surface contours produced by peel ply preparation retained some fractured adhesive; it can be evidence that the adhesive impregnation on the adherend (after curing) was uniform, as shown in Figure 13.37b.

After 84 cycles, the joint failures showed a variation in the fracture pattern, and the full separation was not occurred, as shown in Figure 13.38. The adhesive layer showed a plastic behaviour that induced the adherents to remain connected.

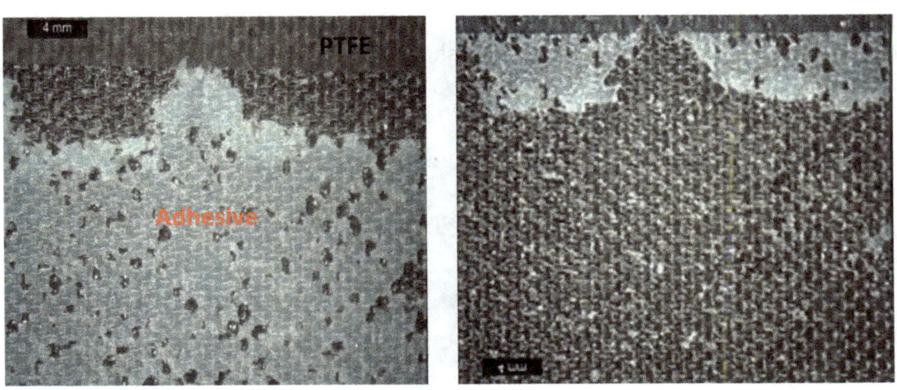

Figure 13.36 Typical surface failure

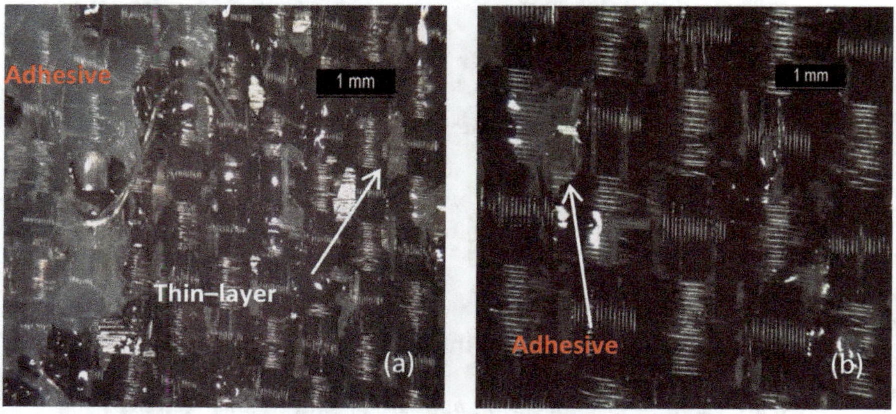

Figure 13.37 Photograph of failure surface unaged and aged SLJ: (a) fracture failure and (b) adhesive in contours

13.4.3.1.2 Fatigue test

To determine the SLJ fatigue life (for unaged and aged specimens), fatigue tests were carried out under a constant amplitude, ratio $R = 0.1$ and frequency of 5 Hz. The testing ran up to 106 cycles and followed the procedure described in section 13.3.7.1.2.

The fatigue life has been normalized with respect to the ultimate static strength of the unaged joints. The fatigue life was studied for 0, 84 and 336 cycles. The load (L) against the number of cycles to failure (N) plots of unaged and aged SLJs are shown in Figure 13.39 . The arrows plotted above some data refer that the specimen didn't fail before 106 cycles.

In order to describe the fatigue behaviour, the fatigue cycles have been grouped into three regions: low cycle fatigue below 10,000 cycles, high cycle fatigue between 10,000 and 10^6 cycles and infinite life after 10^6 cycles.

Unaged joints were tested at four different load levels (40, 45, 50 and 55% of ultimate strength). Unaged joints at lower load (40% ultimate load) supported 10^6 cycles without failure. The unaged joints showed a high cycle fatigue life of around 25,000 cycles at 55% maximum load.

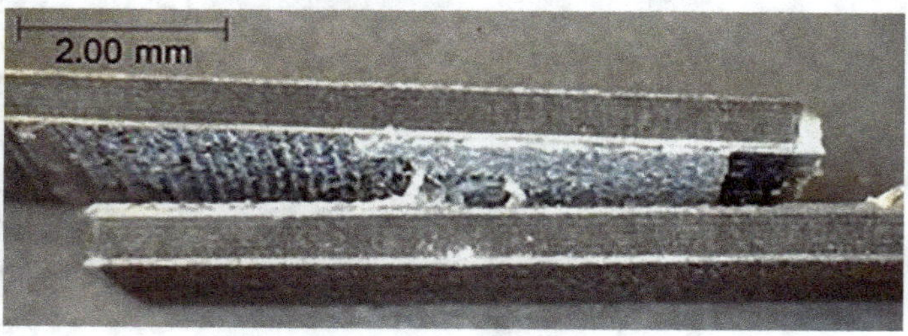

Figure 13.38 Fracture failure after 84 cycles

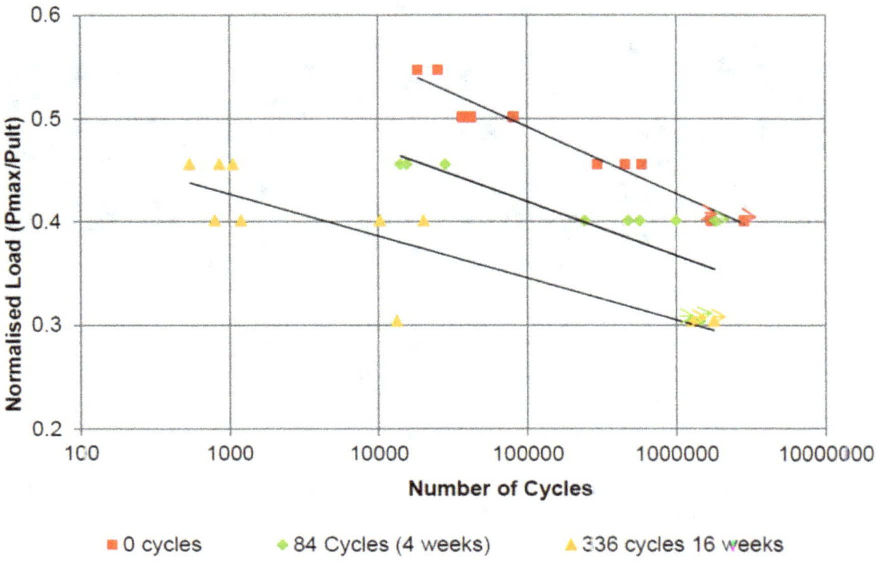

Figure 13.39 *L–N normalized curves of SLJ unaged and aged (R = 0.1, f = 5 Hz)*

Aged Joints at 84 hygrothermal cycles were tested at 30, 40 and 45% of quasi-static strength (unaged). *S–N* curve normalized (84 cycles) shows a reduction of fatigue life although there was a high spread of data. At lower load level (30%), the fatigue life was over 10^6 cycles. At 40% load level, there was a high spread of fatigue life cycles; two specimens presented infinite life while three joints failed at high fatigue cycles (244,456, 478,768 and 572,448 life cycles).

This could be due to the difference of time after their removal from the environmental chamber. Aged joints tested at the first or second week after removal from the chamber failed; additionally, it was observed that the fatigue life increased as the amount of time out of the chamber increased. In contrast to those joints, SLJs evaluated after 3 weeks out of the environmental chamber were not broken.

At a higher load level (45%), there can be seen a significant reduction of the life cycles compared with unaged joints. Aged joints at 336 cycles were tested at the same load levels as aged joints at 84 cycles. At a lower load, two of the joints exceeded the 10^6 life cycles without failing, but one joint failed unexpectedly at 13,296 cycles. At 40% load level, fatigue life showed high and low life cycles; two specimens failed over 10,000 cycles and two joints failed around 1,000 cycles. At a higher load level (45%), the fatigue life decreased by a factor of 10 compared with conditioned joints at 84 cycles.

Overall, the fatigue life decreased with increasing hygrothermal cycles. The life cycles showed a horizontal shift toward the left for each load tested. A similar trend

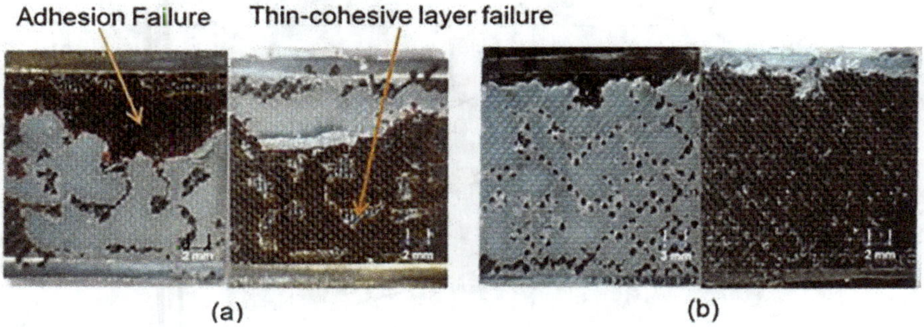

Figure 13.40 Failure surface SLJ tested under fatigue: (a) mode failure – typical failure pattern and (b) failure pattern aged joint 336 cycles at 30% load

has been reported by Katnam *et al.* [88] for metal bonded joints immersed in water at 50°C for 1 and 2 years.

Post-failure analysis

A microscopy evaluation was performed after the fatigue test to identify the failure mode of the aged and unaged joints. A microscope with 400x magnification was used to inspect the failure surface.

A mixed failure mode of adhesion failure and thin-layer cohesive failure can be identified on the fracture surface, as shown in Figure 13.40a. All joints showed failure occurred along the adherend–adhesive interface with a predominant adhesion failure type. The pattern of failure was similar in most of the specimens; two cracks initiated from the ends of the overlap and propagated towards the centre of the overlap until the final failure.

However, the joint aged at 336 cycles and tested at 30% load level and which presented a premature failure showed a slight change of fracture pattern; only one crack propagated and the failure occurred near to one end of the overlap, as illustrated in b.

13.4.3.2 Non-destructive inspection

13.4.3.2.1 Visual inspection-microscopy evaluation

To characterize the bondline with respect to the voids or moisture effects optical microscopy was used under reflected light. A control sample was observed (previous and post-ageing cycles) with a NIKON Optiphot microscope at different magnification lens (5X, 10X, 20X, 40X).

A side view of the overlap region from the unaged specimen can be seen in Figure 13.41. A great void is seen in the bondline, and the PTFE in the free edge was highlighted with red colour. The adherends apparently are free of defects, as shown in Figure 13.41a. A significant amount of voids was found along all bond layers.

The measurement of the bondline thickness and the void height was taken as 0.3 and 0.19 mm, respectively. This void has occupied 64% of the bondline height,

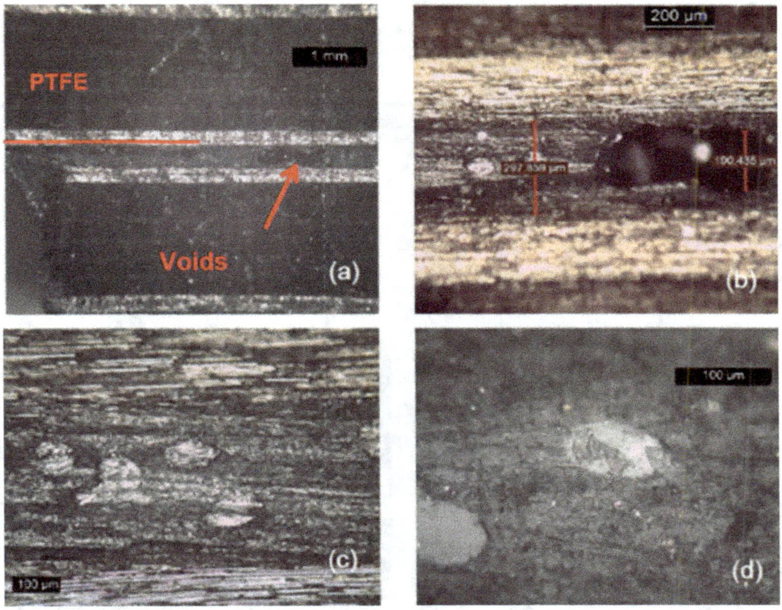

Figure 13.41 *Photographs of side view of unaged adhesive bondline: (a) 5X. (b)10X, (c) 20X and (d) 40X*

as seen in Figure 13.41b. Debonding or micro-cracks haven't been identified in the bond area. There were spots with a white contour along the bond area as shown in Figure 13.41c and d, which seemed to moisture trapped in the bondline.

Figure 13.41 shows the microscopy images after 252 ageing cycles. Void content in the bondline is seen in Figure 13.42a. The bondline thickness measured was 0.3 mm. Evidence of lack of bonding, cracks or degradation was not found. The apparent moisture seen before ageing did not show changes of amount or dimensions, as shown in Figure 13.42c and d.

Evidence of degradation resulting from the ageing cannot be seen. There was not found swelling or trapped moisture in the bondline after the ageing. The white contours cannot be assumed as trapped moisture. But, there was a high amount of voids identified along the bond edge.

13.4.3.2.2 Ultrasonic inspection

Ultrasonic C-scan 5 MHz with a gain of 31 dB was used to evaluate the quality of the bonded joints (unconditioned and conditioned) to hygrothermal cycles in an environmental chamber. A control SLJ was evaluated pre and post ageing process. This joint was removed from the environmental chamber at 84,

168 and 252 cycles and inspected through the ultrasonic C-scan as described in section 13.3.4.2.

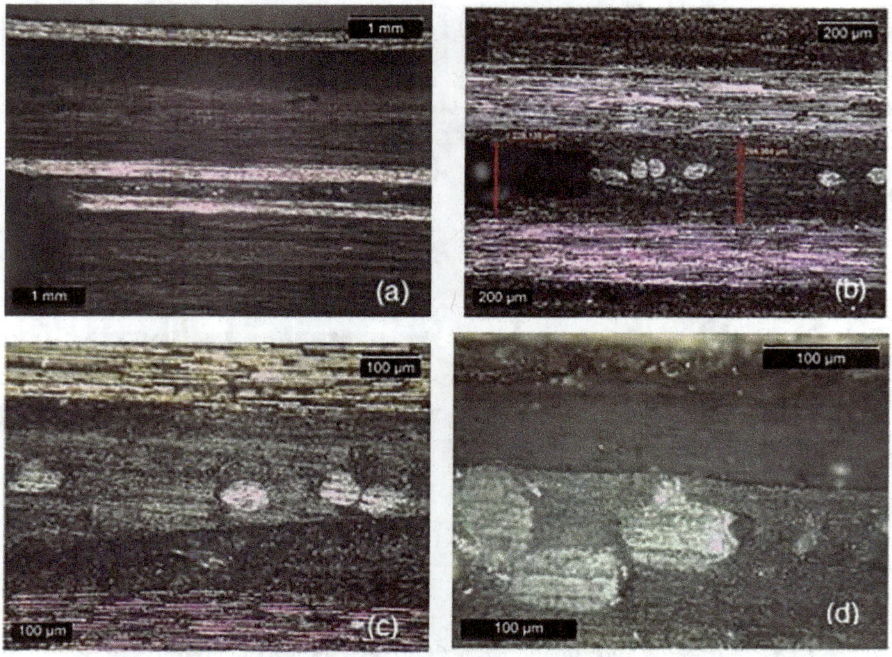

Figure 13.42 *Photographs of side view of aged bondline (252 cycles) bondline (A) 5X, (B)10X, (C) 20X, (D) 40X*

A high amplitude energy transmitted through the bond line is interpreted as good quality bond (100% amplitude). The amplitude variation of the reflected signal passing through the (adherend-adhesive) interface represents attenuations associated with discontinuities in the adhesive layer. Hence, the measurement of ultrasonic energy going through the healthy bond is greater due to less attenuation [103].

The C-scanned images of the joint (unaged and aged) and the signal amplitude are shown in Figure 13.43 . Note that the good bond quality is taken as red colour contours and the lack of bond quality is taken as blue/white colour contours.

The unaged image shows a significant variation of the amplitude. A good bond quality is seen in the middle of the joint. A slight attenuation can be seen as a yellow colour contour (60% signal amplitude), which can be interpreted as differences of adhesion. Also, the lowest amplitude signal can be appreciated in end of the overlap (blue colour contour), particularly on the right side. This lack of adherence can be associated with the PTFE crack initiators, as shown in Figure 13.43a.

After 84 hygrothermal cycles, the C-scan image showed a reduction of the amplitude. The attenuation was greater around all of the bond area (green and yellow colour contours), which can be interpreted as bond deficiencies. At 168 cycles, the amplitude mapped did not show changes compared with that of 84 cycles, as shown in Figure 13.43b and c.

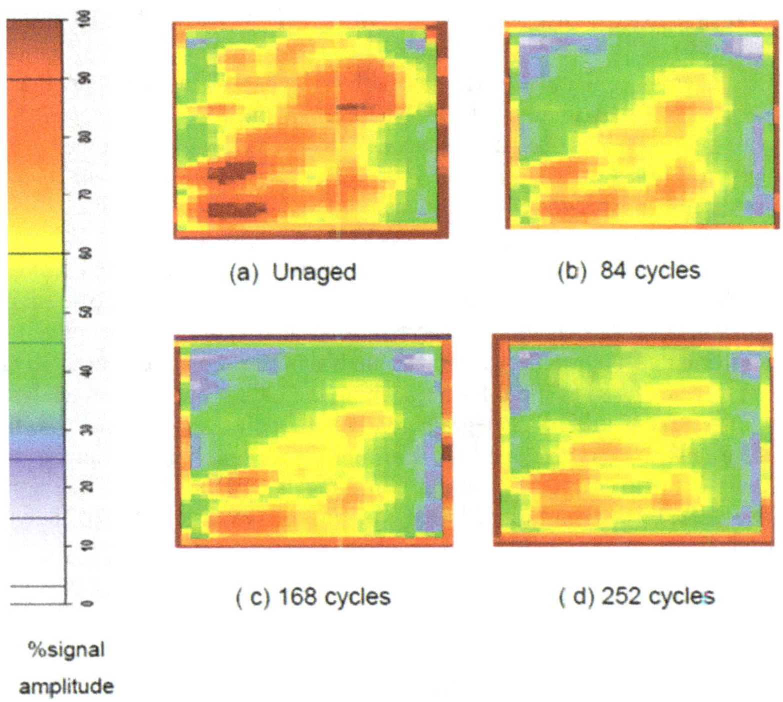

Figure 13.43 *C-scan images of the bond region obtained at different ageing states*

And at 252 cycles, the spectra of the ultrasonic signal showed a small change compared to 168 ageing cycles. There was a moderate rise of amplitude signal in some parts of the overlap, as can be seen in Figure 13.43d.

Overall, the signal penetrated the bond area and provided information about the condition of the bond. The joint inspected before exposure to environmental attack presented bond discontinuities along the overlap. This lack of homogeneity can be associated with bond defects, such as porosity or lack of adhesion. In conclusion, the joint was not a healthy joint before ageing.

Inspection of the joint after hygrothermal ageing showed changes in the spectrum of the ultrasonic signal. This increment of the mismatched areas can indicate degradation of the bond compared to the unaged condition.

By comparing the ageing images (84,168,252 cycles), slight changes of the ultrasonic signal were found. However, they cannot be taken as evidence of incremental mismatching of the bond due to any presence of water being able to affect the sensitivity of the incident ultrasonic wave [110].

13.5 Degradation modelling of adhesively bonded SLJ

To predict the environmental effects on the adhesive joints, a methodology that couples experimental data with numerical modelling approach has been proposed.

The methodology has been divided principally into three steps; in the first step, theoretical model was carried out to identify the diffusion coefficient and the moisture concentration at a different number of cycles based on existing constitutive expressions, as shown previously in section 13.4.

The second step consisted of developing a numerical model, which was established to determine the effect of the moisture concentration on the elastic modulus of the bulk adhesive. The adhesive properties as a function of moisture content were evaluated through a thermal-displacement analysis. The moisture characterization of the first step was used as input for performing the bulk finite-element analysis.

An adhesively bonded SLJ degradation was modelled in the third step based on cohesive law. It was developed to determine the progressing of damage in the bondline area due to moisture absorption and mechanical loading. The degraded parameters of SLJ were determined based on the joint's experiments and analysis at the bulk adhesive.

13.5.1 Finite-element analysis

Two numerical analyses were carried out using Abaqus 3D models: a bulk adhesive model, which used a moisture diffusion displacement method in order to achieve degraded properties, and a CZM of the joint to predict the damage caused by hygrothermal cycles and mechanical load. The analysis was carried out in Abaqus Standard using the implicit solver. The implicit finite-element technique is used to solve linear and non-linear problems, and the implicit solver requires iterate solutions with each time increment. It has been adequate for brittle adhesive behaviour under tensile loading [76]. With a larger time step size, can be achieved more numerical stability and accuracy.

13.5.1.1 Bulk adhesive

To determine the adhesive elasticity modulus degradation due to moisture content, a moisture displacement analysis was performed using an analogy between heat transfer (Fourier's law) and moisture diffusion laws (Fick's law) [111].

Heat transfer expression is presented in (13.20):

$$\frac{\partial T}{\partial t} = \left(\frac{k}{cp}\right) \nabla^2 T \tag{13.19}$$

where T is the temperature, t is the time, k is the thermal conductivity, c is the specific heat, ρ is the density and $\nabla^2 T$ is the temperature gradient.

Moisture diffusion law is shown in (13.21):

$$\frac{\partial C}{\partial t} = D\nabla^2 C \tag{13.20}$$

Table 13.9 FM94 properties

Properties	Unit	Value
Elastic modulus (E)	MPa	3,000
Poisson's ratio (v)		0.35
Shear modulus (G)	MPa	1,111
Diffusion coefficient (D)	mm^2/s	1.64E-06

where C is the moisture concentration, D is the diffusion coefficient and $\nabla^2 T$ is the concentration gradient.

Using the equations, the diffusion is modelled using the material model of heat/transfer built in Abaqus, using the diffusion coefficient as thermal conductivity and concentration as temperature.

For such analysis, density and specific heat can be taken as unity as proposed in Reference [64]. The coefficient of moisture expansion used was 0.0016/(%) [49, 90]. The diffusion coefficient was taken from the experimental results of FM94 (section 13.4.1.1.1). The FM94 properties used to model the bulk are shown in Table 13.9.

13.5.1.1.1 Geometry and boundary conditions

The bulk adhesive geometry is shown in Figure 13.44. To simulate tensile loading, an encastre constraint was assigned to one edge of the bulk specimen, while the other edge was subjected to displacement.

To perform the moisture diffusion-displacement analysis, two coupled temperature-displacement steps were implemented: In the first step, the condition before moisture diffusion was defined. A load (type body heat flux) was applied uniformly distributed with a magnitude of 0.0001 (assumed as the lowest moisture concentration); this magnitude was taken as the moisture concentration at time zero. In the second step, a load (type body heat flux) with the final moisture concentration

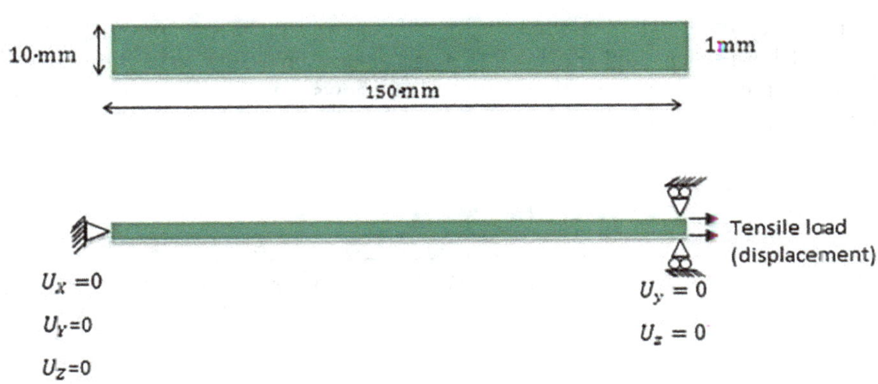

Figure 13.44 Configuration of bulk adhesive

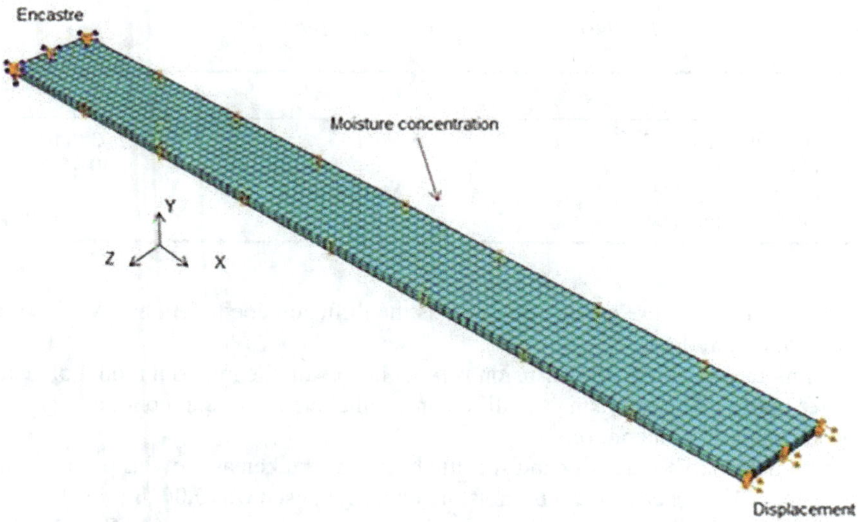

Figure 13.45 *Mesh bulk adhesive*

(for each number of cycles studied: 1.49, 1.51, 1.54) and a displacement (10 mm) at the edge of the specimen were applied. The magnitude of the load was given as the percentage of moisture concentration for any time interval.

Moisture concentration in the bulk was assumed to be equal to the percentage of moisture concentration determined in the experiments for each time interval. The hygrothermal degradation of the adhesive was assumed to be only a function of the moisture concentration. The temperature cycle effect in the properties was considered implicit in the moisture concentration.

The bulk adhesive was modelled as continuum elastic plastic with 2,000 quadratic hexahedral thermal-displacement-coupled elements (C3D20T). The mesh configuration can be seen in Figure 13.45.

The modulus (E) was predicted at each different moisture concentration (experimentally taken at each number of cycles), it was obtained from the curve stress–strain at different moisture concentration. Table 13.10 shows the E modulus predicted for the moisture concentration. A reduction in the elastic modulus can be seen along

Table 13.10 *Finite-element results*

Number cycles	Number of weeks	Wt(%)	Numerical E modulus (MPa)
0	0	0	3,000
42	2	1.49	2,432
84	4	1.51	2,422
168	8	1.54	2,414

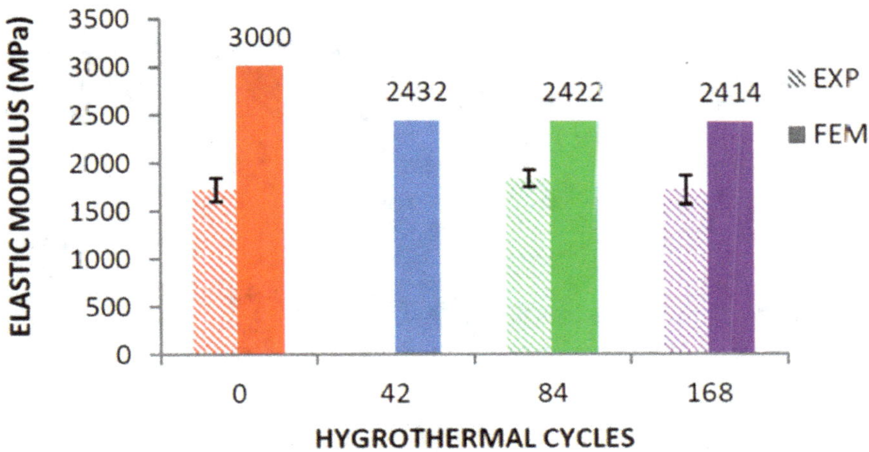

Figure 13.46 *FM94 elastic modulus experimental and numerical; the value of 3,000 MPa was set*

the increment of moisture. The analysis was limited to moisture concentration in the saturation state (1.54%). This concentration was defined for 168 hygrothermal cycles, based on experiments, and then the elasticity prediction was not performed for longer than this time.

E was predicted, and experimental results and number of hygrothermal cycles are plotted in Figure 13.46 . The modulus decreased linearly until 42 cycles, dropping up to 19% of the initial E modulus. Later, the reduction trend levelled off showing slight changes reaching a value of 2,414 MPa at the maximum moisture concentration.

The 3D numerical model results were compared with the experiment results. By contrast, in the experiments, the modulus did not present variations after exposing the adhesive to hygrothermal cycles, as described in section 13.4.2.1.2. However, elastic modulus degradation has been reported by other authors [90, 112]: Sugiman *et al.* [49] reported a reduction of 38% after 1 year of immersion in water at 50°C and achieved a saturation of 5.5 wt%. Also, Mubashar studied the moisture cycling effects on adhesive properties. He found that major reduction occurred during the first absorption cycle and the trend of reduction was stabilized after 2 wt% of moisture concentration [64].

The elastic modulus predicted for the hygrothermal cycles has been used in the degradation model of the SLJ to determine stiffness degradation of CZM.

13.5.1.2 SLJ model

The SLJ degradation was predicted using CZM to simulate the composite-adhesive interface mechanical response. To predict the damage in the SLJs exposed to environmental conditions, the cohesive parameters were degraded based on numerical

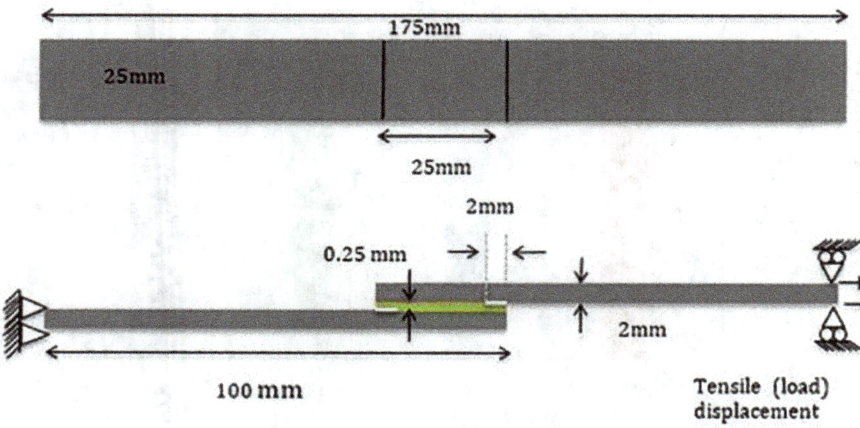

Figure 13.47 SLJ configuration and boundary conditions

methods and experimental data. An empirical equation was defined from the lap shear experiments to calibrate the cohesive strength (tipping stress) in the model.

13.5.1.2.1 Geometry and boundary conditions

SLJ geometry configuration and boundary conditions are shown in Figure 13.47. To simulate the shear load test condition, one edge was fixed and the opposite edge was subjected to a tensile displacement load in the *X*-direction.

Composite adherends were simulated as an orthotropic material. Engineering constants were determined via a laminate analysis (NPL's composite design analysis CoDA), and results are shown in Table 13.11. Adherends were modelled using 3D continuum elements with reduced integration (C3D8R). Reduced elements integration refers to a rule order less than the full rule used to integrate the element's internal forces and stiffness [113].

The adhesive was modelled using 3D cohesive elements layer (COH3D) to represent cohesive crack propagation path at the bond [114]. Two millimetres length was reduced at each end of bondline elements to simulate the crack initiators. A

Table 13.11 Composite laminate properties T800/M21 [UD 0°/45°/90°/–45°]S

	E1	**50,921**
	E2	**50,921**
Elastic modulus (MPa)	**E3**	**9,732**
Poisson's ratio	v12	0.32
	v23 = v13	0.36
Shear modulus (MPa)	G12	19,326
	G13 = G23	3,134

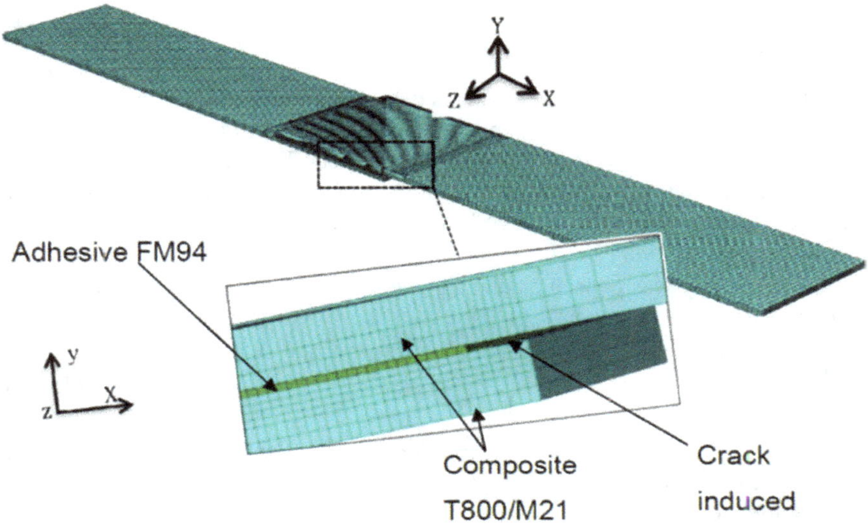

Figure 13.48 Meshing of the SLJ

refined mesh was applied near to the overlap to achieve better accuracy with the smallest element size of 0.26 mm. The meshing is shown in Figure 13.48.

A maximum cohesive element size (*le*) of 0.26 mm was defined to satisfy the condition of minimum three elements in the length of the cohesive zone (*lcz*). The *lcz* is the value for the distance from the crack tip to the point where maximum cohesive traction occurs (refer) . The values for *lcz* for mode I and mode II were calculated following the (13.21), where *E* and *G* are tension and shear modulus, respectively [115]:

$$l_{cz}^{I} = \frac{9\pi}{32}E\left(\frac{G_{Ic}}{T_n^{0^2}}\right) \text{ and } l_{cz}^{II} = \frac{9\pi}{32}E\left(\frac{G_{IIc}}{T_s^{0^2}}\right) \tag{13.21}$$

As discussed in section 13.2.6.3, the bi-linear law has been shown to provide simplicity and adequate accuracy; it is the configuration with less convergence difficulties with the SLJ modelling [76].

13.5.1.2.2 Damage criterion

The CZM has three steps damage initiation criterion, a propagation governed by cohesion law and the stiffness degradation down to failure. At final failure, all elements are removed [64, 116].

Damage initiation can adopt four criteria: maximum stress, maximum strain, quadratic stress, and a quadratic strain criterion. For this model, a quadratic nominal stress criterion (QUADS-in Abaqus) has been adopted, illustrated in (13.22) [115, 116]:

$$\left\{\frac{\langle T_n\rangle}{T_n^0}\right\}^2+\left\{\frac{T_s}{T_s^0}\right\}^2+\left\{\frac{T_t}{T_t^0}\right\}^2=1 \tag{13.22}$$

A QUADS criterion assumes that all stress directions are connected by a quadratic relationship. The values of Tn^0, Ts^0 and Tt^0 traction represent the maximum permissible nominal stresses in normal, first and second shear directions, respectively. *Tn*, *Ts* and *Tt* are tractions in each direction and for any time increment. The McCauley bracket ($\langle \cdots \rangle$) affirms that the damage is not initiated by a purely compression normal stress or deformation [76], i.e. if $Tn < 0$, $\langle Tn\rangle = 0$.

Fracture energy was chosen as damage evolution criterion with a linear power behaviour as predicted by (13.23) [117].

$$\left(\frac{G_{Ic}}{G_I}\right)^2+\left(\frac{G_{IIc}}{G_{II}}\right)^2=1 \tag{13.23}$$

here G_{IC} and G_{IIC} are work done by the traction in mode I and mode II respectively. G_I and G_{II} are the critical fracture energy of the adhesive, and the empirical value of α was assumed to be one [118].

13.5.1.2.3 Cohesive parameters

In this model, the cohesive traction (*T*) and the cohesive fracture energy (G_c) are the major properties for achieving an accurate response [72]. To calibrate the parameters, the fracture energy was taken from experimental studies and the cohesive traction was calibrated via comparison of simulations and the experiment results [117].

The stiffness parameters (K_n and $K_s = K_t$) were obtained by dividing the Young's modulus (*E*) and Shear modulus (*G*) by the adhesive thickness [116, 119], e.g. $K_n = \frac{E}{t}$ and $K_s = K_t = \frac{G}{2}$.

Values of $G_{IC} = 1.7$ and $G_{IIC} = 2.5$ N/mm have been reported by Zavatta [107]. They were adopted in this study. Some tests were carried out to identify the adequate cohesive traction for SLJ (under initial conditions). The parameters used are shown in Table 13.12.

Initially, values of $T_n^0 = 50$ MPa and $T_s^0 = T_t^0 = 30$ MPa were chosen assuming that T_n^0 is equal to ultimate stress (σ_{ul}) reported in the literature for FM94 [107].

Table 13.12 CZM calibration

CZM parameters cohesive						Results: Stiffness		Results: Failure load	
K_n	$K_s = K_t$ (N/mm³)	T_n	$T_s = T_t$ (MPa)	G_I	GcII (N/mm)	FEM	Exp Avg (N/mm³)	FEM (kN)	Exp Avg
12,000	4,444	50	30	1.7	2.5	71	56.6	8.44	6.58
		50	21	1.7	1.75	70		7.23	
		36	15	1.7	1.75	66		6.84	

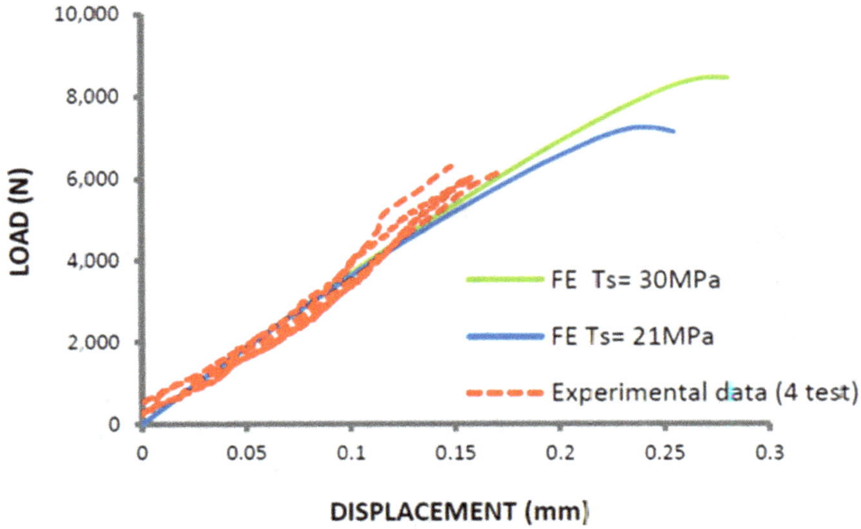

Figure 13.49 *Comparison load–displacement behaviour of SLJ with predicted using Tn = 50 MPa, Ts = 30 MPa and Tn = 50 MPa, Ts = 21 MPa*

Using that ultimate stress taken from the literature, a failure load of 8.4 kN was predicted. The results overestimated the experimental data of 6.58 by 28%.

Regarding the post-mortem evaluation shown in section 13.4.3.2, the failure mode was mainly adhesion failure, evidencing that the adhesive/adherend interface strength was less than that estimated. Consequently, the approach suggested by Belnoue *et al.* (adhesion failure approach) [120], who modelled adhesion failure with CZM, was implemented and compared with the previous findings.

The approach consisted of a modification of the cohesive parameters based on the effective mode II properties when the interface is under through-thickness compressive stresses. The new strength (T_0') and fracture toughness (G_{IIc}') for mode II were calculated using (13.24) and (13.25) :

$$T_s^{0'} = T_s^0 + \eta\sigma_{33} \tag{13.24}$$

$$G_{IIc}' = G_{IIc}\left(\frac{T_s^{0'}}{T_s^0}\right) \tag{13.25}$$

where σ_{33} is the through-thickness stress, which is equal to the maximum traction in shear, and η is an empirically derived enhancement factor analogous, with a value of 0.3 [120].

(T_s^0) mode I traction was assumed to be 50 MPa and mode II traction to be 21 MPa. The new fracture energy in mode II was reduced to 1.75 N/mm [120].

Figure 13.49 shows load–displacement curve for the SLJ unconditioned compared with the FEM results for *Tn* = 50 MPa with *Ts* = 30 MPa and *Ts* = 21 MPa

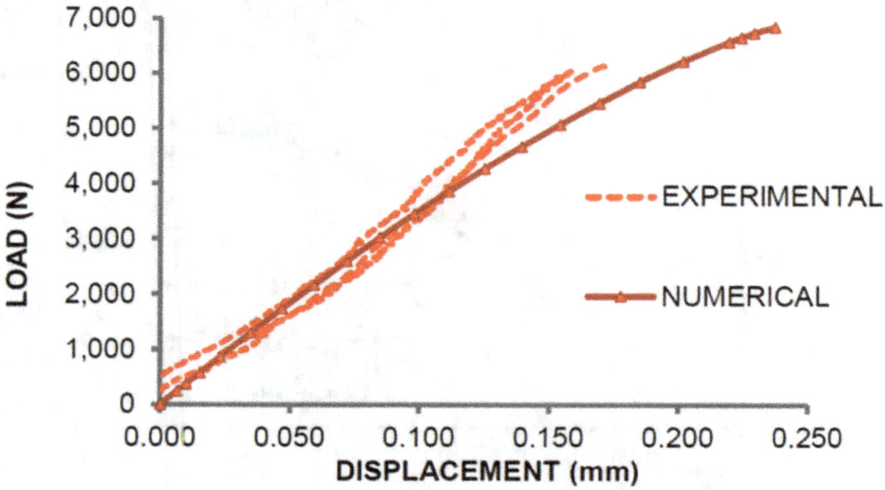

Figure 13.50 Load–displacement behaviour of unconditioned SLJ

(adhesion failure approach). The change of displacement was measured in the overlap area, as evaluated in the shear lap test.

Using the reduction proposed for adhesion failure (T_s=21 MPa), the peak failure predicted was 7.2 kN, hence a significant reduction of peak load was found by applying the reduction in shear strength and fracture energy. The failure load predicted was 9% higher than the average experimental data. The maximum displacement did not show noticeable changes. Results are shown in Table 13.12. Calibration of CZM parameters based on unaged bulk experiments has been carried out for modelling adhesive bonded joints [112, 119]. Hence, based on experimental data, the adhesive tensile strength of 36 MPa (section 13.4.2.2) was used as T_n. Maximum traction in mode II (T_s^0) was calculated considering the ratio of (50/30) taken from theoretical data. The traction mode II ($T_s^{0\prime}$) – adhesion failure approach – was obtained from the (T_s^0), as described earlier.

Figure 13.50 shows load–displacement numerical and experimental curves for unconditioned joints. The model predicted a failure load of 6.8 kN for unconditioned joints, in excellent agreement with the actual data, i.e. 3.8% greater than the average experimental peak load.

The initial stiffness predicted was also in close agreement slightly higher than the stiffness in the experiments. The maximum displacement achieved was 0.24 mm, larger than the experimental displacement average (0.16 mm). Results can be seen in Table 13.13.

Figure 13.51 illustrates the damage evolution in different stages of the deformation process during tensile loading. The legend shows the damage variable (SDEG-Abaqus) contours. The material point (element) is considered undamaged when SDEG is equal to 0, and fully damaged when the value is equal to 1. The adhesive layer prior to damage is shown in Figure 13.51a, the damage initiated in the middle

Table 13.13 Comparison static response SLJ under hygrothermal ageing conditions

Condition	Stiffness		Failure load		Displacement	
Hygrothermal Cycles	**FEM**	**Exp**	**FEM**	**Exp Avg.**	**FEM**	**Exp Avg**
	(N/mm^3)		**(kN)**		**(mm)**	
0	68.71	54.40	6.84	6.58	0.24	0.16
42	67.05	54.20	5.59	4.92	0.19	0.15
84	66.46	51.23	5.27	4.22	0.18	0.16
168	65.47	57.22	4.56	4.60	0.16	1.14
252	64.44	52.29	4.02	4.20	0.15	0.15
714	64.5	52.16	4.34	3.82	0.15	0.13

of one end of the overlap and the growth progressed in +*Z*-direction until reached the edges, as shown in Figure 13.51b. Later, a second crack initiated at the opposite end and both cracks propagated in direction *X*, as seen in Figure 13.51c. The failure occurred when both cracks met in the middle of the overlap, as shown in Figure 13.51d.

The maximum stress principal distribution in the SLJ is shown in Figure 13.51.e. The highest stress concentrations occurred in the substrates near to overlap edge. Peel and shear stresses achieved their maximum value. Additionally, damage variation before and after failure are highlighted.

13.5.1.2.4 Degradation parameters

SLJ ageing under hygrothermal cycles and load was modelled using CZM to predict damage caused by the degradation. The numerical model of the bulk adhesive was carried out to predict the elastic modulus degraded as a function of the moisture cycles. Based on the SLJ test results, an empirical equation was defined to identify the cohesive traction degradation in mode I.

The bilinear traction-separation law is illustrated in Figure 13.52. The tipping traction (T^o) is the maximum traction before damage initiates. The critical value of displacement δ^o refers to the damage initiation and δf refers to the displacement at ultimate failure. The stiffness before damage in the elastic regime is *K*, the slope of the curve. The area under the curve gives the fracture energy (G_c).

To determine the degraded traction as a function of the hygrothermal cycles, the gradient of reduction (from the lap shear strength results) was used as a factor of degradation. Hence, the empirical equation (13.26) was defined to predict the traction in tension (T_n^{deg}) as a function of number of cycles (*h*):

$$T_n^{deg} = T_n^0 \left(2e^{-6}h^2 - 0.0018h + 0.8626\right) \tag{13.26}$$

As fracture energies were reduced proportionally to the cohesive strength reduction, then the fracture energy mode I was obtained using the (13.27), where the

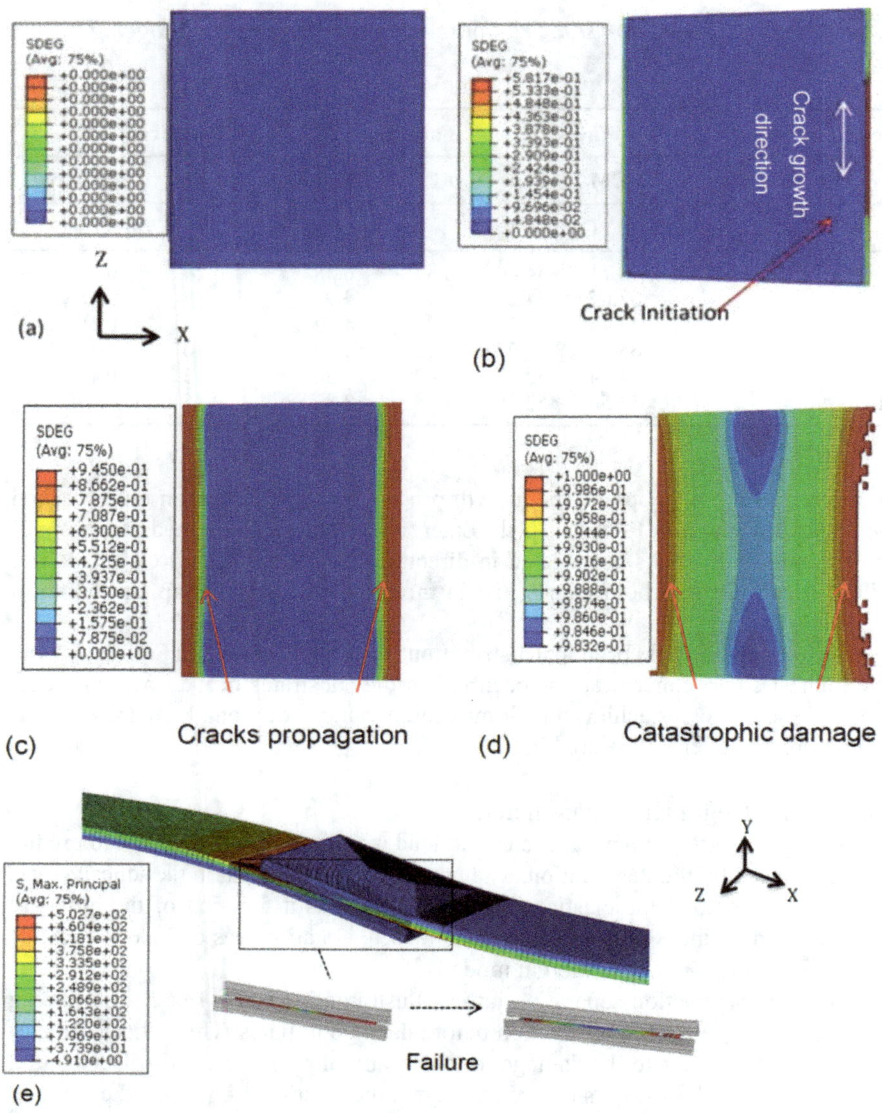

Figure 13.51 Failure process in SLJ; (a) cohesive elements before damage, (b) crack initiation in the cohesive elements, (c) crack propagation and (d) cohesive elements full damage

δ_f calculated for the unconditioned state has remained constant for all ageing conditions:

$$G_{Ic}^{deg} = T_n^{deg}/\delta f \tag{13.27}$$

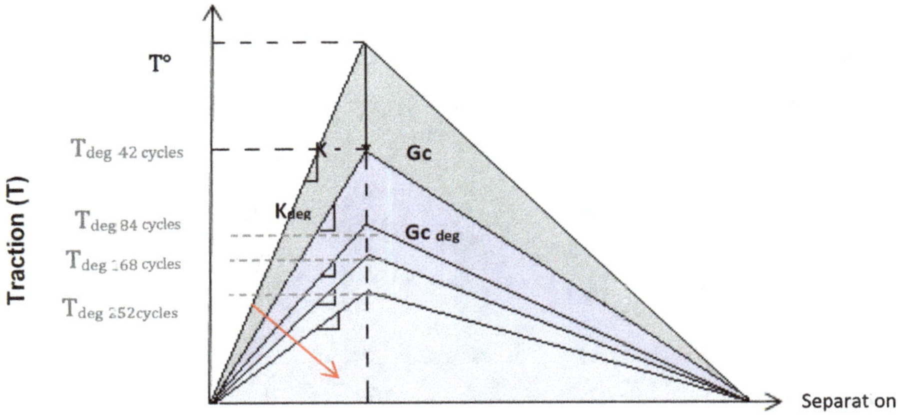

Figure 13.52 *Schematic degradation process for bilinear traction–separation curve. The degraded stiffness K_{deg} for each number of cycles was obtained by dividing the E numerical modulus of adhesive bulk (Table 13.10) by the adhesive thickness, e.g. $K_{deg} = E_{deg}/t$.*

As was defined early, a cohesive strength of 36 MPa was defined as tipping traction in mode I for the non-conditioned state. The procedure to calculate the parameters in case of adhesion failure was followed. Table 13.14 summarizes the CZM parameters for ageing SLJ.

Model validation

The calibrated parameters were applied to predict the single lab joint exposed to hygrothermal cycles. The 3D model has been validated with the experimental results. The predicted and experimental failure load for SLJ ageing under hygrothermal cycles are shown in Figure 13.53. The error bars refer to the standard deviation calculated for each data of cycles.

Table 13.14 *Properties of aged adhesive and cohesive used in the CZM of SLJ*

Adhesive	**Cohesive**						
Hygrothermal Cycles	***E***	K_n	$K_s = K_t$	T_n	$T_s = T_t$	G_{Ic}	G_{IIc}
	(Mpa)	**(N/mm³)**		**(Mpa)**		**(N/mm)**	
0	3,000	12,000	4,444	36	15	1.7	1.75
42	2,432	9,728	3,602	28	12	1.32	1.36
84	2,422	9,688	3,588	26	11	1.23	1.26
168	2,414	9,656	3,576	22	9	1.04	1.07
252	2,414	9,656	3,576	19	8	0.9	0.92
714	2,414	9,656	3,576	21	8.8	0.99	1.02

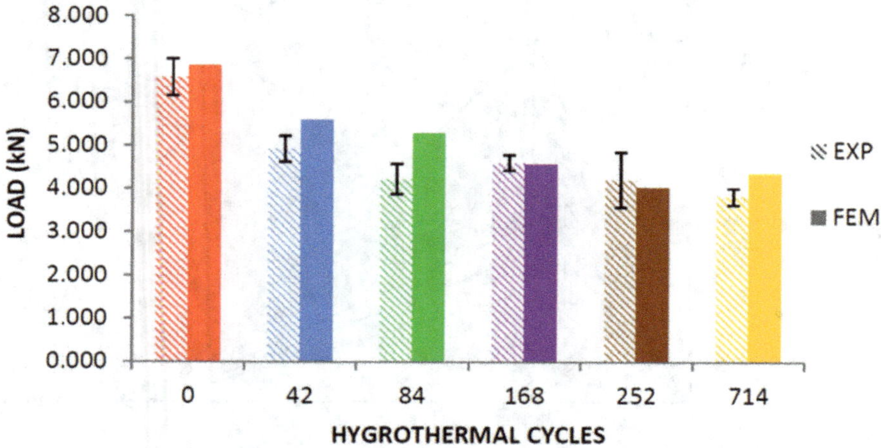

Figure 13.53 *Variation of strength of SLJ*

The variation of the static strength of SLJ, exposed to hygrothermal conditions, has demonstrated strength degradation with the increase in the number of cycles. The strength decreased noticeably in the first 42 cycles. Then a gradual reduction continued until 252 cycles reached a failure load of 4.02 kN. Later, the joint strength rose marginally at 714 cycles. Table 13.13. summarizes the experimental and numerical results.

The failure load for zero cycles was slightly over predicted, being within the standard deviation of the experimental results. The numerical stiffness was found to be around 20% higher than the experimental joint stiffness. Likewise, the displacement predicted was about 50% greater than the measured failure displacement average.

At 42 cycles, the maximum predicted load was 5.59 kN, an overprediction of almost 13%. The displacement predicted was 0.19 mm, 21% greater than the experiments. The numerical stiffness was 23% higher than in the experiments.

At 84 cycles, the overprediction of the failure load was at its highest: 5.27 kN. Likewise, the stiffness was 29% above the experimental results, while the displacement was well predicted with a value of 0.18 mm.

At 168 and 252 cycles, the model closely predicted the failure load and the displacement with 1% being the difference between the experiments and the numerical value. Nevertheless, the predicted stiffness maintained a difference of 20% with the results of the test. Finally, the prediction for 714 cycles was greater by 13% for both the maximum load and displacement.

Figure 13.54 shows the load against displacement curves for numerical and experimental results. It is noted that the model is capable of predicting the degradation of the joint strength. In summarizing, the prediction showed closer alignment with the results as the cycles increased. However, the observed experimental strength reduction at 714 cycles was not predicted by the model. The empirical

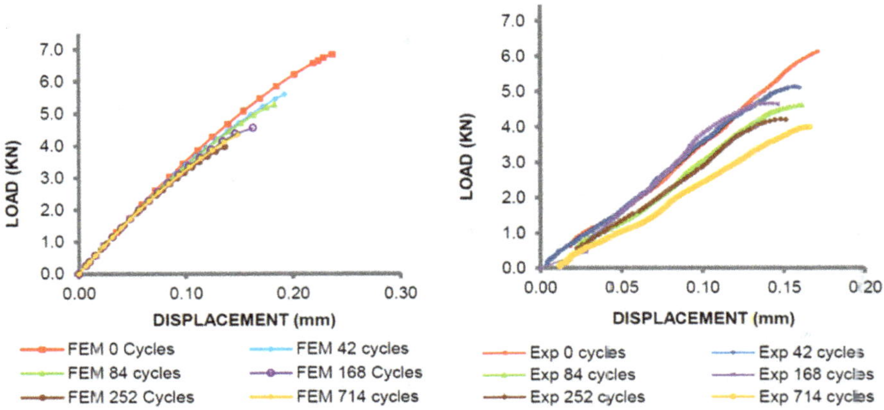

Figure 13.54 Numerical and experimental load–displacement curves FM94

equation, defined to predict the normal degraded traction, was influenced by the errors within the experimental results.

The CZM modelling developed in this study proved to be consistent with the static strength degradation that occurred in the SLJ expose to hygrothermal cycles in presence of mechanical damage (crack initiator). Also, it has been feasible to predict the damage propagation.

13.6 Discussion

13.6.1 Introduction

The aim of this work was to evaluate the second law behaviour second law behaviour structural strength of the bonded joints exposed to temperature and moisture cycles and load. Mechanical tests were performed to determine the static response of bulk adhesive and the static and fatigue response of SLJs after the hygrothermal cycles (ageing). The mechanical test results that were presented in section 13.4 are discussed in this section.

13.6.2 Moisture absorption

Moisture diffusion in both FM94 bulk and composite T800/M21 exposed to the hygrothermal cycles has been fitted to Fick's second law behaviour equation.

Table 13.15 summarizes moisture diffusion properties found in the literature and the experiments from the present study.

Fick's law has been the main model used to describe the moisture absorption moisture in adhesives, as can be seen in Table 13.15. Mainly alternative studies have focused on the absorption properties under static conditions (humidity and temperature).

Table 13.15 Comparison of moisture diffusion properties found for FM94 with other results reported

Authors	Material	Expose condition	Dh (m^2/h)	Mm (%)	Model
Aschcroft et al (2003)	Film Adhesive	Humid 90°C/97 environment % RH 12	7.20E-12	1.50	Fickian
Wahab et al (2002)	FM73	Immersion water at 60°C	1.5-2E-13		Fickian
Crocombe et al (2006)	AV119	Humid environment 50 °C/ 81.2 %RH	3.85E-13	3.06	Fickian
Liljedahl (2006)	FM73	Humid environment 70°C/79.5 %RH	7.90E-12	1.20	Fickian
Mubashar (2010) [65]	FM73	Immersion water at 50°C (cycle absorption and desorption)	9.72E-13 4.75E-14	1.78 1.78	Dual
Liu (2016)	SY14	Immersion water (60 hours) at 90°C	6.28E-12	6.77	Fickian
Experimental results	FM94	Hygrothermal cycles (70°C 85%RH and – 20°C/0 %RH), gravimetric test after removing at hot wet conditions	1.52E-12	1.54	Fickian

Nevertheless, Fick's behaviour has not been always reported for epoxy adhesives as was mentioned in section 13.2.6.2. A dual Fick model predicted the absorption diffusion of FM73 bulk adhesive exposed to absorption–desorption cycles at two different temperatures (immersed in water at 50 and 70°C), with a better agreement than Fick's second law for absorption [64].

As mentioned before in section 13.3.3.1, both FM73 and FM94 are manufactured by Cytec; FM94 shows better performance at high temperature than FM73 [94].

Comparing the experimental results of the FM94 with the absorption at dynamic conditions reported by Mubashar (absorption–desorption cycles at 50°C) [64], the rate of diffusion of the FM94 (under the hygrothermal cycles) was higher than the FM73, but the maximum moisture content of FM94 was 13% smaller than FM73. These differences of diffusion and saturation agree with

Table 13.16 Comparison of moisture diffusion properties found for CFRP T800/M21 with other results reported (PF: parallel to fibre; TF: transverse to fibre)

Authors	Material	Expose condition	Dh(m2/s)	Mm(%)	Model
Aschcroft et al (2003) [87]	CFRP UD	Humid environment 90°C/ 97 %RH	7.90E- 12	0.8	Fickian
Liljedahl (2006) [68]	IM7/8552	Humid enviroment 70°C/ 79.5 %RH	1.40E- 12 PF 5.00E- 13 TF	0.8	Fickian
Liu et al (2016) [90]	CFRP T800/ 5228E	Immersion water (60 hours) 90°C	5.60E- 13 PF 2.90E- 13 TF	1.5	Fickian
Guermazi et al (2015) [43]	CF G814	Immersion water 70°C	4.85E- 13	1.55	Fickian
Experimental results	CFRP T800/ M21 UD	Hygrothermal cycles (70°C 85 %RH / –20°C) Removed at hot wet condition (70°C 85 %RH) for gravimetric test	4.69E- 13	0.68	Fickian

Reference 105, which affirms that the speed to reach the saturation depends on the temperature, but the moisture saturation is only driven by the surrounding humidity.

On the other hand, comparing FM94 results of absorption with the results reported by Liljedahl (FM73 exposed to 70°C/79.5% RH) [68], it was noted that the diffusion coefficient was five times slower. And the maximum moisture content achieved was slightly bigger than FM73.

It indicated that the diffusion into FM94 has been affected by the dynamic conditions, which can reduce the rate of diffusion [71], and cooling after saturation can lead to more water absorption [55]. The analysis allows confirmation that, under the hygrothermal cycles studied, either temperature or relative humidity surrounding has driven the maximum moisture content in the adhesive FM94. This coincides with the results presented by Mubashar who found that the temperature increased the diffusivity and the saturated moisture content.

In case of CFRP, some moisture diffusion properties reported for CFRP laminate are summarized in Table 13.16. The diffusion coefficient obtained for FM94 was in agreement with results for CFRP found in the literature. The moisture absorption in CFRP can also be studied considering the anisotropy of the material. In this approach, it is determined the diffusion in the direction of the fibre and transverse to fibre as presented by Liu *et al*. and Liljedahl [68, 121].

The highest diffusion was shown by Ashcroft *et al.* [87] for CFRP UD at 90°C/97% RH of 7.20 E-12 m^2/s; as it has been discussed earlier, the temperature leads directly to the rate of diffusion. But the difference of maximum moisture content compared with T800/M21 (subject to hygrothermal cycles) was only 0.12%.

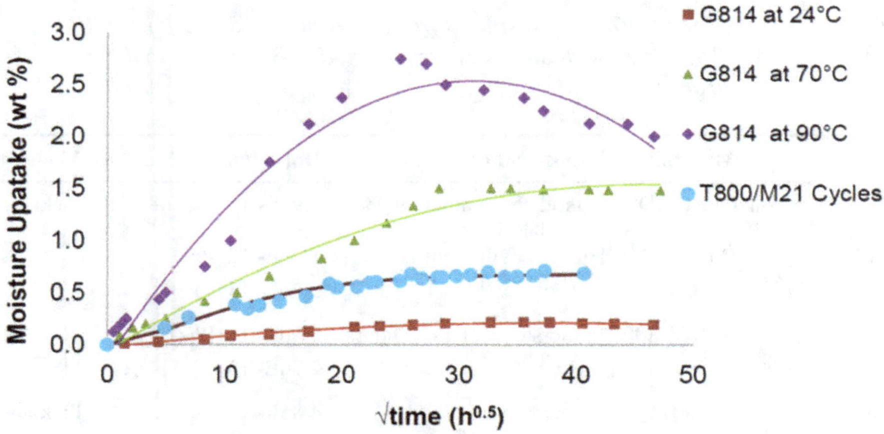

Figure 13.55 *Moisture absorption CFRP G814 (4 mm thickness) exposed to water at different temperatures reported by [43] and T800/M21 (2 mm thickness) exposed to hygrothermal cycles*

The moisture content found for T800/5228E [90] and for CF G814 [43] is significantly greater; it can be caused by the higher thickness (3.2 and 4 mm) because the maximum content increases with the laminated thickness [44].

In general, the fibre reinforced composites are capable of absorbing relatively small amounts of water; it can be seen that the maximum content of the majority of CFRP (2 mm thickness) does not exceed 1%.

A study of T300 graphite/epoxy composite immersed in water at 25, 60 and 80°C demonstrated that under 25 and 60°C, moisture absorption showed a Fickian diffusion behaviour. However, at 80°C, diffusion occurred with non-Fickian behaviour. This was attributed to the high temperature; it increases the molecular chain relaxation, decreasing the Tg and bonding strength and then the water flows into space easier [44].

The moisture absorption found for T800/M21 (2 mm thickness) under hygrothermal cycles has been compared with the moisture absorption of carbon-epoxy G814 (4 mm thickness) exposed to water at 24, 70 and 90°C reported by Guermazi *et al.* [43], as can be seen in Figure 13.55. Analysing Figure 13.55, it can be stated that the maximum for moisture content and the diffusion rate of the CFRP G814 increase with the temperature. Second, the absorption diffusion and moisture content determined for the T800/M21(under hygrothermal cycles) are lower compared with the diffusion of G814 (immersed in water at 70°C), the reduction can be attributed to the differences in thickness and the effect of cyclic conditions (temperature/moisture) suffered by T800/M21.

In conclusion, the lower diffusivity shown by T800/M21 compared with the G814 and IM7/8552 can be due to the polymer relaxation caused by the changes in humidity and moisture concentration [45].

Analysing swelling and desorption results, it can be stated that the FM94 or the T800/M21 exposed to hygrothermal cycles has not shown a molecular chemical interaction and the main absorption occurred through diffusion into the free volume. This is supported by the statement that any gain of volume in the adhesive depends on the chemical structure of the polymer, the amount of water within it and the interaction between both [29].

Modes of water ingress into the joint can take the following ways: travel through the adhesive, travel along the adhesive/adherend interface or by capillary action through crack and crazes in the adhesive, or by diffusing into the adherend [71]. The moisture permeability of the CFRP allows a shorter moisture path through the interface adhesive/adherend [122]. Hence, the moisture diffusion is faster through the interface adhesive/adherend than through the adhesive

13.6.3 Moisture effect adhesive mechanical properties

According to these results (section 13.4.2.5), moisture uptake during the hygrothermal cycles apparently doesn't affect the mechanical properties of the FM94. However, as discussed in section 2.4.1, the moisture ingress in the bulk adhesive can cause changes of the strength, stiffness and strain [108].

The results are different from the results reported for FM73 bulk adhesive samples by Sugiman *et al.* [49], who reported a degradation of 38% of tensile strength and 24% of *E* modulus, after immersion in water at 50°C over 2 years with 5.5 wt% moisture absorption. Mubashar [64] reported for the same adhesive a reduction by 39% tensile stress and 37% *E* modulus, after moisture cycles at 70°C and a moisture uptake of 2.5 wt%. Han *et al.* [112] found a reduction of E modulus by 15% for FM73, immersed in water at 50°C with a moisture content of 3.75%.

According to Cytec, FM94 offers high moisture resistance and a better performance at high temperatures than FM73 [94]. Hence, the lack of strength degradation of the FM94 exposed to hygrothermal cycles can be driven by low affinity of this polymer with the water [45].

However, the moisture content reached by FM94 after the hygrothermal cycles (1.54 wt%) should cause an effect in the plasticity due to the hygrothermal history [123]. Hence, other factors could affect the results, such as the high amount of voids in the bulk adhesives samples (as was mentioned in section 13.3.3.1) and the freezing cycles. The voids could trap the major amount of water and avoid the interaction between moisture with the polymer molecules.

The effect of freeze-thaw cycles was studied in two adhesive epoxies (sikadur 330 and sikadur 30 [52]. After exposure at 100% RH at 38°C for 8 hours followed by freezing at –18°C for 16 hours, there was evidence of differences in the mechanical responses. Sikadur 330 and sikadur 30 showed a reduction of elastic modulus by 19 and 14%, respectively. But, in the case of tensile strength, sikadur 30 did not show any important changes.

Hence, the effect of cycles at low temperature on the mechanical properties of the FM94 should be more deeply studied in the future.

13.6.4 Mechanical test results

13.6.4.1 Lap shear test

As mentioned in section 13.4.3.1, the maximum strength expected for unconditioned joint was 13.8 kN, but the joint only had a strength of 6.58 KN. The reduced strength can be accounted for by the inclusion of tip crack initiator and voids in the bondline.

The crack initiators of PTFE films were included in the bond area to try and provide consistent failure results. However, they avoided the effect of the fillet, causing an increment of the peel stress and reducing the joint strength [124].

Compared to the work by Bhanushali *et al.* [125], who evaluated SLJ of T800/M21 bonded with FM94, peel ply surface preparation and 25 mm × 25 mm overlap, the joints with tip crack were 61% weaker than joints without crack initiators [125].

The visual inspection showed a high number of voids in the bondline previous to the ageing (refer to section 13.4.3.2.1). Pre-bond moisture trapped in the adhesive combined with a vacuum bag curing can lead to void formation and cause an important effect on the bond strength [126]. According to MIL-HDBK-17, a level of 0.2% pre-bond moisture in the material can lead to a strength reduction by as much as 80% (Department of Defence from Reference 17).

The pre-bond moisture can be caused by first the hygroscopic nature of the adhesive films, as the moisture can be absorbed before use (condensation during thaw when removed from the freezer) and second the moisture can be absorbed by the T800/M21 laminate before bonding (Parker, 1983). Although the laminates were dried before bonding, the drying could not be sufficient to remove the moisture trapped.

Other causes of low failure strength of the joints can be the surface preparation mode; the literature indicates that the surface preparation treatment affects the bond strength, and the use of peel ply has shown in preliminary studies smaller strength bond compared with grit blasting and plasma treatment in bonded joints without ageing [32, 33].

The cycles of moisture and temperature in the adhesive joints, employed in this study, have resulted in a degradation of the mechanical properties. The bond strength decreased with the number of exposed cycles. The main strength degradation was achieved during the first 42 cycles of ageing (2 weeks), and then the gradient of reduction tended to level off; this polynomial behaviour has also been observed by other authors and attributed to saturation of moisture in the adhesive [21, 49, 112].

A long-term exposure is needed to reduce strength under hygrothermal static ageing; composite SLJ joints exposed to 82 °C-85% RH over 2 years were mechanically degraded by 30% and changed the failure mode [21]. Aluminium SLJs bonded with FM73 and immersed in water at 50°C were degraded by 22.1% strength after 1 year of exposure [49].

The hygrothermal cycles have caused a higher impact on the joint strength (about 40% reduction at 8 weeks) than ageing under static hot-wet conditions as it was expected and mentioned in section 13.2.4.2. Thermal cycling of the sample containing 'trapped' water molecules acts to accelerate any degradation process. However, the mechanism for this observed degradation is not yet fully understood.

Additionally, studies have shown that under cold conditions, the bond degradation does not always act in the same form [47]; Agarwal *et al.* [52] tested steel-CFRP joints bonded with two different epoxy adhesives (sikadur 330 and sikadur 30), reporting strength reduction of 27.5 and 18%, respectively, after 40 freeze-thaw cycles (6 weeks of cycles); in the case of sikadur 30, the degradation didn't level off [52]. Hu *et al.* [22] reported 5.61% strength degradation of a CFRP-steel SLJ, conditioned under temperature cycles (maximum 80°C and minimum temperature –40°C) and 20% RH for 28 days [22]. Kim [52] reported an increase of 36% after freeze-thaw cycles, attributed to additional curing of the epoxy adhesive in moisture conditioning and contrasted with the stiffness degradation.

Regarding the non-affected FM94 bulk adhesive properties after the hygrothermal cycles, discussed before, the lap joint results indicate that the main degradation occurred in the adhesive/adherend interface rather than the adhesive. Moreover, the interfacial failure mode (evidenced by broken joints) confirms the weakness of the composite/adhesive interface. As stated by Parker (1986) for bonded joints, with a tendency towards interfacial failure mode, the composite/adhesive interface may be weakened with the exposure time under moisture [48].

These findings are contrary to the findings of Banea and da Silva [42], who argue that in epoxy/FRP composite interface the thermodynamic work of adhesion remains positive after ageing in water, reducing the probability of interfacial failure [42].

Under the hygrothermal cycle, the adhesive/adherend interface degradation could have been affected by the expansion of any water present (at low temperature stage) inside the bondline; this can cause cracks and voids that leads to more moisture absorption and causes plasticization and debonding, as suggested in Reference 54. As was illustrated in , the plasticity increased with the number of cycles.

Although the plasticity was observed (as shown in), the effect of hygrothermal cycles in the strain was not clearly evidenced as expected [58], probably because the strain monitoring was not accurate enough. Moreover, the low strain effect might be due to the stress concentration caused by the voids that reduce the resistance to the crack propagation and also the presence of the crack initiators.

13.6.4.2 Fatigue test

The static strength and the fatigue life of the composite bonded joint were reduced as hygrothermal cycles increased. But, the static strength remained nearly constant after 168 hygrothermal ageing cycles, whereas the fatigue life continues its reduction.

The hygrothermal cycling effect increases with the load level; as can be seen at 45% quasi-static strength, the joint life time is reduced by a factor of roughly 1,000 compared with unaged joint. Also, can be identified the 'safe-life' at load up 30% static strength and no longer than one million fatigue cycles.

The results coincide with the reduction of fatigue reported in References 88 and 57 which tested adhesive joints aged under static hygrothermal conditions. In the same way, Ferreira *et al.* [127] evaluated the fatigue life of glass fibre reinforced bonded joints, after exposure to static hygrothermal conditions (immersed water at

20, 40 and 70°C), finding that the reduction of the number of cycles is greater and faster at higher temperature, as seen in].

Similarly, Abdo and Aglan [59] reported a reduction of 47% of the joint lifetime after 6 weeks of hygrothermal cycles condition. As with the present study, joints subjected to temperature and humidity cyclic conditions were similarly affected by the combined effect of temperature/humidity, as well as a hygrothermal expansion phenomenon that aggravated the harsh conditions.

The fatigue cycles suffered by a joint of any aircraft must be obtained by data collection or through computer algorithms. However, assuming that the joint is part of the lower wing root panel of a transport aircraft [128], 1 hour flight on average is equivalent to 100 load cycles. Hence, the SLJ life was reduced from 10.000 flight hours to 10 flight hours, after 336 hygrothermal cycles at 40% quasi-static strength.

Not many studies have shown fatigue studies of composite bonded joints, and the effect on fatigue life. Although findings have revealed that the durability of the joint has decreased with the hygrothermal cyclic attack, further evaluation of crack growth and strain variation caused by the thermal cycle is needed.

13.6.5 Non-destructive inspection adhesive bonded joints

A high amount of voids was identified in the bondline of the joints. These voids in the adhesive bond can result from the evolution of gases during the heat curing. The gases have three main sources: volatiles present in the bond process, water absorbed by the adhesive and water absorbed by the adhered [126]. Also, FM94 has been found to have a tendency to produce more voids than FM73 [106].

Using vacuum bag curing could have increased the size and number of voids because the lower pressure achieved (compared to autoclave) was unable to dissolve the bubbles and the volatiles, which remained trapped in the adhesive [126, 129]. However, voids have also been shown in the bondline of composite bonded joints cured in an autoclave [21].

The joint manufacture in the autoclave is not always possible, especially when the repairs are made *in situ*; then the quality of bond needs to be monitored as processes are developed to minimize defects and voids in the joint [6].

The ultrasonic inspection has shown indications of the bond quality; some zones with lack of adhesion were detected pre-ageing. After the hygrothermal cycles, the defects were greater; however, these results therefore need to be interpreted with caution because the inspection is limited to the technician skills dependence and the difficulties in calibration for each specimen (thickness variations causing uncertainty).

13.6.6 Degradation modelling

A coupled analysis (experimental-numerical) was implemented to model the mechanical behaviour and to predict the failure load of the adhesive SLJ aged (under hygrothermal cycles). First, the adhesive degraded properties were determined using a moisture-diffusion displacement analysis, after a CZM (set with the degraded parameters) was carried out to predict the damage for each number of hygrothermal cycles.

As mentioned in the literature review, moisture-diffusion analysis has been performed by other authors [89, 130] to identify moisture ingress process along the bondline. Then, using the moisture concentration distribution, a simultaneous stress-diffusion analysis has been performed for each element to define the moisture-stress dependence on each one. However, the present study differs from previous research because a moisture concentration value was assumed as homogeneous along the adhesive layer, at each hygrothermal cycle.

The results of the bulk adhesive moisture-diffusion displacement analysis showed a reduction on the E modulus by 20% after 42 hygrothermal cycles (1.49% moisture content) and with the increment of moisture, the elasticity remained nearly constant until the saturation (1.54% moisture content). These findings, contrary to the FM94 tensile results are consistent with other researchers [64, 131] and with the plasticity variation observed in the joints post failure evaluation. For that reason, the E degraded values were used to calculate the joint stiffness degraded as a parameter in the CZM.

Some studies have focused on the hygrothermal expansion of both adherend and adhesive, which may introduce residual stress and strain in the joint [49, 90, 131]. Thermal and swelling residual strain has been included in the FEM by means of subroutines to make more realistic simulations. However, Liljedahl et al. (2005) [131] observed that not including residual strain in the analysis doesn't necessarily affect the accuracy of the numerical models of mixed mode flexure and notched coating adhesion specimens.

The unaged CZM was performed to simulate the progressive interfacial failure, and the cohesive parameters were modified on mode II to drive the adherend/adhesive interface strength reduction. Additionally, a mechanical damage (induced crack initiators) was included in the cohesive elements to simulate the realistic SLJ test conditions. The outcomes showed a good correlation with the experiments, predicting the unaged joint strength and damage propagation, as shown in .

The degradation modelling focused on the degraded properties of the adhesive/adherend interface, and the cohesive parameters (traction and fracture energy) were reduced as the degradation rate showed in the joints after the hygrothermal cycles. Hence, the cohesive traction (for each number of hygrothermal cycles) was calculated using an empirical equation, and these values then became the parameter in the CZM. The predicted static strength showed a very good consistency with the experimental data.

The most interesting finding was that the model developed can be used as a tool that rapidly predicts the degradation of SLJs subject to hygrothermal cycles and mechanical damage.

13.7 Conclusions and contribution

The results obtained in this study show moisture absorption of both FM94 adhesive and T800/M21 laminated exposed to hygrothermal cycles followed Fick's law. The moisture absorption mechanism showed a dynamic behaviour, with a higher absorption rate during hot-wet stages as opposed to a lower rate of absorption during cold-dry stages. FM94 reached the moisture saturation of 1.54% wt at 120 hours of exposure while T800/M21 achieved 0.68 %wt at 672 hours.

The adhesive FM94 exposed to hygrothermal cycles was numerically and experimentally evaluated. Experimentally, the *E* modulus and the tensile strength did not evidence changes after the exposure. Numerically, a diffusion moisture-displacement analysis was carried out to evaluate the moisture effect on the elasticity, but by contrast, the elasticity dropped by 20% after 1.4% moisture uptake (42 hygrothermal cycles).

According to the static test, the shear strength of the SLJ showed reductions of about 42% after 714 hygrothermal cycles, results compared with the control specimen. The major reduction took place at the initial exposure, and then after 84 hygrothermal cycles, the reduction trend levelled off. Despite the reduction in the displacement to failure, an increment of adhesive plasticity was observed in the post-failure analysis.

The results show that under fatigue load conditions, there is an additional mechanism caused by temperature and humidity cycles that combined may lead to a continuous degradation. The fatigue life was reduced with the increase of the ageing cycles, with percentages greater than 45% (quasi-static strength), the life cycles diminished from 500,000 to 1,000 cycles.

The static and fatigue failure obtained from unaged and aged joints revealed principally an adhesion failure pattern, although some thin cohesive failure mode was observed in unaged joints.

A combined experimental–numerical approach was employed to predict the strength and crack propagation in aged joints. The model includes harsh and realistic conditions, such as mechanical damage (induced crack initiator) and exposition to humidity and temperature cycles. This conservative CZM model allows the behaviour of SLJ to be predicted with a good degree of accuracy. The predicted static strength was accurate to 4% for zero ageing cycles, 1% for 252 hygrothermal cycles and 13% for 714 hygrothermal cycles.

The model now developed could be employed by engineers to rapidly predict the reduction of the resistance of composite SLJs exposed to hygrothermal cycles.

The use of ultrasonic inspection to evaluate the joints at pre- and post-exposure helped identify manufacture defects and identify changes in the bond quality. The inspection with C-scan with 5 MHz, 31 dB and normal incidence technique revealed changes of amplitude interpreted as bond quality issues in unaged joints, as well as the integrity of the joint changed due to temperature and humidity cycles.

Few studies exist showing consistent results for composite joint evaluation subject to ageing cycles. The results of this investigation demonstrated a degradation tendency under static and fatigue conditions. Nevertheless, this investigation was limited by time constrains to the extent that deeper research of the individual effects of thermal expansion and moisture expansion effect in the joint is advisable and deserves to be further explored.

The study findings showed the extent of damage, strength reduction and fatigue life of the composite SLJ, providing data to validate the predicted CZM.

References

[1] Ashcroft I.A., Wahab M.M.A., Crocombe A.D., Hughes D.J., Shaw S.J. 'The effect of environment on the fatigue of bonded composite joints. part 1: testing and fractography'. *Composites Part A: Applied Science and Manufacturing*. 2001;32(1):45–58.

[2] Li J., Yan Y., Zhang T., Liang Z. 'Experimental study of adhesively bonded CFRP joints subjected to tensile loads'. *International Journal of Adhesion and Adhesives*. 2015;57(1–2):95–104.

[3] Katnam B., Silva L.F.M.D., Young T.M. 'Bonded repair of composite aircraft structures: A review of scientific challenges and opportunities'. *Progress in Aerospace Sciences*. 2013;61:26–42.

[4] Gay D., Hoa S.V. *Composite materials: design and applications*. Second. USA: CRC Press; 2007.

[5] Escarpita A., Elizalde H., Ramirez R.A. 'Caracterizacion de mecanismos de falla Y propagacion de daño en materiales compuestos parte I-revision del estado del arte'. 2007.

[6] Federal Aviation Administration FAA 'DOT/FAA/TC-14/39 literature review of weak adhesive bond fabrication and nondestructive inspection for strength measurement'. 2015.

[7] Sugiman S., Crocombe A.D., Aschroft I.A. 'The fatigue response of environmentally degraded adhesively bonded aluminium structures'. *International Journal of Adhesion and Adhesives*. 2013;41(2):80–91.

[8] Ashcroft I.A., Shaw S.J. 'Mode I fracture of epoxy bonded composite joints 2. fatigue loading'. *International Journal of Adhesion and Adhesives*. 2002;22(2):151–67.

[9] Ekh J. *Multi-fastener single-lap joints in composite structures*; 2006.

[10] Bardis J., Kedward K. 'Effects of surface preparation on long-term durability of composite adhesive bonds'. *Report No. DOT/FAA/AR-01/08*. 2001;23.

[11] Ameli A., Azari S., Papini M., Spelt J.K. 'Characterization and prediction of fracture properties in hygrothermally degraded adhesive joints: an open-faced approach'. *Journal of Adhesion Science and Technology*. 2013;27(10):1080–103.

[12] NPL National Physical Laboratory Materials Centre 'NPL manual design and testing of bonded and bolted joints'. 2007.

[13] Matthews T.T.T., Tester T.T. 'The influence of stacking sequence on the strength of bonded CFRP single LAP joints'. *International Journal of Adhesion and Adhesives*. 1985;5(1):13–18.

[14] Hart Smith L.J. 'Adhesive bonded single LAP joints'. 1973.

[15] Adams R., Wake W.C. *Structural adhesive joints in engineering*. Essex: ELSEVIER APPLIED SCIENCE; 1984.

[16] Kafkalidis M.S., Thouless M.D. 'The effects of geometry and material properties on the fracture of single lap-shear joints'. *International Journal of Solids and Structures*. 2002;39(17):4367–83.

[17] Department of Defense from United States of America 'Composite materials handbook, volume 3: polymer matrix composites materials usage'. *Design and Analysis*. 2002;3.

[18] Taib A.A., Boukhili R., Achiou S., Gordon S., Boukehili H. 'Bonded joints with composite adherends. part I. effect of specimen configuration, adhesive thickness, spew fillet and adherend stiffness on fracture'. *International Journal of Adhesion and Adhesives*. 2006;26(4):226–36.

[19] Song M.-G., Kweon J.-H., Choi J.-H, *et al.* 'Effect of manufacturing methods on the shear strength of composite single-lap bonded joints'. *Compos Struct*. 2010;92(9):2194–202.

[20] Tsai M.Y., Morton J., Matthews F.L. 'Experimental and numerical studies of a laminated composite single-lap adhesive joint'. *Journal of Composite Materials*. 1995;29(9):1254–75.

[21] G. aK, Hou T.H., M. aB, Palmieri F.L., Wohl C.J., Connell J.W. 'Hygrothermal aging of composite single LAP shear specimens comprised of AF-555M adhesive and T800H/3900-2 adherends'. *International Journal of Adhesion & Adhesives*. 2012;39:1–7.

[22] Hu P., Shi Z.W., Wang X.X., Li W.D., Zhou S.G., Han X. 'Strength degradation of adhesively bonded single-lap joints in a cyclic-temperature environment using a cohesive zone model'. *The Journal of Adhesion*. 2015;91(8):587–603.

[23] Hong J.-H., Han M.-G., Chang S.-H. 'The effects of environmental temperature and additional pressure on the bonding characteristics of co-cured carbon/epoxy-aluminum single LAP joints'. *The Journal of Adhesion*. 2015;91(3):197–213.

[24] Zhai Y., Li D., Li X., Wang L., Yin Y. 'An experimental study on the effect of bolt-hole clearance and bolt torque on single-lap, countersunk composite joints'. *Compos Struct*. 2015;127(1–4):411–19.

[25] Ataş A., Soutis C. 'Application of cohesive zone elements in damage analysis of composites: strength prediction of a single-bolted joint in CFRP laminates'. *International Journal of Non-Linear Mechanics*. 2014;66:96–104.

[26] Bodjona K., Raju K., Lim G.-H., Lessard L. 'Load sharing in single-lap bonded/bolted composite joints. part I: model development and validation'. *Compos Struct*. 2015;129(7):268–75.

[27] Saleem M., Zitoune R., El Sawi I., Bougherara H. 'Role of the surface quality on the mechanical behavior of CFRP bolted composite joints'. *International Journal of Fatigue*. 2015;80(11):246–56.

[28] Abdel Wahab M.M. 'Fatigue in adhesively bonded joints: a review'. *ISRN Materials Science*. 2012;2012:1–25.

[29] ASM International Handbook Commitee *Adhesives and selants. engineered materials handbook*. Vol. 3. ASM International Handbook Commitee; 1990.

[30] Schmid Fuertes T.A., Kruse T., Körwien T., Geistbeck M. 'Bonding of CFRP primary aerospace structures – discussion of the certification boundary conditions and related technology fields addressing the needs for development'. *Composite Interfaces*. 2015;22(8):795–808.

[31] Delmdahl R., Gabler F. 'Excimer lasers clean cfrps for optimised adhesive bonding'. *The Laser User*. 2013:72.

[32] Kumar Singh R. 'Identification and evaluation of QA tools for composite surface for adhesive bonding'. 2015.

[33] Gude M.R., Prolongo S.G., Ureña A. 'Hygrothermal ageing of adhesive joints with nanoreinforced adhesives and different surface treatments of carbon fibre/epoxy substrates'. *International Journal of Adhesion and Adhesives*. 2013;40(250):179–87.

[34] Vaananen R., Heikkila P., Tuominen M., Kuusipalo J., Harlin A. 'Fast and efficient surface treatment for nonwoven materials by atmospheric pressure plasma'. *AUTEX Research Journal*. 2010;10:8–13.

[35] Serrano J.S., Urena A., Lazcano Urena S., Blanco Verala T. 'New approach to surface preparation for adhesive bonding of aeronautical composites: atmospheric pressure plasma. studies on the pretreatment lifetime and durability of the bondline'. *Composite Interfaces*. 2015;6440:22–26.

[36] Hart-Smith L.J., Redmon G., Davis M. 'The curse of the nylon peel ply'. 1996.

[37] Madge M., Brousseau E., Eaton M., Fetherston C., Williams E. 'Materia prima'. *Adhesive Bonding of Laser Processed CFRP Panels*. 2014;21(1).

[38] Gomatam R.R., Sancaktar E. 'Effects of various adherend surface treatments on fatigue behavior of joints bonded with a silver-filled electronically conductive adhesive'. *Journal of Adhesion Science and Technology*. 2005;19(8):659–78.

[39] Encinas N., Oakley B.R., Belcher M.A, *et al*. 'Surface modification of aircraft used composites for adhesive bonding'. *International Journal of Adhesion and Adhesives*. 2014;50(5):157–63.

[40] Kumar B.G., Singh R.P., Nakamura T. 'Degradation of carbon fiber-reinforced'. *Journal of Composite Materials*. 2002;36(24):2713–33.

[41] Pantelakis S., Tserpes K.I. 'Adhesive bonding of composite aircraft structures: challenges and recent developments'. *Science China Physics, Mechanics and Astronomy*. 2014;57(1):2–11.

[42] Banea M.D., da Silva L.F.M. 'Adhesively bonded joints in composite materials: an overview'. *Proceedings of the Institution of Mechanical Engineers, Part L: Journal of Materials: Design and Applications*. 2009;223(1):1–18.

[43] Guermazi N., Ben Tarjem A., Ksouri I., Ayedi H.F. 'On the durability of FRP composites for aircraft structures in hygrothermal conditioning'. *Composites Part B: Engineering*. 2016;85(Special issue-2):294–304.

[44] Wong 'Moisture absorption characteristics and effects on mechanical behaviour of carbon/epoxy composite: application to bonded patch repairs of composite structures'. 2013.

[45] Vanlandingham M.R., Eduljee R.F., Gillespie J.W. 'Moisture diffusion in epoxy systems'. *Journal of Applied Polymer Science*. 1999;71(5):787–98.
[46] Maxwell aS, Broughton W.R., Dean G., Sims G.D. 'Review of accelerated ageing methods'. 2005;84.
[47] Heshmati M., Haghani R., Al-Emrani M. 'Environmental durability of adhesively bonded FRP/steel joints in civil engineering applications: state of the art'. *Composites Part B: Engineering*. 2015;81:259–75.
[48] Parker B.M. 'Some effects of moisture on adhesive-bonded CFRP-CFRP joints'. *Compos Struct*. 1986;6(1–3):123–39.
[49] Sugiman S., Crocombe A.D., Aschroft I.A. 'Experimental and numerical investigation of the static response of environmentally aged adhesively bonded joints'. *International Journal of Adhesion and Adhesives*. 2013;40(3):224–37.
[50] Korta J., Mlyniec A., Uhl T. 'Experimental and numerical study on the effect of humidity-temperature cycling on structural multi-material adhesive joints'. *Composites Part B: Engineering*. 2015;79(1):621–30.
[51] DeVries K.L. 'Environmental effects' in Dostal, Ed C. (ed.). *Engineered materials handbook. volume 3 adhesives and sealants*. United States of America: ASM International Handbook Committee; 1990. pp. 613–15.
[52] Agarwal A., Foster S.J., Hamed E., Ng T.S. 'Influence of freeze–thaw cycling on the bond strength of steel–FRP LAP joints'. *Composites Part B: Engineering*. 2014;60(4):178–85.
[53] Jedidi J., Jacquemin F., Vautrin A. 'Design of accelerated hygrothermal cycles on polymer matrix composites in the case of a supersonic aircraft'. *Compos Struct*. 2005;68(4):429–37.
[54] Wang J., GangaRao H., Liang R., Liu W. 'Durability and prediction models of fiber-reinforced polymer composites under various environmental conditions: a critical review'. *Journal of Reinforced Plastics and Composites*. 2016;35(3):179–211.
[55] Powers D. 'Interaction of water with epoxy'. 2009.
[56] Abdel Wahab M.M. 'Fatigue in adhesively bonded joints: a review'. *ISRN Materials Science*. 2012;2012:1–25.
[57] Broughton R.D. 'Report no. 12 cyclic fatigue testing of adhesive joints environmental effects'. 1999.
[58] Costa M., Viana G., da Silva L.F.M., Campilho R.D.S.G. 'Environmental effect on the fatigue degradation of adhesive joints: a review'. *The Journal of Adhesion*. 2017;93(1–2):127–46.
[59] Abdo Z., Aglan H. 'Analysis of aircraft adhesive joints under combined thermal and mechanical cyclic loadings'. *Journal of Adhesion Science and Technology*. 1997;11(7):941–56.
[60] Federal Aviation Administration FAA 'Bonded joints and structures techniques issues and certification considerations; PS-ACE100-2005- 1003'. 2005.
[61] Kapadia A. 'Best practice guide non-destructive testing of composite materials'. 2006.

[62] Ehrhart B., Valeske B., Muller C.-E., Bockenheimer C. 'Methods for the quality assessment of adhesive bonded CFRP structures - A resumé'. *2nd International Symposium on NDT in Aerospace 2010*; 2010. pp. 1–9.
[63] Australian R., Heslehurst D.F.A. 'Optical NDT of adhesively bonded joints'. *Materials Eevaluation*. 2009;67(7):837–42.
[64] Mubashar A. 'Modelling degradation in adhesive joints subjected to fluctuating service conditions'. 2010.
[65] Broughton B. *Measurement good practice guide no.28 durability performance of adhesive joints*. Teddington Middlesex Uk; 2000.
[66] Bank L.C., Gentry T.R., Barkatt A. 'Accelerated test methods to determine the long-term behavior of FRP composite structures: environmental effects'. *Journal of Reinforced Plastics and Composites*. 1995;14(6):559–87.
[67] Wahab M.M.A., Ashcroft I.A., Crocombe A.D., Shaw S.J. 'Diffusion of moisture in adhesively bonded joints'. *The Journal of Adhesion*. 2001;77(1):43–80.
[68] Liljedahl C.D.M. 'Modelling the interfacial degradation in adhesively bonded joints'. 2006.
[69] Crank J. 'The mathematics of diffusion'. 1975.
[70] Ameli A., Datla N.V., Papini M., Spelt J.K. 'Hygrothermal properties of highly toughened epoxy adhesives'. *The Journal of Adhesion*. 2010;86(7):698–725.
[71] Kinloch A.J. *Durability of structural adhesives*. London: Applied Science Publishers Ltd; 1983.
[72] Katnam K.B., Crocombe A.D., Khoramishad H., Ashcroft I.A, *et al.* 'The static failure of adhesively bonded metal laminate structures: a cohesive zone approach'. *Journal of Adhesion Science and Technology*. 2011;25(10):1131–57.
[73] Högberg J.L. 'Mixed mode cohesive law'. *International Journal of Fracture*. 2006;141(3–4):549–59.
[74] Hallett S.R., Harper P.W. 'Modelling delamination with cohesive interface elements'. *Numerical Modelling of Failure in Advanced Composite Materials*. 2015:55–72.
[75] Roshdestwensky K.N., Mroginski J.L., Morel C.A., D'Ambra R.J.B., Podestá J.M. 'Modelado numérico del comportamiento de la interfase en materiales compuestos'. *Argentine Association for Computational*. 2013;XXXII:1653–70.
[76] da Silva L.F.M., R. D. S. G.C. 'Advances in numerical modeling of adhesive joints'. 2012.
[77] Högberg J.L., Salomonsson K. 'Simulation of an adhesive layer using a novel mixed mode cohesive law'. 2005;46.
[78] Eliasson S., Lundberg A. 'Investigation and comparison of cohesive zone models for simulation of crack propagation'. 2015::1–5.
[79] van den Bosch M.J., Schreurs P.J.G., Geers M.G.D. 'An improved description of the exponential xu and needleman cohesive zone law for mixed-mode decohesion'. *Engineering Fracture Mechanics*. 2006;73(9):1220–34.

[80] Needleman A. 'A continuum model for void nucleation by inclusion debonding'. *Journal of Applied Mechanics*. 1987;54(3):525–31.

[81] Tvergaard V. 'Effect of fibre debonding in a whisker-reinforced metal'. *Materials Science and Engineering: A*. 1990;125(2):203–13.

[82] Tvergaard V., Hutchinson J.W. 'The relation between crack growth resistance and fracture process parameters in elastic-plastic solids'. *Journal of the Mechanics and Physics of Solids*. 1992;40(6):1377–97.

[83] Xu X.-P., Needleman A. 'Void nucleation by inclusion debonding in a crystal matrix'. *Modelling and Simulation in Materials Science and Engineering*. 1993;1(2):111–32.

[84] Scheider I., Brocks W. 'Simulation of cup–cone fracture using the cohesive model'. *Engineering Fracture Mechanics*. 2003;70(14):1943–61.

[85] Camacho G.T., Ortiz M. 'Computational modelling of impact damage in brittle materials'. *International Journal of Solids and Structures*. 1996;**33**(20):2899–938.

[86] Pinto A.M.G., Magalhães A.G., Campilho R.D.S.G., *et al*. 'Single-lap joints of similar and dissimilar adherends bonded with an acrylic adhesive'. *The Journal of Adhesion*. 2009;85(6):351–76.

[87] I. A., Wahab M.M.A., Crocombe D. 'Predicting degradation in bonded composite joints using a semi-coupled finite- element method'. *Mechanics of Advanced Materials and Structures*. 2003;10(3):227–48.

[88] Katnam K.B., Crocombe A.D., Sugiman H., Khoramishad H., Ashcroft I.A. 'Static and fatigue failures of adhesively bonded laminate joints in moist environments'. *International Journal of Damage Mechanics*. 2011;20(8):1217–42.

[89] Han X., Crocombe A.D., Anwar S.N.R., Hu P., Li W.D. 'The effect of a hot–wet environment on adhesively bonded joints under a sustained load'. *The Journal of Adhesion*. 2014;**90**(5–6):420–36.

[90] Liu S., Cheng X., Zhang Q., Zhang J., Bao J., Guo X. 'An investigation of hygrothermal effects on adhesive materials and double LAP shear joints of CFRP composite laminates'. *Composites Part B: Engineering*. 2016;91(1):431–40.

[91] Broughton W.R., Crocker L.E., Gower M.R.L. 'Design requirements for bonded and bolted composite structures'. 2002;46.

[92] Soden P.D., Hinton M.J., Kaddour A.S., Properties "Lamina 'Lay-up configurations and loading conditions for a range of fibre-reinforced composite laminates'. 1998;58.

[93] Hexcel 'Hexply M21 product data'. *Product Data*. 2010::1–6.

[94] Industries Cytec. 'Aerospace materials FM ® 94 adhesive film'. 2015::1–9.

[95] standard A. 'ASTM D 5229– 14 – standard test method for moisture absorption properties and equilibrium conditioning of polymer matrix composite materials'. *Annual B. ASTM Stand*. 2014;92:1–13.

[96] British Standards Institution *BS EN ISO 62:2008 plastics-determination of water absorption*. London: British Standards; 2008.

[97] Christian S.J. *Mechanical characterization and structural assessment of biocomposites for construction*. Stanford University; 2009.

[98] Marín L., Trias D., Badalló P., Rus G., Mayugo J.A. 'Optimization of composite stiffened panels under mechanical and hygrothermal loads using neural networks and genetic algorithms'. *Compos Struct*. 2012;**94**(11):3321–26.

[99] British Standards Institutions 'BS EN ISO 527:2012 plastics— determination of tensile properties. part 2: test conditions for moulding and extrusion plastics'. 2012;44.

[100] Institution B.S. 'BS EN ISO 527-3 plastics — determination of tensile properties- part 3: test conditions for films and sheets'. *Part*. 2004;1:527–1.

[101] Broughton W.R., Dean G.D. *Measurement Good Practice Guide No. 104: Fatigue and Creep Testing of Adhesives and Thermoplastic Joined Systems*. 2007;104.

[102] Palumbo D., Tamborrino R., Galietti U., Aversa P., Tatì A., Luprano V.A.M. 'Ultrasonic analysis and lock-in thermography for debonding evaluation of composite adhesive joints'. *NDT & E International*. 2016;78(4):1–9.

[103] Vijayakumar R.L., Bhat M.R., Murthy C.R.L. 'Non destructive evaluation of adhesively bonded carbon fiber reinforced composite lap joints with varied bond quality'. *AIP Conference Proceedings*. 2012;1430(31):1276–83.

[104] standard A. 'ASTM D 5229– 92 – standard test method for moisture absorption properties and equilibrium conditioning of polymer matrix composite materials'. *Annu. B. ASTM Stand*. 2010;92:1–13.

[105] Seneviratne W., Tomblin J., Kittur M. *Fatigue and fracture of adhesively-bonded composite joints*. Elsevier; 2015.

[106] Roh C.T.S. 'The strength of double strap joint with brittle and ductile adhesives'. *Failure in Composites*. 2013;4:231–39.

[107] Zavatta N. *Influence of adhesive thickness on adhesively bonded joints under fatigue loading*. Universita di Bologna; 2015.

[108] Viana G., Costa M., Banea M., da Silva L. 'A review on the temperature and moisture degradation of adhesive joints'. *Proceedings of the Institution of Mechanical Engineers, Part L Journal of Materials Design and Applications*. 2016:1–14.

[109] Park Y.-B., Song M.-G., Kim J.-J., Kweon J.-H., Choi J.-H. 'Strength of carbon/epoxy composite single-lap bonded joints in various environmental conditions'. *Compos Struct*. 2010;92(9):2173–80.

[110] Vijaya Kumar R.L., Bhat M.R., Murthy C.R.L. 'Some studies on evaluation of degradation in composite adhesive joints using ultrasonic techniques'. *Ultrasonics*. 2013;53(6):1150–62.

[111] Szekeres A. 'Analogy between heat and moisture'. *Computers & Structures*. 2000;76(1–3):145–52.

[112] Han X., Crocombe A.D., Anwar S.N.R., Hu P. 'The strength prediction of adhesive single LAP joints exposed to long term loading in a hostile environment'. *International Journal of Adhesion and Adhesives*. 2014;**55**(5–6):1–11.

[113] Fallis A., Techniques D., Dassault Systèmes Simulia 'ABAQUS documentation'. *Abaqus 6.12*. 2013;53(9):1689–99.

[114] O'Mahoney D.C., Katnam K.B. 'Taguchi analysis of bonded composite single-lap joints using a combined interface-adhesive damage model'. *International Journal of Adhesion and Adhesives*. 2013;40:168–78.

[115] Song K., Davila C., Rose C. 'Guidelines and parameter selection for the simulation of progressive delamination'. *2008 ABAQUS User's Conference*; 2008. pp. 1–15.

[116] Fernández-Cañadas L.M., Iváñez I., Sanchez-Saez S. 'Influence of the cohesive law shape on the composite adhesively-bonded patch repair behaviour'. *Composites Part B: Engineering*. 2016;91(1):414–21.

[117] Jimenez S., Duddu R. 'On the parametric sensitivity of cohesive zone models for high-cycle fatigue delamination of composites'. *International Journal of Solids and Structures*. 2016;82(2):111–24.

[118] Xu W., Wei Y. 'Strength and interface failure mechanism of adhesive joints'. *International Journal of Adhesion and Adhesives*. 2012;34:80–92.

[119] Campilho R.D.S.G., Moura M.F.S.F., Domingues J.M. 'Using a cohesive damage model to predict the tensile behaviour of CFRP single-strap repairs'. *International Journal of Solids and Structures*. 2008;45:1497–512.

[120] Belnoue J.P.-H., Giannis S., Dawson M., Hallett S.R. 'Cohesive/adhesive failure interaction in ductile adhesive joints part II: quasi-static and fatigue analysis of double lap-joint specimens subjected to through-thickness compressive loading'. *International Journal of Adhesion and Adhesives*. 2016;68:369–78.

[121] Li M., Liu H., Gu Y., Li Y., Zhang Z. 'Effects of carbon fiber surface characteristics on interfacial bonding of epoxy resin composite subjected to hygrothermal treatments'. *Applied Surface Science*. 2014;288:666–72.

[122] Heshmati M., Haghani R., Al-Emrani M. 'Environmental durability of adhesively bonded FRP/steel joints in civil engineering applications: state of the art'. *Composites Part B: Engineering*. 2015;81:259–75.

[123] Chen X., Lin Y.C., Chen X. 'Moisture sorption-desorption-resorption characteristics and its effect on the mechanical behavior of the epoxy'. 2014;2005.

[124] Kim K.-S., Yoo J.-S., Yi Y.-M., Kim C.-G. 'Failure mode and strength of uni-directional composite single LAP bonded joints with different bonding methods'. *Compos Struct*. 2006;**72**(4):477–85.

[125] Bhanushali R., Ayre D., Nezhad H.Y. 'Tensile response of adhesively bonded composite-to-composite single-lap joints in the presence of bond deficiency'. *Procedia CIRP*. 2017;59(2):139–43.

[126] Pearce P.J., Arnott D.R., Camilleri A., Kindermann M.R., Mathys G.I., Wilson A.R. 'Cause and effect of void formation during vacuum bag curing of epoxy film adhesives'. *Journal of Adhesion Science and Technology*. 1998;12(6):567–84.

[127] Ferreira J.A.M., Reis P.N., Costa J.D.M., Richardson M.O.W. 'Fatigue behaviour of composite adhesive LAP joints'. *Composites Science and Technology*. 2002;62(10):1373–79.

[128] Lowak D., De Jonge H., Franz J.B., Schutz J. 'The transport wing standard programme (twist)'. 1979.

[129] da Silva L.F.M., Adams R.D., Gibbs M. 'Manufacture of adhesive joints and bulk specimens with high-temperature adhesives'. *International Journal of Adhesion and Adhesives*. 2004;24(1):69–83.

[130] Katnam K.B., Sargent J.P., Crocombe A.D., Khoramishad H., Ashcroft I.A. 'Characterisation of moisture-dependent cohesive zone properties for adhesively bonded joints'. *Engineering Fracture Mechanics*. 2010;77(16):3105–19.

[131] Liljedahl C.D.M., Crocombe A.D., Wahab M.A., Ashcroft I.A. 'The effect of residual strains on the progressive damage modelling of environmentally degraded adhesive joints'. *Journal of Adhesion Science and Technology*. 2005;19(7):525–47.

[132] Kregting R. 'Cohesive zone models'. 2005::1–31.

Chapter 14

Performance of aerospace composites in the presence of process-induced defects

Dimosthenis Stratakis[1], David Ayre[1], and Hamed Yazdani Nezhad[1,2]

This particular research constitutes a research and development project; a novel technique for the embedment of highly thin (nearly zero-thickness) defects in aerospace grade adhesively bonded composite joints is the dominant subject. These defects consist a representative of poor polymer processing and impact induced defects, while this project studies the mechanical response of the joints in the presence of such defects. This was assessed using composite-to-composite single-lap joint (SLJ) specimens having different width and defect type. With respect to this project, two categories of defects with two different widths were recognised for laboratory testing; therefore, four configurations were considered. The joints were manufactured via manual lay-up of carbon fibre-reinforced composite prepreg and autoclave curing, followed by non-destructive inspection (ultrasonic C-scan, pulsed-echo thermography and infrared thermography) as an attempt for detecting embedded zero-thickness defects. The joints were then subjected to mechanical tension loading to determine the ultimate failure load and displacement. Finally, the fracture area of the failed joints in the bond area was observed with optical microscopy to visualise and determine the failure mechanisms, i.e. cohesive and adhesion failure.

14.1 Introduction

During previous years, the most generally known technique for joining in aircraft manufacturing was the mechanical fastening. However, over the last 50 years, the application of adhesively bonded joints has been increased in the manufacture of the primary aircraft fuselage and wing structures [1]. Using the traditional mechanical fastenings, i.e. bolts or rivets, by drilling holes in the composite structure, can

[1]Enhanced Composites and Structures Centre, School of Aerospace, Transport and Manufacturing, Cranfield University, Bedford, UK

[2]Aeronautics and Aerospace Research Centre, Department of Mechanical Engineering and Aeronautics, University of London, London, UK

damage the material and lead to an early and unexpected failure. This is because these types of joints are prone to large stress concentrators around the fastening area [2]. Adhesive bonding can provide more uniform load transfer; furthermore, it can eliminate the stress concentrators due to mechanical fastenings [3].

Moreover, the strength, the load transfer corrosion resistance and the improved durability of the structural adhesives have been significantly improved. This has allowed the replacement of mechanical fastening joints with adhesively bonded joints [4]. For example, adhesively bonded joints are extensively used on aircraft, including the Boeing 787 or Airbus A380 [5]. More specific, adhesives are used to bond the stringers to fuselage and wing skins to stiffen the structures against buckling [1].

To allow adhesive joints to be used more widely in primary structures, one must ensure their reliability [2]. A full understanding of the type and likelihood of defects occurance that may affect the joint efficiency must be established [6]. Therefore, a non-destructive evaluation must be developed, which also can detect types of defects that can occur in adhesive bonded joints [2].

In addition, missing damage tolerance behaviour of bonded joints in size scaling is a major problem. The assessment of damage tolerance is needed to ensure that fatigue, intrinsic/discrete damage, large area manufacturing flaws or severe accidental damage can occur within the operational life of the aircraft [7], the remaining structure will resist to loads without the failure of excessive structural deformation until the damage is detected [8]. According to Robert [8], the key factor for the categorisation of the damage is first of all the identification of the damage itself. It has been noted that approximately 80–90% of the damage has been inspected visually [8]. Therefore, it is necessary for the diagnostic tools to characterise and categorise the damage, more particular, the internal damage in composites. It can be observed in Figure 14.1 that test pyramid indicates the development of a strategy for experimental testing, which is important for a more strategic overall design.

14.1.1 Aim

The aim of this particular research was to investigate a novel technique, which researches the damage behaviour of embedding relatively thin defects, with different geometries, in aerospace grade adhesively bonded composite joints produced with different widths.

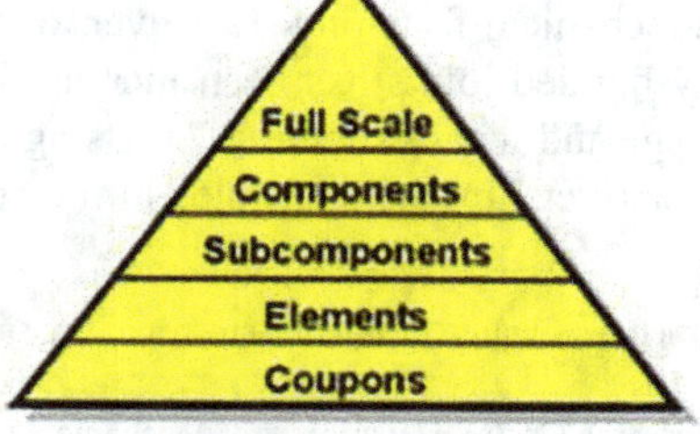

Figure 14.1 Test pyramid [8]

14.1.2 Objectives

1. Manufacture the adhesively bonded joints with two different types of defects (through-the-width and semicircular).
2. Investigate the possibility of detecting the embedded zero-thickness defects by utilising non-destructive inspection techniques (ultrasonic C-scan, pulsed-echo thermography and infrared thermography).
3. Determine the ultimate failure load and displacement by subjected the joints mechanical tension loading.
4. Determine the failure mechanisms, i.e. cohesive and adhesive failure, of the failed joints by observing the fractured areas with optical microscopy.

14.2 Materials and manufacturing

14.2.1 Materials

14.2.1.1 Adherend

The CFRP was manufactured by using the unidirectional prepreg HexPly® 8552/268GSM/IM7. HexPly® 8552 is a high-performance tough epoxy matrix used in primary aerospace structures. It exhibits good impact resistance and damage tolerance for a wide range of applications was developed to operate in environments up to 121°C [9]. HexTow® IM7 carbon fibre is a continuous, high-performance, intermediate modulus, PAN-based fibre [10].

14.2.1.2 Pretreatment product

Oxford Advanced Surfaces has developed Onto™ SB 1050, a functional surface treatment that improves the adhesive bond strength in bonded systems [11]. This product is part of the SB 1000 series of unique adhesion promoters. The treatment is designed to promote adhesion of polyurethane and epoxy adhesives. It is a practical solution to the surface preparation that is simple to use and requires no specialist coating equipment [12].

14.2.1.3 Adhesive

The adhesive material was the adhesive film Cytec ® 94. It is a modified epoxy film adhesive designed for bonding metallic and composite structures. It offers high-temperature performance (service temperature is at 104°C), toughness and moisture resistance. Moreover, FM® 94 exhibits superior elongation, toughness and shear strength properties for composites and metals. The controlled flow characteristics allow for its use in co-cured applications as well as bonding pre-cured structures.

14.2.2 Manufacturing

14.2.2.1 Test plan

In this research, the experiments were designed for static test, more specific shear-tensile, to test the shear strength and the maximum displacement of the adhesive joints. Hence, the SLJ was selected as the preferred method, because of their cost effectiveness, ease of manufacturing. In addition, SLJ are the most simplified method for competitive study.

With respect to this project, two different specimens were manufactured with two different widths and defect types, therefore, four configurations were considered. Specifically, the first configuration was labelled as Category 0. The width of this category was 50 mm, moreover, Categories II and III had the same width. The difference among these categories is that Category 0 did not have any defect. On the other hand, Categories II and III had a semicircular (thumb nail) defect type. However, Categories II and III had a remarkable difference. Category II had one layer of adhesive, while Category III had two layers, as shown in Figure 14.2. In contrast, the width of Category I was 25 mm and the type of the embedded defect was through-the-width, as illustrated in Figure 14.3.

The ASTM standard used for this project was D3165. The dimensions of the specimens for this ASTM were 114.3 (length) × 25.4 mm (width). However, in this project, the study focused more on wider specimens with double the width, because in this research wide specimens were compared with narrow ones. The reason was that specimens with the same length (114 mm) and different width (50 mm) were manufactured. The overlap length was chosen according to ASTM D5868 and not D3165, because, for this standard the overlap area is square.

Table 14.1 summarises the different categories with their experimental labelled, and Table 14.2 shows the total number of manufactured specimens.

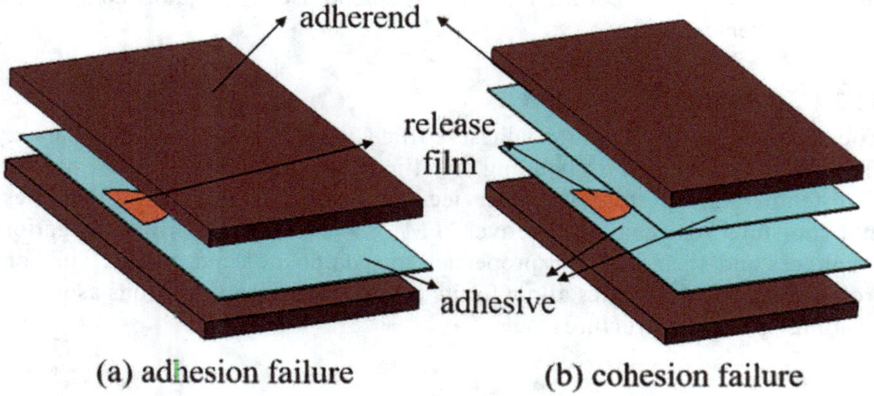

Figure 14.2 Schematic of Category II (a) and Category III (b)

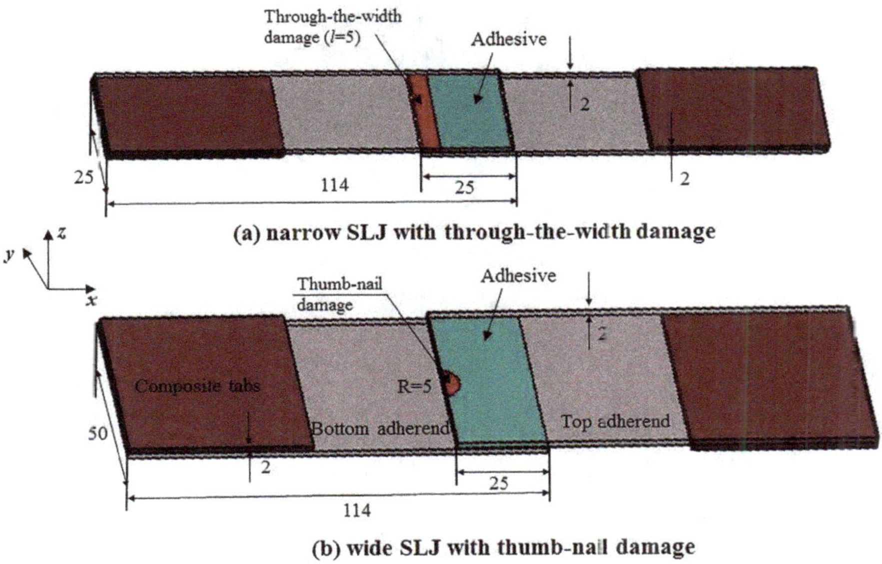

Figure 14.3 *Schematic of specimens based on the different width and defect type*

14.2.2.2 Manufacturing of adherend material

As aforementioned in the subsection 14.1.1, the plates were manufactured using the HexPly® 8552/268GSM/IM7, via manual lay-up of prepreg carbon fibre-reinforced composite and following by autoclave curing. The dimensions of the plates were

Table 14.1 *Categories of the specimens*

Category	Type of width and defect	Experimental label
0	Wide without defect	W-W
I	Narrow with through-the-width defect	N-D
II	Wide with semicircular defect (adhesion failure)	W-DA
III	Wide with semicircular defect (cohesion failure)	W-DC

Table 14.2 *Total number of manufactured specimens*

Specimens label	Number of specimens for manufacturing
W-W	3
N-D	3
W-DA	3
W-DC	3

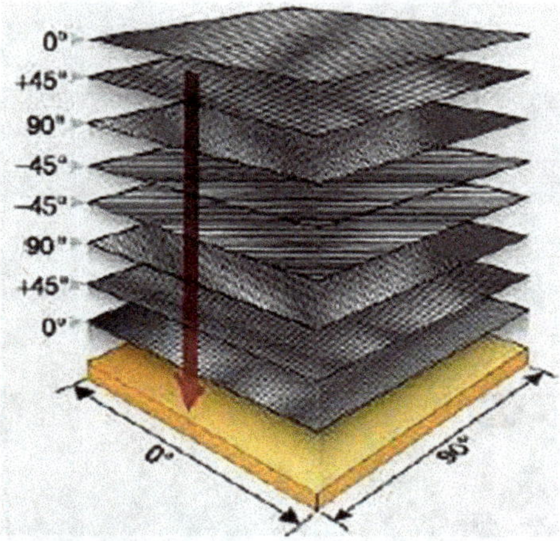

Figure 14.4 Stacking sequence

600 × 600 mm. To achieve uniform properties in all directions, quasi-isotropic lay-up was chosen. The laminate sequence was $[0/+45/90/-45]_s$, as it can be seen in Figure 14.4. The thickness of each ply was 0.25 mm. In this project, eight plies were used in order to manufacture the plates. Therefore, the total thickness of the plates was 0.25 × 8 = 2 mm (minimum).

14.2.2.3 Curing of adherend material

The curing of HexPly® 8552/268GSM/IM7 was held on in an autoclave (Table 14.3). The autoclave was selected to reduce the preparation time of the samples. The curing specifications were employed according to the data sheet of HexPly®8552 .

The selected autoclave was Aero-Form Ltd. (Figure 14.5.). The maximum operating temperature and pressure were 200°C and 100 PSI, respectively, which meets the prepreg curing requirements.

14.2.2.4 Surface treatment

As it has been mentioned above in the literature review, the role of the surface is highly important during the bonding process because it is a crucial factor in order to ensure the quality of the adhesive bonding [13]. For this project, two types of surface

Table 14.3 Curing specification for 8552

Heating rate	**1–3°C/min**
Cure temperature	180°C ± 5
Pressure	7 bars
Vacuum	1 bar

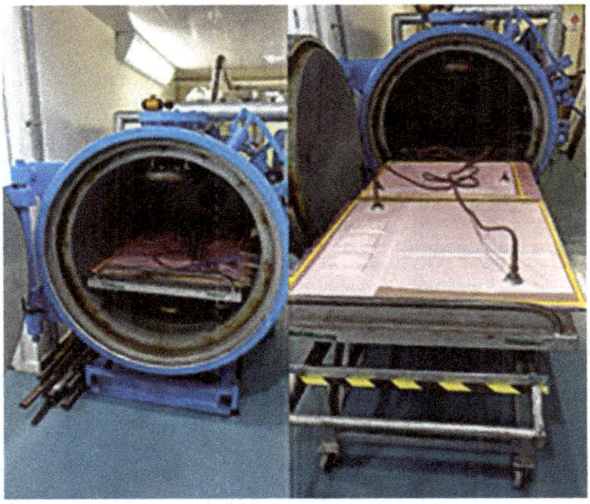

Figure 14.5 Preparation of the plate for the autoclave

treatment were utilised: peel ply and the pretreatment product of Advanced Oxford Surfaces Onto™ SB 1050.

In general, peel ply is the most widely used technique for surface treatment since it ensures continue worthiness, is very rapid and cost-effective [14, 15]. However, peel ply is not an entirely reliable technique for bonding [16] as opposed to techniques such as plasma treatment. This is the reason that the Onto™ SB 1050 had been introduced. The application of the pretreatment Onto™ SB 1050 product was held just before the adhesive bonding. After the completion of the curing process, the peel ply was removed from the cured plate.

14.2.2.5 Cutting

After the curing of the plate in the autoclave, the plate was ready to be cut into coupons. The plates were cut according to ASTM D3165 [17]. Figure 14.6. shows the exact dimensions of the cutting. The cutting was held with a table wheel saw in the laboratory, as shown in Figure 14.7. According to the standard, the coupons had to be cut in 114.3 × 25.4 mm^2; however, to simplify the procedure of cutting, the final dimensions of coupons were 114 × 25 mm^2 (length × width). Moreover, the dimensions of some coupons (Figure 14.8.) were 114 × 50 mm^2.

14.2.2.6 Embedding of the defects

The embedding of the defects was the most challenging stage of this project. This was because the embedment of the highly thin (nearly zero-thickness) semicircular (thumb nail) defect was a novel technique. Moreover, the creation of this kind of defect was quite challenging. On the other hand, the introducing of the through-the-width defect had some difficulties, however, it was not so challenging.

The first stage was the creation of the two different types of defects. The defect in both types was a release film, with the thickness of 25 μm. An A4 paper sheet had

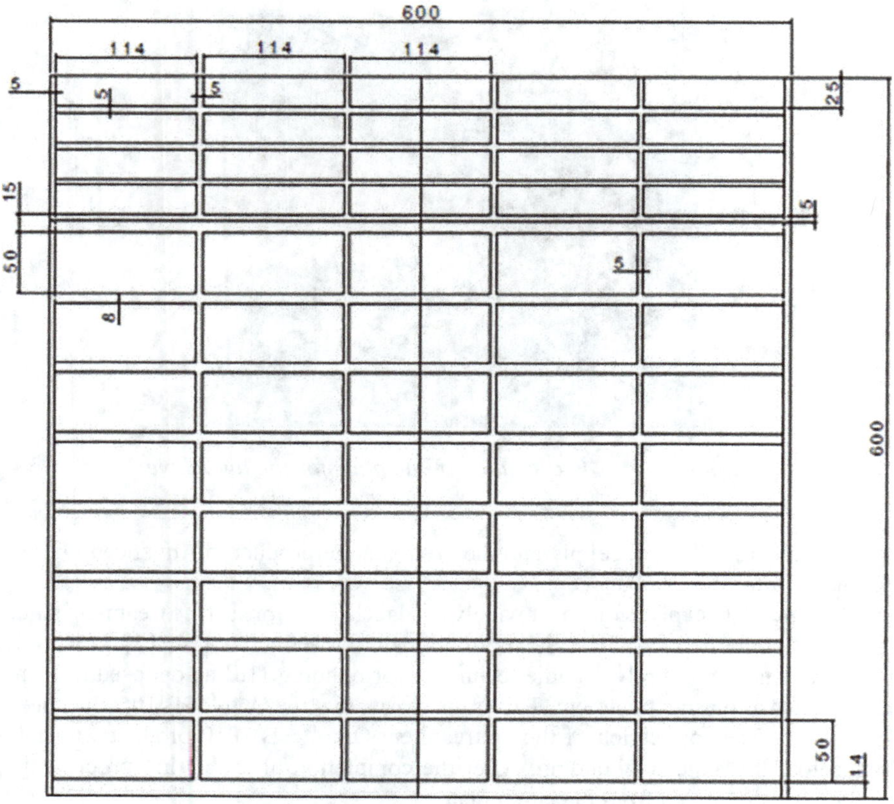

Figure 14.6 *Cutting plan of the plate*

Figure 14.7 *Table wheel saw*

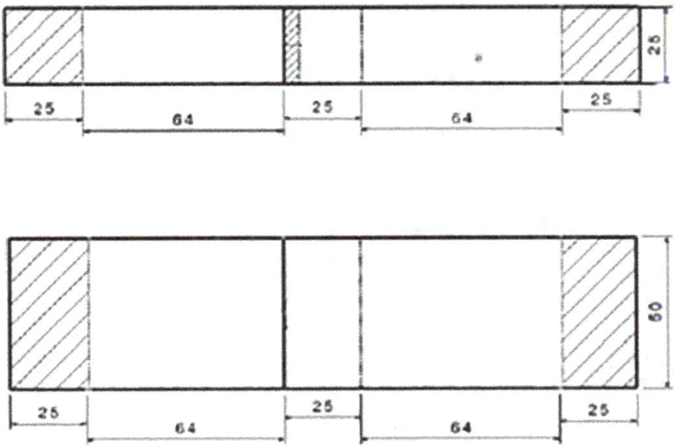

Figure 14.8 Dimensions of wide (w = 50 mm) and narrow (w = 25 mm) coupons

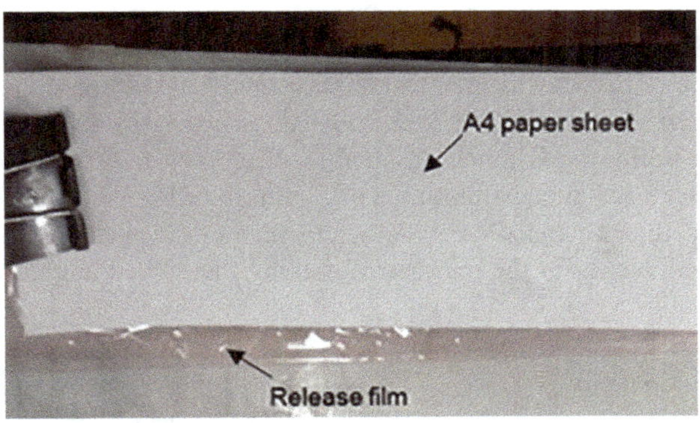

Figure 14.9 Cutting of the release film

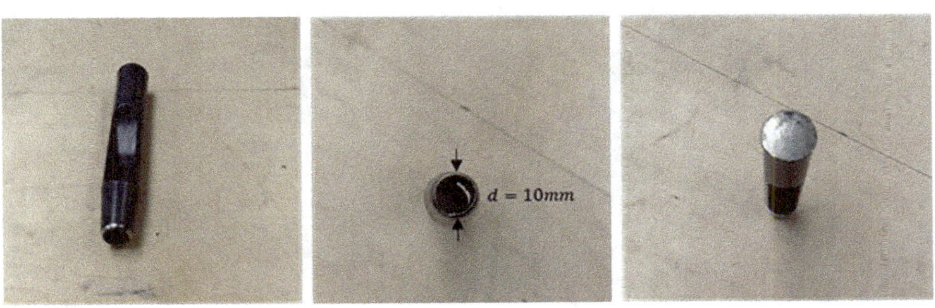

Figure 14.10 Hollow steel punch

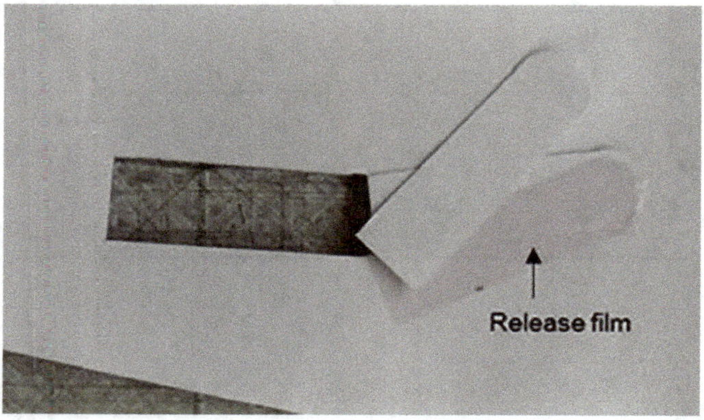

Figure 14.11 The final shape of thumb nail defect

been used to make the cutting more feasible. Figure 14.9. can illustrate the release film, which was placed between two A4 size paper sheets in order to make the cutting more ease and precise.

For the semicircular defects, the radius was 5 mm. As it has been aforementioned, this type of defect was the most challenging. For the creation of thumb nail defects, a hollow steel punch had been utilised, as observed in Figure 14.10.. The diameter of this hollow steel punch was 10 mm. Therefore, the challenge was the creation of a perfect semi-cycle with radius of 5 mm. In order to achieve this, the steel punch was placed in the paper sheet and with a hammer the semicircular defect was carefully created. Next, for the purpose of making the embedding easier, the shape

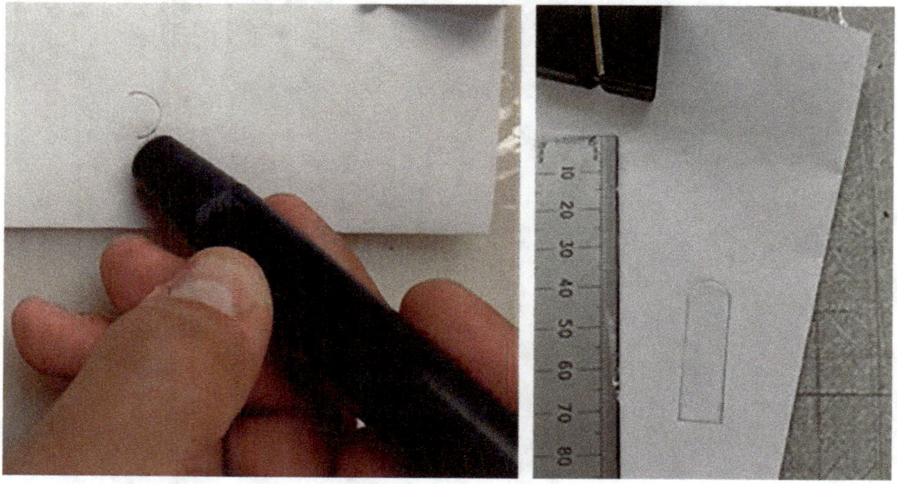

Figure 14.12 The two stages of the semicircular defect

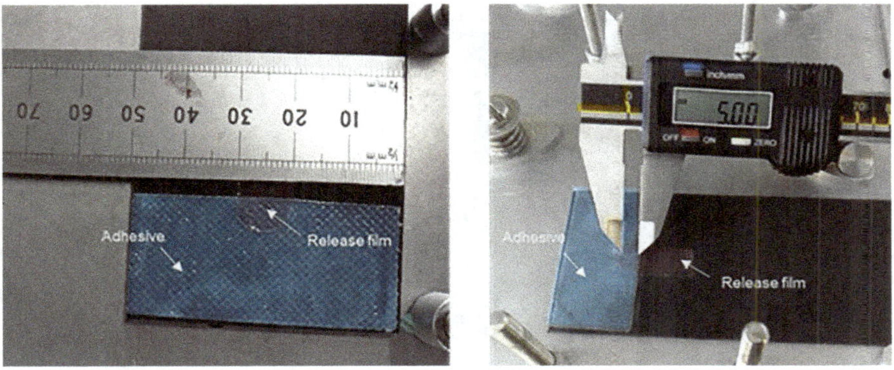

Figure 14.13 Applying the semicircular defect

of the final defect can be shown in Figure 14.10. Figure 14.11. and Figure 14.12. illustrate the two stages of the semicircular defect.

In contrast, the creation of the through-the-width defects was not so complicated. The width of the defect was 5 mm. The defect was cut in the above-mentioned dimensions. Figure 14.31 shows the trough-the-width defect.

The final stage was the embedding of the two types of defects. The defects were applied onto the adhesive (the following subsection provides a more detailed description of the adhesive, i.e. dimensions, size, etc.). Additionally, the application of the semicircular defect was more difficult. This was because the defects had to be placed exactly in the middle of the coupons. Figure 14.13. illustrates the introduction of this type of defect.

In Figure 14.14., the embedding of the through-the-width defect can be seen.

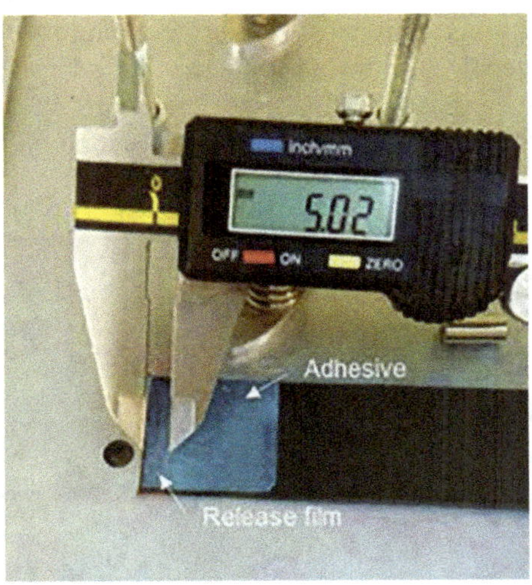

Figure 14.14 Embedding of the through-the-width defect

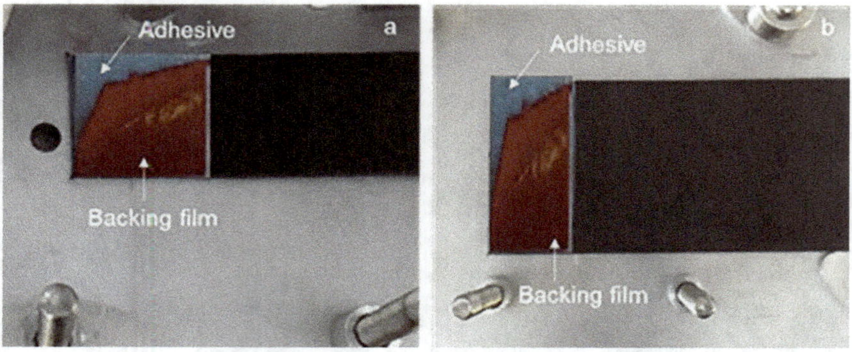

Figure 14.15 *The removal of the backing film: (a) narrow adherend and (b) wide adherend*

14.2.2.7 Adhesive bonding

The bonded area of the joints was 25 mm, therefore, for the narrow specimens was 25 × 25 mm² and for the wide ones was 25 × 50 mm². The adhesive film FM® 94 had to be cut in those two dimensions.

To begin with, the adhesive had to be put out for the freezer to defrost in order to make cutting easier. After approximately 15 minutes, the adhesive was ready to be cut to the aforementioned dimensions. The backing paper of the adhesive was removed just before the application on top of the adherend surface, as it can be seen in Figure 14.15.

The thickness of the adhesive was 0.25 mm. The pretreatment product Onto™ SB 1050 was applied just before the bonding. The bonding process was held by using bonding fixture (jigs). This bonding fixture was necessary because the SLJ had an accurate bond length, proper alignment and uniform bond line thickness. Moreover, this fixture could not prevent spillage of the adhesive, i.e. adhesive fillet. One cycle can adhesively bond up to six specimens with 25 mm width. However, only two specimens with 50 mm width could be bonded per cycle. The required

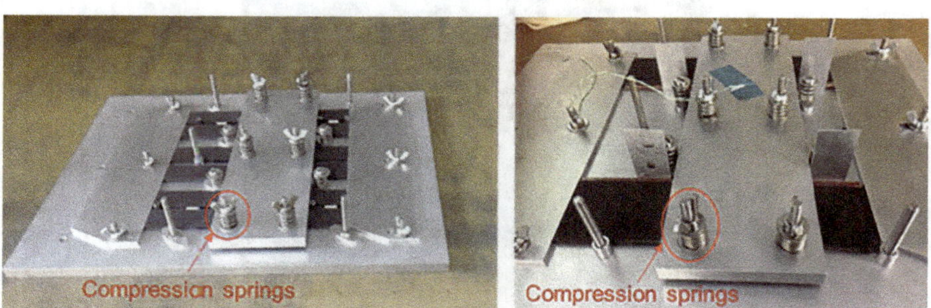

Figure 14.16 *Apply pressure with compression springs*

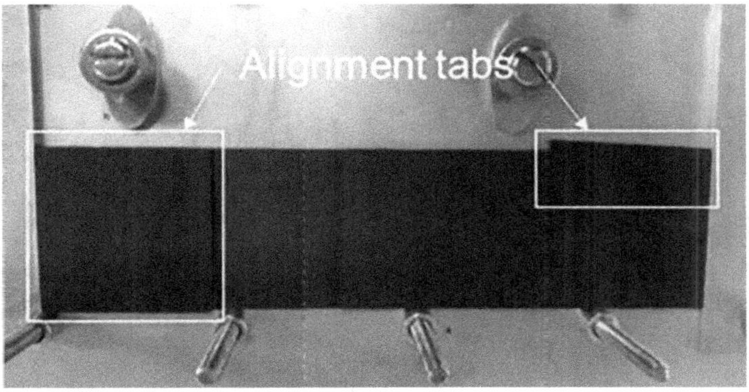

Figure 14.17 Alignment tabs

pressure was calculated based on the required force for six specimens, which was the maximum number per cycle for this particular fixture.

$$Pressure\ (MPa) = Force\ (N)/Bonded\ Area\ (mm^2)$$

$$Force = Pressure \times Bonded\ Area$$

$$Force = Pressure \times Bonded\ Length \times Bonded\ Width$$

$$Force = 0.28\ MPa \times 25\ mm \times 25\ mm \tag{14.2}$$

$$F = 175N$$

For width = 50 mm

$$F = 350N$$

This was the force for one specimen. Therefore, for six, which was the maximum number, $F = 175 \times 6 = 1050$ N ($w = 25$ mm) and F $= 350 \times 6 = 2100$ N ($w = 50$ mm). The calculated force was maintained by using heavy plates on the bonded area fixed by compression springs, as illustrated in Figure 14.16. The forces of 1050 and 2100 N were achieved by adjusting the length of the compression springs to 16

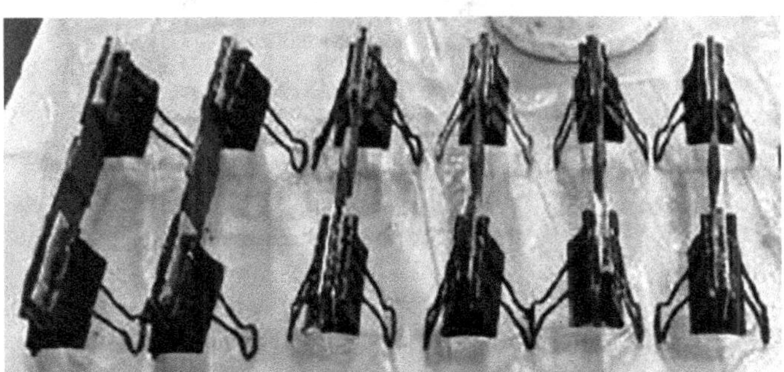

Figure 14.18 Tabbing process

Figure 14.19 *Tabbed single-lap joints*

and 12 mm, respectively. A vernier calliper was utilised for the measurement of the springs' length.

It is worth mentioning that for better alignment of the joints, two alignment tabs (25 × 25 mm and 25 × 50 mm) were placed at the end of the joints: one beneath and one above (Figure 14.17.). The reason was to improve the non-uniform thickness of the joints.

14.2.2.8 Curing of the adhesive

The following stage was to place the whole assembly into the oven for the purpose of curing the adhesive. The first step of this procedure was to heat up the plate up to 40°C, simultaneously, the oven was calibrated to 40°C. Moreover, for better

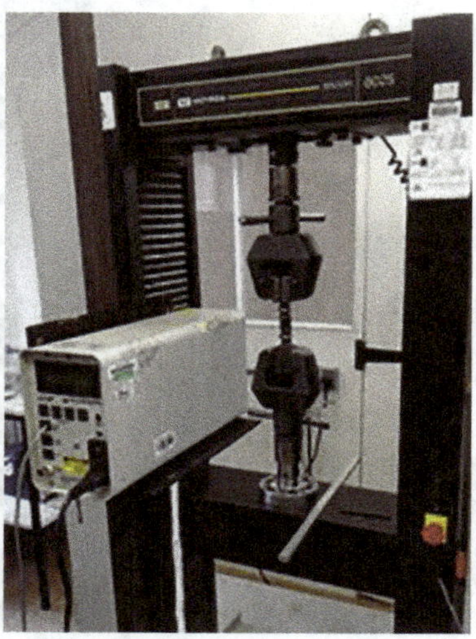

Figure 14.20 *Laser extensometer unit*

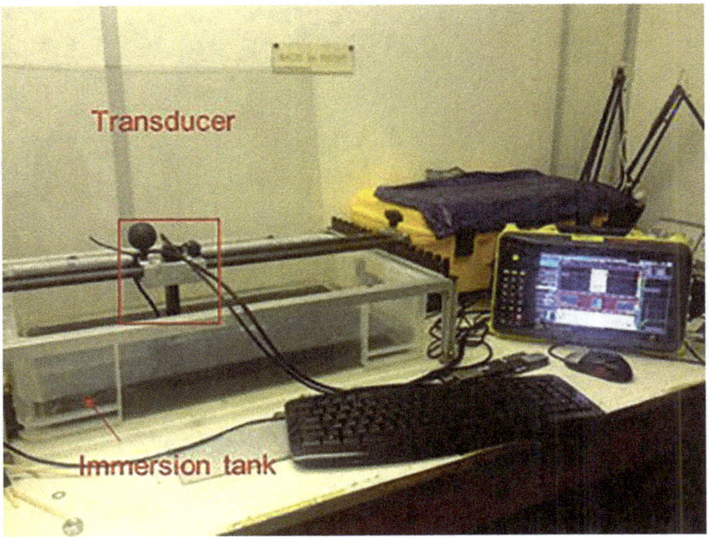

Figure 14.21 *Sonatest apparatus*

monitoring of the curing process, a thermocouple was attached on the top of the plate. The oven was set to have a step of 1°C/min until it reached the temperature of 121°C.

After reaching the 121°C, the entire assembly was held for this temperature for one hour. The final step of the curing process was the cooling of entire assembly to room temperature. First, the jig was gradually cooled to 60°C inside the oven.

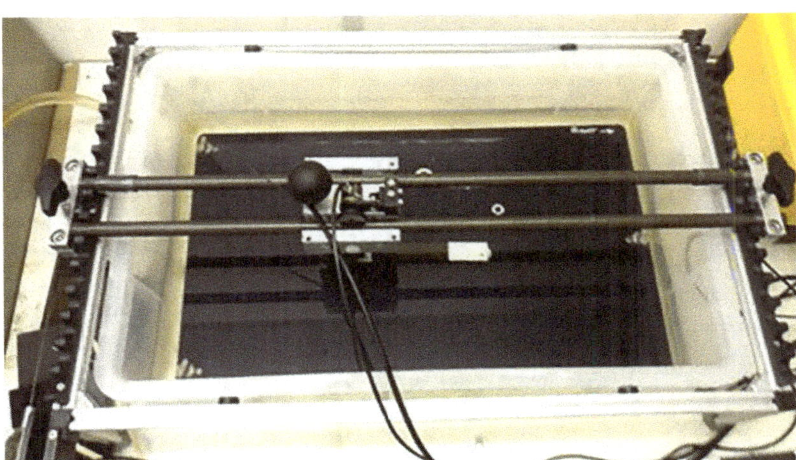

Figure 14.22 *Calibrated sample*

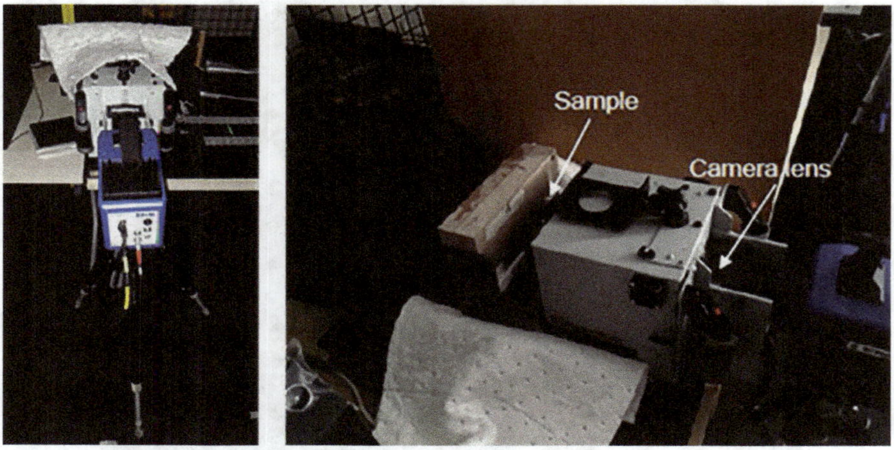

Figure 14.23 Pulsed thermography equipment

Subsequently, the whole assembly was taken out from the oven and with the implementation of a fan, the jig was completely cooled in room temperature.

14.2.2.9 Tabbing

The final stage of the manufacturing process was the tabbing of the SLJ. The alignment tabs were bonded on the two opposite sides of the single-lap joints by using the Araldite 420 adhesive. The dimensions of these tabs were 25 × 25 mm for the narrow joints and 25 × 50 mm for the wide ones. The tabbing process was necessary because

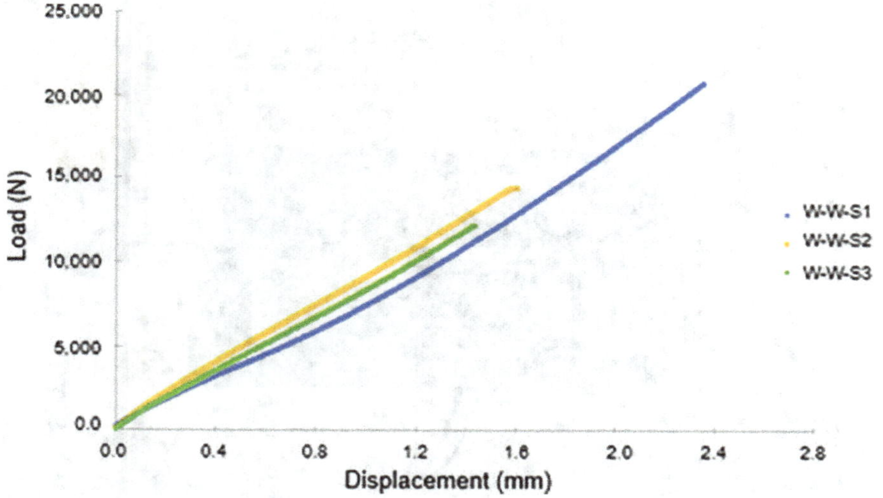

Figure 14.24 Table 14.4 Load vs. displacement of W-W

Table 14.4 *Summarised data for W-W*

Label	Failure load (N)	Displacement (mm)	Shear strength (MPa)
W-W-S1	20 690	2.354	16.55
W-W-S2	14 405	1.608	11.52
W-W-S3	12 194	1.438	9.75

it reduced the effect of the load path eccentricity in SLJ. Furthermore, additional aluminium tabs were important to be bonded in order to ensure good grip during the lap shear test. These tabs were bonded by using the same adhesive, specifically, Araldite 420 (Figure 14.18.). The curing of this adhesive held in room temperature for two-day period (Figure 14.19.).

14.3 Testing

14.3.1 Lap shear test

The lap shear test of the adhesively bonded joints was carried out on an INSTRON 5500R testing machine. The experiment was held in order to determine the ultimate failure load and displacement of the subjected joints. The specimens were mounted in self-tightening jaws, a load cell capacity of 100 kN was chosen to equip with the machine for conducting the test, since the expected failure load was 29.8 kN, for the wide ones, the loading rate was 1 mm/min at room temperature.

The expected failure load for the wide specimens (w = 50 mm) was

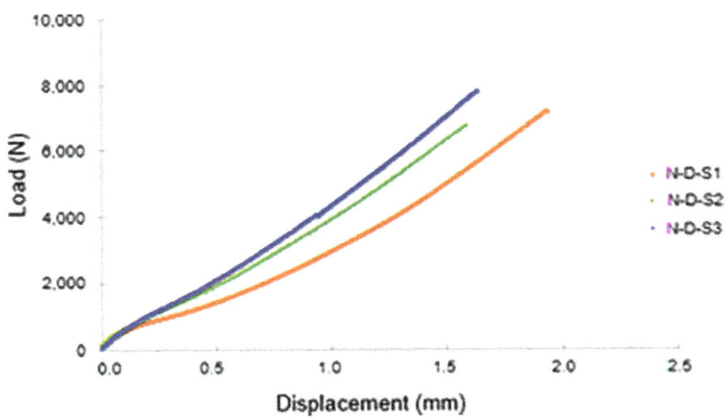

Figure 14.25 *Load vs. displacement of N-D*

Table 14.5 Summarised data for N-D

Label	Failure load (N)	Displacement (mm)	Shear strength (MPa)
N-D-S1	7120	1.937	11.40
N-D-S2	6701	1.589	10.72
N-D-S3	7799	1.633	12.48

$$Shear\ strength\ of\ the\ adhesive = 23.8\ MPa$$
$$F = shearstress\ \times bondedarea$$
$$F = 23.8\ MPa\ \times 25\ mm\ \ \times 50\ mm \quad (14.3)$$
$$F = 29750N$$
$$Failureload = 29.8kN$$

For narrow specimens (w = 25 mm) was

$$F = 23.8MPa\ \ \times 25\ mm\ \times 25\ mm$$
$$F = 14875N \quad (14.4)$$
$$F = 14.9kN$$

To begin with, the appropriate load cell and jaws had to be placed on the machine. The next step was to mount the specimens to the jaws and to set the parameter of this particular experiment to the software, i.e. loading rate, thickness of each joint, etc.

A laser extensometer unit (Figure 14.20.) was implemented in order to focus more on the area of interest, which was the bonded area of the joints.

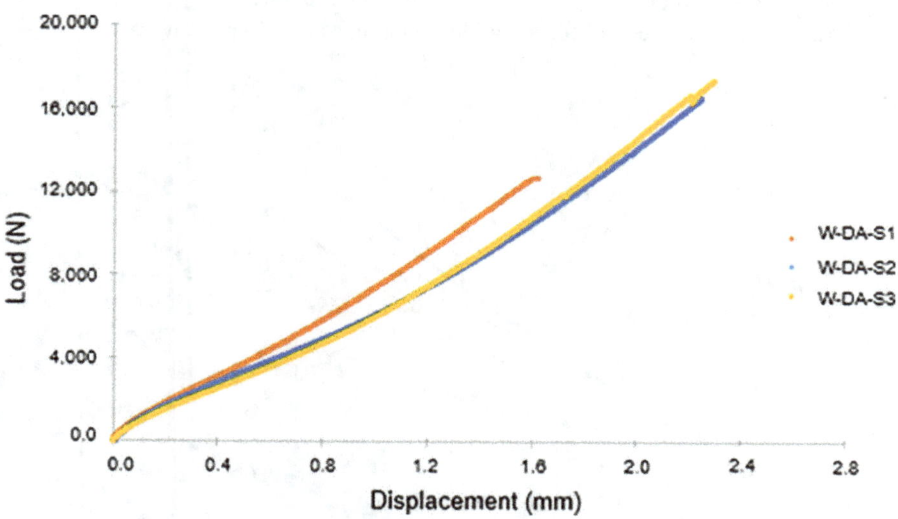

Figure 14.26 Load vs. displacement of W-DA

Table 14.6 Summarised data for W-DA

Label	Failure load (N)	Displacement (mm)	Shear strength (MPa)
W-DA-S1	12 733	1.619	10.19
W-DA-S2	16 527	2.252	13.22
W-DA-S3	17 358	2.296	13.89

After the completion of the test, the data were recorded on the computer and plotted on graphs load vs. displacement for the different joint configurations.

14.3.2 NDI

14.3.2.1 Ultrasonic C-scan

The specimens were subjected to high-frequency sound waves by implementing the Sonatest apparatus (Figure 14.21.). Ultrasonic C-scan was used to investigate the possibility of detecting the embedded defects. The set-up of this machine consisted of an immersion tank with water and a pulse-echo transducer. The scanning speed was 30 mm/s, the gain was 20 db and the frequency of the transducer was 5MHz. Only one gate was set, which was focused to the overlapped region of the specimens.

First, the samples were placed and calibrated inside the immersed tank, as illustrated in Figure 14.22. Subsequently, the transducer scanned the samples manually and the data were recorded on the computer. Whilst the transducer scanned the samples, the amplitude was being recorded and the time transmitted signal at regular intervals. The strength of the signals was illustrated in the form of different colour. Therefore, with a presence of the defect, the strength of the signals was being reduced and as a result of that, the illustrated colour changed compared to the rest

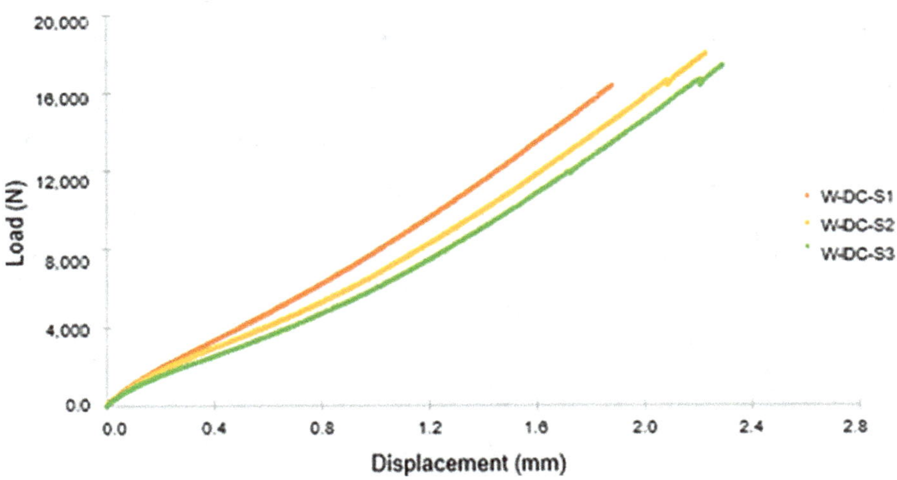

Figure 14.27 Load vs. displacement of W-DC

Table 14.7 Summarised data for W-DC

Label	Failure load (N)	Displacement (mm)	Shear strength (MPa)
W-DC-S1	16 299	1.882	13.40
W-DC-S2	17 942	2.233	14.25
W-DC-S3	17 337	2.298	13.87

of the sample colouration. The recorded images were illustrated by using the UTS Studio software.

14.3.2.2 Pulsed thermography

The adhesively bonded joints were subjected to a short yet, high-energy pulsed light by implementing the Thermoscope II system of Imaging Inc (Figure 14.23.). As in the ultrasonic C-scan, the reason of using pulsed thermography was to investigate the possibility of detecting the embedded defects. The possibility of detecting the defects was pulse heated by two Xenon flash lamps delivering 25 kJ. The distribution of temperature was recorder for 15 seconds as an obtained rate of 50 Hz by utilising an FLIR SC 7600 MB series infrared camera. The frames of the recorded images before the flash were 10 and after 760. The spatial resolution of the camera was 640 × 512 pixels.

In order for this experiment to be performed, the samples were mounted in a sealed black box (because there are interactions between the light and the flashlight of the camera) at 270 mm in front of the camera's lens. The capturing of the data was held by using the MOSAIQ 4.0 acquisition system.

Table 14.8 Summarised data of every category

Specimens label		Failure displacement (mm)	Failure load (N)	Average displacement (mm)	Average load (N)	Average shear stress (MPa)
W-W	S1	2.354	20 690	2.354	15 763	12.60
	S2	1.608	14 405			
	S3	1.438	12 194			
N-D	S1	1.937	7120	1.720	7207	11.53
	S2	1.589	6701			
	S3	1.633	7799			
W-DA	S1	1.619	12 733	2.056	15 540	12.43
	S2	2.252	16 527			
	S3	2.296	17 358			
W-DC	S1	1.882	16 299	2.138	17 640	13.87
	S2	2.233	17 942			
	S3	2.298	17 337			

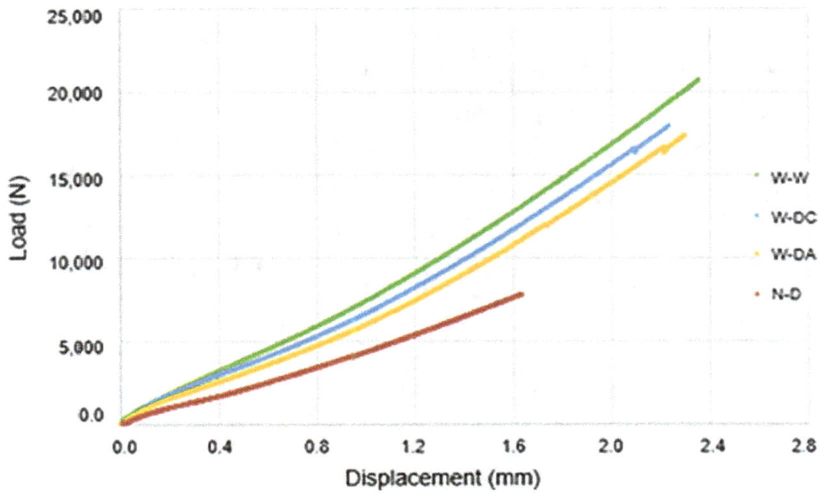

Figure 14.28 *Load vs. displacement for the four categories*

14.4 Results and discussion

14.4.1 Lap shear test

The data obtained for each category were plotted in load vs. displacement. For this experiment the values of the displacement were between 1.4 and 2.3 mm, hence, the measurement of the displacement had to be very precise.

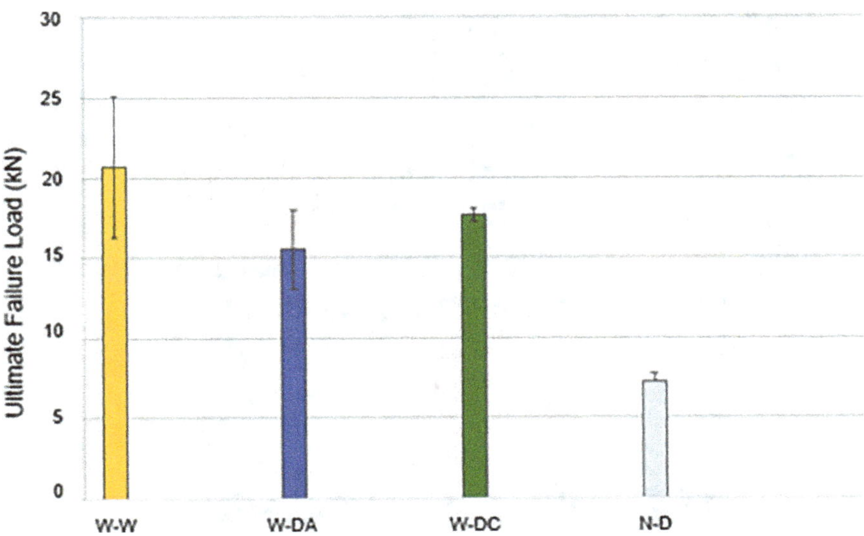

Figure 14.29 *Failure load vs. the four categories*

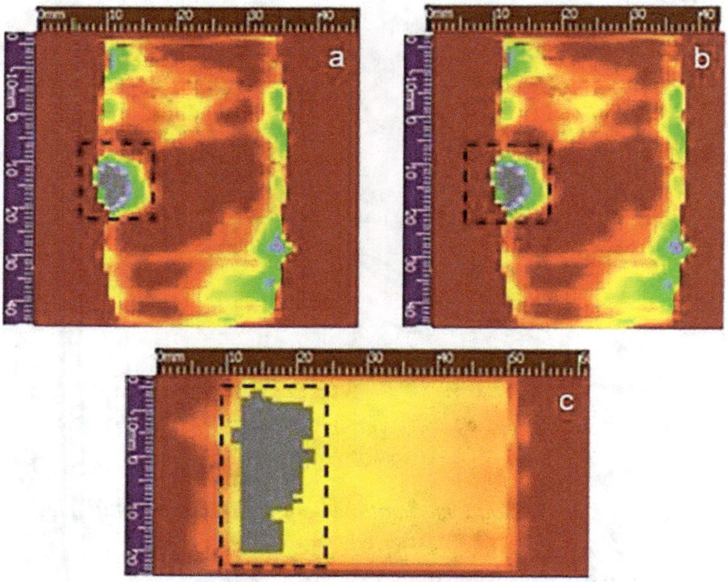

Figure 14.30 Ultrasonic images: (a) W-DA, (b) W-DC and (c) N-D

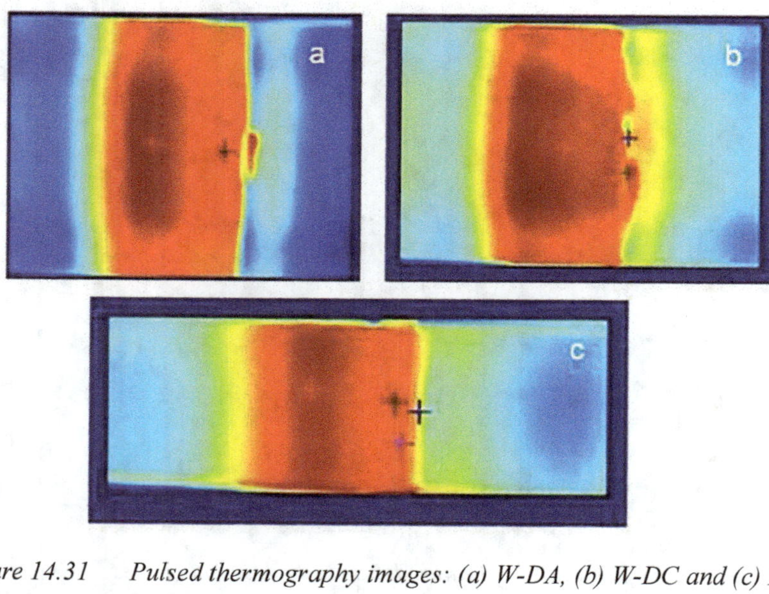

Figure 14.31 Pulsed thermography images: (a) W-DA, (b) W-DC and (c) N-D

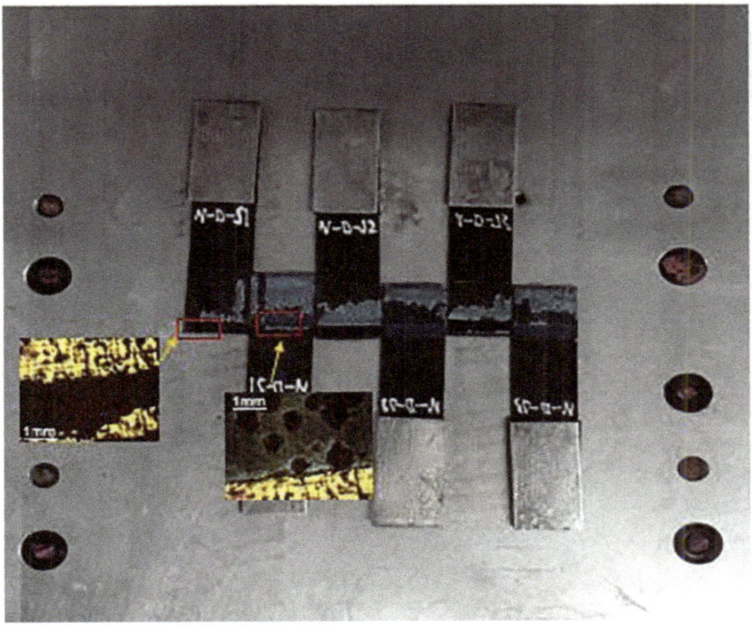

Figure 14.32 Failed surfaces for W-W

Figure 14.24. shows the load vs. displacement curves for Category 0 (W-W). It can be seen that in Table 14.4 that the failure load of the W-W-S1 was 20 690 N, greater than the failure loads for the W-W-S2 and W-W-S3, 14 405 and 12 194 N, respectively. Category 0 was the reference category for comparison, without any defects embedded.

Figure 14.25. illustrates the curves of Category I, i.e. with no defect (N-D). The maximum load was 7820 N (N-D-S3) and the minimum was 6720 N (N-D-S2), as observed in Table 14.5.

Figure 14.26. shows the curves of Category II (N-DA). It can be seen in Table 14.6 that the maximum failure load was 17 358 N and the minimum was 12 733 N.

Last but not least, the load vs. displacement graph for Category III (W-DC) can be illustrated in Figure 14.27.. As observed in Table 14.7, the maximum failure load was 17 942 and the minimum was 16 299.

Observing carefully all the above graphs, it can be noticed that the adhesively bonded joints exhibit an elastic behaviour until failure. It can indicate that the adhesive exhibits brittle behaviour. According to the datasheet for FM® 94, this adhesive has been characterised as a toughened epoxy. Another contributing factor of acting like a brittle material was the thin thickness 0.25 mm.

For the manufacturing of every specimen, exactly the same procedures were followed as aforementioned in 'Materials and Manufacturing'. Nevertheless, the reasons for this big difference in the failure load are not clear. Some of the reason could

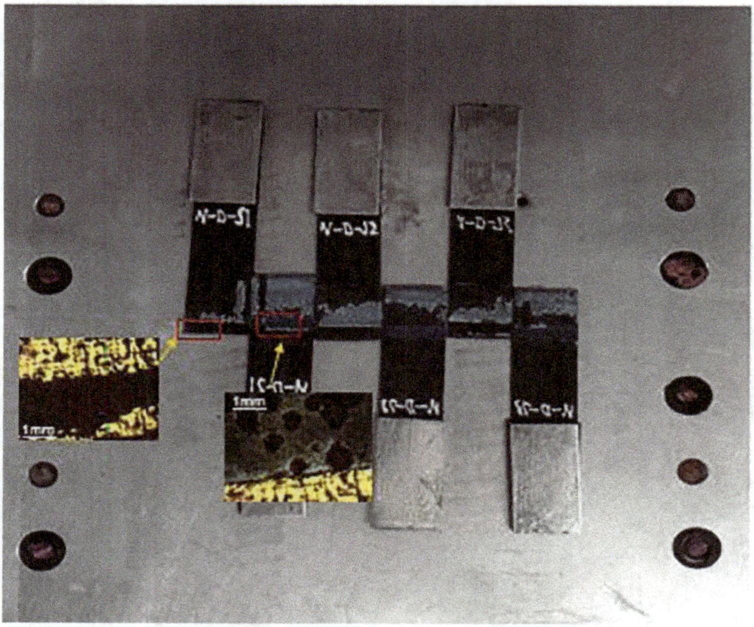

Figure 14.33 Failed surfaces for N-D

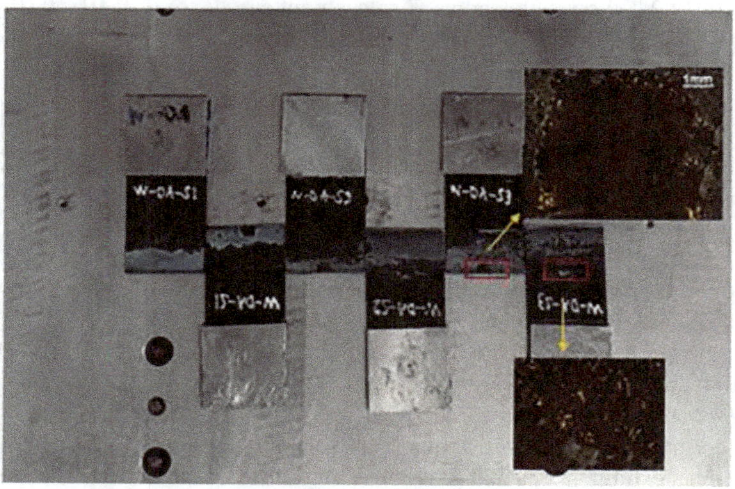

Figure 14.34 Failed surfaces for W-DA

be possible contamination during the application of Onto™ SB 1050, the integrity of the Onto™ SB 1050 itself for this specific prepreg, deficiencies during the curing process of the adhesive or the difference between the specimens itself, i.e. bond line

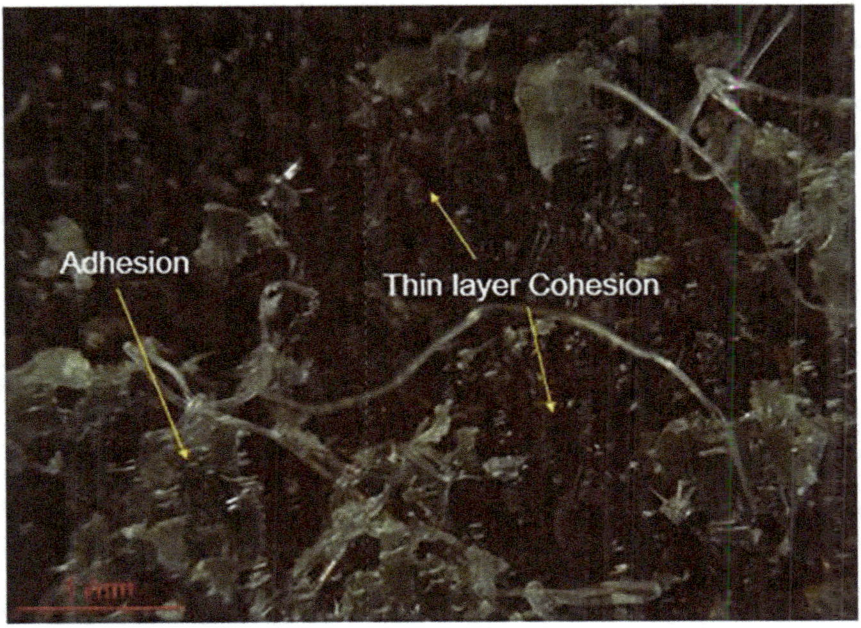

Figure 14.35 Failure Mechanism of W-DA

Figure 14.36 Failed surfaces for W-DC

thickness, overlap area, applied pressure, size of the fillet, etc. It is worth mentioning that all the specimens were from the same plate; moreover, the adhesively bonded W-W-S2 and W-W-S3 were cured in the same cycle. Therefore, this was the most predominant hypothesis, however, additional investigation and more experiments had to be undertaken. A similar incident can be observed in Category II as well. Specifically, the W-DA-S1 exhibited 12 733 N failure load compared to W-DA-S2 and W-DA-S3 with 16 527 and 17 358 N, respectively. This also indicates deficiencies during the manufacturing or curing process.

Table 14.8 summarises the data for every category, and Figure 14.28. illustrates the load vs. displacement curves of the four different categories.

It can be seen that there is a difference in the failure load among the specimens. The maximum failure load of W-W was 20 690 N, 17 358 N for W-DA and 17 942 N for the W-DC. For the narrow specimens were 7799 N. The width of this category was 25 mm compared to 50 mm of the other three. Hence, half of the width means half the failure load. Therefore, considering twice the value of the width, the maximum failure load of this category was 15 598 N, which, value of the failure load, was close the other categories. In addition, the failure loads of Categories II and III had similar values, even if, Category III had two layers of adhesive compared to one layer of Category II.

It can be noticed that for the W-DA, there was a 24.9% reduction of the failure load and for W-DC 14.7% compared to W-W. The reduction of the failure load for the N-D was 65.2%; however, the percentage was high because the width of this category was 25 mm.

The failure load for the four categories is plotted in Figure 14.29.. The maximum failure load for W-W was 20.7 kN and for N-D was 7.8 kN. Therefore, the maximum shear stress was 16.55 MPa and 12.48 MPa. It can be noticed that there was a big difference in the shear stress. For N-D, the difference was almost 30%, and for W-W, it was 48%.

14.4.2 NDI

14.4.2.1 Ultrasonic C-scan

Figure 14.30. illustrates the obtained images for ultrasonic C-scan. It can be observed that the signal changes (different colour) in the centre for the specimen with the semicircular defect and through the width from the specimens with a through-the-width defect. This clearly indicated a change in the material, adherend, adhesive and release film (defect). It is clear that C-scan successfully detects the embedded defects. Moreover, there is an indication that the bond strength was not uniform in the overlap area. Specifically, on the edges of the bonded area, the strength of the bond was reduced (green and blue colour).

Overall, ultrasonic C-scan was successfully detected in the two types of defects. Moreover, this technique is quite sensitive to the detection of the bond's quality [19]. The results identify the location and the size of the defect. However, there is no established correlation between the strength of the bond and the received signal.

Therefore, it is difficult to quantify the integrity and strength of the bond by utilising ultrasonic C-scan.

14.4.2.2 Pulsed thermography

Figure 14.31. shows the obtained images by utilising pulsed thermography. In contrast with C-scan, pulsed thermography was not able to detect the embedded defects. The images are self-explanatory. The identification of the defects of the categories W-DA, W-DC and N-D was not successful. According to Bhanushali [19], the thermographic images for the different bond (good, kissing and weak) appeared to be similar to each other, hence, it was impossible to detect the defects. Thermography is effective for finding voids. In this particular case, the thickness of the release film was so thin (25 μm) that it was not able to be illustrated for the obtained images by implementing PT. Moreover, PT is a suitable NDT for detecting impact damage or delamination in composites. This NDT relies on the accurate position of the pixels, which means that the images had to be captured in the same location at the same time in order to be compared.

14.4.3 Optical microscopy

In Figure 14.32., it can be visualised that adhesion failure was the failure mechanism of Category 0. However, it can be noticed that the W-WS2 and W-W-S3 joints were failed like the adhesive had never bonded properly. It seemed that the adhesive peeled off completely, in addition, the adhesive acts like a brittle material. Moreover, these two joints exhibited adherend delamination. This clearly indicates deficiencies during the curing process. This can explain the difference in the failure load of the three joints.

Figure 14.33. illustrates the obtained images for N-D. The failure mode of this category is adhesion failure. As in the two specimens of the previous category (W-W-S2 and S3), the adhesive exhibits brittle behaviour.

Figure 14.34. shows the obtained images of Category II. In this category, one layer of adhesive was applied in order to cause adhesion failure. The incident with the brittle behaviour of the adhesive can be observed in the W-DA-S1 (Figure 14.35.). This specimen had the lowest failure load of this category. With the naked eye, it can be seen that adhesion failure is the dominant failure mechanism. However, utilising the microscope, it can be observed that thin cohesion failure has been exhibited.

Finally, Figure 14.36. visualises the specimens of Category III. In this category, two layers of adhesive film were applied to cause cohesion failure; because the release film was embedded between the two layers of adhesive. The caused failure mode was a mixed-mode failure (adhesion and cohesion failure). Although in W-DC-S3 (Figure 14.37.), there was a delamination, the integrity of the bond did not degrade.

After examining the failed surface and processing the result, it was difficult to correlate the failure mechanisms with the type of the bond defect [19]. Table 14.9 summarises the results of the failure mechanisms of the failed joints.

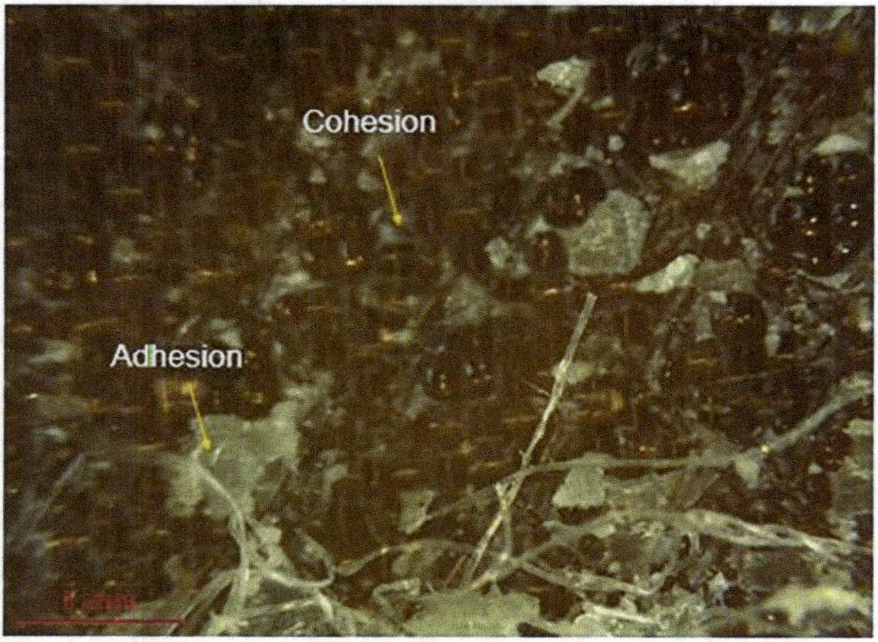

Figure 14.37 Failure mechanism of W-DC

Table 14.9 Failure modes of the different specimens

Label of specimens	Failure mode
W-W	Adhesion
N-D	Adhesion
W-DA	Mostly adhesion (partiality thin cohesion)
W-DC	Mixed-mode (adhesion and cohesion)

14.5 Conclusions

The lap shear test indicated that there was a difference in the shear strength among the four categories. Specifically, Category II exhibited 24.9% less failure load for the reference, Categories 0 and III exhibited 14.7% and Category I 65.2%. The wide specimens did not show so much degradation on failure load. However, considering the size scaling, defects such as those, i.e. in the laminates, might affect the integrity of the whole structure; hence, their consideration is important. According to the literature, a kissing bond exhibits less than 20% of the nominal strength. Therefore, these two types of defects can be characterised as kissing defects, for specimens of

Category III. On the other hand, for Categories II and III, it cannot be characterised. These defects are partially kissing defects. Hence, an exact definition of weak and kissing bonds needs to be defined. In addition, for a more comprehensible explanation, further investigation is required. The strength of the joints between the two different types of the defects was affected, however not considerably. The two types of the defects were exhibited almost the values in the failure load. Hence, a better approach and additional investigation are needed.

NDI was held to investigate the possibility of detecting the embedded defects. Ultrasonic C-scan showed promising results. The defects were clearly illustrated from the obtained images. However, it was not possible to identify the integrity and degradation of the bond quality. In contrast, pulsed thermography showed limited feasibility for the detection of the defects. This NDT is suitable for detecting impact defects and delamination.

Lastly, cohesion failure was detected on the failed surfaces, although with a naked eye it seemed to be adhesion. However, it was not possible the correlation of the failure mode with the defect types.

References

[1] Higgins A. 'Adhesive bonding of aircraft structures'. *International Journal of Adhesion and Adhesives*. 2000;20(5):367–76.

[2] Waugh R.C. *Development of infrared techniques for practical defect identification in bonded joints*; 2016:. pp. 77–95.

[3] Flom Y., Arsenault R.J. ' interfacial bond strength in an aluminium alloy 6061—sic composite '. *Materials Science and Engineering*. 1986;**77**(11):191–97.

[4] Roach D., Rackow K., Duvall R. ' innovative use of adhesive interface characteristics to nondestructively quantify the strength of bonded joints '. *10th European Conference on Non-Destructive Testing*; Moscow, 2010.

[5] Tracey A.C., Belcher M.A.T., Blohowiak K.Y., Flinn B.D. *' improving adhesive bonding through surface characterization: reverse the curse of the nylon peel ply? '* Joint Advanced Materials and Structures- AMTAS Meeting, Seattle, WA; 2014.

[6] da Costa Mattos H.S., Sampaio E.M., Monteiro A.H. 'Static failure analysis of adhesive single LAP joints'. *International Journal of Adhesion and Adhesives*. 2011;31(6):446–54.

[7] Mikkor K.M., Thomson R.S., Herszberg I., Weller T., Mouritz A.P. ' finite element modelling of impact on preloaded composite panels '. *Composite Structrues*. 2006;**75**(1–4):501–13.

[8] Sierakowski R.L. ' ' aircraft damage tolerance '. *16th International Conference on Composite Materials*; Kyoto, 2007.

[9] 'Hexcel-8552 hexply-8552'. 2016.

[10] Data P 'HexTow ® IM7'. 2016.

[11] Adhesives E 'OAS'. n.d.

[12] Description P. ' onto TM SB1050 evaluation pack onto TM SB1050 evaluation pack application note '. 2017:2–4.

[13] Baldan A. 'Adhesively-bonded joints and repairs in metallic alloys, polymers and composite materials: adhesives, adhesion theories and surface pretreatment'. *Journal of Materials Science*. 2004;39(1):1–49.

[14] Bénard Q., Fois M., Grisel M. 'Peel ply surface treatment for composite assemblies: chemistry and morphology effects'. *Composites Part A: Applied Science and Manufacturing*. 2005;36(11):1562–8.

[15] Flinn B.D., Clark B.K., Satterwhite J., Van Voast P.J. ' influence of peel ply type on adhesive bonding of composites '. *Presented at the 2007 Society for the Advancement OfMaterial and Process Engineering (SAMPE) Conference, Baltimore, MD, Jun 2007*. 2008:1–26.

[16] Palmieri L.F., Belcher A.M., Wohl J.C., Blohowiak Y.K., Connell W.J. ' laser ablation surface preparation of carbon fiber reinforced epoxy composites for adhesive bonding ' in *International SAMPE technical conference*. Vol. 68; 2013. pp. 554–66.

[17] ASTM D3165-07 ' standard test method for strength properties of adhesives in shear by tension loading of single-lap-joint laminated assemblies ' '. *Order -A Journal Theory Ordered Sets Its Applications*. 2002;**96**:3–5.

[18] Hart-Smith L' the curse of nylon peel ply '. *41st SAMPE International Symposium and Exhibition*; Anaheim, CA, 1996.

[19] Bhanushali R., Cranfield University 2016.

Chapter 15
Interleaving in composites

Mehdi Yasaee[1]

15.1 Introduction

Fibre-reinforced polymer (FRP) composite laminates have absence of fibre reinforcements in their thickness direction. This means their through thickness stiffness, strength and toughness are weak, relative to their in-plane properties. In multi-directional FRP laminates, first failures in the form of cracks tend to initiate and propagate between the plies of different fibre orientations. This type of cracking is commonly known as delamination, which can be considered a critical weakness of FRP composites. The resistance to delamination can be improved by introducing some form of composite toughening to improve the interlaminar fracture toughness of the material. These toughening concepts can be introduced in the form of intrinsic or extrinsic toughening.

Intrinsic toughening is the improvement of the fracture toughness of the bulk material, either through utilising tougher resin systems or improving the resin fibre interface bond strength. Intrinsic toughening is now part of most commercial composite resin systems, where the brittle epoxy resins have been enhanced with the addition of modifiers such as liquid rubber (e.g., rubber-carboxy terminated butadiene nitrile (CTBN)) which generates regions of finely dispersed phase separations during cure, see Figure 15.1. These phase separations thus produce more than double the fracture toughness relative to the unmodified resin systems [1].

Extrinsic toughening is the improvement of fracture toughness enhancement through the addition of foreign materials. Through thickness reinforcements (TTR) [2] is a type of extrinsic toughening, as this involves the addition of TTR materials in form of pins or fibres.

The procedure to achieve TTR will depend on the manufacturing process of the FRP composite. For resin transfer moulding, dry fibre preforms can be stitched or tufted with thin fibre threads, before resin infusion and cure. For prepreg laminates, small, rigid, FRP or metallic pins (Z-pins) are inserted in the thickness direction of an uncured laminate stack [3]. The FRP composites with

[1]School of Aerospace, Transport and Manufacturing, University of Cranfield, Cranfield, UK

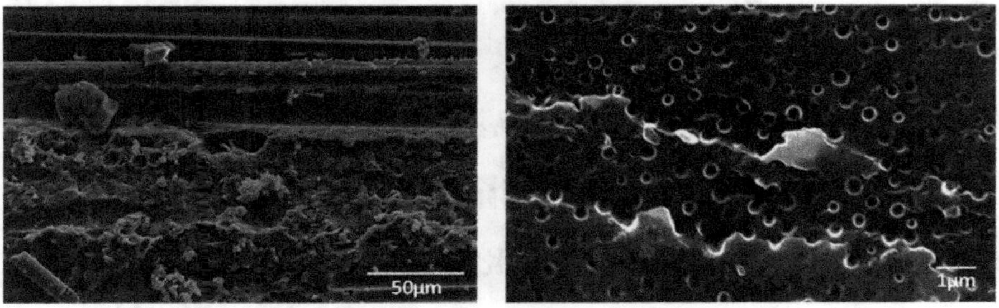

Figure 15.1 *Hexcel Redux 312 (thermoset modified epoxy adhesive)*

TTR demonstrate significant enhancement in their interlaminar fracture toughness. Resulting in significant improvement in both impact and delamination fatigue performance of FRP composite structures, expanding their use to even the most demanding applications. A number of industries in the aerospace and automotive sectors have successfully utilised TTR technologies, including Z-pinning process [3].

TTR technologies are effective; however, they have few downsides. In terms of mechanical performance, the improvements in delamination resistance are countered with reduction in the in-plane strength. Another important downside is the complexity of manufacturing with TTR. These technologies require bespoke manufacturing equipment, expensive raw materials and necessitate expertise on the side of the manufacturer to attain the high-quality components. Meaning, cost of producing composites with TTR is quite expensive which presents a form of high risk for the user deciding whether to utilise them for their product development.

A far more cost effective and widely available extrinsic toughening is interleaving. Rather than modify the toughness of the entire composite laminate volume, interleaving concept is to focus the modification to the weakest regions, here being the interface between dissimilar plies. By placing interleaving materials at the interface between plies during the layup process (Figure 15.2) the interlaminar fracture toughness of laminated composite will be improved.

The purpose of this chapter is to review this concept and present examples of interleaved material types and show the physical mechanisms that contribute to the improvement in the delamination resistance of the bulk composite. In section 15.2, a summary

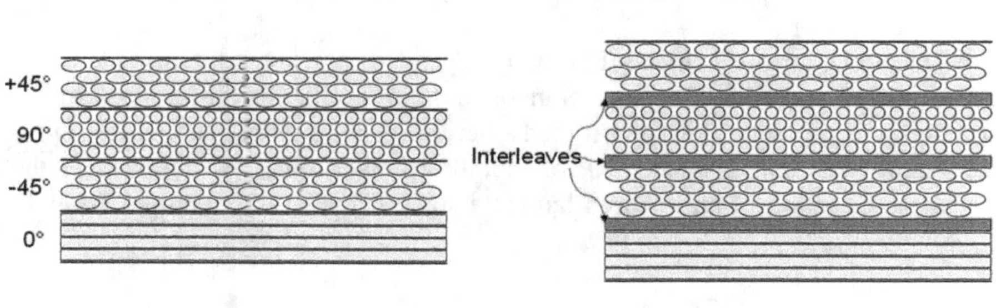

Figure 15.2 *Non interleaved (left) and interleaved laminate stack (right)*

of the measurement methods available to identify the delamination resistance of interleaving is given highlighting important properties to observe. In section 15.3, different interleaving material types and techniques are reviewed. In section 15.4, application of interleaving to composite stiffeners and joints is presented and in section 15.5 the concept of selective interleaving is discussed.

15.2 Measuring interleaving properties

Interlaminar fracture toughness testing is the most appropriate procedure to measure interleaved composite resistance to delamination propagation. Ideally, the fracture toughness test procedures should produce stable delamination propagation, since the crack can propagate into and through the interleaved interface, providing useful insight into the mechanism that produces the fracture toughness improvements.

For Mode I fracture toughness testing, double cantilever beam test (ASTM D5528 or similar) can be used. For Mode II fracture toughness testing, recommendation is to use either end loaded split (ELS) test (ISO 15114:2014) or four point end notched flexure testing. Manufacturing the interleaved specimen is relatively straight forward. For fracture toughness testing, the interleaved layer will only need to be placed on the midplane of the panel (Figure 15.3a). Placement of the release film relative to the interleave may have influence on the measurement, therefore, consideration on this will need to be made when analysing the Modes I and II fracture toughness (G_{IC} and G_{IIC}) properties.

It is essential to conduct cross-sectional microscopy of one interleaved sample in order to identify the thickness of the cured interleave (Figure 15.3b) and gain insight into the distribution of the fibres around the interleaved materials. Adding thick interleaving materials at every interface within a large stacked composite laminate will mean the bulk thickness of the component will increase. This will make the effective mechanical properties of the composite system, such as stiffness and strength to reduce because the fibre volume fraction will decrease.

Interleaved materials may not always produce continuous fracture toughness enhancement. Thereby with a plot of the resistance curve (plot of the G_I vs. crack

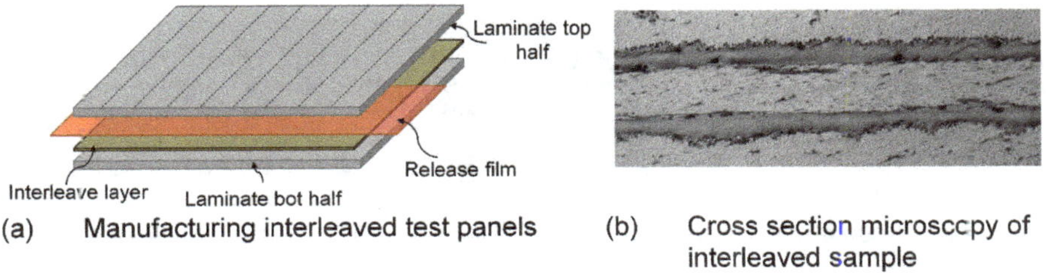

Figure 15.3 (a) Illustration of typical interleaved composite panel manufacturing for fracture toughness testing, (b) essential cross-sectional microscopy for measuring interleaved thickness after cure

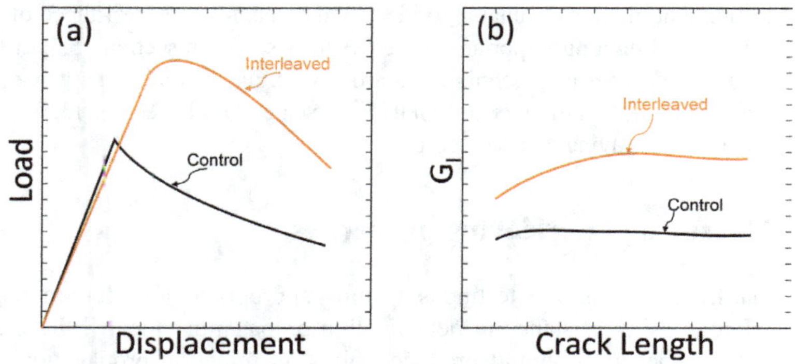

Figure 15.4 Illustration of typical experimental data from Double Cantilever Beam (DCB)/End Loaded Split (ELS). (a) Load-displacement plot (b) Resistance curve, G_I-crack length.

length, Figure 15.4) does one get to understand the behaviour of the interleaved interface with clarity.

A resistance curve that shows a rise in the G_I indicates that the interleaved interface produces a bridging zone behind the crack tip that contributes to the fracture toughness measure. In section 1.2, this concept is described in detail. Other factors that influence the crack propagation are important to observe. Consider the delamination propagation path along the sample (Figure 15.5). The difference between the measure of fracture toughness when the delamination propagates at the interleave/composite interface may be much lower than that which propagates by undulating through the interleave material. The interleave material, cured thickness and arrangement within the composite stack may all contribute to change the crack propagation path and thereby change the overall response. To observe this, one can conduct either scanning electron microscopy (SEM) of the fracture surface or cross-sectional microscopy of the tested sample.

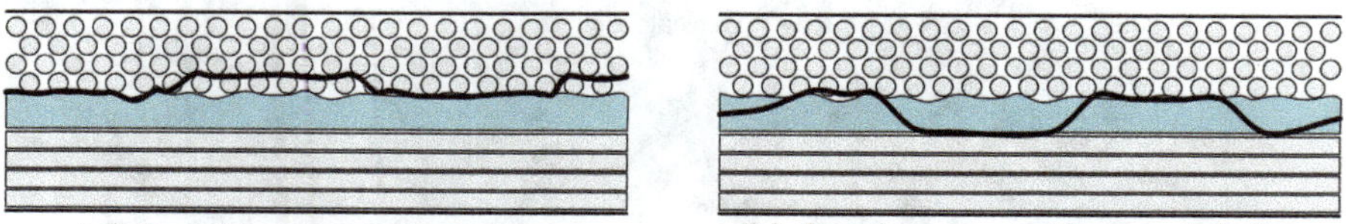

Figure 15.5 Illustration of delamination crack propagation that will produce different fracture toughness measurements

In summary, when characterising the performance gains of interleaved material, it is important to conduct:

1. Fracture toughness tests that produce stable delamination propagation
2. Measure the cured interleaved thickness and observe changes to the fibre dispersions around the interleaves
3. Observe the crack propagation paths through or around the interleaves using cross-sectional microscopy or fracture surface analysis using SEM imaging.

15.3 Interleaving types and performance

15.3.1 Self-same matrix resin

Delamination fracture occurs by breaking the matrix resin as well as the matrix and fibre interface. In essence, the fracture toughness of the resin and the fibre/resin interface strength become the key properties that contribute to the interlaminar fracture toughness of the composite laminate. The inherent toughness of the matrix resin tends to be much higher than that of the fibre/resin interface. Thereby the amount of resin that fractures relative to the fibre debonding, can change the measured fracture toughness.

By introducing extra matrix resin at the interlaminar interface, the resin rich region can be increased, as shown in Figure 15.6. This was achieved by interleaving the laminate with resin films made from the same resin of the composite laminate, called self-same resin [4]. With correct manufacturing conditions to contain the extra resin and prevent the resin bleed, the interlayer resin thickness can be changed.

The measured fracture toughness of the composite laminate with samples of increasing self-same resin interleaving shows rise in the Modes I and II interlaminar

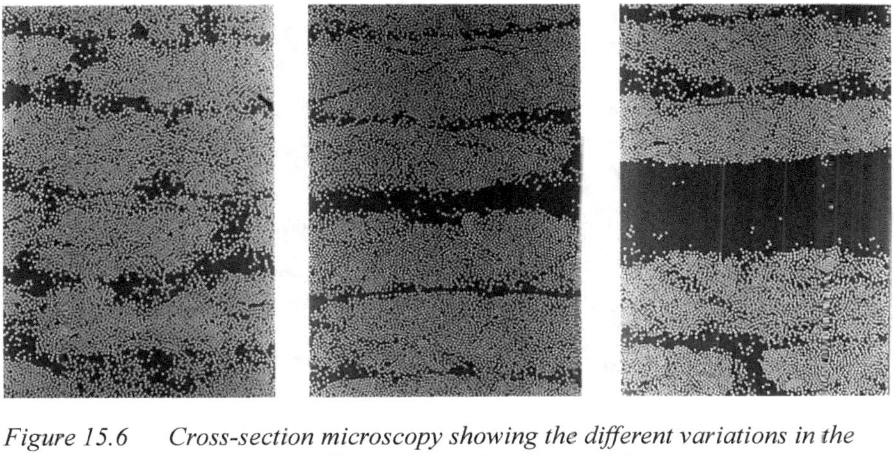

Figure 15.6 Cross-section microscopy showing the different variations in the thickness of self-same epoxy interleaves [4]

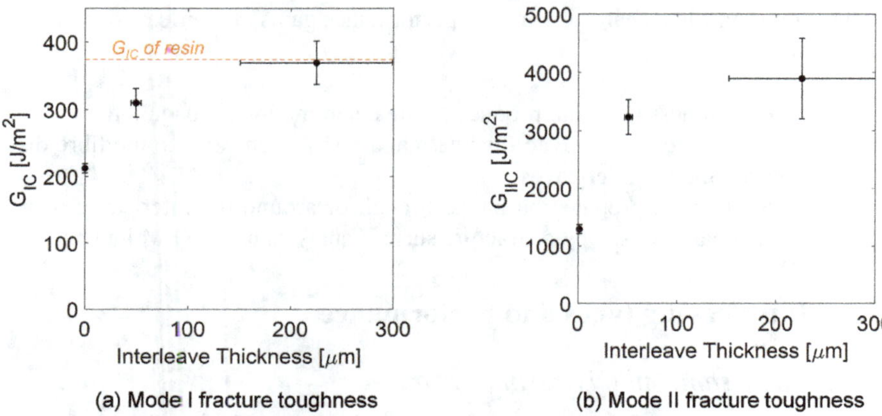

Figure 15.7 Delamination resistance variation with self-same epoxy interleave thickness (data replotted from [5])

fracture toughness(Figure 15.7). The data show non-linear increase in fracture toughness relative to the interleave thickness that tends to plateau at the fracture toughness of the neat resin.

In Figure 15.8, the Modes I and II resistance curves of samples interleaved with 50 μm self-same epoxy are shown. This extra resin layer clearly contribute to the increase of the composite laminate fracture toughness.

The mechanism, which causes this rise, can be explained based on the size of the process zone (plastic yield zone) ahead of the crack tip, as shown in Figure 15.9. Cracks propagate along a path of least resistance that falls inside the process zone. When the resin rich layer thickness is small, the fibre interfaces fall within the

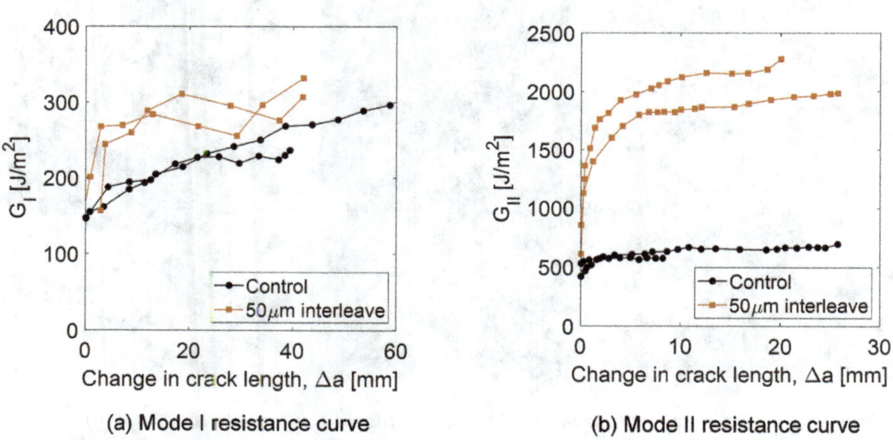

Figure 15.8 Delamination resistance curves of control and 50 μm self-same epoxy interleave laminates (data replotted from [6])

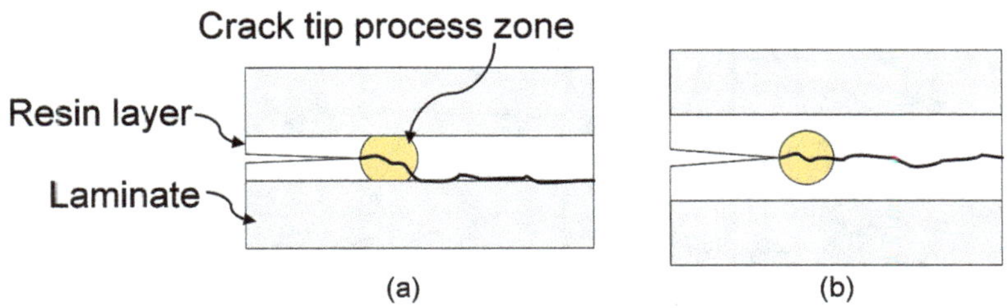

Figure 15.9 *Illustration of plastic yield zone interaction with interleaf boundaries for different thicknesses (a) adhesive failure of the resin and fibre, (b) cohesive failure of the resin*

process zone thus the crack will propagate into the resin/fibre interfaces, given that it is the weakest of the surrounding material to break apart. However, when the resin rich layer increases, the process zone effectively covers only the resin. Crack then propagates within the resin layer, effectively producing a cohesive crack that only breaks the resin, producing fracture toughness same as neat resin. This physical phenomenon is the primary mechanism behind the increased fracture toughness which most interleaved laminates undergo. Of course, using tougher resin, particles or fibres instead of neat self-same resin can further enhance the measure toughness which will be discussed in following sections.

15.3.1.1 Thermoset resin

On par with the self-same resin interleaving, is thermosetting resin interleaving. The thermosets will be of different formulation to the laminate matrix resin and often derived in the form of adhesive films. Such films are widely available and have been designed with excellent toughness properties. Meaning they will contribute to the increased toughness both from the higher toughness of the adhesive film as well as the cohesive fracture made possible due to the resin rich thickness increase.

In Figure 15.10, adhesive film Cytec FM300 were used as interleaves to improve the carbon fibre-reinforced plastic (CFRP) fracture toughness [7]. Evidently, in these results, the peak toughness enhancement was achieved with interleave thickness of larger than 0.2 mm. In Figure 15.11, Yasaee *et al.* [8] show use of Hexcel Redux 312 adhesive film interleave which exhibits more than 200% increase in apparent G_{IIC} of the interface. The diffused distribution of the resin interlayer can be seen in the colour difference in Figure 15.11a. The fracture surface of these test samples is shown in Figure 15.11. This example highlights how the addition of the interleave did not significantly change the resin rich layer in between the fibres. The low viscosity of thermoset resins results in consolidation and diffusion through the composite system during cure. Keeping the resin rich region thickness to a minimum means the average fibre volume fraction of the material will not be reduced, maintaining the inherent strength and stiffness properties of the laminate, while still gaining the fracture toughness enhancements. Many different examples of thermosetting

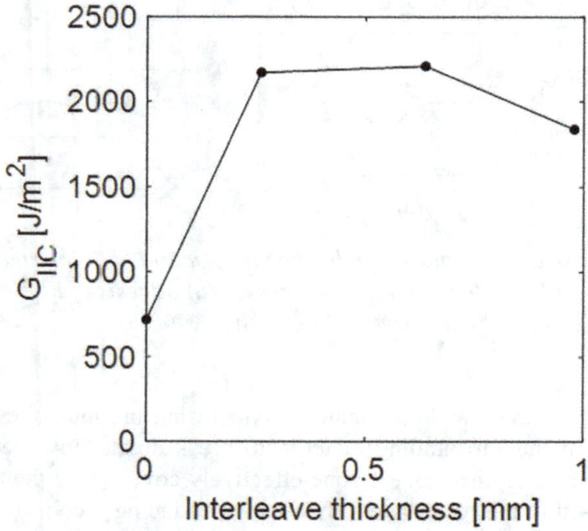

Figure 15.10 *Mode II fracture toughness variation of CFRP composite with change in thermoset adhesive film interleave thickness (Cytec FM300 epoxy) (data replotted from [7])*

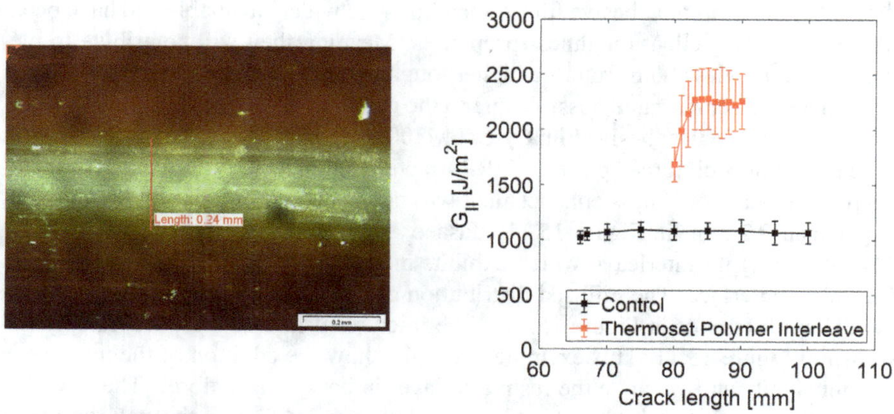

(a) Hexcel Redux 312 Adhesive Interleave

(b) Mode II resistance curve

Figure 15.11 *Thermoset interleave (Hexcel Redux 312 adhesive film). (a) Cured interleave thickness (b) Mode II resistance curve (data replotted from [8]).*

film interleaving have been demonstrated with significant improvements to fracture toughness reported [7–10].

15.3.1.2 Thermoplastic

Thermoplastic polymers on average have higher toughness properties than thermoset polymers. This makes them a suitable candidate for use as interleaving materials. Many examples of thermoplastic film interleaves have been reported to improve fracture toughness of the composite system [11–15].

Not all thermoplastic films can be utilised for interleaving directly. In some cases thermoplastics such as Kapton, Upilex or polytetrafluoroethylene (PTFE) do not chemically bond with standard epoxies, therefore, not suitable as interleaves. In fact, these materials may often be used as release films to produce 'artificial' cracks in composites for experimental testing purposes or to protect tooling to easily separate the composites after cure. Armstrong-Carroll *et al.* [16] polymerised allylamine monomer in a plasma atmosphere and condensed them onto Kapton and PTFE films. This bond enhancement as a result allowed these materials to be used as high-performing interleaves. Yasaee *et al.* [8, 10] surface treated Upilex films by first coarse sanding the surface with 200 grit paper then submerge the films in diethylenetriamine (D93856, Sigma–Aldrich Co.) for 24 hours. This improved the

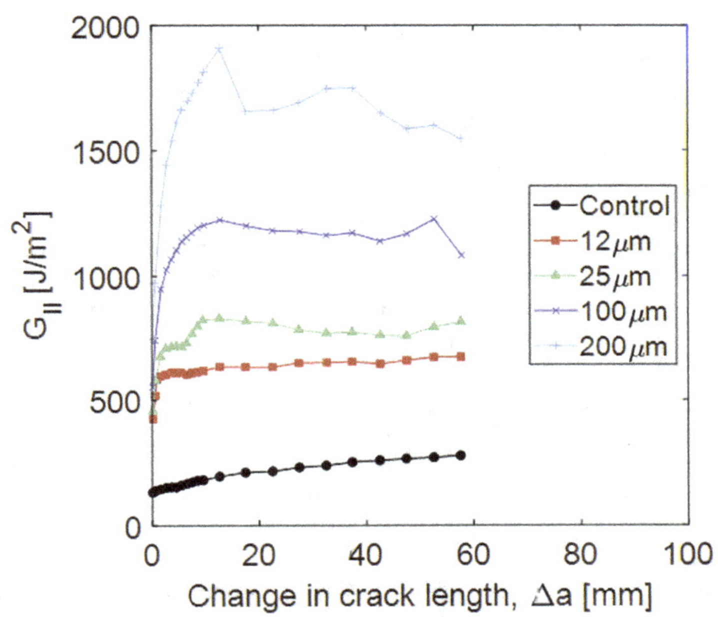

Figure 15.12 Resistance curve of thermoplastic interleaved CFRP composites of increasing film thickness (partially ionized EMAA) (data replotted from [14])

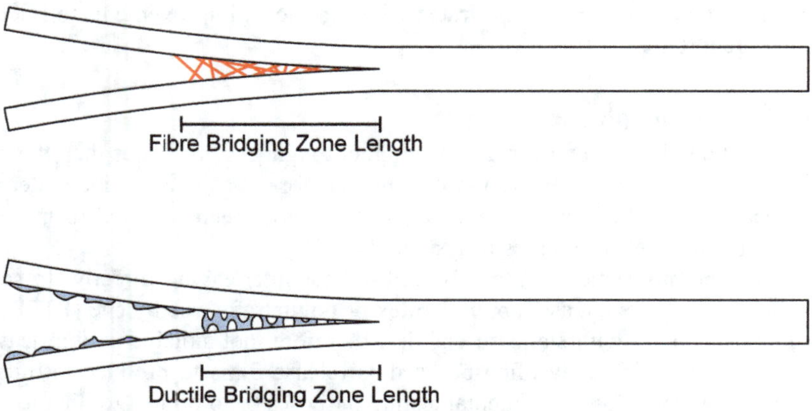

Figure 15.13 *Material bridging contributing to fracture toughness increase*

adhesion of the films to the epoxy matrix and thereby performed well as an interleave material.

Examples of partially ionised ethylene-methacrylic acid (EMAA) films of different thicknesses used as interleaves are shown in Figure 15.12. The increase in G_{IIC} is significant; however, an interesting behaviour is worth noting. All the interleaved samples appear to have similar G_{IIC} initiation values before gradually increasing as the crack length is increased. This rise in the G_{II} with crack length is associated with extrinsic bridging effect which is a result of materials that remain attached to the fracture surface behind the crack tip. This material bridging requires extra strain energy to break apart evidenced from the increase in the G_I curve.

Comparable phenomenon is often seen in composites when tested in Mode I fracture, where fibres remain attached behind the crack contributing to the rise of the resistance curve before plateauing, (see Figure 15.4). This is called 'fibre bridging' illustrated in Figure 15.13. When materials have a ductile interface, similar bridging effect occurs, caused by the bits of ductile materials that remain attached, see Figure 15.13. This is termed here as 'ductile bridging' and is a mechanism that causes the rise in the resistance curves of Figure 15.12. The region behind the crack with bridging still active is called the 'bridging zone length' and can be directly measured from the resistance curves as the distance between the initiation and the plateau of the curves. Knowing the bridging zone length is important as it implies the length cracks need to grow before utilising the full fracture toughness potential of the interleaves.

An advantage of using thermoplastic polymers as interleaves is their inherent capability to transition between solid and liquid states. This means a fractured thermoplastic interleaved interface can be re-bonded by re-heating the sample, effectively gaining back the composite interface integrity [17].

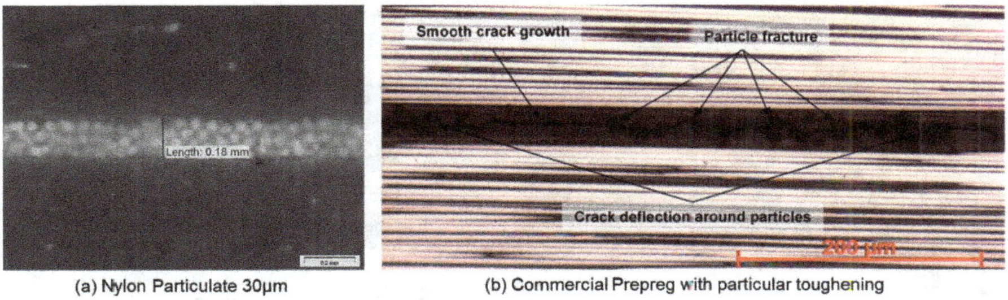

Figure 15.14 Particulate interleaving (a) Nylon particulates. (b) Particulate toughened Hexcel Hexply M21 laminate [26].

15.3.1.3 Particle

Particulates used as interleaves have been one of the most popular methods to enhance the fracture toughness of laminates. Examples of particulates include nylon [8, 10, 18–21], rubber [22, 23], poly-acrylonitrile-butadiene-styrene [24] and silicon carbide [25]. Particle enhanced interlayers can produce anywhere up to 70% to 350% increase in G_{IC} and G_{IIC}, respectively. The mechanisms in which these particles contribute to the enhanced fracture toughness have been argued to be due to the particles' capability to sustain thicker interlayer regions, as shown in Figure 15.14. Particle interleaving is the only procedure that has successfully been commercialised by major pre-preg manufacturers. The supplied pre-preg materials have these particles dispersed evenly across one of their surfaces. In Figure 15.14b, we can see examples of Hexcel Hexply M21 showcasing cohesive crack growth through the interlayer [26].

15.3.1.4 Fibre

The use of additives such as short fibres, nano-fibres or non-woven fibre mats is a popular for use as interleaves. Fibre interleaves must be much thinner than the laminate layers since their purpose is to provide that random fibre bridging effect of the resin rich interlayer region. Various names are often given to such layers, such as non-woven tissue, non-woven mats, veils or whiskers. Many different fibre materials have been used, ranging from standard glass fibres [10, 27] or aramid fibres [10, 27–29] to the bespoke polymer-based fibres such as polyester [27], nylon [30], polysulfone [31] and polyphenylene sulfide (PPS) [32–34]. In all cases the fibre interleaves contribute to significant increase in the interlaminar fracture toughness of the composite system.

Example of polymer fibre (PPS) interleaves arresting delamination cracks after low velocity impact tests is shown in Figure 15.15. The fracture surface of the PPS interleave after Mode I delamination test is shown in Figure 15.16b. The difference between non-interleaved fracture surfaces (Figure 15.15a), is quite evident. The interleaved surface exhibits extensive loose PPS fibre which confirm, fibre bridging to be the key mechanism for the enhanced fracture toughness achieved by fibre-based interleaves.

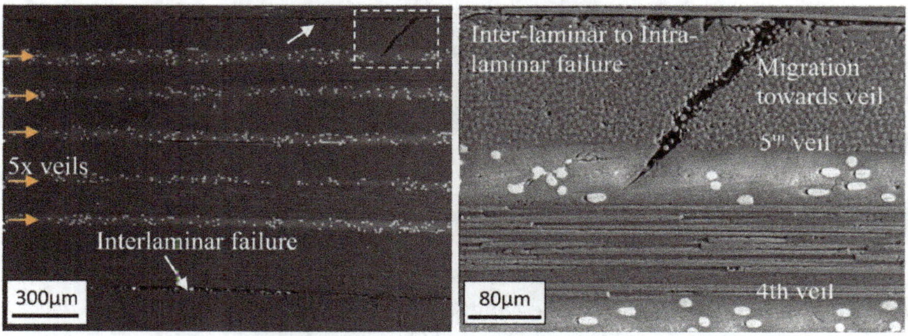

Figure 15.15 Cross-section microscopy of woven CFRP laminate interleaved with non-woven PPS veil (Technical Fibre Products Ltd) after low velocity impact test [32]

Other forms of fibre interleaves, include short nano-fibres such as carbon nanotubes (CNT) many of which has been specifically created either to be randomly dispersed [35, 36] or aligned [37, 38] in the thickness direction. However, experimental tests on randomly distributed CNTs reported only a minor improvement in fracture toughness (up to 0.8 kJ/m^2 in Mode I [39]). Aligned CNTs on the other hand have been shown to be quite effective [37] with improvements of >3 kJ/m^2 in Mode I reported; however, contradicting lower improvements is also reported 4041[40, 41]. Aligned CNT using electric fields has also been demonstrated but only capable of up to 0.75 kJ/m^2 in Mode I toughness [42]. Disadvantage with CNTs is high cost of the raw material, upwards of >£500 per kg in discontinuous powder form [43]. The improvement gains are not superior to other fibre interleaving thereby not justifying the higher cost and complexity to produce these composite systems.

15.3.2 Summary

In summary, the effectiveness of interleaves to increase the interlaminar fracture toughness of composite laminate system is clear. The interleaves provide significant

Figure 15.16 Fractured interface of woven laminates (a) without veil (b) 10 gm^{-2} non-woven PPS veil [32]

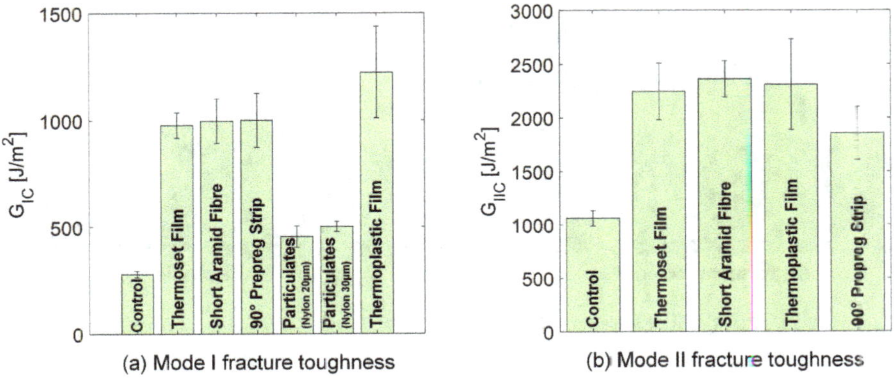

Figure 15.17 *Comparison of different interleaving materials and their influence on the (a) Mode I and (b) Mode II fracture toughness of the interlaminar interface (data replotted from [8, 10])*

improvement to resist delamination initiation and growth. The result of which translates to various applications including reducing low velocity impact damage area and hence improving compression after impact (CAI) strength of the composite system [32]. Yasaee *et al.* [8, 10] explored comparison of various interleaving material types under the same test and base material condition. The results summarised in Figure 15.17 show how short fibre and thermoset films perform very well in both Modes I and II with the thermoplastic film producing the highest fracture toughness enhancement overall.

Interleaved materials in general are a low-cost and straightforward process to improve damage tolerance of composite laminates. There are many potential material types that could be used for the purpose of interleaving and most methods will produce the performance enhancement necessary. However, care must be taken to observe the influence of interleaved materials on other mechanical properties of composite laminate system, such as the in-plane stiffness, strength and overall weight, which can be affected as a result of extra materials and potential for lower fibre volume fraction it generates [44].

15.3.3 *Interleaving applied to stiffeners and adhesive joints*

Some examples of interleaving applied to stiffeners and adhesive joint is shown in this section. The failure mechanism of T-joint stiffeners starts with initiation of cracks in the 'noodle' section of the configuration. The inability of the laminate plies to bend around sharp corners is what results in the resin rich 'noodle' (Figure 15.18). General practice to improve the T-joint performance is to fill the 'noodle' with fibre material to help reinforce the resin and resist crack initiation and propagation. The secondary damage mechanism following the noodle cracking is the delamination that forms at the stiffener-panel interface. Two examples of interleaving on T-stiffeners have been investigated by Sapi *et al.* [45] and Wang and Soutis [46]. In the work

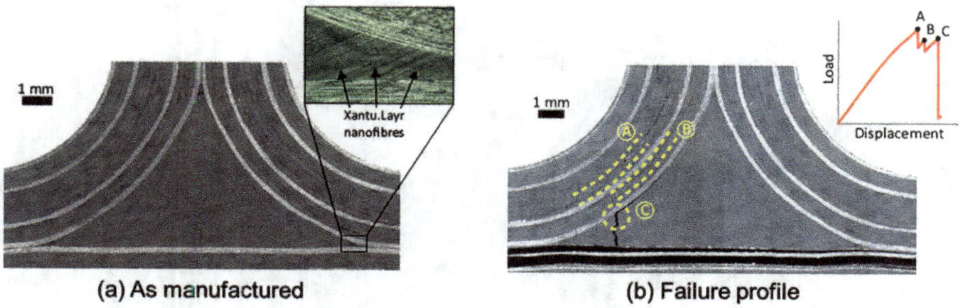

Figure 15.18 *Cross-section imaging of CFRP T-joint with nano-fibre interleaves (Xantu.Layr from Revolution Fibres Ltd) [45]*

of Sapi *et al.* [45] nano-fibre interleaves (Xantu.Layr) from Revolution Fibres Ltd were used on every ply interface of the T-stiffener, as shown in Figure 15.18. This increased the maximum T-stiffener pull-out load by 6% (Figure 15.19), specifically resisting the secondary damage mechanism which is the delamination of the interfaces helping to raise the load carrying capability following the initial load drop.

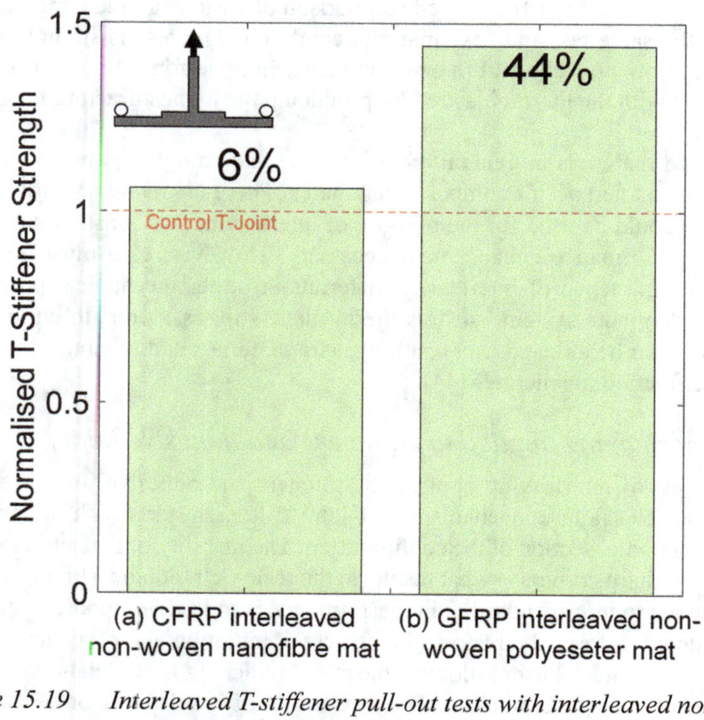

Figure 15.19 *Interleaved T-stiffener pull-out tests with interleaved non-woven mats (a) Sapi* et al. *[45] (b) Wang and Soutis [46]*

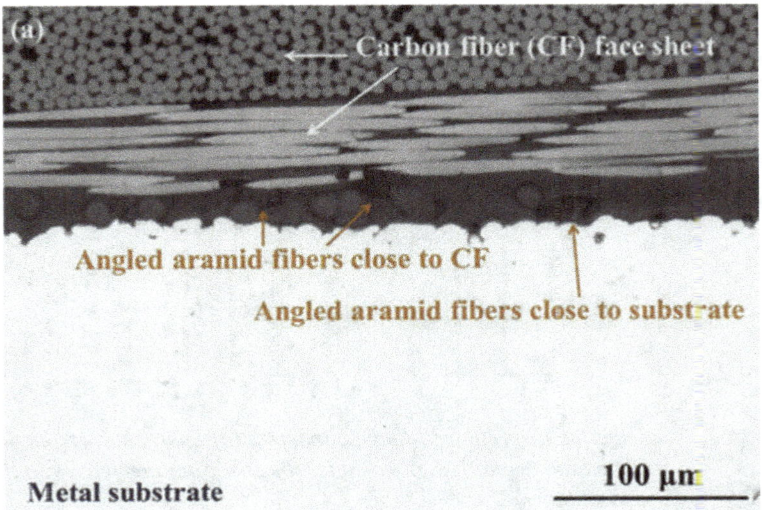

Figure 15.20 Short aramid fibre used as interleaves for co-cured bond between composite and metal substrates [49]

In the example of Wang and Soutis [46], the stiffener and the skin were manufactured with mixture of Uni-Directional (UD) and woven glass fibre laminates. A polyester veil was added in the interface between the flange and the skin. The resulting peak load was increased by 44%, showcasing a major enhancement of the interleaving process to composite stiffener performance.

Many studies have demonstrated the use of interleaving fibres to improve adhesive joint strength between composite and metal substrates. Use of glass-fibre mats [47], nylon 6.6 fibre mats [48], aramid fibre mats [49, 50] has all demonstrated improvements to the joint strength. Example of the specimen cross-section is shown in Figure 15.20. The mechanism that contributes to the improved strength is both the increased adhesive thickness and the fibre bridging resulting from the added fibre mats.

In similar fashion the use of fibre mats can also be quite effective at improving the interface between a composite skin and a honeycomb or foam cores in a sandwich structure with fibre bridging being the dominant mechanism that contributes to increased skin/core interface strength [51].

15.3.4 Selective interface interleaving

Although interleaving produces improvements to delamination resistance, the liberal use of these materials across every interface and entire interface area in a laminate stack may result in lower global stiffness, strength and adverse effects on the global fibre volume fraction of the composite [9]. However, selective use of interleaving not only at critical interfaces but at critical regions may be an interesting approach to enhance composite damage tolerance without severely affecting

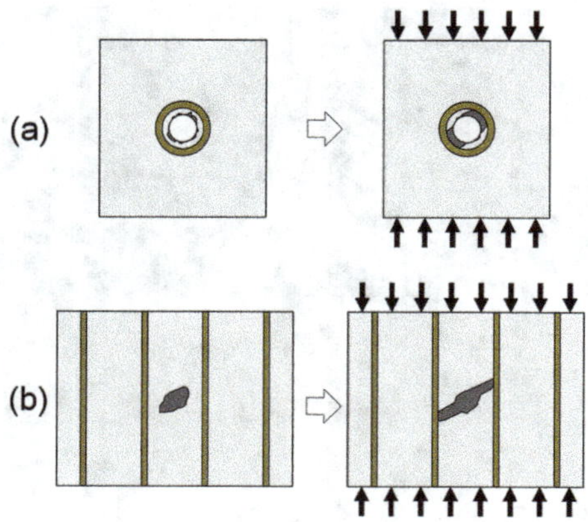

Figure 15.21 *Selective discrete interleave concepts. (a) Discrete interleaves embedded around an open-hole. (b) Discrete interleaves in linear grid pattern.*

the global properties and reducing manufacturing and material costs in the process. The concept of selective interleaving to 'compartmentalise' delamination damage is illustrated in Figure 15.21.

Investigation by Yasaee *et al.* [52] demonstrates the selective interleaving concept to control the impact damage pattern. Thermoplastic interleaves in shape of a ring were used to contain impact damage by only using the interleaves at specific interfaces, see Figure 15.22. In Figure 15.23, , an example of the interleaved rings arresting impact damage is shown. The discrete interleaves provide negligible increase in the composite mass and minimal disturbance to the composite fibres while still capable to contain majority of the damage reducing the impact damage area by 36% and increase the CAI strength of the laminate by 16%.

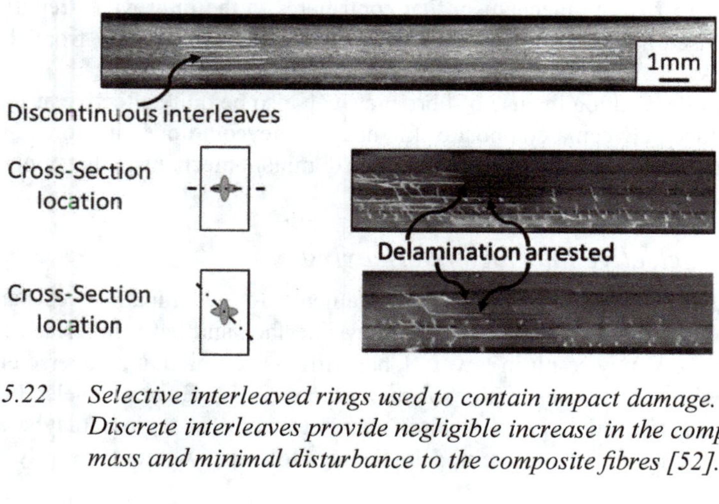

Figure 15.22 *Selective interleaved rings used to contain impact damage. Discrete interleaves provide negligible increase in the composite mass and minimal disturbance to the composite fibres [52].*

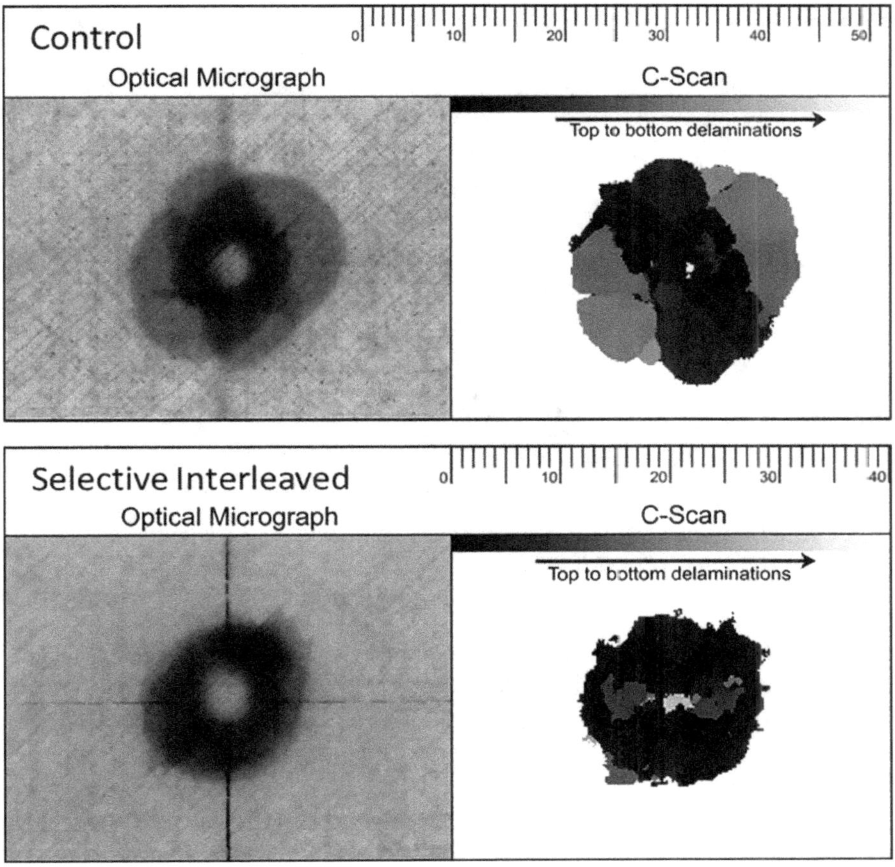

Figure 15.23 *Selective interleaves reducing the impact damage area by up to 38% [52]*

Selective interleaving has also been shown to control the propagation of delamination crack during a fatigue loading. Yasaee *et al.* [15] showed by placing interleaves at critical location within the interface can effectively control the growth of delamination towards a particular direction. Figure 15.24 shows how the selective use of the interleaves helps contain the initial impact damage. Without any interleaving, impact damage would grow evenly in all directions of the laminate until failure. In the interleaved samples, the horizontal interleaves prevent the growth of the delamination to top and bottom of the laminate while the vertical interleaves prevent growth to the right. By choice, no vertical interleaves were used in the left of the impact point. Thereby during fatigue loading, impact damage grew gradually to the left of the laminate. Not only can effective control of delamination growth be achieved, significant increase in the fatigue limit of a CAI panel is also possible. Another example of selective interface interleaving was demonstrated by Ramji *et al.* [32].

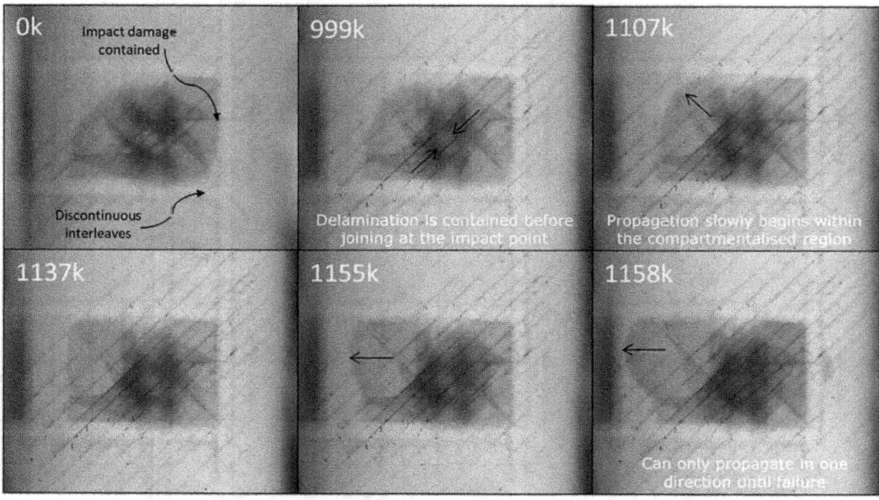

Figure 15.24 Selective interleaving used to control fatigue delamination growth in CAI panels [15]

In this study interleaves used at selective interfaces helped to improve the impact strength of laminate panels without having to implement them at all interfaces, see Figure 15.15.

15.3.5 Summary

Delamination damage is a critical failure mechanism in most composite structures due to the lack of fibre reinforcement in the thickness direction of the laminate. To overcome delamination damage interleaving has been shown to be quite effective. Interleaving materials at the interface where delamination occurs, help to increase the interlaminar facture toughness of the interface. Having a thicker interface has inherent benefit in terms of fracture toughness increase due to the natural phenomena of cohesive fracture of resins being tougher than fracture of both resin and resin/fibre interface. Added to this, inclusion of fibres or particles as interleaves will provide additional toughness enhancement. Thermoplastic materials used as interleaves provide another avenue for interleaving, having higher toughness by default, these materials can significantly increase interlaminar fracture toughness of epoxy laminates.

Due to the advantages of interleaving, commercial suppliers of pre-pregs have begun to provide pre-preg materials with particle interleaves as default. Thereby interleaves has seen many applications across different components from improving the damage tolerance of impact damaged panels to increasing the stiffener or adhesive joints of composites. Finally, use of interleaves discretely within a composite component maybe more beneficial and a method to reduce cost of the structure. Concept of selective interleaving has been shown to be quite effective to compartmentalise delamination damage, contain and even manipulate to make the failure of the structure in a manner that may be beneficial.

References

[1] Johnson W., Mangalgiri P. *Influence of the resin on interlaminar mixed-mode fracture*; 1985.

[2] Mouritz A.P. 'Review of z-pinned composite laminates'. *Composites Part A: Applied Science and Manufacturing*. 2007;38(12):2383–97.

[3] Partridge I.K., Hallett S.R. 'Use of microfasteners to produce damage tolerant composite structures'. *Philosophical Transactions of the Royal Society A: Mathematical, Physical and Engineering Sciences*. 2016;374(2071):20150277.

[4] Partridge I.K., Cartié D.D.R. 'Suppression of initiation of delamination cracking in unidirectional composites by self-same resin interleaving' in Moore D.R. (ed.). *Application of fracture mechanics to polymers, adhesives and composites*. Vol. 33. ESIS Publication Oxford: Elsevier; 2003.

[5] Singh S., Partridge I.K. 'Mixed-mode fracture in an interleaved carbon-fibre/epoxy composite'. *Composites Science and Technology*. 1995;55(4):319–27.

[6] Hojo M., Ando T., Tanaka M., Adachi T., Ochiai S., Endo Y. 'Modes I and II interlaminar fracture toughness and fatigue delamination of CF/epoxy laminates with self-same epoxy interleaf'. *International Journal of Fatigue*. 2006;28(10):1154–65.

[7] Gibson R.F., Chen Y., Zhao H. 'Improvement of vibration damping capacity and fracture toughness in composite laminates by the use of polymeric interleaves'. *Journal of Engineering Materials and Technology*. 2001;123(3):309–14.

[8] Yasaee M., Bond I.P., Trask R.S., Greenhalgh E.S. 'Mode II interfacial toughening through discontinuous interleaves for damage suppression and control'. *Composites Part A: Applied Science and Manufacturing*. 2012;43(1):121–8.

[9] Ozdil F., Carlsson L.A. 'Mode I interlaminar fracture of interleaved graphite/epoxy'. *Journal of Composite Materials*. 1992;26(3):432–59.

[10] Yasaee M., Bond I.P., Trask R.S., Greenhalgh E.S. 'Mode I interfacial toughening through discontinuous interleaves for damage suppression and control'. *Composites Part A: Applied Science and Manufacturing*. 2012;43(1):198–207.

[11] Chen S.F., Jang B.Z. 'Fracture behaviour of interleaved fiber-resin composites'. *Composites Science and Technology*. 1991;41(1):77–97.

[12] Li L., Lee-Sullivan P., Liew K.M. 'The influence of thermoplastic film interleaving on the interlaminar shear strength and mode I fracture of laminated composites'. *Journal of Engineering Materials and Technology*. 1996;118(3):302–09.

[13] Jiang W., Tjong S.C., Chu P.K., Li R.K.Y., Kim J.K., Mai Y.W. 'Interlaminar fracture properties of carbon fibre/epoxy matrix composites interleaved with polyethylene terephthalate (PET) films'. *Polymers and Polymer Composites*. 2001;9(2):141–45.

[14] Matsuda S., Hojo M., Ochiai S., Murakami A., Akimoto H., Ando M. 'Effect of ionomer thickness on mode i interlaminar fracture toughness for ionomer toughened cfrp'. *Composites Part A*. 1999;30(11):1311–19.

[15] Yasaee M., Killock C., Hartley J., Bond I.P. 'Control of compressive fatigue delamination propagation of impact damaged composites using discrete thermoplastic interleaves'. *Applied Composite Materials*. 2015;22(5):559–72.

[16] Armstrong-Carroll E., Iskandarani B., Kamel I., Donnellan T.M. 'The influence of interleaf deformation behavior and film-resin adhesion on the fracture toughness of interleaved composites'. *Composite Materials: Fatigue and Fracture*. 1993;4:299–317.

[17] Pingkarawat K., Wang C.H., Varley R.J., Mouritz A.P. 'Self-healing of delamination cracks in mendable epoxy matrix laminates using poly[ethylene-co-(methacrylic acid)] thermoplastic'. *Composites Part A: Applied Science and Manufacturing*. 2012;43(8):1301–7.

[18] Gao F., Jiao G., Lu Z., Ning R. 'Mode II delamination and damage resistance of carbon/epoxy composite laminates interleaved with thermoplastic particles'. *Journal of Composite Materials*. 2007;41(1):111–23.

[19] Groleau M.R., Shi Y.-B., Yee A.F., Bertram J.L., Sue H.J., Yang P.C. 'Mode II fracture of composites interlayered with nylon particles'. *Composites Science and Technology*. 1996;56(11):1223–40.

[20] Heness G., Stevens M., Dossett B. 'The effect of surface treatment on delamination for a nylon interleaving material'. *Materials Forum*. 2007;31:90–95.

[21] Wang W.-T., Yu H., Potter K., Kim B.C. 'Effect of the characteristics of nylon microparticles on mode-I interlaminar fracture toughness of carbon-fibre/epoxy composites'. *Composites Part A*. 2020;138:106073.

[22] Gilbert E N., Hayes B.S., Seferis J.C. 'Interlayer toughened unidirectional carbon prepreg systems: effect of preformed particle morphology'. *Composites Part A: Applied Science and Manufacturing*. 2003;34(3):245–52.

[23] Malkin R., Trask R.S., Bond I.P. 'Control of unstable crack propagation through bio-inspired interface modification'. *Composites Part A: Applied Science and Manufacturing*. 2013;46(1):122–30.

[24] Stevanovic D., Kalyanasundaram S., Lowe A., PYB J. 'Mode I and mode II delamination properties of glass/vinyl-ester composite toughened by particulate modified interlayers'. *Composites Science and Technology*. 2003;63(13):1949–64.

[25] Alsaadi M., Erkliğ A. 'A comparative study on mode I and mode II interlaminar behavior of borax and sic particles toughened S-glass fabric/epoxy composite'. *Arabian Journal for Science and Engineering*. 2017;42(11):4759–69.

[26] Hunt C., Kratz J., Partridge I.K. 'Cure path dependency of mode I fracture toughness in thermoplastic particle interleaf toughened prepreg laminates'. *Composites Part A: Applied Science and Manufacturing*. 2016;87(4):109–14.

[27] Lee S.-H., Lee J.-H., Cheong S.-K., Noguchi H. 'A toughening and strengthening technique of hybrid composites with non-woven tissue'. *Journal of Materials Processing Technology*. 2008;207(1–3):21–9.

[28] Yadav S.N., Kumar V., Verma S.K. 'Fracture toughness behaviour of carbon fibre epoxy composite with Kevlar reinforced interleave'. *Materials Science and Engineering: B*. 2006;132(1-2):108–12.

[29] Sohn M.-S., Hu X.-Z. 'Comparative study of dynamic and static delamination behaviour of carbon fibre/epoxy composite laminates'. *Composites*. 1995;26(12):849–58.

[30] Akangah P., Lingaiah S., Shivakumar K. 'Effect of Nylon-66 nano-fiber interleaving on impact damage resistance of epoxy/carbon fiber composite laminates'. *Composite Structures*. 2010;92(6):1432–9.

[31] Li G., Li P., Yu Y., *et al*. 'Novel carbon fiber/epoxy composite toughened by electrospun polysulfone nanofibers'. *Materials Letters*. 2008;62(3):511–4.

[32] Ramji A., Xu Y., Yasaee M., Grasso M., Webb P. 'Influence of veil interleave distribution on the delamination resistance of cross-ply CFRP laminates under low velocity impact'. *International Journal of Impact Engineering*. 2021;157:103997.

[33] Ramji A., Xu Y., Yasaee M., Grasso M., Webb P. 'Effect of interfacial fibre orientation and PPS veils on fracture toughness of woven CFRP laminates under mode I loading'. *Composites. Part B, Engineering*. 2019.

[34] Ramji A., Xu Y., Yasaee M., Grasso M., Webb P. 'Delamination migration in CFRP laminates under mode I loading'. *Composites Science and Technology*. 2020;190:108067.

[35] Arai M., Noro Y., Sugimoto K.-ichi., Endo M. 'Mode I and mode II interlaminar fracture toughness of CFRP laminates toughened by carbon nanofiber interlayer'. *Composites Science and Technology*. 2008;68(2):516–25.

[36] Warren G.L., Sun L., Hadjiev V.G., Davis D., Lagoudas D., Sue H.-J. 'B-staged epoxy/single-walled carbon nanotube nanocomposite thin films for composite reinforcement'. *Journal of Applied Polymer Science*. 2009;112(1):290–8.

[37] Wicks S.S., de Villoria R.G., Wardle B.L. 'Interlaminar and intralaminar reinforcement of composite laminates with aligned carbon nanotubes'. *Composites Science and Technology*. 2010;70(1):20–8.

[38] Garcia E.J., Wardle B.L., John Hart A. 'Joining prepreg composite interfaces with aligned carbon nanotubes'. *Composites Part A: Applied Science and Manufacturing*. 2008;39(6):1065–70.

[39] Qian H., Bismarck A., Greenhalgh E.S., Kalinka G., Shaffer M.S.P. 'Hierarchical composites reinforced with carbon nanotube grafted fibers: the potential assessed at the single fiber level'. *Chemistry of Materials*. 2008;20(5):1862–9.

[40] Falzon B.G., Hawkins S.C., Huynh C.P., Radjef R., Brown C. 'An investigation of mode I and mode II fracture toughness enhancement using aligned carbon nanotubes forests at the crack interface'. *Composite Structures*. 2013;106(3):65–73.

[41] Wicks S.S., Wang W., Williams M.R., Wardle B.L. 'Multi-scale interlaminar fracture mechanisms in woven composite laminates reinforced with aligned carbon nanotubes'. *Composites Science and Technology*. 2014;100(12):128–35.

[42] Wu S., Ladani R.B., Ravindran A.R., *et al.* 'Aligning carbon nanofibres in glass-fibre/epoxy composites to improve interlaminar toughness and crack-detection capability'. *Composites Science and Technology*. 2017;152:46–56.

[43] Cheap Tubes *Welcome to cheap tubes [online]*. Available from www.cheaptubes.com [Accessed Sep 2017].

[44] Sela N., Ishai O. 'Interlaminar fracture toughness and toughening of laminated composite materials: a review'. *Composites*. 1989;20(5):423–35.

[45] Sápi Z., Hutchins S., Butler R., Rhead A. 'Novel filler materials for composite out-of-plane joints'. *Composite Structures*. 2019;229:111382.

[46] Wang Y. Soutis C. 'Fatigue behaviour of composite T-Joints in wind turbine blade applications'. *Applied Composite Materials*. 2017;24(2):461–75.

[47] Moreira Arouche M., Teixeira de Freitas S., de Barros S. 'On the influence of glass fiber mat on the mixed-mode fracture of composite-to-metal bonded joints'. *Composite Structures*. 2021;256(3):113109.

[48] Ulus H., Burak Kaybal H., Erdem Berber N., *et al.* 'An experimental evaluation on the dynamic response of water aged composite/aluminium adhesive joints: influence of electrospun nanofibers interleaving'. *Composite Structures*. 2022;280(17):114852.

[49] Sun Z., Shi S., Hu X., Guo X., Chen J., Chen H. 'Short-aramid-fiber toughening of epoxy adhesive joint between carbon fiber composites and metal substrates with different surface morphology'. *Composites Part B: Engineering*. 2015;77:38–45.

[50] Ekrem M., Avcı A. 'Effects of polyvinyl alcohol nanofiber mats on the adhesion strength and fracture toughness of epoxy adhesive joints'. *Composites Part B: Engineering*. 2018;138(4):256–64.

[51] Shi S., Sun Z., Hu X., Chen H. 'Carbon-fiber and aluminum-honeycomb sandwich composites with and without kevlar-fiber interfacial toughening'. *Composites Part A*. 2014;67:102–10.

[52] Yasaee M., Bond I.P., Trask R.S., Greenhalgh E.S. 'Damage control using discrete thermoplastic film inserts'. *Composites Part A: Applied Science and Manufacturing*. 2012;43(6):978–89.

Chapter 16

A deep learning-based tool to predict delamination induced interlaminar stresses in composite structures

Marco Zacaria Wahaba[1], Hamed Yazdani Nezhad[1], Yazhini Chitra Pradeep[2], and Sathiskumar Anusuya Ponnusami[1]

A Neural Network (NN) algorithm was developed to predict the interlaminar stresses in a multi-layered composite panel, given the ply orientations and the 2-D in-plane stresses as sole inputs, usually obtained from classical plate or shell analyses. In particular, the developed tool eliminates the need for any three-dimensional (3-D) analysis or similar approaches to obtain the interlaminar stresses. For training the deep learning NN algorithm, a large dataset was generated to extract the interlaminar stresses in a composite shell with a hole, through time-intensive shell-cohesive analysis using commercial FEA Software (ABAQUS). Automated scripting interface using Python was used to automatically generate the training, validation, and testing datasets. Upon completion of the model training, stress distribution, performance measure, error distribution and regression plots were analysed and discussed. The results showed an overall good prediction of the interlaminar stresses in composite panels with a mean accuracy of 95%. Insights on the model accuracy and the predictions, and possible directions of further improvement were suggested. The resulting tool can be quite handy for composites analysis and design engineers for quickly obtaining interlaminar stresses from a 2-D plate/shell analysis, without resorting to time-intensive 3-D finite element analysis.

16.1 Introduction

Delamination is one of the major damage modes in laminated composite materials. It is a phenomenon that occurs as separation of the laminas near the free edges of the laminate, where the greatest concentration of interlaminar stresses occurs. These

[1]Aeronautics and Aerospace Research Centre, University of London, London, UK
[2]Humanising Autonomy Ltd, London, UK

stresses can cause laminated composites to delaminate and fail at loads that are significantly lower than the failure loads predicted by traditional lamination theories. It is consequently critical to accurately determine the 3D stress state in the boundary-layer areas of laminated plates and shells in order to correctly describe the laminates' behaviour and prevent early failure [1]. For several years, the interlaminar stress distribution in composite laminates has been studied. With the advancement of computational power and high-fidelity numerical models that accurately depict the response and failure of composite structures, alternatives for generating design allowables using sophisticated FEA have also been investigated in order to decrease certification expenses. These solutions, however, are still time intensive and computationally costly, especially when uncertainty is included in. Therefore, an appropriate evaluation of the interlaminar stresses, and their integration into a ML model, can lead to an accurate prediction of failure mechanisms within composite structures.

The computing speed superiority of AI and ML techniques, compared to physics-based modelling tools and finite element methods (FEM), enables them to aid in solving new composite design problems hence generating optimal designs of composite materials [2]. Furtado *et al.* [3] employed ML to predict first-ply failure strength and ultimate strength of multi-directional composite laminates and their statistical distribution, related to the material and geometrical properties. Pathan *et al* [4] developed a supervised ML model combined with Principal Component Analysis (PCA) to predict the elasto-plastic responses such as elastic moduli and yield strengths (E22, E33, G23, σ_{22}, σ_{33}), of a unidirectional fibre composite loaded in the transverse plane (2–3), at negligible computational cost. Predictions were obtained from images of the material microstructure, as well as knowledge of the constitutive models for fibres and matrix, without performing physically based calculations. Mozaffar *et al.* [5] employed deep learning to predict path-dependent plasticity within 0.5% error, aimed at describing the yield loci and work hardening of a material under different deformation states. Register *et al.* [6] employed ML for the purpose of minimising the destructive aeroelastic effect of flutter. An NN algorithm was used to model and predict the effects of the lamina orientation on flutter induced vibrational modes in a simplified composite wing-box composite. The NN algorithm demonstrated to be able to accurately predict the vibrational frequencies in much less time compared the FEA solver Abaqus that initially predicted them.

The purpose of this work is to develop a ML-based algorithm that enables prediction of interlaminar stresses between composite layers, given the material orientations and the in-plane stresses (σ_{11}, σ_{22}, σ_{12}). This will be achieved by employing FEA software to perform shell-cohesive analysis, aimed at retrieving interlaminar stresses between layers. The resulting data will be used to train an existing ML model for prediction of interlaminar stresses, given material orientations and in-plane stresses as sole inputs. Such aim will be accomplished via the following objectives:

1. Design of a multi-layered (10) composite panel in Abaqus software and development of a Python code to autonomously generate multiple simulations aimed

at characterising the mechanical behaviour of the panel under varying input conditions.

2. Development of a database pairing each set of input parameters (descriptors) with corresponding output parameters, and subdivision of the dataset into training, validation and testing datasets.
3. Development of an ML algorithm from which a model can be derived using the created dataset. The model must enable prediction of the interlaminar shear stresses of the composite structural panel, given input parameters.

16.2 Methodology

With the objectives discussed above, the adopted methodology consists of the following phases:

1. Design and execution of FEA experiments.
2. Development of a database with the simulation results.
3. Employment of a ML algorithm to predict interlaminar stresses, followed by validation and testing.

16.2.1 Design of experiments

Design of experiments consists of synthesising different configurations of the composite structural panel, with the purpose of characterising mechanical properties such as in-plane and out-plane stresses. The range of possible composite configurations (design matrix) is identified as a finite set of descriptors or features [7][8]. For this project two types of descriptors are used:

- Ply orientations of the lamina (microstructural descriptors).
- Magnitude of external load conditions (boundary descriptors).

These descriptors were arranged within column vectors (known as examples), each of same size. Ply orientations had 9 examples of size 10, whilst boundary descriptors had 10 examples of size 1. To evaluate the interlaminar stresses of a composite structure, a shell-cohesive analysis must be performed. It consists of introducing zero thickness layers of cohesive elements between the composite layers.(Figure 16.1) The cohesive element can then be used to simulate the initiation and growth of delamination between stacked layers of shell elements [9]. These types of simulations are computationally intensive, hence employment of ML limits time and costs required for the design optimisation process. The flowchart in Figure 16.2 illustrates the Abaqus simulation setup.

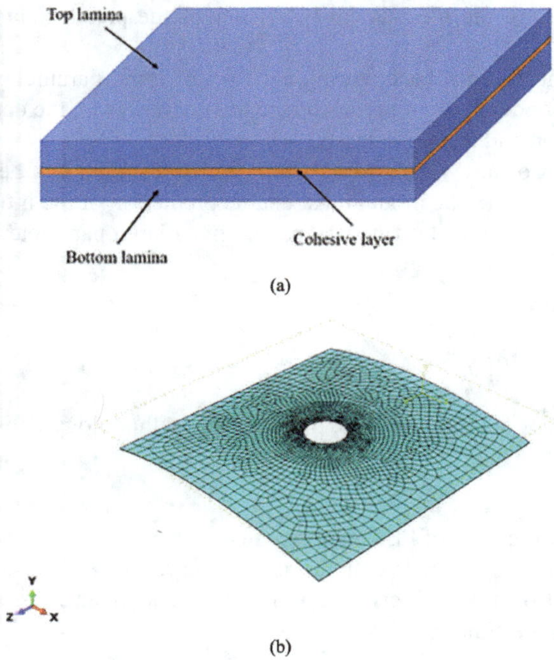

Figure 16.1 Schematic of cohesive layer between laminas meshed panel

Figure 16.2 Panel with load and boundary conditions applied

16.2.1.1 Simulation setup

16.2.1.1.1 Geometry creation

A solid curved panel 1 m × 1 m with a hole in the middle was created using the FEA software Abaqus (Figure 16.3). The panel was partitioned in a way to obtain 10 layers to which assign an elastic material behaviour, and nine thin layers with cohesive material properties. As an input for the database development, at first the FEA simulation was performed with a material orientation of (0, –45, 30, 60, 90 – symmetric) that was assigned to the 10 composite layers. Mechanical properties assigned for the

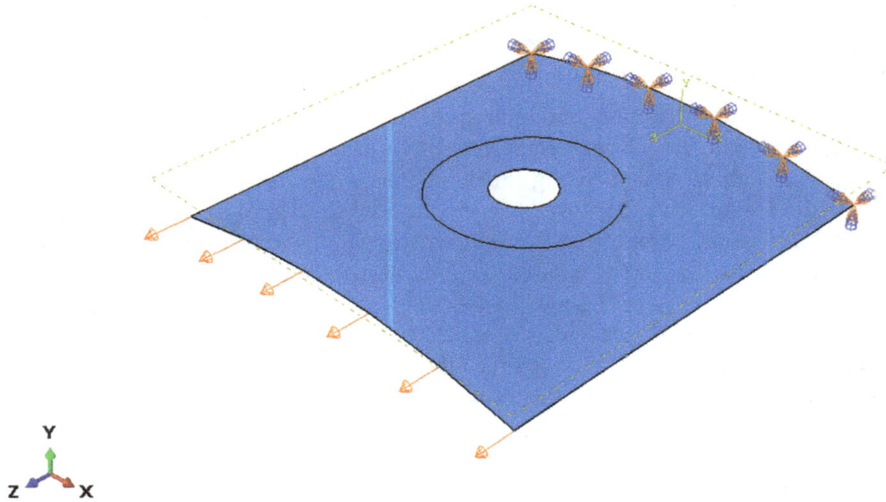

Figure 16.3 *Panel (1 m × 1 m) with axial load and BC*

lamina are: $E_1 = 139 \times 10^9$ Pa, $E_2 = E_3 = 54 \times 10^9$ Pa, $G_{12} = G_{13} = G_{23} = 5.4 \times 10^9$ and Pa, and $v_{12} = v_{13} = v_{23} = 0.05$. A simple uniaxial displacement load in the z direction was applied on one curved edge of the panel while the other was fixed in three directions.

16.2.1.1.2 Meshing and material assignment

S4R elements were used to mesh the structural panel, being the most suitable for a 3D shell model of this type. Around the hole, a circular area was partitioned in which a finer mesh was set. A finer mesh would have smaller spacing between each computed point thus having a more accurate and realistic representation of the stress distribution. However, having a grid that is far too refined increases the computational cost drastically without having a great impact on the results. A continuum shell element type SC8R and a cohesive element type COH3D8 were set for the lamina and cohesive layers, respectively. Upon simulation completion, input, output and journal files are generated as an outcome of the simulation.

16.2.2 Database development

After experiments design, a database with parameters of interest q (interlaminar stresses), corresponding to the macroscopic response (mechanical properties) of the material for each input set x (ply orientation and in-plane stresses) was developed. (Figure 16.4) This database constituted the

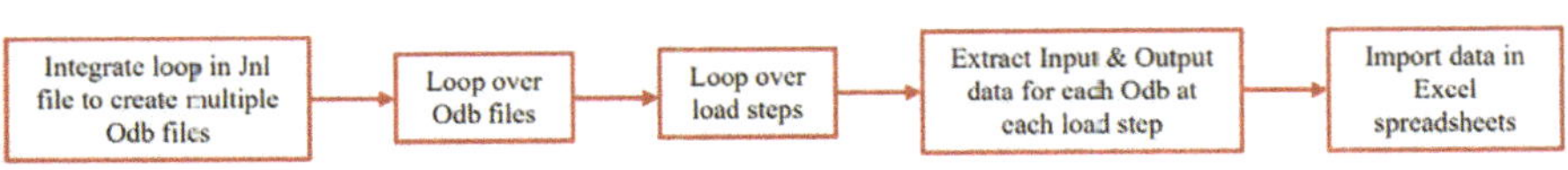

Figure 16.4 *Database development with Python/Abaqus script interface*

training and validation dataset to which a machine learning algorithm was applied. $\left\{\left(x_j^{(1)}, q_i^{(1)}\right), \left(x_j^{(2)}, q_i^{(2)}\right), \ldots, \left(x_j^{(S)}, q_i^{(S)}\right)\right\}$, for $j = 1, \ldots, n$ and $i = 1, \ldots, m$. Where m corresponds to the number of output quantities of interest q. Determination of the parameters of interest for each configuration of the structural panel via individual Abaqus simulations is demanding timewise. Thanks to Abaqus scripting interface with Python, loops were employed to reproduce the stress analysis multiple times to obtain a wide range of input variables and corresponding quantities of interest to use as the training and validation datasets for the following machine learning step of the project. The python script was able to generate automatically and sequentially, nine different stacking sequences with a 10-time step load condition each. Datapoint vectors from the completed runs were saved to an excel spreadsheet upon completion. The flowchart below illustrates the steps of the database development with Python/ Abaqus script interface.

16.2.3 Machine learning modelling

Following completion of the database, a model that captures the influence of the input variables on output mechanical properties was developed through ML methods. The purpose of the model is to predict the interlaminar stresses of the composite lamina configuration that the structural panel will be subjected to.

A ML algorithm is an algorithm that is able to learn from data. Mitchell *et al.* (1997) provide the following definition: 'A computer program is said to learn from experience E with respect to some class of tasks T and performance measure P, if its performance at tasks in T, as measured by P, improves with experience E'.

- In this problem, the ML model is required to output a function $f: R^n \rightarrow R^m$ in order to predict certain numerical values given some inputs. Therefore, the type of task is a regression.
- A quantitative measure of the model performance must be designed to assess the ML algorithm's capabilities on unknown data that has not seen before. Therefore, an independent test dataset is employed to evaluate the performance of the algorithm by using the MSE performance function, which is defined as the following:

$$MSE = \frac{1}{N}\sum_{i=1}^{N}(y_i - \hat{y})^2 \tag{16.1}$$

here:

N is the total number of input vectors in the validation dataset;
i is the example considered;
y is the target value;
$\hat{y}$ is the model predicted value of y.

- The dataset developed, contains eight features to which three targets are associated, therefore the algorithm employed is based on a supervised learning experience [10].

16.2.3.1 Training, validation, testing sets and hyper-parameters

The dataset was split prior to the learning process into training set (90% of the total dataset), validation (5% dataset) and testing sets (5% dataset). The training set is used to train the selected model, whilst the validation set is employed to train the hyper-parameters and estimate the generalisation error during or after training. The testing set is used to test the model accuracy. The following hyper-parameters were employed:

- Number of neurons
- Number of network layers
- Activation functions
- Learning rate.

16.2.3.2 Neural Networks

Within the wide variety of model development methods, NN seems to be the most suitable. They have multiple layers, resulting from the assembly of neurons into a specific network architecture. NNs, assume that each output parameter of interest q results from applying a transformation function to a linear combination of the input variables x (neurons). Equation (16.2) represents the relationship between input and output parameters

$$q_i = f\left(w_{ij}x_j + b_i\right) \tag{16.2}$$

where w_{ij} are the weight to be determined, b_i are the bias or offset parameter which are unknown, and f is the transfer or activation function which will be chosen [7]. The flowchart in Figure 16.5 illustrates the ML algorithm development steps.

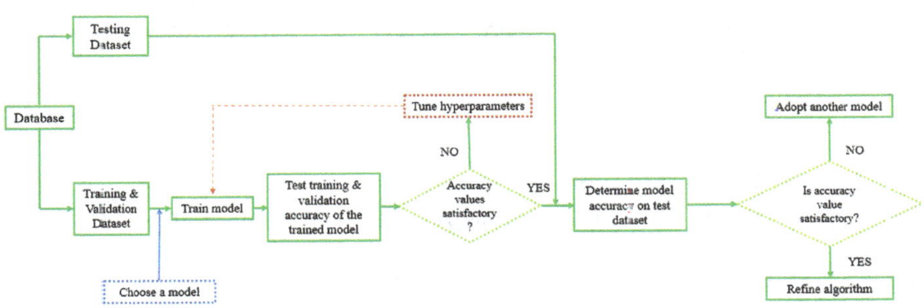

Figure 16.5 ML algorithm development steps

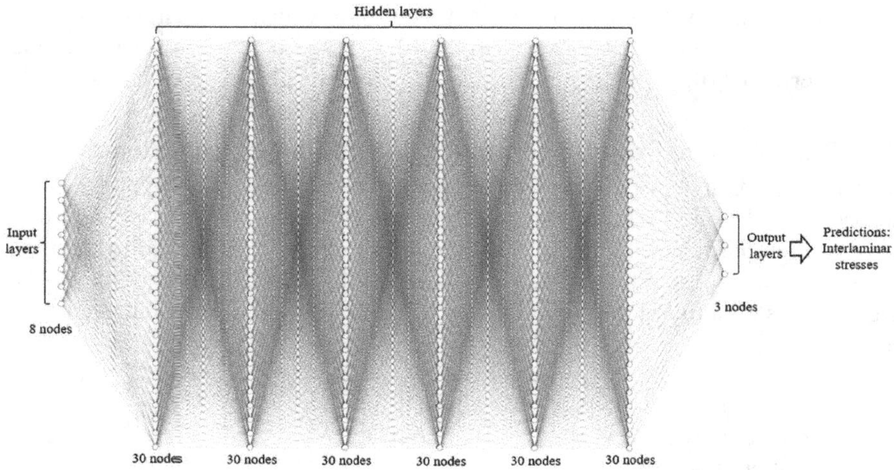

Figure 16.6 NN Architecture designed for the current study

An NN model (Figure 16.6) was created in MATLAB to process the developed FEA dataset, it consisted of eight fully connected layers; one input layer, six hidden layers and one output layer. Figure 16.7 shows the NN architecture.

16.2.3.3 Network type

The type of network used is a feed-forward backpropagation, which proceeds backwards from the final layer to the first layer of the network and computes the gradient of the error function for a single weight by the chain rule. In contrast to the native direct approach of calculating the gradient of each layer separately, this backwards flow of error information permits to compute efficiently the gradient one layer at a time. This choice was based on its flexibility as it neither requires prior knowledge about the network or parameters to be tuned, apart from the numbers of the network input. Figure 16.8 illustrates the backpropagation (right to left) of the gradients at each node using the chain rule and computed values [11].

16.2.3.4 Transfer function

The hyperbolic tangent sigmoid function (TanSigmoid) was utilised as the transfer function in the hidden layers, whereas an identity transfer function was used in the output layer. Transfer functions compute a layer's output from its net input. The TanSigmoid (M) function takes as input the matrix $M \rightarrow SxQ$ (output) of net input vectors and returns each element of M squeezed between –1 and 1, therefore layer output is centred at zero [10].

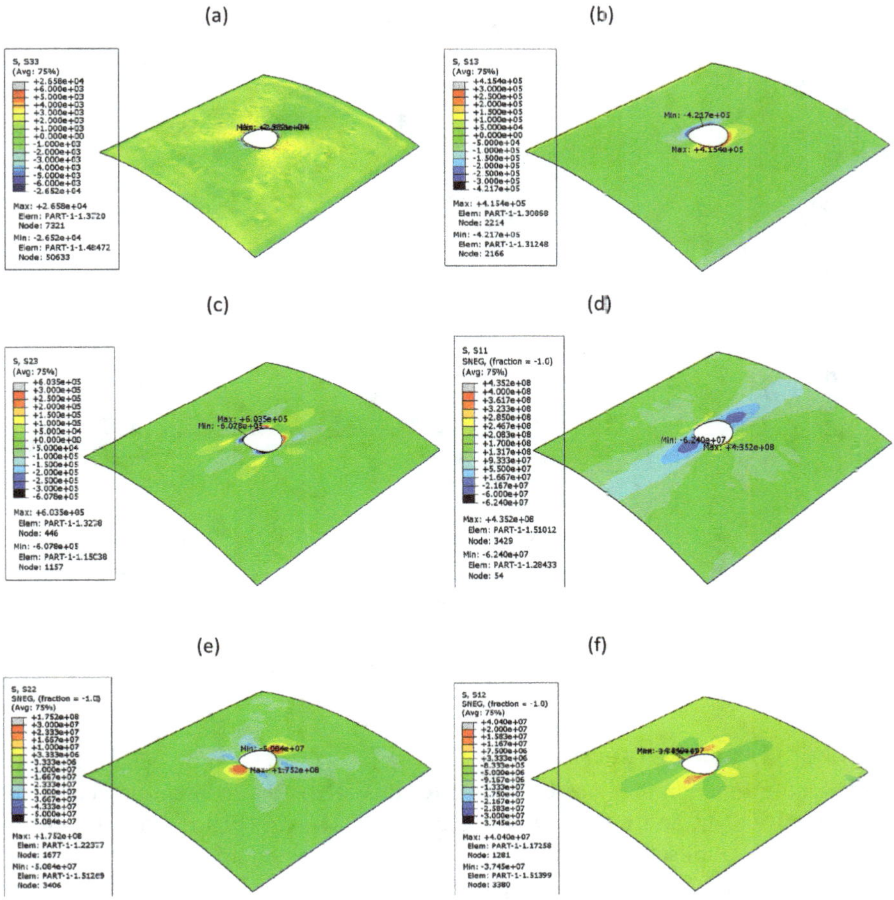

Figure 16.7 *(a) S33 stress in cohesive layers, (b) S13 stress in cohesive layers, (c) S23 stress in cohesive layers, (d) S11 stress in composite layers, (e) S22 stress in composite layers, (f) S12 stress in composite layers*

16.2.3.5 Training function

Training function was based on the Levenberg–Marquardt (LM) algorithm because it is stable and needs less time and epochs for convergence. LM algorithm provides a numerical solution to the problem of minimising a nonlinear function. The LM algorithm is an interpolation between Gauss–Newton method and the steepest descent algorithm, combining the speed advantage of the former with the stability of the latter. The main concept behind the LM algorithm is to perform a combined training process where: the steepest descent algorithm is used around areas with a complex curvature until the curvature is suitable enough to perform a quadratic approximation (Gauss Newton Algorithm) [12].

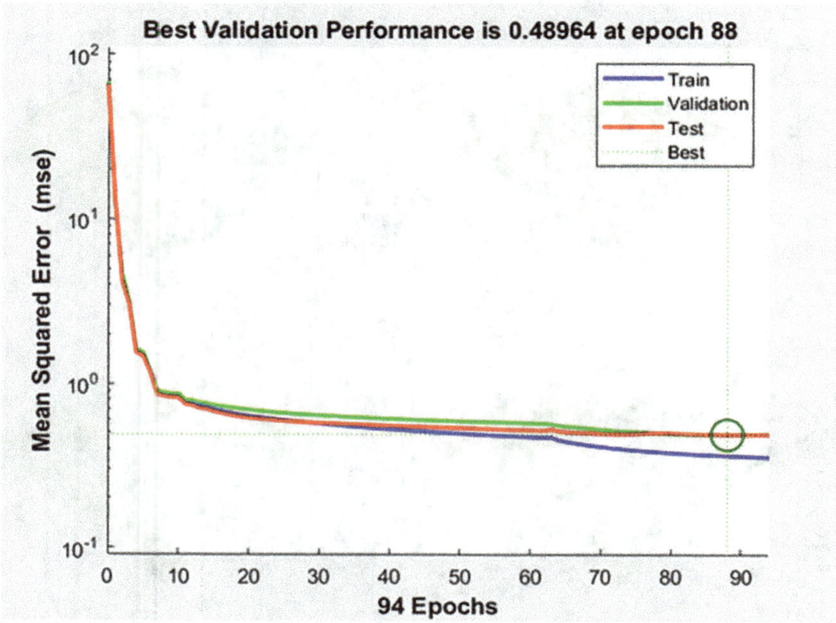

Figure 16.8 Performance plot

16.2.3.6 Learning function

The learning function that was employed for this NN is the gradient descent with momentum weight and bias learning (LEARNGDM), e.g. Figure 16.9. Gradient descent is an optimisation algorithm that locates the minimum value of the target function by searching the negative gradient of the function. In a search space with a high quantity of curvatures, the gradient descent algorithm can get 'stuck' in flat

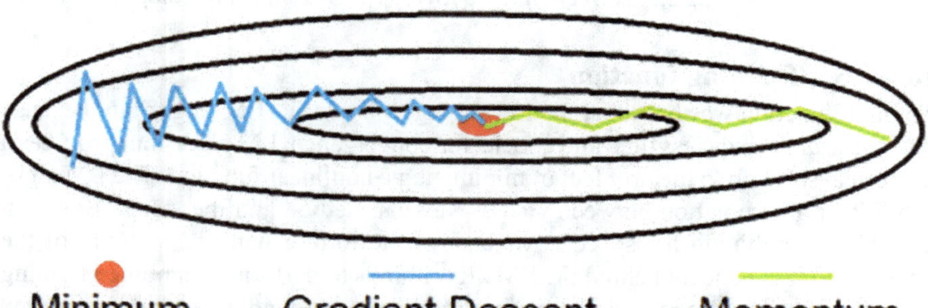

Figure 16.9 Comparison between gradient descent and momentum approaches [10].

regions (with no gradient). Momentum algorithm is a variation that accelerates the optimisation process by overcoming oscillations caused by the presence of noisy gradients in the search space.

LEARNGDM computes the weight variation *dw* with the following formulae:

$$dw = MC * dw_{prev} + (1 - MC) * lr * gw \tag{16.3}$$

where:

w is the weight or bias;
MC is the momentum constant;
lr is the learning rate;
dw_{prev} is the previous weight change stored from the learning state [10].

16.3 Results

16.3.1 *Stress distributions within the composite panel*

Data for in-plane and out-plane stresses were produced respectively for composite and cohesive layers of the panel. The figures below highlight the aforementioned stresses distribution and their respective values in the expected regions of the panel, according to the experimental results encountered in the literature review . From the outcome of the simulation, only the cohesive layers present the interlaminar stresses (S33, S13, S23), whilst the in-plane stresses (S11, S22, S12) can be visualised only in the composite layers. It can be observed that maximum and minimum stresses are always located in the regions in proximity of the hole. The maximum stress values of around 4e+08 N/m^2 are well within acceptable range, implying that the composite panel is a good model of a working real-world counterpart.

16.3.2 *Performance measure plot*

The performance of the NN is displayed in Figure 16.8. It shows that the algorithm is performing well overall. Since train, validation and test lines are all beyond the best (dotted) line, set at 88 epochs; convergence is successfully achieved. Since there is no evidence of the test line rising significantly before the validation line, then there is no evidence of overfitting occurring. The MSE at 88 epochs is 0.49 which is an acceptable value, in range with values encountered within some articles in the literature review.

16.3.3 *Training state plot*

The training state plot illustrates the progress of the training parameters, such as the gradient magnitude, the momentum update (Mu) and the number of validations checks versus the epochs. Figure 16.10 shows that error reduces after more epochs of training and the validation error reaches its lowest at 88 epochs due to training process termination after six subsequent validation checks. Mu is the adaptation parameter used by the LM optimisation process when computing the update of the parameters (weight update) to avoid the problem of local minimum. The Mu plot

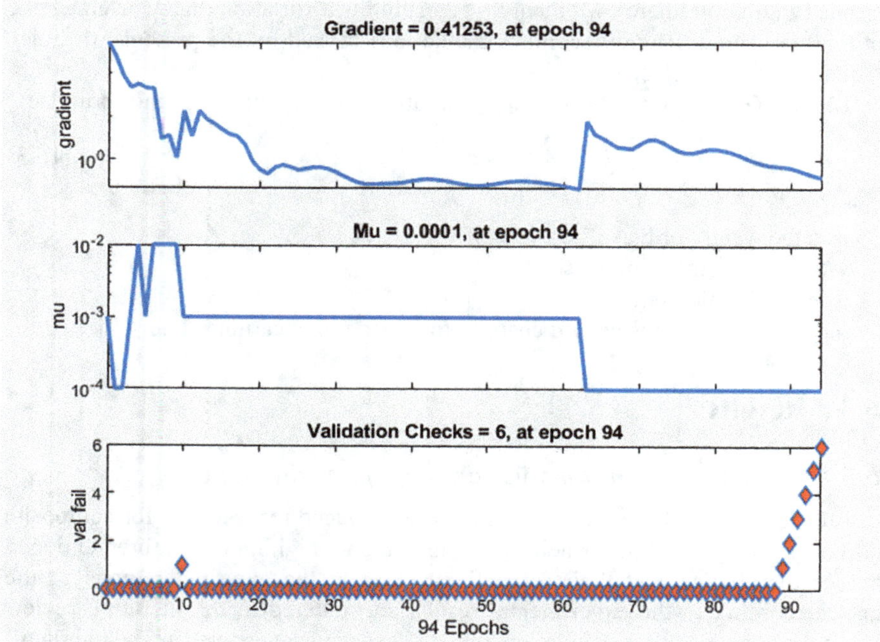

Figure 16.10 Training state plot

versus Epochs indicates very low values of Mu at epoch 94. Training, testing, and validation datasets present a good correlation since the training state plot values are in line with values encountered throughout literature review articles.

16.3.4 Error histogram

The error histogram in Figure 16.11 shows errors between target values and predicted values after training the NN. The error values indicate how the predicted values differ from the target values. The vertical axis (instances) represents the number of dataset samples that lie in a certain bar. It is evident that very few dataset samples are located in regions with high errors (from –3.087 to –1.054 and from 1.486 to 1.994). The vertical zero-error line corresponds to the zero-error value on the x-axis. It is important to note that, the zero-error line is on the most populated bar with centre –0.03823. Therefore, there is a good correlation between prediction and target. The error histogram proves that the pre-processing operations performed on the dataset alongside the training model chosen were well suited for the problem considered.

16.3.5 ML predictions and regression plot

The training, validation, and testing regression plots displayed below (Figure 16.12) represent the correlation between the outputs of the network and the targets. *R* is the

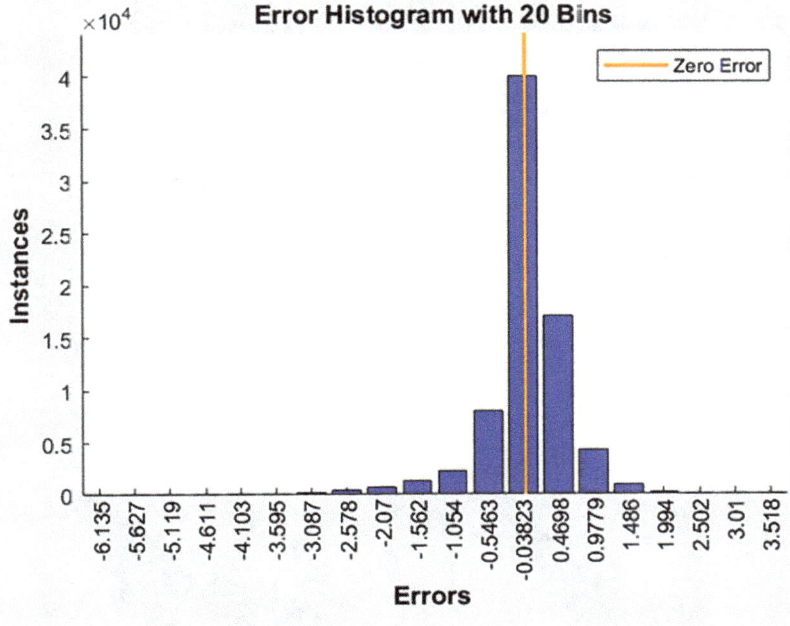

Figure 16.11 Error histogram plot

correlation coefficient. The dashed line indicates the ideal result when outputs are equal to targets (R=1 exact linear relationship). The solid line indicates the best fit linear regression between outputs and target values. Training, validation and testing correlation coefficients (R) are all above 0.92 which is in line with values encountered throughout literature review articles.

16.4 Discussion

According to the findings, ML-based procedures are highly suited for predicting the interlaminar stresses in composite structures. The NN algorithm developed, presented promising results as highlighted by the performance measure, error histogram and regression plots mentioned above. In addition, a comparison of the simulation durations between the FEA solver (Abaqus) and the NN algorithm demonstrated the actual advantage of ML; with 1-minute necessary for the NN to match the dataset which is a fraction of the time required by the FEA Software. When it comes to resources and computational time, it made sense to focus on a simplified panel rather than a complex one. A complex geometry poses additional challenges to a problem which might compromise solution outcome, if tackled incorrectly. This is why a simplified curved panel was chosen. To tackle challenging real-world applications, a trained model that has been exposed to gradual augmentation of levels

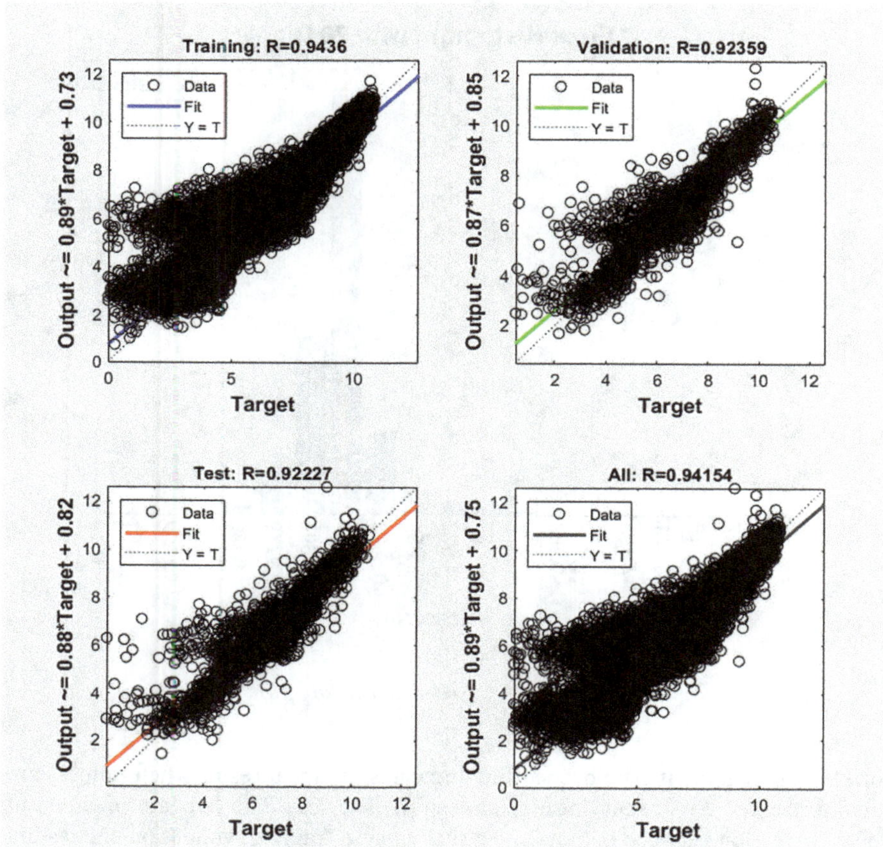

Figure 16.12 Regression plot

of complexity, should be adopted instead. The results of the FEA study (in-plane and out-plane stresses) were consistent with what was expected from the theory of mechanics of composite materials. Although the errors obtained for the NN setup are very low, it is clear that the results generated by the NN do not completely replicate the behaviour of the FEA as R is overall 0.94. Therefore, changes ought to be made to the network architecture and the training parameters. The process of finding an optimal NN depends on many variables. For example, the number of layers and neurons, affects if the network is underfitting or overfitting. Moreover, the dataset size and the hyperparameters used in training also need to be fine-tuned. Hence, the next step would be to modify further the NN architecture and investigate other suitable models for comparison. The regression plot reveals that certain data points have poor fits, since they are placed further from the trendline. Noise within the dataset should be minimised during the pre-processing of the data. This is to ensure incorrect behaviours are not incorporated during the learning process of training, thus increasing the overall accuracy of the end results. This can be achieved by removing outliers in the dataset and filtering anomalous results. The scripts were constructed in such a manner that they can be easily modified and reprogrammed, with the ability to tune values and parameters. It is possible to explore a wide range of ply orientations, thicknesses, materials, and number of plies using the Python script employed to create the panel and automate its parametric modifications. Due to the integrated structures of the Abaqus, Python, and MATLAB environments, little modifications would need to be made to the MATLAB NN model in order to accommodate such changes. Using Bessa's (2016) framework of three steps proved to be an appropriate structure for this study, with the non-specific character of the framework being easily transferable to the examined topic, as was theorised. As a result, ML techniques might be used to investigate a wide range of engineering issues with complicated and parameter-rich design spaces. The study described above offers as evidence of this, as well as offering insight into the particular application of such methodologies to composite structure modelling, the Python and MATLAB scripts developed provide a foundation for additional research into the impacts of interlaminar stresses.

16.5 Conclusions

A shell-cohesive analysis was performed on Abaqus to extract the in-plane and out-plane stresses of a multi-layered composite panel. A Python code was designed to automatically generate a multitude of FEA simulations with different layer configurations in order to develop a database of stresses and ply angles that was used to train and test a NN model in MATLAB. The developed tool based on NNs is shown to predict the interlaminar stress with good accuracy, showing the relevance of ML approaches to composite delamination studies. In particular, circumventing the need for complex 3D FEA to obtain 3D interlaminar stress from simpler plate/shell analysis saves significant time and effort for engineers in the areas of composite analysis, design and repair.

References

[1] Ealias J., Mattam J.J. Study of Inter-laminar Shear stress of composite structures. Undefined. 2013. Available from https://www.semanticscholar.org/paper/Study-of-Inter-laminar-Shear-stress-of-composite-Ealias-Mattam.

[2] Chen C.-T., Gu G.X. 'Machine learning for composite materials'. *MRS Communications*. 2019;9(2):556–66.

[3] Furtado C., Pereira L.F., Tavares R.P., *et al.* 'A methodology to generate design allowables of composite laminates using machine learning'. *International Journal of Solids and Structures*. 2021;233(6):111095.

[4] Pathan M.V., Ponnusami S.A., Pathan J., *et al.* 'Predictions of the mechanical properties of unidirectional fibre composites by supervised machine learning'. *Scientific Reports*. 2019;9(1):13964.

[5] Mozaffar M., Bostanabad R., Chen W., Ehmann K., Cao J., Bessa M.A. 'Deep learning predicts path-dependent plasticity'. *Proceedings of the National Academy of Sciences*. 2019;116(52):26414–20.

[6] Nezhad H.Y., Register J. *Artificial intelligence based aerospace composite design*. City, University of London; 2021.

[7] Bessa M.A., Bostanabad R., Liu Z., *et al*. 'A framework for data-driven analysis of materials under uncertainty: Countering the curse of dimensionality'. *Computer Methods in Applied Mechanics and Engineering*. 2017;320(1):633–67.

[8] Bessa M.A., Glowacki P., Houlder M. 'Bayesian machine learning in Metamaterial design: fragile becomes Supercompressible'. *Advanced Materials*. 2019;31(48):1904845.

[9] Dávila C., Camanho P., Turon A. Cohesive elements for shells. 2007. Available from https://ntrs.nasa.gov/api/citations/20070018344/downloads/20070018344.pdf.

[10] Goodfellow I., Bengio Y., Courville A. Deep learning. [online] Deeplearningbook.org. 2016. Available from https://www.deeplearningbook.org/.

[11] Johnson D. Back propagation neural network: What is backpropagation algorithm in machine learning? [online] www.guru99.com. Available from https://www.guru99.com/backpropogation-neural-network.html#6.

[12] Yu H., Wilamowski B.M. 'Levenberg–marquardt training'. *The Industrial Electronics Handbook*. CRC Press; 2011. Available from https://www.routledgehandbooks.com/doi/10.1201/b10604-15.

Chapter 17

Damage assessment of composites based on non-destructive pulsed thermographic inspection

Yifan Zhao[1], Weixiang Du[1], Lawrence Tinsley[1], Sri Addepalli[1], Jörn Mehnen[1], and Rajkumar Roy[2]

Thermography is a relatively new technique for non-destructive testing (NDT), which has been gathering increasing interest due to its relatively low-cost hardware and extremely fast data acquisition properties. This technique is especially promising in the area of rapid automated damage detection and quantification. This study introduces a coefficient clustering analysis (CCA) method to detect and quantitatively measure damage occurring in composite materials using non-destructive pulsed thermographic inspection. This method is based on fitting a low-order polynomial model for temperature decay curves, which a) provides an enhanced visual confirmation and size measurement of the damage, b) provides the reference point for sound material for further damage depth measurement and c) reduces the burden in computational time. The performance of the proposed method is evaluated through a practical case study with carbon fibre-reinforced polymer (CFRP) laminates, which were subjected to a drop impact test with varying energy levels.

17.1 Introduction

Composite materials are well known for their high strength-to-weight ratios, low density and corrosion-resistant properties in comparison with traditional metallic components. As such, they are applied in a wide variety of contexts, increasingly in automotive and aerospace sectors; where there is a huge requirement to improve system performance through weight reduction. With the rising price of aviation fuel and attitude towards environmental issues, modern aircraft manufacturers are looking out for innovative solutions that can offer better performance without compromising the structural integrity

[1]School of Aerospace, Transport and Manufacturing, Cranfield University, Cranfield, UK
[2]School of Mathematics, Computer Science and Engineering, University of London, London, UK

and safety features of the aircraft. Thus, the current generation of aircraft is seeing a large introduction of composite components that constitute about 50% of weight of the aircraft in parts such as engine casings, wing sections, tailplane, control structures and fuselage [1]. However, composites are also well known for their vulnerability to impact damage and their difficulty to repair compared to metal-based components. An impact or strike on the surface may cause structural damage that may be exhibited with only a small surface visual profile – this is known as barely visible impact damage (BVID) [2]. Even though the damage is barely visible on the surface, the damage to the structure could severely affect its properties and performance, which may not be apparent from the surface profile of the impact.

A variety of impact sources exist, such as stones, hail, bird strikes and even accidental drops of workmen's tools during maintenance that can cause impact damage. Literature suggests that bird strikes account for up to 80% of service damage to composites in the aerospace sector [3]. While the surface may appear sound, there may be significant damage hidden in the internal structure, and may not be appreciated on the surface because of a difficult relationship between the appearance of surface features and the structural integrity of the part [4, 5]. When the composite structure is subjected to minor impact damage that is barely visible on the surface, the damage even at the microscale can progress to significant structural damage that affects the strength, durability and stability of the composite laminates [6].

In the aviation industry, multiple non-destructive testing or NDT methods are employed, ranging from direct visual inspection, dye-penetrant, magnetic particle, eddy current and radiography to advanced methods such as 3D computed tomography, ultrasound and thermography to capture the health and structural integrity of the component without creating or intensifying any further damage to the component that is being inspected. This diversity of inspection methods requires a range of skills and expertise, providing results with differing margins of error between them.

Thermography has been attracting increasing attention over recent decades as the method involves a rapid, robust, non-contact and non-invasive inspection. Thermography can be divided into two modes: passive and active. The passive mode applies where deviations from normal operation exhibit a change in thermal contrast to be observed by an infrared (IR) imager, while active thermography involves the input of an external heat that generates a measurable thermal contrast. This particularly applies where an inspected part is not in use and is in thermal equilibrium with its surrounding environment, when detection of sub-surface damage and defects is sought, or for the measurement of material thermo-physical properties. To generate a thermal contrast, heat input is designed to highlight damages and defects either using them to generate the heat signal, as in the case of Vibro-thermography, or as an obstruction or conduit to heat flow. Thermography in its varied forms has found applications in many contexts such as condition monitoring of electrical equipment [7], mechanical equipment [8], welds [9], structures [10], and aerospace composites [11, 12] and through-life engineering [13].

In IR thermography, various image processing methods are in common use, from the basics of dealing with fixed pattern noise, vignetting, bad pixels and spatial noise smoothing [14], to thermal contrast algorithms [15] which have been further developed over the years [16]. The Thermographic Signal Reconstruction (TSR) [17] algorithm was a landmark development in pulsed thermography that dramatically increased spatial and temporal resolution of a thermogram sequence and opened up the opportunity for new ways of processing pulsed thermography data. Others have applied Principal Component Analysis (PCA) to thermography, also referred to as Principal Component Thermography (PCT), on TSR coefficients that improved the results from typical averaging filters, with comparable results to Pulse-Phase Thermography [18]. PCA analysis has also been applied to a TSR-type treatment of per-pixel signal data [19] to differentiate different delamination sizes and lengths, with sensitivity to delamination opening. Additional study has involved the use of skewness parameter [20], and high-order statistics, with somewhat consistent performance in signal-to-noise ratio for defects of different size and depth [21]. TSR-based analysis processes have previously been explored with exciting developments, involving the transformation of the thermography data into a series of RGB colour images synthesised from TSR polynomial coefficients [22]. The data thus obtained may be plotted for each colour shade to estimate contrast emergence times together with depth scaling and has been applied to both artificial and real damage features [23]. This specific application has been proven to be quite powerful at extracting multiple features into a single image.

This chapter is limited to focus on damage detection and corresponding sizing measurement with an application in assessing degradation caused by drop impact. Damage detection is important because a number of commonly used feature depth measurement methods, such as Peak Temperature-Contrast [15] and Peak Temperature-Contrast Slope [24], often require a reference point that is known on a sound material. Ringermacher [24] used the average temperature from the entire surface before the flash as a reference. This can work well only when the defective region is small and the surface is uniformly illuminated. Curve fitting-based methods, such as Shepard's Peak second derivative method [25] and Sun's Least-Squares Fitting method [26, 27], require fitting either a high-order polynomial model or a complex heat diffusion model. A high-order polynomial model can experience the over-fitting problem when the model has too many parameters relative to the number of observations, especially when the data are noisy. Although fitting based on a physical model (the model structure being known) reduces the sensitivity to noise, it requires multiple unknown parameters to be estimated simultaneously using optimisation techniques. However, this can be very time-consuming and sometimes only locally optimal solutions are produced rather than globally optimal solutions. Developing a fast, automatic and reliable technique with high robustness against noise for damage detection is, therefore, a key goal of the community.

17.2 Experimental data

17.2.1 Specimen

Specimens were produced with the dimension of 150 mm × 100 mm × 4 mm, which were made of unidirectional Toray 800 carbon fibres pre-impregnated with Hexcel M21 epoxy resin. The laminates were subjected to a drop impact test with predefined

(a) (b)

Figure 17.1 (a) The weight-drop machine used for impact generation and (b) specimen support fixture

energy levels using a semi-spherical 16 mm diameter weight drop machine which employed a drop weight of 2.281 kg, as illustrated in Figure 17.1(a). The support used to hold the sample in place was designed by following the instructions given by the standard BS ISO 18352, as shown in Figure 17.1(b).

The weight impact energy is equivalent to $m \times g \times h$, where m refers to impact mass, here 2.281 kg was used, $g = 9.8 m/s^2$ is gravitational acceleration and h is the drop height. Impact energy is adjusted by changing the height of the drop-weight, details of which are shown in Table 17.1. The specimens were subjected to represent impact energies of 5, 10, 15, 20, 25 and 30 J respectively. As shown in Figure 17.2(a), in all samples, the damage is visible from the impacted side, but they are hidden or less obvious from the rear surface, as shown in Figure 17.2(b).

Table 17.1 Drop height and impact energy level

Specimen no.	Drop height (m)	Energy level (J)
#1	0.22	5
#2	0.45	10
#3	0.67	15
#4	0.89	20
#5	1.12	25
#6	1.34	30

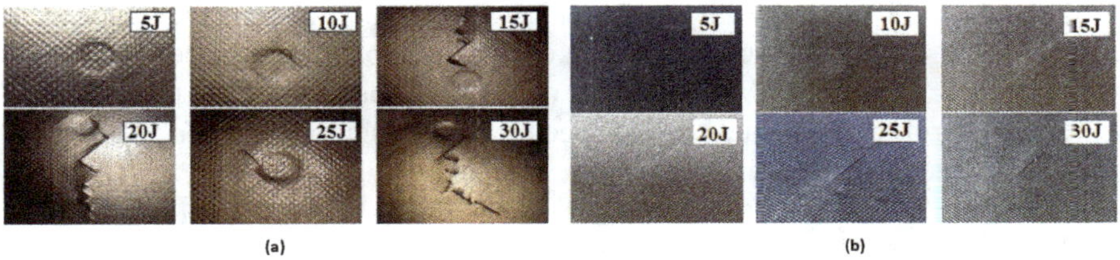

Figure 17.2 *Snapshot of six studied specimens. (a) Impact damage is visible from the impacted side of the specimens; (b) the rear surface of the samples presents invisible or subtle evidence of damage.*

17.2.2 Data collection

As illustrated in Figure 17.3, during the pulsed thermographic inspection working under the reflection mode, a short- and high-energy light pulse from a flash lamp is projected onto the sample surface. Heat conduction then takes place from the heated surface to the interior of the sample, leading to a continuous decay in the surface temperature. An IR radiometer controlled by a computer captures the time-dependent response of the sample surface temperature (decay profile). In areas of the sample surface above a defect (see point two in Figure 17.3(a)) the transient flow of heat from the surface into the sample bulk is wholly or partially separated, thus causing a temperature deviation from the sound areas, where the sample continues to cool until it reaches the sample back wall (see point one in Figure 17.3). Most of the defect detection methods are based on the analysis of these temperature decay curves. The experiment was conducted with the Thermoscope® II, a proprietary pulsed-active thermography system from Thermal Wave Imaging Inc. This system comprises two Xenon flash lamps mounted on an internally reflective hood with a capacitor bank providing power and a desktop PC to capture and store data [13, 28]. A FLIR SC7000 series IR radiometer was used, which has an Indium

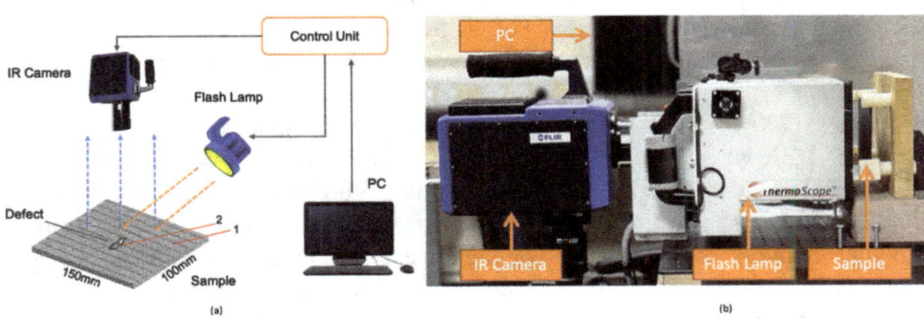

Figure 17.3 *(a) Experimental configuration of the pPulsed thermographic inspection under the reflection mode, where point 1 denotes a sound area on the sample surface and point 2 denotes a position with defects underneath; (b) A photo of the experimental setup*

Antimonide (InSb) sensor with a spectral range of 3–5.1 μm. The radiometer has a full spatial resolution of 640 × 512 pixels. The samples were placed with their surface perpendicular to the camera's line of sight at a distance of 300 mm from the lens. Considering the thickness of the specimens and their low thermal diffusivity, a sampling rate of 25 Hz was used.

17.2.3 Temperature decay curve

In pulse thermography, a short- and high-energy light pulse is projected onto the sample surface. The surface absorbs the light energy and its temperature increases. This heat is conducted through the sample, propagating inside the material causing a decrease in surface temperature. The surface temperature for a plate with a defect at a depth L is given by Lau *et al.* [29]

$$T(t) = \frac{Q}{\sqrt{\pi \rho c k t}}\left[1 + 2\sum_{n=1}^{\infty} R^n \exp\left(\frac{-n^2 L^2}{\alpha t}\right)\right] \tag{17.1}$$

where $T(t)$ is the temperature of the surface at time t, Q is pulse energy, ρ is density, c is heat capacity, k is the thermal conductivity of the material, α is thermal diffusivity and R is the thermal reflection coefficient of the air gap interface. For a semi-infinite medium, the temperature difference has a linear relation with time in the logarithmic domain with a slope of –0.5 [30].

The temperature response of any change in thermal material property from structure, damage or defect will derivate from the linear response. Shepard [17] proposed a Thermal Signal Reconstruction (TSR) technique to reduce temporal noise using a high-order polynomial model to fit the temperature cooling curve. The model can be written as

$$\ln(T(t)) = \sum_{i=0}^{N} a_i (\ln(t))^i \tag{17.2}$$

where N is the model order and a_i are coefficients to be estimated. Normally, N is larger than 5. Once the unknown coefficients a_i are estimated by the least square method, the temperature behaviour can be reconstructed to replace the raw data. Noise is significantly reduced in TSR data and damages can then be better visualised. The first and second derivatives of the fitting curve can be easily calculated by using the coefficients directly, and they have been proposed to determine the damage depth.

Figure 17.4(a) shows an IR image of the laminate that was subjected to an impact of 30 J, where the 'jet' colour map was used. Three pixels from different areas were considered and corresponding temperature decay curves in the logarithmic scale are shown in Figure 17.4(b). It is shown that when the heat diffuses through the sound area (represented by the red curve and marked by the red cross in Figure 17.4(a)) the temperature decay characteristics are uniform for a slope of –0.5 as established in Reference [30]. However, it can be inferred from the thermogram that there is the area represented by the green marker that shows a different temperature decay profile in comparison with the area where the material is not supposed to have any damage. This profile when plotted over time is represented by the green curve in Figure 17.4(b), which confirms this deviation from the time-temperature

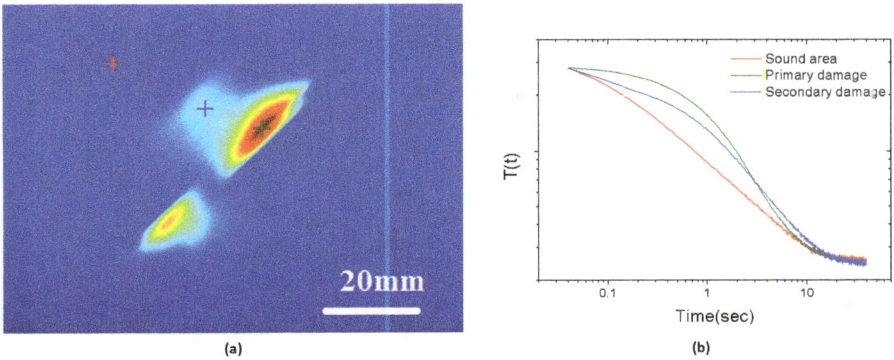

Figure 17.4 *(a) Infrared image of the laminate as shown in Figure 17.2 for 30 J impact, inspected from the non-impact side of the laminate. This snapshot was sampled 1 second after the flash, where the 'jet' colour map was used. (b) The logarithmic time-temperature curve was plotted for three selected pixels in (a) with the marker colours indicating the respective curves.*

profile of a sound area, confirming the presence of near and sub-surface damage as established in the literature [11, 12, 14, 31]. From Figure 17.4(a), it can also be inferred that there is secondary damage that has also been caused by the impact but has much lower contrast. It can be inferred that the deviation of the curve, now represented by the blue curve, confirms the presence of additional damage, which in this case may be referred to as secondary damage. The results obtained thus provide validation ensuring that the trend obtained is in line with what has been established in other researches.

17.3 Coefficient clustering analysis

Fitting a high-order polynomial model can be time-consuming depending on the spatial resolution of images and the number of frames to be analysed. It also can cause over-fitting when the polynomial model describes noise instead of the underlying relationship [32]. Overfitting generally occurs when a model is excessively complex, such as having too many parameters relative to the number of observations. If a relatively low number of data are sampled, over-fitting can affect the performance of a high-order model, and this problem will be further amplified when using the first and second derivatives. To address this issue, this study proposes a new method to rapidly detect damaged areas using a second-order polynomial model, which can be written as

$$\ln\left(T(t) - T_0\right) = a_0 + a_1 \ln(t) + a_2 \ln(t)^2 \tag{17.3}$$

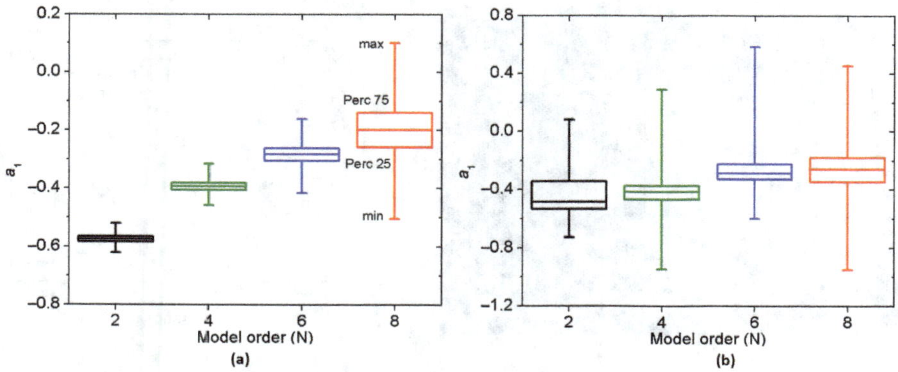

Figure 17.5 *Box chart for a_1 with different model orders estimated from a damage-free specimen (left) and a damaged specimen of 30 J (right)*

Although the model fitting is not as close as a high-order model, the coefficients are much less sensitive to noise and more consistent for pixels from sound areas. This fact is evidenced by Figure 17.5, where Figure 17.5(a) shows the box chart for the polynomial coefficients a_1 with different model orders estimated from a damage-free specimen, and Figure 17.5(b) shows the box chart for a_1 with different model orders estimated from a damaged specimen (30 J). It can be observed from Figure 17.5(a) that the variance of a_1 indicates the sensitivity to variation of pixels from the damage-free specimen and its significant increase following the increment of the model order. As before, Figure 17.5(b) represents the variation a_1 for a damaged area, indicating the sensitivity to variation of pixels. No significant difference has been observed between different model orders.

A novel CCA is introduced in this study to explore the inner relationship between coefficients to characterise and classify the thermal behaviour of each pixel of captured thermal images. Considering a damage-free specimen, an area of 100 × 100 pixels was sampled and the model (4) was fitted for each pixel. Figure 17.6 shows the distribution of the first-order coefficient a_1 and the second-order coefficient a_2 with corresponding Gaussian fitting, respectively. It is clearly shown that both coefficients are located within a narrow range with an approximately Gaussian distribution. The standard deviation of the fitting, σ, is 0.01328 and 0.00166, respectively, which is relatively very small compared with the mean of the distribution, μ, which is –0.57482 and 0.01799, respectively. The value of R-square, as the indicator of approximation level, is 0.99314 and 0.99422, respectively.

To investigate the variance of coefficients across different samples, the above step was repeated for all specimens, and the mean and standard deviation of Gaussian fitting for both coefficients of sound areas from each sample are shown in Table 17.2. The statistics results of the mean of a_1 (-0.5633 ± 0.0107) and a_2 (0.01385 ± 0.00349) demonstrate the consistency of thermal behaviour for sound areas. On assessing sound area data, it has been observed that the first-order

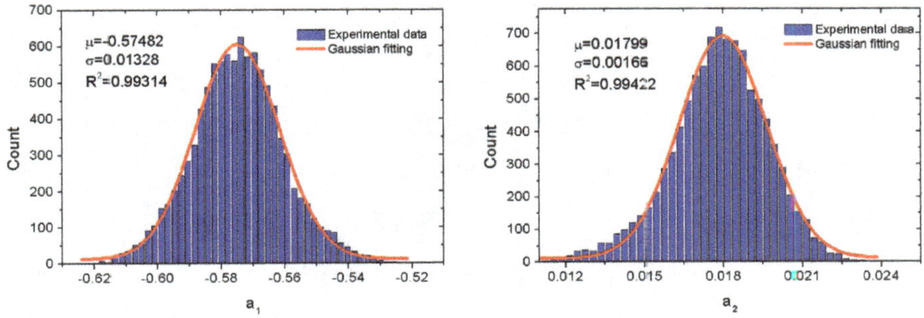

Figure 17.6 *The distribution of the coefficients for the damage-free specimen with corresponding Gaussian fitting*

coefficient increases following the increment of impact energy, which indicates that the impact event may have led to an overall change in the thermal behaviour of the sound area. All these observations reveal the potential to use both coefficients as the feature to classify sound and damaged pixels. This study proposes to use the clustering between a_1 and a_2 as the base to perform classification.

Figure 17.7 shows the scatter plot, illustrated by blue dots, for pixels sampled from the damage-free specimen. The plot exhibits a spindle-shaped profile which indicates a strong linear relationship between these two coefficients. The linear relationship can be described as

$$a_2 = b_0 + b_1 a_1 \tag{17.4}$$

where b_0 and b_1 are intercept and slope to be estimated, respectively. This model along with the boundary condition defines the constraint of a_1 and a_2 ($a_1 \in [\tau_1, \tau_2]$) of pixels from sound areas. The classification process is illustrated in Figure 17.8, where the red line segment denotes the model (4). Considering a pixel p (x, y), the first step is to estimate the first- and second -order coefficient of the model (3), and

Table 17.2 *The parameters of the Gaussian fitting for the distribution of the first- and second-order coefficients for the six specimens, reflecting data extracted from 'sound' areas*

	a_1		a_2	
Specimen (J)	**Mean**	**Standard deviation**	**Mean**	**Standard deviation**
5	–0.57466	0.01241	0.01807	0.00146
10	–0.56968	0.01422	0.01813	0.00196
15	–0.56692	0.01546	0.01090	0.00103
20	–0.56713	0.01566	0.01090	0.00102
25	–0.55683	0.01573	0.01397	0.00180
30	–0.54458	0.01163	0.01116	0.00177

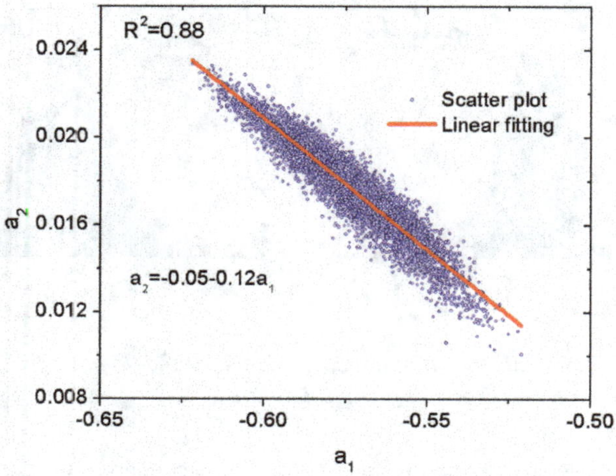

Figure 17.7 *The plot of the second-order coefficient a_2 as a function of the first-order coefficient a_1 for one damage-free specimen*

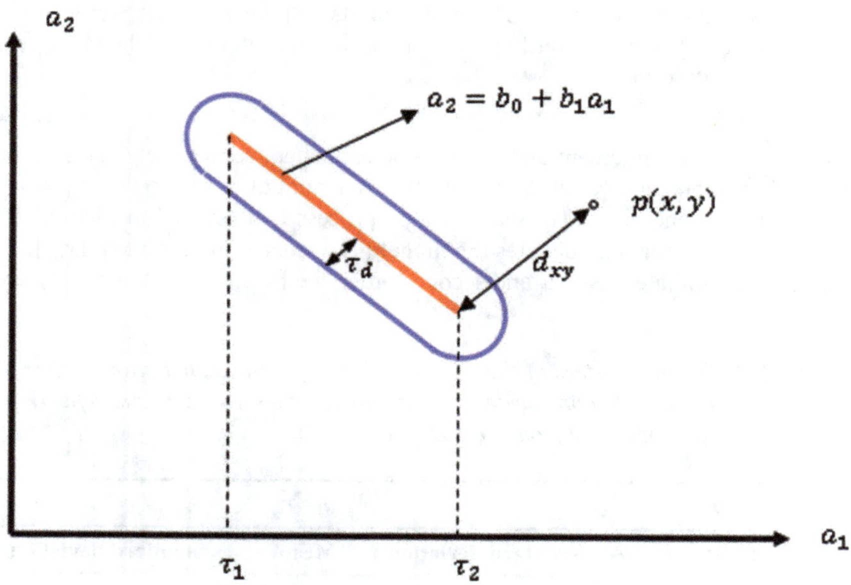

Figure 17.8 *Illustration of the classification process, where the red line segment denotes the baseline. Any pixel whose scatter of a_1 vs a_2 locates inside the region surrounded by the blue curve is determined as a pixel from the sound area, or it is determined as a pixel from the damaged area.*

then calculate the shortest distance from the coefficient coordinate $(a_1^{x,y}, a_2^{x,y})$ to the line segment, denoted by d_{xy}. If d_{xy} is smaller than a preset tolerance τ_d, this pixel is determined as a pixel coming from a sound area of the component, or the pixel is determined as a damaged pixel. The value of d_{xy} indicates the level of difference of thermal behaviour with the damage-free area. The larger the value of d_{xy}, the more complex the thermal behaviour is, which could directly relate to the level of the structure change. The value of τ_d is chosen based on the standard derivation of the coefficients calculated from the damage-free specimen. Through a large amount of testing and performance comparison, the study proposes to select τ_d in

$$\tau_d = 2.5\sqrt{\sigma_1^2 + \sigma_2^2} \tag{17.5}$$

where σ_1 and σ_1 denotes the standard deviation of a_1 and a_2, respectively.

Note the values of the parameters depend on the material and thickness of specimens. The identified model is applicable for components with the same material and thickness, any change of which requires re-calibrating the parameters using a reference sample or an area known as damage-free.

The steps of the proposed CCA method can be summarised as:

1. Select a region of a damage-free specimen and estimate the baseline coefficients of the model (3): a_0, a_1 and a_2;
2. Fit the model (4) to estimate b_0, b_1 based on the scatter plot of a_1 and a_2, as shown in Figure 17.7; Determine the boundary condition τ_1, τ_2 and tolerance τ_d by estimating the distribution of the coefficients, as shown in Figure 17.6;
3. Select a region of interest for a testing specimen, and estimate the coefficients of the model (3): a_0, a_1 and a_2;
4. Calculate the shortest distance from the coefficient coordinate $(a_1^{x,y}, a_2^{x,y})$ of each pixel to the baseline, as shown in Figure 17.8. If d_{xy} is smaller than τ_d, this pixel is determined as a pixel coming from a sound area, or the pixel is determined as a damaged pixel.
5. Visualise the damage by reversing pixels to space domain in a black-white form or colour form.

17.4 Results and discussion

17.4.1 Damage detection

A region with a size of 200 × 200 pixels around the centre of the specimen was sampled for all specimens. The unknown parameters for the model (4) were estimated from the reference sample and the model is finally expressed as

$$a_2 = -0.05 - 0.12a_1 \tag{17.6}$$

As shown in Figure 17.7, the red plot illustrates the model (6) with $R^2 = 0.88$. The boundary condition of a_1 is determined by 99% quantile of distribution, which is $[-0.62, -0.53]$. The distance tolerance τ_d is chosen as 0.033 based on (5).

By applying the developed CCA method with the estimated parameters to the thermal decay data, the scatter plots of a_1 vs a_2 are shown in Figure 17.9. The green scatters were classified as the pixels from sound areas, and the red scatters were classified as the pixels from damaged areas based on the process shown in Figure 17.8.

There are almost no damaged pixels detected for the specimen at 5 J. As expected, the number of damaged pixels detected increases following the increment of impact energy, which indicates a growing area of damage. The change in the contour of clustering indicates that a higher energy impact causes more complex thermal behaviour indicating potentially a more complex structural damage, which adds to deviation in the thermal behaviour exhibited on the surface in comparison to sound areas. To explore more information, the left figure of Figure 17.10 illustrates the overlap of coefficient clustering for pixels from defective areas for the specimen 10–30 J (5 J is not included due to no damage detected). It has been observed that the scatters grow towards the bottom-right corner following the increase of impact energy. This indicates that the linearity of the temperature decay curve reduces because a_1 is closer to zero, and more complicated thermal behaviours have been observed because a_2 is further away from zero. The set of figures on the right in Figure 17.10 shows the box charts of both coefficients for pixels from damaged areas. It has been observed that following the increase of impact energy, the range between the maximum and the minimum for both coefficients increases which indicates that a higher energy impact potentially causes more complex structural damage. The minimums of a_1 are similar for all tests, but the maximum is closer to zero; the maximums of a_2 are similar for all tests, but the minimum is further away from zero. The variation of the box size for all tests is relatively small, which indicates that most damaged pixels have similar thermal behaviour.

17.4.2 Damage measurement

Figure 17.11 shows the visualisation of the detected damage area (black colour) in the binary form by converting the classification results into the spatial domain. To quantify the measurement, Table 17.3 presents the calculated properties of the detected damage area for each specimen, where the compactness measures the shape. Compactness was calculated by

$$compactness = \frac{4\pi \times area}{perimeter^2} \tag{17.7}$$

The value of the compactness is not greater than 1. If the shape of the area is a circle, the value of compactness arrives at its maximum of 1. Table 17.3 shows that the higher the impact energy, the larger the area and perimeter of the detected damage is. It can be observed that the damaged area of the specimen 30 J is smaller than 25 J, but the perimeter is larger than 25 J, which indicates a more irregular shape. This observation is confirmed by inspection of the compactness. The value of compactness for 10, 15, 20 and 25 J is

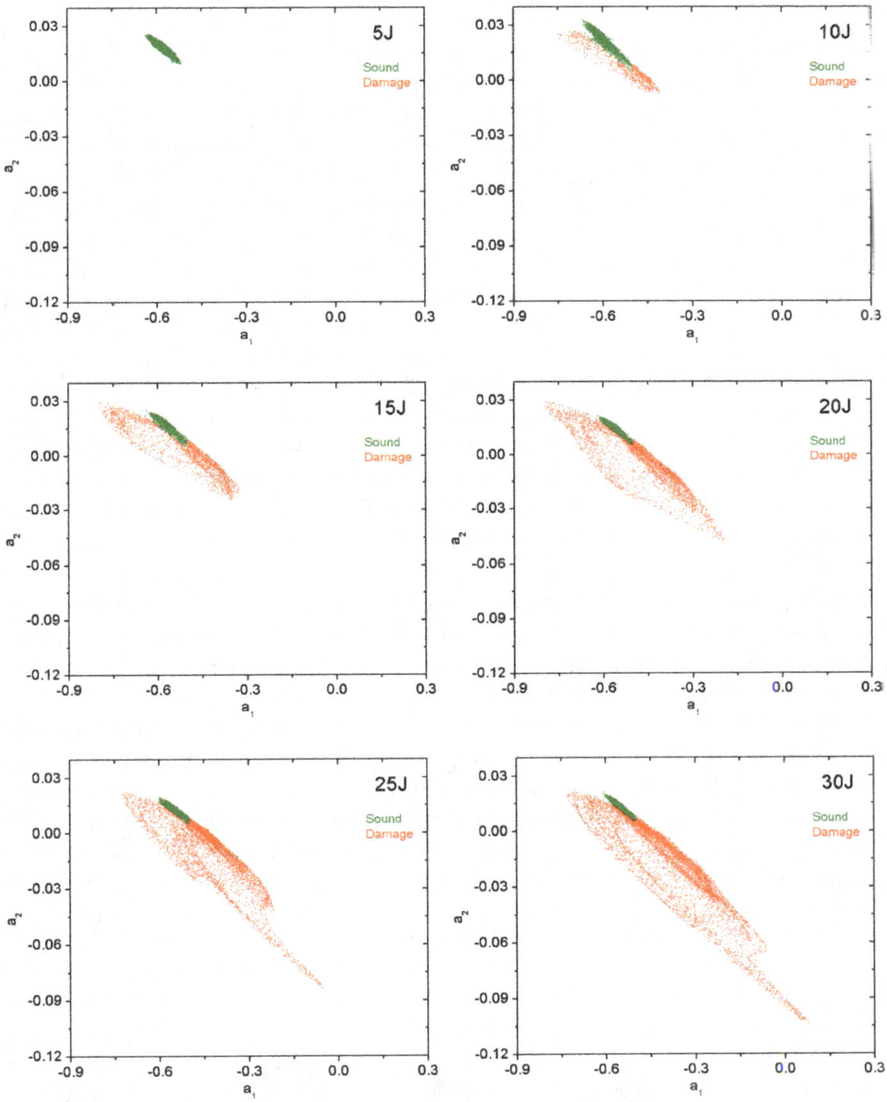

Figure 17.9 *The classification results are based on the scattering of the first-order coefficient and the second-order coefficient for all specimens, where the green colour denotes the detected sound pixels and the red colour denotes the detected damaged pixels*

similar (mean: 0.634 and standard derivation: 0.043). The compactness of 30 J is significantly smaller with the value of 0.489, which indicates a more irregular shape. A plot of perimeter as a function of impact energy, shown in Figure 17.12, indicates a highly linear

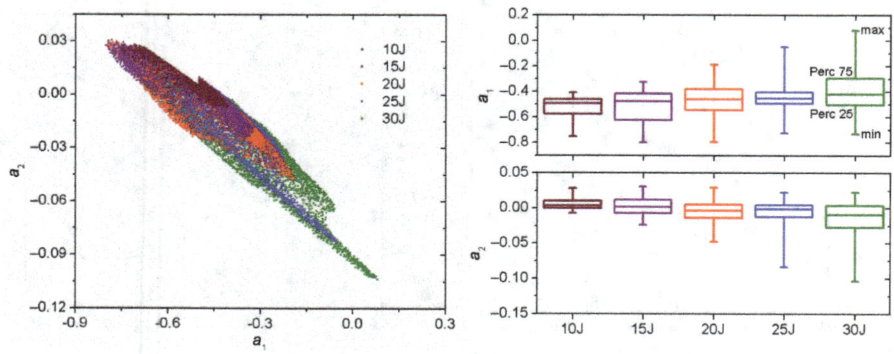

Figure 17.10 *Properties for pixels from defective areas for the specimen 10–30 J. Left: coefficient clustering overlay; top-right: box chart of the first-order coefficient; bottom-right: box chart of the second-order coefficient, which shows the minimum, maximum, 25% percentile and 75% percentile of all of the data.*

relationship. A line is fitted to the plot by the least square method with R^2 (coefficient of determination) = 0.97. For the specimen with 30 J impact, the high impact energy causes a potentially more complex damage structure, rather than an increased damage area in comparison with that of the 15 J specimen.

Once the raw data sequence is converted into polynomial coefficients and has undergone the coefficient clustering step transforming them into d_{xy}, these values can then be plotted back in the spatial domain, providing a new and alternative visualisation of the detected damage areas, as can be seen in Figure 17.13. It has been observed that there are consistently two areas (top right and bottom left) that have high values of $\mathbf{d_{xy}}$ for all specimens with impact damage. This phenomenon may be caused by the mechanism of the weight-drop machine itself.

Figure 17.11 *Visualisation of the detected defect area for all specimens in the binary form*

Table 17.3 *The property measures for the detected damage areas for all specimens*

Specimen (J)	Area (mm^2)	Perimeter (mm)	Compactness
10	154.92	58.19	0.575
15	397.08	86.60	0.665
20	648.84	111.18	0.667
25	1347.36	163.85	0.630
30	1283.04	181.51	0.489

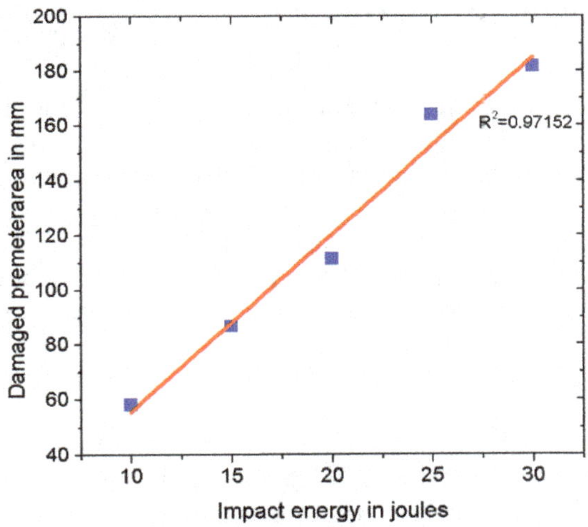

Figure 17.12 *Perimeter of the detected damaged area as a function of impact energy*

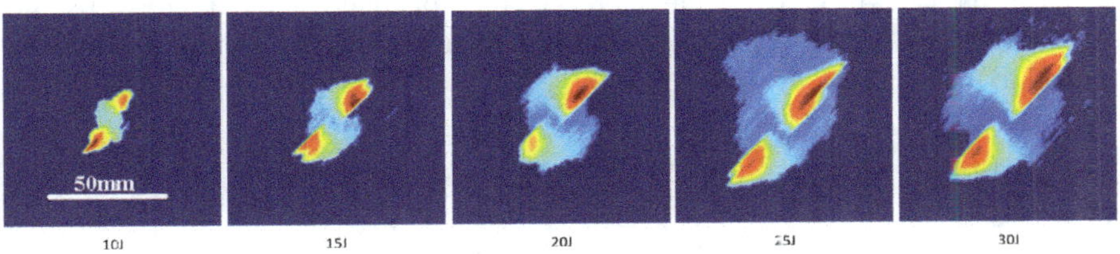

Figure 17.13 *Reconstructed data visualisation of damage across a range of samples based on the* dxy *values from the CCA method*

17.5 Conclusions

Quantitative measurement of defective areas by pulse thermography has been an important research topic over the last decade. A high-order polynomial model is normally used in existing methods to better fit the experimental data. However, overfitting generally occurs when the number of observations is limited or the model is complex. To avoid this problem, this study has developed a novel CCA method based on a second-order polynomial fitting for the temperature decay curve. It has been observed that the estimated coefficients of sound pixels exhibit an approximately Gaussian distribution, and the variance of both coefficients for different testing specimens is relatively small. The clustering of coefficients is therefore selected as a feature to represent the thermal behaviour of each pixel. A simple linear function describing the relationship between two coefficients with a corresponding boundary condition has been developed to classify pixels from damaged and sound areas of the specimens.

The evaluation of the performance and accuracy of the proposed CCA method is based on experimental data from carbon fibre laminates that suffered different impact energies. Clustering results reveal that the higher the impact energy, the wider the distribution of the scatter plot of damaged pixels indicating a more complex thermal behaviour. Both binary and colour visualisations show that the higher the impact energy, the more complex the thermal behaviours it causes. It has been observed that the specimen with 30 J impact has a more irregular shape than the other five specimens. The perimeter of the detected area and the impact energy exhibits a highly linear relationship. Analysis of the results reveals the clear potential of the new CCA method.

It should be inferred from the reverted spatial images that they don't just represent a single frame from the dataset, as represented by other existing methods but represent the reconstruction of the entire dataset compressed into a single image. This technique not only provides an enhanced visual confirmation of the damage but also reduces the burden on the operator in post-processing the data.

The present study can only detect and measure damage size. Future research will be focussed on the development of an efficient coarse-to-fine damage assessment routine, which would perform a rapid inspection, initially identifying the damaged areas based on CCA to provide the reference point for sound material. From this, the region of interest can be deduced, allowing depth measurement techniques to then be applied to the identified areas.

Acknowledgements

The authors would like to thank Professor Peter Foote of Cranfield University for providing the CFRP specimens. This work was supported by the EPSRC Centre for Innovative Manufacturing in Through-life Engineering Services (Grant number EP/I033246/1).

References

[1] Boeing. *Boeing 787 from the ground up*. BOEING; 2006.

[2] Polimeno U., Meo M. 'Detecting barely visible impact damage detection on aircraft composites structures'. *Composite Structures*. 2009;91(4):398–402.

[3] CAP 562: Civil Aircraft Airworthiness Information and Procedures (CAAIP) *Issue 4, Ammendment 3. Authority Civil Aviation*. 2020.

[4] Newaz G., Sierakowski R. *Damage tolerance in advanced composites*. Boca Raton, FL: CRC Press; 1995.

[5] Sanchez-Saez S., Barbero E., Zaera R., Navarro C. 'Compression after impact of thin composite laminates'. *Composites Science and Technology*. 2005;**65**(13):1911–19.

[6] de Freitas M., Reis L. 'Failure mechanisms on composite specimens subjected to compression after impact'. *Composite Structures*. 1998;42(4):365–73.

[7] Jadin M.S., Taib S. 'Recent progress in diagnosing the reliability of electrical equipment by using infrared thermography'. *Infrared Physics & Technology*. 2012;55(4):236–45.

[8] Bagavathiappan S., Saravanan T., George N.P., Philip J., Jayakumar T., Raj B. 'Condition monitoring of exhaust system blowers using infrared thermography'. *Insight Non-Destructive Testing and Condition Monitoring*. 2008;**50**(9):512–15.

[9] Meola C., Carlomagno G.M., Squillace A., Giorleo G. 'The use of infrared thermography for nondestructive evaluation of joints'. *Infrared Physics & Technology*. 2004;46(1-2):93–9.

[10] Grinzato E., Bressan C., Marinetti S., Bison P.G., Bonacina C. 'Monitoring of the Scrovegni chapel by IR thermography: Giotto at infrared'. *Infrared Physics & Technology*. 2002;43(3-5):165–9.

[11] Vavilov V., Maldague X., Dufort B., Robitaille F., Picard J. 'Thermal nondestructive testing of carbon epoxy composites: detailed analysis and data processing'. *NDT & E International*. 1993;26(2):85–95.

[12] Ball R.J., Almond D.P. 'The detection and measurement of impact damage in thick carbon fibre reinforced laminates by transient thermography'. *NDT & E International*. 1998;31(3):165–73.

[13] Mehnen J., Tinsley L., Roy R. 'Automated in-service damage identification'. *CIRP Annals*. 2014;63(1):33–6.

[14] Ibarra-Castanedo C., Bendada A., Maldague X.P.V. 'Thermographic image processing for NDT'. *IV Conferencia Panamericana de END*; 2007. pp. 1–12.

[15] Maldague X. *Theory and Practice of Infrared Technology for Nondestructive Testing*. New York: Wiley; 2001.

[16] Ibarra-Castanedo C., Bendada A., Maldague X. 'Image and signal processing techniques in pulsed thermography'. *GESTS Int'l Trans. Computer Science and Engineering*. 2005;22(1):89–100.

[17] Shepard S.M. 'Temporal noise reduction, compression and analysis of thermographic image data sequences'. *US Patent 6516084 B2*. 2003.

[18] Rajic N. 'Principal component thermography for flaw contrast enhancement and flaw depth characterisation in composite structures'. *Composite Structures*. 2002;58(4):521–8.

[19] Schmutzler H., Garcia A., Sato N, *et al.* 'Influence of delamination characteristics in carbon fibre/epoxy laminates on signal features of pulse thermography'. *Journal of Nondestructive Evaluation*. 2014;**34**(1).

[20] Madruga F.J., Ibarra-Castanedo C., Conde O., López-Higuera J., Maldague X. 'Automatic data processing based on the skewness statistic parameter for subsurface defect detection by active infrared thermography'. *9th International Conference on Quantitative InfraRed Thermography*; Krakow, Poland; 2008.

[21] Madruga F.J., Ibarra-Castanedo C., Conde O.M., López-Higuera J.M., Maldague X. 'Infrared thermography processing based on higher-order statistics'. *NDT & E International*. 2010;43(8):661–6.

[22] Roche J., Leroy F., Balageas D. 'Information condensation in defect detection using TSR coefficients images'. *12th International Conference on Quantitative InfraRed Thermography*; Mahabalipuram, India 2015.

[23] Roche J.-M., Passilly F., Balageas D. 'A TSR-based quantitative processing procedure to synthesize thermal D-scans of real-life damage in composite structures'. *Journal of Nondestructive Evaluation*. 2015;**34**(4):1–15.

[24] Ringermacher H.I., Archacki R.J., Veronesi W.A. 'Nondestructive testing: transient depth thermography'. *US Patent 5711603*. 1998.

[25] Shepard S.M. 'Automated binary processing of thermographic sequence data'. *US Patent 8449176 B2*. 2003.

[26] Sun J.G. 'Analysis of pulsed thermography methods for defect depth prediction'. *Journal of Heat Transfer*. 2006;128(4):329–38.

[27] Sun J.G. 'Method for determining defect depth using thermal imaging'. *U.S. Patent 6,542,849*. 2003.

[28] Widjanarko T., Tinsley L., Roy R., Mehnen J. 'Characterisation and performance assessment of a pulsed-thermography camera system for component degradation inspection'. *Proceedings of the 1st International Conference on Through-life Engineering Services*; Cranfield, UK; 2012.

[29] Lau S.K., Almond D.P., Milne J.M. 'A quantitative analysis of pulsed video thermography'. *NDT & E International*. 1991;24(4):195–202.

[30] Shepard S.M., Hou J., Lhota J.R., Golden J.M. 'Automated processing of thermographic derivatives for quality assurance'. *Optical Engineering*. 2007;46(5):051008.

[31] Shepard S.M. 'Advances in pulsed thermography'. *Processings of SPIE*. 2011;4360:511–5.

[32] Chan K.Y., Kwong C.K., Dillon T.S., Tsim Y.C. 'Reducing overfitting in manufacturing process modeling using a backward elimination based genetic programming'. *Applied Soft Computing*. 2011;11(2):1648–56.

Chapter 18
Augmented reality-equipped composite monitoring

Idriss Ben Jamaa[1] and Hamed Yazdani Nezhad[1]

This project has been partially based on a work produced by Dr. Hamed Yazdani Nezhad and his partners, entitled "Numerical Analysis of Low Velocity Rigid Body Impact Response of Composite Panels." This report has gone through the reverse finite element analysis (FEA), simulating an experimental process involving measuring the damage caused by a body impacting a composite panel at low velocity. It has shown that a numerical simulation of an experimental procedure involving composite materials can produce an accurate representation/prediction of the experimental results. Using this assumption, the following project has been carried out.

Presuming that a composite laminate's structural integrity may be difficult to monitor without using invasive methods, a procedure requiring only visual surface data would be a major advancement in the structural health monitoring (SHM) of composite panels. Therefore, the purpose of this research was to explore the use of virtual technologies like AR or XR and question the validity of their use in the SHM of composite structures. Using photogrammetry and light detection and ranging (LiDAR), a composite panel's three-point bend deformation was measured, through which a three-dimensional (3D) model was created and analysed.

This study has shown that the use of LIDAR is superior to that of photogrammetry for capturing surface data of a flat composite panel. Although a MATLAB® code computing the stresses and strains of the panel from the surface deformation data was unable to be implemented, an Abaqus simulation was used to provide the required relationships. The latter was used to be visualized through AR using the Unity 3D Engine. If explored further, the use of surface scanning and AR visualisation of computed data from said surface scanning can indeed revolutionise the SHM of composites.

[1]Aeronautics and Aerospace Research Centre, University of London, London, UK

18.1 Introduction

The main aim of this project was to explore the use of augmented reality (AR) and virtual technologies for the SHM of composite panels. As mentioned, existing monitoring systems may not always be an ideal solution, depending on the context. Certain methods may have compromises that have been made. Such examples will be cited in the following sections of this report. Noninvasive solutions for SHM are sought out in the composite material industry. To restrict the scope of this research, composite materials typically used in the aeronautical and aerospace industry were considered. This has led to ask the following: using only visual data acquired through a camera or a set of 2D images, can a virtual 3D model be reconstructed accordingly, and carry out accurate simulations? To accomplish this aim, the following objectives were considered.

Creating a 3D model using 2D images requires one of two possible methods: LIDAR and photogrammetry. Therefore, one or both methods (depending on the difficulties each may have) were used to collect the necessary visual data to reconstruct a 3D model of the specimen. This is the crucial milestone that needs to be completed before being able to proceed to the 3D analysis of the images. The issue remains in applying these techniques for the specific purpose of this project. Therefore, existing codes, algorithms and functions cannot be used as provided, and need to be adapted. Under normal circumstances, laboratory equipment or university-provided hardware would be used to carry out these measurements. Under the COVID-19 restrictions, however, readily available tools were used.

After processing the visual data, a point cloud of the observed specimen can be generated. This leads to the next objective, where a software suited for FEA will be used to predict trends from the 3D model. Such trends can be damage within the composite panel, or stress–strain propagation.

Next, creating a virtual mock-up using AR is a focus point. Through this virtual mock-up, the predicted trends from the FEA can be visualized by superposing the virtual model onto or next to the real-life panel. This provides a virtual representation of the performed experiment's results.

Finally, validating the simulated results with manually measured data is an important detail. Only by comparing manually measured deformations with calculated deformations (from the generated point cloud) can a general idea of the accuracy of the system be given. Ideally, strain gauges would be placed onto the panel to compared measured and calculated strain. Again, due to the pandemic, such equipment is not available. Therefore, if this project is to be used as a reference for a similar experiment, the latter is advised. Through these four [1] objectives, this study aims to answer the question of whether surface data analysis through cameras and AR are viable options for SHM. It is important to note that the aim for this project is not to discredit existing SHM techniques for composite materials, nor is it an attempt to find a definitive alternative to existing methods. This report does, however, serve to highlight the attributes of combining computer vision/photogrammetry and AR in the engineering industry. Reducing the scope to a specific topic, such as the SHM of composites used in the aerospace and aeronautical industry, is important to achieve a more in-depth understanding of the possible contributions this project may have in a larger scale. Only by understanding and dealing with a problem in a specific context can an answer be adapted to a more generalist application.

This report will first discuss the research and motivation behind exploring this subject. The literature review will consist of defining the main definitions, through the research of some accredited academics. The difficulty in SHM of composites will also be discussed in this section, as it is an essential aspect of this project. By the end of this report, an alternative to existing methodologies can hopefully be presented. The accuracy of simulations and the techniques used will also be explored briefly in the theory section. After which, a general outline on how to extrapolate stress–strain relationships from surface deformation data will be explained. Said surface deformation is calculated from the 3D reconstruction of the observed panel. This aspect will also be explored.

Once the underlying theory has been explained, and the reasons behind the experiments have been justified, the methodology section exclusively deals with the process behind acquiring the 3D data. Finally, the results and discussion section is reserved to conclude on the findings, and propose further study ideas that this report does not or was unable to address.

18.2 Theory: measurement, simulation, and visualization

18.2.1 Review of literature and research

18.2.1.1 What are computer vision, photogrammetry, LIDAR, and AR?

Computer vision is the process of achieving a data-extracting capability equalling that of a human [2]. In other words, the computer is trained to recognise shapes, forms, and objects relative to their environment. The relationship that the shapes have with its environment is processed in the same manner that an individual may process that shape. Computer vision can be used to recognise faces in live or recorded videos or identify cars moving in traffic. By recognising certain key features and comparing them frame by frame, computer vision's aims are to recognise objects, provide navigation to an automated vehicle, and model the object, among other examples.

Photogrammetry may share common attributes to computer vision, which is why certain people may easily confuse both together. It is therefore important to clarify the misunderstanding before tackling the main core of this project. Although both fields involve a computer processing image data through feature recognition, they differ in their goals. Photogrammetry's main aims are to accurately map topographical features, both from afar (e.g., mapping terrain) and from close range (e.g., architecture, anthropometrics). Application of this technology may range from anywhere between archaeology to game-world designing.

LIDAR is a technology that can be used in the 3D reconstruction of a topography through a camera. The main difference being that LIDAR camera uses lasers to carry out the measurements. Like SONAR, the time that the light wave takes to go from the emitter, to bounce off of the subject and into the receiver is calculated, giving the exact distance between the camera and the subject. A very accurate topography can therefore be recreated.

AR is the process of overlaying virtual models onto real life through an interface. Similarly, AR and virtual reality (VR) may be two technologies than some may confuse one from another. The main difference between AR and VR is that VR immerses the viewer into a virtual world, whereas AR enhances it [3]. Both, however, fall under that category of extended reality (XR). The student conducting this project has had experience in the application of AR in the engineering industry, having worked as an in intern at Lumiscaphe (a French company specialised in AR). It is indeed a powerful tool that helps visualise designs and their prototypes. Proof of concepts can be tested full-scale in a noninvasive manner. For example, during a session at Lumiscaphe, a robotic arm's range of motion was displayed using AR, without the need of building a prototype, nor relying solely on an on-screen animation.

AR is used in different engineering industries. Smart Augmented Reality Tool, e.g., serves as an aid to aircraft assembly, quality control, and inspection by overlaying a virtual model onto the components being inspected. Any construction or assembly defects compared to the virtual model can be detected [4].

18.2.1.2 Composite structures' maintenance and testing

Composite materials are hard to monitor without using noninvasive methods. The layers of material that constitute the laminates react differently to the same load or strain applied. Each individual layers not only have different characteristics but may also have their fibres oriented at different angles. This makes some layers resistant to directional loads or stresses, while others may be prone to damage more easily. Damages can occur within the inner layers without necessarily being visible on the outer layers. Other characteristics such as thermal resistance also differentiate each layers' behaviours from each other [5]. Therefore, it is difficult to predict a composite material's failure mechanism, without using invasive methods. This leads to products being over-designed [6]. Damage to the laminate can be under the form of (but not limited to) fibre breakage, delamination, and matrix cracking [7].

Conventional methods of monitoring composite laminates involve the use of X-ray, thermography, and eddy currents, among others. This, however, requires the sample to be tested off-site in some cases (i.e., to test the structural integrity of a laminate, it needs to be disassembled/removed) [8]. This therefore would imply a higher maintenance cost. Furthermore, non-destructive testing (NDT) seems to be adopted for detecting existing damages, as opposed to predict whether a damage will occur [5].

Each SHM method comes with its own set of advantages and disadvantages. Some methods used to monitor laminates in the aerospace and automobile industries may be limited to these domains only. More importantly, each method is restricted to detecting only certain types of damage (e.g., using a particular method may detect delamination and cracks between layers, but it may not detect adhesive defects) [9].

This report is an attempt to address this issue, by questioning whether the degradation of a composite structure can be predicted through observation and simulation, rather than using conventional monitoring techniques. Again, this report is not attempt to discredit said SHM techniques, but rather is an attempt to highlight the attributes of combining AR and photogrammetry in the engineering industry.

18.2.1.3 How accurate can the simulations be?

As mentioned, photogrammetry is the process of mapping out in detail the topography of a given subject. An structure-for-motion (SfM) algorithm tracks features from one image to another and matches them accordingly. SfM is the process through which a 3D reconstruction can be possible using 2D images. The accuracy of photogrammetry is proportional to the quality of the camera. Seeing as modern smartphones have cameras that can go up to 50 Megapixels [1], even those can be considered suitable to carry out an SfM reconstruction with a great deal of accuracy. Furthermore, the increasing use of LIDAR may lead to the obsolescence of photogrammetry, especially when it comes to the mapping of an object (as opposed to terrain or buildings). The facial recognition on some smartphones uses LIDAR to accurately detect facial features of a particular individual. Therefore, the tools used to measure the laminate plates' deformation can be deemed very accurate for the purposes of this research.

This can be further shown in a study where photogrammetry was used to carry out similar measurements on a deformed beam. The recreation of 3D point clouds was found to be accurate on the submillimeter scale [10].

When it comes to simulating experimental procedures through a numerical FEA, as long as the material properties of a composite panel, and the boundary conditions of the experiments are well defined within the simulation, then the numerical analysis can be used as an accurate predictive tool [11].

Thus, both the measuring equipment/technique and the numerical analysis methods have been shown to be accurate. Through this study, the reliability of the combination of both fields will be explored.

18.2.2 Stress and strain calculation through surface data

To understand how the simulated experiments will work, it is important to go through the following theory concerning stress–strain relationship in a composite material. The relationship can be given by the following equation:

Equation 1: strain-deformation relation

$$\epsilon = z \cdot \frac{d^2w}{dx^2} \tag{18.1}$$

The double-derivative of "*w*" is called the "Bending Curvature" of the bent laminate in the "*w*" direction (in world coordinates, its equivalent is the "*y*" direction). "*z*" is the distance from the centre-point of the panel's thickness, and the point of interest (at which the strain will be calculated). Although only an approximation, this would serve to give a good enough estimation of the strains.

Maximum strains (both compressive and tensile) would happen at the top-most and bottom-most point of the laminate. Therefore, "*z*" is equal to the thickness of the panel divided by 2. Calculating the double derivative of "*w*" can be done by taking three points and going through the following operation:

Equation 2: finding the double-derivative [6]

$$\left(\frac{dw}{dx}\right)_{@1,2} = \frac{w_2 - w_1}{x_2 - x_1} = \frac{W_{2,1}}{X_{2,1}} = \theta_A \tag{18.2}$$

Equation 3: finding the double-derivative [12]

$$\left(\frac{dw}{dx}\right)_{@2,3} = \frac{w_3 - w_2}{x_3 - x_2} = \frac{W_{3,2}}{X_{3,2}} = \theta_B \tag{18.3}$$

Therefore:

Equation 4: finding the double-derivative [10]

$$\frac{d^2w}{dx^2} = \frac{\theta_B - \theta_A}{x_3 - x_1} \tag{18.4}$$

Through this value, the strain at each individual point can be calculated. After which, the stress at these same points can also be calculated. The stress–strain relation can be given by the following equation:

Equation 5: stress–strain relationship in a composite panel [6]

$$[\sigma] = [\bar{Q}]\{\epsilon\} \tag{18.5}$$

Equation 6: stress–strain relationship in a composite panel [12]

$$\begin{Bmatrix} \sigma_x \\ \sigma_y \\ \tau_{xy} \end{Bmatrix} = [\bar{Q}] \begin{Bmatrix} \epsilon_x \\ \epsilon_y \\ \gamma_{xy} \end{Bmatrix} \tag{18.6}$$

In the "Methodology" section of this report, details about the composite panel's deformation will be given. In short, however, the panel will only be bent in one direction. No twisting curvature nor "xy" deformation is caused. Therefore, the other values in the above equation can therefore be equated to zero, giving the following relationship:

Equation 7: stress–strain relationship in a composite panel [10]

$$\{\sigma_x\} = \begin{bmatrix} \bar{Q}_{11} & \bar{Q}_{21} & \bar{Q}_{31} \end{bmatrix}^T \cdot \{\epsilon_x\} \tag{18.7}$$

To calculate $\bar{Q}$:

Equation 8: how to calculate $\bar{Q}$

$$[\bar{Q}|_{11}]|_\theta = m^4 Q_{11} + n^4 Q_{22} + 2m^2 n^2 Q_{12} + 4m^2 n^2 Q_{33}$$
$$[\bar{Q}|_{12}]|_\theta = [\bar{Q}|_{21}]|_\theta = n^4 Q_{11} + m^4 Q_{22} + 2m^2 n^2 Q_{12} + 4m^2 n^2 Q_{33} \tag{18.8}$$
$$[\bar{Q}|_{13}]|_\theta = [\bar{Q}|_{31}]|_\theta = m^2 n^2 Q_{11} + m^2 Q_{22} + 2m^2 n^2 Q_{12} + (m^2 - n^2)^2 Q_{33}$$

where θ: fibre direction of the composite panel's ply, at which the stress is calculated

$$m = \cos(\theta)$$
$$n = \sin(\theta)$$
$$[Q] = \begin{bmatrix} Q_{11} & Q_{12} & 0 \\ Q_{21} & Q_{22} & 0 \\ 0 & 0 & Q_{33} \end{bmatrix}$$
$$Q_{11} = \frac{E_1}{1 - \nu_{12}\nu_{21}}$$
$$Q_{12} = Q_{21} = \frac{\nu_{21}E_1}{1 - \nu_{12}\nu_{21}}$$
$$Q_{22} = \frac{E_2}{1 - \nu_{12}\nu_{21}}$$
$$Q_{33} = G_{12}$$
$$\frac{\nu_{12}}{\nu_{21}} = \frac{E_1}{E_2}$$

The above equations serve to calculate stress and strain on a composite panel, based off the relative positions of points on the material's surface. In short, if the coordinates of a point on a panel and/or its displacement are known, then the stresses and strains on said surface can be found through the above relationships. This method only works, however, if the material properties of the composite material are known. More importantly, the points chosen for the analysis must be close together to minimise error.

18.2.3 Image data processing and 3D modelling

18.2.3.1 Hausdorff distance equation

Let two sets of Points A and B in a Euclidian 3D space, where

$$A = \{a_1, a_2, a_3, \ldots, a_n\}$$
$$B = \{b_1, b_2, b_3, \ldots, b_n\}$$

The Hausdorff distance equation can therefore be defined as:
Equation 9: Hausdorff equationsoftware is very useful for several point

$$\tilde{\delta}_H(A, B) = \max_{a \in A} \min_{b \in B} \|a - b\| \tag{18.9}$$

In other words, the absolute distance between two sets of points can be calculated by taking the maximum distance between one point from a given set, to its nearest neighbour on another set. Comparing one set of points (or a point cloud) to another is referred to as C2C for short. The main issue with the basic given formula is that several points from one set can be mistakenly associated multiple times to one same point in the other set. Calculating the Hausdorff distance between two point clouds therefore may not be as straightforward. Different algorithms may be used to compute the C2C Hausdorff distance, some having benefits that others do not.

Variations of this formula exist to compute the M3C2 of two point clouds, for example. Although computation time is longer, this version offers more a more detailed distance calculation. All in all, creating and writing a Hausdorff distance algorithm from scratch is a subject that requires its own separate study. It is a delicate field in mathematics that can be used in several different contexts. For example, comparing proteins' geometric structures in biology, or motion detection in computer vision, or even evaluating magnetic resonance volume segmentation. Existing software will therefore be used to carry out the above operation.

18.2.3.2 AR and marker image libraries

Marker-based AR uses a visual predefined marker that guides the projection of the virtual model. In other words, the virtual model attaches itself onto an image target, therefore locking the scale and position of the model. This method of projecting an AR model is useful as it uses a pre-existing and pre-defined relationship between the model and an element of the real world, which leads to a simpler (user interface) UI. The scale and orientation of the model would therefore depend entirely on the marker used and can be adjusted to the users' convenience.

Using physical image markers may also become useful if one decides to visualize multiple models at once. Switching between markers would consequently switch the projected models, giving the user a full visual experience without complicating the programmers' tasks. Open-source AR libraries exist for those aiming to implement AR into a project. These libraries contain the necessary code to actively track features in real time, allowing for the AR model to lock on to real-world features. Specifically, open-source image marker libraries also exist, allowing the programmer to implement an already-existing algorithm into their project, involving the use of marker-based AR.

18.3 Methodology

18.3.1 Deformation type

The type of deformation imposed onto the composite panel is a three-point bend. Having no access to laboratory equipment, the experiments should rely on mechanisms that are both simplistic, and reliable by scientific standards. For an experiment to be considered valid, said process should be repeatable, controlled, and the dependent and independent variables are known. Using a three-point bending rig, the panel's deformation is solely dependent on the central force applied. With this configuration, the panel's deformation is restricted along a single axis. For theoretical analysis, this reduces the panel's FBD from a 3D to a 2D system. Therefore, restricting the deformation to a single axis simplifies the overall operation, and

Figure 18.1 *Front view of three-point bending rig*

allows for the stress–strain equations from the "Theory" section to be applied. It is important to remember that these equations were adapted for a unidirectional bend. Any deformation along any other axis would require the use of more complicated and detailed equations. ASTM E290 is a standard used to test for the ductility of materials by using a three-point bending rig. It proposes the use of a minimum of 2:1 width-to-length ratio, and ideally an 8:1 ratio for thin materials. Under ideal conditions, these specifics would be followed. In the case of this research, however, ASTM E290 would serve only as a general guideline for the three-point bending specifics.

The bending rig was first built with available and cheap material before deciding on the dimensions of the composite panel. The three circular dowels acting as the pressure points are equally distanced from each other by 125 ± 2 mm (this distance is measured from the centre of each dowel). Both points at the extremities are fixed to the rig, whereas the central point is free to move up and down. This configuration allows for the dowels at the extremities to act as the points applying pressure downwards, whereas the central dowel applies a force upwards. The movement of the moving point is restricted by the notches carved into the wood, guiding the nylon strings, allowing only for an up-and-down motion. The full specific of this rig is available in **Appendix A**, along with this choice in configuration.

Figure 18.1 represents the three-point bending rig constructed for the purposes of the experiments. The tall frame allows for measurements to be taken and allows for sufficient leeway for the panel to bend upwards. Certain unknown factors,

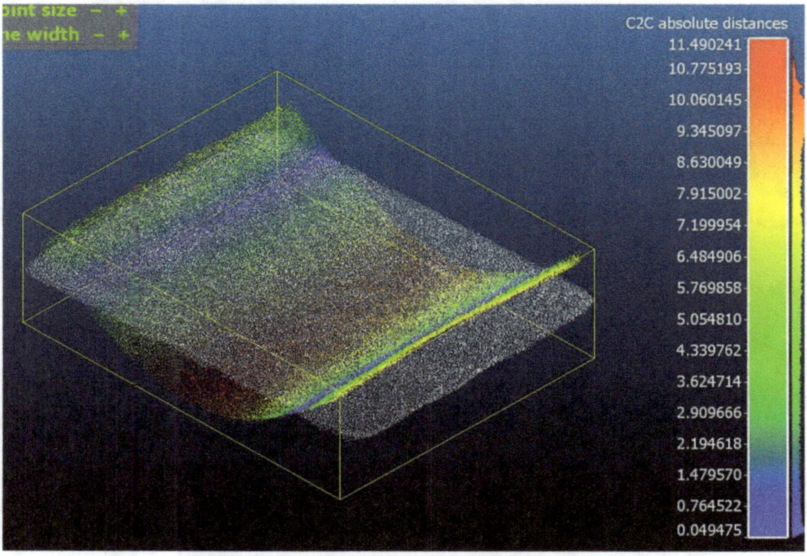

Figure 18.2 C2C analysis between an unbent and bent playing card

however, remain in this experiment and need to be taken into consideration when concluding this report. The force applied onto the panel is not known. The ductility of the composite panel is not quantitatively known, or in other words, the maximum deformation before catastrophic failure that the composite panel can undergo is not known.

The experimental procedure is as follows: the first step is to lock the composite panel into place by tightening the winch, assuring that the central dowel is placed at the span-wise mid-point of the laminate. Once locked into place, the first set of measurements can take place.

Several photos are taken using both the regular camera, and the LIDAR camera (more details below). These measurements serve as the "datum" points of the laminate. Then, the winch pulls the central dowel upwards bending the panel. A gear tooth of the winch is marked to keep track of the induced displacement per panel. Although this measure is too vague to calculate the amount of force applied per panel, it does serve as a marker that can be used for the following experiments. This ensures that the experiments are repeatable. Once the composite laminate is bent, it is locked into place by the winch. This allows for the surface measurements to take place. Similarly, a set of photographs are taken with both regular and LIDAR cameras. The procedure is repeated for all laminates. The "unbent" position is measured, then the "bent" position is measured (Figure 18.2). From there, the point clouds are reconstructed using the appropriate software, as detailed below. To ensure that the experiments are repeatable, markings are drawn on the winch's gear. This assures that a panel can be bent into the same position multiple times, keeping the same amount of deformation. During the experiments, the mechanism is locked into two

Table 18.1 Composite panels' specifics

Panel name	Number of plies	Overall thickness	Fibrer orientation
1	10	1.92 mm	Unidirectional
2	10	1.92 mm	[0/90] and alternating
3	40	7.69 mm	Unidirectional
4	40	7.68 mm	[0/90] and alternating

positions. In other words, two different displacements are induced onto each composite panel. The approximate measured displacements are as follows for a ten-ply laminate:

Both displacements in the *y*-direction shown in the figures are the reference displacements that will be used to quantify the accuracy of LIDAR and SfM measurements.

18.3.2 Composite laminates specifications

As mentioned in the introduction, the scope of these experiments was reduced to composite materials typically used in the aeronautics and aerospace field: CFRP. Originally, four carbon fibre composite panels were (Table 18.1) intended to be used for the experiments, to give variety to the results. Upon experimenting, however, two among the four purchased were too thick to be bent using the wooden rig.

The purchased CFRP composites were manufactured by Uchida, a composite material manufacturer and subcontractor of FRP Services & Co. Using the ASTM E290 standard specified above as a guideline, the panel's dimensions were chosen to be 360 × 45mm. These dimensions correspond to the 8:1 recommended length-to-width ratio. All four purchased laminates share these dimensions, and only differ in thickness, and in ply orientation. One ply of carbon fibre is 0.192 mm thick, with a Tensile Modulus of 235 GPa. Each composite panel were pre-impregnated and cured at 130°C for 1 hour.

The lamina properties for the above material are

The composite panels are placed into the rig, making sure that it is symmetrically positioned. The winch is turned as much as manually possible without damaging the rig, keeping the attention focused on the sound the entire wooden frame is making. Although the frame itself can sustain more than 100 kg in vertical force, the mechanical strain applied by the winch may easily cause damage on certain parts of the rig. Therefore, although four composite panels are available, only the two thinner ones will be tested in these sets of experiments. The two thicker panels only bent the wooden dowels.

18.3.3 Photogrammetry measurements

Using photogrammetry to recreate a 3D subject requires very specific conditions. The lighting must remain constant and should not shine directly onto the measured object. If an object were to be measured outdoors, e.g., the ideal weather would be an overcast or cloudy one, where the sun does not shine directly. If indoors, although having a strong lamp to highlight some details obscured by light shadows may seem favourable, it is advised that said lamp does not shine directly onto the measured object. A strong light source directly shining onto something may lead to unwanted glare, reflections, and an uneven distribution of colour when photographing the object. To prevent that, the three-point bending rig and the panel were photographed in the corner of the room next to a white wall. The light from the lamp shining off the white wall assures an even distribution of light without relying on a direct light source.

Similarly, the object being measured should not be glossy. A reflective surface is very difficult to reconstruct through SfM, as each frame may have distinct differences because of the refractive surface. Ideally, matte surfaces are what one should be looking for when photographing an object for photogrammetry. The carbon fibre composite panels being reflective, a piece of paper cut into the appropriate shape is stuck onto the top surface of the panel, thus assuring that no reflection happens without affecting the structural integrity of the panel. A grid-pattern is printed onto the pieces of paper covering the panel, which will aid as a marker that can be easily recognised by the SfM software.

The photos were taken using the camera of an iPhone X, which has a maximum resolution of 12MP. By default, this smartphone automatically adjusts lighting and focus. To avoid an automatic readjustment between each photos, the "manual" camera mode can be set by selecting a surface within the frame of the camera (like a wall or a tabletop), and by holding the finger on the screen for 5 seconds until the words "AE/AF Lock" appear. This indicates that the phone camera has been set to manual mode, and that the lens focus and lighting adjustment has been locked. In other words, the camera intrinsic will not change from one image to another. This is an important element in photogrammetry, along with the lighting conditions. The focal length of the camera lens must remain the same between each image. By default, the majority of SfM algorithms use the first image of a given image set for a scene reconstruction as reference. That is, the first image fed into the algorithm is used to estimate camera parameters that serve as the boundary conditions for the SfM algorithm, if the latter are not specified by the user beforehand. If these camera intrinsic were to change from one image to another, then the SfM algorithm will apply faulty boundary conditions on those images, resulting in faulty 3D camera position reconstruction (which is a crucial element in photogrammetry). Furthermore, to ensure that the images taken are of highest quality, the "HDR" option is turned on. This creates one image from multiple images taken at once by the iPhone, allowing for a detailed image with higher contrasts.

Each picture is taken side by side continuously, making sure that the camera position follows a specific unbroken path. It is important that each photos are taken side-by-side. The SfM relies on matching features from one frame to the next. If two consecutive frames are taken from two different locations, no features can be reliably matched. For each experimental configuration (unbent and bent for both ten ply panels), an image set of between 40 and 50 photos was created.

Finally, the image sets are fed into **Regard 3D**, a free-to-use photogrammetry software that can reconstruct a 3D scenery from the image set. Although other free-to-use software such as **Meshroom** exist, Regard 3D allows for more control over the output. In Meshroom, the 3D reconstruction takes place after pushing one button. After which, the program runs for several hours depending on the size of the image set. Through Regard 3D, the program pauses after each task, and prompts the user to select the parameters for the following task before continuing. Although this may indicate that Regard 3D is less user-friendly and less intuitive (and the UI is very clearly less elegant than Meshroom), the output can be directly controlled by the user. This leads to Regard 3D outputting a more precise and detailed 3D reconstruction than Meshroom, even on its default values. Generally, Meshroom is considered to be an ideal tool for artistic reasons, whereas Regard3D can be seen as a more professional tool. In these experiments, all parameters between tasks were set to default, except for the final task, which is the point densification phase. This parameter is set higher than its default value, which creates a denser point cloud allowing for more details. The default value creates a very "blobby" result. The figure below shows the difference between a "low-density" reconstruction and a "high-density" reconstruction.

The resulting point cloud is then exported into a .OBJ file. The figure below compares the output between a "low point density" surface reconstruction, and a "high point density" surface reconstruction.

18.3.4 LIDAR measurements

In a similar manner, the LIDAR measurements are taken, while the three-point bending rig is under both "unbent" and "bent" configuration (Figure 18.3). The front-facing camera of the iPhone X is equipped with a LIDAR camera used for the

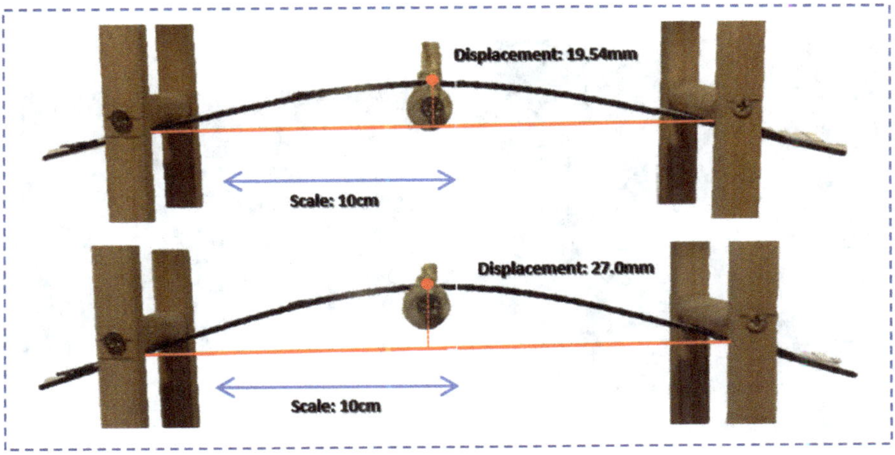

Figure 18.3 Displacement settings on the three-point bending rig

facial recognition. Newer versions of the iPhone are equipped with such a camera on the back side, which render the measurements easier.

The conditions within which this camera can be used to take reliable data are less restraining than the conditions for photogrammetry. Ideally, the surface being measured not being glossy is one similarity. The most important aspect of LIDAR photography, however, is that nothing that is unimportant should be within range of the camera. This includes a wall behind the object, or any feature on the sides. Appropriate lighting is not as much of an issue as with photogrammetry.

Once the panels are in position, the LIDAR camera is used to take a continuous view of the whole panel, making sure to rotate the camera accordingly to take every possible angle of the bending rig. The iOS App used for this operation is called "Capture." Each measurement only takes no more than 30 seconds. The point clouds are then exported and saved in the .OBJ format.

18.3.5 Point cloud denoising and distance measurements

The next necessary step is to scale, denoise, and convert the point clouds into meshes. **Blender** can be used to do so, as it compatible with a number of 3D file formats. Blender is also typically used by photogrammetry users to clean up their models. Point clouds can be converted into meshes through a straightforward UI, and the mesh itself can be edited to sharpen or smooth surfaces.

Such functions, however, are not useful for the purposes of this project. Editing the surface features of the converted mesh would be counter intuitive, as the aims of this project are to use the direct measurements from SfM and LIDAR. Blender is a very useful free-to-use tool for artistic renditions of an object captured through photogrammetry or LIDAR. For this project, **MeshLab®** is used (Figure 18.4).

The .OBJ files are imported into MeshLab, and the model is first reoriented accordingly along the global coordinates using the "Manipulator" tool. The "Measuring"

Figure 18.4 "Before" and "after" point cloud denoising

tool is used to measure the distance between two points; in this case, any known distance is measured such as the composite panel's length or width used to calculate the scaling ratio. Using these measurements, the whole model is scaled by applying a "scale filter." Finally, any unwanted surrounding points are selected and deleted from the model. The point cloud is now finally exported in .XYZ file format for the following step. The images below are an example of the "before" and "after" conditions of the point clouds.

The application used for the Hausdorff Distance measurement is **CloudCompare**. This software is very useful for several point cloud comparison operations. Two of them are the more interesting functions, which are the C2C comparison and M3C2. Although the M3C2 add-on will not be used for these experiments, it may be used in a future project. M3C2 is a more robust algorithm than C2C, offering more detailed estimations of movement from one point cloud to another. For the purposes of these experiments, however, the C2C algorithm will be used to calculate the required distances. Point clouds of both "unbent" and "bent" configurations for each laminate are imported and superposed, their bounding box centres superposed over each other. If the previous step was completed properly, then both point clouds should have the same size. Using the "unbent" configurations as the reference point cloud, the Hausdorff distance is calculated. The following image is the computation's output from a test simulation, where the deformation of a playing card is measured.

18.3.6 Abaqus and MATLAB® simulation

A replica of the panels is reproduced in **Abaqus** for the FEA. The dimensions, ply numbers, and material properties from the real-life composites are used for the simulations. Parallelepipeds with a circular head are acting as the medium for the pressure points of the three-point bending are also added in the simulations, where the two shapes at the extremities act as stationary objects, and the one in middle acts as the moving object.

The boundary conditions of the simulation are set to replicate the experimental setup as accurately as possible. This includes setting the displacement of the middle point. The latter's displacement is set to be the displacement measured during the C2C analysis. The results from the Abaqus simulations are recorded to be used for the final step of the project. The main interest in this situation will be the Von Mises Stress displayed on default by Abaqus.

On Abaqus, the composite panel is modelled by first drawing a rectangle with the appropriate dimensions. This part is set as a deformable, "extrusion type" shell. The pressure points on the other hand are modelled as discrete rigid, "extrusion type" solids. Knowing that two of these parts are placed at 60 mm from the edges, and one in the centre, reference lines were drawn onto the composite panel in Abaqus accordingly. A new material is defined, where elastic lamina properties, and the material properties outlined in Table 18.2 are used. Next, the composite sample is given a conventional shell, and the newly defined material is applied to each ply. All remaining fields that require being filled out to define the panel's properties on Abaqus are mentioned in the "Composite Laminates Specifications" section.

Table 18.2 Lamina properties of the composite panels

E_1	**: 170,000**
E_2	: 8,500
ν_{12}	: 0.35
G_{12}	: 4,500
G_{13}	: 4,500
G_{23}	: 5,000

The interaction between both the panel and the three pressure points is set as a "contact type" penalty tangential behaviour (with a friction coefficient of 0.1), in addition to a normal behaviour where separation is allowed after contact. The three static pressure points' boundary conditions are set as an "Encastre" type. Only the centre pressure point's conditions vary, as it is the only one given a displacement with uniform distribution, whereas the others remain static. Finally, the meshes' element types are set as discrete rigid elements.

Concerning the **MATLAB®** simulation, the latter is set to use the displacement data acquired by the C2C analysis through a .TXT file. This .TXT file was exported from CloudCompare after the Hausdorff distances were computed. The main issue with running the code would be to optimise it due to the number of points being displayed. Therefore, the number of points is arbitrarily reduced to a more manageable amount. It is important to remember that the .TXT contains not only the position of the points from the reconstructed 3D point clouds but also the displacement data (or absolute distance) for each point from the "bent configuration" of the point clouds. This is the surface data required to perform the stress–strain analysis. Seeing as the only force applied to the panel is in the y-direction, it is safe to assume that said measured absolute distances only represent displacement in the y-direction.

The points' displacement coordinates from the .TXT file are then sorted in the x-direction, before applying the stress–strain estimation formulas discussed in the "Theory" section of this report. The stress or strains computed from the equations are then displayed with a colour scheme appropriate to each points' corresponding stress/strain values. The values computed from this surface data analysis are then compared with the Abaqus simulations' data, and an estimation of the degree of accuracy can be made.

The aim of this step is to compare both the Abaqus simulation's results with the MATLAB code's results. The Abaqus simulation would carry a direct FEA of the stresses and strains using the maximum displacement from the C2C computations, whereas the MATLAB code would calculate the stresses and strains from the point clouds' displacement data using the equations discussed in the "Theory" section.

18.3.7 AR Visualization

As discussed previously, AR is a visualisation method that involves overlaying a virtual image over real-life, whether in real-time or in post-processing. Because the process requires a 3D model to be actively animated with regard to the visual inputs from a camera, a game-developing engine is best suited for creating an AR interface. In this project, **Unity 3D** was used to do so. Although **Blender** has the advantage of having an easy and intuitive UI and modelling tools already incorporated, Unity excels in designing game mechanics. Blender and Unity, however, are both used hand-in-hand by artists and developers, where Blender is used to animate and create objects through its modelling tools, and said objects are imported into Unity to integrate user-to-interface interaction-based mechanics for a game.

Unity 3D is a software used for game development at any scale. The **Vuforia** add-on is downloaded and activated, which allows for the use of AR-related libraries. Creating a developer account is necessary for using the Vuforia libraries. Through its website, marker images can be uploaded to be processed and used in Unity. The objective of this section is to develop a smartphone application which will serve as the AR interface. Developing an app for iPhone being an exceedingly demanding operation, an app for Android devices is instead developed. Therefore, it is important to set the settings of Unity to be compatible with Android products, which may include downloading additional data.

Making sure that the appropriate licenses for using Vuforia are added into Unity, the marker images are added into the editing window, and are set as the parents of any desired models. In this case, two markers serve as the parents of two separate models: one that will visualize the Von Mises Stresses from the Abaqus simulations on a virtual analogue of the composite panels, and one visualizing the stacking sequence of said composite panels. **Pixilart** is the website used to create the colour gradient required for the stress distribution visualisation.

Finally, the project is saved, processed, and exported as a file that can be opened by an Android device. Although the results of this section will be discussed in the conclusion of this report, the results are best seen in the form of a video, accessible below. Manually switching the smartphone's camera from one marker to another will change what is being visualised, according to the user's preference.

18.4 Results and discussion

Before concluding and comparing results, it is important to reaffirm the important points of these sets of experiments. Photogrammetry has been established to be an accurate way to carry out measurements. How precise is it, however, when it comes to using the reconstructed point clouds for further analysis? The same question can be asked about LIDAR, whether the surface data acquired from this technology can be relied upon to conduct a more in-depth structure analysis. Furthermore, through the research of Dr. Hamed Yazdani Nezhad and his colleagues, the use of reverse-FEA for analysing the structural health of composite panels has been shown to be accurate enough to be used as a predictive tool for composites' SHM.

Using this knowledge, two composite panels of same thickness but different stacking sequences were fixed into a homemade three-point bending rig. Each panel is bent until two different displacements, and for each displacement both LIDAR and photogrammetry was used to measure the surface deformation.

From the LIDAR and SfM measurements, a 3D point cloud is reconstructed, all of which are scaled, reoriented and denoised after hand. The C2C displacements are calculated using the Hausdorff equations and fed into Abaqus and MATLAB®.

In Abaqus, a finite element analysis (FEA) of the three-point bending is simulated while respecting the appropriate boundary conditions, and the displacement of the panel's mid-point is set to the same displacement calculated from the C2C analysis. The stresses and strains within the composite panels are therefore calculated.

In MATLAB®, the displacements from the C2C analysis are used to compute the stresses and strain within the composite panel, using a relationship discussed in the "Theory" section. These stress and strain values would therefore be computed using solely the surface data acquired from the 3D reconstructions. In an ideal situation, the MATLAB® and Abaqus results would be compared to each other. The MATLAB® simulation would provide an idea of the degree of accuracy in composite structural analysis from surface data, compared to the more robust FEA from Abaqus.

The final important aspect of these experiments is simulating the stresses and strains from the analyses using AR. The importance of this part is crucial in visually explaining the importance that virtual technology may have on the SHM of composite materials. How easy, however, is it to create this AR? what is the range of use?, and can it be designed to be responsive to real-time inputs?

18.4.1 Photogrammetry results

Although it has been established by a previous study that photogrammetry can be used to reliably measure a panel's deformation by a three-point bending rig, such an operation involves a very specific set of parameters. Figure 18.5 is among the best photogrammetry results created, and therefore does not require any further results to be shown. One main issue stems from the fact that the conditions in which the panel was photographed were not ideal. The lack of access to laboratories or a studio

Figure 18.5 Comparison between low-density and high-density point surface reconstruction

with decent homogenous lighting due the pandemic led the experiments to be taken place at home. The shadow of the photographer may have obscured certain areas of the images, notably on the floor and vertical beams of the main rig. This may have led the SfM algorithm to faultily matching certain features, and ineffectively scaling certain dimensions during the reconstruction phase.

Although the panel is still in its datum position (i.e., flat and unbent), the surface of the panel has been reconstructed as bumpy. Another definite source of error for this result is the dimensions of the panel. As mentioned previously, thin surfaces are difficult objects to be created through photogrammetry. Even the vertical beams were unable to be properly reconstructed, despite their relative thickness to the composite panel.

Does this indicate that photogrammetry cannot be an option for the SHM of composite materials? The reasons stated above are not enough to support this assertion. Under ideal lighting conditions (e.g., in a laboratory or proper testing facilities) certain types of composites may be photographed as long as their thickness does not impede the SfM feature-matching algorithm. The results of these experiments, however, do raise a new set of requirements and hindrances, if photogrammetry should be used in SHM of any type of structures, including composite materials.

Including the already-mentioned lighting conditions, and camera quality recommendations, the type of deformation or damage sustained by a structure has to be visible with the naked eye (or in this case, a high-quality camera). Any visible change in the structure should be recognisable by the SfM algorithm as a feature that has 3D depth. In other words, scratches or dents may be recognised as a texture, instead of being recognised as a 3D geometry that is part of the environment. Deformations on this scale may be recognised only with difficulty on the macroscopic level.

Small-scale damages or deformations may be recognised as 3D geometrical features using photogrammetry, if microscopes were to be used as the medium through which the image sets can be created. In this situation, scratches, dents, or small tears can be subjects of photogrammetry-based analysis. The deformations should also remain at the same configuration for long enough for the photogrammetry measurements to take place. Any structure that deteriorates through time, either because of imposed loadings, or temperature, or chemical reactions, cannot be a subject of photogrammetry-based analysis. Whenever this type of analysis is intended to be used, a model of both "before" and "after" configurations should exist and should be taken from unmoving structures.

One main issue with photogrammetry that needs to be highlighted is the time factor. Assuming that photogrammetry is a viable and accurate option for surface deformation measurements (as stated in Reference 3), reconstructing said point cloud is extremely time consuming. The figure shown above is the result of a 2-hour long analysis for an image set of 45 pictures. Higher definition images, and a higher number of images will lead to an even longer processing time. Even with a powerful computer, it is safe to assume that photogrammetry may not be a viable option for a real-time (or at the very least, as close to real-time as possible) structural health analysis of a composite panel.

The figure below is the photogrammetry result of a bent laminate. Unlike the previous situation, the white surface of the panel was not registered. This is due to a lack of features discernible on the surface of the panel. Because of such results, a grid pattern was printed onto the surface, which allowed for the SfM algorithm to pin-point features on the panel's surface. This leads to the following issue with photogrammetry-based analysis: any surface that lacks discernible features, like colour contrasts, or geometric patterns, are unlikely to be registered by the point cloud reconstruction software. This therefore needs to be considered in addition to the previously mentioned aspects.

In conclusion to this section, photogrammetry may be a good tool for simple deformations (such as bending, stretching, torsions, folding, etc.), and can reconstruct an accurate 3D point cloud of said deformations for further analysis, only if the conditions are met, and if real-time results are not a prerequisite. Any other types of surface damage that a structure can sustain during testing or during use (scratching, denting, stretching, or bending on the millimetre scale, etc.) are more difficult to monitor through image analysis alone, and is not recommended.

18.4.2 C2C from LIDAR and Abaqus results

As explained in the "Methodology" section of this report, the LIDAR measurements were used to reconstruct 3D point clouds of the undeformed and deformed panels (Table 18.3). The values below are the results of the experiments. The LIDAR measurements took no more than 15 seconds per panel, and the manual denoising and scaling of the point clouds each took approximately 3 minutes. The imposed displacement in Abaqus is what the C2C analysis has calculated (Figures 18.6–18.13).

Unidirectional panel, lesser displacement, LIDAR measurement

Unidirectional panel, higher displacement, LIDAR measurement

0°/90° panel, lesser displacement, LIDAR measurement

0°/90° panel, higher displacement, LIDAR measurement

First and foremost, the deformation error for both unidirectional composite panels seems to be below the 5% threshold, indicating that the LIDAR measurements for both deformations of the first panel were accurate (Table 18.4). This also leads to the assumption that the FEA results are accurate representations of the experiments, as it has been established by Dr. Yazdani Nezhad that FEAs can be used as a predictive tool for composite materials' SHM. Ideally, the Abaqus results would be compared to those computed using MATLAB®. A working code, however, was unable to be developed because of time constraints among other issues. The main discrepancies that were identified will be enumerated in the following section.

The two experiments of the first panel therefore indicate that the LIDAR measurements are accurate. The second panel, however, shows a noticeable error. Upon closer inspection, it seems that the miscalculation is due to the bent panel not being properly aligned with the unbent panel, when superposing both point clouds. Superposing both point clouds is done in CloudCompare by matching the bounding boxes' centres from one point cloud to the other. The bounding box can be thought of as parallelepiped-shaped volume enclosing the entire point cloud. Overlaying

Table 18.3 LIDAR-measured deformed panels' Hausdorff distances

Panel type	Actual deformation	LIDAR-measured deformation	Deformation error (%)	Max. Von Mises stress (FEA)	Max. Abs. principal strain
Unidirectional	19.7 mm	20.45 mm	3.8	6.07 e2	0.885 e−4
Unidirectional	27.0 mm	26.64 mm	2.96	7.93 e2	1.165 e−4
$0^\circ/90^\circ$ and alternating	19.57 mm	27.8 mm	42.05	7.42 e2	0.104 e−4
$0^\circ/90^\circ$ and alternating	27.0 mm	30.58 mm	13.26	8.17 e2	0.116 e−4

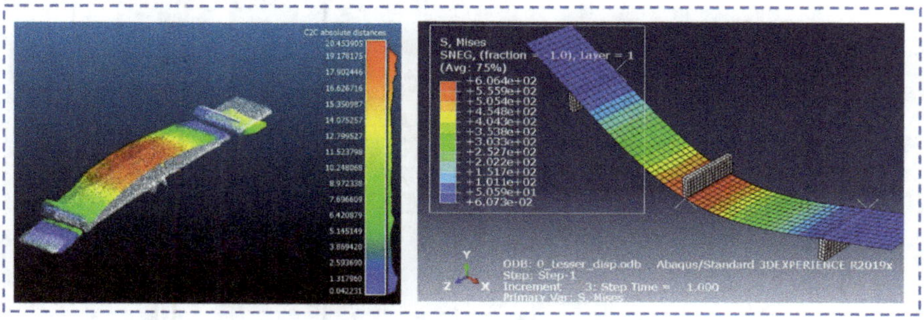

Figure 18.6 Unidirectional panel, lesser displacement, LIDAR measurement

the geometrical centres of both volumes with each other is the only way of automatically setting the panels' reconstructions in the desired configuration for the Hausdorff distance measurements. In doing so, however, the y-position of the bent panel may not be in the desired position, as the bounding box for this mesh is larger than that of the unbent panel. The larger bounding box leads to its centroid being further placed away from the point clouds, whereas the unbent panel's bounding box centroid perfectly matches the centroid of the point cloud.

This therefore leads to faulty distance measurements. The reason why such error was not an issue in the first two experiments is unknown. It can be safe to assume that the two last experiments' results were due to human error, rather than software. If the two last panels were properly scaled during the denoising phase, and properly superposed in the C2C measurement phase, then the error would have been similar to the first two experiments. The error should not, however, be put on the LIDAR measurements themselves.

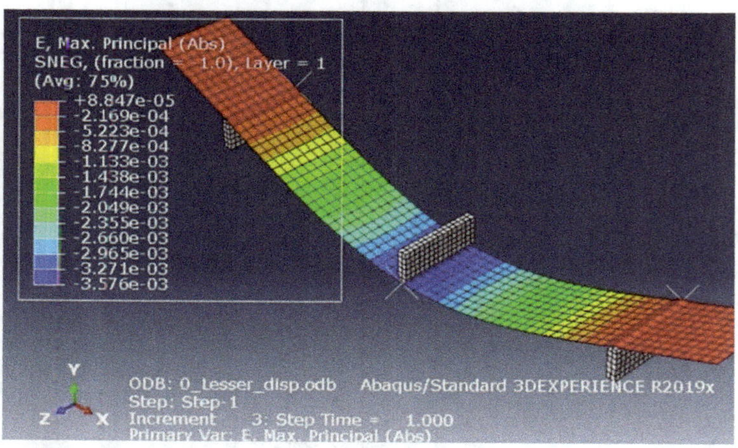

Figure 18.7 Abaqus' strain calculation from C2C measurement, unidirectional panel, lesser displacement

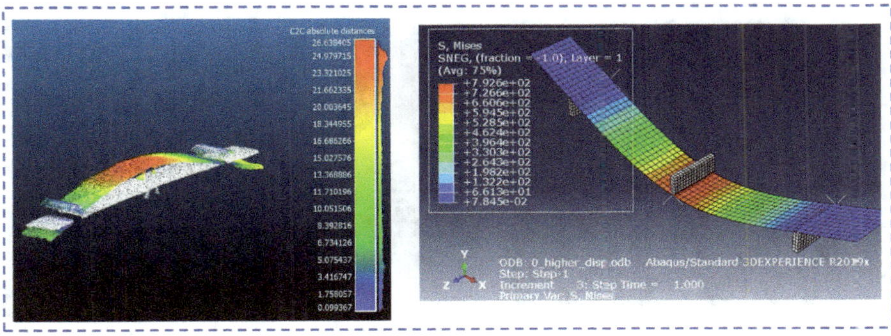

Figure 18.8 *Unidirectional panel, higher displacement, LIDAR measurement*

As mentioned, the FEA results can be assumed to accurately depict the effect that the three-point bending had on the panels, because of the measured displacement. Such statement, however, cannot be entirely validated within the scope of this report, because of the difficult access to laboratories. In an ideal situation, gauges would have been placed on the panels' surface to validate the FEA from LIDAR-measured displacements.

18.4.3 Difficulties with MATLAB® code

By comparing the Abaqus and MATLAB® results, one can have a general idea of how accurate a standalone code can compute stresses and strains from surface data, without the use of a powerful software such as Abaqus. To respond, the following descriptions associated with Figures 18.14–18.17 have been provided:

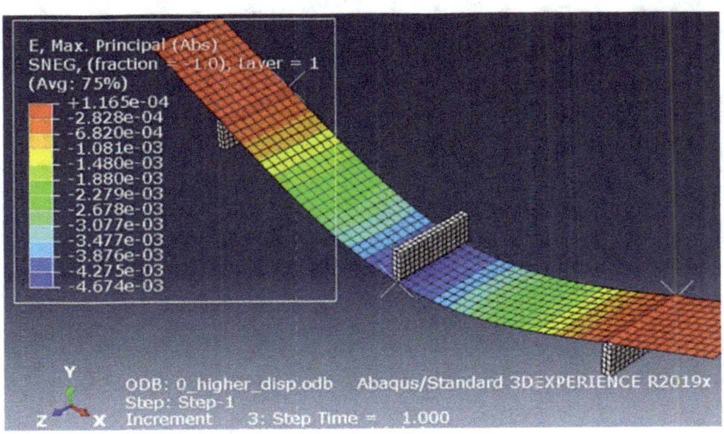

Figure 18.9 *Abaqus' strain calculation from C2C measurement, unidirectional panel, higher displacement*

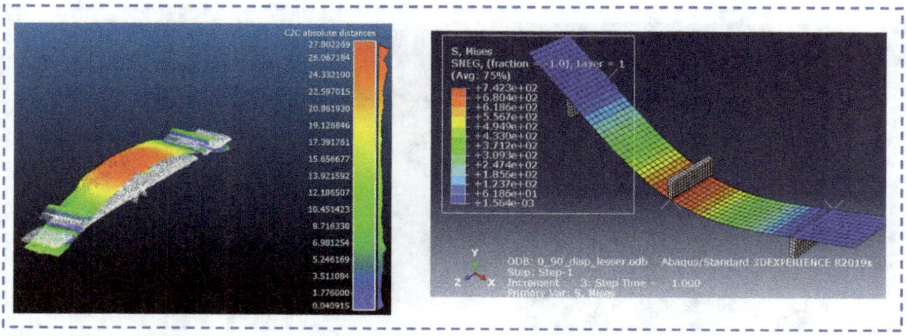

Figure 18.10 [0/90] panel, lesser displacement, LIDAR measurement

If the MATLAB® code can be finalised in a future project, and yield results like those computed by Abaqus, it would show that using surface deformation data in the form of a 3D point cloud could be used for SHM of structures. Using such a code for an almost real-time analysis of composite structures would be more flexible than using designated FEA software. Applications like Abaqus require the user to model the structures and perform a simulation from that, whereas the aim of the MATLAB® code was to predefine a set of parameters and equations that will carry out the operation. The only input required for the MATLAB® code would be the surface deformation data from the C2C analysis, from which the equations discussed in the "Theory" section are applied.

One main issue in creating the MATLAB® code was the size of the point cloud files. MATLAB® does not seem to be suited for carrying out matrix operations on the scale of the provided deformation data. The latter provides information on every

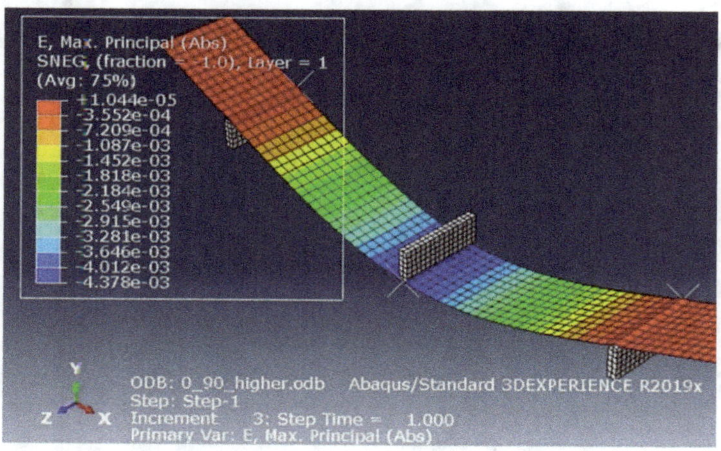

Figure 18.11 Abaqus' strain calculation from C2C measurement, [0/90] panel, lesser displacement

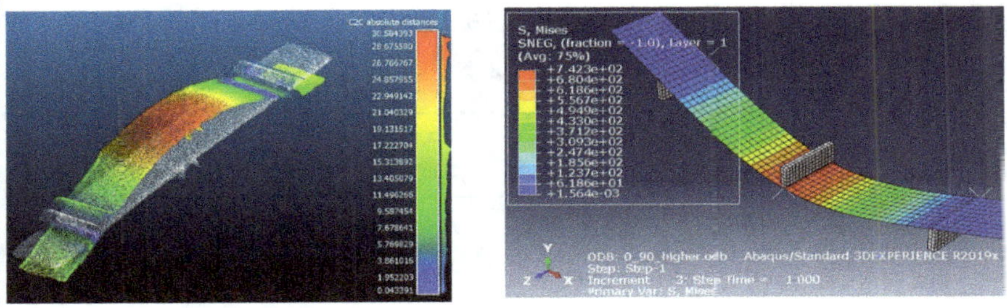

Figure 18.12 *[0/90] panel, higher displacement, LIDAR measurement*

points' location and displacement from its "original" position. Therefore, the number of points that can be analysed needed to be cut down.

The next issue is in the sorting of the points' coordinates and data. The .TXT file containing the deformation data from the C2C analysis does not arrange the points' coordinates in any order. In the "Theory" section, it is shown that the main equation relies on finding the double-derivative of a point, with regard to the points immediately next to it. Such operation is difficult if the point clouds' coordinates are not set in a sequence. Further work in finalising the code would therefore require a sorting algorithm. An attempt to address this issue was made by sorting all points' coordinates in an ascending order in the x-axis. The output, which consists of displaying the original point cloud and labelling each point with a colour scheme indicating the calculated stress, was not satisfactory even after this alteration.

Unidirectional, lesser displacement

Unidirectional, higher displacement

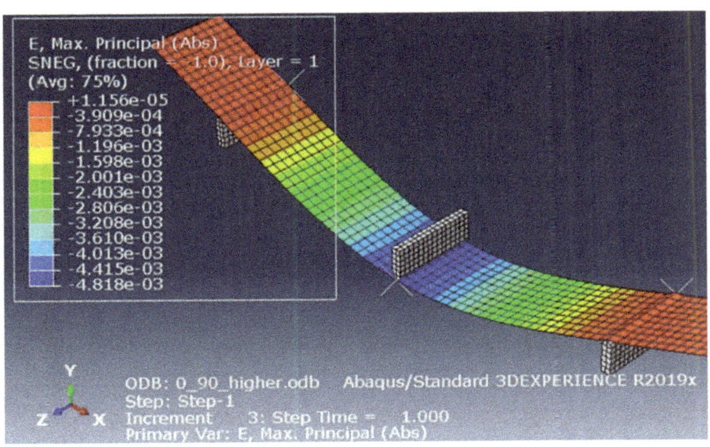

Figure 18.13 *Abaqus' strain calculation from C2C measurement, [0/90] panel, higher displacement*

Table 18.4 *Comparison between MATLAB's strain and Abaqus' strain calculations*

Panel type	Deformation distance	FEA max. strain	MATLAB max. strain	Calibration/ scaling factor
Unidirectional	Lesser	0.885 e−4	4.14 e5	2.14 e−10
Unidirectional	Higher	1.165 e−4	5.7 e5	2.04 e−10
[0/90] and alternating	Lesser	0.104 e−4	7.5 e5	0.14 e−10
[0/90] and alternating	Higher	0.116 e−4	5.74 e5	0.2 e−10

0°/90°, lesser displacement

0°/90°, higher displacement

The above images are the computed absolute maximum principal strains from both MATLAB® and Abaqus. As expected, the centremost area from the MATLAB® simulations seems to indicate a near-zero strain, which is expected through the Abaqus results. The maximum strain on the edges, however, indicates that there is a scaling issue with regard to the panel's dimensions in the MATLAB® calculations. By reading the colour bar on the right of the figures, it enforces this statement. The values, however, seem consistent. The first two figures are from the

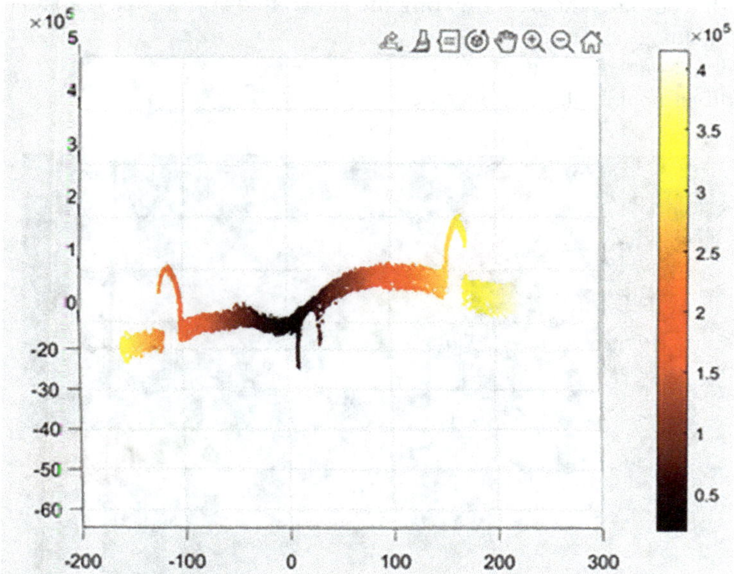

Figure 18.14 *MATLAB's strain calculation from C2C results, unidirectional panel, lesser displacement*

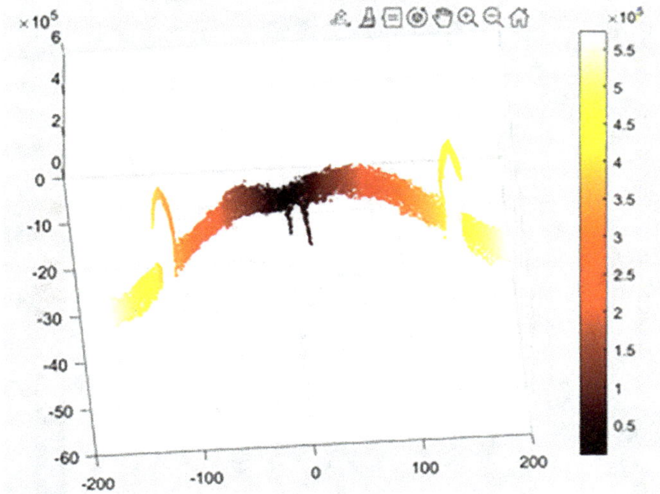

Figure 18.15 MATLAB's strain calculation from C2C results, [0/90] panel, higher displacement

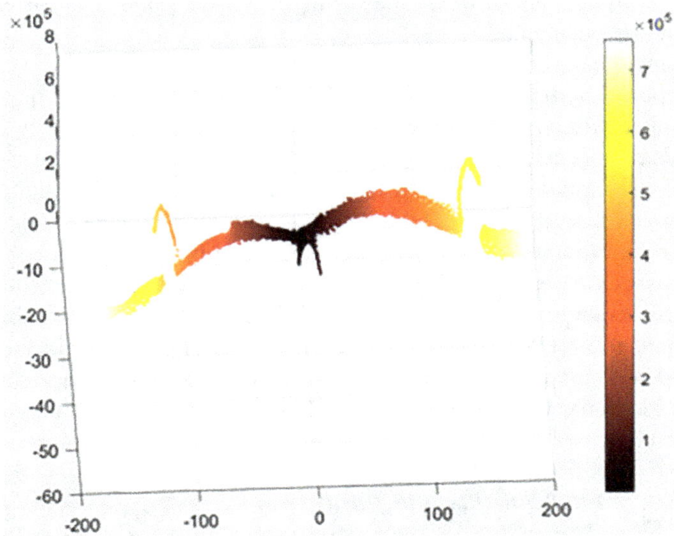

Figure 18.16 MATLAB's strain calculation from C2C results, [0/90] panel, lesser displacement

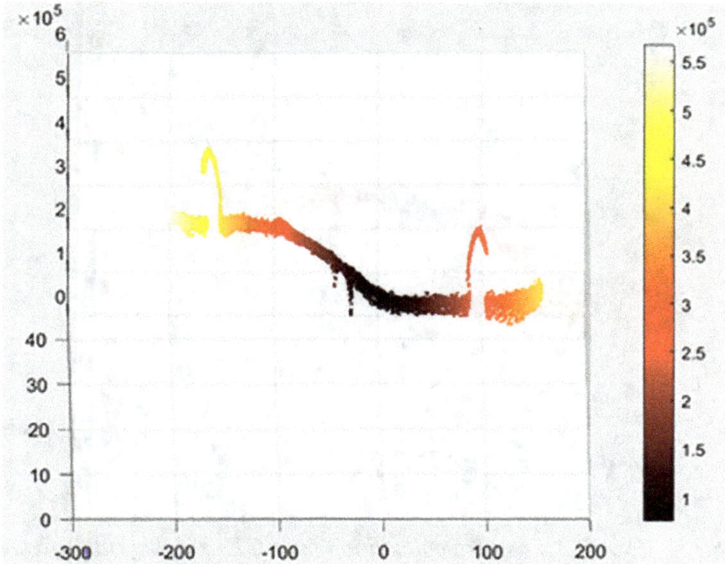

Figure 18.17 *MATLAB's strain calculation from C2C results, unidirectional panel, higher displacement*

same panel, with two different deformations. The scaling factor between the Abaqus-computed values and MATLAB®-computed values is very similar. The same can be said for the [0/90] panels, where the scaling factor for strain is consistent despite the different displacement.

Observing the severe distortion that the panels' point clouds have sustained in the data output, in addition to the strain order of magnitude, indicates that there is a scaling problem within the code. In other words, the global coordinates versus the local coordinates of each individual point clouds do not seem to have been translated properly when computing the strain through MATLAB®. Indeed, the calculated scaling factors (as seen in the table) are consistent with each panel, indicating that the code can be calibrated accordingly for each panel type within a certain margin of error. No specific method to fix this issue can be proposed, as this requires more in-depth knowledge in data processing, which is beyond the scope of this project. It is important to note that these results do not dismiss the possibility of using a MATLAB® code to carry out a stress–strain calculation from surface data. It only shows that additional work is required to calibrate the results until they match the more robust FEA.

18.4.4 Recommendations and future work

The main question that was posed for this report was the following: using only visual data acquired through a camera or a set of 2D images, can a virtual 3D model be reconstructed accordingly, and carry out accurate simulations?

The surface data of composite panels were acquired through both photogrammetry and LIDAR. Photogrammetry is a reliable method of reconstructing topography from 2D images, which is used in many fields. In these sets of experiments, however, the 3D reconstruction from the SfM of 2D images did not yield useable results. Such was due to the lighting, and the topology of the panel. Thin surfaces are generally difficult to capture, and although initially the panels were deemed to be thick enough to be registered, the results proved otherwise. These experiments do not conclusively affirm that photogrammetry is an unsuitable tool for the SHM of composite materials, but rather it is unsuitable for thin glossy surfaces. Under laboratory conditions, however, with proper lighting, proper cameras, and proper testing equipment, photogrammetry may be used as a viable option in some cases.

The main downfall of using this technique, however, remains in its time-consuming use. Traditionally, photogrammetry uses a large set of images taken manually, which may take a certain amount of time. Some may argue that video frames can be used. This poses the problem, however, that certain frames may be prone to motion blur unless the cameraman takes great care, or if the video frames are manually sorted to eliminate blurry images. Furthermore, in addition to the conditions highlighted previously, a detailed 3D reconstruction requires a large image set; and a large image set requires a large amount of time to be processed. In these experiments, an average of 50 images requires almost 2–3 hours of processing.

Photogrammetry therefore may be a useful tool for SHM, only if time is not a factor. Editing the photogrammetry results by smoothing and sharpening certain features, in addition to denoising and rescaling, is also recommended. The aim of this report, however, is to attempt to combine surface measurements from 3D reconstructions, as well as use AR to visualise results. This implies that the SfM reconstruction must be done within a timely manner, and ideally in real-time, which cannot be done using photogrammetry. This therefore makes LIDAR a more interesting tool.

The LIDAR measurements have shown to be quite accurate in half of the cases. The other half of the experiments encountered difficulties in properly measuring the results. This does not, however, entirely dismiss LIDAR as a viable option for surface data measurement for SHM or NDT. Rather, the problems encountered are avoidable ones if this project is to continue further in a future work. For instance, implementing a more robust C2C Hausdorff distance calculator.

Combining surface deformation measurement through LIDAR, with an element of machine learning can be a possible subject of study for future work. Machine learning can be used to pinpoint visual markers on a given subject, therefore automatically denoising, rescaling, and reorienting a model directly after having taken the LIDAR measurements.

The MATLAB® code needs to be refined, however, it does show consistent values despite the compute order of magnitude. This indicates that after the appropriate changes have been made, calculating stresses and strains through purely surface data is possible. To be able to calibrate these results and validate them, using strain gauges attached to the panels is recommended.

Finally, concerning AR, Unity 3D uses C# as its main language. Therefore, a fully functional code can be written to automatically import C2C results and stress/

strain analysis results into the projected model. The corresponding colour schemes and stress/strain values can be added to predefined meshes and objects in the AR world-building window.

In conclusion to these experiments, we believe that we have managed to establish the necessary building blocks for further exploration in the field. The methods used in these experiments were found to potentially be complimentary to existing methods and can serve as an additional SHM of composite materials in addition to existing methods.

18.4.5 Scope of application

AR is a tool that prove to be very useful in the engineering industry. Through personal experience by working as an intern in Lumiscaphe (a French AR company that visualises clients' concept models), this statement can be affirmed. Using AR as a means of visualising the internal stresses and strains in real-time of a structure can bring a novel way for manufacturing structures and training workers.

If a structure's deformation's effects can be visualised using AR, industrial scale maintenance of certain structures can be a possibility. Noninvasive testing and in-situ testing can also be an application of this technology. For instance, in the aeronautics industry, if stress/strain propagation in a plane's wing can be visualised through AR after a simple LIDAR measurement.

The analysis of dynamic components may also be an option by using high-speed cameras, and using individual frames to reconstruct a 3D model through photogrammetry. Finally, training employees and novel engineers in a manufacturing plant through real-time AR simulation is a plausible prospect than can simplify tasks and enhance the learning experience.

Appendix A: wooden three-point bending rig

Two flat 300 × 150 mm wooden boards serve as the base of the structure, and as the supporting platform for the winch.

Four 400 mm long wooden bars with a 15 mm square cross section serve as the four main supportive beams. These are fixed to both flat boards using angled brackets screwed at the extremities.

At 250 mm from the base board, a wooden dowel is cut to size, and fitted between two of the rig's supporting legs. Before doing so, the centers of the dowels are properly marked and aligned so as to keep them perfectly parallel to the ground when drilling the screws.

The central dowel is attached with nylon strings tied at the extremities where a screw is affixed. On one end, the nylon string is looped around the screw, and tied using a hangman's noose on the other end. The nylon string is passed through a hole drilled in the center of the winch's axe.

Finally, the winch' gear is carved out of a 1 cm thick piece of plywood. Its axe has a square cross section for maximum friction between it and the gear. A triangular piece of wood is wedged and hammered between the square axel and the gear, ensuring a tight fit. The blocking mechanism is made from a metal screw placed perpendicularly between two wooden slats. It is manually pushed under or away from the gear whenever tightening or loosening the winch.

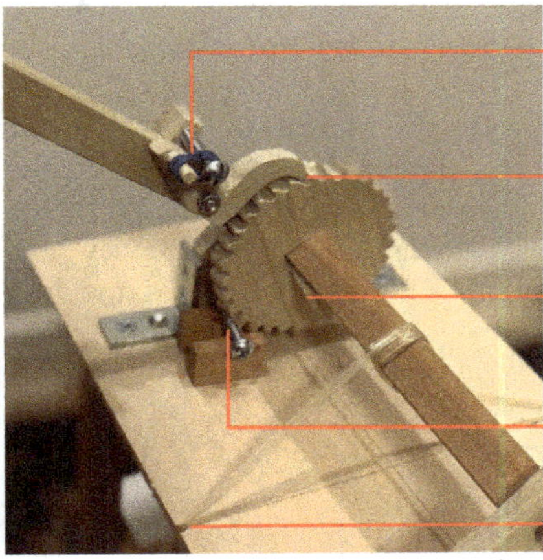

References

[1] Kronfli B., Peckham J., Rogerson J. *Best camera phones 2021: our picks for the best smartphone cameras right now [online]*. TechRadar; 2020. Available from https://www.techradar.com/news/best-cameraphone

[2] Mundy J.L. *The relationship between photogrammetry and computer vision, schenectady [online]*. New York: GE Corporate Research And Development; Available from https://citeseerx.ist.psu.edu/viewdoc/download?doi=10.1.1.69.537&rep=rep1&type=pdf

[3] Greenwald W. Augmented reality (AR) vs. virtual reality (VR): what's the difference? [online]. 2018. Available from https://www.pcmag.com/news/augmented-reality-ar-vs-virtual-reality-vr-whats-the-difference

[4] Howard C.E. Spirit AeroSystems deploys augmented reality to speed inspections, quality control [online]. 2016. Available from https://www.intelligent-aerospace.com/satcom/article/16539007/spirit-aerosystems-deploys-augmented-reality-to-speed-inspections-quality-control

[5] Cai J., Qiu L., Liang D., Yuan S., Liu P. Structural health monitoring for composite materials [online]. 2012. Available from https://www.intechopen.com/books/composites-and-their-applications/structural-health-monitoring-for-composite-materials

[6] Awad Z.K., Aravinthan T., Zhuge Y., Gonzalez F. 'A review of optimization techniques used in the design of fibre composite structures for civil engineering applications'. *Materials & Design*. 2012;33:534–44.

[7] Davis M.J., McGregor A. Adhesive bond damage tolerance and failure assessment. Course material of CIRJ [online]. 2010. Available from https://

scholar.google.co.jp/scholar?q=Adhesive+Bond+Damage+Tolerance+and+Failure+Assessment&hl=en&as_sdt=0&as_vis=1&oi=scholart

[8] Qing X.P., Chan H.-L., Beard S.J., Kumar A. 'An active diagnostic system for structural health monitoring of rocket Engines'. *Journal of intelligent material systems and structures*. 2006;17(7):619–28.

[9] David-West O.S., Haritors G., Amafabia D. -aM, Montalvao D. A review of structural health monitoring techniques as applied to composite structures. SDHM structural durability and health monitoring [online]. 2017. Available from https://www.researchgate.net/publication/320729996_A_Review_of_Structural_Health_Monitoring_Techniques_as_Applied_to_Composite_Structures

[10] Moghaddame-Jafari B. *Deflection measurement through 3D point cloud analysis*. Fairfax, VA: George Mason University; 2016.

[11] Yazdani Nezhad H., Merwick F., Frizzell R.M., McCarthy C.T. 'Numerial analysis of low-velocity rigid-body impact response of composite panels'. *International Journal of Crashworthiness*. 2014;20(1):27–43.

[12] GSP *GSP 216, introduction to remote sensing*. Arcata, CA: Humboldt State University; 2017.

Chapter 19

Energy harvesting and self-sensing multi-functional polymer composites

Tasnuva Khaleque[1], Hamed Yazdani Nezhad[1], and Marcin Słoma[2]

Frequently, the technical standards for the properties of materials to be utilised in products surpass the capabilities of a single material class, such as polymers, metals or ceramics. Such limits may sometimes be circumvented by mixing two types of materials in a single product, i.e., by creating a composite. In such composites, one of the two material classes oversees supplying one set of desired features, while the other material is in charge of giving a separate set of desirable (but related) traits [1].

Because of their combination of high strength, high stiffness and low density, advanced polymer composite materials are gaining popularity for applications need high strength, stiffness and low density, such as automobile and aerospace industries [2]. While the main role of these composites has historically been to support certain loads, there is a rising tendency in research to include various secondary functions into composite structures. There is a growing need for materials that combine required functional qualities like load sensing, energy harvesting, temperature monitoring, actuation, etc. with mechanical performance [1]. Moreover, aviation safety standards and assuring structural integrity dependability are critical elements that drive the usage of such materials in aircraft structures. To meet both requirements, non-destructive inspection of the structure is performed, which is more difficult and expensive than inspecting metallic structures. As a result, new in situ health monitoring approaches must be developed [3].

In the quest to decrease maintenance costs, remote monitoring of components in systems and structures has become vital. Typically, a health monitoring module with a sensor, wireless telemetry and processing capacity is spread across complicated structures and must function remotely utilising battery power. A battery can only provide power for a certain amount of time before it must be recharged or replaced, resulting in a cost of maintenance. A wireless data transmission module, sensors

[1]Aeronautics and Aerospace Research Centre, Department of Mechanical Engineering and Aeronautics, City, University of London, London, UK
[2]Faculty of Mechatronics, Warsaw University of Technology, Warsaw, Poland

and an energy harvesting device are often included in a self-powered health monitoring module. Systems have been extensively researched to convert energy from widely accessible sources such as vibration, human motion, water, air, heat, light and chemical processes into usable electrical energy.

In this chapter, we briefly describe energy harvesting and self-sensing polymer composites.

19.1 Energy harvesting and self-sensing

Composite materials are often used in harsh environments. Their usefulness is mostly dictated by unintentional damage, ageing and corrosion. Fracture is the most hazardous flaw in composite materials because it is invariably preceded by cracking and crack growth. Designing composite components and modelling their behaviour continue to be difficult [4], and there is insufficient data for experimental structural solutions. As a result, diagnostic testing for composite structural materials is assuming a greater significance [5–7]. Removing and reinserting the test component consume time and energy that can be saved with a well-designed testing procedure that has a built-in sensor, and the problem may be foreseen and avoided. Because the sensor is integrated into the component, it can continually provide a signal during operation, and the data can be processed to conduct continuous structural health monitoring. There are pre-configured systems for structural health monitoring that use optical sensors (e.g., fibre Bragg grating sensors that provide exact temperature and deformation measurements [8, 9]). The piezoresistive effect of materials may be exploited to create strain-sensitive 'smart' materials capable of self-monitoring their own strain and internal damage condition [10]. The coupling phenomenon between a mechanical stimulus (strain or stress) and a change in electrical resistance is known as piezoresistivity [11].

Energy harvesting is the technique of extracting energy from external sources and storing it for later use in a variety of applications. To satisfy future energy needs, the development of renewable, environmentally friendly and cost-effective energy sources has become more crucial [12]. Our environment has a variety of renewable energy sources, ranging from kinetic energy to solar energy and biofuels. Kinetic energy harvesting based on three basic transduction mechanisms, electromagnetic, piezoelectric and triboelectric, has been the focus of recent study. The benefits and drawbacks of various energy collecting techniques have been extensively studied in the literature with most of the research on piezoelectric transduction has been inspired by its higher power densities, high-energy conversion efficiency, simple topologies and great scalability [13–15]. Piezoelectric materials are often employed in energy harvesters as electromechanical transducers to transform ambient vibrations into electrical energy that may be stored in batteries or utilised to power small electronic devices. The advantages of piezoelectric energy harvesting include the use of a simple structure without the need for multiple additional components, the absence of moving parts or mechanical constraints, portability, integration with other renewable energies, the absence of a need for an external voltage source,

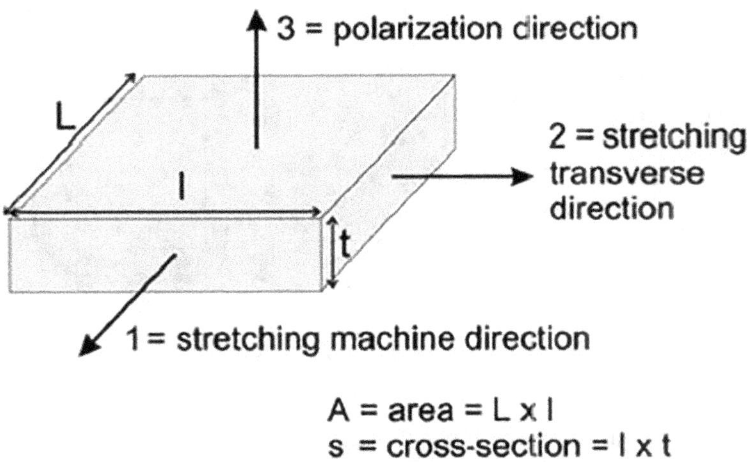

Figure 19.1 *Axis orientation of a piezoelement [30]*

compatibility with microelectromechanical systems (MEMS), ease of fabrication with microelectronic devices and a reasonable output power density. As a result, piezomaterials are a great contender for powering macro- to nanoscale electronic devices in lieu of batteries, which have a limited lifetime (piezoelectric energy harvesting). Therefore, piezoelectric materials are now widely used in a variety of sectors, including sensors [16–19], actuators [20–22], nanogenerators [23–26], MEMS devices, portable electronics [26, 27] and biomedicine [28, 29]. In this chapter, we describe the piezoelectric energy harvesting and piezoresistive self-sensing polymer composites.

19.2 Piezoelectric effect

In 1880, Pierre and Jacques Curie discovered the phenomena of piezoelectricity while performing experiments on quartz, tourmaline and Rochelle salt crystals [30]. The elastic energy of a piezoelectric material may be linked to the dielectric energy of the material. As a result, a relationship between the material's mechanical deformation and an electrical change may be formed.

Because of their unique crystallographic structure, piezoelectric materials are extremely anisotropic, meaning that their electrical characteristics change depending on the orientation of the crystals [31]. As a result, the properties observed must be oriented. The axis orientation of a piezoelement is shown in Figure 19.1.

In the figure, the axis along which the piezoelectric material is created to enhance its piezoelectricity capacity is known as the stretching machine direction (1). In the case of a piezoelectric polymer, it is also the principal orientation of the polymer chain. The axis along which the electrical dipole forms in the piezoelectric

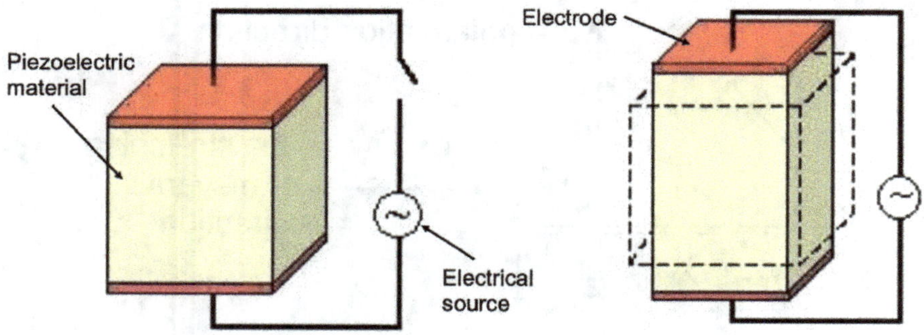

Figure 19.2 *Reverse piezoelectric effect of a piezoelectric block [31]*

material is known as the polarisation direction (3). The third direction is stretching transversely (2), which is 90° from the other two axes.

The direct and inverse piezoelectric effects are the two types of piezoelectric effects. A material is polarised and generates voltage under tensile or compressive stress in the direct piezoelectric action. The application of electric potential causes mechanical displacement in the material in the inverse effect (see Figure 19.2). When the material extends (tension) in response to an electrical stimulus, it is 'positive', and when it contracts (compression), it is 'negative' [30]. The distribution of ions in the crystalline structure of some materials causes the piezoelectric effect. There is a steady-state balance between the positive and negative electric charges in the material in the absence of an external force, therefore it stays neutral.

The direct piezoelectric effect is used in energy harvesting, where an applied force generates an electrical charge. In the case of piezoelectric materials, there are two types of electrical circuits that may be used to receive the electrical current created by the material when it is deformed: sandwich electrodes (shown in Figure 19.3) and interdigitated electrodes (IDE) (shown in Figure 19.4).

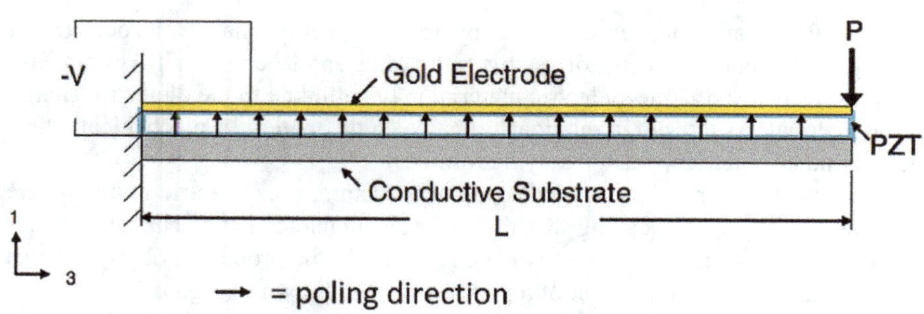

Figure 19.3 *Cross-sectional view of a uniform d_{31} beam, V is the electrical voltage, L is the length of the beam, P is the load applied, PZT is the piezoelectric material [32]*

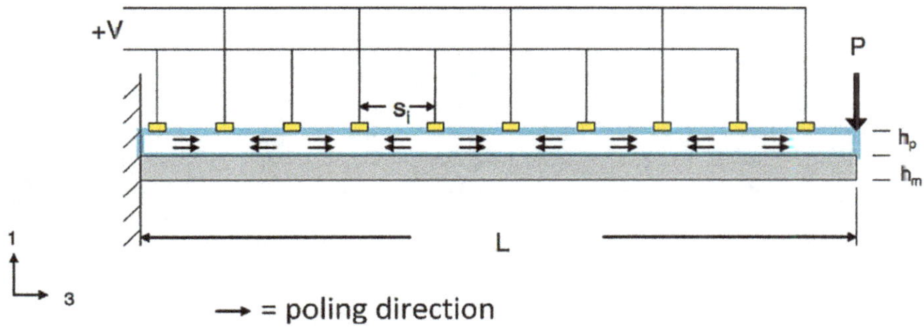

Figure 19.4 *V is the electrical voltage, L is the length of the beam, P is the load applied, h_p is the thickness of piezoelectric layer, h_m is the thickness of the neutral beam, s_i is the spacing between positive and negative electrode fingers in a cross-sectional view of an interdigitated unimorph d_{33} beam [32]*

It is necessary to pick the optimum electrode setting to optimise the sensitivity of a piezoelectric sensor. The ratio of S^E_{33}/S^E_{11} changes depending on the piezoelectric material, where S^E_{33} is the piezoelectric material elastic compliance in the d_{33} mode, and S^E_{11} is the piezoelectric material elastic compliance in the d_{11} mode. The 'sandwich' electrode (d_{31} unimorph) should be used if the ratio is larger than 1 [32].

When the piezoelectric material is thick enough to stand alone and prevent contact between the two upper and lower electrodes, which would ultimately lead to a short circuit, this d_{31} kind of electrical circuit is simple and straightforward to produce.

This d_{33} electrode type is more complicated, but it has the benefit of only standing on one side of the piezoelectric material. It may be applied to very thin or non-hermetic surfaces. This technology is especially suitable for MEMS applications [32].

The non-uniform voltage in the piezoelectric material is caused by the alternating placement of fingers from the positive and negative electrodes, as shown in Figure 19.5.

An electrical field is created between fingers when this piezoelectric material is bent, as seen in Figure 19.6.

The electrode configuration must be optimised to maximise the device's electrical output. Several settings may be altered: the distance between fingers, s, finger width, w_e, IDE width, w and the thickness of the piezoelectric material, h_p. The IDE pattern alters because of the mechanical deformation delivered to the piezoelectric material. The space between the positive and negative electrode fingers increases by lengthening the section along direction 3 (shown in Figure 19.7) The electrical resistance rises because of the IDE's dimensional shift [33]. As a result, a link between mechanical deformation and electrical output change may be formed.

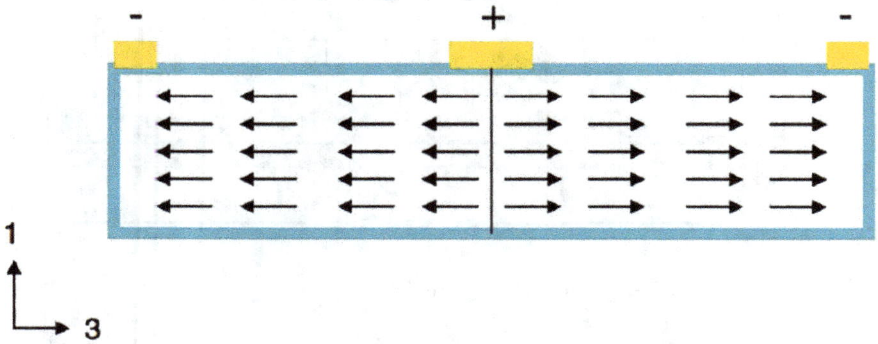

Figure 19.5 *The piezoelectric layer of the IDE d_{33} beam in magnified cross section [32]*

19.2.1 Piezoelectric materials

The use of piezoelectric nanostructured materials for energy harvesting applications has grown dramatically in the recent decade [34, 35]. In actuating and sensing applications, the piezoelectric strain and piezoelectric voltage constants are appropriate criteria. However, the electromechanical coupling factor, power density, mechanical stiffness, mechanical strength, manufacturability and quality factor are the most important factors for energy harvesting. Furthermore, the operating temperature is important in material selection [36]. Based on their structural features, piezoelectric materials are classified into four categories: ceramics, single crystals, polymers and composites. Li *et al.* [15] discussed the fundamental properties of

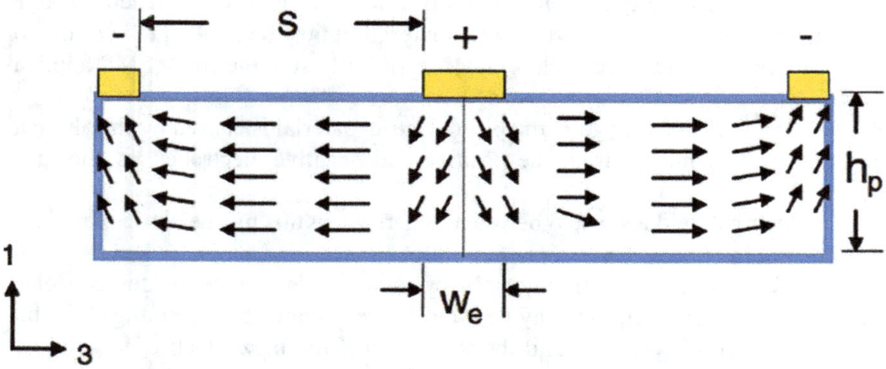

Figure 19.6 *h_p is the thickness of piezoelectric layer, s_i is the spacing between positive and negative electrode fingers, w_e is the width of an electrode finger in a cross-sectional view of a non-uniform electric field within the piezoelectric layer under mechanical deformation [32]*

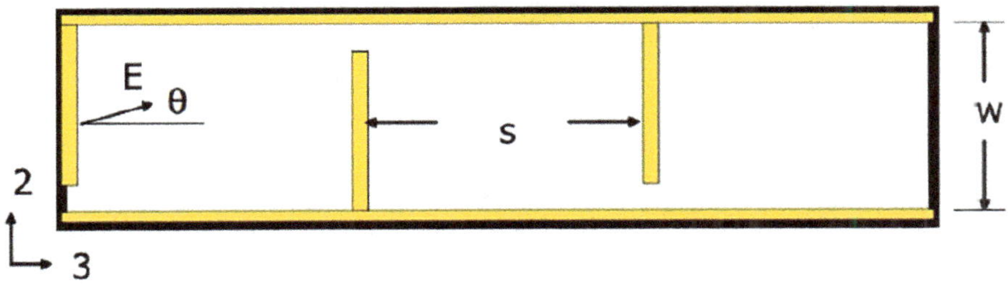

Figure 19.7 *Top view of IDE layout, w is the width of the IDE layout, s is the spacing between positive and negative electrode fingers, E and q show the orientation of the IDE layout with the orientation of the piezoelectric fibres [32]*

these four piezomaterial categories and compared several of the most promising candidate materials in terms of piezoelectric strain constant, piezoelectric voltage constant, electromechanical coupling factor, mechanical quality factor and dielectric constant. Due to the high polarisations present in their crystalline formations, piezoelectric ceramics and single crystals exhibit much superior piezoelectric characteristics versus piezoelectric polymers. In comparison with piezoelectric polymers, piezoelectric ceramics and single crystals are stiffer and brittle. Both piezoelectric and mechanical qualities play a role in determining which piezoelectric material is best suited for a particular piezoelectric harvesting application. Other critical characteristics to consider when choosing appropriate materials include the frequency of application, the accessible volume and the way mechanical energy is delivered into the system. To maximise energy harvesting, the piezoelectric energy harvester should be operated at its resonance frequency. However, in many circumstances, such as low-frequency applications, it is impossible to match the resonance frequency of piezoelectric to the input frequency of the host structure. For low-frequency applications in non-resonant situations, the piezoelectric element may be approximated by a parallel plate capacitor, and the product of the piezoelectric strain constant and piezoelectric voltage constant should be large to collect more electric energy. For near-resonance circumstances, on the other hand, the optimal output power of the harvester is independent of the piezoelectric characteristics of the piezoelement, while the maximum output voltage is dependent on the piezoelectric strain constant. Obviously, the selection of an appropriate piezomaterial for a piezoharvester is conditional on the working circumstances, which complicates the process of selecting a piezomaterial. Natural crystals such as quartz and synthetic ceramics such as PZT (lead zirconate titanate) as well as polymers such as PVDF (polyvinylidene difluoride) all exhibit the piezoelectric effect. Zinc oxide (ZnO) nanostructures form readily at low temperatures and are crystallographically aligned; hence, the bulk of the research has been done on this material [37]. Other materials with strong piezoelectric coefficients, such as PZT, barium titanate ($BaTiO_3$) and lead magnesium niobate–lead titanate, have recently been investigated

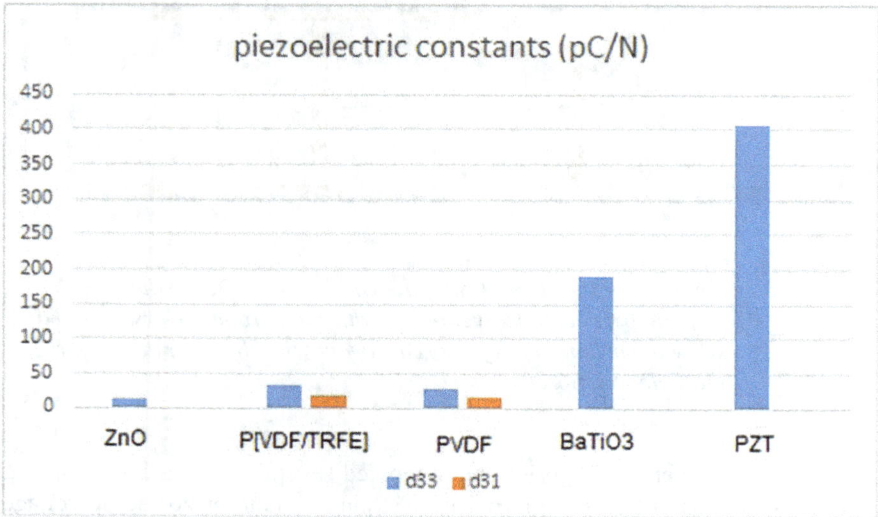

Figure 19.8 Piezoelectric coefficient range for some piezoelectric materials [38]

for nanostructured energy harvesters. Narita and Fox [38] classified piezoelectric materials depending on their piezoelectric coefficients (see Figure 19.8).

Polymers are carbon-based materials that, due to their molecular structure and orientation, have a piezoelectric effect. They are substantially softer than ceramics, and their strain coefficients (d_{33}) and voltage coefficients (g_{33}) are moderate. Their unique characteristics, including as design flexibility, low density and ease of processing, make them suitable for a variety of energy harvesting applications. Microscopic crystals are randomly scattered within the amorphous bulk of semi-crystalline polymers. PVDF, fluoride trifluoroethylene P(VDF-TrFE) [39], liquid crystal polymers, polyamides [40], cellulose and its derivatives [41], etc. are among them. When dipoles are present in the molecular structure of amorphous or non-crystalline polymers, piezoelectricity is found. At temperatures above the polymer's glass transition temperature (T_g), the dipoles are aligned via poling. Polyimide, nylons, polyurea, polyurethanes and other polymers fall within this category.

19.2.1.1 Polyvinylidene fluoride

The fluoropolymer thermoplastic PVDF is made by the polymerisation of vinylidene difluoride (VDF) [31] (see Figure 19.9). PVDF is chemically stable, preventing chemical reactions when it is put into a system, and it has excellent oxidation and UV radiation resistance [43]. The electroactive polymer family includes this naturally piezoelectric polymer [32].

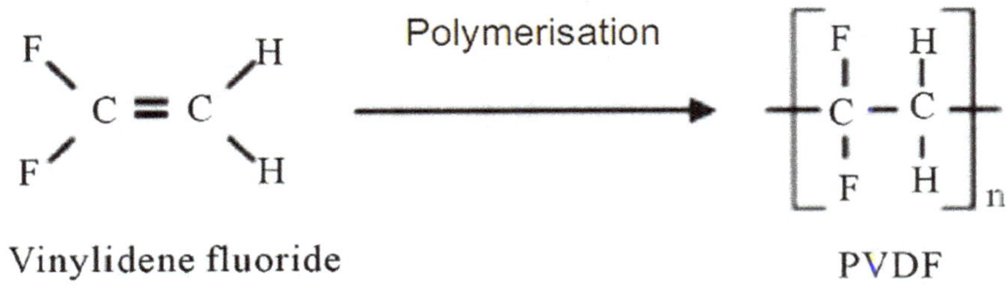

Figure 19.9 Vinylidene fluoride polimerisation [42]

PVDF has a polymorphic semi-crystalline structure with five phases α, β, γ, δ and ε. The first three stages, as shown in Figure 19.10, are the most often employed [44].

The α-phase features a monoclinic unit cell with an alternating trans-gauche–trans-gauche (TG+ TG−) conformation (a = 4.96 Å, b = 9.64 Å, c = 4.62 Å, β = 90°) [45]. Out of the five phases, the α-crystal phase is the most thermodynamically stable [44]. The β-phase exhibits an all-trans (TTT) planar zigzag conformation with an orthorhombic unit cell (a = 8.58 Å, b = 4.91 Å, c = 2.56 Å) and excellent electrical characteristics [44, 45]. The γ-phase features an orthorhombic unit cell and is T3G+ T3G− [46]. The polar and anti-polar equivalents of the α- and γ-phases are polar and anti-polar of δ and ε, respectively [46]. The centroid of the negative charges differs from the centroid of the positive charges in a polar molecule.

Kawai discovered the piezoelectric property of PVDF in 1969 after seeing that it had much higher d_{31} and g_{31} coefficients than other polymers [47]. The electrical charge density (C/m^2), which happens when a mechanical stress of 1 N/m^2 is applied, is used

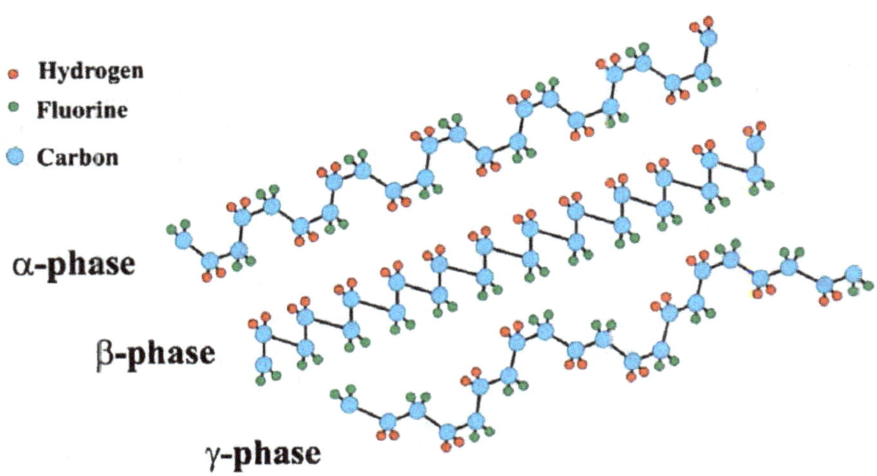

Figure 19.10 PVDF most used configurations [31]

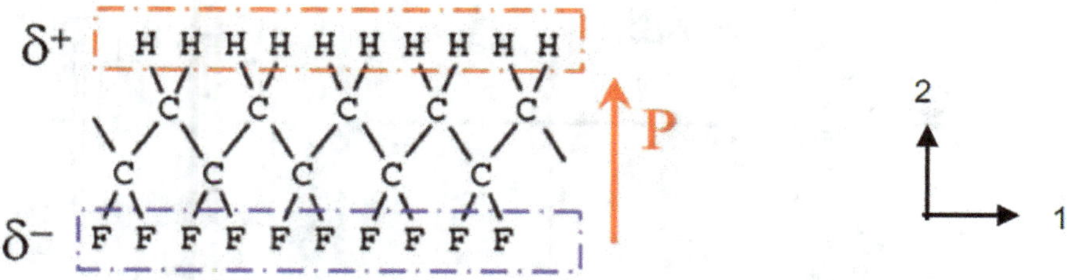

Figure 19.11 β-Phase conformation and its electric dipole moment [52]

to calculate the 'd' coefficient. The value is then expressed as a C/N ratio [48]. When a force of 1 Pa is applied along the axis 1, the electrical field g_{31} (in V·m/N) is formed. $g_{31} = d_{31}/4$ is the relationship between the g and d coefficients. PVDF materials have a g_{33} value of 200–300 mV·m/N [48]. Bergman *et al.* found the pyroelectricity of PVDF just 2 years later [49].

The PVDF polymer chain is made up of two sorts of groups: the CF_2 fluorine group (which is electrically negative) and the CH_2 hydrogen group, which is electrically positive. An electrical dipole moment is formed (5–8 × 10^{-30} C·m) depending on the location of these groups along the chain. As a result, its use is suitable for sensors and actuators [50]. Because the α-phase is non-polar, it lacks electrically interesting features for piezoelectric applications. However, both β- and γ-phases are polar. As a result, it is necessary to regulate the material quality utilised in a sensor or actuator application to shift the fraction of α-phase to γ- and β-phases, which has the largest residual polarisation (8 × 10^{-30} C·m) [50]. The PVDF β-phase exhibits exceptional electrical characteristics because of the presence of all trans C–C bonds: alternating CH_2 and CF_2 groups [51], as seen in Figure 19.11.

According to Figure 19.1, the piezoelectric effect of PVDF is larger throughout the length of the fibres (the polymer chain axis, axis 1) than along the cross section (axis 2) [31]. For piezoelectric, pyroelectric and ferroelectric applications, this β-crystal form is highly regarded. Despite the fact that the γ-structure has less electrical properties than the β-structure, its polar characteristics are still interesting in comparison with other polymers and have the advantage of a higher curing temperature, which makes the γ-structure more resistant to harsh environmental conditions [51]. As mentioned before, it is needed to employ PVDF material in a β-phase conformation or at the very least a γ-phase for sensor or actuator applications. After synthesis, the PVDF powder is mostly in the α-conformation; hence, a transformation is necessary to enhance the fraction of β-phase. For example, this α-to-β transition may be affected by mechanical stretching, solution growth, metal salt or nanoparticle (NP) addition, melt crystallisation under high pressure or application of a strong electrical field [44].

Thus, the electrospinning technique, which will be discussed in detail later, is often employed to manufacture PVDF nanofibres. Figure 19.12 illustrates the links between the α, β and γ structures, as well as the processes discussed above.

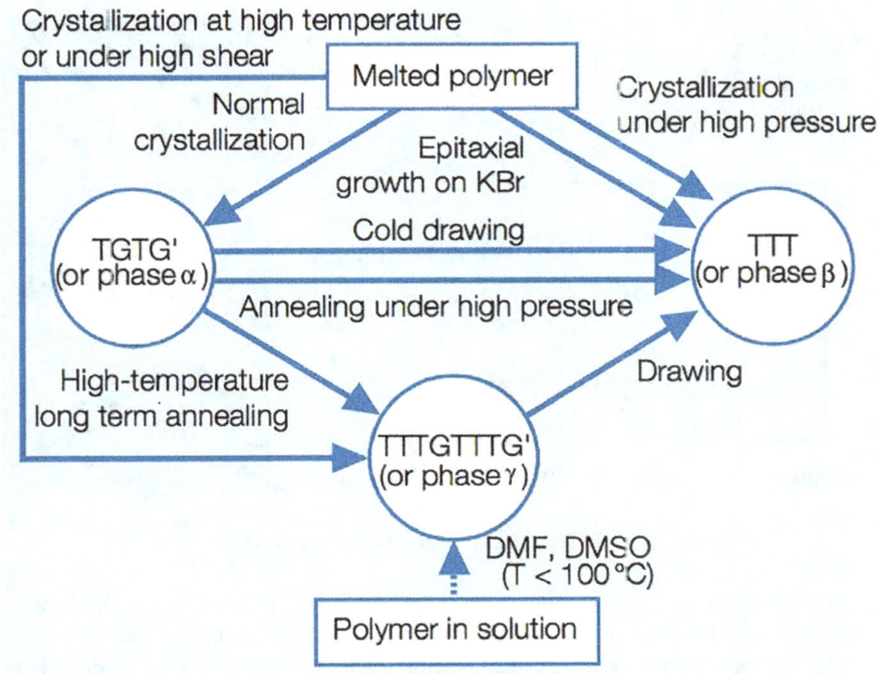

Figure 19.12 Phase transformation and processes of PVDF [53]

Rayleigh studied the electrospinning method in 1882 and Formhals patented it in 1934. Taylor (1964) [54] named the electrospinning cone after his studies on the deformation of the droplet [55]. The fundamental concept is shown in Figure 19.13.

Both molten PVDF material and powdered PVDF material in a gel-like solvent may be employed. This selected raw material is inserted in a syringe. The material is gradually forced into the rotating nozzle tip, where it first forms a droplet.

The droplet is directed and stretched to a nanoscale by the application of a strong electrical field between the nozzle tip and the collection. This distortion influences the droplet's surface energy. The process is continuous, which implies that after the first phase, no suitable droplet can form, while a Taylor cone is continually supplied with material. Additionally, the collector's form and movement dictate the ultimate quality of the electrospun material (e.g., mat or oriented fibres) [57]. Depending on the material and desired outcome, the following parameters may be modified in particular [55]:

- Viscosity, conductivity, molecular weight, surface tension of the solution, etc.
- Procedural information: electric field, distance, rate, etc.
- Ambient: relative humidity, temperature, etc.

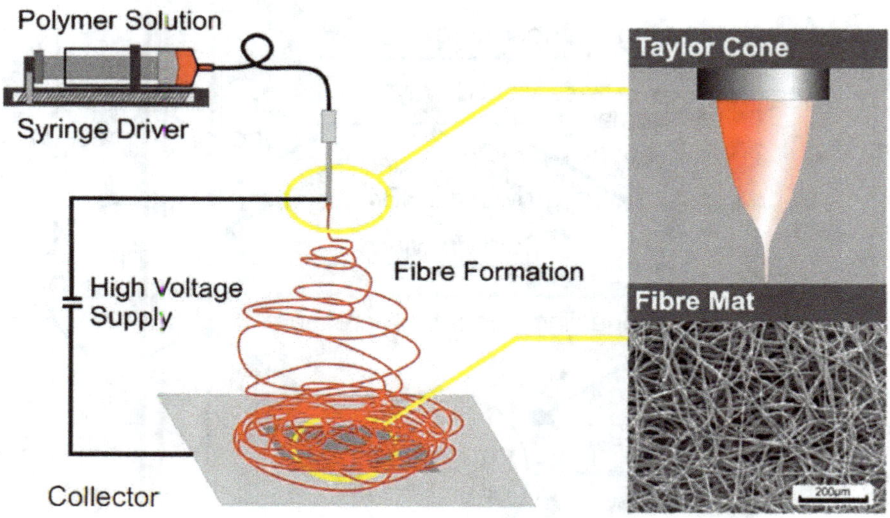

Figure 19.13 Electrospinning process principle [56]

According to research conducted on stretched PVDF films, the stretching ratio seems to have a greater effect on phase change than the stretching temperature [58].

The Taylor cone generated during the electrospinning process enables deformation of the bulk material up to the necking stage, at which the crystal transition (from α- to β-phase) begins [59, 60].

Indeed, when subjected to strain, the polymer chains in the α-phase gradually deform and arrange themselves to create the β-phase, as shown in Figure 19.14 [32].

19.2.1.2 P(VDF-TrFE) copolymers

P(VDF-TrFE) is a semi-crystalline thermoplastic copolymer in which VDF copolymerises with TrFE in any amount [61, 62] (see Figure 19.15a). TrFE incorporation into PVDF at any concentration results in an α- to β-phase transition and a conformational shift from trans-gauche to all-trans [64, 65]. P(VDF-TrFE) exhibits piezoelectric capabilities below its Curie temperature because of the orientation of crystalline phase under a strong poling field. It possesses strong piezoelectric coefficients (d_{33} = −30 to −40 pC N^{-1} and d_{31} = 25 pC N^{-1}) and has been investigated for use in a variety of energy harvesting applications [66].

Pi *et al.* measured a V_{oc} of 7 V, an I_{sc} of 58 nA and a j of 0.56 μA cm^{-2} from a piezoelectric nanogenerator (PENG) utilising a spin-coated P(VDF-TrFE) (75/25 mol/cent) film [67]. The effect of strain rate change on the electrical output of PENG was investigated, and an analytical model was developed that corresponded well with the experimental data. Bae and Chang demonstrated that a series of 'annealing, and cooling-pressing-poling' treatments resulted in a maximum F(β) of 66.33% and a voltage of 1.10 V for P(VDF-TrFE) films [68]. Oliveira *et al.* demonstrated a 19% increase in crystallinity after 48 hours of annealing at 130°C [69].

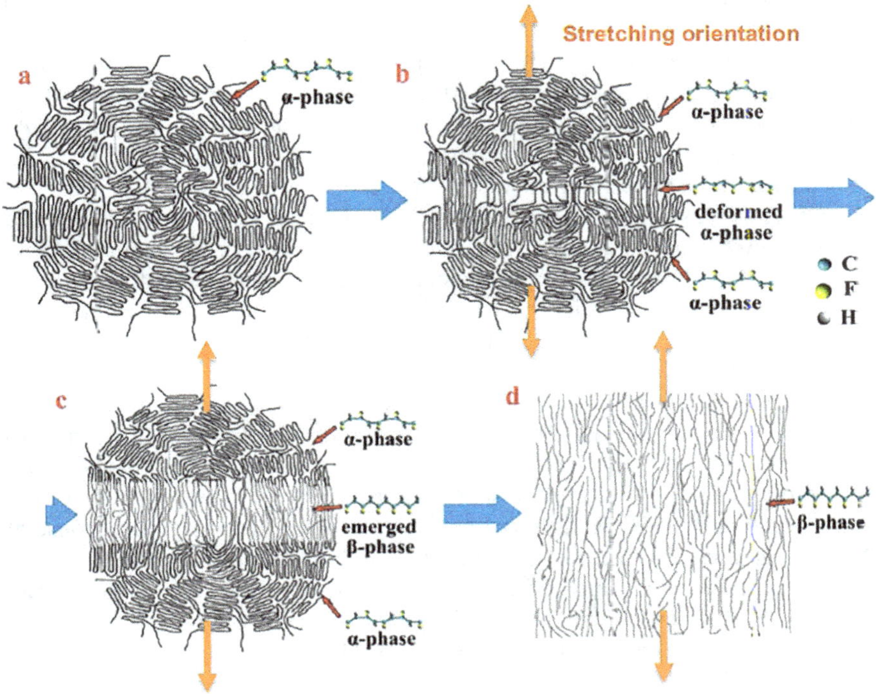

Figure 19.14 Phase change with stretching of PVDF material [31]

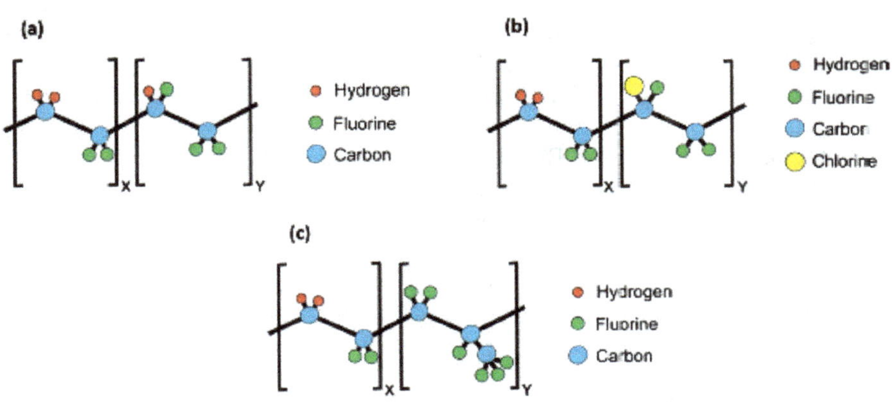

Figure 19.15 Schematic representation of (a) P(VDF-TrFE), (b) P(VDF-HFP) and (c) P(VDF-CTFE) repeat units [63]

After annealing, the shift from α- to β-phase occurred, resulting in a F(β) of 94% and an increase in d_{33} from 13 to 18 pC N^{-1}.

Lee *et al.* created pyramid- and trigonal line-shaped morphologies on P(VDF-TrFE) by micropatterning [70]. Under vertical compressive stress, the micropatterned P(VDF-TrFE)-based PENG produced a greater output (4.4 V and 3.4 μA) than the flat film PENG (1 V and 0.1 μA). The increased output was caused by the increased strain created in micropatterned films when subjected to the same force. Cauda *et al.* synthesised PVDF and P(VDF-TrFE) nanowires in a 70/30 ratio via template wetting into 200 nm sized channels of anodic porous alumina (APA) membranes [71]. Nanoconfinement crystallised the β-phase in both polymers, while the reference PVDF thin film crystallised in the non-polar α-phase. P(VDF-TrFE)- and PVDF-templated nanowires had highest d_{33} values of 8.2 and 6.5 pC N^{-1}, respectively. The non-poled PVDF film had a d_{33} value of 15 pC N^{-1}, but the poled and bulk PVDF and P(VDF-TrFE) films had d_{33} values ranging from 20 to 30 pC N^{-1}. Similarly, Lutkenhaus *et al.* observed a considerable shift in the crystallisation behaviour of P(VDF-TrFE) when it was restricted inside a 40 nm diameter APA template [72]. β-Phase crystallisation was accelerated while α-phase development was inhibited. This results in an enhancement in the piezoelectric response of P(VDF-TrFE) due to the directed crystallisation of ferroelectric domains. Bhavanasi *et al.* demonstrated that nanoconfinement effect boosted the piezoelectric characteristics of P(VDF-TrFE) nanotubes [73]. Poled 200-nm nanotubes with a 40% decreased poling field had a higher d_{33} value of 44 pC N^{-1} than completely poled films with a value of 20 pC N^{-1}. Under compressive pressure of 0.075 MPa at a frequency of 1 Hz, the P(VDF-TrFE) nanotubes outperformed bulk films (voltage = 4.8 V, power = 2.2 W cm^2). Chen *et al.* synthesised P(VDF-TrFE) (75/25) films with programmable microporous architectures and a maximum d_{33} value of 25.4 pC N^{-1} [74]. Under 0.5-n periodic bending stress, porous architectures boosted power production fivefold as compared with solid films. Aliane *et al.* demonstrated that with a 72.2/27.8 mol/cent P(VDF-TrFE) composition, a greater β-phase was achieved with a shorter polymer chain length and a bigger grain size [75]. Additionally, by lowering the molecular weight of P(VDF-TrFE) to 470 kg mol^{-1}, the d_{33} value was enhanced to 58 pC N^{-1}. This study established that the composition of TrFE and the crystallisation state of P(VDF-TrFE) have a significant effect on the piezoelectric properties.

Additionally, P(VDF-TrFE) fibre nanogenerators have shown potential. Electrospun P(VDF-TrFE) (70/30) nanofibre-based piezoelectric energy harvesters (PEH) yielded a maximum output voltage of 1 V and a power output of 0.6 nW [76]. Persano *et al.* reported ultrahigh sensitivity for monitoring pressures as low as 0.1 Pa using P(VDF-TrFE) nanofibres (75/25 weight/cent). Under bending circumstances, the PENG demonstrated a voltage of 1.5 V and currents up to 40 nA [77].

19.2.1.3 P(VDF-CTFE) copolymers

P(VDF-CTFE) is formed by the insertion of chlorotrifluoroethylene (CTFE) into PVDF chains ($[(CH_2CF_2)\times(CF_2CFCl)\gamma]n$) [78, 79] (Figure 19.15b). P(VDF-CTFE) copolymers containing minimal VDF are semi-crystalline, but those containing 25–70 mol% VDF are amorphous [80]. At higher VDF concentrations, thermoplastic copolymers with a melting point more than 70% are formed, referred to as flexible PVDF [81]. This copolymer has a greater strain response than PVDF but needs a stronger electric field to operate. When poled at a 70 MV m^{-1} electric field, the copolymer displays a more than 5% electrostrictive strain response and a d_{33} value of 140 pC N^{-1} [78].

19.2.1.4 P(VDF-HFP) copolymers

When hexafluoropropylene (HFP) is added to PVDF, a P(VDF-HFP) copolymer with varying characteristics is formed (Figure 19.15c). At low HFP concentrations of 15–19 mol/cent, the copolymer displays thermoplastic capabilities; at higher HFP concentrations (more than 19–20 mol/cent), the copolymer exhibits elastomeric qualities [82]. For uniaxially drawn P(VDF-HFP) films, a high d_{31} value of 30 pC N^{-1} has been reported [83].

Another research showed a d_{31} value of 43.1 pC N^{-1} and a k_{31} value of 0.187 for P(VDF-HFP) at a concentration of 10% [84]. Extruded P(VDF-HFP) films with simultaneous stretching and poling demonstrated a maximum d_{33} value of 24 pC N^{-1} [85].

19.3 Energy harvesting polymer composites

A filler with a high piezoelectricity is impregnated in a polymer matrix in piezoelectric polymer composites. This allows the benefits of both materials to be combined, such as the high piezoelectric constant and coupling factor of the fillers and the flexibility of the polymers. Fillers have an influence on the overall performance of the piezoelectric polymer composites.

19.3.1 Energy harvesting fibre-reinforced polymer composites

Sensors linked to the structure are commonly used for in situ damage detection of fibre-reinforced composites. Geller *et al.* investigated the process integration of piezoelectric sensors in the form of short fibres inside glass fibre–reinforced polymer (GFRP) composites as an example of endeavour to include piezoelectric materials into advanced composites [86]. Weder *et al.* used a multi-fibre injection spray technique to automate the fabrication of intelligent composite components with tiny piezoelectric fibre components implanted in a GFRP composite, with the potential for use in structural health monitoring [87]. Bowen *et al.* showed the possibility of integrating piezoelectric fibres in a carbon fibre–reinforced polymer (CFRP) for actuation or sensing; however, processing issues arose due to the embedding of individual components and electrodes rather than a stand-alone device [88]. Malakooti *et al.* proposed a new method based on the development of piezoelectric ZnO nanowires on aramid fibres [89]. The embedding of entire piezoelectric devices has also been suggested to ease the production process. Hufenbach *et al.* investigated the development of thermoplastic compatible piezoelectric modules for use in fibre-reinforced composites [90]. Yang *et al.* also produced a smart layer module with polyimide polymer layers and a piezoelectric, which solved the issues of decreasing electrical short circuits and enhancing the ease of electrical connection to the device with the modular system [91]. By embedding the system in a GFRP composite, the mechanical characteristics and potential for vibration suppression were investigated. Tang *et al.* used off-the-shelf piezoelectric transducers implanted in GFRP to generate Lamb waves for damage detection and structural health monitoring [92]. The electrically insulating property of glass fibres, as opposed to more conductive carbon fibres, was one motivation for embedding in GFRP. The effectiveness of embedded piezoceramic transducers for use as Lamb wave generators was explored via embedding dense piezoelectrics inside a CFRP composite to enhance the efficiency of the process. Konka used fibre-based piezoelectric transducers and incorporated macro-fibre composites (MFCs) in a GFRP with no discernible effect on fatigue parameters [93]. Finally, Hu *et al.* investigated the possibility of embedding sub-surface piezoelectric sensors into composites to recognise impact loads and calculate impact force history, though the technique of embedding the components was not discussed in detail [94].

The use of coupled bend–twist oscillations in the vibration analysis of aircraft wings and spinning turbine blades has attracted the attention of the aeronautical engineering community for decades [95]. By exploiting coupled oscillations that create favourable strain mode distributions, the energy harvesting capabilities of resonant harvesting devices, such as piezoelectric cantilever beams, can be enhanced. Xie *et al.* [95] demonstrated for the first time the utilisation of a laminated CFRP to induce linked bending–twisting oscillations in cantilever beams for energy harvesting applications. To fully utilise the strain distribution throughout the beam, piezoelectric layers operating in bending and shear modes are mounted to the bend–twist-coupled beam surface at places of highest bending and torsional stresses in the first mode of vibration. The cantilever beam employed in this study was constructed entirely of unidirectional CFRP plies (Cytec MTM 49–3/T800HB-140–36%). Plies were placed down manually in the stacking process, with vacuum debulking performed after every fourth ply to eliminate air gaps. After laying all plies, the laminate was cured for 2 hours in an autoclave at 125°C and 7 bar pressure using a vacuum bag moulding procedure. To create smooth outside surfaces, a stiff aluminium plate was put on top of the laminate, and a flexible 3-mm thick silicone rubber sheet was employed at the bottom and top. Figure 19.16 illustrates the cantilever beam schematically, along with its size and the location of the piezoelectric patches.

The diversity of bend and torsional modes of the harvester may be used to construct a harvester that works over a broader frequency range, and such device topologies give a unique method for tuning the resonant frequency response of the harvesting system [95]. The suggested concept has the benefit of being able to be modified to provide a broader band energy harvester by altering natural frequencies. Additionally, torsional modes often have lower displacement amplitudes than bending modes, allowing for the possibility of reduced volume device topologies. By utilising a laminate architecture that inherently exhibits bend–twist coupling, the complexity of the system is reduced because no eccentric proof mass or asymmetric tip mass is required to achieve torsion, and the numerous possible bend–twist

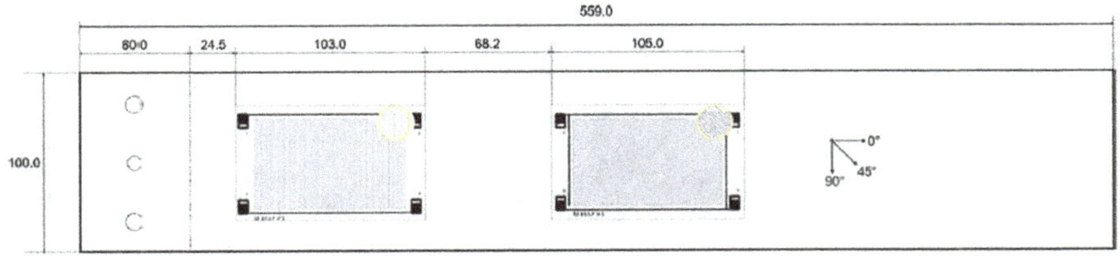

Figure 19.16 Schematic of the cantilever beam and position of piezopatches [95]

coupling lay-ups provide a large design space for optimising the frequency response of the cantilever.

For complicated component geometries or in hard-to-reach areas, this might make periodic inspection problematic. To address these issues, Yan *et al.* [96] investigated the utilisation of piezoelectric materials incorporated in CFRP composites in the form of MFC (see Figure 19.17). Damage detection, sensing and energy harvesting are all possible with such a multi-material system in place. The piezoelectric devices are implanted between the carbon fibre pre-preg and heat treated at high temperatures in this study, allowing for total piezoelectric integration into the structure. They examined the effect of processing temperature on the characteristics of MFC, with a focus on the Curie temperature of the embedded ferroelectric. They also determined the influence of the piezoelectric on tensile strength, the mechanical characteristics of CFRP composites. The ability of embedded piezoelectric devices

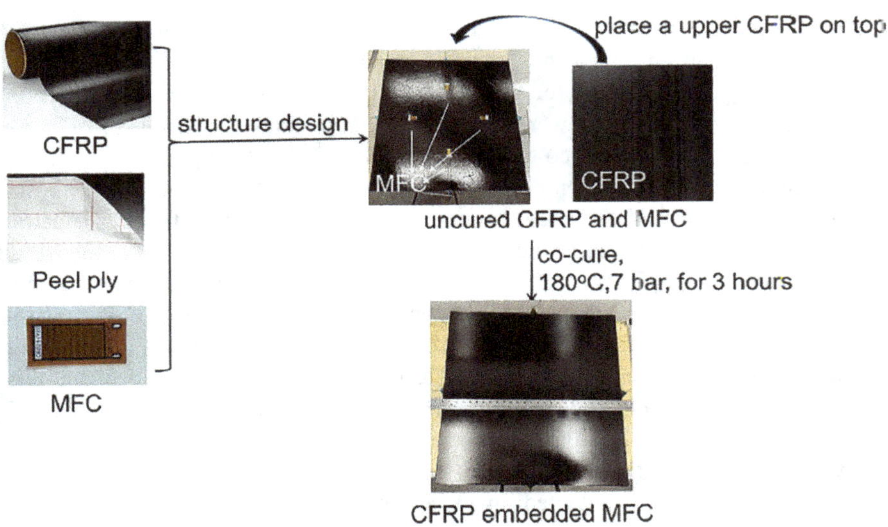

Figure 19.17 CFRP processing with embedded piezoelecrric for a multi-functional composite [96]

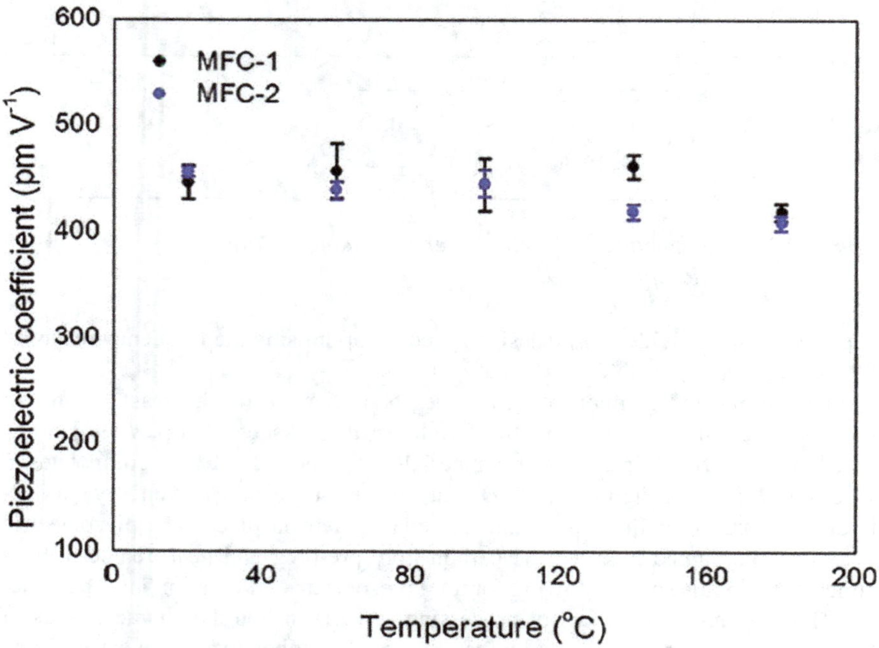

Figure 19.18 *Variation of longitudinal piezoelectric coefficient (d_{33}) of MFC with temperature [96]*

to send and receive ultrasonic signals, as well as the possibility of harvesting power from mechanical strain for self-powered systems was assessed.

Yan *et al.* reported that for a period of 1 hour, no significant changes in piezoelectric coefficient (d_{33}) were found with temperature range 20–180°C (see Figure 19.18), suggesting that co-curing MFCs with CFRP at temperatures up to the target composite processing temperature of 180°C is possible. Figure 19.18 shows that curing the MFC at 180°C does not have a substantial impact on the piezoelectric d_{33} coefficient. This is owing to the Curie temperature (T_c) of PZT-5A of the MFC being relatively high (T_c ~335°C) and the residual polarisation of the material not being lowered during processing [88]. In general, working temperature of a piezoelectric material is restricted to around one-half of its T_c (in °C) [97]. To prevent depolarisation, a curing temperature of 180°C (~0.53 T_c) for 1 hour is adequate. It is worth noting that piezoelectric devices based on a lower Curie temperature PZT-5H (Tc ~215°C) [14] were totally depolarised (d_{33} ~0 pC/N) after being exposed to 180°C (~0.84 T_c) over the same time.

Alsaadi *et al.* [98] conducted an experimental and computational investigation that presents a novel method for harvesting energy from ambient vibrations. Using the co-curing process, a piezoelectric energy harvester was incorporated onto a CFRP laminate structure (see Figure 19.19). The integrated composite with the

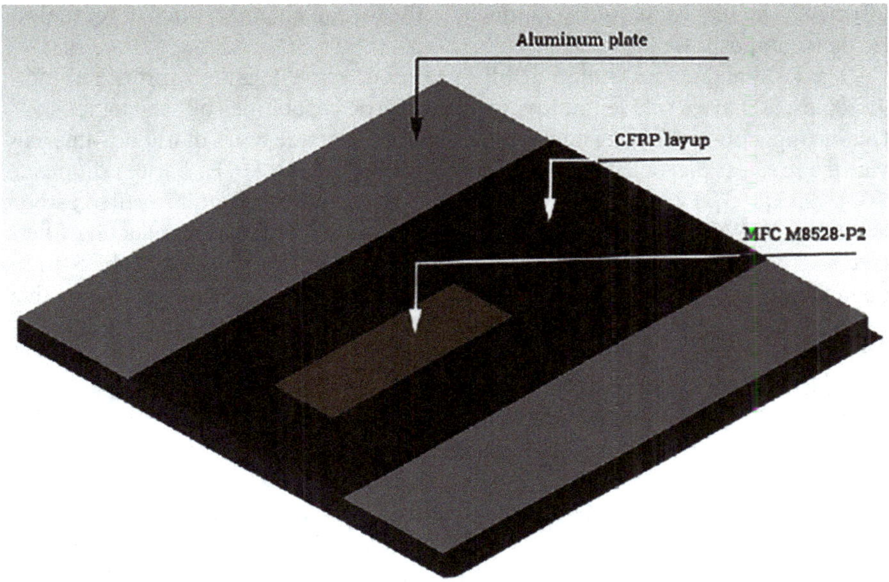

Figure 19.19 Integration process of MFC onto a CFRP composite [98]

energy harvester was lightweight, flexible and produced consistent and dependable energy results.

Shi *et al.* [99] investigated an integrated smart composite made up of an MFC and a carbon/epoxy pre-preg co-cured MFC that can harvest ambient vibrational energy. The resulting vibration energy harvesting-enabled carbon fibre plate produced a normalised power density. The comparatively high-power output of the thin profile harvester enables the integration of micro-electronics and sensors onto the composite stack.

Wearable Internet of Things devices for sports necessitate an innovative design of a self-powered technology capable of wirelessly transmitting and receiving signal (force, ambient condition) from a body wearing item, relying on vibration/impact energy harvesting materials that are lightweight but have excellent resistance to intense exercise [100]. Narita *et al.* [100] developed piezoelectric hybrid CFRP composite laminates by implanting a potassium sodium niobate (KNN) NP-filled epoxy interlayer, and the composite samples are effectively polarised. The output voltage of the samples was monitored as a function of the impact load. The findings indicate the generation of a voltage greater than 50 mV.

19.3.2 Energy harvesting polymer nanocomposites

A high piezoelectricity filler is injected in a polymer matrix in piezoelectric polymer composites. This enables the benefits of both materials to be combined, namely the high piezoelectric constant and coupling factor of fillers and the flexibility of polymers. The inclusion of fillers influences the overall characteristics of piezoelectric nanocomposites.

The kind and concentration of fillers utilised, as well as their forms and morphologies, all have a role in improving piezoelectric capabilities of nanocomposites. Conducting, non-conducting and hybrid fillers are the three types of fillers. Materials with the greatest piezoelectric coefficients include $BaTiO_3$ [101], barium dititanate [102], ZnO [103], PZT [104], KNN [105], etc. Graphene [106], multi-walled carbon nanotubes (MWCNTs) [107], carbon black (CB) [108] and other conducting fillers have also been employed extensively. At the nanometre/micron scale, hybrid fillers are made up of two or more components, one of which is conducting and the other non-conducting. $NaNbO_3$-reduced graphene oxide (rGO), titanium oxide (TiO_2)-rGO nanotubes, manganese oxide (MnO_2)/graphene/MWCNT hybrid and MnO_2/graphene/MWCNT hybrid are all examples of hybrid fillers [109].

Ahn *et al.* [46] investigated the influence of altering MWCNT concentration (0.05–1 wt%) on the piezoelectric characteristics of PVDF/MWCNT nanocomposites, as well as the use of drawing and poling [46]. Because of overstretching the PVDF jets, the electrospun nanocomposites displayed a high phase amount. During the crystallisation process, the 0.2 wt% long MWCNTs functioned as nuclei, and poling caused charge aggregation at the PVDF/MWCNT [110] interfaces. As illustrated in Figure 19.19, both mechanisms facilitated phase changeover [111]. External stresses were applied by drawing, which led in interactions between the functional groups contained in MWCNTs and the CF_2 dipoles of PVDF, increasing the β-phase content (Figure 19.20). Kim *et al.* [112] conducted a similar work, finding increased phase content in PVDF/MWCNT composites after drawing and poling when MWCNT concentration was less than 0.2 wt%.

PVDF/MWCNT composite nanofibres produced by near-field electrospinning were examined by Liu *et al.* [113]. Compared with a 600 V mm^{-1} electric field, a greater electric field of 1,200 V mm^{-1} enhanced the crystallinity of PVDF fibres by

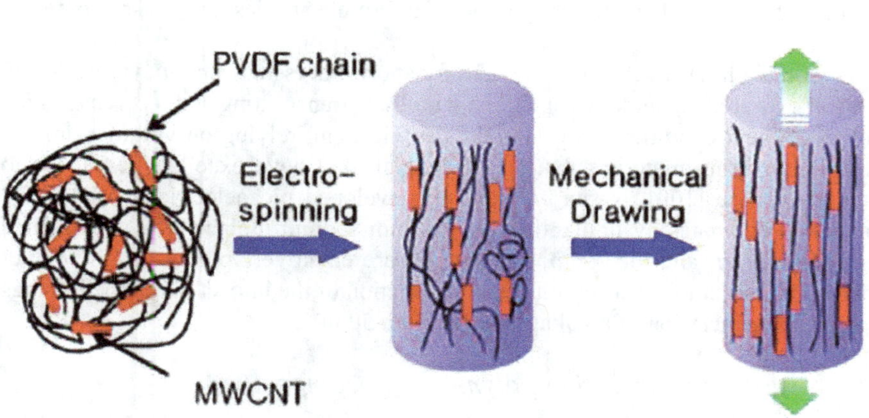

Figure 19.20 The mechanism of chain extension in PVDF induced by electrospinning and mechanical drawing [46]

nearly 53%. PVDF/MWCNT 0.03 wt% nanofibres had a d_{33} value of 57.6 pC N^{-1} [114], compared with 15 pC N^{-1} for PVDF thin films [115]. Yu *et al.* developed an electrospun PVDF/MWCNTs-based PENG that did not need poling [116]. At an ideal MWCNT concentration of 5 wt%, the output voltage produced by the piezoelectric nanogenerator was 6 V, which is 200% greater than plain PVDF nanofibres. The explanation for this is because the inclusion of MWCNTs increased the surface conductivity of the PVDF nanofibres.

Wu *et al.* investigated the effect of carbon nanotubes (CNTs) on the piezoelectric performance of aligned PVDF/CNT nanofibres and adjusted electrospinning settings [110]. β was 88%, and d_{33} was 27.4 pC N^{-1} in electrospun PVDF fibres; 18 wt% CNT loading boosted β to 89% and d_{33} to 31.3 pC N^{-1}. In bending testing, a non-linear trend in output voltage was seen in response to bending angle, with the greatest value of 1.89 V at a 100° bent. Ning *et al.* made PVDF/MWCNT nanocomposites (0–0.3 wt%) by solution casting, then drawing to 400–500% elongation and stepwise poling to produce β-phase [107]. The output voltage of the nanocomposite film containing 0.05 wt% MWCNT was double that of the pure PVDF film.

Melt mixing was used by Ke *et al.* to make PVDF nanocomposites with various surface functionalised MWCNTs [55]. MWCNTs measured 1 μm in length and had a 9.5 nm outer diameter. Non-functionalised CNTs were dubbed u-CNT, a-CNT, c-CNT and h-CNT, whereas those containing amino, carboxyl and hydroxyl groups were dubbed a-CNT, c-CNT and h-CNT. These four kinds of CNTs were disseminated in PVDF at 1.5, 3 and 5 wt% loadings, respectively. a-CNT nanocomposites had the highest β-phase (17.4%), followed by h-CNTs (11.6%), u-CNTs (9.4%) and finally c-CNT nanocomposites (4.7%). The interactions between the impacts of MWCNT dispersion in PVDF and their interactions resulted in the creation of β-phase. Excessive melt shearing, as well as chemical bonding between moieties on the surface of CNTs and PVDF, generated these interactions, as seen in Figure 19.21a. Because F_2 atoms had a higher electronegativity than C and H_2 atoms, the $-CF_2$ dipoles readily connected with the CNT interface, which has a lot of electrons, and created PVDF chains with TT conformation. PVDF dispersion of the u-CNTs, a-CNTs and h-CNTs was excellent, resulting in PVDF adsorption along the CNT surface (Figure 19.21b).

Sharma *et al.* found that both carboxyl functionalised MWCNTs and Ag-decorated CNTs aided the transition to phase in PVDF [118]. The synthesis process had a significant impact on β-phase production, with electrospun fibres having a greater β-phase than melt-mixed samples. This was attributed to electrospinning's high electrostatic field, which aligned the PVDF dipoles in one direction. The d_{33} value of Ag-CNTs/PVDF fibres was 54 pC N^{-1}, compared with 35 pC N^{-1} for PVDF/CNT fibres and 30 pC N^{-1} for clean PVDF.

With the inclusion of different weight percentages of CNTs and the surfactant Triton X-100, Dognan *et al.* [10] evaluate the efficacy of CNTs/epoxy nanocomposites for strain sensing through electrical resistance measurement. Using injection floating catalyst chemical vapour deposition (ICCVD), MWCNTs were synthesised that were sufficiently conductive to be used in the construction of CNTs/epoxy nanocomposites that could detect tensile deformation by electrical resistance change

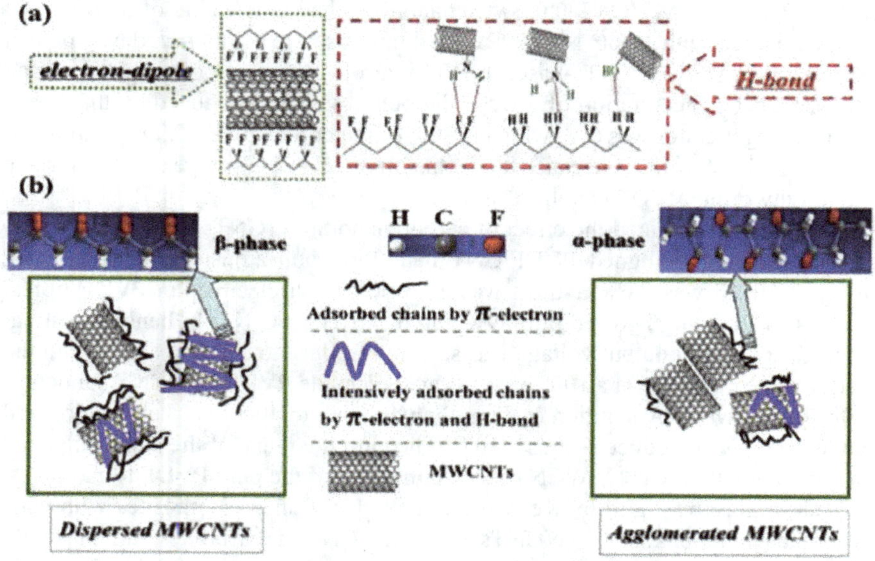

Figure 19.21 *(a) The chemical bonding of functionalised CNTs to PVDF chains, (b) the dispersion of CNTs affects the adsorbed PVDF chains on the CNT surface [117]*

measurements. The fabrication procedure of nanocomposite samples was meant to be simple and doable using laboratory chemicals. To make the samples, a combination of epoxy resin and CNTs was mixed, then cast into clean glass moulds and dried under vacuum to guarantee that no voids or air bubbles were trapped. The samples were subsequently reduced to size using a fine bench saw. Finally, the electrical connection and insulation tabs were glued to the samples so that the test machine could measure electric resistance and electrical insulation, respectively. An aerospace-grade epoxy resin system, Araldite 1564 and Aradur 3486 (hardener) were employed to create the CNTs/epoxy nanocomposite. Because of the viscosity of the resin (1,200–1,400 mPa·s) and the hardener (10–20 mPa·s), acetone (99%) was used to reduce the viscosity of the resin system and make it easier to handle. The epoxy modified with 0.1 wt% CNTs exhibited sufficient piezoresistivity for the resistance measurements, and those with 0.01 wt% CNTs did not show sufficiently measurable conductivity. It was reported that, with 0.1 wt% CNTs, adding 0.5% content of the surfactant improved the GF.

PVDF composites containing Buckminsterfullerenes (C_{60}) and single-walled carbon nanotubes (SWCNTs) were created by Baur *et al.* in a variety of compositions [119]. C_{60} fullerenes have a sp^2 hybridised structure that looks like CNTs but is spherical rather than tubular-like CNTs [120]. At a concentration of 0.05 and 0.2 wt%, the SWCNT and C_{60} composites had a maximum d_{31} of 65 and 63 pC N^{-1}, respectively. The d_{31} value of pristine PVDF films was 32 pC N^{-1}, showing that composites had a d_{31} value twice that of pure PVDF. This was accomplished by a mixture of elements including crystalline site nucleation, internal charge generation, enhanced polarisation and Young's modulus and dielectric constant optimisation.

P(VDF-TrFE)/graphene composites were developed by Wu *et al.*, who found that 0.15 wt% graphene provided a maximum V_{oc} of 12.43 V, about twice that of pure P(VDF-TrFE) films (6.10 V) [121]. In the same way, power density increased by 302% for AC circuits and 359% for DC circuits. The addition of a little quantity of graphene had no effect on phase development during early crystallisation. Stretching was also a major role in enhancing the degree of crystallinity of the composite films; crystallinity was 40% for untreated films and 60% for stretched films. Abolhasani *et al.* used electrospinning to make graphene/PVDF composite nanofibres and found that a tiny amount of graphene boosted β substantially [122]. With pristine PVDF nanofibres, the V_{oc} and I_{sc} were 3.8 V and 2 A, respectively, which rose to 7.9 V and 4.5 A with 0.1 wt% graphene/PVDF nanofibres. The addition of graphene to the randomly oriented nanofibres below 0.1 wt% reduced both β and voltage output. The output voltage of nanofibres was shown to have a convoluted connection with numerous parameters such as β, crystallinity, diameter and conductivity. According to Bidsorkhi *et al.*, adding 2 wt% graphene nanoplatelets to the PVDF matrix guaranteed excellent adherence and dispersion as well as the maximum sensitivity to produce strain [123].

Sinha *et al.* made a PENG out of graphene–silver (GAg)-doped PVDF nanocomposites, which they made using a one-pot process [124]. By integrating Ag NPs into graphene, GAg was created, which resulted in n-type doping of graphene owing to electrostatic contact. These interactions resulted in the production of polar GAg, which caused PVDF self-polarisation and favoured β-phase nucleation.

Because photo-sensitive plasmonic GAg was present, the PENG was able to detect visible light, resulting in an increased value of up to 46.6%. These findings indicated that the PENG might be employed in flexible optoelectronics for self-powered visible light sensors. Using atom transfer radical polymerisation, Layek *et al.* created polymethyl methacrylate (PMMA)-functionalised graphene (MG) and integrated it into PVDF [125]. Due to MMA polymerisation at the basal and side planes, PMMA raised the height of the graphene sheet from 1.2 to 2 and 4.5 nm. At a loading of 5%, MG sheets nucleated β-PVDF crystals, resulting in a decrease in the α-phase and an increase in the β-phase in the nanocomposites. Electrospun nanocomposite fibres were made by Barstugan *et al.* using a solution comprising polybenzoxazole (PBO), PVDF and graphene [126]. Adding 1 wt% PBO and 0.3 wt% graphenes to PVDF enhanced the piezoelectric effect and thermal strength of the material. When the PBO/graphene/PVDF fibres were thinner, exposed to greater pressure and electrospun on copper plate instead of aluminium foil, the voltage produced by them increased. The greatest voltage produced by the 0.02- and 0.06-mm diameter fibres was 60 V and 9.68 V, respectively.

Amorphous CB was chosen as a nanofiller by Wu *et al.* because it has a much higher surface area than CNTs and is simpler to disperse than graphene [108]. A 0.5 wt% CB concentration in P(VDF-HFP)/CB nanocomposite decreased crystal size and increased the number of crystals. In the process of stretching, this facilitated the creation of elongated and aligned crystal morphologies, resulting in extremely effective poling. The calibrated V_{oc} of the nanocomposite films was 104% of pristine P at an optimal CB concentration of 0.5 wt% (VDF-HFP). Pure P(VDF-HFP) films had a power density of 6.1 W m^{-3}, but the P(VDFHFP)/CB system had a power density of 28.3 W m^{-3}, which was 464% higher than pure P(VDF-HFP) films (VDF-HFP).

Cai *et al.* made P(VDF-HFP) nanocomposites with CB and few-layer graphene (FLG), demonstrating that changing the amounts of CB and FLG enhanced the β-phase percentage considerably. After poling and stretching, the β-phase fraction increased from 27% for pure P(VDF-HFP) to 98% for CB-0.3/FLG-0.02 wt% and 99% for CB-0.5/FLG-0.02 wt% films. Agglomeration of CB was observed in the films that exclusively contained CB. FLG minimised CB agglomeration and resulted in a consistent distribution of the hybrid fillers in the nanocomposites. CB-0.5/FLG-0.02 wt% films had a maximum harvested power density of 51.9 W m^{-3}, which was much greater than 15.8 W m^{-3} for P(VDF-HFP) films. CB-0.3/FLG-0.02 wt% films yielded the second-maximum power density, which is consistent with open-circuit voltage and β-phase fraction. Piezoelectricity has been improved.

Kim *et al.* employed three-dimensional (3D)-printed nanocomposites including varying quantities of MWCNTs and $BaTiO_3$ NPs for flexible sensing applications using the fused deposition modelling method [127]. MWCNTs were included into the PVDF matrix to provide a stress-reinforcing network and electron conduction channel, as well as to function as a dispersion for $BaTiO_3$ NPs. At 0.4 wt% MWCNTs/18 wt% $BaTiO_3$, the greatest d_{31} value was 0.13 pC N^{-1}. The output voltage was 120 mV when bent and 435 mV when the nanocomposite sensor was pressed with human fingers. Additionally, increasing the MWCNT content resulted in increased production owing to the stress-reinforcing effect of $BaTiO_3$ and PVDF.

Yang *et al.* constructed a 3D nanostructure of manganese dioxide (MnO_2)/graphene/MWCNTs with varying MnO_2 loadings to improve the piezoelectricity of PVDF [104]. The synthesis of MnO_2/graphene/MWCNTs hybrids is shown in Figure 19.22. Two hybrids containing 23 and 66 wt% MnO_2 were synthesised. In comparison with pure MWCNTs, hybrids had a higher surface area and roughness due to the presence of a significant number of connected –OH groups, which resulted in excellent compatibility with PVDF. After poling in a 50–80 MVm^{-1} electric field, the nanocomposites with 0.03–0.07 wt% and 0.1–0.5 wt% hybrids loading demonstrated a d_{33} value of

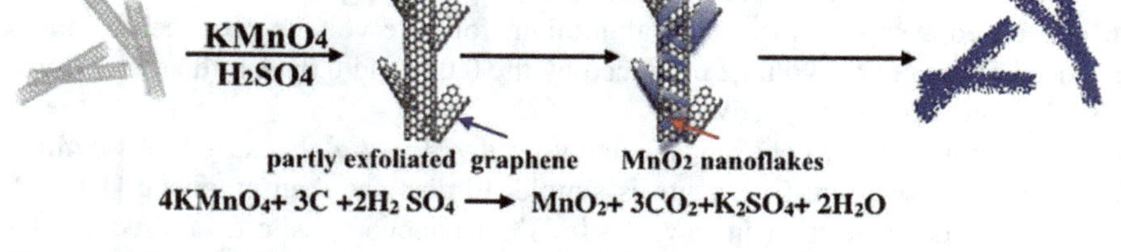

Figure 19.22 The synthesis route of the MWCNTs/graphene/MnO$_2$ hybrid [128]

17–33 pC N^{-1}, while a field greater than 100 MV m^{-1} was needed to create comparable piezoelectric performance in pure PVDF.

19.4 Self-sensing and health monitoring polymer composites

The application of carbon-based polymer nanocomposites in the development of smart piezoresistive materials that can sense their own elastic strain, plasticity and damage has been studied extensively [129]. When polymer composites attain percolation, they exhibit an electrical conductivity whose variations as a function of strain may be used for self-sensing of strain and damage. Insulating polymers packed with conductive (carbon) nanostructures are not generated by continuous conductors running the length of the material, but by the accumulation of discrete (micron-scale) components integrated to form continuous paths/networks. Due to the non-conductive matrix in these materials, the only way for current to flow through them is by the development of a network of conductive fillers (physically connected or not) that contain delocalised electrons. This is referred to as electrical percolation, and the precise filler concentration at which such a network forms is referred to as the percolation threshold [130]. There are many factors that influence the concentration of carbon fillers required to achieve this percolation threshold, including the morphology of the filler, its own conductivity, its state of dispersion within the matrix and the rheology of the polymer matrix, and it can range from 0.01 weight/cent to less than 10 weight/cent [131, 132]. The composite becomes electroconductive once the percolation threshold is achieved, and the change in its effective electrical resistance because of loading may be exploited as a piezoresistive feature to self-sense deformation and damage. It is now well acknowledged that the electromechanical (piezoresistive) sensitivity of these composites is greater at low filler concentrations (just above electrical percolation) than at high filler concentrations (i.e., composites with high conductivity) [133]. This is because for composites with a high filler content (a large number of conductive paths, i.e., a densely packed network), the interruption or change of a single conductive channel (or a few of them) has only a minimal impact on the composite's effective electrical resistance [133, 134]. Operating at or near the electrical percolation threshold results in a noisy electrical signal and significant instrumentation limitations, which presently preclude the adoption of this very low filler concentration regime. To circumvent this practical constraint, the majority of piezoresistive work on polymer composites uses filler concentrations much beyond the percolation threshold, where electrical resistances in the kΩ–MΩ range are frequently attained and readily managed.

The gauge (or gauge) factor, which is a measure of linear piezoresistive sensitivity, may be used to quantify linear piezoresistivity. The gauge factor (GF) is defined as the slope of the linear connection between fractional change in electrical resistance and uniaxial strain applied. GFs of typical metallic strain gauges used in materials testing range from 2.1 to 2.3. On the other hand, the intrinsic piezoresistivity of carbon nanostructures may be at least two orders of magnitude greater than when they are embedded in polymer composites. To some extent, the imparted strain is

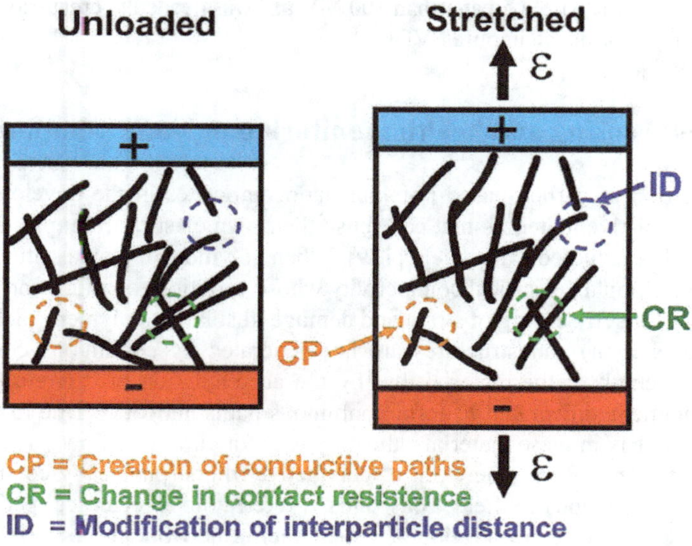

Figure 19.23 *Piezoresistive processes in carbon nanostructured polymer composites [136]*

not transmitted to the filler material. Thus, the intrinsic piezoresistivity of the carbon fillers and the change in their size following loading contribute only a minor amount to the observed piezoresistivity in these types of nanocomposites [135].

In polymer composites with low filler levels, piezoresistivity is influenced by the configuration of the filler network because of mechanical stress. The key underlying processes governing the change in electrical resistance in response to loading are roughly shown in Figure 19.23. They may be classified into three major effects of strain, namely rebuilding of the conductive network (formation and destruction of conductive routes), changes in the contact resistance between conductive fillers and changes in the filler-to-filler interparticle distance [137–140]. Straining the composite results in the creation of new conductive routes and the destruction of others (Figure 19.23); the filler-to-filler contact resistance also varies as the filler-to-filler contact area and number of contact points vary. Additionally, there is a critical non-contact situation that is determined by the interparticle distance. It is well acknowledged that when electrons are excited enough energetically (voltage, temperature), they may tunnel or hop across short distances across a thin polymeric layer; this (quantum) tunnelling resistance exponentially increases with the interparticle distance [141, 142]. The maximum tunnelling distance in polymers packed with carbon nanostructures varies according to the nanostructure and polymer type; theoretically, the maximum tunnelling distance is close to 2 nm [139, 142], but some other research explored somewhat larger tunnelling lengths (5 nm) [143].

Due to the electrical–mechanical properties of continuous carbon fibre, it is an ideal choice for strain sensing in fibre-reinforced polymer composites. They may be used to analogise the usual strain gauge approach, which measures changes in electrical resistance to indicate strain [144]. Conor and Owston [145] were the first to describe the electrical–mechanical behaviour of seven distinct signal carbon fibres, demonstrating that electrical characteristics varied according to fibre type. Blazewicz *et al.* [146] studied the influence of piezoresistance on several kinds of fibres and hypothesised a relationship between the microstructure of the fibres, the piezoresistivity and the coefficient of strain sensitivity. Gordon *et al.* [147] discovered an unusual phenomenon in which the resistivity rose somewhat with longitudinal tensile strain in the single-lamina composite but reduced with strain in the two-lamina composite. Huang and Wu [148, 149] investigated the electrical sensing characteristics of CFRP strips for the detection of sub-millimetre strain. A continuous carbon fibre grid may likewise be used to monitor the two-dimensional strain fields [150].

CNT-modified polymer composites are the most extensively researched polymer composites among all the carbon-modified polymer composites. In comparison with other thermosetting polymers such as vinyl ester or polyester, epoxy thermosettings filled with SWCNTs [151] and MWCNTs [10, 152, 153] are among the most thoroughly examined polymers in terms of piezoresistivity of polymer composites. Due to their low processing viscosity, these polymer composites are distinguished by their ability to achieve electrical percolation at CNT concentrations substantially lower than those seen in thermoplastics or elastomers. While linear piezoresistivity is quite common for CNT thermosetting composites loaded in tension with low ultimate strains, non-linear piezoresistivity can also occur for thermosettings with relatively large ultimate strains [153]. In this latter instance, an inversion (from positive to negative) of the piezoresistive trend may occur for certain CNT weight concentrations when subjected to relatively significant deformations, likely due to the inelastic behaviour of polymer [153]. Negative piezoresistivity indicates that the effective (3D) distance is reduced, which may be produced by matrix plasticity, Poisson's effect-induced lateral CNT motion and/or 3D CNT network rearrangement at high strain [153]. As the polymer is axially stretched, the distance between CNTs decreases. The weight concentration of CNTs also influences the piezoresistive response, and it is widely known that lower CNT concentrations result in larger piezoresistive sensitivity [154, 155]. This is because lower CNT concentrations result in fewer redundant routes creating the CNT network, and the tunnelling effect is predicted to be more pronounced at CNT concentrations around the percolation threshold [156]. Although tensile loading is by far the most often explored loading mode, a few studies have also examined compression [154] and bending [157].

Piezoresistivity has been expanded to greater strain sensing capacities for CNT composites utilising thermoplastic matrices. A wide range of thermoplastic polymers have been modified with CNTs to achieve piezoresistive strain sensing capabilities, including PMMA [134, 158], PVDF [159] , polycarbonate [160], polyethylene oxide [161], polysulphone [162] and polyethylene terephthalate. The GFs observed for CNT/thermoplastic composites in the elastic regime are comparable with those reported for thermosettings. In the plastic domain, however, greater GFs have been reported for CNT/thermoplastic composites than for CNT/thermosetting composites because to their wider deformation capacities [136].

CB (a black powder formed by the incomplete combustion of carbonaceous or petroleum products) is the oldest carbon filler that has been widely employed to make strain sensing polymer composites with electroconductive properties. Piezoresistive applications need a high concentration of CB, compared with other carbon nanostructures like CNTs or graphene flakes. Because spherical particles need greater filler concentrations to achieve electrical percolation than nanostructures with a high aspect ratio, this is a geometrical consideration. The piezoresistive response of CB/polymer composites is non-linear.

Graphene (a monolayer) has exceptional physico-chemical characteristics, including a very high elastic modulus, good electrical and thermal conductivity and a large specific surface area [163]. Single-layer graphene has piezoresistive characteristics, making it ideal for the development of electromechanical devices and strain sensing applications [164, 165]. The GF of suspended single-layer graphene has been reported to be as high as 151 [166], while the GFs of FLG films with various thicknesses and conductivities have been reported to be anywhere between 10 and 300 [167]. When a family of graphenic sheets (GSs) is deposited on substrates or employed as fillers in polymer composites, the high GFs found for individual graphene or FLG sheets are considerably reduced. Numerous polymers have been employed to generate piezoresistive nanocomposites based on GSs, ranging from thermosettings such as epoxy [168] and vinyl ester [169] to elastomers such as PDMS [170] and ethylene-butene copolymer (EB) [171]. Uniaxial tension has been the most extensively researched loading mode for this class of polymer nanocomposites, followed by bending tests and, less often, direct uniaxial compression. In case of polymer nanocomposites filled with GSs, the sensitivity of the nanocomposites is dependent on the filler concentration, with a greater sensitivity for a lower filler content [172].

19.5 Conclusions

Multi-functional applications (energy harvesting and self-sensing) of polymer composites are reviewed in this chapter. Carbon nanomaterial-based composites are ideal for sensor and energy harvesting applications because of their unique dimensions, electrical, thermal and mechanical characteristics. For sensors and energy harvesting devices, there has been a great deal of improvement in the selection of diverse carbon nanomaterials and matrices, as well as unique structural design and chemical treatment.

Piezoelectric materials are often used as electromechanical transducers in energy harvesters to convert ambient vibrations to electrical energy that may be stored in batteries or used to power tiny electronic devices. The advantages of piezoelectric energy harvesting include the use of a simple structure without the need for multiple additional components, the absence of moving parts or mechanical constraints, portability, integration with other renewable energy sources, the elimination of the need for an external voltage source, compatibility with MEMS, ease of fabrication with microelectronic devices and a reasonable output power density. Therefore, piezomaterials are an excellent candidate for powering macro- to nanoscale electronic devices instead of batteries, which have a finite life (piezoelectric energy harvesting). As a result, piezoelectric materials are now extensively employed in several industries, including sensors [16–19], actuators [20–22], nanogenerators [23–26], MEMS devices, portable electronics and biomedicine [26, 27].

Carbon compounds such as graphite, CB, CNTs and, more recently, single- and multi-layer graphene are excellent candidates for use as polymer fillers in self-sensing polymer composites. The coupled electromechanical response of such polymer nanocomposites is dependent on a few factors, including the nanostructure architecture, dimensions, layer count, functional groups of the fillers, dielectric constant, processing viscosity of the polymer matrix and the filler dispersion within the matrix. From the geometrical perspective of the filler, the excluded volume is a critical parameter affecting the piezoresistive sensitivity of the polymer composites. The present uses of thermosetting and thermoplastic piezoresistive materials are projected to enable the development of the next generation of smart materials for the automotive, aerospace, transportation and energy sectors. The applications of these nanocomposites, particularly for thermosetting polymers, are rapidly evolving toward self-sensing of strain and damage in fibre-reinforced multi-scale hierarchical composites, for virtually any industrial application requiring high load bearing capacity and self-sensing composites. On the other hand, the applications of carbon-filled elastomeric materials are concentrated on sensors for human motion, soft skins, smart wearable sensors and robotics, all of which need substantial deformations. However, non-linearity, cycle repeatability, undesirable viscoelastic response and hysteresis are all practical difficulties that must be addressed before commercial devices based on these materials can be developed further.

References

[1] van der Zwaag S., van den Ende D.A., Groen W.A.P. 'Sensing and energy harvesting novel polymer composites'. *Materials experience*. 2014:221–34.

[2] Hull D.C.T.W. 'An introduction to composite materials' in *Cambridge solid state science series*; 1996.

[3] 'Giraudmaillet, C. 2017 'Health monitoring of polymer composite structures using piezoelectric nanofibres'. Cranfield University.

[4] Hegedus G., Sarkadi T., Czigany T. 'Self-sensing polymer composite: white-light-illuminated reinforcing fibreglass bundle for deformation monitoring'. *Sensors*. 2019;19(7):1745.

[5] Speranzini E., Marsili R., Moretti M., Rossi G. 'Image analysis technique for material behavior evaluation in civil structures'. *Materials*. 2017;10 (7):77008.

[6] Castori G., Speranzini E. 'Structural analysis of failure behavior of laminated glass'. *Composites Part B: Engineering*. 2017;125(4):89–99.

[7] Corradi M., Borri A., Righetti L., Speranzini E. 'Uncertainty analysis of FRP reinforced timber beams'. *Composites Part B: Engineering*. 2017;113:174–84.

[8] Konstantopoulos S., Fauster E., Schledjewski R. 'Monitoring the production of FRP composites: a review of in-line sensing methods'. *Express Polymer Letters*. 2014;8(11):823–40.

[9] Ma Z., Chen X. 'Fiber Bragg gratings sensors for aircraft wing shape measurement: recent applications and technical analysis'. *Sensors*. 2019;19(1):55.

[10] An D., Nourry J., Gharavian S, *et al.* 'Strain self-sensing tailoring in functionalised carbon nanotubes/epoxy nanocomposites in response to electrical resistance change measurement'. *TESConf 2020 - 9th International Conference on Through-life Engineering Services*; Cranfield, UK, 2020.

[11] Kobayashi A.S. *Handbook on experimental mechanics*. Englewood Cliffs, NJ: Prentice-Hall; 1987.

[12] Heard B.P., Brook B.W., Wigley T.M.L., Bradshaw C.J.A. 'Burden of proof: a comprehensive review of the feasibility of 100% renewable-electricity systems'. *Renewable and Sustainable Energy Reviews*. 2017;76:1122–33.

[13] Covaci C., Gontean A. 'Piezoelectric energy harvesting solutions: a review'. *Sensors*. 2020;20(12):E3512):3512:.

[14] Bowen C.R., Kim H.A., Weaver P.M., Dunn S. 'Piezoelectric and ferroelectric materials and structures for energy harvesting applications'. *Energy & Environmental Science*. 2014;7(1):25–44.

[15] Li H., Tian C., Deng Z.D. 'Energy harvesting from low frequency applications using piezoelectric materials'. *Applied Physics Reviews*. 2014;1(4):041301.

[16] Tressler J.F., Alkoy S., Newnham R.E. 'Piezoelectric sensors and sensor materials'. *Journal of Electroceramics*. 1998;2(4):257–72.

[17] Lang S.B., Muensit S. 'Review of some lesser-known applications of piezoelectric and pyroelectric polymers'. *Applied Physics A*. 2006;85(2):125–34.

[18] Khan S., Dang W., Lorenzelli L., Dahiya R. 'Flexible pressure sensors based on screen-printed P(VDF-TrFE) and P(VDF-TrFE)/MWCNTs'. *IEEE Transactions on Semiconductor Manufacturing*. 2015;28(4):486–93.

[19] Song K., Ma N., Mishra Y.K., Adelung R., Yang Y. 'Achieving light-induced ultrahigh pyroelectric charge density toward self-powered UV light detection'. *Advanced Electronic Materials*. 2019;5(1):1800413.

[20] Periasamy C., Chakrabarti P. 'Time-dependent degradation of PT/ZNO nanoneedle rectifying contact based piezoelectric nanogenerator'. *Journal of Applied Physics*. 2011;**109**(5):054306.

[21] Francis L.F., Polla D.L. 'Processing and characterization of piezoelectric materials and integration into microelectromechanical systems'. *Annu. Rev. Mater. Sci*. 1998;**28**:563–97.

[22] Adriaens H.J.M.T.S., Koning W.L.D., Banning R. 'Modeling piezoelectric actuators'. *IEEE/ASME Transactions on Mechatronics*. 2000;5(4): 331–41.

[23] Ahmad Z., Prasad A., Prasad K. 'A comparative approach to predicting effective dielectric, piezoelectric and elastic properties of PZT/PVDF composites'. *Physica B: Condensed Matter*. 2009;404(20):3637–44.

[24] Tamang A., Ghosh S.K., Garain S, *et al.* 'DNA-assisted β-phase nucleation and alignment of molecular dipoles in PVDF film: a realization of self-poled bioinspired flexible polymer nanogenerator for portable electronic devices'. *ACS Applied Materials & Interfaces*. 2015;7(30):16143–47.

[25] Vatansever D., Hadimani R.L., Shah T., Siores E. 'An investigation of energy harvesting from renewable sources with PVDF and PZT'. *Smart Materials and Structures*. 2011;20(5):055019.

[26] Priya S., Song H.-C., Zhou Y, *et al.* 'A review on piezoelectric energy harvesting: materials, methods, and circuits. energy harvesting and systems'. *Energy Harvesting and Systems*. 2017;4(1):3–39.

[27] Cook-Chennault K.A., Thambi N., Sastry A.M. 'Powering MEMS portable devices—a review of non-regenerative and regenerative power supply systems with special emphasis on piezoelectric energy harvesting systems'. *Smart Materials and Structures*. 2008;17(4):043C01.

[28] Slowing I.I., Vivero-Escoto J.L., Wu C.-W., Lin V.S.-Y. 'Mesoporous silica nanoparticles as controlled release drug delivery and gene transfection carriers'. *Advanced Drug Delivery Reviews*. 2008;60(11):1278–88.

[29] Motamedi A.S., Mirzadeh H., Hajiesmaeilbaigi F., Bagheri-Khoulenjani S., Shokrgozar M.A. 'Piezoelectric electrospun nanocomposite comprising Au NPs/PVDF for nerve tissue engineering'. *Journal of Biomedical Materials Research Part A*. 2017;105(7):1984–93.

[30] Yang R., Qin Y., Li C., Zhu G., Wang Z.L. 'Converting biomechanical energy into electricity by a muscle-movement-driven nanogenerator'. *Nano Letters*. 2009;9(3):1201–5.

[31] Gusarov B. 'PVDF piezoelectric polymers: characterization and application to thermal energy harvesting'. Université Grenoble Alpes; 2015.

[32] Knight R.R., Mo C., Clark W.W. 'MEMS interdigitated electrode pattern optimization for a unimorph piezoelectric beam'. *Journal of Electroceramics*. 2011;26(1):14–22.

[33] Guillot-Ferriols M., Rodríguez-Hernández J.C., Correia D.M., *et al.* 'Poly(vinylidene) fluoride membranes coated by heparin/collagen layer-by-layer, smart biomimetic approaches for mesenchymal stem cell culture'. *Materials Science and Engineering: C*. 2020;117 111281.

[34] Hatch S.M., Briscoe J., Dunn S. 'Improved CuSCN–ZnO diode performance with spray deposited CuSCN'. *Thin Solid Films*. 2013;531:404–7.

[35] Dunn S. 'Strain behavior of thin film PbZr0.3Ti0.7O3 (30/70) examined through piezoforce microscopy'. *Journal of Applied Physics*. 2003;94(9): 5964–8.

[36] Bedekar V., Oliver J., Zhang S., Priya S., *et al*. 'Comparative study of energy harvesting from high temperature piezoelectric single crystals'. *Japanese Journal of Applied Physics*. 2009;48(9):091406.

[37] Kaps S., Bhowmick S., Gröttrup J, *et al*. 'Piezoresistive response of quasi-one-dimensional ZNO nanowires using an in situ electromechanical device'. *ACS Omega*. 2017;**2**(6):2985–93.

[38] Narita F., Fox M. 'A review on piezoelectric, magnetostrictive, and magnetoelectric materials and device technologies for energy harvesting applications'. *Advanced Engineering Materials*. 2018;20(5):1700743.

[39] Zhang Q.M., Bharti V., Zhao X. 'Giant electrostriction and relaxor ferroelectric behavior in electron-irradiated poly(vinylidene fluoride-trifluoroethylene) copolymer'. *Science*. 1998;280(5372):2101–4.

[40] Fukada E. 'History and recent progress in piezoelectric polymers'. *IEEE Transactions on Ultrasonics, Ferroelectrics and Frequency Control*. 2000;47(6):1277–90.

[41] Kim J., Yun S., Ounaies Z. 'Discovery of cellulose as a smart material'. *Macromolecules*. 2006;39(12):4202–06.

[42] Hartono A., Djamal M., Satira S. 'Preparation of PVDF film using deep coating method for biosensor transducer applied'. 3rd International Conference on Instrumentation, Communications, Information Technology and Biomedical Engineering (ICICI-BME); IEEE, 2013.

[43] Humphrey J., Amin-Sanayei R. 'Vinylidene fluoride polymers'. *Encyclopedia of Polymer Science and Technology*. 2002;4.

[44] Mohamadi S. 'Preparation and characterization of PVDF/PMMA/graphene polymer blend nanocomposites by using ATR-FTIR technique'. *Infrared Spectroscopy-Materials Science, Engineering and Technology*. 2012;1.

[45] Satapathy S., Gupta P., Pawar S., Varma K.B.R. 'Crystallization of beta-phase poly (vinylidene fluoride) films using dimethyl sulfoxide (DMSO) solvent and at suitable annealing condition'. *ArXiv: Materials Science*. 2008.

[46] Ahn Y., Lim J.Y., Hong S.M, *et al*. 'Enhanced piezoelectric properties of electrospun poly(vinylidene fluoride)/multiwalled carbon nanotube composites due to high β-phase formation in poly(vinylidene fluoride'. *The Journal of Physical Chemistry C*. 2013;117(22):11791–99.

[47] Kawai H. 'The piezoelectricity of poly (vinylidene fluoride)'. *Japanese Journal of Applied Physics*. 1969;8(7):975–76.

[48] Ueberschlag P. 'PVDF piezoelectric polymer'. *Sensor Review*. 2001;21(2):118–26.

[49] Bergman J.G., McFee J.H., Crane G.R. 'Pyroelectricity and optical second harmonic generation in polyvinylidene fluoride films'. *Applied Physics Letters*. 1971;18(5):203–5.

[50] Correia H.M.G., Ramos M.M.D. 'Quantum modelling of poly(vinylidene fluoride'. *Computational Materials Science*. 2005;33(1):224–9.

[51] Kanik M., Aktas O., Sen H.S., Durgun E., Bayindir M. 'Spontaneous high piezoelectricity in poly(vinylidene fluoride) nanoribbons produced by iterative thermal size reduction technique'. *ACS Nano*. 2014;8(9):9311–23.

[52] Han H., Nakagawa Y., Takai Y., Kikuchi K., Tsuchitani S., Kosimoto Y. 'Microstructure fabrication on a β-phase PVDF film by wet and dry etching technology'. *Journal of Micromechanics and Microengineering*. 2012;22(8):085030.

[53] Solvay® 'Solef® PVDF: design & processing guide, Bollate'. *Solvay® Speciality Polymer*. 2015.

[54] Taylor G.I., McEwan A.D. 'The stability of a horizontal fluid interface in a vertical electric field'. *Journal of Fluid Mechanics*. 1965;22(1):1.

[55] Ke P., Jiao X.-N., Ge X.-H., Xiao W.-M., Yu B. 'From macro to micro: structural biomimetic materials by electrospinning'. *RSC Advances*. 2014;4(75):39704–24.

[56] Laux E. *Electrospinning: technology and application*. Berne: Swiss MNT Network; 2006.

[57] Leach M.K., Feng Z.-Q., Tuck S.J., Corey J.M. 'Electrospinning fundamentals: optimizing solution and apparatus parameters'. *Journal of Visualized Experiments*. 2011;47:249421.

[58] Salimi A., Yousefi A.A. 'Analysis method: FTIR studies of β-phase crystal formation in stretched PVDF films'. *Polymer Testing*. 2003;22(6):699–704.

[59] Matsushige K., Nagata K., Imada S., Takemura T. 'The II-I crystal transformation of poly(vinylidene fluoride) under tensile and compressional stresses'. *Polymer*. 1980;21(12):1391–7.

[60] Sencadas V., Gregorio R., Lanceros-Méndez S. 'α to β phase transformation and microestructural changes of PVDF films induced by uniaxial stretch'. *Journal of Macromolecular Science, Part B*. 2009;48(3):514–25.

[61] Yagi T., Tatemoto M. 'A fluorine-19 NMR study of the microstructure of vinylidene fluoride–trifluoroethylene copolymers'. *Polymer Journal*. 1979;11(6):429–36.

[62] Aimi K., Ando S., Avalle P., Harris R.K. 'Solid-state 19F MAS and 1H→19F CP/MAS NMR study of the phase transition behavior of vinylidene fluoride–trifluoroethylene copolymers: 1. Uniaxially drawn films of VDF 75% copolymer'. *Polymer*. 2004;45(7):2281–90.

[63] Martins P., Lopes A.C., Lanceros-Mendez S. 'Electroactive phases of poly(vinylidene fluoride): determination, processing and applications'. *Progress in Polymer Science*. 2014;39(4):683–706.

[64] Yagi T., Higashihata Y., Fukuyama K., Sako J. 'Piezoelectric properties of rolled vinylidene fluoride and trifluoroethylene copolymer'. *Ferroelectrics*. 1984;**57**(1):327–35.

[65] Tashiro K., Tadokoro H., Kobayashi M. 'Structure and piezoelectricity of poly(vinylidene fluoride)'. *Ferroelectrics*. 1981;32(1):167–75.

[66] Zhang L., Oh S.R., Wong T.C., Tan C.Y., Yao K. 'Piezoelectric polymer multilayer on flexible substrate for energy harvesting'. *IEEE Transactions on Ultrasonics, Ferroelectrics, and Frequency Control*. 2013;60(9):2013–20.

[67] Pi Z., Zhang J., Wen C., Zhang Z.-bin., Wu D. 'Flexible piezoelectric nanogenerator made of poly(vinylidenefluoride-co-trifluoroethylene) (PVDF-TrFE) thin film'. *Nano Energy*. 2014;7:33–41.

[68] Bae J.-H., Chang S.-H. 'Characterization of an electroactive polymer (PVDF-TrFE) film-type sensor for health monitoring of composite structures'. *Compos Struct*. 2015;131(1):1090–98.

[69] Oliveira F., Leterrier Y., Månson J.-A, *et al*. 'Process influences on the structure, piezoelectric, and gas-barrier properties of PVDF-TrFE copolymer'. *Journal of Polymer Science Part B: Polymer Physics*. 2014;52(7):496–506.

[70] Lee K.Y., Kim D., Lee J.-H., Kim T.Y., Gupta M.K., Kim S.-W. 'Unidirectional high-power generation via stress-induced dipole alignment from $ZnSnO_3$ nanocubes/polymer hybrid piezoelectric nanogenerator'. *Advanced Functional Materials*. 2014;**24**(1):37–43.

[71] Cauda V., Stassi S., Bejtka K., Canavese G. 'Nanoconfinement: an effective way to enhance PVDF piezoelectric properties'. *ACS Applied Materials & Interfaces*. 2013;5(13):6430–37.

[72] Lutkenhaus J.L., McEnnis K., Serghei A., Russell T.P. 'Confinement effects on crystallization and curie transitions of poly (vinylidene fluoride-co-trifluoroethylene)' in *Macromolecules*. Vol. 43; 2010;8(43):. pp. 3844–50.

[73] Bhavanasi V., Kusuma D.Y., Lee P.S., Orientation P. 'Polarization orientation, piezoelectricity, and energy harvesting performance of ferroelectric PVDF-TrFE nanotubes synthesized by nanoconfinement'. *Advanced Energy Materials*. 2014;4(16):1400723.

[74] Chen D., Chen K., Brown K., Hang A., Zhang J.X.J. 'Liquid-phase tuning of porous PVDF-TrFE film on flexible substrate for energy harvesting'. *Applied Physics Letters*. 2017;110(15):153902.

[75] Aliane A., Benwadih M., Bouthinon B., Coppard R., Domingues-Dos Santos F., Daami A. 'Impact of crystallization on ferro-, piezo- and pyro-electric characteristics in thin film P(VDF–trfe)'. *Organic Electronics*. 2015;25(10):92–98.

[76] Wang A., Hu M., Zhou L., Qiang X. 'Self-powered well-aligned P(VDF-TrFE) piezoelectric nanofiber nanogenerator for modulating an exact electrical stimulation and enhancing the proliferation of preosteoblasts'. *Nanomaterials*. 2019;9(3):349.

[77] Persano L., Dagdeviren C., Su Y, *et al*. 'High performance piezoelectric devices based on aligned arrays of nanofibers of poly(vinylidenefluoride-co-trifluoroethylene'. *Nature Communications*. 2013;4(1):1633.

[78] Safari A.A.E.K. 'Piezoelectric and acoustic materials for transducer applications'. 2008.

[79] Ameduri B. 'From vinylidene fluoride (VDF) to the applications of VDF-containing polymers and copolymers: recent developments and future trends'. *Chemical Reviews*. 2009;109(12):6632–86.

[80] Kalfoglou N.K., Williams H.L. 'Mechanical relaxations of poly(vinylidene fluoride) and some of its copolymers'. *Journal of Applied Polymer Science*. 1973;17(11):3367–73.

[81] Latour M., Dorra H.A. 'Structural study of a vinylidene fluoride chlorotrifluorethylene copolymer by × ray and infrared spectroscopy-piezoelectric properties'. *Ferroelectrics*. 1982;44(1):197–203.

[82] Améduri B., Boutevin B., Kostov G. 'Fluoroelastomers: synthesis, properties and applications'. *Progress in Polymer Science*. 2001;26(1):105–87.

[83] Künstler W., Wegener M., Seiß M., Gerhard-Multhaupt R. 'Preparation and assessment of piezo- and pyroelectric poly (vinylidene fluoride-hexafluoropropylene) copolymer films'. *Applied Physics A: Materials Science & Processing*. 2001;73(5):641–5.

[84] Neese B., Wang Y., Chu B., *et al*. 'Piezoelectric responses in poly(vinylidene fluoride/hexafluoropropylene) copolymers'. *Applied Physics Letters*. 2007;90(24):242917.

[85] Huan Y., Liu Y., Yang Y., Wu Y. 'Influence of extrusion, stretching and Poling on the structural and piezoelectric properties of poly (vinylidene fluoride-hexafluoropropylene) copolymer films'. *Journal of Applied Polymer Science*. 2007;104(2):858–62.

[86] Geller S., Tyczynski T., Gude M., Sauer S., Fischer W.-J. 'Glass fiber-reinforced polyurethane composite structures with integrated piezoelectric sensor elements and corresponding electronics'. *Advanced Engineering Materials*. 2018;20(12):1800447.

[87] Weder A., Geller S., Heinig A., Tyczynski T., Hufenbach W., Fischer W.-J. 'A novel technology for the high-volume production of intelligent composite structures with integrated piezoceramic sensors and electronic components'. *Sensors and Actuators A: Physical*. 2013;202:106–10.

[88] Bowen C.R., Watson M., Betts D.N., Harris P., Bertin M., Kim H.A. 'Piezoelectric fibres integrated into structural composites'. *Ferroelectrics*. 2014;466(1):14–20.

[89] Malakooti M.H., Patterson B.A., Hwang H.-S., Sodano H.A. 'ZnO nanowire interfaces for high strength multifunctional composites with embedded energy harvesting'. *Energy & Environmental Science*. 2016;9(2):634–43.

[90] Hufenbach W., Gude M., Heber T. 'Design and testing of novel piezoceramic modules for adaptive thermoplastic composite structures'. *Smart Materials and Structures*. 2009;18(4):045012.

[91] Yang S.M., Hung C.C., Chen K.H. 'Design and fabrication of a smart layer module in composite laminated structures'. *Smart Materials and Structures*. 2005;14(2):315–20.

[92] Tang H.-Y., Winkelmann C., Lestari W., La Saponara V. 'Composite structural health monitoring through use of embedded PZT sensors'. *Journal of Intelligent Material Systems and Structures*. 2011;22(8):739–55.

[93] Konka H.P. 'Embedded piezoelectric fiber composite sensors for applications in composite structures'. [Louisiana State University].

[94] Hu N., Fukunaga H., Matsumoto S., Yan B., Peng X.H. 'An efficient approach for identifying impact force using embedded piezoelectric sensors'. *International Journal of Impact Engineering*. 2007;34(7):1258–71.

[95] Xie M., Zhang Y., Kraśny M.J., Rhead A., Bowen C., Arafa M. 'Energy harvesting from coupled bending-twisting oscillations in carbon-fibre reinforced polymer laminates'. *Mechanical Systems and Signal Processing*. 2018;107(4):429–38.

[96] Yan X., Courtney C.R.P., Bowen C.R, *et al.* 'In situ fabrication of carbon fibre–reinforced polymer composites with embedded piezoelectrics for inspection and energy harvesting applications'. *Journal of Intelligent Material Systems and Structures*. 2020;31(16):1910–19.

[97] Zeng W., Zhou C., Xiao J., Ma J. 'Correlation between temperature-dependent permittivity dispersion and depolarization behaviours in zr4+-modified bifeo$_3$–batio$_3$ piezoelectric ceramics'. *Bulletin of Materials Science*. 2015;38(7):1737–41.

[98] Alsaadi A., Shi Y., Pan L., Tao J., Jia Y. 'Vibration energy harvesting of multifunctional carbon fibre composite laminate structures'. *Composites Science and Technology*. 2019;178(12):1–10.

[99] Shi Y., Piao C., El Fadlaoui D., Al-Saadi A., Jia Y. 'Integration and characterisation of piezoelectric macro-fibre composite on carbon fibre composite for vibration energy harvesting'. *Journal of Physics: Conference Series*. 2019;1407(1):012080.

[100] Narita F., Nagaoka H., Wang Z. 'Fabrication and impact output voltage characteristics of carbon fiber reinforced polymer composites with lead-free piezoelectric nano-particles'. *Materials Letters*. 2019;236:487–90.

[101] Wang F., Mai Y.-W., Wang D., Ding R., Shi W. 'High quality barium titanate nanofibers for flexible piezoelectric device applications'. *Sensors and Actuators A: Physical*. 2015;233:195–201.

[102] Fu J., Hou Y., Gao X., Zheng M., Zhu M. 'Highly durable piezoelectric energy harvester based on a PVDF flexible nanocomposite filled with oriented BaTi 2 O 5 nanorods with high power density'. *Nano Energy*. 2018;**52**:391–401.

[103] Meyers F.N., Loh K.J., Dodds J.S., Baltazar A. 'Active sensing and damage detection using piezoelectric zinc oxide-based nanocomposites'. *Nanotechnology*. 2013;24(18):185501.

[104] Chen H., Dong X., Zeng T., Zhou Z., Yang H. 'The mechanical and electric properties of infiltrated PZT/polymer composites'. *Ceramics International*. 2007;33(7):1369–74.

[105] Wu B., Wu H., Wu J., Xiao D., Zhu J., Pennycook S.J. 'Giant piezoelectricity and high curie temperature in nanostructured alkali niobate lead-free piezoceramics through phase coexistence'. *Journal of the American Chemical Society*. 2016;138(47):15459–64.

[106] Abzan N., Kharaziha M., Labbaf S. 'Development of three-dimensional piezoelectric polyvinylidene fluoride-graphene oxide scaffold by non-solvent induced phase separation method for nerve tissue engineering'. *Materials & Design*. 2019;167(9):107636.

[107] Ning H.M., Hu N., Kamata T., *et al.* 'Improved piezoelectric properties of poly(vinylidene fluoride) nanocomposites containing multi-walled carbon nanotubes'. *Smart Materials and Structures*. 2013;22(6):065011.

[108] Wu L., Yuan W., Hu N., *et al.* 'Improved piezoelectricity of PVDF-HFP/carbon black composite films'. *Journal of Physics D: Applied Physics*. 2014;47(13):135302.

[109] Mahapatra S.D., Mohapatra P.C., Aria A.I, *et al.* 'Piezoelectric materials for energy harvesting and sensing applications: roadmap for future smart materials'. *Advanced Science*. 2021;8(17):2100864.

[110] Wu C.-M., Chou M.-H., Zeng W.-Y. 'Piezoelectric response of aligned electrospun polyvinylidene fluoride/carbon nanotube nanofibrous membranes'. *Nanomaterials*. 2018;8(6):420.

[111] Yu S., Zheng W., Yu W., Zhang Y., Jiang Q., Zhao Z. 'Formation mechanism of β-phase in PVDF/CNT composite prepared by the sonication method'. *Macromolecules*. 2009;42(22):8870–74.

[112] Kim G.H., Hong S.M., Seo Y. 'Piezoelectric properties of poly(vinylidene fluoride) and carbon nanotube blends: beta-phase development'. *Physical Chemistry Chemical Physics*. 2009;11(44):10506–12.

[113] Liu Z.H., Pan C.T., Lin L.W., Lai H.W. 'Piezoelectric properties of PVDF/MWCNT nanofiber using near-field electrospinning'. *Sensors and Actuators A: Physical*. 2013;193(6):13–24.

[114] Pu J., Yan X., Jiang Y., Chang C., Lin L. 'Piezoelectric actuation of direct-write electrospun fibers'. *Sensors and Actuators A: Physical*. 2010;164(1-2):131–6.

[115] Mahapatra S.D., Mohapatra P.C., Aria A.I, *et al.* 'Piezoelectric materials for energy harvesting and sensing applications: roadmap for future smart materials'. *Advanced Science*. 2021;8(17):2100864–e2100864.

[116] Yu H., Huang T., Lu M., Mao M., Zhang Q., Wang H. 'Enhanced power output of an electrospun PVDF/MWCNTs-based nanogenerator by tuning its conductivity'. *Nanotechnology*. 2013;24(40):405401.

[117] Ke K., Pötschke P., Jehnichen D., Fischer D., Voit B. 'Achieving β-phase poly(vinylidene fluoride) from melt cooling: effect of surface functionalized carbon nanotubes'. *Polymer*. 2014;55(2):611–19.

[118] Sharma M., Srinivas V., Madras G., Bose S. 'Outstanding dielectric constant and piezoelectric coefficient in electrospun nanofiber mats of PVDF containing silver decorated multiwall carbon nanotubes: assessing through piezoresponse force microscopy'. *RSC Advances* 2016;6(8):6251–8.

[119] Qin S., Yu Z., Zhang X., Yang S., Peng W., Zhao F. 'Calibration of polyvinylidene fluoride (PVDF) stress gauges under high-impact dynamic compression by machine learning'. *Journal of Applied Physics*. 2022;131(2):024502.

[120] Kroto H.W., Heath J.R., O'Brien S.C., Curl R.F., Smalley R.E. 'C60: Buckminsterfullerene'. *Nature*. 1985;318(6042):162–3.

[121] Wu L., Jing M., Liu Y., *et al.* 'Power generation by PVDF-TrFE/graphene nanocomposite films'. *Composites Part B: Engineering*. 2019;164(7):703–9.

[122] Abolhasani M.M., Shirvanimoghaddam K., Naebe M. 'PVDF/graphene composite nanofibers with enhanced piezoelectric performance for development of robust nanogenerators'. *Composites Science and Technology*. 2017;138:49–56.

[123] Cheraghi Bidsorkhi H., D'Aloia A.G., Tamburrano A., De Bellis G., Sarto M.S. 'Waterproof graphene-PVDF wearable strain sensors for movement detection in smart gloves'. *Sensors*. 2021;21(16):5277.

[124] Sinha T.K., Ghosh S.K., Maiti R, *et al.* 'Graphene-silver-induced self-polarized PVDF-based flexible plasmonic nanogenerator toward the realization for new class of self powered optical sensor'. *ACS Applied Materials & Interfaces*. 2016;8(24):14986–93.

[125] Layek R.K., Samanta S., Chatterjee D.P., Nandi A.K. 'Physical and mechanical properties of poly(methyl methacrylate) -functionalized graphene/poly(vinylidine fluoride) nanocomposites: Piezoelectric β polymorph formation'. *Polymer*. 2010;51(24):5846–56.

[126] Barstugan R., Barstugan M., Ozaytekin I. 'PBO/graphene added β-PVDF piezoelectric composite nanofiber production'. *Composites Part B: Engineering*. 2019;158(8):141–8.

[127] Kim H., Islam M.T., Md Didarul I., *et al.* 'Increased piezoelectric response in functional nanocomposites through multiwall carbon nanotube interface and fused-deposition modeling three-dimensional printing'. *MRS Communications*. 2017;7(4):960–6.

[128] Yang L., Cheng M., Lyu W, *et al.* 'Tunable piezoelectric performance of flexible PVDF based nanocomposites from MWCNTS/graphene/MnO2 three-dimensional architectures under low poling electric fields' in *Composites part A: applied science and manufacturing*. Vol. 107; 2018. pp. 536–44.

[129] Avilés F., Oliva-Avilés A.I., Cen-Puc M. 'Piezoresistivity, strain, and damage self-sensing of polymer composites filled with carbon nanostructures'. *Advanced Engineering Materials*. 2018;20(7):1701159.

[130] Pike G.E., Seager C.H. 'Percolation and conductivity: a computer study. I'. *Physical Review B*. 1974;10(4):1421–34.

[131] Bauhofer W., Kovacs J.Z. 'A review and analysis of electrical percolation in carbon nanotube polymer composites'. *Composites Science and Technology*. 2009;69(10):1486–98.

[132] Watters A.L., Palmese G.R. 'Ultralow percolation threshold of single walled carbon nanotube-epoxy composites synthesized via an ionic liquid dispersant/initiator'. *Materials Research Express*. 2014;1(3):035013.

[133] Hu N., Masuda Z., Yan C., Yamamoto G., Fukunaga H., Hashida T. 'The electrical properties of polymer nanocomposites with carbon nanotube fillers'. *Nanotechnology*. 2008;19(21):215701.

[134] Pham G.T., Park Y.-B., Liang Z., Zhang C., Wang B. 'Processing and modeling of conductive thermoplastic/carbon nanotube films for strain sensing'. *Composites Part B: Engineering*. 2008;39(1):209–16.

[135] Ren X., Seidel G.D. 'Computational micromechanics modeling of inherent piezoresistivity in carbon nanotube–polymer nanocomposites'. *Journal of Intelligent Material Systems and Structures*. 2013;24(12):1459–83.

[136] Avilés F., Oliva-Avilés A.I., Cen-Puc M, Piezoresistivity S. 'Piezoresistivity, strain, and damage Self-Sensing of polymer composites filled with carbon nanostructures'. *Advanced Engineering Materials*. 2018;20(7):1701159.

[137] Flandin L., Chang A., Nazarenko S., Hiltner A., Baer E. 'Effect of strain on the properties of an ethylene-octene elastomer with conductive carbon fillers'. *Journal of Applied Polymer Science*. 2000;76(6):894–905.

[138] Theodosiou T.C., Saravanos D.A. 'Numerical investigation of mechanisms affecting the piezoresistive properties of CNT-doped polymers using multi-scale models'. *Composites Science and Technology*. 2010;70(9):1312–20.

[139] Mohiuddin M., Hoa S.V. 'Estimation of contact resistance and its effect on electrical conductivity of CNT/PEEK composites'. *Composites Science and Technology*. 2013;79(15):42–8.

[140] Fang W., Leung S.N. 'Elucidation of structure-to-property relationships of piezoresistive polymer-carbon nanotube nanocomposites'. *Journal of Applied Physics*. 2015;118(4):044907.

[141] Simmons J.G. 'Electric tunnel effect between dissimilar electrodes separated by a thin insulating film'. *Journal of Applied Physics*. 1963;34(9): 2581–90.

[142] Hu N., Karube Y., Yan C., Masuda Z., Fukunaga H. 'Tunneling effect in a polymer/carbon nanotube nanocomposite strain sensor'. *Acta Materialia*. 2008;56(13):2929–36.

[143] Chaurasia A.K., Seidel G.D. 'Computational micromechanics analysis of electron-hopping-induced conductive paths and associated macroscale piezoresistive response in carbon nanotube–polymer nanocomposites'. *Journal of Intelligent Material Systems and Structures*. 2014;25(17):2141–64.

[144] Luan C., Movva S., Wang K., Yao X., Zhang C., Wang B. 'Towards next-generation fiber-reinforced polymer composites: a perspective on multifunctionality'. *Functional Composites and Structures*. 2019;1(4):042002.

[145] Conor P.C., Owston C.N. 'Electrical resistance of single carbon fibres'. *Nature*. 1969;223(5211):1146–7.

[146] Blazewicz S., Patalita B., Touzain P. 'Study of piezoresistance effect in carbon fibers'. *Carbon*. 1997;35(10):1613–18.

[147] Gordon D.A., Wang S., Chung D.D.L. 'Piezoresistivity in unidirectional continuous carbon fiber polymer-matrix composites: single-lamina composite versus two-lamina composite'. *Composite Interfaces*. 2004;11(1):95–103.

[148] Huang H., Yang C., Wu Z. 'Electrical sensing properties of carbon fiber reinforced plastic strips for detecting low-level strains'. *Smart Materials and Structures*. 2012;21(3):035013.

[149] Huang H., Wu Z.S. 'Static and dynamic measurement of low-level strains with carbon fibers'. *Sensors and Actuators A: Physical*. 2012;183:140–7.

[150] Matzies T., Christner C., Horoschenkoff A., Rapp H. 'Carbon fibre sensor meshes: simulation and experiment'. *Computational Materials Science*. 2012;64(10-11):122–5.

[151] Mol Menamparambath M., Arabale G., Nikolaev P., Baik S., Arepalli S. 'Near-infrared fluorescent single walled carbon nanotube-chitosan composite: interfacial strain transfer efficiency assessment'. *Applied Physics Letters*. 2013;102(17):171903.

[152] Panozzo F., Zappalorto M., Carraro P.A., Quaresimin M. 'Electrical resistance change vs damage state in cracked symmetric biaxial laminates: a closed form solution'. *Composite structures*. 2017;184.

[153] Wichmann M.H.G., Buschhorn S.T., Gehrmann J., Schulte K. 'Piezoresistive response of epoxy composites with carbon nanoparticles under tensile load'. *Physical Review B*. 80 12; 2009. p. 245437.

[154] Ku-Herrera J.J., Avilés F., Seidel G.D. 'Self-sensing of elastic strain, matrix yielding and plasticity in multiwall carbon nanotube/vinyl ester composites'. *Smart Materials and Structures*. 2013;22(8):085003.

[155] Yin G., Karube Y., Liu Y., *et al*. 'A carbon nanotube/polymer strain sensor with linear and anti-symmetric piezoresistivity'. *Journal of composite materials*. 2011;45(12):1315–23.

[156] Rahman R., Servati P. 'Effects of inter-tube distance and alignment on tunnelling resistance and strain sensitivity of nanotube/polymer composite films'. *Nanotechnology*. 2012;23(5):055703.

[157] Vertuccio L., Guadagno L., Spinelli G., Lamberti P., Tucci V., Russo S. 'Piezoresistive properties of resin reinforced with carbon nanotubes for health-monitoring of aircraft primary structures'. *Composites Part B: Engineering*. 2016;107(055001):192–202.

[158] Kang I., Schulz M.J., Kim J.H., Shanov V., Shi D. 'A carbon nanotube strain sensor for structural health monitoring'. *Smart Materials and Structures*. 2006;15(3):737–48.

[159] Ferrreira A., Rocha J.G., Ansón-Casaos A., Martínez M.T., Vaz F., Lanceros-Mendez S. 'Electromechanical performance of poly(vinylidene fluoride)/ carbon nanotube composites for strain sensor applications'. *Sensors and Actuators A: Physical*. 2012;178(6):10–16.

[160] Zhang W., Suhr J., Koratkar N. 'Carbon nanotube/polycarbonate composites as multifunctional strain sensors'. *Journal of Nanoscience and Nanotechnology*. 2006;6(4):960–4.

[161] Park M., Kim H., Youngblood J.P. 'Strain-dependent electrical resistance of multi-walled carbon nanotube/polymer composite films'. *Nanotechnology*. 2008;19(5):055705.

[162] Bautista-Quijano J.R., Avilés F., Aguilar J.O., Tapia A. 'Strain sensing capabilities of a piezoresistive MWCNT-polysulfone film'. *Sensors and Actuators A: Physical*. 2010;159(2):135–40.

[163] Aoki H., Dresselhaus M.S. '*Physics of Graphene*'. *Springer Science & Business Media*. 2013.

[164] Huang M., Pascal T.A., Kim H., Goddard W.A., Greer J.R. 'Electronic-mechanical coupling in graphene from in situ nanoindentation experiments and multiscale atomistic simulations'. *Nano Letters*. 2011;11(3):1241–46.

[165] Zhang Y., Lei C., Kim W.S. *Design optimized membrane-based flexible paper accelerometer with silver nano ink*. Applied Physics Letters; 2013;6(8).

[166] Fu X.-W., Liao Z.-M., Zhou Y.-B, *et al*. 'Graphene/zno nanowire/graphene vertical structure based fast-response ultraviolet photodetector'. *Applied Physics Letters*. 2012;100(22):223114.

[167] Zhao J., He C., Yang R, *et al*. 'Ultra-sensitive strain sensors based on piezoresistive nanographene films'. *Applied Physics Letters*. 2012;101(6):063112.

[168] Zha J.-W., Zhang B., Li R.K.Y., Dang Z.-M. 'High-performance strain sensors based on functionalized graphene nanoplates for damage monitoring' in *Composites science and technology*; 2016;123:. pp. 32–38.

[169] Acquarelli C., Paliotta L., Tamburrano A., De Bellis G., Sarto M. 'Electromechanical properties of multilayer graphene-based polymeric composite obtained through a capillary rise method'. *Sensors*. ;16(11).n.d. E1780.

[170] Chen Z., Ren W., Gao L., Liu B., Pei S., Cheng H.-M. 'Three-dimensional flexible and conductive interconnected graphene networks grown by chemical vapour deposition'. *Nature Materials*. 2011;10(6):424–28.

[171] Palza H., Garzon C., Rojas M. 'Elastomeric ethylene copolymers with carbon nanostructures having tailored strain sensor behavior and their interpretation based on the excluded volume theory'. *Polymer International*. 2016;65(12):1441–8.

[172] Shang S., Yue Y., Wang X. 'Piezoresistive strain sensing of carbon black /silicone composites above percolation threshold'. *Review of Scientific Instruments*. 2016;87(12):123910.

Chapter 20

Tailoring thermo-mechanical properties of hybrid composite-metal bonded joint

Xiaolong Zhang[1], Vijay Kumar Thakur[2], Adrianus Indrat Aria[3], and Hamed Yazdani Nezhad[1,4]

The metallic substrates and polymer adhesive in composite-metal joints have a large coefficient of thermal expansion (CTE) mismatch, which is a barrier in the growing market of electric vehicles and their battery structures. It is known that adding carbon nanotubes (CNTs) to the adhesive reduces the CTE of the CNT-adhesive multi-material system, therefore, when used in adhesively bonded joints it would, theoretically, result in low CTE mismatch in the joint system, and thus would reduce the risk of joint failure at high temperatures. In this research, the influence of two specific mass ratios of CNTs on the CTE of the polymer was evaluated. A laser extensometer was utilised to measure the thermal strains over the surface of thin specimens representative of joint's substrates (composite and metal) subjected to uniform and gradually increasing temperature from 30 to 120. The CTEs of the specimens were calculated, compared and analysed.

The results indicated that the addition of 1.00 wt% and 2.68 wt% of multi-walled CNTs (MWCNTs) in the adhesive reduce the CTE of the polymer adhesive from 7.5×10^{-5} °C^{-1} (pristine level) to 5.87×10^{-5} °C^{-1} and $4.43 \times 10^{-5} \times$ °C^{-1}, respectively by 22% and 41% reduction. Theoretical calculations showed that the CTE should have been reduced to 3.6×10^{-5} °C^{-1} (52% reduction) and 1.4×10^{-5} °C^{-1} (81%). The Raman spectroscopy was used to evaluate the quality of the MWCNTs in the lab and the result showed there were many defects in the raw material, which may attribute to the difference between theory and experiment in the CTE estimation. Moreover, scanning electron microscopy (SEM) was used to evaluate the dispersion of MWCNTs on the surface and cross-section of the specimens, from which

[1]Enhanced Composites and Structures Centre, School of Aerospace, Transport and Manufacturing, Cranfield University, Cranfield, UK

[2]Biorefining and Advanced Materials Research Centre, SRUC, Edinburgh, UK

[3]Surface Engineering and Precision Centre, School of Aerospace, Transport and Manufacturing, Cranfield University, Cranfield, UK

[4]Aeronautics and Aerospace Research Centre, University of London, London, UK

Figure 20.1 Adhesive applied to the battery pack system [3]

agglomeration of MWCNTs were observed distributed over a large area of the specimens. Therefore, the enhancement of MWCNTs on the properties of polymer adhesive was impaired by previous two negative factors.

20.1 Introduction

Due to its advantages of energy saving and environmental protection, high-performance electric vehicles are under rapid development [1]. The structure adhesive bonding technology, as a new type of lightweight structure connection technology nowadays, has a good performance to connect heterogeneous material and solves the technical bottlenecks in traditional connection process. Besides, it has not only ideal heterogeneous connection performance but also good sealing and fatigue resistance compared with traditional welding and other connection methods. Because of these advantages, this technology has been applied in the assembling process of the high-performance electric vehicles [2]. For example, the adhesive bonding technology is applied on the battery pack system, as shown in Figure 20.1, which is one of the core components of the electric vehicles and determines the mileage and speed of the electric vehicles [4].

The main problem of this battery pack system is that the batteries discharge at different rates during use and generate a lot of heat at different rates of heat generation. Besides, the accumulation of time and space effect will accumulate a lot of heat. The temperature in the battery pack area can be up to 200 [5]. The high temperature will lead to thermal expansion and residual stress in the battery compartments, resulting in failure of composite-metal bonded joints. The joint failures in the battery pack system will result in some serious problems, such as cracks appear in joints and moisture in the air enters the battery [2], reducing the performance of the battery, and even leading to the failure of entire battery pack system, leading to traffic accidents

The main factor impacting the failure of the bonding joint is the coefficient of thermal expansion (CTE) mismatch among components of the joint, the metal and polymer adhesives. In some studies, it is said that the carbon nanotubes (CNTs) can be added to the epoxy-based adhesive to adjust the CTE of the adhesive, therefore, the adhesives with different CTEs can be produced for various requirements [6].

20.1.1 Aim

As the CTE mismatch is the main source inducing the internal shear stress and resulting in the joint failure at a high temperature. This project is aimed to analyse the relationship between the mass ratio of CNTs in the adhesive and the CTE of the adhesive, verifying the proper mass ratio of CNTs in the adhesive to optimise the CTE mismatch between adhesive and the metal in the joint, reducing the risk of joint failure at a high temperature.

20.1.2 Objectives

The following objectives were defined with the intention of successfully accomplishing the aim of the project:

1. Analytical calculation
   Calculating the potential weight fraction of MWCNTs in polymer by analytical calculations. The calculation result will be adopted in the specimens manufacturing.
2. Dispersion method of high mass ratio MWCNTs in polymer
   The MWCNTs with a high mass ratio are not easy to disperse well in the polymer because of severe agglomeration of MWCNTs, high viscosity and poor flowability of the polymer. So, a high-efficient dispersion method is necessary to ensure the final property of the composite material.
3. Specimen manufacturing
   The polymer specimens with CNTs and polymer-metal bonded specimens with CNTs are prepared. The MWCNTs used in the manufacturing process are produced by a PHD student in a lab. The specimens are manufactured according to the procedures regarding surface treatments, curing conditions, adhesive application etc.
4. Thermal strain measurement
   The thermal strains of specimens are measured with relevant equipment at a high temperature to assess the influence of CNTs on the CTE of the polymer used as adhesive in the joints.
5. Microscopy
   The surface and cross-section of specimens will be analysed with SEM to assess the dispersion quality of MWCNTs in polymer and Raman spectroscopy will be applied to assess the quality of CNTs.

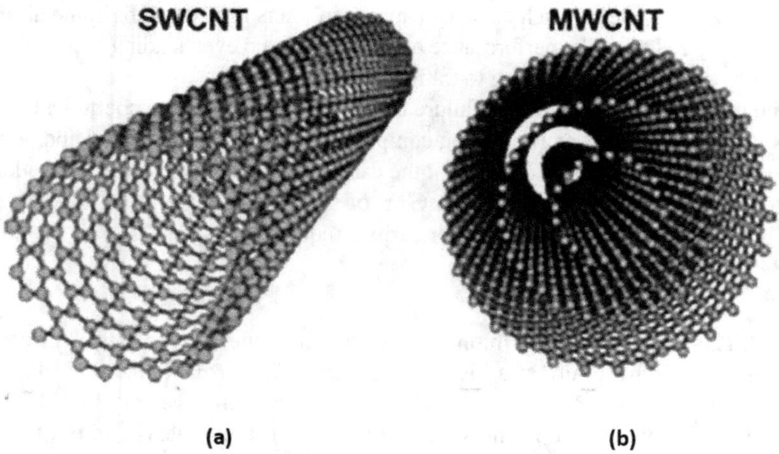

Figure 20.2 Diagram of (a) SWCNT (b) MWCNT [8]

20.2 Literature review

20.2.1 Carbon nanotube

20.2.1.1 Introduction of carbon nanotube

CNTs are tubular structures consisted of carbon atoms, having a nanoscale diameter and length is micron-sized [7]. The CNT has two forms of structures. (1) Single-walled carbon nanotube (SWCNT): regarding as a single layer of graphene curled to form a hollow cylinder, as seen in Figure 20.2(a). (2) Multi-walled carbon nanotube (MWCNT): consisting of two or more nested nanotubes, as seen in Figure 20.2(b).

Because of their nanostructure and the strength of the bonds between atoms, these two materials have excellent mechanical stiffness and tensile strength over 100GPa [6]. They also have other valuable properties like chemical stability, high electrical conductivity and thermal conductivity, which can be applied in many manufacturing industries [9].

20.2.1.2 Production of CNTs in multicomp laboratory

Chemical vapor deposition (CVD) is a widely used method to produce CNTs nowadays because it has advantages like low set-up cost and ease of mass-production [10]. The technology applied in production is injection floating catalyst CVD (IFCVD). Table 20.1 indicates the parameters and raw materials. The working procedure of IFCVD [11] is shown in Figure 20.3.

The MWCNTs produced are shown in Figure 20.4, the MWCNTs exist as lumps with different dimensions. The average diameter of MWCNTs produced is around 60 nm (from 20 to 100 nm) measured with SEM, as shown in Figure 20.5.

Table 20.1 Materials and parameters for IFCVD in the lab

Hydrocarbon	Toluene
Catalyst	Ferrocene
Temperature	750°C
Pressure	Normal atmosphere
Gas flow rate	6.5 mL/h
Deposition time	3.5 hour

20.2.2 Composite-metal adhesively bonded joint

Adhesive bonding is a brave new bonding technology besides conventional jointing technologies such as rivets and welding [13]. Adhesive bonding is more efficient in structural bonding compared to conventional joining due to its advantages like weight reduction and elevated stiffness and this technology has been widely applied in manufacturing of vehicle and aviation industry [14].

Adhesive bonding joins are consisted of adhesives and adherents. The adhesives can be epoxies, polyesters, polyurethanes, methacrylate, and so on [15]. The epoxy-based adhesives are the most common types due to their high performance and easy application [14]. The adherents can be similar materials like metal-metal, composite plate-composite plate. Additionally, dissimilar materials can be bonded together like metal-composite plate, which is the greatest advantage of the adhesive bonding.

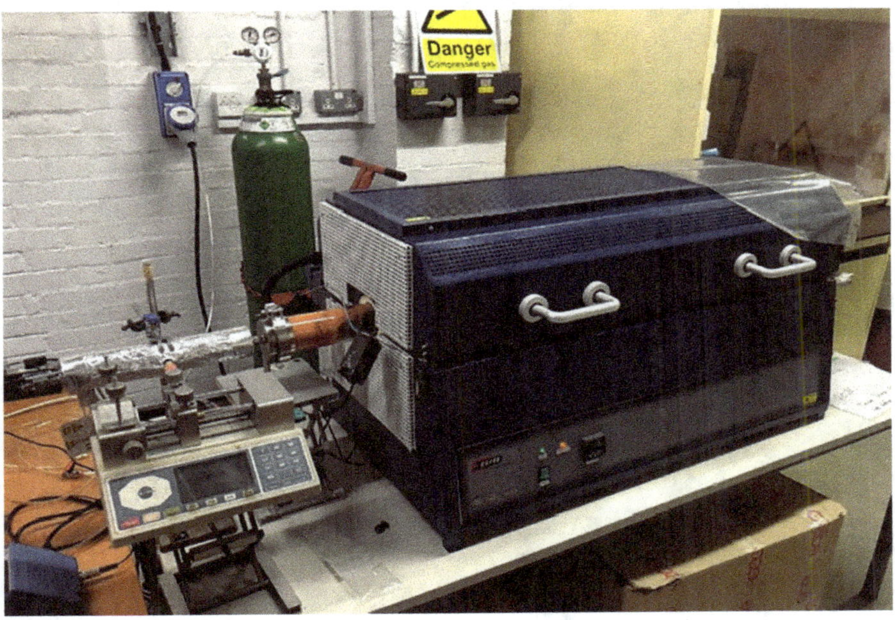

Figure 20.3 Equipment for IFCVD

Figure 20.4 MWCNTs produced in the lab

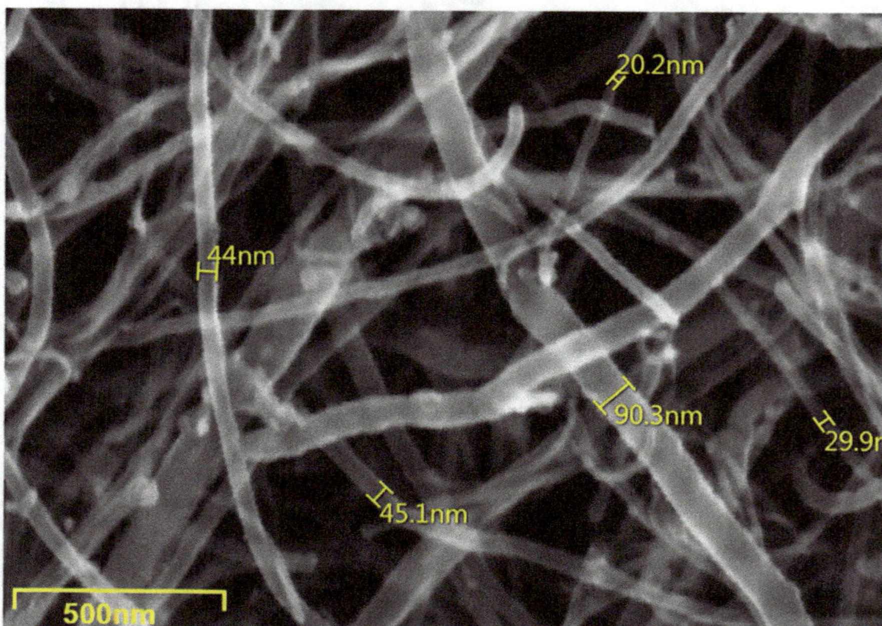

Figure 20.5 SEM image of MWCNTs [12]

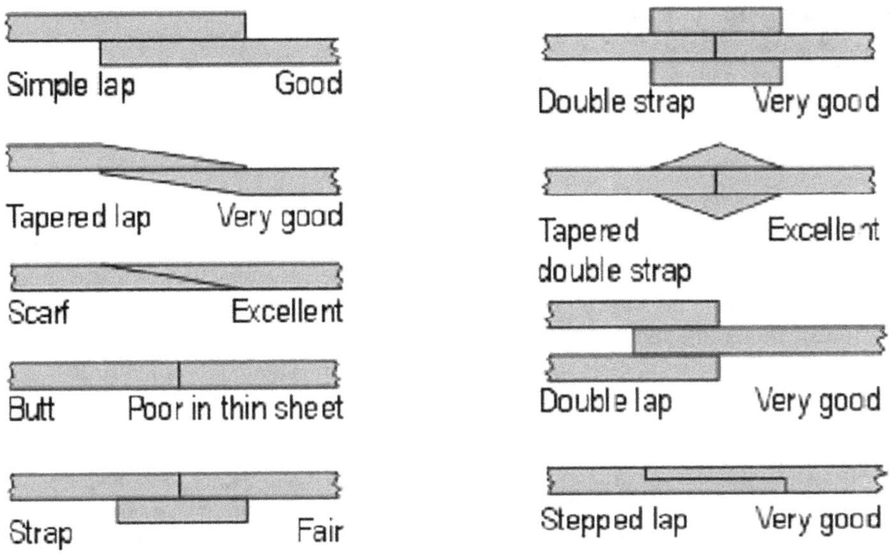

Figure 20.6 Various configurations of adhesive bonding joints [13]

20.2.2.1 Types of joints

There are various configurations that adhesive bonding joints can be made, as shown in Figure 20.6.

20.2.2.2 Advantages of adhesive joint

Adhesively bonded joints have numerous advantages in the growing composite market such as:

- Dissimilar materials can be bonded together.
- The stiffness and strength of joint are better compared to conventional joints.
- There are fewer stress concentrations and the load is distributed over the whole bonding area.
- There is no need to change the adherents physically, like cutting or drilling.
- The weight of joints is reduced without nuts and bolts.

The bonding operation can be automatic in mess production [13].

20.2.2.3 Disadvantages of adhesive joint

Although so many advantages have been mentioned, the adhesive joint also has some disadvantages, such as:

- The joint strength is sensitive to variations in service temperature, humidity and more influenced by the surface treatments to the adherents.

- Lacking a non-destructive testing (NDT) technique for quality evaluation of adhesive joints.
- The joints cannot be disassembled to check or replace the broken parts.
- Thick joints are susceptible to high peel stress [13].

20.2.3 Effects of CNTs on thermo-mechanical performance

A few studies demonstrated that CNTs have some good mechanical properties like high tensile strength and characteristics like low density and high surface area. Which make CNTs become a good filler, improving the properties of epoxy-based composite materials, such as thermal and mechanical properties [16]. Also, the CTE of the epoxy-based composite materials will change by adding CNTs inside [6].

20.2.3.1 Mechanical properties

CNTs have excellent mechanical properties, like high Young's modulus and tensile strength. The high aspect ratio of CNTs can improve the load transmission between the CNTs and epoxy to enhance the mechanical properties. Large interfacial interactions between CNTs and polymer, which benefit from high surface area effectively reinforce all the mechanical properties [17].

A research has been done on the fatigue strengths of a composite-to-aluminium joints with 2 wt% CNTs. The result showed that the static strengths of joints decreased by 36.6%, the fatigue strengths increased by 12.8%. Also, the endurance limit of fatigue lives was 12.8% higher [13]. In another research, it has been approved that CNTs with higher aspect ratios could enhance the impact resistance and fracture toughness of nanocomposites more effectively [17].

20.2.3.2 Thermal properties

The effects of CNTs on thermal conductivity are influenced by the content, aspect ratio, dispersion quality of CNTs and their interfacial interactions with the polymer. The CNTs have good thermal conductivity and thermal stability, so CNTs can be applied to improve the thermal properties of polymers.

A large number of experiments were conducted to explore the enhancement of CNTs on thermal conductivity of polymer and researchers observed some strengthening effects. Besides the thermal conducting properties, the thermal stability and flammability of polymer can also be enhanced adding CNTs [17].

20.2.3.3 CTE of polymer

The structure of composite materials determines the value of CTE. The CTE depends on the interaction between phases and properties of each phase in the material. Pure epoxy resins have a higher CTE, generally at $7–7.5 \times 10^{-5} K^{-1}$, but MWCNTs show a negative CTE [6]. When MWCNTs are added to the epoxies, the CTE of MWCNTs/epoxy composite material is lower than that of pure epoxies. Good dispersion quality of MWCNTs in the epoxies also decreases the CTE of composite material.

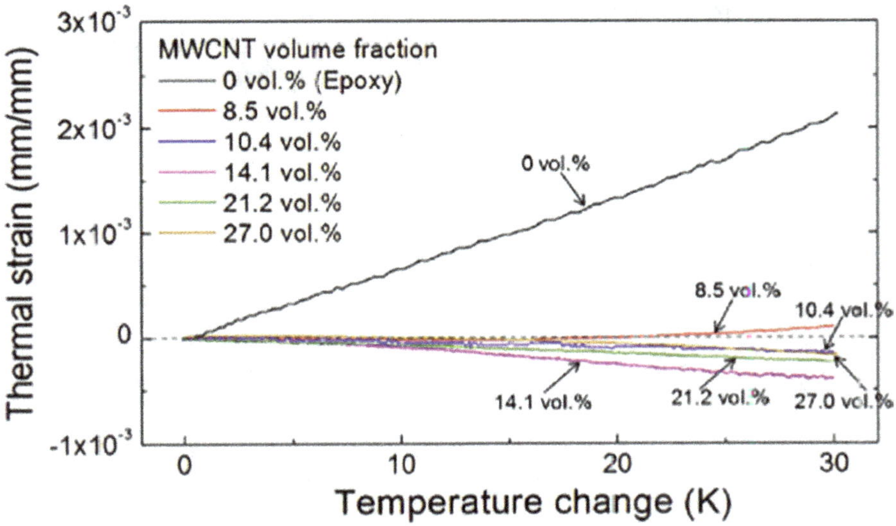

Figure 20.7 *Thermal strain of composites with various volume fractions of MWCNT [6]*

The numerical results in some studies indicated that the thermal strain of the composite with 8.5 vol% MWCNTs decreased rapidly compared to the pure epoxy. If there was more than 10.4 vol% MWCNTs, the thermal contraction would occur on the specimens, as shown in Figure 20.7.

The function between volume fraction and CTE of composites was expressed by the following equation [6]. There are only six accurate data from the experiments in the literature, in order to get the linear relationship of these two variates, the least-squares regression analysis was employed to provide the regression line in Figure 20.8.

$$\alpha_C = \frac{\alpha_m \left(1 - V_f\right) E_m + \alpha_f V_f E_f}{\left(1 - V_f\right) E_m + V_f E_f} \tag{20.1}$$

α_c: CTE of the composites in the direction of the MWCNT alignment
E_m: Young's modulus of the epoxy $E_m = 2.5$ GPa
E_f: Young's modulus of the MWCNT's $E_f = 180$ GPa
α_m: CTE of the epoxy $\alpha_m = -7.0 \times 10^{-5} K^{-1}$
α_f: CTE of the MWCNTs $\alpha_f = -1.2 \times 10^{-5} K^{-1}$
V_f: MWCNT volume fraction

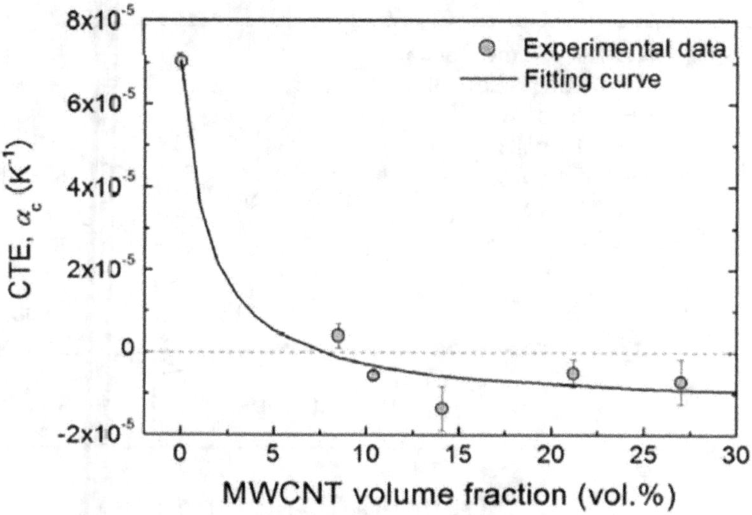

Figure 20.8 CTEs of composites with different MWCNT volumes [6]

20.2.4 CNTs dispersion

As described in section 20.2.3, the dispersion quality of CNT in the polymer is important and affects the properties of CNTs/polymer composite materials. A good dispersion of CNTs provided more surface area to bond with matrix, and it reduced the amount of aggregated CNTs (shown in Figure 20.9). The stress concentration occurred on the agglomeration of CNTs with external load. There are various methods have been applied for CNT dispersion, and there are some problems to solve [17].

20.2.4.1 CNTs agglomeration

The agglomeration of CNTs (shown in Figure 20.10) is still the most serious problem for dispersing CNTs in epoxy, and the following factors result in this problem.

20.2.4.1.1 Size of CNTs

It has been well recognized that CNTs have a nanometer-scale diameter. While they have a large aspect ratio, can be over 1,000, inducing the large surface area. So the problem of dispersing CNTs is quite different from other conventional materials, such as carbon fibres. The CNTs can be considered as nano-sized ropes and they are quite easy to entangle to bundles after the growing process, as shown in Figure 20.11, inducing a serious problem in dispersion.

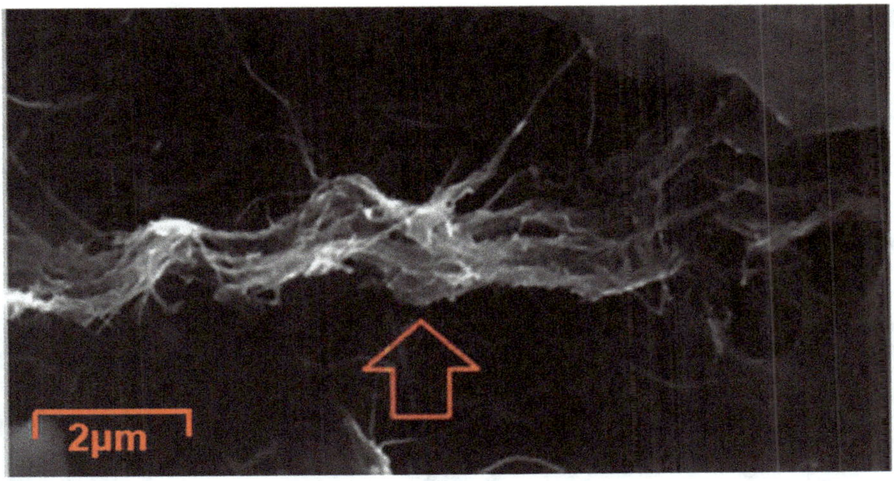

Figure 20.9 CNTs agglomeration [11]

20.2.4.1.2 Interfacial interaction

Interfacial interaction is quite important to CNTs/polymer composites, as all the CNTs have a large surface area. There is the strong electrostatic interaction and van der Waals force between the CNTs [18]. Which is the main reason why CNTs appear as entanglements, increasing the viscosity of the CNT/polymer systems. The composites will have a poor mechanical performance with a high viscosity.

20.2.4.1.3 Content of CNTs

The content of CNTs in the polymer is also a key issue in CNTs dispersion of because it will make the previous two problems worse as the increase of the content [17]. Which means that although CNTs can improve the properties of the epoxy resin, there is a critical volume fraction of CNTs in the polymer. When the volume fraction of CNTs is under the critical content, the CNTs will have a good enhancement on the properties of CNT/epoxy composites. While if there are too many CNTs, the CNTs

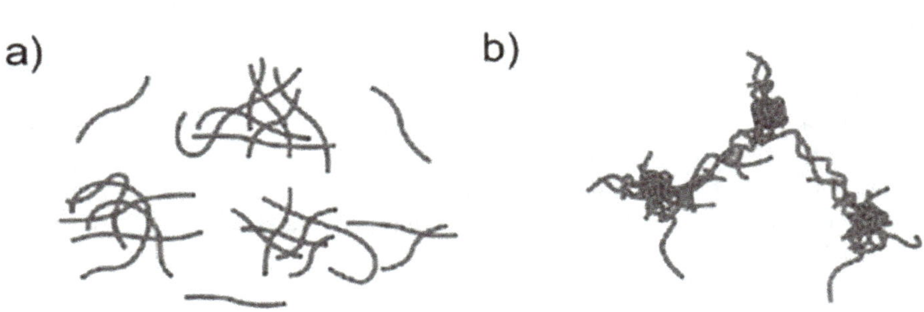

Figure 20.10 (a) Light and (b) heavy agglomeration of CNTs [11]

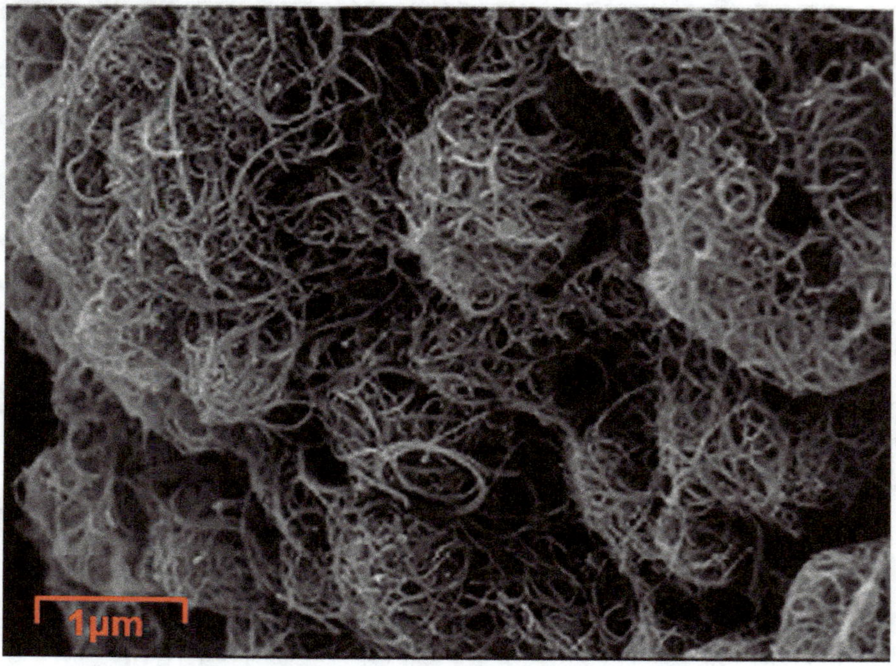

Figure 20.11 Entangled MWCNTs [15]

will have negative effects on the properties of CNTs/epoxy composites, and the properties will be even lower than the properties of pure epoxy resin in some cases.

20.2.4.2 Mechanical dispersion methods of CNTs

CNTs with a small volume contain an extremely large number of particles and these particles have the large surface area, resulting in the difficult dispersion in the polymer matrix. In order to give full play to improve the enhancement of CNTs on the polymer, an efficient particle dispersion technology is of great necessity [17]. In other words, the CNTs dispersion technology greatly affects the performance of CNTs/polymer nanocomposites in all aspects.

20.2.4.2.1 Ultrasonication

Ultrasonication is a widely used small-scale method in the laboratory. Ultrasound energy is used to stir the particles in the solvent during ultrasonication process. In the lab, ultrasonication process can be conducted with a sonicator, and the sonicator has different modes, e.g., the ultrasonic bath and the ultrasonic probe, as shown in Figure 20.12.

During the sonication process, the attenuated waves are produced by the compressed ultrasound. The individual particles or agglomerates which located on the surface of the bundles will peel off because of the shock from these waves and

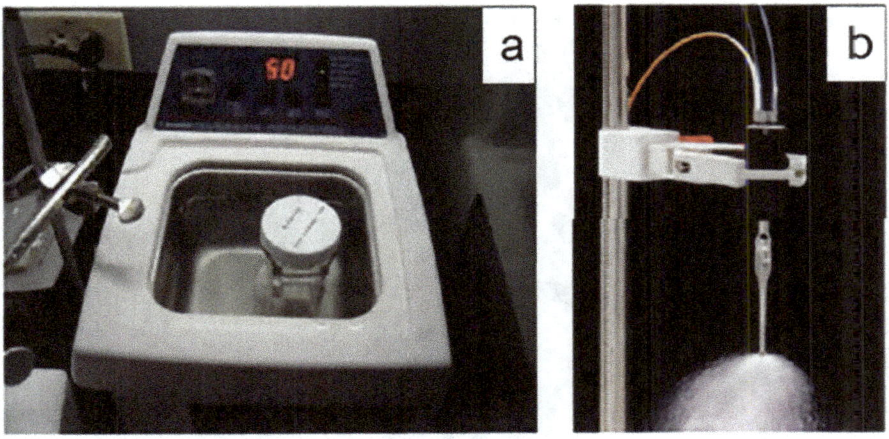

Figure 20.12 (a) Ultrasonic bath. (b) Ultrasonic probe [15].

further disperse in the matrix until uniformity of CNTs meets the requirement of the test. Ultrasonication is an efficient method to disperse CNTs in low-viscosity liquids [19].

CNTs can be seriously damaged if the sonication treatment lasts too much time. Many defects occurred on CNTs by applying an ultrasonic probe as atomic structures of CNTs are destroyed easily [20].

20.2.4.2.2 Stirring

Stirring is another technology to disperse materials in the liquid in the laboratory, and it can be applied to disperse CNTs in the polymer. In contrary to the Ultrasonication, this technology has no damage to the CNTs during the mixing process. The dispersion of particles is affected by some factors, such as the dimension of the specific equipment while mixing speed is the main factor for this mixing process. Epoxy resins usually have a high viscosity, so a higher mixing speed is necessary to achieve a good dispersion quality of CNTs.

20.2.4.2.3 High shear mixer

The CNTs could reagglomerate during the stirring process because of the static conduction, and this has been approved by some researches [21]. While this problem can be overcome with a high-speed shear mixer. The rotation speed of mixer can be up to 1,500 rpm, while a higher speed will induce many defects in CNTs [17]. A high-shear mixer uses a rotating impeller to stir the fluid and create flow and shear in the liquid. The fluid at the edge of the container flows faster than in the middle. The shear occurs because of the difference in the speed of the liquid shear [22]. The high shear mixer in the lab is as shown in Figure 20.13.

Figure 20.13 High shear mixer in the lab

20.2.4.3 Functionalization of CNTs

The properties of composites are influenced by the dispersion quality of the CNTs in the polymer and the interfacial interaction between the CNTs and epoxy. A lot of research has been done on the methods of modifying the surface properties of CNTs [18]. These methods can be conveniently divided into chemical functionalization methods and physical functionalization methods. The difference between these two methods is the interaction between the molecules and carbon atoms of CNTs.

20.2.4.3.1 Chemical functionalization

Chemical functionalization is a bond of functional entities on carbon nanotube-based carbon scaffolds. It occurs mainly on the sidewall of the CNTs. The covalent sidewall functionalization is related to the changes in sp2 hybridization to sp3 and simultaneous loss of the p-conjugated system [23].

Defect functionalization is another available method. By oxidation reaction with strong acids or with strong oxidizing agents, defects can be produced on the surface of CNTs [9]. By combining these defects with carboxylic acid (-COOH) or hydroxyl group (-OH), functional groups with rich chemical properties can be formed [24]. The CNTs functionalized in this way can be well dispersed in many organic solvents. The chemical functionalization method has two disadvantages. The defects occur during functionalization reaction can lead to many damages on the CNTs. Besides, this method is not as good for the environment and has a great threat to personal safety because of the strong acids and oxidants involved.

20.2.4.3.2 Physical functionalization

Non-covalent functionalization is a physical method to modify the interfacial properties of CNTs. The CNTs suspended in the polymer and the polymer wrapped around the CNTs. The wrapping process is induced by the interface interactions between the CNTs and polymer [25].

Besides, the surfactant can also be used for the functionalization of CNTs. For example, Triton X-100 is a popular surfactant in many fields [26].

The surfactant is attached on the CNTs by physical adsorption, changing the surface state of the CNTs and the surface tension is reduced. Consequently, the agglomeration of CNTs reduces rapidly. The excellent properties of surfactant and polymer highly contribute to the performance of the functionalization process.

20.2.5 Mechanism of joint failure by CTE mismatch

The constitution of a composite-metal bonded joint before deformation is shown in Figure 20.14(a). The CTE of polymer-based adhesive is around 7–7.5 × 10^{-5} K^{-1}, while the CTE of the metal is much smaller, e.g., the CTE of steel is only 1–2 × 10^{-5} K^{-1}, there is a big gap between these two CTEs [27]. When the temperature increases, the thermal strain of the two materials is quite different. The adhesive tends to expand as the dotted line shown in Figure 20.14(b), but it is still connected with the substrate before failure, shear stress occurs on the interface between the metal and the adhesive and the direction of stress is shown in Figure 20.14(b). The joint will fail if the shear stress is larger than the adhesion bond strength of adhesive to the steel surface. If there are some defects in the cured adhesive, such as delamination, cracks, porosity or debonding, this shear stress will expedite the failure process of the adhesive joint [13].

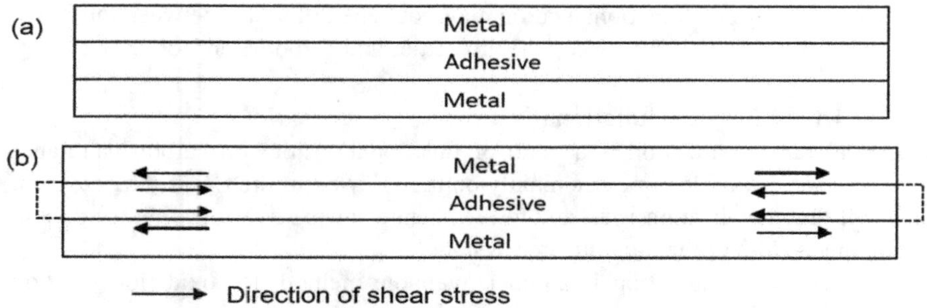

Figure 20.14 (a) Constitution of joint at a low temperature and (b) constitution of joint at a high temperature

20.2.6 Analytical calculations

The volume fraction of the fillers is important for the properties of composites. And in this project, the volume fraction of CNTs is involved in CTE calculation process. However, during the preparing progress of CNTs/polymer composites in the laboratory, it is more convenient to measure the weight of the CNTs compared to the volume of CNTs. We need to transform the mass ratio of CNTs to volume fraction for the calculation. An equation [28] used for the transformation between the mass ratio and volume fraction was found as follows:

$$V_{CNT} = \left(\frac{\rho_{CNT}}{\rho_{polymer}} \times \frac{1}{wt._{CNT}} - \frac{\rho_{CNT}}{\rho_{polymer}} + 1 \right)^{-1} \tag{20.2}$$

VCNT: the volume of CNTs

ρCNT: the density of CNTs

ρPolymer: the density of the polymer

wtCNT: the weight fraction of CNTs

The equation is based on the densities of the constituents in the composites. A specific weight fraction (wt%) can be calculated from the equation with a specific volume fraction of CNTs.

For MWCNTs with an average diameter of 39 nm (15–57 nm), it is possible to reduce the CTE of polymer from 7×10^{-5} K^{-1} to 0.4×10^{-5} K^{-1} using 8.5 vol% of MWCNTs. The fitting curve in Figure 20.15 indicates that the volume fraction of CNTs would be 2.2–3.6% to decrease the CTE of polymer to the CTE range of steel $1–2\times10^{-5}$ K^{-1}. So, use of up to 3 vol% CNTs should satisfy our needs.

Now, the MWCNTs produced have a higher average diameter 60 nm (20 –100 nm, shown in Figure 20.5), 1.5 times radially larger than the MWCNTs with a diameter of 37 nm. As CNTs are generally hollow large molecules, this means that a specific weight fraction would give a higher volume fraction. If the average diameter of CNTs increases by 1.5 times, the volume density of CNTs decreases by 1.5 times. So, we can regard the density of MWCNTs we have as 1.00 g/cm^3. The density of

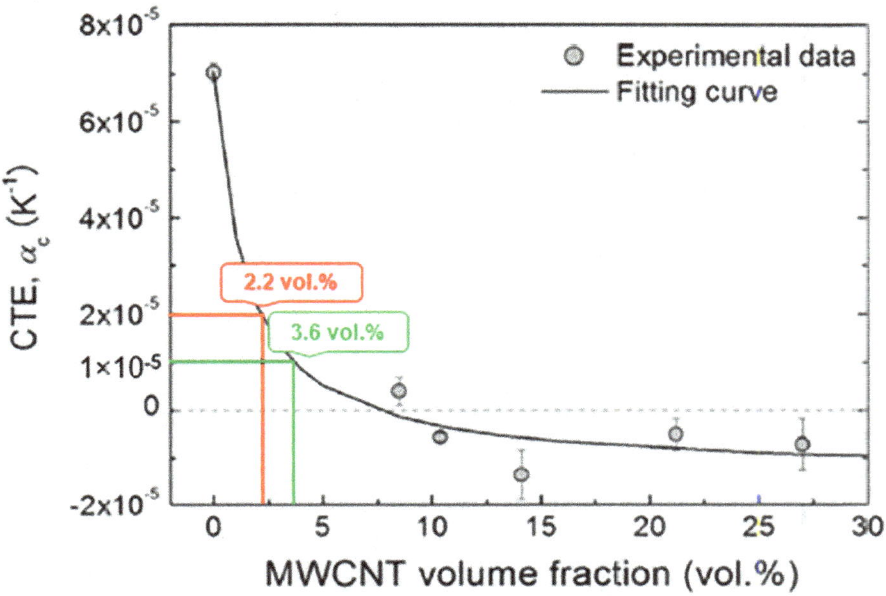

Figure 20.15 Range of volume fraction of CNTs required [6]

the polymer used in this project is 1.1.5 g/cm³ [29]. So, the weight fraction which satisfies our need is 2.68% calculated in (20.2).

20.2.7 Methods of thermal strain measurement

20.2.7.1 Strain gauges

Strain gauges, known as conventional strain sensors, are resistive transducers. Strain gauges are usually attached on a flat surface of a component, the electrical resistance of strain gauges will be changed because of the tensile or compressive stress from the strain of the component, and the change of the electrical resistance corresponds to the specific strain of the component [30].

20.2.7.2 Digital image correlation strain analysis

Digital image correlation (DIC), also known as digital speckle correlation, is used to obtain deformation information of the region from two digital images before and after the deformation of the specimens. The principle of the method is meshing the regions of the image before deformation and for each sub-region, a calculation is carried out according to the defined function through a certain search method, as shown in Figure 20.16. The displacement can be obtained by finding the specific subregion in the deformed image. This method has the advantages of wide-field measurement and high measurement accuracy.

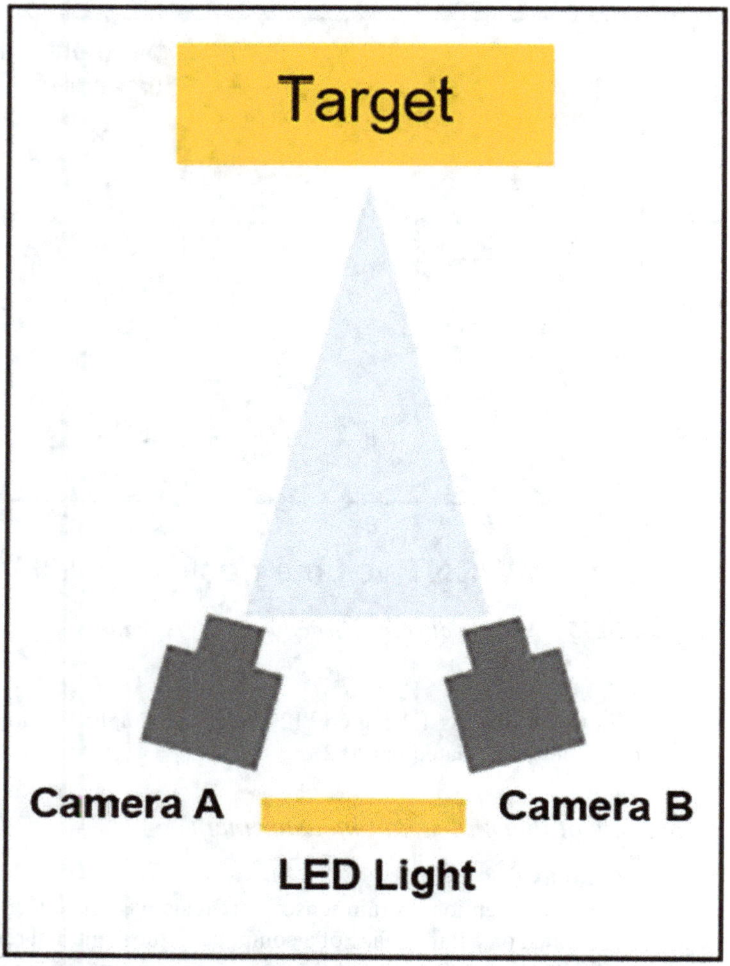

Figure 20.16 Working principle of DIC [31]

20.2.7.3 Laser extensometer

The laser extensometer is also a non-contacting strain measuring method for materials testing. A laser scanner is applied to receive the laser signal and measure the distance between the lasers reflected from the specimens. The extensometer shows the actual measured gauge length, the difference value of the gauge lengths indicates the strain of the specimen. The laser extensometer also has the advantages of high accuracy, easy to use, self-contained and portable [32]. The working principle of the laser extensometer is as shown in Figure 20.17.

In this project, the laser extensometer was selected after the comparison of three measuring methods. There were no strain gauges in the lab. The DIC and

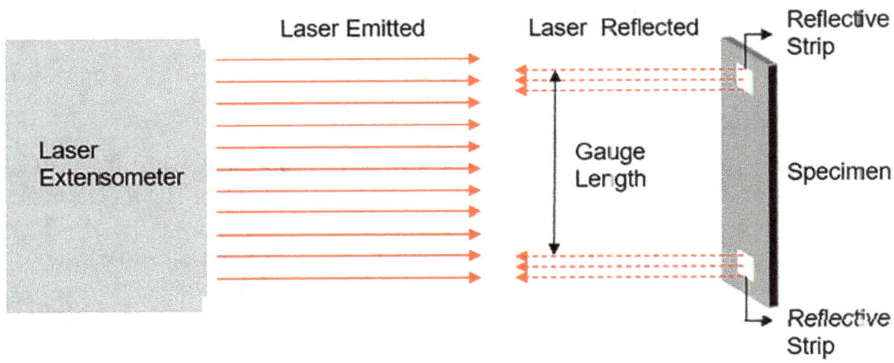

Figure 20.17 Working principle of laser extensometer

laser extensometer equipment were both available, while the specimens needed to be heated during the measurement, it was difficult to set up a hot plate with DIC. The laser extensometer was selected for its portability and high accuracy.

20.3 Specimens manufacturing

20.3.1 Material

20.3.1.1 MWCNTs

The mass ratios of MWCNTs used in the project is 2.68 wt% and 1 wt%. The mass ratio 2.68 wt% comes from the analytical calculation in section 20.2.6. The mass ratio 1 wt% comes from the cost control requirements in the large-scale production process of the factory, as the price of CNTs on sale varies from 15 to 20 pounds/gram [33], more expensive than the polymer.

20.3.1.2 Resin and curing agent

The resin applied in the project is Araldite LY 1564 and the curing agent is Aradur 3486 from HUNTSMAN. The laminating system has a low viscosity, contributing to the dispersion of MWCNTs inside. The mix ratio of Araldite LY 1564 and Aradur 3486 is 100:34 by weight. The recommended working temperature is up to 120. The pot life of the mixture is up to 10 hours which facilitates the mass production of industrial parts [29].

20.3.1.3 Surfactant for CNTs

The Triton X-100 is one of the most popular surfactants used during the dispersion of the MWCNTs. The CMC of Triton X-100 is from 0.22 to 0.24 mM [34] and the molecular weight is 625. So, the CMC is 0.14 mg/mL by weights. When the concentration of Triton X-100 is higher than the CMC, micelles of Triton X-100 occur in the solution and they further contribute to the dispersion quality of CNTs [35]. The concentration of Triton X-100 used in the test is ten times of CMC.

20.3.1.4 Substrate

In this project, the original substrate supposed to be the steel plate, while the steel is replaced by the aluminium plate as there is a lack of steel in the lab. The replacement of aluminium has no negative impacts on the project.

The material of substrates used for specimens is aluminium 6063-T6, widely used in the architectural field and manufacturing industry. It has the advantages of anti-corrosion and good strength, important for ensuring a good working performance at a high temperature.

20.3.2 Specimens manufacturing

20.3.2.1 Schematic of specimens

There are five types of specimens in the test plan, the constitutions and the dimensions of the specimens are shown as follows (Figure 20.18).

20.3.2.2 Surface treatment

The first step in specimens manufacturing is the surface treatment, and this is a very important step, the quality of the surface treatment will have a great impact on the final properties of the specimens. The entire surface of the specimen is the bonding area, which is sanded with 100-grit sandpaper to make the surface chemically active (Figure 20.19). Within 1 hour before gluing, 400-grit sandpaper was used for sanding again to avoid the potential oxidation on the bonding area, and acetone was used to wipe the bonding area to thoroughly clean any remaining contaminants on the surface, so as to avoid the negative effects on the bonding quality and the final performance of the specimens.

20.3.2.3 Mould

The moulds for the manufacturing of specimens were manufactured, as shown in Figures 20.20–20.21. The moulds are treated with release film to avoid the residue of the polymer after curing. The surface treatment with acetone and release agent is also needed right before the application of the adhesive.

The mould 1 is used to manufacture specimen a and specimen b. The thickness of the chamber in mould 1 is 1 mm, an aluminium substrate with the thickness of 0.5 mm is put inside to produce a bonding thickness of 0.5 mm.

The mould 2 is used to manufacture specimen c and specimen d, the thickness of the chamber in mould is 0.5 mm so that the thickness of the specimens will be 0.5 mm as required.

The length and the width of the chambers in the moulds are 110 and 35 mm, a little bit bigger than the specimens, to keep the dimensions of the specimens after trimming the rough edges.

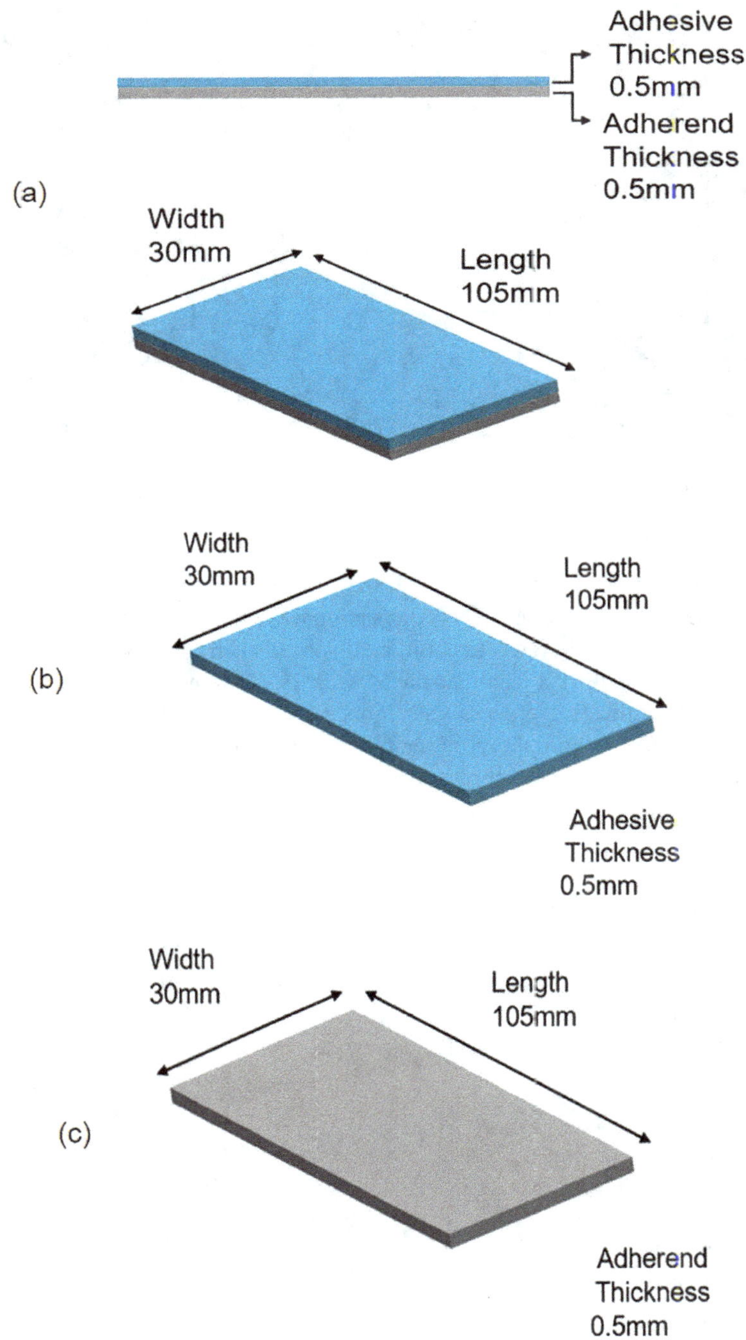

Figure 20.18 *Schematics of specimens[36]*

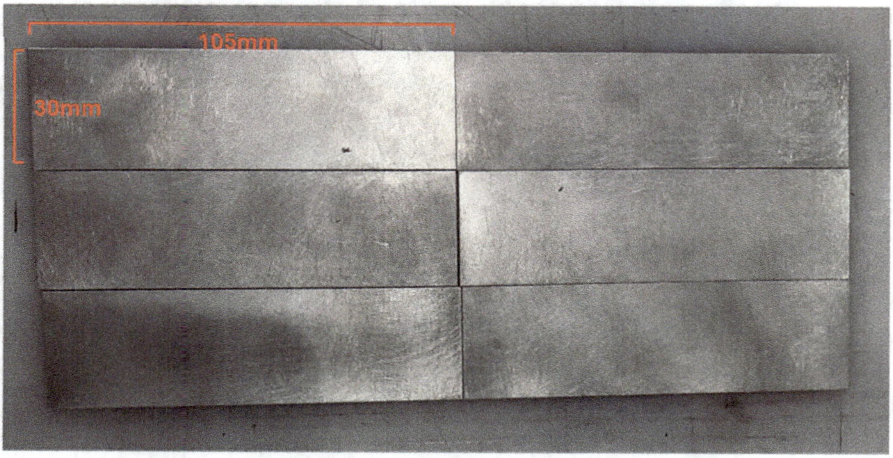

Figure 20.19 *Aluminium plates sanded with 100-grit sandpaper*

20.3.2.4 Adhesive preparation and application

One of the main steps of the MWCNTs dispersion was sonication (see Figure 20.22a), while only 0.1 gram of MWCNTs could be disposed each time by the limitation of the 50 mL beaker on the sonicator, as the MWCNTs would not even flow in the acetone during the sonicating process with more MWCNTs inside. All the MWCNTs were gathered in the high shear mixer (see Figure 20.13), and further treatment lasted for 20 minutes after the sonication process.

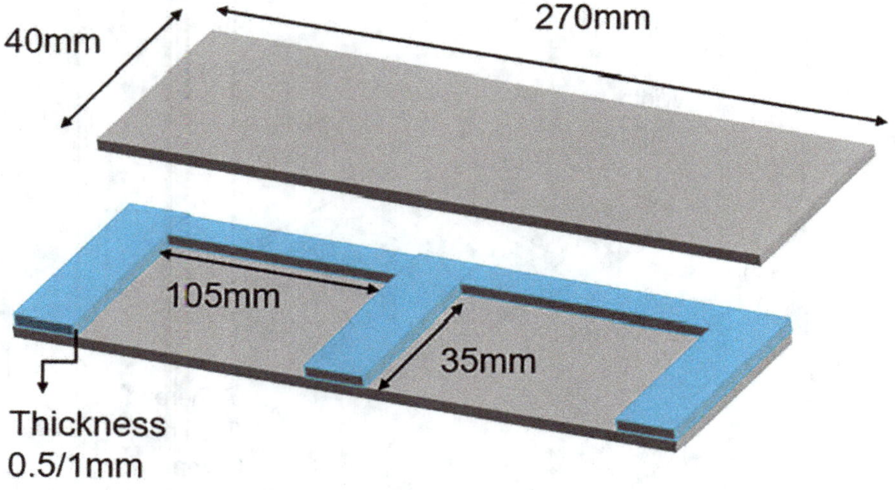

Figure 20.20 *Schematic of the mould*

Figure 20.21 *Image of the moulds*

The resin (Table 20.1) was put in the beaker with MWCNTs on the heating magnetic stirrer, as shown in Figure 20.22(b). The flowability of the resin increased by temperature increasing. The dispersion quality of CNTs increases with a good resin flowability.

The detailed steps for the dispersion process:

- Weighing, 0.1 gram of CNTs each time.
- Adding 40 mL acetone to MWCNTs as the solvent.
- Sonicating for 30 minutes with 80% amplitude.
- Adding Triton X-100 with 56 mg.

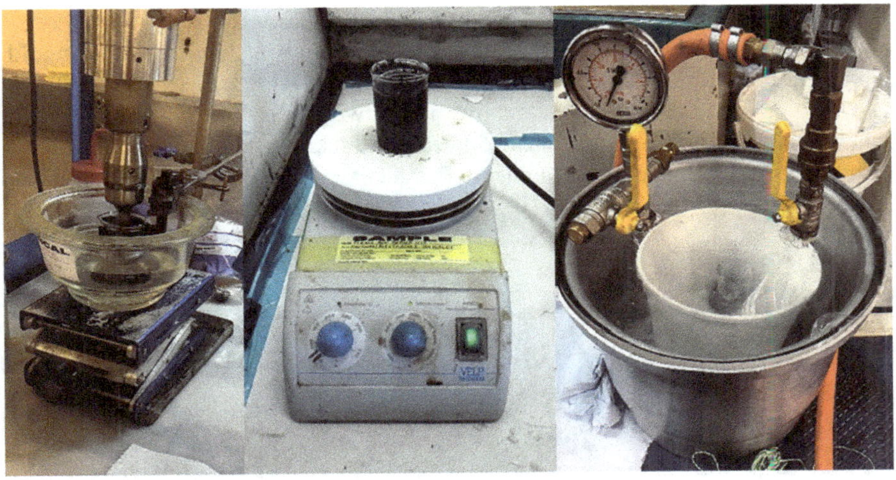

(a) (b) (c)

Figure 20.22 *(a) Sonication (b) Heating magnetic stirring (c) Degassing*

Table 20.2 Quantity of materials for specimens

Material	**Specimen a and c**	**Specimen b and d**	**Specimen e**
Aluminium	105 mm × 30 mm × 0.5 mm	105 mm × 30 mm × 0.5 mm	105 mm × 30 mm × 0.5 mm
Plate	quantity: 2	quantity: 2	quantity: 2
MWCNTs	(2.68 wt%) 0.6 g	(1 wt%) 0.2 g	N/A
Resin and hardener	Araldite LY 1564: 16.7 g Aradur 3486: 5.68 g	Araldite LY 1564: 14.9 g N/A Aradur 3486: 5.1 g	
Surfactant	Triton X-100: 336 mg	Triton X-100: N/A 112 mg	

- Sonicating for 30 minutes with 80% amplitude.
- Gathering all MWCNTs and treating with high shear mixer for 20 minutes.
- Adding resin in the solvent.
- Mixing and evaporating acetone for 5 hours at 80.

When all acetone has been evaporated, adding the curing agent inside the epoxy resin, the required curing agent is shown in Table 20.2. Mixing the adhesive for 30 minutes by hand, scraping the sides and bottom of the vessel frequently. Intense stirring was avoided to minimize air being introduced into the adhesive [37]. Degassing the adhesive (as shown in Figure 20.22c) for 30 minutes to discharge the bubbles in the adhesive. Then the adhesive was applied on the mould and the aluminium substrate with a spatula by hand, as shown in Figure 20.23. The last step of manufacturing is that the moulds were fixed with clamps, the extra adhesive was extruded and cleaned to keep the thickness of the specimens. Two were made for each specimen to ensure the accuracy and the repeatability of the results.

Figure 20.23 Application of adhesive on the mould

Figure 20.24 Specimens after trimming.

20.3.2.5 Curing

The moulds were then moved into a preheated oven to cure the specimens. The curing process lasted for 5 hours at 100 [38] in the oven with the intention to prevent any undercuring.

20.3.2.6 Trimming

The specimens were taken from the moulds carefully after curing. The specimens were trimmed with a cutter knife to remove the residuary side fillets and make sure all the specimens have the same dimension, as shown in Figure 20.24.

20.3.2.7 Final manufactured specimens

As a laser extensometer would be applied in the test, the reflective strips on the specimens are necessary because of the measuring principle of the laser extensometer. Two reflective strips with a length of 10 mm were attached on the surface of the specimen and the gap between two strips is about 50 mm to avoid the big defects in the specimen. The thermocouple was attached on the surface of the specimen to monitor the temperature. Figure 20.25 shows one of the final manufactured specimens.

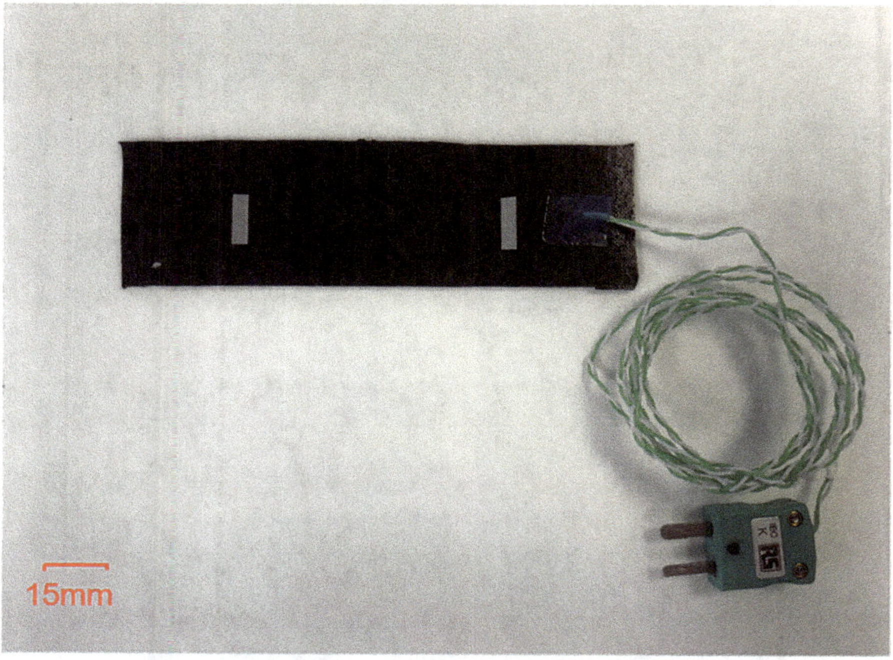

Figure 20.25 Final manufactured specimen

20.4 Testing

20.4.1 Thermal strain measurement

20.4.1.1 Equipment

1. Fume hood

   The measurement was conducted in a fume hood, as the specimens would be heated up to 120 and might be overheated to over 120 if the hot plate did not stop heating in time, the specimens had a possibility to degrade or burn at the high temperature. So, a fume hood was applied for safety purpose.

2. Hot plate

   A hot plate was used as a rig to fix the specimens on the plate, vertically to the laser from the laser extensometer. Also, it was applied to heat the specimens to 120. The heating rate is 5/minute, it should be low enough to allow sufficient time for specimens to produce the thermal strain during the heating process. The heating rate for aluminium specimens can be faster than that of polymer specimen.

3. Laser extensometer

   Laser extensometer was applied to test the deformation on the specimen during the heating process. The difference value of the gauge lengths at the different temperatures indicates the thermal expansion of the specimens.

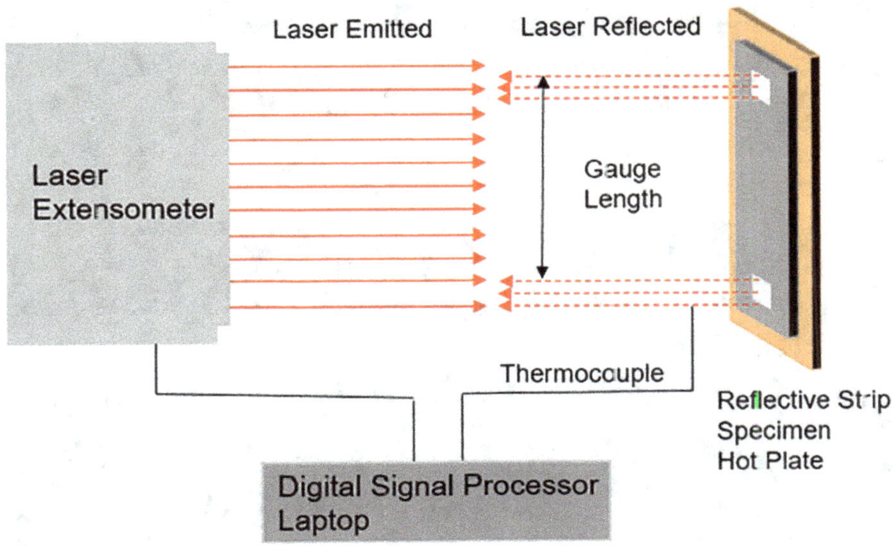

Figure 20.26 *Schematic of thermal strain measurement system*

4. Laptop and digital signal processor

   The digital signal processor was used to collect the signals from the thermocouple and the laser extensometer and then transmit the signals to the laptop. A LABVIEW software in the laptop was used to record all the data, including the time, temperature and gauge lengths for the final analysis.

20.4.1.2 Test scheme

The schematic of the thermal strain measuring system applied in the project is as shown in Figure 20.26. When the test started, the laser emitted from the laser extensometer vertically aimed at the surface of the specimen. And the laser extensometer started to measure and indicate the gauge length between the two reflective strips by the laser reflected from the reflective strips, the gauge length increased by the increasing temperature. The thermocouple monitors the temperature of the specimen during the test and the measurement ended when the temperature rose up to 120. The digital signal processor transmits the data to the laptop and the laptop recorded all the data of the time, gauge length and time automatically for the further analysis as shown in Figure 20.27. The CTE of the specimens is calculated in the following equation:

$$\alpha = \frac{L_{end} - L_{start}}{L_{start} \times \left(T_{end} - T_{start}\right)} \tag{20.1}$$

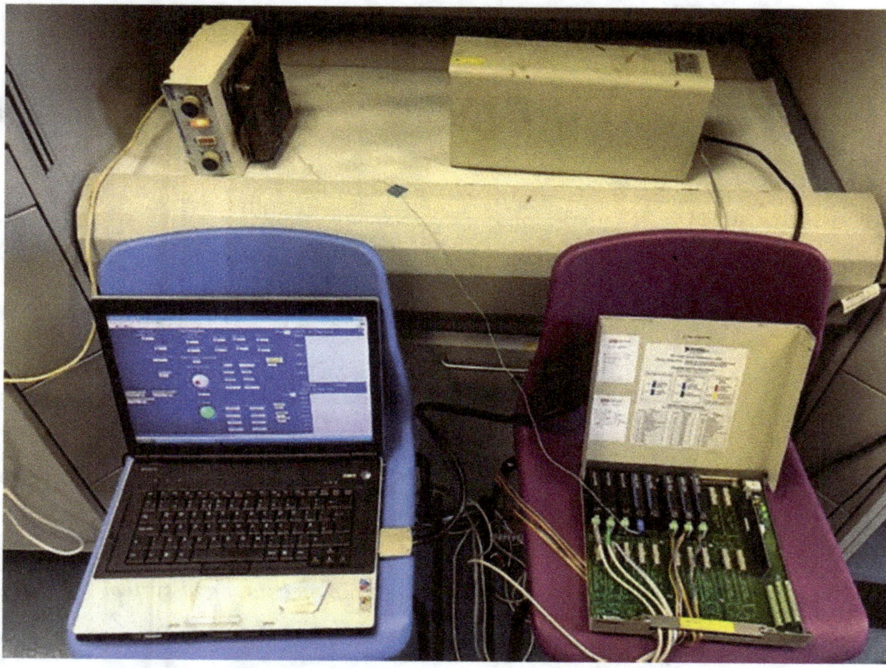

Figure 20.27 Setup of the measuring system

20.4.1.3 Procedure of measurement

The measurement was conducted in a fume hood for safety purpose, as the polymer may degrade or burn at a high temperature. The specimen was attached on the hot plate (Figure 20.28). The laser extensometer was then positioned at a distance of 305 mm from the specimen to the laser panel under a slight angle (~5°) to avoid the interference from the possible spurious reflections from other reflective surfaces [39]. The next step was running the LABVIEW program on the laptop and recording all the data. The last step was opening the hot plate and heating the specimen with a heating rate of 5/minute. The measurement stopped by closing the hot plate manually when the program indicated that the temperature of the specimen has risen to 120 °C.

20.4.2 Scanning electron microscopy

SEM is applied to analyse the dispersion quality of CNTs in the adhesive, the agglomeration of CNTs can be observed in the SEM images. The SEM instrument used was TESCAN LYRA3, as shown in Figure 20.29.

Four samples from specimen c and specimen d were analysed. One of the two samples from each specimen was analysed for the dispersion of CNTs on the surface of the specimen, the other one was analysed for the dispersion of CNTs on the cross-section of the specimen. Before analysis a 10 nm gold coating was sputtered

Figure 20.28 Setup of the specimen on the hot plate

on to minimize the specimen charging and ensure the high-quality SEM images from analysis [40].

20.4.3 Raman spectroscopy

The Raman spectroscopy is applied to assess the quality of CNTs [41]. The number of defective structures and complete molecular structures will be measured, a comparison of the intensity of these two structures gives an evaluation of the quality of the CNTs used as the raw material in this project. The MWCNTs was dispersed and mixed with silicon oil. The mixture was then loaded with glass slide for the Raman measurement.

20.5 Results and discussion

20.5.1 Coefficient of thermal expansion

The gauge lengths of all the specimens were measured with the temperature increasing. The measurement started at 30 and ended by 120, the measuring results of each specimen are presented in Table 20.3. The CTE of each specimen has been calculated in (20.1) and presented in Table 20.4. The comparison of CTEs of different materials, including steel, aluminium, polymer use in the project, and different specimens is shown in Table 20.3.

Figure 20.29 SEM instrument applied in the measurement

The comparison in Figure 20.30 indicated that the CTEs of the specimen a and specimen b, which contain an aluminium substrate, are close to the CTE of the aluminium plate. Which is because the adhesive and aluminium were bonded together

Table 20.3 *Gauge lengths of specimens by temperature in test*

	Gauge lengths of specimens (mm)				
Temperature (°C)	Specimen a	Specimen b	Specimen c	Specimen d	Specimen e
30	54.088949	47.569949	48.660051	54.630237	53.112284
35	54.092191	47.579613	48.674393	54.632009	53.115781
40	54.098447	47.584648	48.680509	54.646085	53.124402
45	54.100876	47.597783	48.692602	54.650078	53.132502
50	54.113108	47.604256	48.714701	54.657071	53.14312
55	54.124107	47.606099	48.72523	54.668083	53.148397
60	54.131635	47.621702	48.737437	54.680519	53.153699
65	54.134394	47.627551	48.737399	54.697495	53.157374
70	54.141909	47.636757	48.749987	54.707146	53.163783
75	54.145203	47.64876	48.759232	54.719671	53.170204
80	54.148458	47.653669	48.761241	54.72744	53.17674
85	54.147758	47.664083	48.766136	54.738859	53.181928
90	54.146932	47.673187	48.779284	54.742597	53.191541
95	54.156748	47.692019	48.779704	54.761696	53.206698
100	54.183973	47.697042	48.791847	54.782791	53.209661
105	54.193484	47.706528	48.806521	54.814618	53.215523
110	54.205411	47.706502	48.90756	54.830271	53.227793
115	54.211197	47.702611	48.911235	54.83949	53.23438
120	54.216385	47.70312	48.917122	54.848048	53.243853

Specimen a: Polymer+ Al+1.00 wt.% CNTs
Specimen b: Polymer+ Al+2.68 wt.% CNTs
Specimen c: Polymer +1.00 wt.% CNTs
Specimen d: Polymer +2.68 wt.% CNTs
Specimen e: Al

Table 20.4 *CTEs of five specimens*

	Specimen a	Specimen b	Specimen c	Specimen d	Specimen e
Length of thermal expansion (mm)	0.127436	0.133171	0.257071	0.217811	0.131569
Interval of Temperature (°C)	90				
CTE ($\times 10^{-5} \times^{-1}$)	2.66	2.87	5.87	4.43	2.75

and the bond strength is stronger than the shear stress in the specimen, the adhesive was forced to expand with the aluminium substrate during the heating process. Besides, due to the CTE mismatch between adhesive and aluminium, the specimen

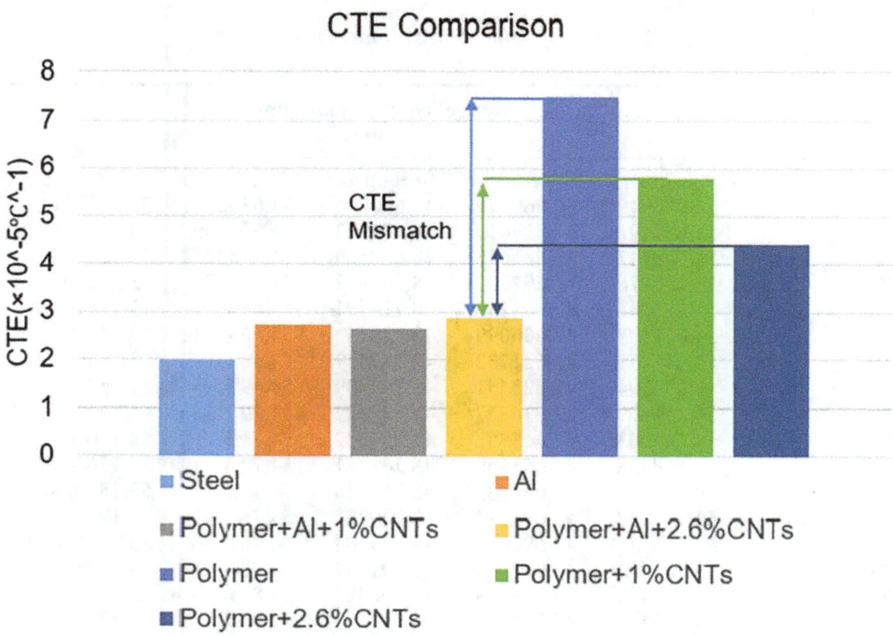

Figure 20.30 CTE comparison of different materials and specimens

may bend as shown in Figure 20.31. While no bending was observed in the test as the thermal strain of the specimen was too small to be observed.

According to the fitting curve in Figure 20.32, the CTE of the specimen c should be 3.6×10^{-5} °C^{-1}, and the CTE of the specimen d should be 1.4×10^{-5} °C^{-1}. While the results in Table 204 indicate that the CTE of specimen c is 5.87×10^{-5} °C^{-1} and the CTE of specimen d is 4.43×10^{-5} °C^{-1}. Which means although we have an achievement proving the addition of MWCNTs inside the polymer will get a smaller level of CTE mismatch between metal and polymer, as shown in Figure 20.30, the enhancement effect of MWCNTs on polymer performance did not meet our expectation well.

This result may come from the following reasons:

1. Defects in MWCNTs

Figure 20.31 Specimen bending with CTE mismatch

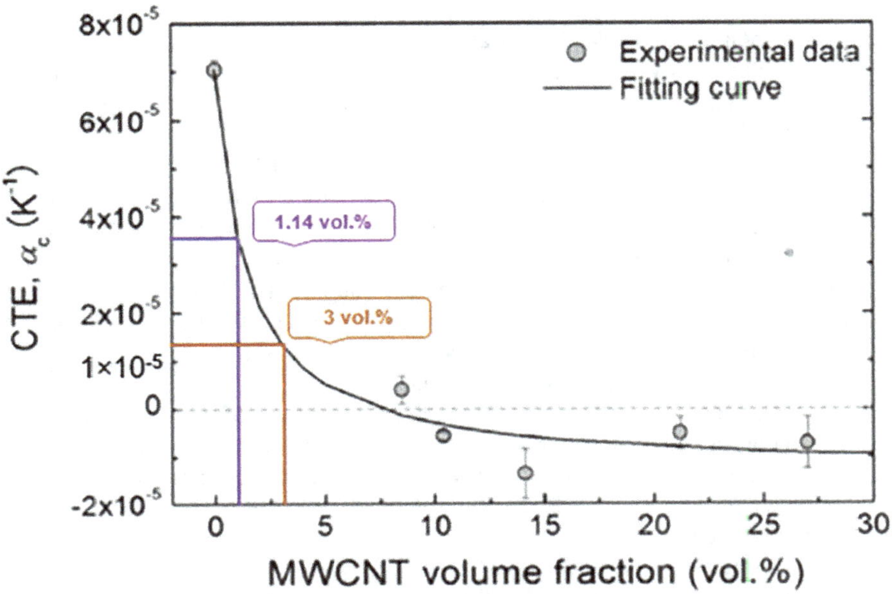

Figure 20.32 *CTE with 1.14 vol% and 3 vol% MWCNTs*

The defects in MWCNTs take a toll on the properties of MWCNTs [18], inducing a weaker enhancement on the performance of the polymer than the MWCNTs with fewer defects. The quality of MWCNTs is as shown in Figure 20.33.

2. Catalyst in MWCNTs

The catalyst used for MWCNTs in the lab is Ferrocene. Because of the production method applied in the lab, it is impossible to separate the catalyst from the MWCNTs, the catalyst still exists in the MWCNTs, and it was ignored in the test. Consequently, the weight of MWCNTs used in specimens is a bit less than expected. The catalyst exists in the MWCNTs did not contribute to the performance of the polymer [42].

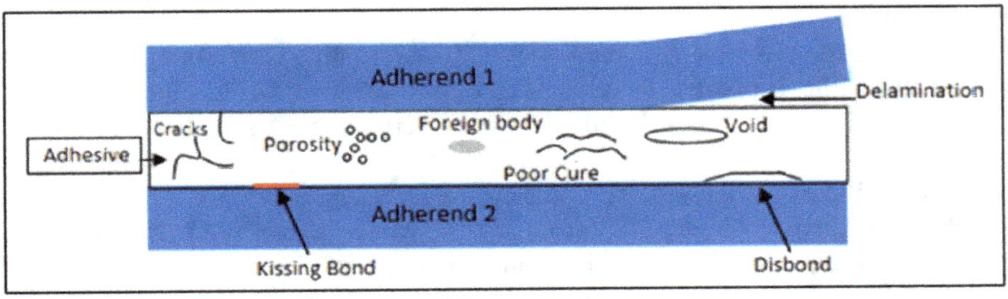

Figure 20.33 *Quality evaluation of MWCNTs*

3. Content of MWCNTs

   The content of MWCNTs was calculated based on the fitting curve in Figure 20.32, only six experimental data were involved in creating this curve, there was a large fluctuation on the data of volume fractions and CTEs adopted in the project.

   Besides, there is a deviation on the density of MWCNTs involved in the analytical calculation, as there is no effective method to measure the density of MWCNTs in our lab, the density of MWCNTs is related to the dimension of CNTs and the quantity of the walls in CNTs [43]. All we know by now is the diameter of the MWCNTs, varies from 20 –100 nm. The density of MWCNTs used in the project might be under-estimated as the density of MWCNTs on the market varies from 1.5 to 2.5 g/cm^3 [44].

4. Reagglomeration of MWCNTs in polymer

   The agglomeration occurs spontaneously because of the static conduction, consequently the reagglomeration happened after curing, as shown in Figure 20.34. CNTs agglomeration will induce a negative influence on the performance of polymer. Agglomerates may work as initiation sites for failure or accelerate the final breakage.

5. Deficiency in the adhesive

   Many deficiencies occurred in the polymer after curing process, as shown in Figure 20.35. The stress transmission was disturbed, and localized stress concentration occurred on the deficiencies, resulting in a different thermal expansion of the specimens.

20.5.2 Scanning electron microscopy

Figure 20.34a, b and Figure 20.35c, d indicate the dispersion quality of MWCNTs both on the surface and cross-section of the different specimens.

In Figure 20.34(a), we can see some highlighted bundles of MWCNTs, which are protruding out of the surface, and an obvious agglomeration of MWCNTs. It seems the dispersion quality of MWCNTs in specimen c was poor, while in Figure 20.35(b), the dispersion quality was better, and only two agglomerations on the surface. This performance may be because of the improper method used for dispersing and mixing MWCNTs in specimen c, more specifically, the sonication process was not long enough to disperse all the MWCNTs separately from bundle state.

In Figure 20.35(c) and (d), many MWCNTs protruding out from the surface of cross-section, this is because the preparation for the sample was too abrasive as the samples were cut with a scissor and the cross-section was not sanded before gold coating.

The MWCNTs dispersed well, and no obvious agglomeration occurred in a small area (shown in Figure 20.35(c) and (d)), while we can see the agglomeration occurred in Figure 20.34 with a lower magnification times. This is because the MWCNTs were not dispersed perfectly before application and reagglomeration happened in the curing process [45]. The high mass-ratio of MWCNTs in adhesive

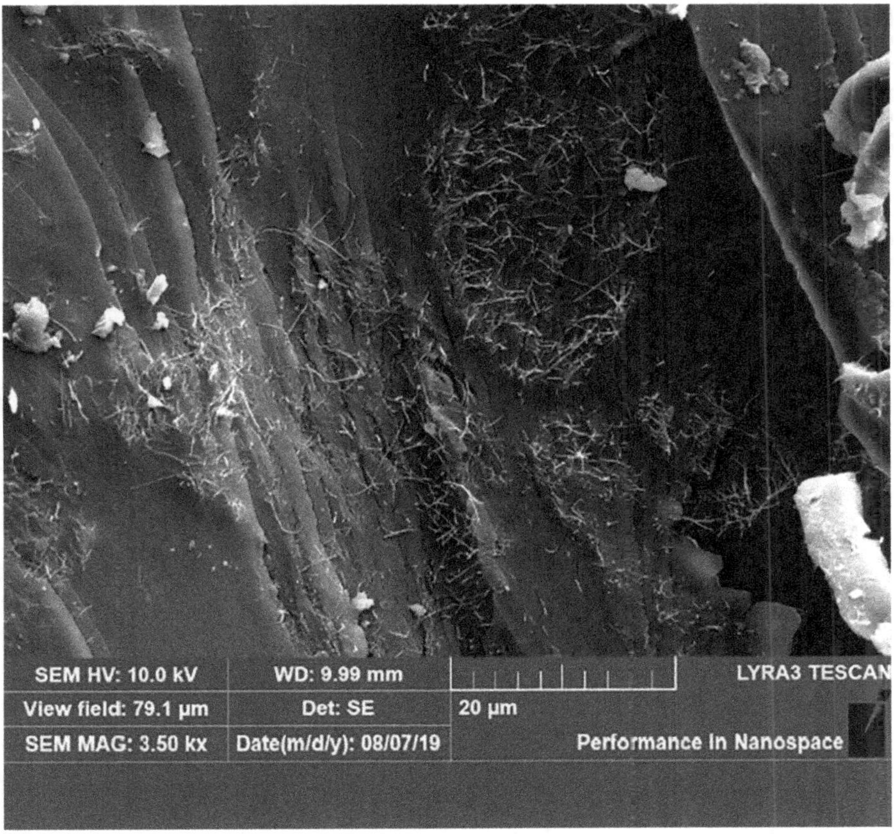

Figure 20.34 Reagglomeration of MWCNTs in specimen d

is the main reason for the agglomeration. Also, a better dispersion method is needed for the high mass-ratio of MWCNTs .

20.5.3 Raman spectroscopy

The MWCNTs used in this project were assessed by Raman spectroscopy. Figure 20.33 indicated the evaluation result. The G band corresponds to the order character, and the D band represents the disorder character in Raman spectrum. The intensity of D band is around 1040 and the intensity of the G band is 1270, the intensity ratio of D and G bands, ID/IG is 0.8189. The high-intensity ratio indicates that there are a lot of defects in the CNT. The defects damage the properties of the MWCNTs. Consequently, the enhancement of MWCNTs on the polymer did not meet the expectation of the project [41].

There are two reasons for the defects occurred in MWCNTs. The first is that the defects occurred during the production process in the lab. The quality of CNTs was affected by lots of parameters, such as catalyst, temperature, pressure and so on, the

(a)

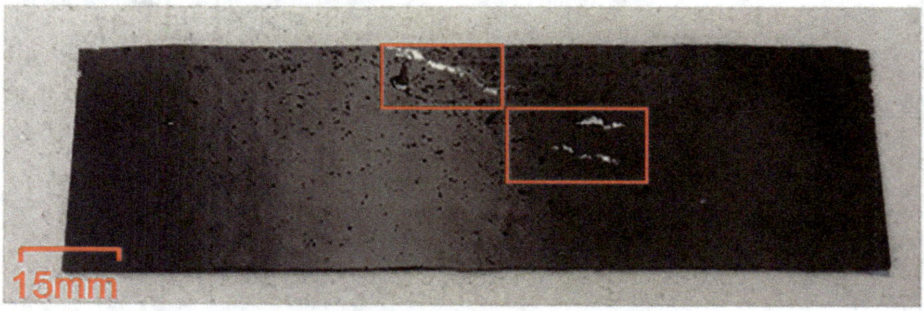

(b)

Figure 20.35 Deficiency on specimen (a) Porosity (b) Void

CNTs grew with many defects inside. Besides, the defects of MWCNTs also can be induced during the gathering process from the quartz substrate. The quality of CNTs can be improved with a better production method of MWCNTs in the lab and a better collection method.

The second reason, which is also the main reason, is that a large number of defects occurred during the dispersion processes, the sonication process and high shear mixing process. The dispersion processes could be too drastic and lasted too long for MWCNTs. The sonication process with lower amplitude and the high shear mixing with a lower rotate speed could prevent defects in MWCNTs. A shorter dispersion processes is also necessary. More researches and experiments are required to find the best parameters for the dispersion processes.

20.5.4 Bond deficiency on specimens

Various deficiencies will be induced with the improper production process and in-service environmental degradation. There are several categories of deficiencies in the adhesive.

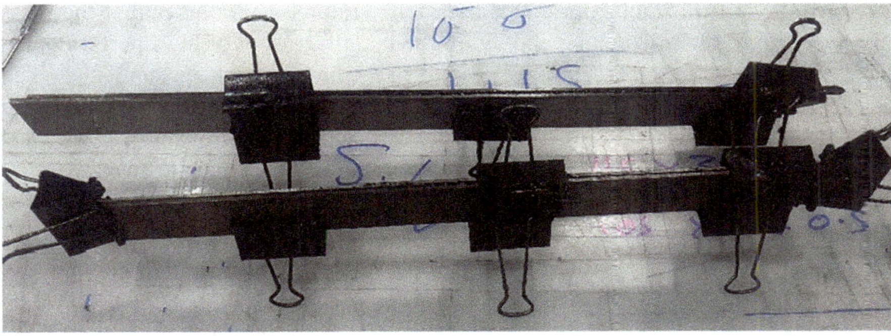

Figure 20.36 *Moulds fixed with clamps*

The deficiencies occurred on the specimens are the porosity and void, as shown in Figure 20.35, other categories of deficiencies cannot be distinguished clearly on the specimens.

The deficiencies in the adhesive will be harmful to the bond strength of the joint, as the localized stress concentration occurs in the deficiencies, the joint will be much easier to fail with the shear stress induced at a high temperature.

There are some reasons for the deficiencies occurred. The first reason is that the flowability of adhesive was poor with a high mass ratio of MWCNTs inside, although the adhesive was degassed for 30 minutes before application [45], the remaining air bubble was not easy to discharge from the bottom of the beaker. The second reason is that a lot of air bubbles were brought into the adhesive during the application process of the adhesive, and the air bubbles were locked in the chamber in the mould, as shown in Figures 20.23, 20.36.

The potential solutions for the deficiencies are as follows:

1. Extending degassing time

   Enough degassing time will be able to discharge most of the air bubble in the adhesive.
2. Using injection moulding for the manufacturing of specimens

   A lot of air bubbles were brought in during the manual application of the adhesive, while the bubbles can be avoided with the injection method in a vacuum environment.
3. Using vacuum oven for the curing process

   The vacuum oven will help discharge the air bubbles in the adhesive during the curing process, contributing to the performance of the adhesive.

20.6 Conclusion and further work

We have an achievement proving the addition of MWCNTs inside the polymer will obtain a smaller level of CTE mismatch between metal and polymer, which is able to reduce the risk of joint failure at a high temperature. This means that the new adhesive with MWCNTs can be applied in the manufacturing of high-performance electric vehicles, e.g., the assembling process of the battery pack system.

However, the test results have shown that the enhancement of the specific mass ratios of MWCNTs used in the test did not meet the objective of the project because of the deficiencies in MWCNTs and the agglomeration occurred in the adhesive, etc. Therefore, some issues need to be further addressed.

- The properties of CNTs are assessed for perfect CNTs ignoring all the defects. It is quite important to develop new methods for the dispersion and functionalization of CNTs, especially for the high mass-ratio CNTs. It takes a long time and complex steps to deal with the CNTs with current dispersion methods, the deficiencies are easily brought in and the result of the research will be away from expectation.
- Test results from SEM cannot fully verify the dispersion quality of CNTs, and the evaluation of the dispersion quality of CNT is not easy because the solvent and the specimens after curing are not transparent with CNTs inside, even a small mass of CNTs has a large number of particles. It is important to develop a new method for quantitative analysis of dispersion quality or agglomeration in a solvent or a polymer.
- The test temperature in the project is only up to 120 for safety purpose, while the working temperature of adhesive in a real situation could be much higher, like 200, further test should be conducted for the simulation close to real situation. Besides, the epoxy resin used in this project cannot work for a long-term with a temperature over 200 degrees. Therefore, a high-temperature resistant resin should be applied in further research.

References

[1] VENEDO *Global EV sales for 2018 - final results [online]*. 2019. Available from http://www.ev-volumes.com/country/total-world-plug-in-vehicle-volumes/ [Accessed May 2019].

[2] Long L. *Research on static strength analysis and strength prediction of adhesively bonded single lap joint for autoclaveresearch on static strength analysis and strength prediction of adhesively bonded single lap joint for autoclave*. China: Ji Lin University; 2011.

[3] Alvarez S. *Tesla patent hints at more reliable batteries through dynamic management system [online]*. 2019. Available from https://www.teslarati.com/tesla- patent-model-s-model-x-model-3-battery-management-system-update/ [Accessed May 2019].

[4] GPC *GPC battery [online]*. 2015. Available from http://www.gpcbattery.com/English/index.aspx [Accessed May 2019].

[5] CBEA. 'Thermal management of electric vehicle'. *SAE Technical Papers*. 2019.

[6] Shirasu K., Yamamoto G., Tamaki I., *et al*. 'Negative axial thermal expansion coefficient of carbon nanotubes: experimental determination based on measurements of coefficient of thermal expansion for aligned carbon nanotube reinforced epoxy composites'. *Carbon*. 2015;95:904–9.

[7] Kumar M., Ando Y. 'Chemical vapor deposition of carbon nanotubes: a review on growth mechanism and mass production'. *Journal of Nanoscience and Nanotechnology*. 2010;10(6):3739–58.

[8] St. Clair B. *Nanotechnology and predicting the future (review of the current state of nanotechnology and making future predictions with a discussion of forecasting methods)*. 1. The Amateur Academic; 2016. p. 10.

[9] Saito T., Matsushige K., Tanaka K. 'Chemical treatment and modification of multi-walled carbon nanotubes'. *Physica B: Condensed Matter*. 2002;323(1–4):280–3.

[10] Aria A.I., Gharib M. 'Physicochemical characteristics and droplet impact dynamics of superhydrophobic carbon nanotube arrays'. *Langmuir: The ACS Journal of Surfaces and Colloids*. 2014;30(23):6780–90.

[11] Roy S., David-Pur M., Hanein Y. 'Carbon nanotube growth inhibition in floating catalyst based chemical vapor deposition and its application in flexible circuit fabrication'. *Carbon*. 2017;116(14):40–9.

[12] An D., Nourry J., Gharavian S, *et al*. 'Strain self-sensing tailoring in functionalised carbon nanotubes/epoxy nanocomposites in response to electrical resistance change measurement'. *SSRN Electronic Journal*. n.d

[13] Bregar T. *Smart adhesive evaluation for structural in- situ strain measurement application msmsc thesis*. Cranfield University; 2017.

[14] Xu B. *Mechanisms and failure criteria of adhesive joints and toughened epoxy adhesives by [PhD]*. Queen Marry University of London; 2010.

[15] Urbina S.R.R. *Mechanical degradation of composite structures subject to environmental effects.Pdf*. Cranfield University; 2015.

[16] Sreekumar T.V., Liu T., Min B.G., *et al*. 'Polyacrylonitrile single-walled carbon nanotube composite fibers'. *Advanced Materials*. 2004;16(1):58–61.

[17] Ma P.-C., Siddiqui N.A., Marom G., Kim J.-K. 'Dispersion and functionalization of carbon nanotubes for polymer-based nanocomposites: a review'. *Composites Part A: Applied Science and Manufacturing*. 2010;41(10):1345–67.

[18] Aria A.I., Lyon B.J., Gharib M. 'Morphology engineering of hollow carbon nanotube pillars by oxygen plasma treatment'. *Carbon*. 2015;81:376–87.

[19] Branson *Sonifier productsc [online]*. 2019. Available from http://www.sonifier.com/en

[20] Lu K.L., Lago R.M., Chen Y.K., Green M.L.H., Harris P.J.F., Tsang S.C. 'Mechanical damage of carbon nanotubes by ultrasound'. *Carbon*. 1996;34(6):814–6.

[21] Schmid C.F., Klingenberg D.J. 'Mechanical flocculation in flowing fiber suspensions'. *Physical Review Letters*. 2000;84(2):290–3.

[22] Ross C. *Ultra high shear mixing technology [online]*. 2019. Available from https://www.mixers.com/whitepapers/Ultra High Shear Mixing Technology 08242007.pdf

[23] Aria A.I., Gharib M. 'Reversible tuning of the wettability of carbon nanotube arrays: the effect of ultraviolet/ozone and vacuum pyrolysis treatments'. *Langmuir: The ACS Journal of Surfaces and Colloids*. 2011;27(14):9005–11.

[24] Lyon B.J., Aria A.I., Gharib M. 'Fabrication of carbon nanotube-polyimide composite hollow microneedles for transdermal drug delivery'. *Biomedical Microdevices*. 2014;16(6):879–86.

[25] Hill D.E., Lin Y., Rao A.M., Allard L.F., Sun Y.-P. 'Functionalization of carbon nanotubes with polystyrene'. *Macromolecules*. 2002;35(25):9466–71.

[26] Geng Y., Liu M.Y., Li J., Shi X.M., Kim J.K. 'Effects of surfactant treatment on mechanical and electrical properties of CNT/epoxy nanocomposites'. *Composites Part A: Applied Science and Manufacturing*. 2008;39(12):1876–83.

[27] Hasheminia S.M., Park B.C., Chun H.-J., Park J.-C., Chang H.S. 'Failure mechanism of bonded joints with similar and dissimilar material'. *Composites Part B: Engineering*. 2019;161(4):702–9.

[28] Loos M. *Carbon Nanotube Reinforced Composites: CNT Polymer Science and Technology*. Elsevier; 2014.

[29] Huntsman *Mechanical properties: araldite ® LY 1564* / aradur ® 3486*/ aradur ® 3487 * * [online]*. 2012. Available from http://www.swiss-composite.ch/pdf/t-Araldite- LY1564-Aradur3486-3487-e.pdf

[30] Kumar K.M., Jagannathan V. *Basic electrical electronics computer and communication engineering [online]*. 2000. Available from http://www.philadelphia.edu.jo/newlibrary/english-books/163-engineering/9929- 13568

[31] Seika *Seika dgital image [online]*. 2019. Available from https://www.seikadi.com/cn/measurement/dic.html%0A

[32] Azom *Non-contacting laser extensometers for strain measurement in materials testing [online]*. 2000. Available from https://www.azom.com/equipment-details.aspx?EquipID=4270

[33] CHEAPTUBES *Multi walled carbon nanotubes [online]*. 2019. Available from https://www.cheaptubes.com/product-category/multi-walled-carbon-nanotubes/

[34] Rojas J.A., Ardila-Rodríguez L.A., Diniz M.F., Gonçalves M., Ribeiro B., Rezende M.C. 'Optimization of triton X-100 removal and ultrasound probe parameters in the preparation of multiwalled carbon nanotube buckypaper'. *Materials & Design*. 2019;166:107612.

[35] Bai Y., Lin D., Wu F., Wang Z., Xing B. 'Adsorption of Triton X-series surfactants and its role in stabilizing multi-walled carbon nanotube suspensions'. *Chemosphere*. 2010;79(4):362–7.

[36] Khaleque T., Zhang X., Kumar Thakur V., Aria A.I., Yazdani Nezhad H. 'Tailoring of thermo-mechanical properties of hybrid composite-metal bonded joints'. *Polymers*. 2021;**13**(2):E170.

[37] HumiSeal *4 tips to successfully apply two-component epoxy adhesives [online]*. 2018. Available from https://blog.humiseal.com/4-tips-to-successfully-apply- 2-component-epoxies

[38] Phillips R.W. *Composite resin systems. annals of the academy of medicine, singapore [online]*. 1986. Available from https://www.huntsman.com/corporate/a/Home

[39] Epsilon *Model LE-05 and LE-15 non-contact laser extensometer [online]*. 2019. Available from https://www.epsilontech.com/wp-content/uploads/product- specs/Laser_Extensometer.pdf

[40] Ikeda H., Iwai M., Nakajima D., *et al.* 'Nanostructural characterization of ordered gold particle arrays fabricated via aluminum anodizing, sputter coating, and dewetting'. *Applied Surface Science*. 2019;4652018:747–53.

[41] Costa S., Borowiak-palen E., Kruszynska M., Bachmatiuk A., R. J. K. *Characterization of Carbon Nanotubes by Raman Spectroscopy*. Materials Science-Poland. Szczecin University; 2008.

[42] Esquenazi G., Brinson B., Barron A. 'Catalytic growth of carbon nanotubes by direct liquid injectioncatalytic growth of carbon nanotubes by direct liquid injection CVD using the nanoclusterusing the nanocluster [$H_x pmo_{12}O_{40}CH_4 mo_{72}fe_{30}(O_2cme)_{15}O_{254}(H_2O)_{98-y}(etoh)_y$'. *Journal of Carbon Research*. 2018;4(1):17.

[43] Laurent C., Flahaut E., Peigney A. 'The weight and density of carbon nanotubes versus the number of walls and diameter'. *Carbon*. 2010;48(10):2994–6.

[44] Mazlan N.B. 'The effect of oxidation treatment of MWCNT on the properties of PDMS nanocomposite'. *PhD. Sains Malaysia University*. 2012.

[45] Davis C.S., Woodcock J.W., Gilman J.W. 'Preparation of nanoscale mutliwalled carbon nanotube dispersions in a polyetheramine epoxy for ecotoxicological assessment'. *NIST Special Publication*. 2015;1200(9).

Chapter 21

High-performance nanocomposites for strain self-sensing applications in composite joints

Jim Nourry[1], Tadej Bregar[2], Marek Burda[3], Vijay Kumar Thakur[4], and Hamed Yazdani Nezhad[1,5]

Abstract

Carbon nanotubes (CNTs) are inherently multifunctional, conductive and possess piezo-resistive characteristics. They are able to enhance various characteristics of everyday materials and even widen the spectrum of tasks these materials can be used for. Aiming at the multi-functionality of materials, nanocomposites made of epoxy resin with embedded CNTs are a promising solution for strain self-sensing applications. A composite structure with an embedded network of conductive CNTs will see its electrical conductivity vary as it is strained during service. A critical parameter to achieve repeatable and reliable measure is the nanotube dispersion state in the resin. Nanocomposites made of CNTs dispersed in aerospace grade epoxy resin were created using a simple process. More specifically, probe sonication and magnetic stirring were used alongside different concentrations of surfactants. To investigate the resistance-based strain gauge abilities of such material, a testing apparatus was set-up and real-time measurement of strain and resistance were recorded. Gauge factor (GF) of around one was calculated when loading in the elastic region. Load versus strain curves for pure cured epoxy were compared to the fabricated nanocomposite, and mechanical properties were investigated. Raman characterisation of CNT sample, as well as scanning electron microscope (SEM) imaging was performed to investigate dispersion and quality of nanotubes. In the end, the feasibility of nanocomposites made with a simple process using CNTs produced was demonstrated.

[1]Enhanced Composites and Structures Centre, School of Aerospace, Transport and Manufacturing, Cranfield University, Bedford, UK

[2]Heating and Process Engineering, Faculty of Mechanical Engineering, University of Ljubljana, Ljubljana, Slovenia

[3]Ampashield Ltd., Cambridge House Camboro Business Park, Cambridge, Cambridgeshire, UK

[4]Biorefining and Advanced Materials Research Centre, SRUC, Edinburgh, UK

[5]Aeronautics and Aerospace Research Centre, Department of Mechanical Engineering and Aeronautics, City, University of London, London, UK

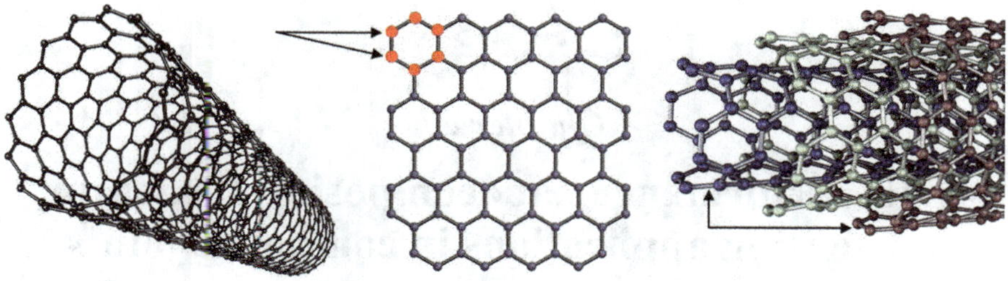

Figure 21.1 Schematic of (A) SWCNT, (B) graphene sheet and (C) MWCNT

21.1 Introduction

21.1.1 Structure of carbon nanotubes

All materials find their properties in their atomic structure and macroscopic arrangement. CNTs are no exception to that rule as most of its high-end properties are directly linked to its atomic structure. A CNT is a seamless tube made of a graphene sheet, composed of carbon atoms forming a hexagonal crystal lattice (Figure 21.1b).

The cylindrical shape of nanotubes induces a change in the orbital structure of the $sp_{\#}$ hybridised atoms of the graphene sheet. The curvature of the graphene creates a mix between σ and π orbitals, which is called rehybridisation [1]. The σ bonds are responsible for strong covalent bonding in lattice direction, whereas the π bonds account for the van der Waals forces in the out of plane direction [2].

The chiral vector defines the way the sheet of graphene is wrapped to create the nanotube and is represented by the indices n and m as follows:

$$C = na_{*} + ma_{\#} \tag{21.1}$$

The indices account for the number of unit vectors a_{*} and $a_{\#}$ in the two directions of the hexagonal lattice. 'Zigzag' is the configuration where $m = 0$, 'armchair' is the one where $n = m$, and, in the other cases, the configuration is called 'chiral' [3]. These differences in chirality impact on the electronic properties of CNT. The armchair configuration presents no band gap and is metallic; zigzag and chiral configurations with $n \neq m$ are semiconducting with some band gap. CNTs can be composed of a single-walled CNTs (SWCNTs) or multiple-walled CNTs (MWCNTs) (Figure 21.1a,c). SWCNTs mainly have diameter of 0.4–3 nm for a length in a micrometre and even centimetre range [4, 5], whereas for MWCNT, the diameter can be from 1.4 to 100 nm at least [6, 7]

This geometry gives the nanotubes an impressive aspect ratio up to 10^{1}. SWCNTs, in particular, exhibit improved electric properties as they can be superior conductors [8], but their quality and production are still challenging [5, 9].

CNTs exhibit various enhanced properties. Their characteristics come from their tubular shape and their inherent graphene atomic structure. In this research, the electromechanical properties are specially of interest and will be described in the following sections.

21.1.2 Electromechanical properties

As mentioned previously, the mechanical properties of CNTs are mainly due to the σ bond occurring between the carbon atoms of the graphene structure. These strong covalent bonds allow for a theoretical Young's modulus of around 1 TPa for SWCNTs [10, 11]. Also, MWCNTs have been reported to show a Young's modulus of around 1.30 TPa [12]. CNTs also express a high tensile strength of the order of hundreds of GPa greatly exceeding steel [13]. The hollow structure combined with the covalent bonds present between carbon atoms gives the CNT extreme energy absorption capabilities, which explains their high mechanical properties. Hence, they have been used as reinforcement for polymer matrices in composites manufacturing [14], dramatically increasing the mechanical properties of the created composite material.

Electrical properties of CNTs are directly linked to the orbital organisation of the graphene structure. As mentioned previously, the π orbital is orthogonal to the lattice formed by the σ bonds, and its delocalised electron is responsible for electric conduction. Typical values for electrical conductivity of CNT is in the order of 10^2–10^3 S/m [15]. Although electric conductivity greatly depends on purity, crystallinity and entanglement of CNT, it still represents an area of great interest for research.

The piezoresistive effect is when mechanical loadings of the material induce a change in the measured resistivity [16]. This effect can arise from direct modification of the CNT band gap [17] or by macroscopic reorganisation of a CNT network. Depending on the radius of the CNTs, their chirality and mode of deformation, and their resistivity may increase or decrease. This effect is the basis of strain sensing applications of CNTs [18].

21.1.3 CNTs for strain sensing

Nanotechnology aims to develop products that integrate nanostructured material to enhance their properties. The intense interest in nano-embedded materials comes mainly from the composite manufacturing industry where the development of nanoscale parts, self-sensing structures and advanced functional properties are highly researched. Where traditional strain sensor lacks of integration abilities and nanoscale sensibility, CNTs offer a solution to develop new sensors [19].

With respect to strain sensing, an important indicator of sensing performance is the GF and is defined as follows:

$$GF = \frac{\Delta R/R}{\Delta L/L} = \frac{\Delta R/R}{\varepsilon} \tag{21.2}$$

where, ΔR is the change in resistance due to strain, R is the initial resistance, and ε is the strain. CNTs GF has proven to be relatively high [20], significantly greater than most known materials [21].

The use of CNTs for strain sensing applications takes different forms. Thin films of SWCNT and MWCNT, buckypapers, have been created and embedded in structures allowing for damage and deformation sensing [22–24]. With this approach, scientists were able to reach high sensibility (GF) [23] with high stability and high repeatability. However, limitations such as fragility and strain transfer from the medium to the buckypapers exist.

As an alternative method, CNTs have been dispersed into polymer matrices to create nanopolymers for which increased mechanical and electric properties have been reported [25–27]. This method allows for better contact between CNTs and the surrounding medium, multidirectional sensing and easier integration to existing manufacturing processes.

Sensibility of such methods to strain seems to be high in the order of 10# for low strain values <10% and tends to have lower sensitivity <5 when design for higher mechanical elongation >30% [23].

The main idea behind these techniques is to create a path for electrons to travel across the material. Achieving an interconnected network of CNTs evenly distributed across the area of interest is crucial and determines the performance of the sensor. This research focuses on the properties of nanopolymers for strain sensing. As opposed to buckypapers, this technique is particularly adapted to composite manufacturing as the use of matrices like epoxy resins is widely spread. Moreover, a uniform distribution of nanoparticles inside a medium is believed to less disturb the mechanical properties compared to more CNT concentrated buckypapers.

Nanopolymers are multi-phased systems where many parameters such as mixing, stabilisation and functionalisation come into play in the properties of the final material. In our case, strain sensing is done through resistivity measurements, taking advantage of the piezoresistivity effect of those materials. In the following section, the occurrence of this effect in CNT nanopolymers along with its parameters is detailed.

21.1.3.1 Piezoresistivity of CNT nanopolymers

The piezoresistivity effect observed in CNT nanopolymers relies on achieving a conducting CNT network in the polymer matrix. The relation between filler addition and electric conductivity follows three main stages. First, upon addition of a low amount of CNT, the conductivity of the polymer starts to change, and it increases slowly due to few connections and mainly tunnelling of electrons across the dispersed CNT. Regarding piezoresistivity, these tunnelling paths are essential and discussed further. Then, as the amount of CNT increases, a great increase in electric conductivity occurs, the volume fraction when this dramatic increase occurs is known as the percolation threshold and simply indicates the formation of a conduction network formed by CNT in contact with each other. Finally, after the percolation threshold more paths for conduction are created as the volume fraction of fillers increases, the electrical conductivity will reach a plateau value, which will be the maximum achievable conductivity.

Creating a conduction network is key to assure high flow of electrons across the polymer. The literature implies that low percolation threshold is ideal because of the low impact on the mechanical properties, weight saving and costs. The relation between CNT loading and strain sensitivity (GF) has not been clearly identified. Increasing the CNT loading does not necessarily increase the GF [28].

It is expected that a filler concentration above the percolation is preferable for piezoresistivity as more contact point and tunnelling points exist between CNTs. Figure 21.2 shows a schematic of the path of electrons across the conductive network inside the polymer matrix.

SWCNTs and MWCNTs have been widely used to create CNT nanopolymers to use in resistance-type strain sensors. When strain is applied, e.g., tensile test is performed, and the conductive network is deformed along with the sample. SWCNTs are known to show a linear response over a larger strain range compared to MWCNTs, which tends to respond non-linearly when larger strain is applied [19, 29]. Within the conductive network, the electrons travel along CNTs, which have an intrinsic resistivity ; they move across CNTs experiencing the contact resistance $R_{=}$, and they tunnel between but not touching CNTs where tunneling resistance exists (Figure 21.2). Piezoresistivity of CNT nanopolymers is mainly dependent on the three resistances. When the conductive network is deformed, the contact resistance increases due to the loss of contact between nanotubes. Also, as the space between CNTs increases, the tunneling one increases [30, 31], and the deformation of the CNTs themselves induces an increase in the intrinsic resistivity. However, the role of the intrinsic resistance in the piezoresistive effect can be neglected for various reasons. The most important one being, the epoxy has a Young's modulus of around 3 GPa, whereas the CNT has a Young's modulus that is 100 times higher. This mismatch added to a poor load transfer between the matrix and the nanotubes leads to believe that the deformation of CNT themselves is highly unlikely.

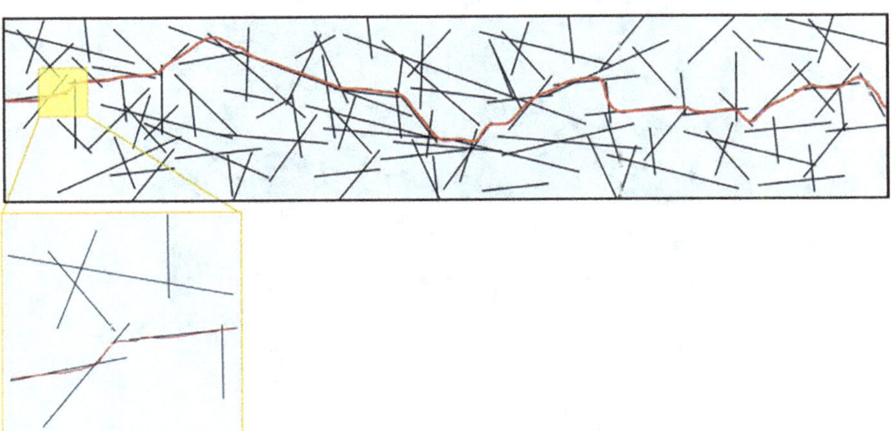

Figure 21.2 Schematic of the path of electrons (red) across the CNT conductive network

21.2 Materials and methods

21.2.1 CNT fabrication

The chemical vapor deposition (CVD) process was used to produce the nanomaterial, and the inside of the CVD tube chamber was used as the growing surface. The details and parameters of the process were as follows.

Argon was used as the purging gas flowing in the chamber at atmospheric pressure. The growth was conducted for 3.5 hours, and ferrocene was used as the catalyst. To source the carbon atoms, toluene gas was used and the concentration of ferrocene was 5% in toluene gas. Approximately, 4 g of CNTs was produced using this method.

Figure 21.3 shows the nanotube flakes after being removed from the reaction chamber.

21.2.1.1 Adhesively bonded joints

21.2.1.1.1 Advantages of adhesive joints

One of the main advantages of using adhesives is that there is no need for physical changes to the substrates, in form of holes, which need to be cut or drilled. Consequently, there are less stress concentrations, and the load is distributed over a larger area [1, 3, 11, 12]. Additionally, dissimilar materials can be bonded together; the joints are lightweight [13]; they are resistant to corrosion; they have excellent insulating properties and high strength-to-weight ratio; superior damping and noise

Figure 21.3 CNT flakes produced by CCVD

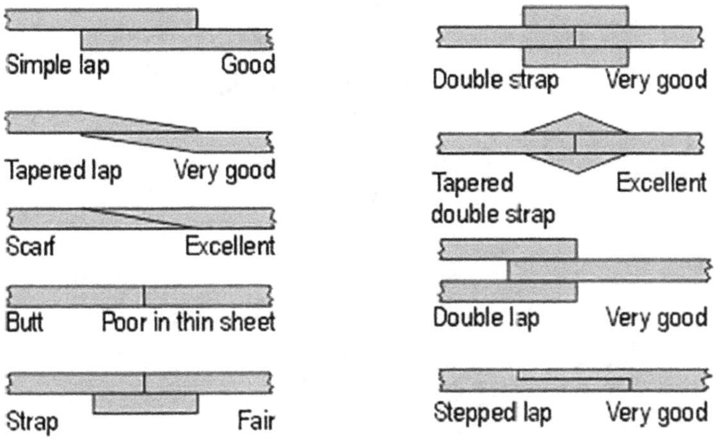

Figure 21.4 Configurations of adhesively bonded joints with an indication of their strength [17]

reduction capability, and the bonding operation can be automated which is cost effective [1].

21.2.1.1.2 Disadvantages of adhesive joints

Adhesives are much more influenced by the environment and are sensitive to variations in service temperature, humidity and also surface treatments [11, 12]. Strength and toughness of adhesives in tension or shear are also relatively low when compared to many metals [1]. They are also more susceptible to creep [14]. Furthermore, their inability to be disassembled for periodic inspections of underlying damage represents another downside, and non-destructive test methods for inspection are currently still limited [1–3].

To overcome these disadvantages, industries and researchers have developed new high strength, high toughness adhesives via inclusion of nano- and microparticles, e.g., carbon or rubber [15].

21.2.1.1.3 Lap-shear joints

Adhesively bonded joints can be made in various configurations (see Figure 21.4), amongst which the most common and simple to manufacture are single lap-shear joints [16].

Adhesive joints generally perform better in compression, shear and tension than in peel and cleavage [17], as seen in Figure 21.5.

In a single lap-shear joint, the most dominant stresses present are high peel stresses and non-uniform shear stresses due to out-of-plane bending moments, which are caused by the eccentricity of the load path [18], as seen in Figure 21.6. This results in peeling stress-driven failure, which means that lap-shear joints provide conservative failure load predictions [19–21].

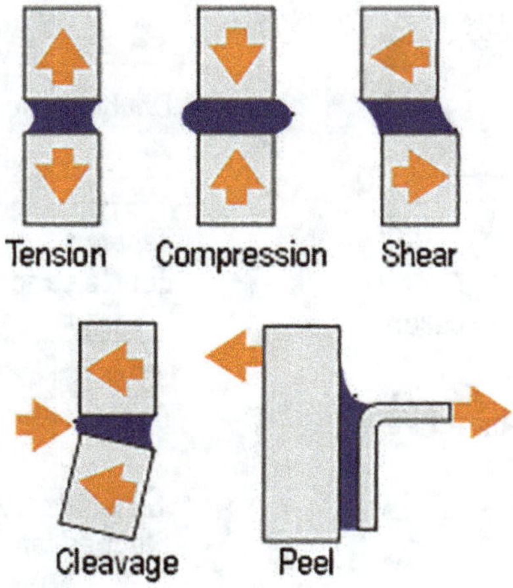

Figure 21.5 Types of loads experienced by adhesively bonded joints [17]

Because maximum failure stresses occurring at the edge of the bond can be much higher than the average failure stress across the bond, it is better to use the ultimate failure load for failure analysis [22]. The failure load itself can be increased proportionally by increasing the joint width [18] and unproportionally by increasing the bond length (see Figure 21.7). A sufficient overlap length is necessary to reduce vulnerability to creep and peel [10]. ASTM D1002 standard for metal–metal lap-shear joints suggests an overlap length of 12.7 mm.

To reduce the eccentricity, end-tabs are bonded to adherends. However, due to variability in thickness of adhesive bonds (of both adherends and end-tabs),

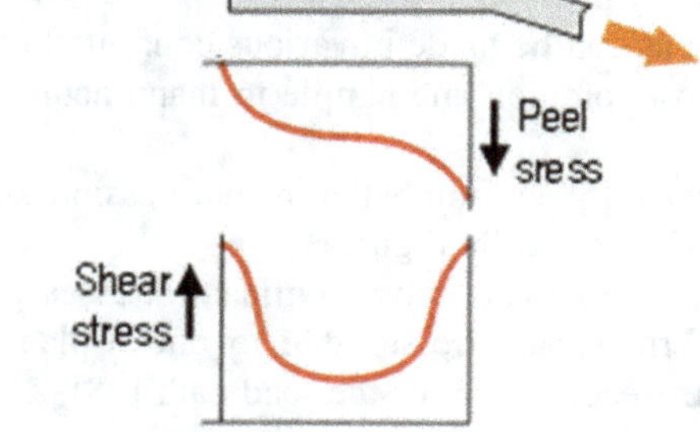

Figure 21.6 Stress distribution in a lap-shear joint [17]

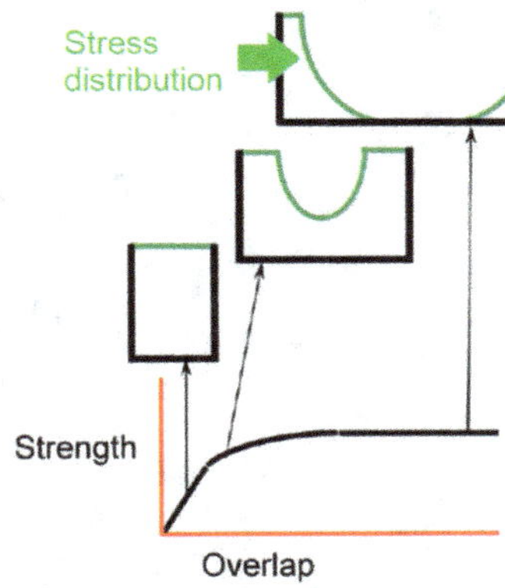

Figure 21.7 Influence of overlap length on strength [10]

complete elimination of eccentricity in a made lap-shear joint is not possible [18]. A representation of this can be seen in Figure 21.8, where for illustration purposes, the thickness of the adhesive in the middle is much higher than the thickness of adhesive at the sides for bonding end-tabs, which are of the same thickness as adherends.

21.2.1.1.4 Conventional strain sensors

Conventional strain sensors, commonly known as strain gauges, are resistive transducers which change their electrical resistance when they are elongated or compressed due to a tensile or compressive stress [23]. The most common type of a

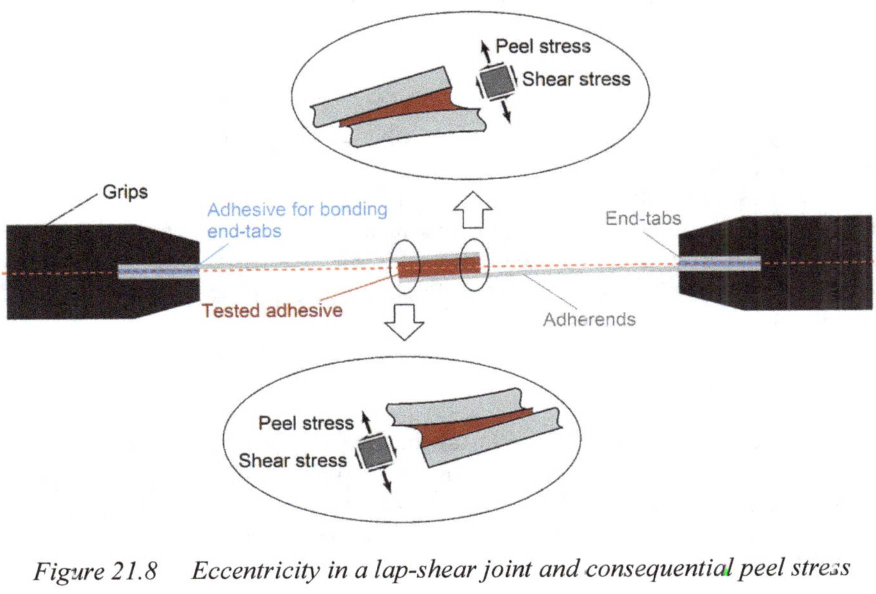

Figure 21.8 Eccentricity in a lap-shear joint and consequential peel stress

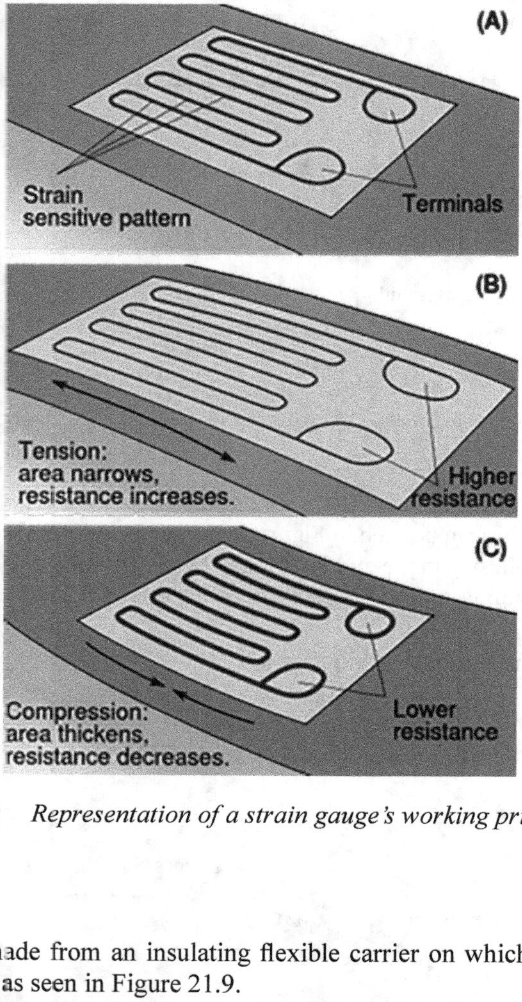

Figure 21.9 Representation of a strain gauge's working principle [24]

strain gauge is made from an insulating flexible carrier on which a metallic wire pattern lays [24], as seen in Figure 21.9.

The wires are positioned in the direction of load. If the gauge is tensile loaded, the wire will become narrower and longer which will in turn increase its electrical resistance and vice versa if the load is compressive [23]. This is represented in equation (21.3), which relates wire's electrical resistance R to its radius r and length l through resistivity ρ:

$$R = \rho \frac{l}{\pi r^2} \tag{21.3}$$

Strain gauges are usually bonded on to the surface of a component of which strain we want to measure and connected to a Wheatstone bridge for measuring very minute resistance changes, which are related to corresponding strain by a quantity known as the GF [25]. GF for such strain sensors is usually equal to 2 [26].

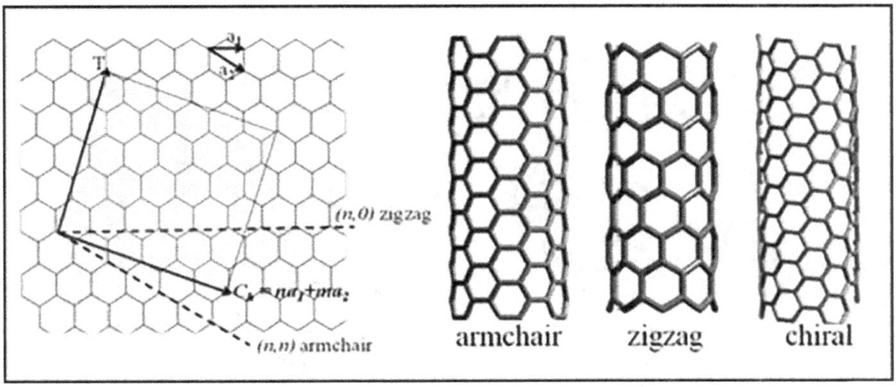

Figure 21.10 Chiral vector and types of SWCNTs [31]

21.2.1.2 CNT–epoxy nanocomposites

21.2.1.2.1 Carbon nanotubes

CNTs can be thought of as single graphene sheets (comprising molecularly bonded carbon atoms) rolled into cylinders. They are of nanoscale size with a high aspect ratio (length/diameter) [3], high specific stiffness and strength [27], and exceptionally high electrical and thermal conductivities [28, 32]. They are categorised as SWCNTs and MWCNTs. There are different types of SWCNTs which can be distinguished one from another by the way a graphene sheet is rolled – this is defined by the Chiral vector, $\vec{c_h}$ [29]. As seen in Figure 21.10, three orientations are possible: armchair, zigzag and chiral. Nature of wrapping (chirality) defines whether a CNT is conductive or semi-conductive. CNTs with an armchair structure are conductive [30]. Meanwhile, other two structures, zigzag and chiral, act as semiconductors and will conduct electric current only when additional energy in form of light or electrical field is applied, which free electrons from carbon atoms [30].

MWCNTs consist of multiple rolled layers of graphene one in another, as shown in Figure 21.11. Based on the type of SWCNTs that form them, they can be either conductive or semi-conductive. They are easier to mass produce and are therefore less-expensive than SWCNTs [34]. In this project, MWCNTs with an armchair structure are used, which makes them conductive.

21.2.1.2.2 Effect of CNTs on epoxy

A research has been done on how the static and fatigue strengths of carbon-epoxy composite-to-aluminium single lap-shear joints (with 2 wt% CNTs) in the paste adhesive are compared to those without [11]. Although the static strengths of adhesive joints with CNTs decreased by 36.6%, the fatigue strengths increased by 12.8%. Also, not only did the fatigue lives increase, the endurance limit was also 12.8% higher. Other researchers found that the presence of CNTs increases bonding strength and durability of epoxy joints in a humid environment, due to high

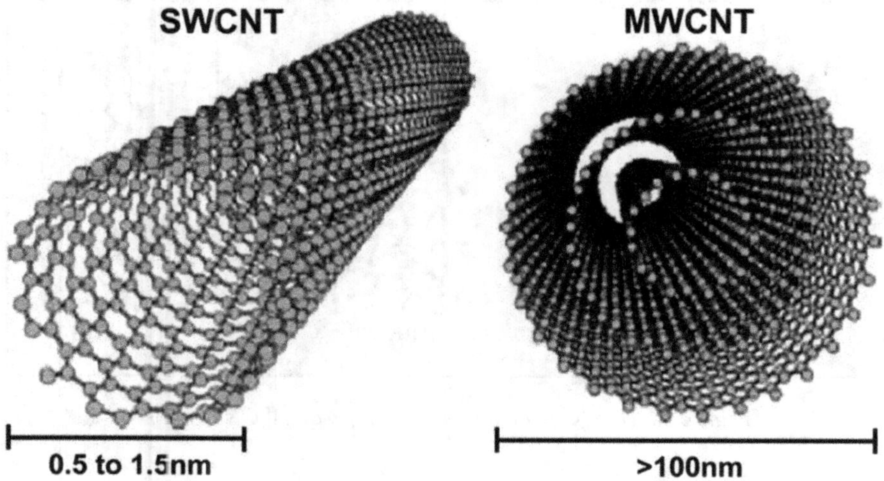

Figure 21.11 SWCNTs and MWCNTs [33]

mechanical strength and hydrophobic nature of CNTs [35]. CNTs in the adhesive also reduce stress concentrations and enhance uniformity of stress distributions [36].

Another research mechanically and electrically tested lap-shear joints with different carbon fractions (0.5, 1 and 2 wt%) present in the epoxy adhesive [2]. Epoxy joints without CNTs exhibited a resistance of 5×10^{11} Ω. Inclusion of 0.5 wt% and 1 wt% of MWCNT decreased the average resistance 7 and 8 orders of magnitude, respectively, as seen in Figure 21.12. Incorporating 2 wt% of MWCNTs, the average resistance reduces 10 orders of magnitude to 63 Ω. When measuring average shear strength of specimens with different carbon fractions, it is apparent from Figure 21.12 that the inclusion of CNTs did not affect specimens' shear strength in a negative way.

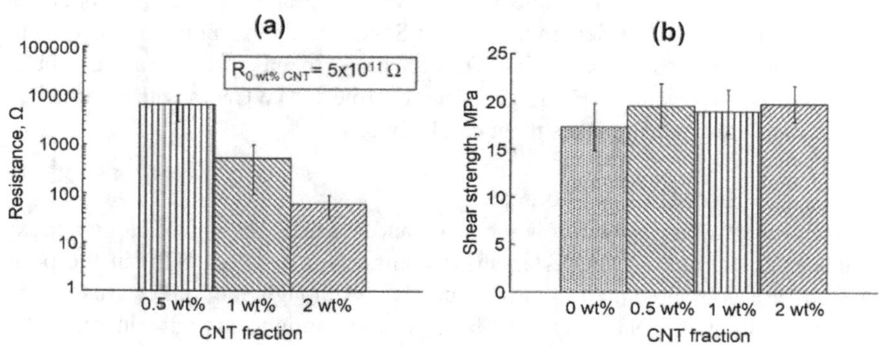

Figure 21.12 (a) Average electrical resistance and (b) average shear strength versus CNT concentrations [2]

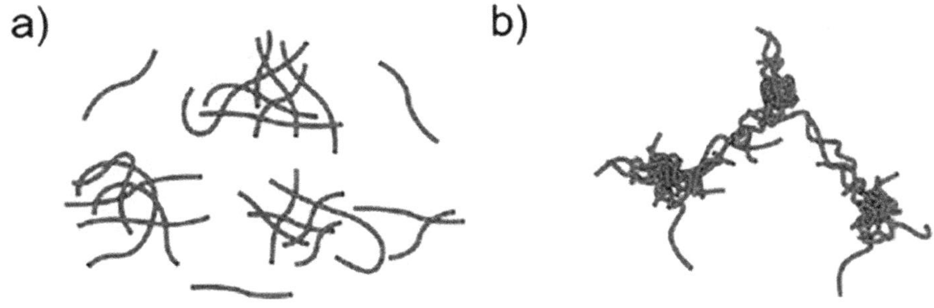

Figure 21.13 Light (a) and heavy (b) CNT agglomeration [38]

21.2.1.2.3 CNT agglomeration

Non-uniform dispersion of CNTs in a polymer matrix currently presents an issue. CNTs interact with each other through van der Waals forces, which are attractive, and this results in an aggregate of CNT ropes [37], which can be seen in Figure 21.13. Electrostatic charges of CNTs counteract that with repulsive forces and stabilise the dispersion; however, the right balance between the two forces is hard to achieve [39]. Agglomeration is especially significant with CNTs produced using CVD, which is the most common technique nowadays [40]. A consequence is reduced electrical conductivity and non-linearity of piezoresistivity [41] 41. The production technique of MWCNTs used in this project is unknown due to being confidential.

21.2.1.2.4 Electrical conductivity of CNT–epoxy nanocomposites

Electrical conductivity of polymers with dispersed CNTs inside depends on the CNT fraction [26], as seen in Figure 21.14. In the first stage (a), there is only a small amount of CNTs inside and the conductivity is similar to polymers alone. By increasing the CNT fraction in the second stage (b), clusters start to form, and conductivity starts to increase. This is mainly due to the tunnelling effect between CNTs and not because of physical contacts between them, as seen in Figure 21.15. In the third stage (c), an electrically conductive pathway for electrons to travel is formed. Conductivity in this stage increases drastically and the CNT fraction represents a percolation threshold. With further increase in the CNT fraction in the fourth stage (d), more electrically conductive pathways are created and with that a conductive network is formed. The conductivity does not increase indefinitely and comes to a constant after a certain fraction and is lower than the conductivity of CNTs alone.

It has been proven that percolating networks of MWCNTs in epoxy can be formed at concentrations below 0.1 wt% [37]. Process parameters used for manufacturing CNT–epoxy nanocomposites have an influence on the percolation threshold and electrical properties. Influence of mixing speed, mixing time, resin temperatures and curing temperatures were investigated [39]. Researchers established that mixing conditions and curing temperature have the biggest influence [42]44. High curing temperatures speed up the diffusion process and with that the mobility of CNTs, which speed up agglomeration and help to form a conductive network with higher

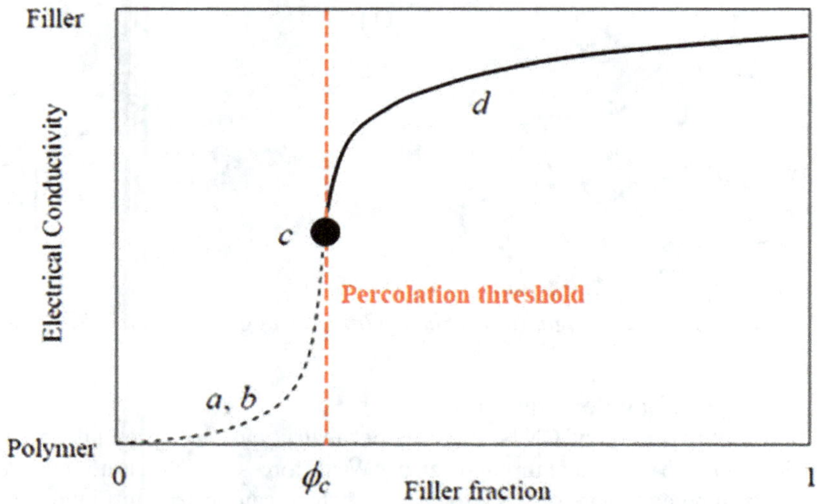

Figure 21.14 Filler (CNT) fraction [26]

conductivity. A mixing process involving modest shear forces and short time was found to be the most suitable for MWCNT–epoxy nanocomposites to avoid both over-dispersion and substantial aggregation; however, this information varies from research to research [26]. Ideal dispersion is meanwhile also not desirable for lowering the percolation threshold of an epoxy with a certain CNT fraction as much as possible. A lightly aggregated state can help in forming the first conductive pathway in a matrix and with that lowers the percolation threshold [26].

21.2.1.2.5 The piezoresistive effect of CNT–epoxy nanocomposites

The piezoresistive effect is a change in the electrical resistance of an object under the effect of mechanical strain [26]. CNTs exhibit such behaviour due to band gap changes in relation to strain changes [43, 44]. A band gap represents an energy range in which no electron state can exist 47. Once CNTs are dispersed in a polymer matrix, they make the matrix piezoresistive not only because CNTs themselves are piezoresistive but also because of additional main two reasons. First, when tensile strain is applied, contact between some CNTs is lost and second, some CNTs change their tunnelling resistance with neighbouring CNTs due to distance change [26]. It is reversed for when compressive strain is applied, as illustrated in Figure 21.16. It has been estimated that the tunnelling resistance increases non-linearly with strain [26]. The effect of CNTs' piezoresistivity is not substantial due to a poor transfer of load from the polymer matrix to CNTs [26].

21.2.1.2.6 CNT strain sensors

Various researches in the development of piezoresistive strain sensors made from CNT–polymer nanocomposites have been done [19][45, 46]. Compared to

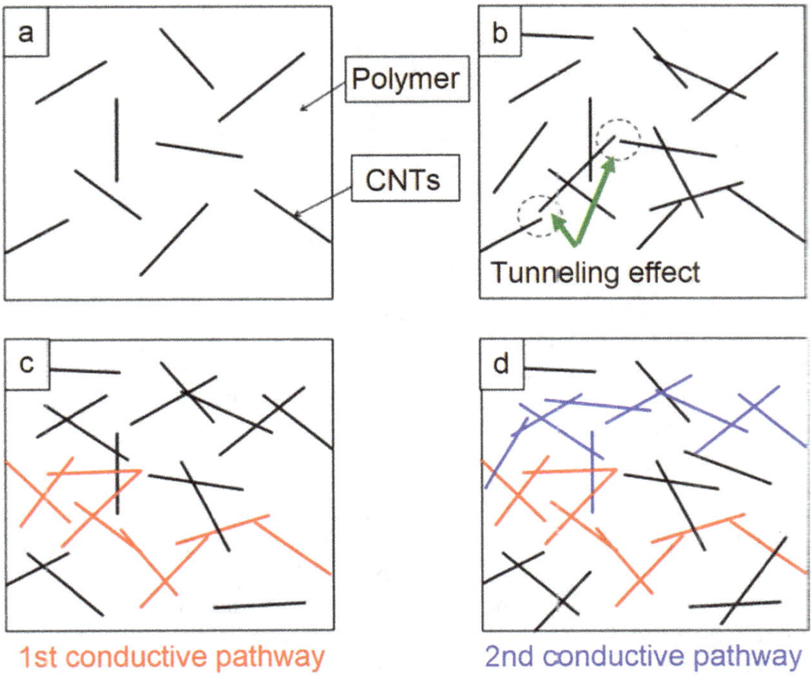

Figure 21.15 Polymer structure with different fractions of CNTs inside [26]

conventional strain gauges, higher sensitivity was observed. For purposes of analysis, piezoresistivity of a sensor is defined as $\Delta R/R_0$, where R_0 represents the starting electrical resistance at zero strain, and ΔR is the change in electrical resistance at a certain strain ε [26]. Sensitivity of a sensor in the elastic region is meanwhile defined with the GF. Studies that have researched such sensors used either a two-probe [30, 47] or a four-probe method [48] for measuring resistance changes. Various researches involving MWCNTs in a polymer matrix to make strain sensors identified that the piezoresistivity was linear at low strains and became non-linear at high strains [30, 31, 45, 49] due to the tunnelling resistance increasing non-linearly [26].

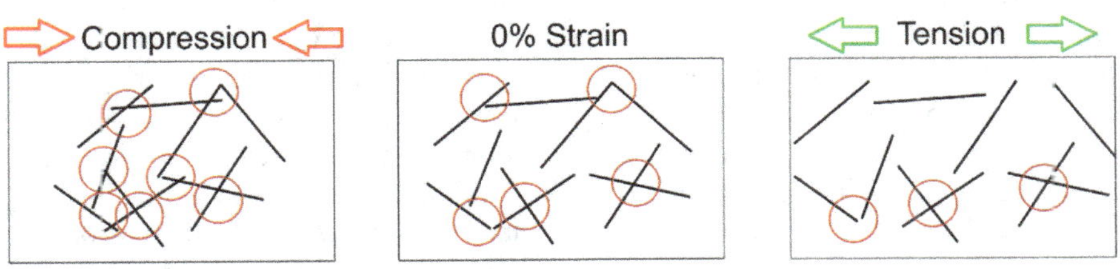

Figure 21.16 Change in number of conductive pathways between CNTs when under compression and tension

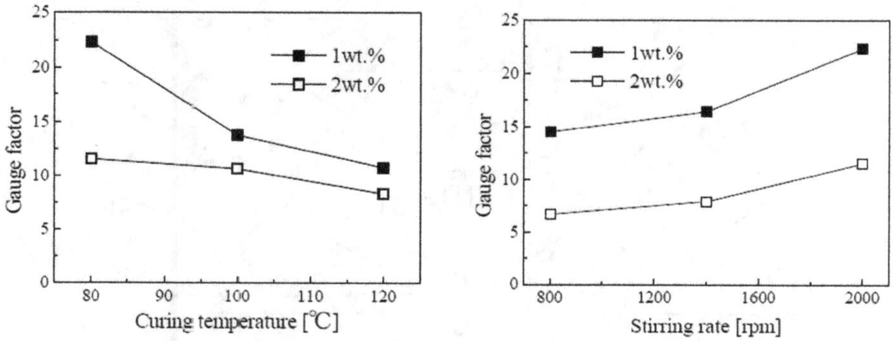

(a) influence of curing temperature

(b) influence of mixing speed

Figure 21.17 Influence of curing temperature and mixing speed on sensor sensitivity (GF) [26]

It has also been established for strain sensors that by using an electric field to align MWCNTs in the polymer matrix during specimen manufacture, sensitivity (GF) is improved and the linear region of piezoresistivity is lengthened [50].

Apart from tensile strains, compressive strains can be measured with such strain sensors as well50. In compression, a MWCNT polymer sensor has high sensitivity and linear piezoresistivity only for a small compressive strain inside the elastic regime. In this regime, the piezoresistivity of compressive strain is anti-symmetric compared to piezoresistivity of tensile strains. Once the strain level increases, the sensitivity eventually falls and saturates due to minimum distances between CNTs reached50.

It has been identified that a low curing temperature and a high mixing speed can result in higher sensor sensitivities (see Figure 21.17), which, based on what was said in the previous Chapter 10, means that sensors are more sensitive when their electrical resistance is higher 50.

21.2.1.3 Defects in adhesively bonded joints

21.2.1.3.1 Types of defects

Adhesively bonded joints can incorporate several different defects [51–58]. Porosity is caused by entrained air or volatiles and is present in most bonds to some degree. Cracks can be formed during monotonic and cyclic loading or due to incorrect curing or thermal shrinkage. Voids are produced by entrapped air or gasses during adhesive preparation and application or due to insufficient amount of adhesive applied. Zero thickness voids (kissing bonds) can occur due to the presence of a contaminant on an adherend, such as grease. Even though both adherends could be touching in areas containing zero thickness voids, minimal or no transfer of load from one to another will be achievable. A less obvious defect can be the adhesive itself and its poor cohesive and adhesive properties due to either incomplete mixing, incorrect formulation or insufficient cure [51] .

21.2.1.3.2 Amine blush

When using two-component epoxy adhesives, amine blush kissing bonds can be produced [59]. Amines tend to react with water and carbon dioxide, and result in the formation of ammonium carbamates, carbonates and bicarbonates . They form a greasy, waxy layer or crystal on the surface of the adhesive, which in some cases may not be visible [59]. This happens during the curing process when moisture condenses on the adhesive's surface or when present moisture causes water soluble compounds to migrate from the body of the adhesive to the surface, which will appear as sticky deposits. A short-time exposure to humid conditions with higher CO_2 concentration after spreading the adhesive on a specimen can already produce significant amine blush. Due to the fact that a portion of amine compounds are used for the creation of these blushes, not all epoxide groups can react with the curing agent, which results in under-cure and a reduced shear strength of the bond [59] 59] [61. Temperature plays a vital role here and as a rule of thumb, for every 10°C reduction in curing temperature, the cure speed will be reduced by a factor of approximately 2, which leaves amines more time to migrate to the surface and undergo alternative reactions . It has been confirmed that blush formation can be observed with Fourier-transform infrared spectroscopy(FTIR) analysis.

21.2.1.3.3 Detection of defects with the impedance method

Defects (except kissing or weak bonds) can be detected with a widely used non-destructive ultrasonic method [51, 60] 12] [57. However, this method cannot detect joint strength degradation due to surface defects [12]. By using a CNT–epoxy nanocomposite adhesive on lap-shear joints, a more novel method for detection of defects was researched. Through changes of resistance, they were able to detect and distinguish between different artificial surface defects in joints, as seen in Figure 21.18.

Additionally, joints with different surface defects gave a different resistance response under loading [12], as seen in Figure 21.19.

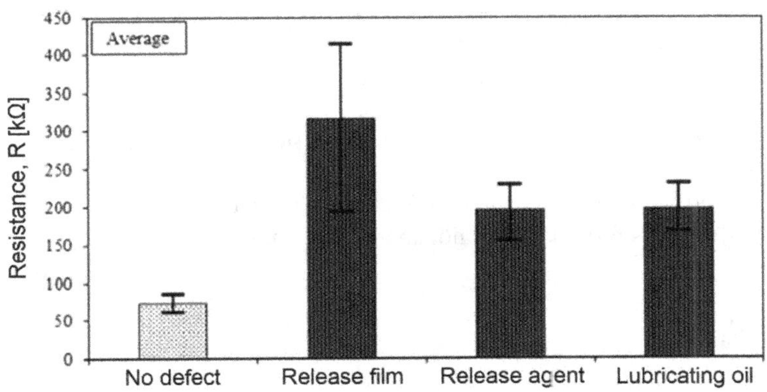

Figure 21.18 Resistance of adhesive joints with and without artificial defects [12]

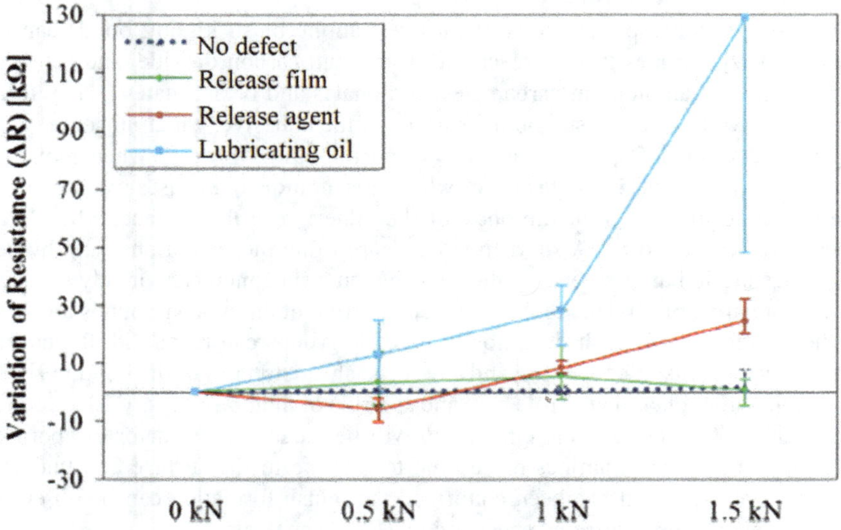

Figure 21.19 Resistance response of adhesive joints with and without artificial defects under loading [12]

21.2.1.3.4 Failure mechanisms of adhesively bonded joints

Adhesively bonded joints can fail in four different ways [62]. These failure mechanisms are as follows:

- Adhesion failure

Adhesion failure is an interfacial failure between the adhesive and the adherends. It can be recognised by the absence of adhesive on one of the bonding surfaces. This can happen due to the following reasons:

- Hydration of the chemical bonds between the adhesive and adherend
- Failure to create a chemically active surface (insufficient or inappropriate surface preparation)
- The adhesive has already cured before being applied
- The adhesive's formulation is not appropriate for the application

- Cohesion failure

Cohesion failure is a bond failure where the adhesive fractures. It can be recognised by the adhesive being present on both bonding surfaces. It happens gradually and is therefore more favourable than adhesion failure, which happens instantly. Failure is usually by shear; however, the following reasons can also be considered:

- Combination of shear and peel stresses or high peel stresses alone
- Insufficient overlap length
- Excessive porosity or void content

- Mixed-mode failure

Mixed-mode failure is a combination of adhesion and cohesion failure. It occurs when a partially degraded bond between the adhesive and the adherend is at some point overloaded and the weakened adhesive's interface fails.

- Adherend failure

Adherend failure occurs when the adhesive has a load capacity which is higher than the adhesive's strength. This is achievable if the adhesive bonds were well fabricated and designed.

21.2.1.4 Non-destructive testing techniques of health monitoring

21.2.1.4.1 Acoustic emission method

Acoustic emission (AE) is a method for monitoring the occurrence of damage [63]. A signal is acquired by the use of surface-mounted sensors when permanent damage is caused in a material which causes energy release and associated generation of an elastic wave [63]. In this way, we can detect the acoustic activity within the material in real time [64]. Other non-destructive testing (NDT) techniques (such as ultrasound and X-ray) focus on recognising damage after it has already occurred. However, enough energy has to be released in order to be able to detect the AE signal [63]. The method can be used not only for damage initiation and propagation monitoring in composite materials [61, 62] and adhesives [3] but also for distinguishing between different damage mechanisms [62, 63].

21.2.1.4.2 Health monitoring of CNT–epoxy adhesives

Substantial amount of research has been done in detecting onset and progression of damage in composites through resistance changes, either by using load-carrying carbon fibres as a conductive pathway [64, 65] or by using CNTs dispersed in the polymer matrix [66–68]. Adhesives containing CNTs have not been researched in the same breadth for NDT applications. Nevertheless, the work that has been done suggests that their usability for such applications is potentially high.

Ability of CNT–epoxy adhesives to distinguish between different types of damage in adhesively bonded lap-shear joints has been evaluated [3]. Under monotonic tensile loading, each failure mechanism possessed a distinct resistance response which corresponded directly with recorded acoustic events, as seen from

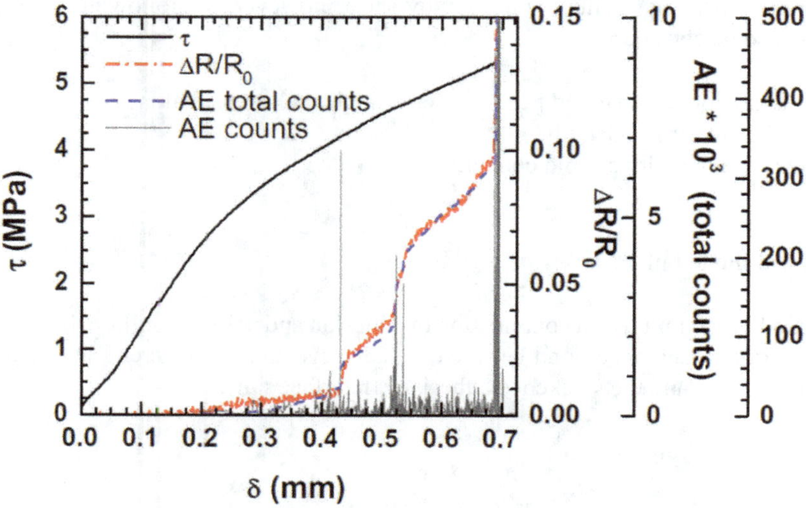

Figure 21.20 Mechanical, electrical and AE response of a specimen failed from adhesion failure [3]

Figure 21.20, showing a specimen undergoing an adhesion failure. The resistance response was able to detect the onset of damage long before the specimen reached its ultimate average shear strength, τ.

Same research performed incremental cyclic tension loading tests to determine the relation of resistance changes to deformation with progressive damage [3]. As shown in Figure 21.21., when permanent damage occurs in the ninth cycle, the baseline resistance noticeably changes and so only a certain part of resistance is reversible if the specimen is unloaded. During the first eight cycles in the elastic region, the baseline resistance also changes; however, due to viscoelastic properties of the polymer adhesive, as most likely not enough time was left between cycles for the adhesive to return to its original condition [83].

The last cycle before failure in Figure 21.21. was plotted separately with associated AE measurements (see Figure 21.22.). The electrical resistance response was non-linear and was described by several researchers [69, 70] [71]. The initial resistance increase is correlated to cracks, which were formed in the previous cycle, being reopened. After the change of slope, the specimen begins to elastically deform. The final linear region is related to new crack formation with a high AE response. Non-linear behaviour of the resistance response is therefore an indication of permanent damage.

21.2.2 Sample manufacturing

The manufacturing process of the test samples was designed to be simple and straightforward, achievable with laboratory chemicals and cheap. To create the samples, a mixture of epoxy resin and CNTs was first prepared, casted into glass

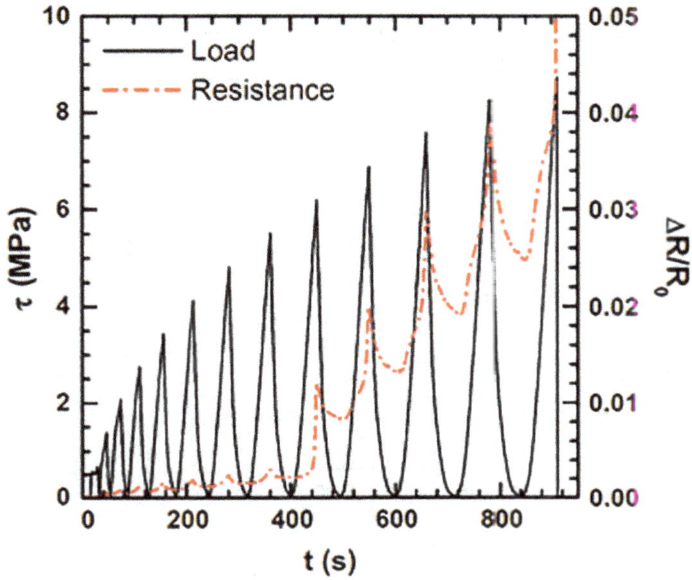

Figure 21.21 *Mechanical and electrical response of a specimen undergoing progressively cyclic loading showing adhesive failure [3]*

moulds and cured, then, they were cut with a bench saw at the appropriate size to create the rectangular-shaped test samples. Finally, the electrical connection and insulation were fabricated on the sample to allow electric resistance measurement and electrical insulation from the test machine. The following sections aim to detail the characteristics, provenance and method of fabrication behind sample manufacturing.

21.2.2.1 Materials

To create the nanocomposite the Araldite 1,564 and Aradur 3,486, an epoxy system was used. This warm curing epoxy system is based on an aerospace grade epoxy resin: Araldyte 1,564 and an amine-based hardener Aradur 3,486. Taking into account the viscosity of the resin (1,200–1,400 mPa.s) and of the hardener (10–20 mPa.s), acetone 99% was provided by the Composite Centre to lower the resin viscosity and allow for better mixing properties.

Triton x-100, in aqueous solution, was used as a non-ionic surfactant and supplied from Sigma-Aldrich. This surfactant was used without further dilution.

To further prepare the sample for tensile and electrical resistance testing, liquid commercial super-glue as well as copper conductive tape and thick kraft paper were bought from a local hardware shop. The conductive paint, RS pro silver paint, was provided by Composite Centre and bought from RS.

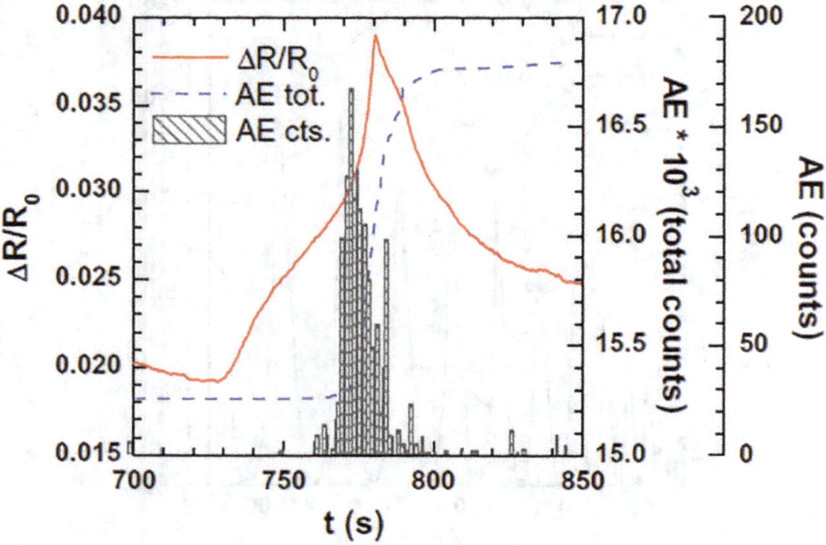

Figure 21.22 *Electrical resistance and AE response during the last cycle before adhesion failure [3]*

21.2.2.2 Sample specifications

Several samples were created to investigate the CNT loading of the epoxy and the effect of surfactant on the dispersion and electrical properties. Table 21.1 shows the details of the different concentrations of CNT and surfactant used in each sample along with the numbering of the sample to allow follow-up of the work in the future. Table 21.1 details the dimensions of each sample that was tested along with numbering.

Table 21.1 *Samples composition details*

CNT wt%	Surfactant %	Number
0	0	7
0.1	0	3
	0.5	11
	1	6
0.01	0	5
	0.2	10
	0.5	9
	1	8

Figure 21.23 *Probe Sonicator and water bath set up*

Each composition was prepared to fill the glass mould. The cured material of each preparation batch allowed for four samples to be produced, the numbering is logically, e.g., 7–1, 7–2, 7–3 and 7–4 for the sample with 0 wt% CNT.

21.2.2.3 Manufacturing process

In order to create the nanocomposite samples, a simple and straightforward method was applied. Using laboratory equipment, 32 test samples were created in a repeatable manner as follows:

First, the CNTs were weighted on a laboratory scale under Local Exhaust Ventilation (LEV), and a certain weight percentage either 0.1 wt% or 0.01 wt% was selected relatively to the weight of epoxy. The appropriate amount of Triton x-100 (Tx100) was then added to the CNT and 20 mL of acetone was poured to the CNT + Tx100 and sonicated using a Branson Digital Sonifier probe Sonicator for 15 minutes at 50% amplitude. During this sonication process, the beaker was held in a water bath to prevent the solution from overheating and the acetone from evaporating, see Figure 21.23..

Second, the dispersion of CNT in acetone was poured in 66.4 g of epoxy resin, hand mixed to incorporate the acetone to the epoxy and lower its viscosity and then sonicated again for 30 minutes at 60% amplitude to disperse the CNT into the epoxy resin. This time the beaker was not held in water and the solution was left to heat up under sonication energy in order to evaporate the acetone present in the solution.

Then, the dispersion of CNT in epoxy was further mixed on a hot plate at 150°C with magnetic stirring at medium speed for 1 hour. This step provided full evaporation of the acetone and further dispersion of the nanotubes in the resin. The mixture was left to cool down to room temperature before the 22 g of hardener was added and mixed using magnetic stirring again for 15 minutes at medium speed.

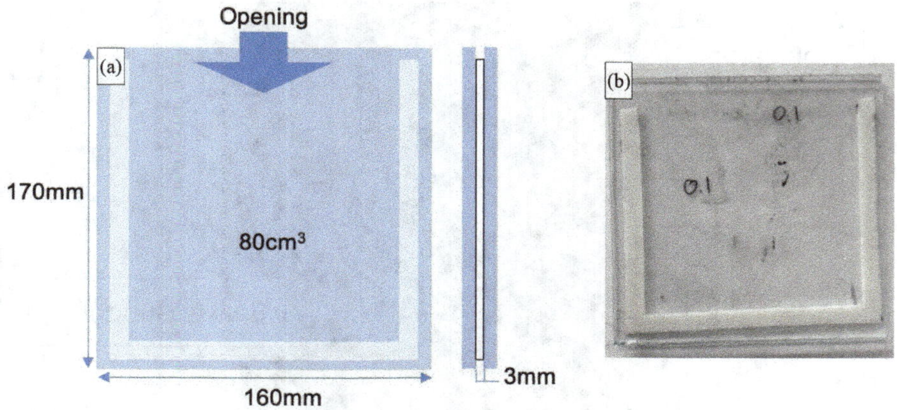

Figure 21.24 Rectangular mould (A) schematic and (B) picture

After degassing in a vacuum chamber for 45 minutes until no bubbles were left, the mixtures were poured into a glass mould with a cavity of 80 cm^3 and thickness of 3 mm, see Figure 21.24..

Finally, the mould was placed in the oven and the resin was left to cure at 80°C for 8 hours following manufacturer specification for Araldyte + Aradur curing time and temperature. When curing was complete, four samples were cut out of the cured material using a bench saw and the sample size was specified as 150 × 25 × 3 (see Figure 21.25.).

21.2.3 Tensile and electrical resistance testing

To assess the ability of the nanocomposite to sense elastic deformation, tensile tests were carried out on the manufactured samples. During those tests, the electrical resistance of the sample was monitored. The following sections aim to detail the equipment and method used for testing the samples.

21.2.3.1 Sample preparation for testing

After curing and cutting the samples, a preparation stage was needed. In this stage, the electrodes for electrical resistance measurements were created as well as the insulated tabs where the tensile machine grips the sample. To create the electric

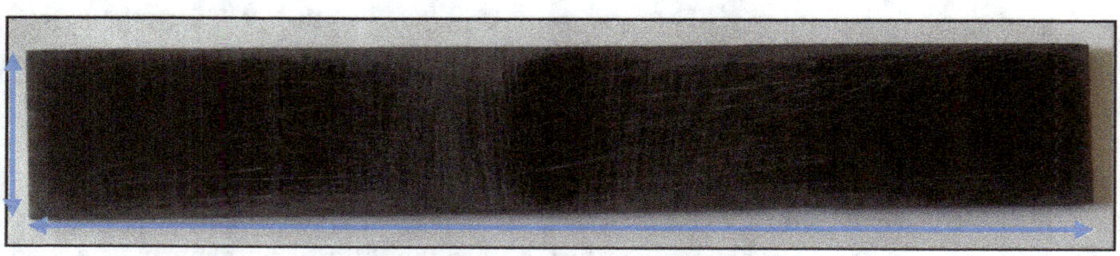

Figure 21.25 Test sample after cutting and surface scratching

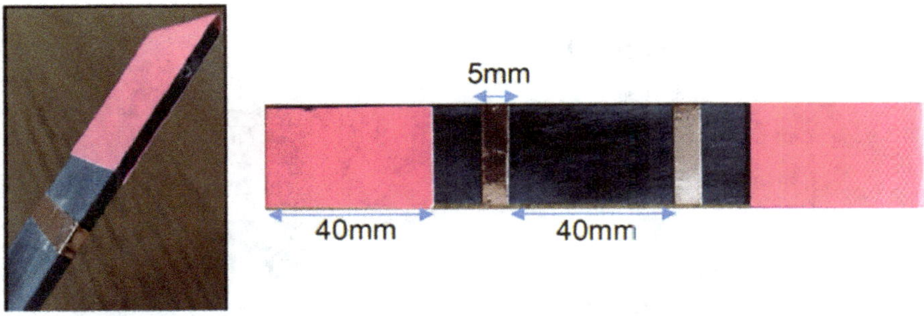

Figure 21.26 Test sample with tabs and electric connections

connection, each sample was first polished with abrasive paper with a grain size of 120 in order to scratch the surface and reveal the CNTs. Then, after washing off the dust with acetone, silver paint was used to create a 5 mm wide path across the width of the sample, and each path was separated by approximately 40 mm. Finally, copper tape was placed on the still fresh silver paint and wrapped three times around the sample. In order to electrically insulate the sample from the test machine and thus forcing electrons to flow through the nanocomposites, thick Kraft paper was used as tabs on each end of the sample. The paper tabs were approximately 40 mm in length and covered the entire width of the sample (see Figure 21.26.).

21.2.3.2 Equipment

To perform the test, a full testing apparatus was modified from a previous MSc project and adapted to use within this research. The testing apparatus consisted of the following:

- Instron tensile machine +5 KN load cell+ controlling computer_A
- Computer_B running a custom LabView acquisition software
- Data Acquisition (DAQ) system
- Keithley Ohmmeter
- Laser extensometer

The Instron tensile machine was linked to the DAQ using Bayonet Neill–Concelman (BNC) cables with machine load on Ch0 and displacement on Ch2. The laser extensometer was also connected to the DAQ on Ch1 using a custom-made connection cable phone plug to BNC. The DAQ serves as a gathering point for all data; it is plugged on Computer_B via USB and is used to input the load, displacement and extension from the laser into the LabView software, which plots and saves the data. The ohmmeter was also plugged to Computer_B via USB and the electrical resistance data were fed to the LabView software. Crocodile Clips were used to connect the ohmmeter to the test sample and two reflective targets were placed

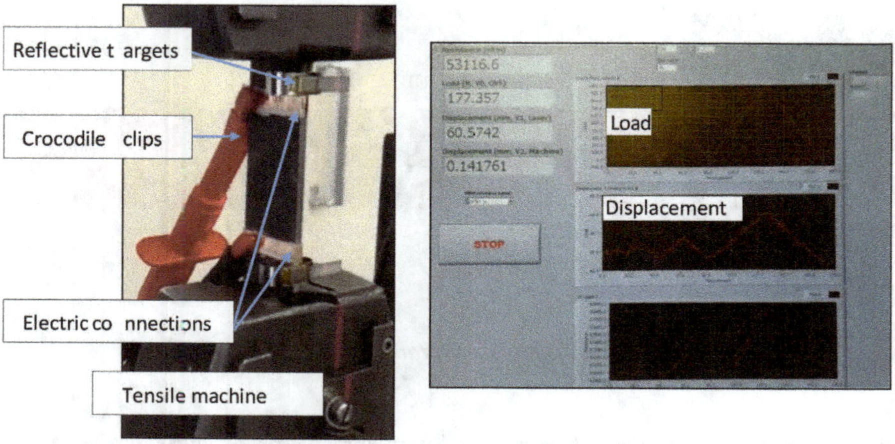

Figure 21.27 Test sample in the tensile machine and LabView software interface

on the sample to allow measurement of displacement with the laser extensometer. Figure 21.27. shows the tensile testing setup and the LabView software interface.

21.2.3.3 Testing plan

Based on manufacturer specification for the resin system, it was defined that the tensile strength of the cured resin will be reached before 5000 KN load. In order to test the electrical resistance and see its evolution when elastic deformation is applied to the nanocomposite it was decided to test the samples as follows.

Each sample was loaded in cyclic manner, and the first load was set to 150 N then the load was set back to 0 N. Thus, the material would go from 0 to 150 N and back to 0 N; this was the first 'loop'. The tensile machine was programmed to perform five loops, 150, 300, 450, 1000 and 2000 N, going back to 0 N between each loop. Doing such a process allowed for investigating materials behaviour and electric resistance evolution under elastic deformation and plastic deformation regime.

During testing, the LabView software displayed three curves, load versus time, displacement versus time and resistance versus time. Each data set was saved into an excel file from which more plots were drawn.

21.2.4 Scanning electron microscopy

In order to investigate the dispersion of CNTs in the epoxy as well as the aspect of the CNTs themselves, the LYRA3 TESCAN Scanning Electron Microscope (SEM) was used. Pieces of each sample were cut from corners and middle parts, polished with 120 grain size abrasive paper and coated with 20 nm of gold. The pieces were then held on aluminium stubs with carbon tape. Gold coating of the sample allowed for improved conductivity of the sample leading to better SEM images, and the carbon tape was used to disperse the electrons to prevent sample charging and bright glow on the SEM images.

21.2.5 Raman spectroscopy

Raman spectroscopy was performed on the nanotubes to further characterise their structure. The nanotubes were dispersed in silicon oil and in water using probe sonication for 15 minutes, then a droplet was laid on a silicon substrate and placed in the Raman spectrometer. The parameters on the machine were set as follows: slit: 300 μm, hole: 400 μm, 10 seconds acquisition time and 5 spectrums to be acquired. A 632.81 nm wavelength red laser at 100% power and the ×100 objective was used.

21.3 Results and discussion

The main aspects of this research are the strain sensing ability of the nanocomposite and the CNTs themselves. Structural characteristics of the nanotubes such as their length and diameter and their chemical characteristics such as their purity result directly in the ability of the nanocomposite to sense strain. Aiming at the perfect SWCNT not always leads to the best sensing results: it is always a matter of finding the right characteristics that fit the application we aim for. These following sections touch on the characteristics of the nanotubes produced, their flaws and advantages with respect to resistive strain sensing application. Then, the sensing capabilities of the nanocomposites are shown and discussed.

21.3.1 Carbon nanotubes characterisation

21.3.1.1 Scanning electron microscopy

Figure 21.28. shows the SEM image of a CNT bundle, without any treatment. The first step towards characterising the CNTs is to look at their length, diameter and entanglement.

From the measurements taken on the SEM microscope, we can see that the apparent diameter is between 20 and 100 nm and we can approximate the length to

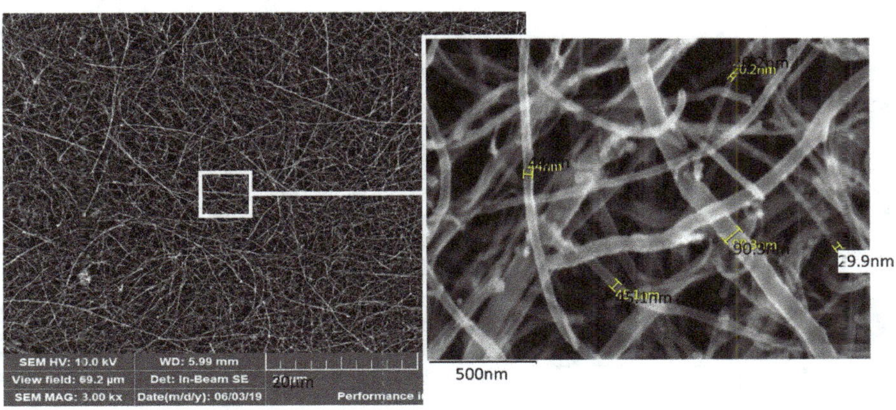

Figure 21.28 SEM image of pristine CNT bundle

40 μm. This first visual investigation seems to indicate the presence of MWCNTs, even if these images cannot exactly give information about the diameter because of angle of view and uneven height level of each nanotubes, it provides a rough idea. On the larger scale image, we can follow one nanotube and appreciate the way it bends and turns in the bundle. In SEM images of SWCNTs and of small diameter MWCNTs, the nanotubes are bent with a greater angle, and the curves are sharper. In this case, this suggests a higher number of walls in the structure of nanotubes.

The length and diameter of the nanotubes are directly related to the manufacturing process. The temperature of the furnace as well as the rate of flow of ferrocene in the chamber and the growth time are reported to affect the morphology of the CNTs [34]. Finer tuning of the manufacturing parameters can be applied to increase the uniformity of the CNTs or lower their diameter. However, the length and diameter observed on these SEM images are consistent with other work [33], but the inner diameter of the tubes was not investigated in this work.

MWCNTs with large diameter are preferable for strain sensing application because of the improved strain transfer between the resin and the nanotubes, repeatability and linearity of the measurements [19]. Furthermore, their electrical conductivity remains comparable to the SWCNTs because most of the electrons are travelling between the layers of the nanotubes [35, 36].

21.3.1.2 Raman spectroscopy

To further characterise the type of nanotubes, their electronic properties and morphology, Raman spectroscopy was performed. Figure 21.29. shows the Raman spectrum of dried CNT on a silicon substrate along with the important peaks relevant to CNTs identification.

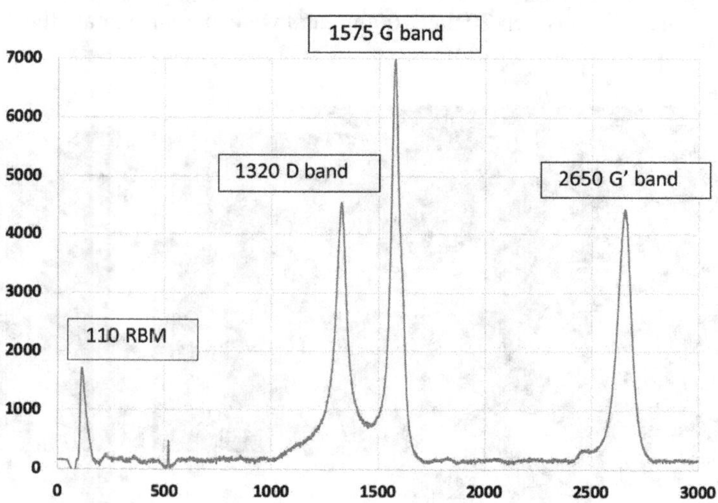

Figure 21.29 Raman spectrum of dried CNTs on a silicon substrate

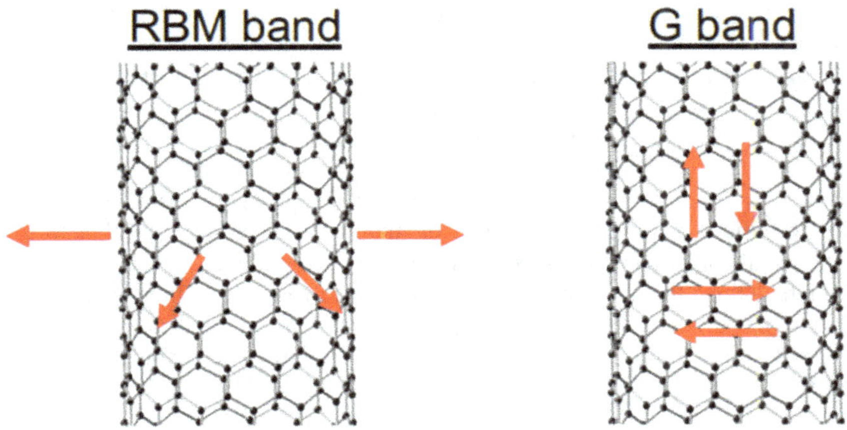

Figure 21.30 Schematic of atoms movement for RBM and G bands

Four major bands are present in the spectrum, the radial breathing mode (RBM) band at 110 cm^{-1}, D band at 1320 cm^{-1}, G band at 1575 cm^{-1} and G' band at 2650 cm^{-1}. The RBM band corresponds to the in-phase movement of all carbon atoms in the radial direction [37]. The D band is related to disturbance in the hexagonal graphitic structure, and it is often evidence of amorphous face or impurity of the nanotubes [39]. The third peak, the G band, is directly related to the graphitic structure and relates to the movement of neighbouring atoms in the opposite direction. Lastly, the G' band which is a second-order harmonic of the D band is representative of a self-annealing pair of phonons and can occur in defect free nanotubes. Figure 21.30. shows a schematic of atoms movement leading to RBM and G bands.

We can see on the spectrum that the RBM band at 100 cm^{-1} is clearly visible and of good intensity. Other works have shown that the RMB band of SWCNT is observable with good intensity, whereas the RMB band of MWCNT is low in intensity and difficult to see [39] [2]. This effect is due to damping of atoms movement due to multiple moving walls on the inside of the tube. The RBM band is linked to the nanotube diameter by the following relation: $\omega_{RBM} = A/d + B$, where d is the diameter of the tube, A and B are constants related to the surroundings of the nanotube. Other works have suggested the value of $A = 248$ and B to be neglected [37]. When applying this relation, we found that the diameter of the nanotube is 2.25 nm. This rough calculation may apply more accurately to SWCNTs but gives an idea about the number of walls of the CNTs.

To assess the quality and the abundance of graphitic structure, the ratio between the intensity of the G and D band is often considered [39]. In our case, we can see that the intensity of the D band is close to the intensity of the G band. This behaviour would suggest high defect and amorphous phase in the CNT as produced, but it is also a good indicator of the presence of MWCNTs. The multiple inner walls of the tubes are prone to defects, by increasing the number of inner tubes and creating

MWCNTs the probability of having defects increases, explaining the intensity of the D band.

Another feature of the G band is its shape, in our case we see that the shape of the G band is thin and sharp. This shape can be linked to the electric properties of the nanotubes, whether they are metallic or semi-conductive. In our case, the shape of the G band is close to the shape of a highly oriented pyrolytic graphite as mentioned in other research [2]. Although the high intensity of the peak would suggest a semi-conductive type of nanotube, further investigation is needed to draw a full picture.

The simultaneous rough analysis of the Raman spectrum and the SEM images led authors to think that the nanotubes are multiwalled with a large number of walls. Their purity and amount of crystalline graphene structure is still to be precisely determined along with a more thorough investigation of peak position related to diameter.

21.3.2 Strain sensing

21.3.2.1 Carbon nanotubes dispersion

In order to understand the nanocomposite's electric behaviour, a link must be made to the dispersion of nanotubes within it. First, a visual characterisation was performed. Each sample was placed on a backlit surface, the epoxy resin being transparent and the nanotubes being black; these images provided a good first look at the CNT distribution. The pictures are shown in Figure 21.31.

First of all, we can clearly separate each group (0.1 wt% CNT and 0.01 wt% CNT), thanks to their opacity. The 0.1 wt% CNT group was the only conductive one, which means that in the 0.01 wt% CNT group, the dispersion of nanotubes did not manage to reach percolation threshold and create a conductive network within the material. The conductive group is opaque, and we can see that without surfactant (0%) the light manages to go through the sample and reveals some areas where the CNTs are agglomerated and some areas with only resin. This is also seen, but to a lesser extent, when 1% surfactant is added, we see that the CNTs are covering more space. When 0.5% is added, the sample is fully opaque and no light comes through the sample, this shows the best level of dispersion.

Regarding the non-conductive group (0.01 wt% CNT), without surfactant the sample shows fine, dust-like, agglomerates. With a high percentage of surfactant, it is clearly visible that the nanotubes formed large agglomerates, and the surfactant in excess formed long micelles that prevented the nanotubes from dispersing in the resin. This effect is seen to a lesser extent for 1% and 0.5% surfactant, where fine agglomerates are still visible. However, the sample with 0.2% surfactant presents a continuous grey colour without visible agglomerates. At this concentration, the surfactant is supposedly under its critical micelle concentration (CMC) and did not form long micelles but surrounded and dispersed the nanotubes.

It is not possible to know precisely the 3D distribution of the CNT in the material, especially within the conductive group with 0.1 wt% CNT. To have a better understanding of the dispersion inside the material, pieces of each sample were observed with the SEM. The purpose was to look at the cross-sectional part, inside

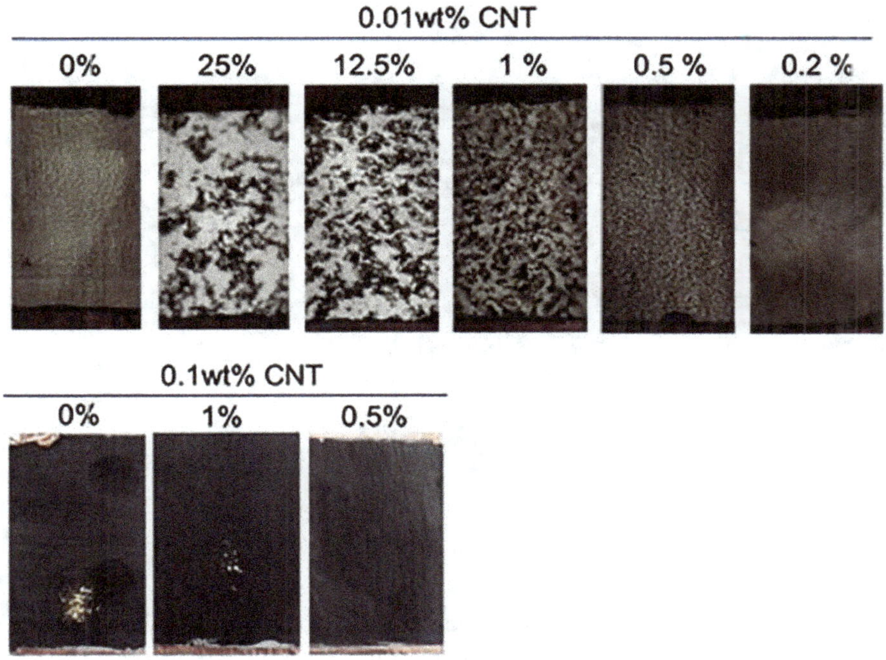

Figure 21.31 *Picture of backlit sample for 0.01 wt% CNT and 0.1 wt% CNT with different surfactant concentrations*

the material but depending on where the SEM samples were cut on the original samples it was difficult to see the CNT in the resin. Figure 21.32. shows SEM image of the 0.1 wt% CNT sample with 1% Tx100. This sample was the only one where the CNTs were clearly visible.

We can see in the images that the CNTs have been shortened by the process. One main reason may be the continuous use of probe sonication for 45 minutes on the CNT during the mixing stage. They do not seem highly entangled, probably because of their short length, and the area is well covered by CNT; However, this image is of an area densely populated by CNT. When scanning the surface with SEM it was seen that many places were empty of CNTs.

21.3.2.2 Tensile tests and resistance measurements

Within the two groups of created samples, with 0.1 wt% and 0.01 wt% of CNT loading, only the one with 0.1 wt% showed electric conductivity when tested with the apparatus. The other one, with 0.01 wt% of CNT, still had the non-conductive behaviour of the epoxy resin. In this section, the evolution of the electrical resistance during tensile testing is shown and compared for different concentrations of surfactant. Only the results for the conductive group (0.1 wt% CNT) is presented, the

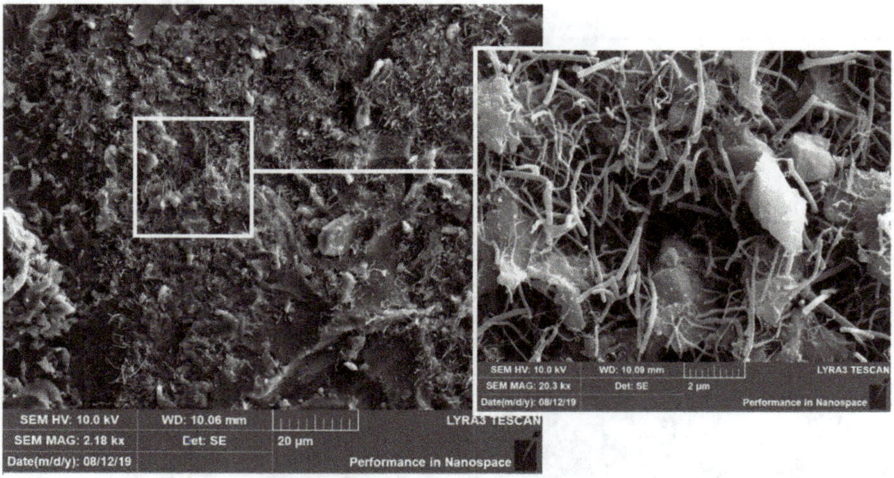

Figure 21.32 Scanning Electron Microscopy Image of Carbon Nanotubes dispersion in Epoxy resin, from 0.1wt% CNT + 1%Tx100

electrical resistance of the other group being too high for the ohmmeter to perceive any changes. The GF is also shown as an indication of sensor performance.

We start by looking at Figure 21.33. showing the change in electrical resistance when tensile strain is applied. Each cycle (1–5) is shown along with the load level. The GF was calculated and plotted on the chart as well.

The starting resistance of the sample 0.1 wt% CNT with no surfactant was 20±5 kOhm. We can see that the tensile strain is in the elastic region of the material

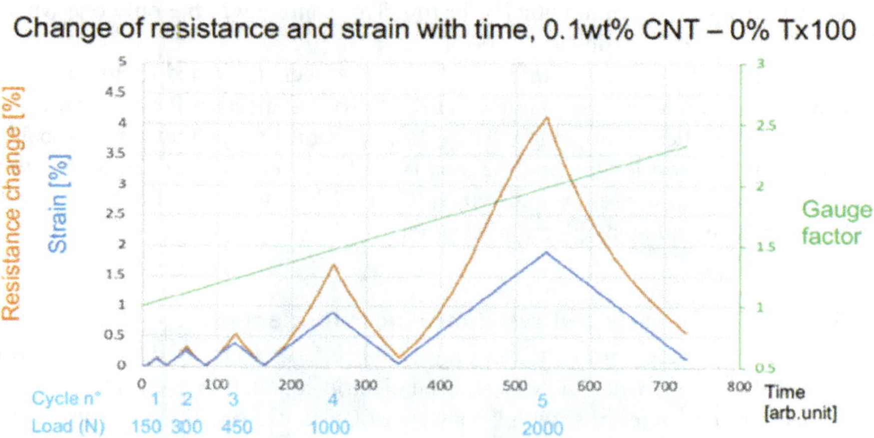

Figure 21.33 Curve of change in resistance and strain versus time for 0.1 wt% CNT and no surfactant

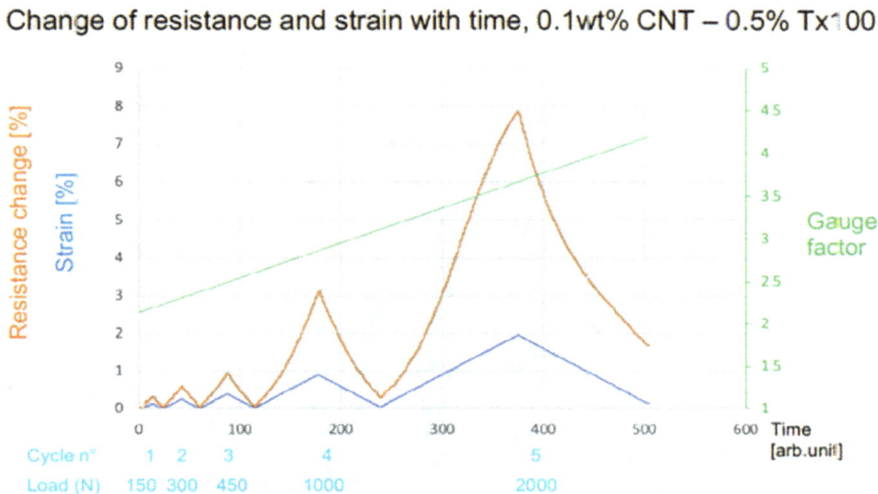

Figure 21.34 Curve of change in resistance and strain versus time for 0.1 wt% CNT and 0.5%Tx100

for the first three cycles, and then plastic deformations are induced. We see that the de-strain does not return to 0 after the fourth cycle. Also, the resistance change is following the strain really accurately during the first three cycle but during the fourth and fifth cycles the resistance changes much more than the strain. This is shown by increasing GF thought the test; the resistance does not change linearly with strain. Moreover, the maximum change in resistance was of approximately 4% of the original value for a strain of 2%. With the addition of 0.5% of surfactant (Tx100), the curve changed drastically. Figure 21.34. shows the change in resistance and strain versus time for 0.1 wt% CNT with 0.5%Tx100.

On that plot, we can see the same trend as without any surfactant, the resistance does not change linearly with strain and the GF increases throughout the test. However, even if the same level of plastic deformation is induced in the material, the resistance response is 100% greater than without surfactant. We can see that when 2% strain is induced, 8% change in electrical resistance occurs. The sensitivity seems to be improved, but the starting resistance of these sample was 3.5±0.1MOhm.

With further increase in the surfactant concentration to 1% Tx100, the materials' properties seem to stabilise. Figure 21.35. shows the change in resistance and strain versus time for 0.1 wt% CNT with 1%Tx100.

We can see that the GF is still increasing but much slower than with no surfactant or 0.5% surfactant and stays, approximately, between 0.75 and 1. The resistance started at 10±1 kOhm, and it is the highest conductivity reached by all samples. Although the sensitivity is lower than before with 0.5% or no surfactant, the resistance change is following the strain in a more linear manner, reaching approximately 2.5% change when strain reaches 1.75%.

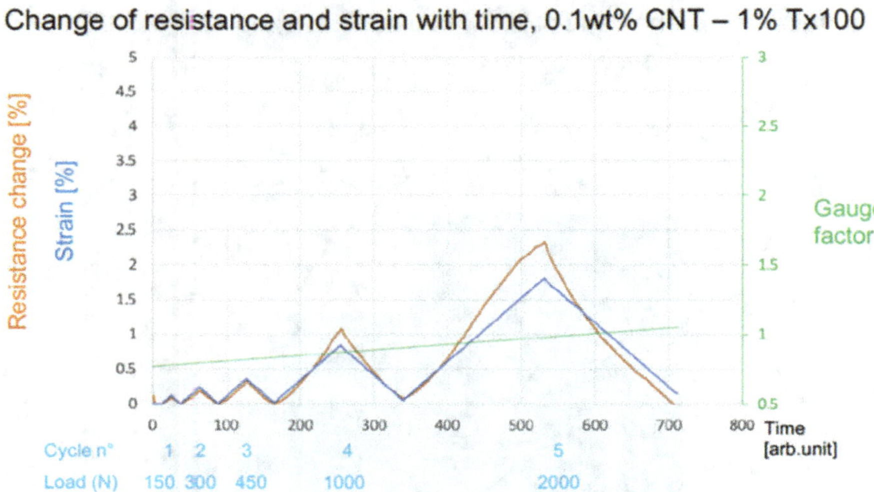

Figure 21.35 *Curve of change in resistance and strain versus time for 0.1 wt% CNT and 1%Tx100.*

21.3.2.3 Mechanical properties

When looking at manufacturing new composite materials, one main objective is to maintain or improve the mechanical behaviour of the base constituents. In our case, the resin system was studied under elastic deformation only. Figure 21.36. shows the load versus displacement curves of the conductive samples (0.1 wt% CNT) for each concentration of surfactant and compared to pure epoxy. In figure panel (A), we can see that the curves match each other, showing that there is neither improvement nor decrease in mechanical properties in that region of tests. However, when 0.5% tx100 is added, the sample seems to weaken and shows more strain at the same load level. More plasticity is induced. When 1%Tx100 is added, the behaviour is opposite, and the mechanical properties are slightly improved.

The hypothesis behind this behaviour is that at 0.5%Tx100 the CMC is nearly reached, and the surfactants are finely dispersing the CNT in the matrix. Because of a poor load transfer from the resin to the CNT, the nanotubes act like defect sites and weaken the epoxy. On the other hand, when the concentration of surfactant is above the CMC, for 1%Tx100, the nanotubes are fully surrounded by surfactant micelles. These micelles provide additional bonding space between the epoxy and the CNT, improving the load transfer and strengthening the epoxy.

21.3.2.4 Effect of surfactant

The purpose of adding surfactant before the mixing stage is to aid the dispersion of nanotubes by unravelling the bundles. A direct link can be made between the concentration of surfactant and the electrical behaviour of the material (i.e., its resistance curve). When 0.5% surfactant is added to the CNT, the measured resistance of

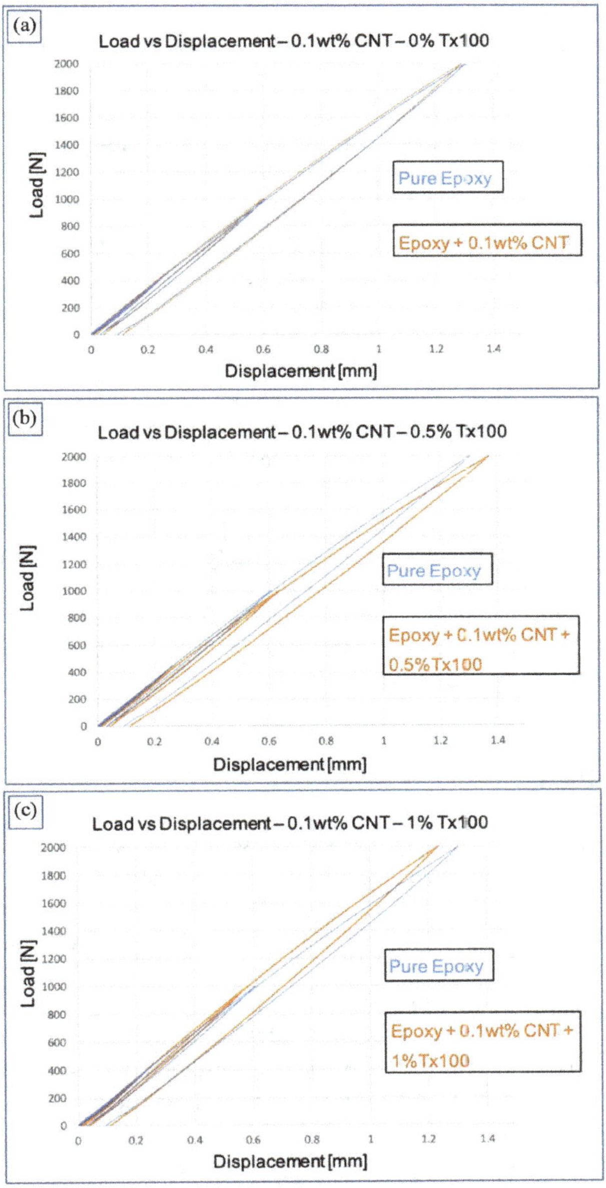

Figure 21.36 Load versus displacement curves for pure epoxy and epoxy +0.1 wt% CNT with no surfactant and 0.5% Tx100 C) 1% Tx100

the sample is around 3.5 MOhm, which was the highest measurable resistance for all samples. This high resistance, reached only with that concentration of Tx100, indicates that the surfactant succeeded to disperse the nanotube and unravel the bundles. As a matter of fact, to obtain high conductivity, it is unwanted to achieve an extra fine dispersion. For example, a matrix where every single nanotube will be separated from its neighbour will exhibit an extra fine dispersion but a high resistance because electrons will be able to travel only through tunnelling effect. It is believed that for 0.1 wt% CNT with 0.5% Tx100, see Figures 21.7, the dispersion was too fine to achieve good conductivity hence the high value of resistance. Moreover, since the dispersion is extra fine and the electrons are travelling through a tunnelling effect, the sensitivity of the sensor is greater than any other sample. The tunnelling effect decreases exponentially with distance between the CNT; a little change induces a large change in resistance, hence the large increase in GF with strain.

Regarding the sample with 0.1 wt% CNT with 1% Tx100, the resistance is around 10 kOhm. It is the lowest resistance reached amongst all samples, see Figures 21.8. The hypothesis behind this behaviour is the following: the surfactant is above its CMC and started to agglomerate the CNT. The conductive network of CNT is denser, with more nanotubes actually touching each other. This would explain the more linear behaviour of the resistance change when strain was applied.

The original idea was to achieve a high level of CNT dispersion, but the results suggest that some level of agglomeration is desirable for the linearity of the measurement, to have detectable resistance while using a low loading of CNT. The other strategy would be to seek the best dispersion possible and increase the loading of CNT to reach a better level of conductivity and linearity.

21.4 Conclusion

This chapter aimed to investigate the feasibility of nanocomposite for strain sensing through electrical resistance measurement. The first concluding point is that feasibility has been demonstrated. Using CNTs produced, it was possible to create a nanocomposite that is able to sense elastic strain through electrical resistance measurements.

The effect of non-ionic surfactant, TrionX100 was investigated and a concentration of 1% in 0.1 wt% CNT was found to yield the best results in terms of linearity of gauche factor and conductivity of the material. Improving the dispersion to a perfect state with all CNTs detached from each other is not desirable as it lowers the electrical conductivity of the sample.

The loading of CNT that was used in this work was found to slightly increase the mechanical properties of the epoxy resin, only when 1% surfactant is added to the composition. The percolation threshold was reached for samples loaded with 0.1 wt% CNT, whereas samples loaded with 0.01 wt% CNT were not conductive.

The change in electrical resistance was mainly attributed to increasing distance between the CNTs due to applied strain. This directly affects the ability of the electron to tunnel from one nanotube to another and increases the resistance of the sample leading to the measurement.

21.5 Future work

Deeper analysis must be carried out on characterising the nanotubes, and the use of Raman spectroscopy must be completed by a full statistical analysis of CNT size distribution using SEM. It might be interesting to investigate the Raman spectrum obtained with another laser wavelength, i.e., green 514 nm [39]. The transmission electron microscope must also be added to the characterisation procedure as it provides a straightforward observation of the nanotube's walls.

In order to obtain more consistent resistance data, better electrical connection between the ohmmeter and the sample must be designed. Suggestions will be to include copper strips or any conductive wire inside the material before curing. The contact between the CNTs and the conductive connection would be improved, and contact resistance would be minimised.

To fully understand the impact of a dispersion of CNTs in the epoxy resin, the tensile test must be performed until failure of the material. Other mechanical tests such as compression or fatigue tests could be implemented to have a more complete picture of the nanocomposite's performances.

On the material side, the author could be interested to test different loadings of CNT close to 0.1 wt%. By doing so, the percolation threshold of the CNT in epoxy using the proposed dispersion method could be defined. Lower concentration of Triton x-100 should also be tried. Also, it seems like a better method to have the mass of CNT rather than the mass of epoxy drive the concentration of surfactant.

Appendix A: Tensile tests

A.1 Sample 1 and 2 for each composition

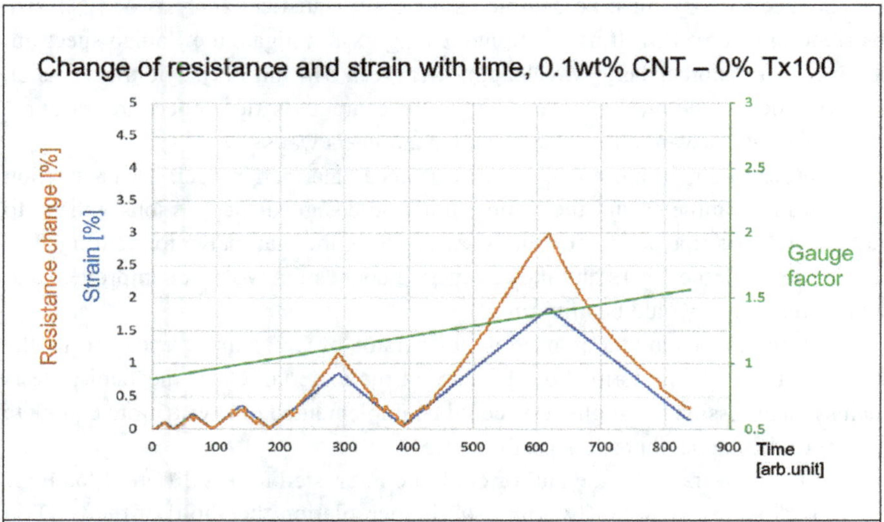

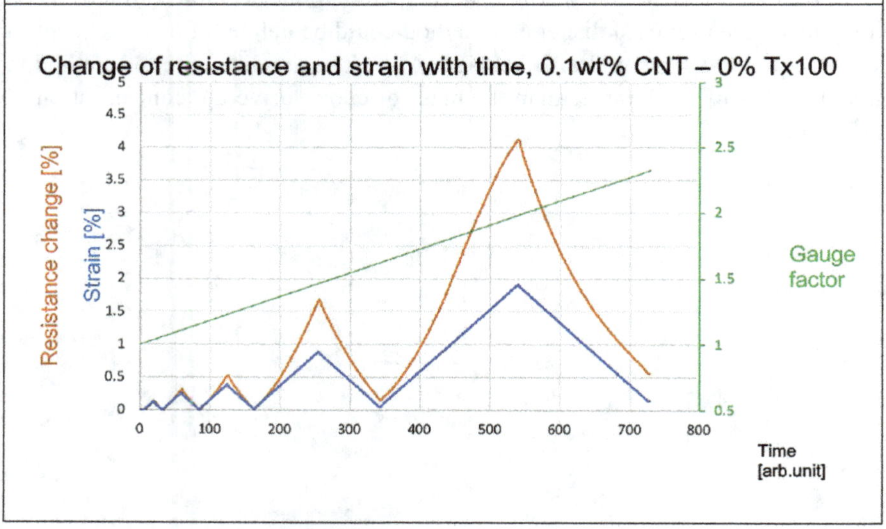

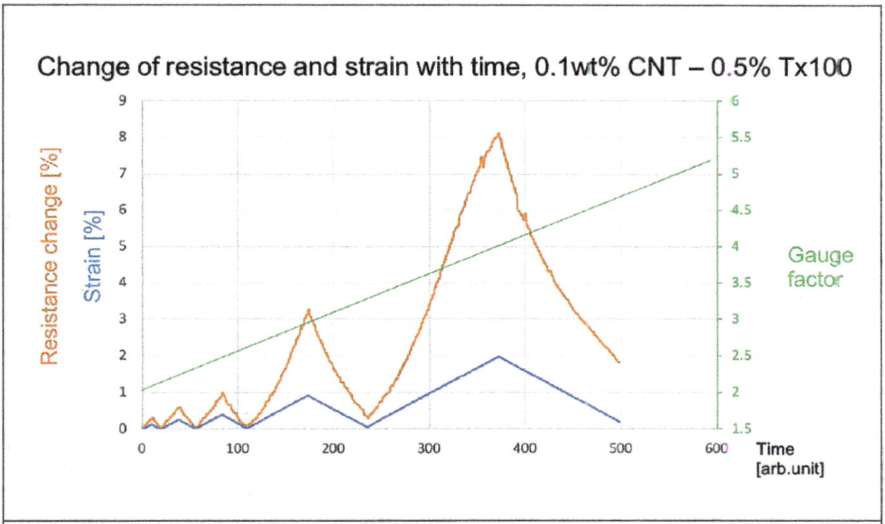
Change of resistance and strain with time, 0.1wt% CNT – 0.5% Tx100
Resistance change [%]
Strain [%]
Gauge factor
Time [arb.unit]
0
100
200
300
400
500
600
1
2
3
4
5
6
7
8
9
1.5
2
2.5
3
3.5
4
4.5
5
5.5
6

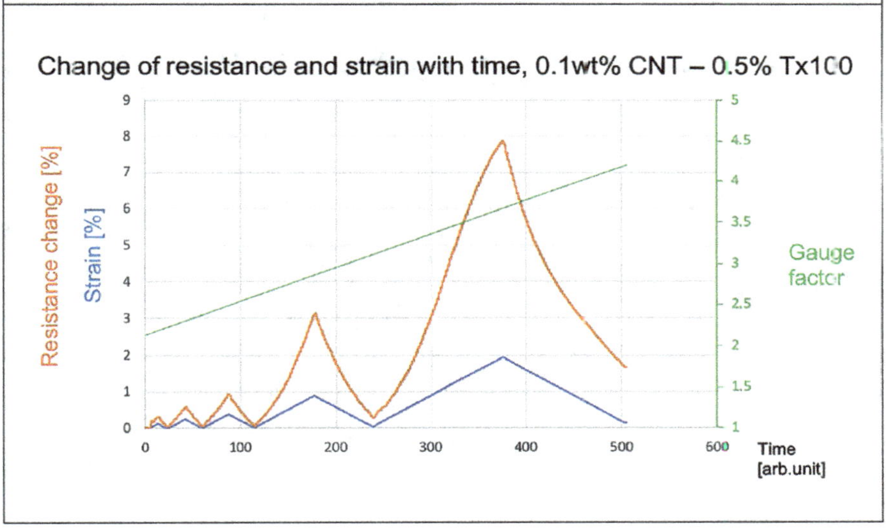
Change of resistance and strain with time, 0.1wt% CNT – 0.5% Tx100
Resistance change [%]
Strain [%]
Gauge factor
Time [arb.unit]
0
100
200
300
400
500
600
1
2
3
4
5
6
7
8
9
1
1.5
2
2.5
3
3.5
4
4.5
5

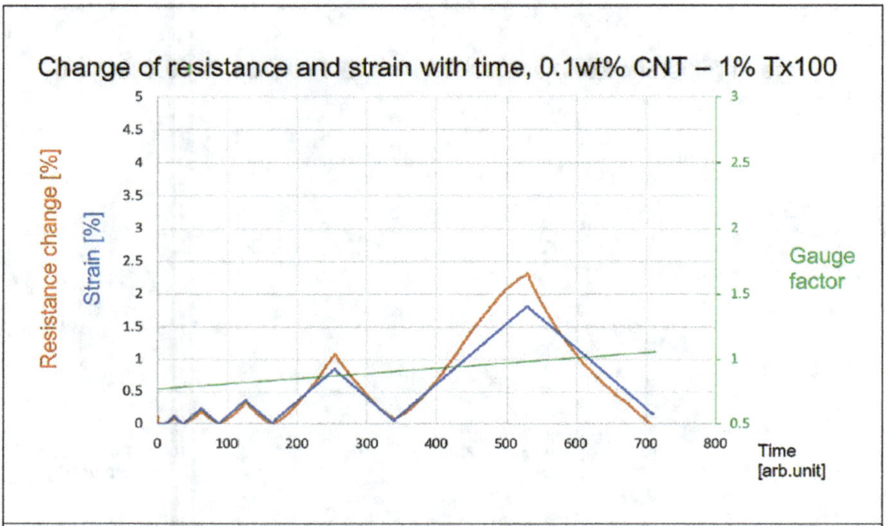
Change of resistance and strain with time, 0.1wt% CNT – 1% Tx100
Resistance change [%]
Strain [%]
Gauge factor
Time [arb.unit]
0
100
200
300
400
500
600
700
800
0.5
1
1.5
2
2.5
3
3.5
4
4.5
5

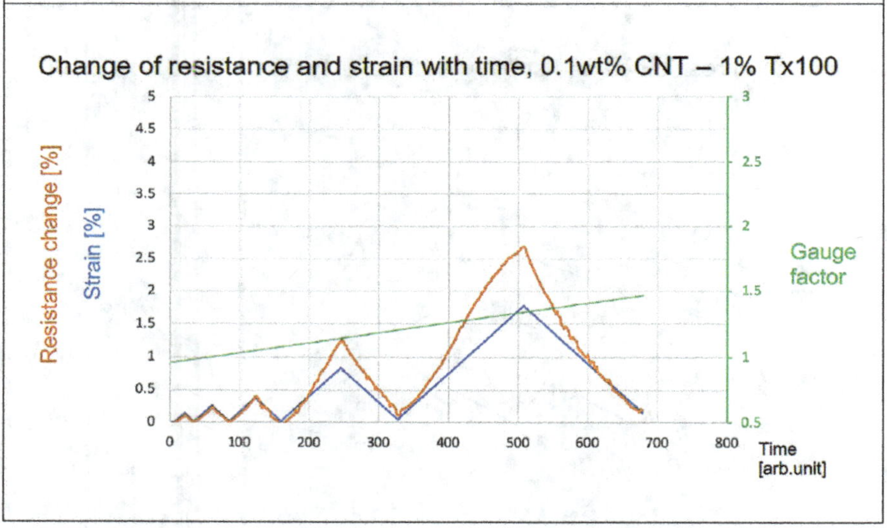
Change of resistance and strain with time, 0.1wt% CNT – 1% Tx100
Resistance change [%]
Strain [%]
Gauge factor
Time [arb.unit]
0
100
200
300
400
500
600
700
800
0.5
1
1.5
2
2.5
3
3.5
4
4.5
5

Appendix B: Load vs displacement curves

B.1 First set of samples with 0%, 0.5% and 1% surfactant

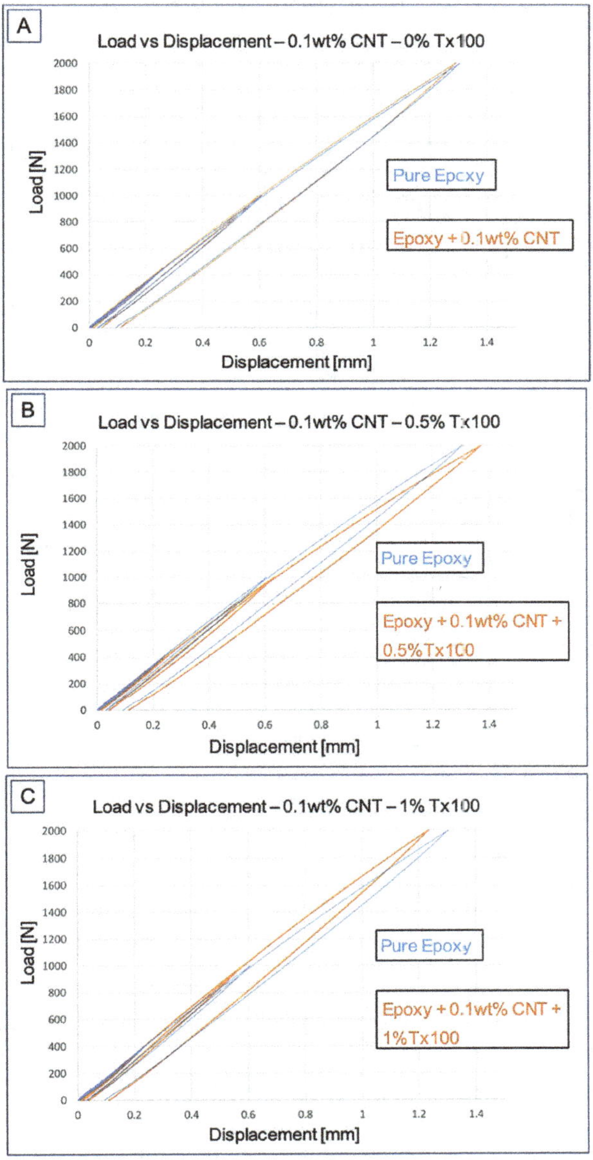

B.2 Second set of samples with 0%, 0.5% and 1% surfactant

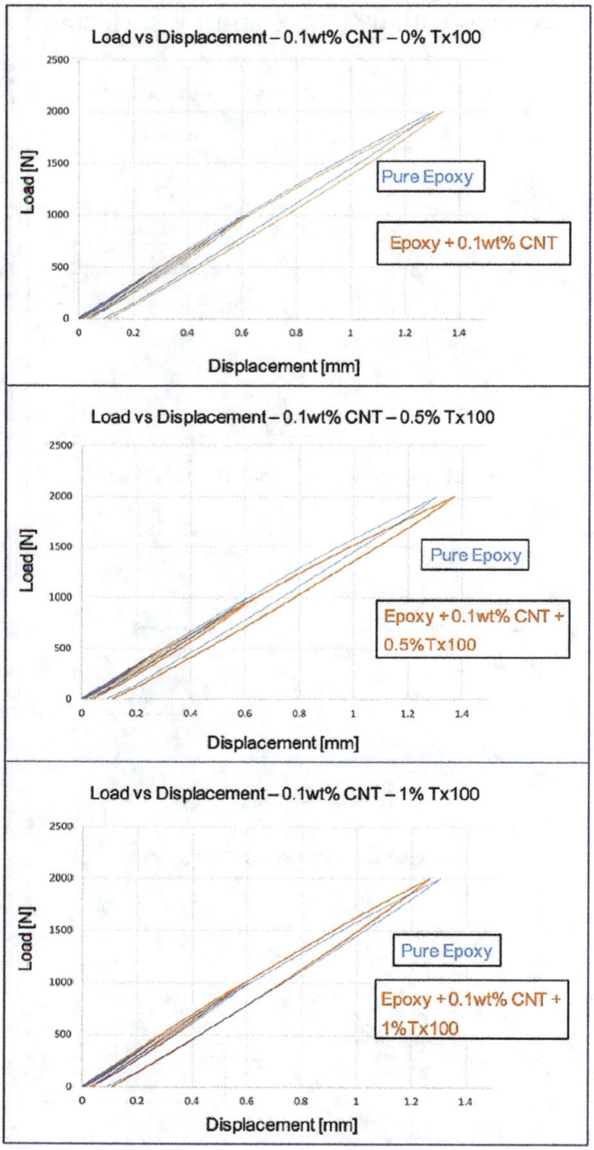

References

[1] Peschel G. *Carbon-carbon bonds: hybridization peschel carbon- carbon bonds: hybridization [online]*. Vol. 2; 2011. pp. 1–6. Available from http://www.physik.fu-berlin.de/einrichtungen/ag/ag-reich/lehre/Archiv/ss2011/docs/Gina_Peschel- Handout.pdf%0Apapers3://publication/uuid/0622B03B-CC86-40E2-9D9E- DC2B08809512

[2] Dresselhaus M.S., Dresselhaus G., Saito R., Jorio A. 'Raman spectroscopy of carbon nanotubes'. *Physics Reports*. 2005;409(2):47–99.

[3] Odom T.W., Huang J.-L., Kim P., Lieber C.M. 'Atomic structure and electronic properties of single-walled carbon nanotubes'. *Nature*. 1998;391(January):62–4.

[4] Grossiord N., Loos J., Regev O., Koning C.E. 'Toolbox for dispersing carbon nanotubes into polymers to get conductive nanocomposites'. *Chemistry of Materials*. 2006;18(5):1089–99.

[5] Rahman G., Najaf Z., Mehmood A, *et al*. 'An overview of the recent progress in the synthesis and applications of carbon nanotubes'. *C Journal of Carbon Research*. 2019;5(1):3.

[6] Mousa M.S. 'Comparison between single-walled CNT, multi-walled CNT, and carbon nanotube-fiber pyrograf III'. *IOP Conference Series: Materials Science and Engineering*. 2018;305(1):012025.

[7] Ding R.G., Lu G.Q., Yan Z.F., Wilson M.A. 'Recent advances in the preparation and utilization of carbon nanotubes for hydrogen storage'. *Journal of Nanoscience and Nanotechnology*. 2001;1(1):7–29.

[8] Aqel A., El-Nour K.M.M.A., Ammar R.A.A., Al-Warthan A. 'Carbon nanotubes, science and technology part (I) structure, synthesis and characterisation'. *Arabian Journal of Chemistry*. 2012;5(1):1–23.

[9] Collins P.G., Avouris P. 'Nanotubes for electronics'. *Scientific American*. 2000;283(6):62–69.

[10] Overney G., Zhong W., Tomnek D., Tománek D. 'Structural rigidity and low frequency vibrational modes of long carbon tubules'. *Zeitschrift Fr Physik D Atoms, Molecules and Clusters*. 1993;27(1):93–96.

[11] Yakobson B., Brabec C.J., Berhnolc J. 'Nanomechanics of carbon tubes: instabilities beyond linear response'. *Physical Review Letters*. 1996;76(14):2511–14.

[12] Wong E.W., Sheehan P.E., Lieber C.M. 'Nanobeam mechanics: elasticity, strength, and toughness of nanorods and nanotubes'. *Science*. 1997;277(5334):1971–75.

[13] Demczyk B.G., Wang Y.M., Cumings J, *et al*. 'Direct mechanical measurement of the tensile strength and elastic modulus of multiwalled carbon nanotubes'. *Materials Science and Engineering: A*. 2002;334(1–2):173–78.

[14] Andrews R., Jacques D., Rao A.M, *et al*. 'Nanotube composite carbon fibers'. *Applied Physics Letters*. 1999;75(9):1329–31.

[15] Jia Choi P., Y.Z. (no date o) *Single, Double, MultiWall Carbon Nanotube Properties & Applications, Sigmaaldrich.com*. Available from https://www.sigmaaldrich.com/technical-documents/articles/materials- science/single-double-multi-walled-carbon-nanotubes.html#ref.

[16] Kleiner A., Eggert S. 'Band gaps of primary metallic carbon nanotubes'. *Physical Review B*. 2001;63(7):1–4.

[17] Anantram M.P., Han J., Lu J.P. 'Band-gap change of carbon nanotubes: effect of small uniaxial and torsional strain'. *Physical Review B - Condensed Matter and Materials Physics*. 1999;60(19):13874–78.

[18] Yang L., Han J. 'Electronic structure of deformed carbon nanotubes'. *Physical Review Letters*. 2000;85(1):154–57.

[19] Kang I., Schulz M.J., Kim J.H., Shanov V., Shi D. 'A carbon nanotube strain sensor for structural health monitoring'. *Smart Mater Struct*. 2006;15(3):737–48.

[20] Obitayo W., Liu T. 'A review: carbon nanotube-based piezoresistive strain sensors A review: carbon nanotube-based piezoresistive strain sensors'. 2017.

[21] Cao J., Wang Q., Dai H. 'Electromechanical properties of metallic, quasimetallic, and semiconducting carbon nanotubes under stretching'. *Physical Review Letters*. 2003;90(15):4.

[22] Rein M.D., Breuer O., Wagner H.D. 'Sensors and sensitivity: carbon nanotube buckypaper films as strain sensing devices'. *Composites Science and Technology*. 2011;**71**(3).

[23] Li C., Zhang D., Deng C, *et al*. 'High performance strain sensor based on buckypaper for full-range detection of human motions'. *Nanoscale*. 2018;10(31):14966–75.

[24] Vemuru S.M., Wahi R., Nagarajaiah S., Ajayan P.M. 'Strain sensing using a multiwalled carbon nanotube film'. *The Journal of Strain Analysis for Engineering Design*. 2009;44(7):555–62.

[25] Ardebili M.K. 'Strain and damage sensing of CNT-EPOXY composite via electrical conductivity'. 2019:1–6.

[26] Hu N., Fukunaga H., Atobe S., Liu Y., Li J. 'Piezoresistive strain sensors made from carbon nanotubes'. *Sensors*. 2011:;**11**:10691–723.

[27] Rams J., Sánchez M., Ureña A., Jiménez-Suárez A., Campo M., Güemes A. 'Use of carbon nanotubes for strain and damage sensing of epoxy-based composites'. *International Journal of Smart and Nano Materials*. 2012;3(2):152–61.

[28] Loh K.J., Lynch J.P., Shim B.S., Kotov N.A. 'Tailoring piezoresistive sensitivity of multilayer carbon nanotube composite strain sensors'. *Journal of Intelligent Material Systems and Structures*. 2008;19(7):747–64.

[29] Yin G., Hu N., Karube Y., Liu Y., Li Y., Fukunaga H. 'A carbon nanotube/polymer strain sensor with linear and anti-symmetric piezoresistivity'. *Journal of composite materials*. 2011;45(12):1315–23.

[30] Yasuoka T., Shimamura Y., Todoroki A. 'Electrical resistance change under strain of CNF/flexible-epoxy composite'. *Advanced Composite Materials*. 2010;19(2):123–38.

[31] Hu N., Karube Y., Arai M, *et al.* 'Investigation on sensitivity of a polymer/carbon nanotube composite strain sensor'. *Carbon.* 2010;48(3):680–87.

[32] Zhao H., Zhang Y., Bradford P.D, *et al.* 'Carbon nanotube yarn strain sensors'. *Nanotechnology.* 2010;21(30):305502.

[33] Singh C., Shaffer M.S.P., Windle A.H. 'Production of controlled architectures of aligned carbon nanotubes by an injection chemical vapour deposition method'. *Carbon.* 2003;41(2):359–68.

[34] Zhang Q., Liu J., Sager R., Dai L., Baur J. 'Hierarchical composites of carbon nanotubes on carbon fiber: influence of growth condition on fiber tensile properties'. *Composites Science and Technology.* 2009;69(5):594–601.

[35] Frank S., Poncharal P., Wang Z.L., Heer W.A. 'Carbon nanotube quantum resistors'. *Science.* 1998;280(5370):1744–46.

[36] Bachtold A., Strunk C., Salvetat J.-P, *et al.* 'Aharonov–bohm oscillations in carbon nanotubes'. *Nature.* 1999;397(6721):673–75.

[37] Dresselhaus M.S., Dresselhaus G., Jorio A., Souza Filho A.G., Saito R. 'Raman spectroscopy on isolated single wall carbon nanotubes'. *Carbon.* 2002;40(12):2043–61.

[38] Hollertz R., Chatterjee S., Gutmann H., Geiger T., Nüesch F.A., Chu B.T.T. 'Improvement of toughness and electrical properties of epoxy composites with carbon nanotubes prepared by industrially relevant processes'. *Nanotechnology.* 2011;22(12):125702.

[39] Kruszynska M. 'Characterization of carbon nanotubes by Raman spectroscopy'. 2014.

[40] Benoit J.M., Buisson J.P., Chauvet O., Godon C., Lefrant S. 'Low-frequency Raman studies of multiwalled carbon nanotubes: Experiments and theory'. *Physical Review B.* 2002;66(7)pp.:1–4.

[41] Gong S., Zhu Z.H., Meguid S.A. 'Carbon nanotube agglomeration effect on piezoresistivity of polymer nanocomposites'. *Polymer.* 2014;55(21):5488–99.

[42] Hu N., Masuda Z., Yamamoto G., Fukunaga H., Hashida T., Qiu J. 'Effect of fabrication process on electrical properties of polymer/multi-wall carbon nanotube nanocomposites'. *Composites Part A.* 2008;39(5):893–903.

[43] Grow R.J., Wang Q., Cao J., Wang D., Dai H. 'Piezoresistance of carbon nanotubes on deformable thin-film membranes'. *Applied Physics Letters.* 2005;86(9):093104.

[44] Tombler T.W., Zhou C., Alexseyev L, *et al.* 'Reversible electromechanical characteristics of carbon nanotubes under local-probe manipulation'. *Nature.* 2000;405(6788):769–72.

[45] Hu N., Karube Y., Yan C., Masuda Z., Fukunaga H. 'Tunneling effect in a polymer/carbon nanotube nanocomposite strain sensor'. *Acta Materialia.* 2008;56(13):2929–36.

[46] Gang Y., Ning H., Karube Y., Yaolu L., YuanL., Fukunaga H. 'A carbon nanotube/polymer strain sensor with linear and anti-symmetric piezoresistivity'. *Journal of Composite Materials.* 2011;45(12):1315–23.

[47] Shimamura Y., Kageyama K., Tohgo K., Fujii T. 'Cyclic behavior of electrical resistance type low stiffness, large strain sensor by using carbon nanofiber/

flexible epoxy composite'. *Key Engineering Materials*. 2011;462–463:1200–05. Available from https://www.scientific.net/KEM.462-463

[48] Dharap P., Li Z., Nagarajaiah S., Barrera E.V. 'Nanotube film based on single-wall carbon nanotubes for strain sensing'. *Nanotechnology*. 2004;15(3):379–82.

[49] Wichmann M.H.G., Buschhorn S.T., Gehrmann J., Schulte K. 'Piezoresistive response of epoxy composites with carbon nanoparticles under tensile load'. *Physical Review B*. 2009;80(24).

[50] Oliva-Avilés A.I., Avilés F., Sosa V. 'Electrical and piezoresistive properties of multi-walled carbon nanotube/polymer composite films aligned by an electric field'. *Carbon*. 2011;49(9):2989–97.

[51] Adams R.D., Drinkwater B.W. 'Nondestructive testing of adhesively-bonded joints'. *NDT & E International*. 1997;30(2):93–98.

[52] Bregar T., An D., Gharavian S, *et al*. 'Carbon nanotube embedded adhesives for real-time monitoring of adhesion failure in high performance adhesively bonded joints'. *Scientific Reports*. 2020;10(1):16833.

[53] Liu Y., Zhang X., Lemanski S., Nezhad H.Y., Ayre D. 'Experimental and numerical study of process-induced defects and their effect on fatigue debonding in composite joints'. *International Journal of Fatigue*. 2019;125:47–57.

[54] Bhanushali R., Ayre D., Nezhad H.Y. 'Tensile response of adhesively bonded composite-to-composite single-lap joints in the presence of bond deficiency'. *Procedia CIRP*. 2017;59:139–43.

[55] Yazdani Nezhad H., Zhao Y., Liddel P.D., Marchante V., Roy R. 'A novel process-linked assembly failure model for adhesively bonded composite structures'. *CIRP Annals*. 2017;66(1):29–32.

[56] Nezhad H.Y., Stratakis D., Ayre D., Addepalli S., Zhao Y. 'Mechanical performance of composite bonded joints in the presence of localised process-induced zero-thickness defects'. *Procedia Manuf*. 2018;16:91–98.

[57] An D., Lotfian S., Mesbah D, *et al*. 'Ultra-thin electrospun nanofibers for development of damage-tolerant composite laminates'. *Materials Today Chemistry*. 2019;14:100202.

[58] Khaleque T., Zhang X., Kumar Thakur V., Aria A.I., Yazdani Nezhad H. 'Tailoring of thermo-mechanical properties of hybrid composite-metal bonded joints'. *Polymers*. 2021;13(2):E170.

[59] Poveromo S.L., Earthman J.C. 'Analysis of "kiss" bonds between composite laminates'. *JOM*. 2014;66(6):970–78.

[60] Kim C.-H., Choi J.-H., Kweon J.-H. 'Defect detection in adhesive joints using the impedance method'. *Composite Structures*. 2015;120:183–88.

[61] Moevus M., Rouby D., Godin N, *et al*. 'Analysis of damage mechanisms and associated acoustic emission in two sic/ [si–B–C] composites exhibiting different tensile behaviours. Part I: damage patterns and acoustic emission activity'. *Composites Science and Technology*. 2008;68(6):1250–57.

[62] Barré S., Benzeggagh M.L. 'On the use of acoustic emission to investigate damage mechanisms in glass-fibre-reinforced polypropylene'. *Composites Science and Technology*. 1994;52(3):369–76.

[63] de Oliveira R., Marques A.T. 'Health monitoring of FRP using acoustic emission and artificial neural networks'. *Computers & Structures*. 2008;86(3–5):367–73.

[64] Wang X., Chung D.D.L. 'Real-time monitoring of fatigue damage and dynamic strain in carbon fiber polymer-matrix composite by electrical resistance measurement'. *Smart Mater Structures*. 1997;6(4):504–08.

[65] Wang S., Chung D.D.L. 'Self-sensing of flexural strain and damage in carbon fiber polymer-matrix composite by electrical resistance measurement'. *Carbon*. 2006;44(13):2739–51.

[66] Thostenson E.T., Chou T.-W. 'Carbon nanotube networks: sensing of distributed strain and damage for life prediction and self healing'. *Advanced Materials*. 2006;18(21):2837–41.

[67] Gao L., Thostenson E.T., Zhang Z., Chou T.-W. 'Coupled carbon nanotube network and acoustic emission monitoring for sensing of damage development in composites'. *Carbon*. 2009;47(5):1381–88.

[68] Lekawa-Raus A., Gizewski T., Patmore J., Kurzepa L., Koziol K.K. 'Electrical transport in carbon nanotube fibres'. *Scripta Materialia*. 2017;131:112–18.

[69] Lim A.S., Melrose Z.R., Thostenson E.T., Chou T.-W. 'Damage sensing of adhesively-bonded hybrid composite/steel joints using carbon nanotubes'. *Composites Science and Technology*. 2011;71(9):1183–89.

[70] Thostenson E.T., Chou T.-W. 'Real-time in situ sensing of damage evolution in advanced fiber composites using carbon nanotube networks'. *Nanotechnology*. 2008;19(21):215713.

[71] Nofar M., Hoa S.V., Pugh M.D. 'Failure detection and monitoring in polymer matrix composites subjected to static and dynamic loads using carbon nanotube networks'. *Composites Science and Technology*. 2009;69(10):1599–606.

Index

Abaqus 607–608, 612–615
Abrasive waterjet machining 137
Absorbing/Breather fabric 215
Acoustic emission 134, 727
Acrylonitrile butadiene styrene (ABS) 65, 66, 68, 70, 74, 75, 77, 231, 232, 240, 251, 265, 271, 273, 274, 275, 277, 289, 292, 300, 308
Active thermography 130, 132, 426, 576
Additive manufacturing (AM) 51, 53–54, 233, 283, 333–334, 364, 403, 407
Additive manufacturing facility (AMF) 289
Adherend materials 509
Adhesive-beam model 398–399
Adhesive bonded SLJ
  degradation modelling 474–487
  finite-element analysis (FEA) 474
    bulk adhesive 474–477
    SLJ model 477–487
  test 463–464
    mechanical testing 464–470
    non-destructive inspection 470–473
Adhesive bonding 12, 385–386, 508, 518–520
  advantages 26
  disadvantages 26
Adhesive composite joints 415
  adhered manufacturing 437
  adhesive bonded SLJ (*see* Adhesive bonded SLJ)
  bulk adhesive specimens manufacturing 439
    preliminary test 440–443
    sheet manufacturing 443
  damage prediction
    cohesive zone modelling (CZM) 429–431
    durability prediction models 427
    environmental degradation models 431–434
    moisture diffusion modelling 427–429
  degradation modelling 494–495
  design
    lamintate lay-up and stacking sequence 418–419
    type of joint 415–417
  durability 421–425
    environmental effects 421–422
    fatigue durability 424–425
    hygrothermal cycle effects 422–424
  experimental results 450
  hygrothermal cycles 439, 442
  manufacture process 419–421
  materials' characterization 444
    diffusion coefficient 446–447
    gravimetric test 444–446
  mechanical properties
    adhesive FM94, 448
    adhesively bonded SLJ tests 448–449
    CFRP 447
    microscopy evaluation 449–450
    non destructive inspection(NDI) 449
    ultrasonic inspection 450
  mechanical test results
    fatigue test 493–494
    lap shear test 492–493
  moisture absorption 487–491
  moisture characterization
    CFRP T800/M21, 460
    FM94, 451–457, 460
    laminate T800/M21, 457–460
    swelling 460
  moisture effect adhesive mechanical properties 491

non-destructive inspection adhesive bonded joints 494
non-destructive testing 425–426
defects, types of 426
non-destructive evaluation (NDE) 426–427
overview 434–435
research methodology 435–436
SLJ manufacturing 437–441
Adhesive joints, interleaving and 549–551
Adhesively bonded joint, composite-metal 671–673
advantages of adhesive joint 673
disadvantages of adhesive joint 673–674
types of joints 673
Adhesively bonded polymer composites
adherend modifications
functionally graded materials (FGMs) 363–365
reinforced adherends 359–363
transverse toughened adherend 359–363
bi-adhesive joints 365–366
double-lap joints 370–372
double-strap joint 372
manufacturing techniques 366–367
plate joints 372–373
single-lap joints (SLJs) 367–370
T-joints 373–374
bonding 356–359
with natural fibres 374
cork particles 375–376
date palm tree fibres 376–377
surface preparation 374–375
overview 355–356
Aerospace composite repair
adhesives 163, 164
aim 120
bolted joints 145–146
bonded patch 145–146
commercial options 159–160
composite adaptable inspection and repair (CAIRe) 151
conventional machining 157
experimental testing
data analysis methodology 168–169
pulsed thermography results 170–182
setup and methodology 165–168
ultrasonic testing 169–170
ultrasonic testing results 182–185
gap analysis 152
hard and soft patch 141–145
in-service damage 125–127
laser system 157–158
machinable area 155–156
material and conditions 140
material removal 135–136, 135–139, 156–157
methodology 121–123
NDI techniques 153–154, 185–186
other 134
shearography 132–134
thermography 130–134
ultrasonic testing (UT) 129–130
on-site curing 163–165
overview 117–119
patch 160–161
design 161–162
production 162–163
post repair inspection 135
process flow diagram 187
pulsed thermography 155
quality control 146
repair certification 149–151
structural health monitoring (SHM) 147–149
structures 123–124
surface preparation 137–138
surface treatments
laser treatment 140
peel-plyA 138–140
plasma treatment 140
surface abrasion/grit blasting 140
typical damages 127–128
3D scanning 158–159
ultrasonic equipment 154
Aerospace composites 507–535
aim 508
manufacturing
adhesive bonding 518–520
curing of adherend material 512
curing of adhesive 520–522

cutting 513
embedding of defects 513–518
of adherend material 511–512
surface treatment 512–513
tabbing 522–523
test plan 510–511
materials
adherend 509
adhesive 509
pretreatment product 509
objectives 509
overview 507–508
results and discussion
lap shear test 527–532
NDI 532–534
optical microscopy 533–534
pulsed thermography 533
ultrasonic C-scan 532–533
testing
lap shear test 523–525
NDI 525–527
pulsed thermography 526–527
ultrasonic C-scan 525–526
Agglomeration 676, 721
Amine blush 725
Anaerobics adhesives 391
Aramid reinforced aluminium laminate (ARALL) 359
Ariane 5 ACY 5400, 15–16
Assembled composite aerostructures
bonded joints 7–8, 10
in aviation industry 20–24
fabrication 10–15
failure modes 8, 11
in space industry 15–20
stress analysis 9–12
fastened joints 6–7
analysis 7
failure modes 6, 9
load distribution 6, 8
overview 1
strength analysis 1–2
structural analysis methods 2–6
Augmented reality (AR) 593–623
composite structures' maintenance and testing 596–597
computer vision 595–596
Hausdorff distance equation 599–600
image data processing and 3D modelling 599–600
LIDAR 595–596
marker image libraries 600
methodology 600–609
Abaqus 607–608
composite laminates specifications 603
deformation type 600–603
distance measurements 606–607
LIDAR measurements 605–606
MATLAB® simulation 607–608
photogrammetry measurements 603–605
point cloud denoising 606–607
visualization 608–609
overview 593–595
photogrammetry 595–596
recommendations for future 620–622
results 609–610
Abaqus 612–615
difficulties with MATLAB® code 615–620
LIDAR 612–615
photogrammetry 610–612
scope of application 621
simulation accuracy 597
stress and strain calculation, through surface data 597–599
Augmented reality (AR), aircraft composite repairing 197
aim 199
in aircraft industry
CAIRe 212, 213
IRIS-integrated AR 211
MiRA technology 211
PlyMatch in ACE 211–212
challenges 224–226
coding language 218
common service-life damages 200–201
composite materials
classification 202
defects and damages in 200
high-energy impact defect 202
intermediate-energy impact defect 202
low-energy impact defect 202
repairing technique 202

concept design
assisting materials and equipment 215–216
bagging layers 214–216
functionality 216–217
panel 214
patches 214
project scope selection 212–213
reparation method 213
target process 213–214
vacuum bagging system 216
developing process 218–221
environment configuration 217–218
gap analysis 198–199
hardware configuration 217
manufacturing defects 200
marker-based augmented reality 210
markerless augmented reality 210
methodology 199–200
objectives 199
overiview 197–198
program validation 221–223
projection-based augmented reality 210
scarf-based repair process 202–210
strengths 224
superimposition-based augmented reality 210
Autoclave curing 34, 437
of carbon fiber-reinforced polymer 39
challenges 39–40
heat transfer modes 32
Average shear stress model 397
Aviation industry, joints in 20–24

Bagging film 215
Barely visible impact damage (BVID) 5–6, 576
Bigelow Expandable Activity Module (BEAM) 289, 290
Bismaleimides 392
Bi-STEM 327
Business challenges 49–51

Carbon fibre reinforced polymer (CFRP) 38, 67, 360–361, 414
Carbon nanofibers 66
Carbon polymer composites 254
Ceramic polymer composites 252–253
Chiral 710
Clearly visible impact damage (CVID) 5–6
Closed mould technique 440
Co-bonding method 11
Co-curing joining mechanism 13–15
Cohesive zone modelling (CZM) 429–431
Co-liquid resin infusion (LRI) 20–22
Collapsible tubular mast (CTM) booms 327–333
design 335–336
cross-section geometry 336–339
CuBe 336
finite element analysis 339–343
maximum length analysis 339–343
Onyx 336
three-point bending test methodology 343–349
manufacturing processes 331–332, 349–353
optimal geometry 330–331
preliminary design parameter calculation 328–330
Complex carbon fiber-reinforced polymer (CFRP) composites
manufacturing method 70–71
Composite adaptable inspection and repair (CAIRe) 151, 212, 213
Composite filament fabrication (CFF) 67–68
Composite-metal bonded joint, thermo-mechanical properties: *See* Thermo-mechanical properties, composite-metal bonded joint
Composites technology 46–49
Composite Technology for Exploration (CTE) 18–20
Continuous carbon fiber (CCF) 68
Continuous fiber-reinforced additively manufactured (CFRAM) printing process 69
Continuous lattice fabrication (CLF) 67
Conventional machining 136–137
Conventional strain sensors 717–719

Copper beryllium (CuBe) boom 321–323, 325, 336
Copper traces 299–300
Cork particles 375–376
CTE Point Design 19
Curing 691
of adherend material 512
of adhesive 520–522
pressure 205, 207
temperature 207
Cyanoacrylates adhesives 391

Damage assessment, composites: *See* Thermography, composite materials and
Damage prediction
adhesive joints
cohesive zone modelling (CZM) 429–431
durability prediction models 427
environmental degradation models 431–434
moisture diffusion modelling 427–429
Data analysis methodology 168–169
Date palm tree fibres 376–377
Degrees of freedom (DOF) 272
Delamination 559–573
discussion about 571–573
methodology 561
database development 563–564
design of experiments 561–562
geometry creation 562–563
hyper-parameters 565
learning function 568–569
machine learning modelling 564–565
meshing and material assignment 563
network type 566
neural networks 565–566
simulation setup 562–563
testing sets 565
training 565
training function 567–568
transfer function 566–567
validation 565
overview 559–561
results
error histogram 570
ML predictions and regression plot 570–571
performance measure plot 569
stress distributions within the composite panel 569
training state plot 569–570
De-orbit mechanism (DOM) 321
aim 324
boom designs 325
CTM booms (*see* Collapsible tubular mast (CTM) booms)
storable tubular extendible member (STEM) boom 326–327
triangular rollable and collapsible (TRAC) boom 325–326
Cranfield University 322–323
objectives 324
overview 322
Deployable boom 325
Digital image correlation (DIC) 683
Digital light processing (DLP) 62, 234, 235
Digital signal processor 693
Dimethylpolysiloxane 291
Direct digital manufacturing (DDM) 61, 62
Direct ink writing (DIW) 291, 305
DLP: *See* Digital light processing (DLP)
Double cantilever beam (DCB) test 396, 397
Double-lap joints 370–372
Double-strap joint 372
Drop-on-demand inkjet printing 304
Dry peel ply 214
Durability prediction models 427

Electrical conductivity 721–722
Electrical discharge machining (EDM) 137
Electromagnetic induction methods 134
Electronic beam melting 62, 233
Electronic speckle pattern interferometry 134

End-notched flexure (ENF) test 396, 397
Energy harvesting 625–653: *See also* Self-sensing
  overview 625–627
  piezoelectric effect 627–630
  piezoelectric materials 630–632
  polymer composites 639
    fibre-reinforced 639–643
  polymer nanocomposites 643–649
  polyvinylidene fluoride 632–636
  P(VDF-HFP) copolymers 639
  P(VDF-TrFE) copolymers 636–639
Environmental degradation models 431–434
Epoxies 391

Fabrication, CNT 714
  adhesively bonded joints 714–719
  advantages of adhesive joints 714–715
  conventional strain sensors 717–719
  disadvantages of adhesive joints 715
  lap-shear joints 715–717
FDM: *See* Fused deposition modeling (FDM)
Fiber-reinforced additive manufacturing (FRAM) 61
  acrylonitrile butadiene styrene (ABS) 65–66
  applications 78–79
  carbon nanofibers 66
  carbon nanotubes 66
  CFRP 67
  challenges 82–85
  classification 62–63
  composite filament fabrication (CFF) 67–68
  continuous carbon fiber (CCF) 68
  continuous lattice fabrication (CLF) 67
  future prospective 82–85
  hydrogels 68
  materials used, mechanical properties of 63–64
  mechanical properties
    creep deformation 76–77
    fatigue strength 75–76
    fracture toughness 77
    machine learning (ML) 77–78
    tensile strength 74–75
    topology optimization (TO) 77
  nanoparticles 66
  polyester 63, 65
  polylactic acid (PLA) 66
  process 63, 65
  sustainability 79–81
Fibre-reinforced polymer (FRP) 61, 537
  printing 69
Fibre 254–259, 547–548
Fibre bridging 546
Fibre metal laminates (FMLs) 359, 362
Fibre-reinforced plastic (FRP) composite, ABJ
  adhesive type 390–392
  advantages 386
  analytical analysis 396
    adhesive-beam model 398–399
    average shear stress model 397
    Goland and Reissner's model 399–400
    shear lag model 398
  characterisation methods 395
    double cantilever beam (DCB) test 396, 397
    end-notched flexure (ENF) test 396, 397
    mixed-mode bending (MMB) test 396, 397
    peel test 396
    shear test 395–396
    tensile test 395
    wedge cleavage test 396
  cost-effective assembly 394
  design 387–388
  design principles 402–406
  disadvantages 386
  environmental parameters 394
  failure modes 389–390
  finite element analysis (FEA) 400–402
  joint configuration 389
  manufacturing considerations 387–388
  oveeview 385–387
  recent developments 402–406
  stress distribution 389–390
  surface preparation 392–394

Fick's law 428
Finite element analysis(FEA) 260
  CTM booms 339–343
First Ply Failure (FPF) 4
FM94
  mechanical properties 461–463
  moisture characterization 451–457, 460
Fume hood 692
Functionally graded materials (FGM) 286–287, 363–365
Fused deposition modeling (FDM) 62, 65, 66–71, 74, 231, 232, 234, 236, 237, 239–242, 256, 261, 262, 276, 277, 287, 305, 333, 334, 336, 343, 349
Fused deposition modelling (FDM) 333–334
  Markforged's FDM printer and materials 334–335
Fused deposition modeling/fused-filament-fabrication (FDM/FFF) method 291

Gauge factor (GF) 649
Geometry creation 562–563
Glass fibre-reinforced polymer (GFRP) composites 69
Goland and Reissner's model 399–400
Granta EduPack 274

Hard patches 208
Hausdorff distance equation 599–600
High shear mixer 679–680
Hot plate 692
Hydrogels 68
Hygrothermal cycle effects 422–424
Hyper-parameters, delamination and 565
Hysol E-20HP 292–293

Industry 4.0
  additive manufacturing 53–54
  augmented reality 55
  autonomous robots 52–53
  big data and analytics 54
  cloud 54
  composites context 44–45
  cybersecurity 54
  horizontal and vertical system integration 52
  industrial Internet of Things 51
  overview 43–44
  simulation 51–52
Inkjet printing 305
In-space manufacturing (ISM) 309
Interfacial interaction 677
Interleaving
  measuring properties 539–541
  overview 537–539
  types and performance
    applied to stiffeners and adhesive joints 549–551
    fibre 547–548
    particle 547
    selective interface interleaving 551–554
    self-same matrix resin 541–543
    thermoplastic 545–547
    thermoset resin 543–545
IRIS-integrated AR 211

Joint assembly 24
  adhesive bonding
    advantages 26
    disadvantages 26
  bond-line control
    need for 25–26
    significance of 25

Laminate stacking sequence (LSS) 418
Lap-shear joints 715–717
Lap shear test 492–493, 523–525, 527–532
Laser extensometer 684–685, 692
Laser machining 137
Laser treatment 140
Learning function, delamination and 568–569
LIDAR 595–596, 605–606, 612–615
Linear elastic fracture mechanic (LEFM) 401

Linear-elastic Vokersen's analysis 417
Liquid acrylic resin-based composites
  vs. conventional thermoset composites 96, 97
  end-of-life opportunities with 103, 106–109
  future opportunities 109–101
  joining 96, 98–102
  manufacturing 94–95
  recyclate matrix 106–107
  repairability 107
  reshapability 107–109
Lithium-ion battery (LIB) technology 304–305
LVA ring/CFRP thrust cone 16–18

Machine learning modelling, delamination and 564–565
Marker-based augmented reality 210
Marker image libraries 600
Markerless augmented reality 210
Markforged's FDM printer 334–335
Material removal 156–157, 203, 206
MATLAB®, 607–608, 615–620
Mechanical fastening 385, 507–508
Meshing and material assignment, delamination and 563
Metallic polymer composites 248–251
Metal-oxide-semiconductor field-effect transistors (MOSFET) 289
Microwave curing 35–36
  of carbon fiber-reinforced polyme 39
  challenges 39–40
  heat transfer modes 32
MiRA technology 211
Mixed-mode bending (MMB) test 396, 397
Moisture diffusion modelling 427–429
Multi-material polymer composite systems 3D printing
  carbon 254
  ceramic 252–253
  discussion 275
  FDM, material science 241–241
  fibres 254–259
  fillers 242–245
  finite element analysis (FEA) 260
  metallic 248–251
  numerical simulation
    boundary conditions 264–272
    mesh discretisation 272–273
    parameters 264–266
    schematic 262–264
    thermo-physical properties 273–274
  objectives 261–262
  overview 231–232
  particulates 245–248
  polymer blends 241–242
  reinforcements 242–245
  results 275–277
  types 233–240
  whiskers 254–259
Multi-walled CNT (MWCNT) 286

Nanoparticles 66
NaOH treatment 374–375
Natural fibres
  composite adhesives with 374
    cork particles 375–376
    date palm tree fibres 376–377
    surface preparation 374–375
Navier-Stokes momentum equation 267
Network type, delamination and 566
Neural network (NN) algorithm 559
Neural networks, delamination and 565–566
Non-destructive evaluation (NDE) 426–427
Non-destructive inspection (NDI) technologies 525–527, 532–534
  damage assessment (*see* Thermography, composite materials and)

Olympus OmniScan MX2 154–155
On-site curing 163–165
Onyx 336
Optical microscopy 533–534

Particle 547
Patches repair 202

Payload attach fitting (PAF) conical structure 18–20
Peel test 396
Peel-plyA 138–140
Perforated release film 215
Performance measure plot, delamination and 569
Phenolic-impregnated carbon ablator (PICA) 284
Phenolics 392
Photogrammetry 595–596, 610–612
Piezoelectric effect 627–630
Piezoelectric materials 630–632
Piezoresistive effect 722
Piezoresistivity 626
Plasma treatment 140
Plate joints 372–373
PlyMatch, in ACE 211–212
PMC: *See* Polymer matrix composites (PMC)
Point cloud denoising 606–607
Polyamide (PA) 240, 241
Polyarylether sulphones (PAES) 308–309
Polycaprolactone (PCL) 240, 241
Polyester 63, 65
Polyetheretherketone (PEEK) 298, 300, 307, 308
Polyetheretherketoneketone (PEEKK) 308
Polyetherimide/PC (PEI/PC) 292
Polyetherketone (PEK) 308
Polyetherketoneetherketoneketone (PEKEKK) 308
Polyetherketoneketone (PEKK) 308
Polyimides 392
Polylactic acid (PLA) 66, 240, 241
Polymer composites 639
  adhesively bonded (*see* Adhesively bonded polymer composites)
  carbon 254
  ceramic 252–253
  fibre-reinforced 639–643
  metallic 248–251
  multi-material systems (*see* Multi-material polymer composite systems 3D printing)
Polymer matrix composites (PMC)
  autoclave curing 31–32, 34, 39–40
    of carbon fiber-reinforced polyme 39
    challenges 39–40
    heat transfer modes 32
  basic principles 31–34
  curing methods 30
  experimental details 36–39
  heating pattern 33–34
  mathematical models
    autoclave curing model 34
    microwave curing model 35–36
  microwave curing 31, 35–36, 39–40
    of carbon fiber-reinforced polyme 39
    challenges 39–40
    heat transfer modes 32
  overview 29–31
Polymer nanocomposites 643–649
Polyurethane 392
Polyvinylidene fluoride (PVDF) 361, 632–636
Pretreatment product 509
Printed circuit board (PCB) 299, 300
Projection-based augmented reality 210
Pulsed thermography 170–182, 526–527, 533
P(VDF-HFP) copolymers 639
P(VDF-TrFE) copolymers 636–638

Raman spectroscopy 695, 701–702, 735
Rapid prototyping 71
Regression plot, delamination and 570–571
Rehybridisation 710
Reinforced carbon-carbon (RCC) 284
Resin injection repair 202
Resonance methods 134

Scanning electron microscopy 694–695, 700–701, 734
Scarf-based repair process 202–210
Selective interface interleaving 551–554
Selective laser melting (SLM) 62, 233, 234
Selective laser sintering (SLS) 62, 65, 71–74, 233, 234

Self-same matrix resin 541–543
Self-sensing 649–652: *See also* Energy harvesting
Shear lag model 398
Shear test 395–396
Shearography
  applicability 132
  benefits 133
  drawbacks 133
  overview 132, 133
  principle 133
  thermography and 133–134
Silicones 392
Simulation setup, delamination and 562–563
Single-event effect 290
Single-lap joints (SLJs) 367–370, 413, 415, 437–441
SIUI Supor 155
SLA: *See* Stereo lithography (SLA)
Slicing 255, 275
SLM: *See* Selective laser melting (SLM)
SLS: *See* Selective laser sintering (SLS)
Soft patches 208
Sonic infrared thermography 134
Space industry, joints in 15–20
Stereo lithography (SLA) 62, 68, 234, 235, 305
Stiffeners, interleaving and 549–551
Stirring 679
Storable tubular extendible member (STEM) boom 326–327
Strain gauges 683
Strain self-sensing applications, high-performance nanocomposites for 709–750
  CNT-epoxy nanocomposites 719–724
    agglomeration 721
    electrical conductivity 721–722
    piezoresistive effect 722
    strain sensors 722–724
  CNTs
    electromechanical properties 711
    for strain sensing 711–712
    structure of 710
  defects in adhesively bonded joints 724–727
    amine blush 725
    detection with the impedance method 725–726
    failure mechanisms of adhesively bonded joints 726–727
    types of defects 724
  fabrication, CNT 714
    adhesively bonded joints 714–719
    advantages of adhesive joints 714–715
    conventional strain sensors 717–719
    disadvantages of adhesive joints 715
    lap-shear joints 715–717
  materials and methods 714–735
  non-destructive testing techniques of health monitoring 727–728
    acoustic emission method 727
    health monitoring of CNT-epoxy adhesives 727–728
  overview 709–713
  piezoresistivity of CNT nanopolymers 712–713
  Raman spectroscopy 735
  sample manufacturing 728–729
    manufacturing process 731–732
    materials 729–730
    specification 730–731
  scanning electron microscopy 734
  tensile and electrical resistance testing 732
    equipment 733–734
    sample preparation for testing 732–733
    testing plan 734
Superimposition-based augmented reality 210
Surface abrasion/grit blasting 140
Surface preparation 203, 205

Tabbing 522–523
Tensile test 395
Thermal blanket 215
Thermal curing techniques 30
Thermography

applicability 131–132
benefits 131
drawbacks 131
overview 130
and shearography 133–134
techniques 130–132
Thermography, composite materials and 575–590
coefficient clustering analysis 581–585
experimental data
data collection 579–580
specimen 577–579
temperatute decay curve 580–581
overview 575–576
results
damage detection 585–586
damage measurement 586–589
Thermo-mechanical properties, composite-metal bonded joint 667–704
aim 669
analytical calculations 682–683
CNTs dispersion 676
agglomeration 676
contents 677–678
interfacial interaction 677
size of CNT 676–677
CNTs, effects on thermo-mechanical performance 674–676
CTE of polymer 674–676
mechanical properties 674
thermal properties 674
CNTs, production in multicomp laboratory 670–671
composite-metal adhesively bonded joint 671–673
advantages of adhesive joint 673
disadvantages of adhesive joint 673–674
types of joints 673
functionalization of CNTs 680–681
chemical functionalization 681
physical functionalization 681
material 685–686
MWCNTs 685
resin and curing agent 685
substrate 686
surfactant for CNTs 685–686
mechanical dispersion methods of CNTs 678
high shear mixer 679–680
stirring 679
ultrasonication 678–679
mechanism of joint failure by CTE mismatch 681–682
objectives 669–670
overview 667–669
results 695–703
bond deficiency on specimens 702–703
coefficient of thermal expansion 695–700
Raman spectroscopy 701–702
scanning electron microscopy 700–701
specimens manufacturing 686–692
adhesive preparation and application 688–691
curing 691
final manufactured specimens 691–692
mould 686–688
schematic of specimens 686
surface treatment 686
trimming 691
testing
digital signal processor 693
fume hood 692
hot plate 692
laser extensometer 692
procedure of measurement 694
Raman spectroscopy 695
scanning electron microscopy 694–695
test scheme 693
thermal strain measurement 692–694
thermal strain measurement, methods of 683–685
digital image correlation strain analysis 683
laser extensometer 684–685
strain gauges 683

Thermoplastic 545–547
Thermoplastic acrylic-matrix fibre-metal laminates 102–105
Thermoplastic acrylic resins 93
Thermoset resin 543–545
Thin-film transistors (TFT) 304
3D modelling, augmented reality and 599–600
3D-printed composites, for spacecraft
  batteries 304–307
  devices for medical and life support purposes 307–309
  electronics
    active components 303–304
    passive components 300–303
    substrates 298–300
    traces 298–300
  overview 283–284
  structures
    heat shields for suborbital flight 284–287
    issues 291–298
    radiation shields in low-earth orbits/ deep space 287–291
3D-printing 62, 71, 283, 306, 406
Three-point bending test methodology
  CTM booms 343–349
T-joints 373–374
Toughened acrylics 391
Training function, delamination and 567–568
Training state plot, delamination and 569–570
Transfer function, delamination and 566–567
Trimming 691

UK composites industry 152–153
ULTEM 9085, 292
ULTEM series 292, 295, 296
Ultra-high-molecular-weight polyethylene (UHMWPE) composites 288
Ultrasonication 678–679
Ultrasonic C-scan 525–526, 532–533
Ultrasonic testing (UT) 169–170
  applicability 130
  overview 129
  results 182–185
  techniques 130

Vacuum-assisted resin transfer moulding (VARTM) 20
Vacuum bagging 208–209
Vat photo-polymerisation technique 235
Vibration-based methods 134
Volumetric feed rate 275

Wedge cleavage test 396
Whiskers 254–259

X-ray 291
X-ray computed tomography 134

www.ingramcontent.com/pod-product-compliance
Lightning Source LLC
LaVergne TN
LVHW011535200926
848484LV00001B/268